Polypharmacology

Polypharmacology

Strategies for Multi-Target Drug Discovery

Edited by
Jens-Uwe Peters

WILEY

Published by John Wiley & Sons, Inc., Hoboken, New Jersey.
Published simultaneously in Canada.

For general information on our other products and services or for technical support, please contact our Customer Care Department within the United States at (800) 762-2974, outside the United States at (317) 572-3993 or fax (317) 572-4002.

Wiley also publishes its books in a variety of electronic formats. Some content that appears in print may not be available in electronic formats. For more information about Wiley products, visit our web site at www.wiley.com.

Library of Congress Cataloging-in-Publication Data applied for:

Hardback ISBN: 9781394182831

Cover Design: Wiley
Cover Image: Courtesy of Balaguru Ravikumar

Set in 9.5/12.5pt STIXTwoText by Straive, Chennai, India

Contents

List of Contributors

Tero Aittokallio
University of Helsinki
Finland/University of Oslo
Norway
tero.aittokallio@helsinki.fi

Alberto Ambesi-Impiombato
PsychoGenics
USA
alberto.ambesi@psychogenics.com

Jürgen Bajorath
Department of Life Science Informatics
B-IT LIMES Unit Chemical Biology and Medicinal Chemistry
University of Bonn
Bonn
Germany
bajorath@bit.uni-bonn.de

Ian M. Bell
Merck & Co.
Discovery Chemistry
USA
ian_bell@merck.com

Eric A.G. Blomme
AbbVie Inc.
USA
eric.blomme@abbvie.com

Maria Laura Bolognesi
University of Bologna
Department of Pharmacy and Biotechnology
Italy
marialaura.bolognesi@unibo.it

Martin Bossart
Sanofi Germany
Integrated Drug Discovery Frankfurt
Germany
martin.bossart@sanofi.com

Daniela Brunner
PsychoGenics
Icahn School of Medicine at Mount Sinai
USA
Daniela.Brunner@psychogenics.com

Helmut Buschmann
Pharma Consulting Aachen
Germany
hbuschmann@gmail.com

María-José Camarasa
Instituto de Química Médica
Spain
mj.camarasa@iqm.csic.es

Laurent R. Chiarelli
University of Pavia
Department of Biology and Biotechnology "Lazzaro Spallanzani"
Italy
laurent.chiarelli@unipv.it

Kelly Chibale
University of Cape Town
South Africa
kelly.chibale@uct.ac.za

Thomas Christoph
Translational R&D, Aachen
Germany
info@t-christoph.de

Anna Cichońska
Harmonic Discovery
Finland
anna@harmonicdiscovery.com

Mario Cocorullo
University of Pavia
Department of Biology and Biotechnology "Lazzaro Spallanzani"
Italy
mario.cocorullo01@universitadipavia.it

Lauren B. Coulson
University of Cape Town
South Africa
lauren.coulson@uct.ac.za

Sonia de Castro
Instituto de Química Médica
Spain
sonia@iqm.csic.es

Maedeh Darsaraee
University of Berne
Department of Chemistry
Biochemistry and Pharmacy
Switzerland
maedeh.darsaraee@unibe.ch

Paul Dent
Virginia Commonwealth University
Department of Biochemistry and Molecular Biology
USA
Paul.Dent@vcuhealth.org

Eleonora Diamanti
University of Bologna
Department of Pharmacy and Biotechnology
Italy
eleonora.diamanti2@unibo.it

Dang Ding
Shandong University
P.R. China
2429895062@qq.com

Attila Egyed
Research Centre for Natural Sciences
Medicinal Chemistry
Drug Innovation Centre
Hungary
egyed.attila@ttk.hu

Shenghua Gao
Shandong University
School of Pharmaceutical Sciences
Department of Medicinal Chemistry
Key Laboratory of Chemical Biology
P.R. China
shh_gao@163.com

Samaneh Goorani
University of Arkansas for Medical Sciences
USA
SGoorani@uams.edu

Jonathon R. Green
AbbVie Inc.
USA
jonathon.green@abbvie.com

Hiroshi Handa
Tokyo Medical University
Department of Molecular Pharmacology
Japan
hhanda@tokyo-med.ac.jp

Norbert Handler
RD&C Research
Development & Consulting GmbH
Austria
n.handler@rdc-concepts.com

Axel Helmstaedter
University of Marburg
Germany
helmstae@staff.uni-marburg.de

Gerhard Hessler
Sanofi Germany
Integrated Drug Discovery Frankfurt
Germany
gerhard.hessler@sanofi.com

John D. Imig
University of Arkansas for Medical Sciences
USA
jimig@uams.edu

Sacha Javor
University of Berne
Department of Chemistry
Biochemistry and Pharmacy
Switzerland
sacha.javor@unibe.ch

György M. Keserű
Research Centre for Natural Sciences
Medicinal Chemistry
Drug Innovation Centre
Hungary
keseru.gyorgy@ttk.hu

Meehyein Kim
Infectious Diseases Therapeutic Research Center
Korea Research Institute of Chemical Technology (KRICT)
Daejeon
Republic of Korea
mkim@krict.re.kr; sirnaworld@yahoo.com

Stephan Kirchner
F. Hoffmann-La Roche Ltd.
Roche Pharma Research and Early Development
Roche Innovation Center Basel
Switzerland
stephan.kirchner@roche.com

Dóra J. Kiss
Research Centre for Natural Sciences
Medicinal Chemistry
Drug Innovation Centre
Hungary
kiss.dora.judit@ttk.hu

Armando A. Lagrutta
Merck & Co.
Nonclinical Drug Safety
USA
armando_lagrutta@merck.com

Xiangqian Li
State Key Laboratory of Microbial Technology
Shandong University
P.R. China
lixiangqian@sdu.edu.cn

Helen L. Lightfoot
F. Hoffmann-La Roche Ltd.
Roche Pharma Research and Early Development
Roche Innovation Center Basel
Switzerland
helen_louise.lightfoot@roche.com

Felix F. Lillich
University of Frankfurt
Germany
lillich@pharmchem.uni-frankfurt.de

Xinyong Liu
Shandong University
School of Pharmaceutical Sciences
Department of Medicinal Chemistry
Key Laboratory of Chemical Biology
P.R. China
xinyongl@sdu.edu.cn

Prathap Kumar S. Mahalingaiah
AbbVie Inc.
USA
prathapkumar.mahalingaiah@abbvie.com

Lee McDermott
PsychoGenics
USA
lee.mcdermott@psychogenics.com

Zeinab Mamdouh
Zagazig University
Department of Pharmacology and Toxicology
Egypt
zmamdouh@ppmlab.net

Cristian Nogales
Maastricht University
Department of Pharmacology and Personalised Medicine
The Netherlands

and

Max Perutz Labs
Vienna Biocenter Campus (VBC)
Austria

and

University of Vienna
Center for Molecular Biology
Department of Structural and Computational Biology
Austria
cristian@menchelab.com

Arsenio Nueda
Almirall, S.A.
Spain
arsenio.nueda@almirall.com

Mayra Pacheco Pachado
Maastricht University
Pharmacology and Personalised Medicine
The Netherlands
mpachado@ppmlab.net
mayra.pachado@gmail.com

Anna M. Pasieka
University of Bologna
Department of Pharmacy and Biotechnology
Italy
pasieka.ann@gmail.com

Alan Lars Pehrson
PsychoGenics
USA
alan.pehrson@psychogenics.com

Jens-Uwe Peters
Skyhawk Therapeutics
Switzerland
Jup1@live.com

Zina Piper
Maastricht University
Pharmacology and Personalised Medicine
The Netherlands
zpiper@ppmlab.net

Andrew Poklepovic
Virginia Commonwealth University
Department of Medicine
USA
andrew.poklepovic@vcuhealth.org

Michal Poznik
RD&C Research
Development & Consulting GmbH
Austria
m.poznik@rdc-concepts.com

Ewgenij Proschak
University of Frankfurt
Germany
proschak@pharmchem.uni-frankfurt.de

Rayees Rahman
Harmonic Discovery
USA
rayees@harmonicdiscovery.com

Balaguru Ravikumar
Harmonic Discovery
Finland
guru@harmonicdiscovery.com

Jean-Louis Reymond
University of Berne
Department of Chemistry
Biochemistry and Pharmacy
Switzerland
jean-louis.reymond@unibe.ch

Sonia Roberts
F. Hoffmann-La Roche Ltd.
Roche Pharma Research and Early Development
Roche Innovation Center Basel
Switzerland
sonia.roberts@roche.com

Navriti Sahni
Harmonic Discovery
USA
navriti@harmonicdiscovery.com

Ana-Rosa San-Félix
Instituto de Química Médica
Spain
anarosa@iqm.csic.es

Tiffany Schwasinger-Schmidt
University of Kansas School of Medicine-Wichita
Department of Internal Medicine
USA
tschwasinger-schmidt@kumc.edu

Dayong Shi
State Key Laboratory of Microbial Technology
Shandong University
P.R. China
shidayong@sdu.edu.cn

Lynn L. Silver
LL Silver Consulting
USA
llsilverconsulting@gmail.com

Letian Song
Shandong University
School of Pharmaceutical Sciences
Department of Medicinal Chemistry
Key Laboratory of Chemical Biology
P.R. China
kevinsong319@163.com

Giovanni Stelitano
University of Pavia
Department of Biology and Biotechnology
"Lazzaro Spallanzani"
Italy
giovanni.stelitano@unipv.it

Kai Tang
Shandong University
P.R. China
13027750680@163.com

Peter L. Toogood
University of Michigan
USA
toogood@umich.edu

Elisa Uliassi
University of Bologna
Department of Pharmacy and Biotechnology
Italy
elisa.uliassi3@unibo.it

Terry R. Van Vleet
AbbVie Inc.
USA
terry.vanvleet@abbvie.com

Andy Vo
AbbVie Inc.
USA
Andy.vo@abbvie.com

Shujing Xu
Shandong University
P.R. China
xu17854111942@163.com

Yuki Yamaguchi
Tokyo Institute of Technology
School of Life Science and Technology
Japan
yyamaguc@bio.titech.ac.jp

Junichi Yamamoto
Tokyo Institute of Technology
School of Life Science and Technology
Japan
yamamoto.j.ab@m.titech.ac.jp

Bing Ye
Shandong University
School of Pharmaceutical Sciences
Department of Medicinal Chemistry
Key Laboratory of Chemical Biology
P.R. China
2860322431@qq.com

Dimitar Yonchev
F. Hoffmann-La Roche Ltd.
Roche Pharma Research and Early Development
Data & Analytics
Roche Innovation Center Basel
Switzerland
dimitar.yonchev@roche.com

Peng Zhan
Shandong University
School of Pharmaceutical Sciences
Department of Medicinal Chemistry
Key Laboratory of Chemical Biology
P.R. China
zhanpeng1982@sdu.edu.cn

Yang Zhou
Shandong University
P.R. China
13503806229@163.com

Preface

Many drugs act on more than one target [1]. This can be necessary for efficacy, but can also lead to adverse effects [2]. For instance, it was discovered in the 1980s that dual $D_{2/3}$ and 5-HT_{2a} receptor antagonism is needed for efficacy in antipsychotic drugs [3]. Today we know that antipsychotics bind to more than 20 targets, some of which contribute to efficacy, but also cause adverse effects [4].

In the early 2000s, the term polypharmacology was introduced to describe this concept of drugs binding to several targets. During this time, it became increasingly recognized that multi-target activity is often needed for efficacy. For instance, the antibiotic research of the 1990s focused on single targets derived from bacterial genomes. These single-targeted approaches were generally fruitless and did not lead to new drugs. Instead, nearly all systemically efficacious antibiotics bind to multiple targets or to targets encoded by multiple genes, so that single mutations do not lead to drug resistance (further discussed in Chapter 10) [5]. It was also recognized that unintended "anti-target" activity leads to adverse effects. Here, the most prominent example is an unusual high number of drugs withdrawn from the market in 1996–2001. These drugs were withdrawn due to adverse effects, which could be traced back to anti-target activity. For instance, half a dozen of drugs was withdrawn due to their potential to cause cardiac arrhythmias caused by unintended blockade of the hERG channel (see Chapter 7) [6]. Thus, polypharmacology can be beneficial or detrimental – these two sides of the polypharmacology coin are further detailed in the introduction.

Following some widely read papers on concepts such as "Network Pharmacology" [7] or "Magic Shotguns" [8], polypharmacology became an increasingly popular research topic. From 2010 onward, Scifinder searches retrieve an ever-increasing number of publications on polypharmacology and related topics, such as multi-target, off-target, and secondary or network pharmacology. A first book on polypharmacology was published in 2012 and became a popular read [9]. This current book is a follow-up with an updated and expanded content.

The book is divided into four parts A–D. Part A discusses undesired polypharmacology, which is often a safety concern. For instance, many drugs bind to "anti-targets" or "off-targets", e.g. to cardiac ion channels. This causes adverse effects such as cardiac arrhythmia. The relevance of such anti-targets for adverse effects will be discussed in a first chapter, followed by chapters on the link between off-targets and adverse drug reactions, on how to screen for off-target activity and how to recognize and optimize compounds with a potential for off-target activity. This is followed by a discussion of kinases and cardiac ion channels, two of the most important classes of anti-targets.

The remainder of the book is dedicated to intended polypharmacology. Part B discusses disease areas, which benefit from polypharmacological approaches. A first chapter outlines the general concept of network pharmacology and multi-target drugs. The following chapters focus on oncology, bacterial and viral infections, CNS diseases, and metabolic diseases, followed by a discussion of the role of polypharmacology in the history of drug discovery.

But how can we discover such multi-target drugs? Part C of the book highlights important approaches, such as compound design, data mining with web-based tools, multi-target peptides, as well as phenotypic screening in cells, tissues, and animal models. A related topic is the Selective Optimization of Side Effects (SOSA) approach to drug discovery, which will be discussed as well.

The final Part D collects case studies on polypharmacological drugs and current research. PROTACs and molecular glues are hot topics in drug discovery, and the first chapter outlines how these originate from the polypharmacology of thalidomide. Next is a story on achieving "selective dual activity" for cyclin-dependent kinase inhibitors. This is followed by a bouquet of topics, from the discovery of cariprazine and tapentadol, to current research on antivirals, malaria, tuberculosis, HIV, and pancreatic cancer.

This book on polypharmacology is intended as a comprehensive resource for industrial drug hunters and academic researchers. It illuminates all facets of polypharmacology, from anti-target screening, to the design of multi-target ligands. A comparison of the current book with the first book from 2012 [9] shows that polypharmacology has certainly come of age. Polypharmacology research has improved the drug discovery process, has delivered ideas for Biotech Startups, and has garnered the attention of the media [10]. Hopefully, this book will inspire readers for new drug discovery projects and will help to mitigate attrition due to safety issues.

I am very grateful to all contributing authors, who invested their time and their expertise into this book. Also, I thank the team at Wiley for proposing this book and for their advice throughout this project: Katherine Wong, Jonathan Rose, Sabeen Aziz, Shwathi Srinivasan, and Keerthana Baskaran.

Enjoy reading!

Jens-Uwe Peters
Basel, Switzerland, December 2024

References

1 Hu, Y. and Bajorath, J. (2013). Compound promiscuity: what can we learn from current data? *Drug Discov. Today* 18 (13/14): 644–650.
2 Peters, J.-U. (2013). Polypharmacology – Foe or Friend? *J. Med. Chem.* 56 (22): 8955–8971.
3 Meltzer, H.Y., Matsubara, S., and Lee, J.C. (1989). Classification of typical and atypical antipsychotic drugs on the basis of dopamine D-1, D-2 and serotonin2 pKi values. *J. Pharmacol. Exp. Ther.* 251 (1): 238–246.
4 Riemer, C. (2012). Antipsychotics. In: *Polypharmacology in Drug Discovery* (ed. J.-U. Peters), 343–362. Hoboken: Wiley.
5 Silver, L.L. (2012). Polypharmacology as an Emerging Trend in Antibacterial Discovery. In: *Polypharmacology in Drug Discovery* (ed. J.-U. Peters), 167–202. Hoboken: Wiley.
6 Bell, I.M., Bilodeau, M.T., and Lagrutta, A.A. (2012, 2012). Activity at Cardiovascular Ion Channels: A Key Issue for Drug Discovery. In: *Polypharmacology in Drug Discovery* (ed. J.-U. Peters), 83–109. Hoboken: Wiley.
7 Hopkins, A.L. (2008). Network pharmacology: the next paradigm in drug discovery. *Nat. Chem. Biol.* 4 (11): 682–690.
8 Roth, B.L., Sheffler, D.J., and Kroeze, W.K. (2004). Magic shotguns versus magic bullets: selectively non-selective drugs for mood disorders and schizophrenia. *Nat. Rev. Drug Discovery* 3 (4): 353–359.
9 Peters, J.-U. (ed.) (2012). *Polypharmacology in Drug Discovery*. Wiley: Hoboken.
10 Kurji, N. (2019). The master key to drug design: multi-target drugs. https://www.forbes.com/sites/forbestechcouncil/2019/09/16/the-master-key-to-drug-design-multi-target-drugs/?sh=54ba77576cfe (accessed 27 September 2024).

1

Introduction

Facets of Polypharmacology – a Janus-Headed Concept for Drug Discovery

Jürgen Bajorath

1.1 Origins

Since the 1980s, target-centric approaches have dominated drug discovery efforts, triggered by the molecular-biology-driven reductionist approach [1] and leading to the "one drug, one target," or "drug specificity" paradigm [2]. Molecular reductionism aimed at "dissecting biological systems into their constituent parts" [1]. Different from the preceding more holistic and pharmacology-oriented era in drug discovery, molecular sciences and the single-target (ST) focus took the centerstage and shaped drug discovery efforts for many years to come [1, 2]. These developments were paralleled by advances in X-ray crystallography and molecular graphics catalyzing a wave of structure-based ("rational") drug design efforts [3, 4], which further emphasized the focus on target-specific compounds in drug discovery.

In the early 2000s, systems biology emerged [5] and also entered the drug discovery arena [6] introducing, for example, network representations of biological systems, pathway modeling, and computational disease models. These developments originating from bioinformatics also altered the view of traditional disciplines such as pharmacology, giving rise to a network perception of physiological processes and increasing the notion of their interdependence [7]. In pharmacological networks, multiple signaling and metabolic pathways establish functional links and dependencies between different target proteins [7, 8]. If pathways in such networks are perturbed or regulatory and control mechanisms compromised, different types of multifactorial diseases might arise, including various forms of cancer, complex diseases of the central nervous system, or metabolic diseases [9–12]. Such diseases could most likely not be effectively treated by therapeutic intervention of individual targets, but required multi-target (MT) engagement instead, thus departing from the target specificity paradigm in drug discovery. MT activity of drugs was not unknown and probably first observed for anti-psychotics and antiepileptics beginning in the late 1980s [12, 13].

In 2006, as a consequence of the increasing notion of pharmacological networks, the concept of polypharmacology was introduced [14], focusing on MT agents for the treatment of multifactorial diseases: "*Contrary to the dogma that the 'rational' way to discover drugs is to design exquisitely selective ligands for single molecular targets, a rival hypothesis proposes polypharmacology or the promiscuous modulation of several molecular targets*" [14]. In 2014, a formal definition of polypharmacology appeared in the US National Library of Medicine (NLM) as "*the design or use of pharmaceutical agents that act on multiple targets or disease pathways.*" Polypharmacology also encompasses the pharmacological effects resulting from the use of MT compounds (MT-CPDs), consistent with the principles of network pharmacology. MT activity of drugs and other bioactive compounds is often also referred to as "promiscuity" (not to be confused with nonspecific compound–protein interactions).

1.2 Pros and Cons

Following its inception, polypharmacology emerged as an alternative to reductionist approaches and rational drug design and further evolved into a multifaceted drug discovery strategy [15–17], albeit "Janus-headedly." In Roman mythology, Janus, the god of the beginnings, passages, and endings, had two opposing faces. Accordingly, the "Janus head" became a symbol of duality and ambivalence – which exactly applied to the polypharmacology concept: on the one hand, MT activity of drugs is a prerequisite for therapeutic efficacy in the treatment of multifactorial diseases; on the other, it is responsible for unwanted (adverse) side effects [15, 18, 19]. While adverse side effects can be elicited by the engagement of a primary target, they are more frequently caused by inhibiting so-called anti-targets

Polypharmacology: Strategies for Multi-Target Drug Discovery, First Edition. Edited by Jens-Uwe Peters.

such as cardiac ion channels (hERG), drug-metabolizing cytochrome P450 isoforms, or G-protein-coupled neurotransmitter receptors [15, 16]. Furthermore, side effects of MT-CPDs might also be caused by interacting with other targets not implicated in immediate toxicity, due to pathway modulations. In the pharmaceutical industry, potential liabilities as a consequence of interactions with anti-targets are a major concern, for example, leading to the assessment of newly identified candidate compounds in various safety screens for activity against such targets. However, not all unexpected side effects are undesired, taking into consideration that MT activity also provides the basis for drug repurposing [20]. Benefits of MT activity of drugs were often discovered post hoc. For example, adenosine triphosphate (ATP)-site-directed kinase inhibitors used in cancer therapy were originally thought to be kinase-selective, based on reductionist assessment, before it was discovered that their clinical efficacy depended on multi-kinase activity and simultaneous interference with multiple deregulated signaling pathways [21]. This also applied to imatinib, the first kinase inhibitor marketed as a drug [21].

Despite the Janus-headed nature of polypharmacology and the risks associated with potential adverse side effects resulting from the MT activity of drugs, the positive impact of polypharmacology on drug discovery and development is undeniable, as demonstrated by the continuous occurrence of MT agents among newly approved drugs. For example, 10 of 49 European Medicines Agency (EMA)-approved drugs marketed in Germany in 2022 were annotated with two or more targets [22]. Of course, despite the strong impact of polypharmacology, the development of compounds with target selectivity or specificity continues to be a pillar of drug discovery and development. For example, for long-term treatment of chronic and non-life-threatening diseases, drug side effects must inevitably be minimized, rendering target-selective compounds highly desirable.

1.3 Discovery and Design

Similar to coincidental findings that side effects of drugs originally thought to be specific were caused by previously unknown secondary targets, new MT-CPDs are often discovered serendipitously, for example, in screening campaigns or target deconvolution of active compounds from phenotypic assays. Given the high interest in compounds with defined MT activity in different therapeutic areas, prospective design of such compounds is also a topical issue in drug discovery [23, 24]. However, consistent with findings that characteristic structural features of MT-CPDs generally depend on target combinations, as further discussed below, the prospective design of MT-CPDs with desired activity is challenging, mostly carried out on a case-by-case basis in medicinal chemistry and far from being routine. For all practical purposes, prospective design of MT-CPDs for polypharmacology is limited to two or at most three targets. To this end, combining or merging target-dependent pharmacophores is a popular knowledge-based approach for MT-CPD design that is readily applicable in the practice of medicinal chemistry and does not require sophisticated computations [23–25]. Pharmacophore fusion attempts can be further extended by screening of test compounds using pharmacophore models for different targets and follow-up analysis of shared hits [26]. As an alternative to pharmacophore modeling, scaffolds isolated from compounds with known activity against different targets can also be used as templates for MT-CPD design, as further discussed below.

In addition to knowledge-based design strategies that are close to practical medicinal chemistry, other ligand- or target-structure-based computational approaches have been applied to identify compounds for polypharmacology [27, 28]. For example, various machine learning (ML) models have been reported to distinguish between compounds with MT activity and corresponding ST activity (typically achieving reasonable to high prediction accuracy). Furthermore, ML models have been used for computational target profiling. Here, test compounds are virtually screened using large numbers of individually derived target-based models to predict MT-CPDs. As a deep learning alternative, multitask models have also been developed to predict compounds with activity against related targets. At the structural level, similarities of binding sites in different targets have been quantified as an indicator of polypharmacology potential at the target level. In addition, parallel docking campaigns or cross-docking screens have been carried out for structure-based target profiling. Furthermore, ligands bound to different proteins have been systematically compared to identify compound pairs with the highest shape similarity to prioritize and evaluate putative cross-target activities [28].

1.4 Structural Data

In addition to its relevance for polypharmacology, the study of MT-CPDs is also of interest from a basic scientific perspective. For example, exploring the mechanisms by which small molecules "multi-specifically" or "pseudo-specifically" interact with different targets helps to better understand these special molecular recognition phenomena. To this end, currently available X-ray structures of complexes formed by MT-CPDs and different proteins provide substantial information. For example, in 2018, we identified 1418 crystallographic MT-CPDs

(>300 Da) in X-ray structures of complexes with different targets available in the Protein Data Bank (PDB) [29, 30]. These MT-CPDs included 702 ligands forming complexes with targets from different protein families (termed multifamily ligands) [30]. Bound conformations of multifamily ligands available in complexes with unrelated targets were compared in detail, revealing a variety of ligand binding modes [31]. In some instances, these ligands conformationally adapted to binding sites having different architectures and chemical features and displayed different binding modes; in others, binding modes were surprisingly conserved in differently shaped active sites. If binding modes of multifamily ligands were conserved, characteristic interaction patterns emerged for targets from a given family that differed from others, hence providing a possible rationale for the conservation of binding modes [31].

As a representative example, Figure 1.1 shows conserved and variable binding modes in different active sites for indomethacin, a nonsteroidal anti-inflammatory drug (NSAID) with known polypharmacology used for the treatment of acute pain and symptoms of osteoarthritis and rheumatoid arthritis.

For 243 of the 702 multifamily ligands, 168 analogue series were detected in the ChEMBL database [32]. These series consisted of a total of 4829 compounds, covered 190 additional targets, and yielded 133 unique analogue series-based scaffolds [30]. Figure 1.2 shows an exemplary scaffold. All analogue series scaffolds were annotated with different target combinations, providing a knowledge base of MT template compounds.

1.5 Activity Data

Rapidly growing volumes of compound activity data provide another information-rich resource for the study of MT-CPDs and polypharmacology. Since the analysis of MT activity is particularly vulnerable to false-positive activity annotations, compound activity data should be carefully curated and potential assay interference compounds [33, 34] or colloidal aggregators [35] should be removed. Indeed, results of MT activity analysis strongly depend on applied data confidence criteria [36], as illustrated in Figure 1.3 for imatinib, suggesting to restrict the assessment of MT-CPDs to high-confidence activity data [36].

There are different facets of MT activity. For instance, it is not very surprising that some active compounds exhibit a tendency to interact with more than one closely related target, such as ATP-site-directed protein kinase inhibitors. By contrast, compounds binding to structurally and functionally unrelated proteins are rather unexpected, but of special interest, from both a basic scientific and a polypharmacology perspective. For example, such compounds might interfere with distinct physiological functions and elicit

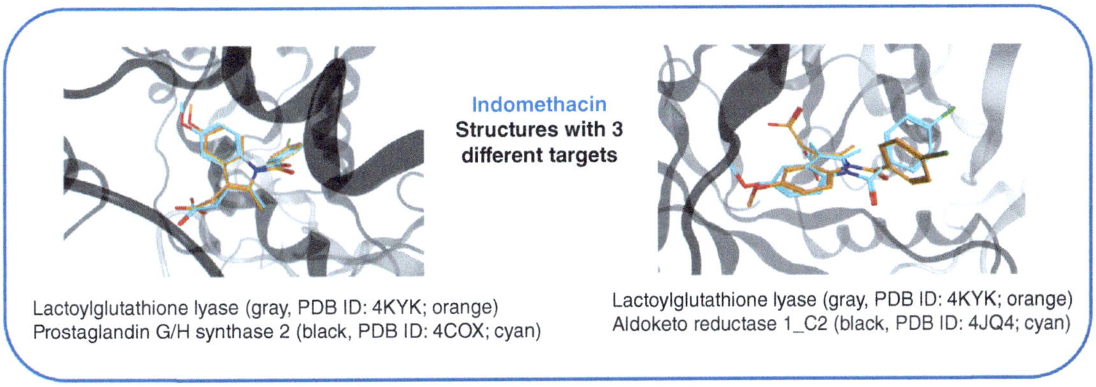

Figure 1.1 X-ray structures of indomethacin in complex with three distinct targets. On the left, and right, pairwise superpositions of bound ligand conformations are shown, revealing conserved (left) and variable binding modes (right) in different protein environments.

Figure 1.2 Scaffold of a multifamily ligand with kinase activity representing an analogue series. For the ligand, crystal structures of complexes with Aurora and TEC kinases were available (PDB) and structural analogues found in ChEMBL were active against additional kinase targets from other families.

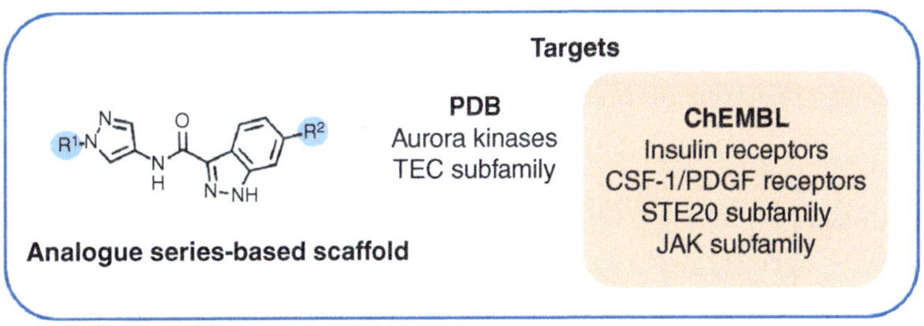

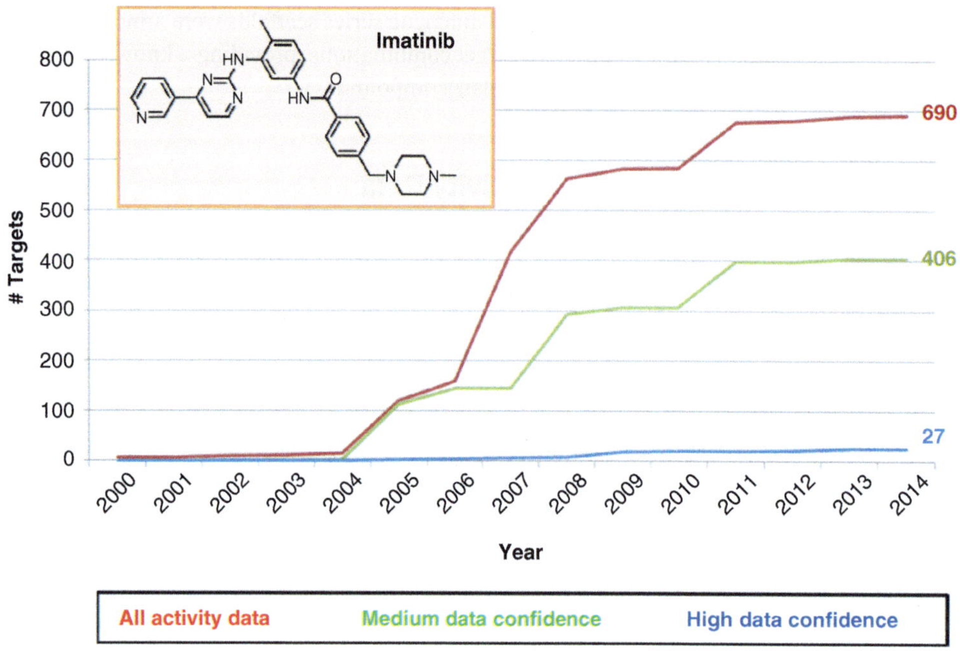

Figure 1.3 Target annotations of imatinib. Based on increasing volumes of activity data from ChEMBL collected over time, the number of targets reported for imatinib is monitored at three different confidence levels: all activity data (no confidence criteria were applied) medium- and high-confidence data. Adapted from Hu and Bajorath [36]. The number of target annotations based on all activity data and medium-confidence data (690 and 406, respectively) is unrealistic.

unexpected pharmacological effects. Systematic analysis of compound activity data helps to estimate the frequency with which MT-CPDs occur and the number of targets they are active against. Especially for candidate compounds and drugs, such estimates are relevant to balance often articulated expectation values that are largely unsubstantiated (e.g., "most drugs bind to 10 or 20 targets …"). In addition, careful analysis of available compounds and activity data also helps to gauge predictions of MT-CPDs and their target numbers, for example, from computational target profiling (*vide supra*). Notably, compound-data-driven analysis principally underestimates MT activity due to data incompleteness, given that not "all compounds have been tested against all targets" (the ultimate goal of chemogenomics). This must be taken into consideration. On the other hand, analysis of the large and rapidly growing volumes of activity data available in the public domain should reveal some statistically sound trends [36]. For instance, in 2019, we carried out a large-scale analysis of biological screening data from PubChem [37] in the search for compounds with activity against targets from different classes [38]. A total of 1063 compounds were identified that were tested in assays for at least 100 human target proteins and were active against at least 10 targets from more than one class [38]. These findings showed that MT-CPDs with activity against distantly or unrelated targets occurred rather frequently.

1.6 Drug Target Estimates

Systematic experimental determination of the targets that drugs are active against is far from being an easy task. Accordingly, insights into drug target numbers are typically confined to case-by-case proteomic analysis or statistics from target panel assays such as kinome screens [39]. However, based on compound data analysis, different estimates of target numbers for drugs and other active compounds have been reported.

Early attempts to predict drug targets used network representations of drug–target interactions [40]. From different databases, drugs, targets, and interaction data were collected and analyzed in drug–target networks. From such network representations, it was estimated that a drug on average interacted with six targets. Depending on the data used, targets per drug ranged from approximately 3 to 13 [40]. Comparable estimates were obtained when approved and experimental drugs taken from DrugBank [41] were mapped to ChEMBL and drug data and targets were monitored over a 15-year period [42]. For bioactive compounds from screening assays, different target numbers were determined. In an early analysis of PubChem [37], MT activities were analyzed on the basis 600+ assays [43]. It was found that approximately 58% of active screening compounds only displayed ST activity in combined primary and confirmatory assays. In addition, based on high-confidence activity data extracted from ChEMBL, it was determined that a bioactive compound on average interacted with only one to two targets, with no significant variations across different families [44, 45]. Different from PubChem data, ChEMBL does not contain test frequencies for compounds taken from the literature (where negative data is typically not reported). Hence, data incompleteness generally plays a greater role in target analysis based on ChEMBL. On the other hand, high-confidence activity data can best be obtained from ChEMBL. In 2016, approximately 430,000 extensively assayed compounds were extracted from PubChem (tested in both primary and confirmatory assays), with mean and median values of 411 and 437 assays per compound, respectively [46]. Most extensively assayed screening hits were on average active against 2.5 targets [46], thus only slightly exceeding the mean number of targets determined for ChEMBL compounds based on high-confidence activity data (in the presence of data sparseness).

Global estimates of drug targets are generally affected by the presence of small subsets of highly promiscuous drugs. For example, during a period of strong growth of public compound activity data (2012–2016), the mean number of targets per drug increased from 5.9 (2012) to 7.3 (2016) for approximately 1600 drugs available in DrugBank. When the top-50 most promiscuous drugs were omitted, the mean (median) value decreased from 7.3 (5.0) to 6.4 (4.5) targets per drug; when the top-100 drugs were omitted, it further decreased to 6.0 (4.0) [36]. Our most recent in-house calculations for drugs with available high-confidence activity data indicate that the mean number of drug targets has essentially remained constant over the past five years, with approximately four targets per drug. Of note, computational predictions of drug targets tend to be higher than data-driven estimates.

1.7 Explainable Machine Learning

In addition to activity or target predictions, ML can also be used for hypothesis testing in polypharmacology research, quasi as a "diagnostic" approach. We have carried out a large-scale ML analysis to address the questions (i) if structural features exist that systematically distinguish MT-CPDs from compounds with corresponding ST activity (ST-CPDs) and (ii) if such structural features might be shared by compounds with activity against different targets. Therefore, a compound test system was generated by systematically searching ChEMBL for compounds with activity against different target pairs (dual-target compounds, DT-CPDs) and corresponding ST-CPDs that were available in sufficient numbers for ML. The search identified 170 target pairs (involving 137 unique targets) for which at least 100 MT- and (50 + 50) ST-CPDs with high-confidence activity annotations were available [47]. Many more target pairs with less than 100 DT-CPDs were detected. These findings also demonstrated that a wealth of DT-CPDs with activity against different targets is publicly available, from which analogue series and representative scaffolds can be isolated and considered as design templates for polypharmacology (*vide supra*).

For testing via ML, the following hypothesis was formulated: if structural features exist that systematically distinguish DT- and corresponding ST-CPDs, then ML models derived based on chemical structures should detect these features and accurately classify MT- and corresponding DT-CPDs; if such structural features do not exist, the models could not possibly be predictive [47]. Accordingly, for each target pair, ML models were derived using different algorithms and the same structural fingerprints. These models were then used to predict test compounds in "native" predictions (i.e., using test compounds of the target pair for which the models were derived) and for all other 169 target pairs in "cross-pair" predictions (i.e., using test compounds of all target pairs for which the models were not derived). Figure 1.4 summarizes the composition of the target-pair-based test system and the ML prediction strategy.

The 170 native predictions using different ML models yielded overall consistently accurate predictions with a median accuracy higher than 80% and only a few target pair datasets with an accuracy higher than 60% (statistical outliers of the distribution). By contrast, the vast majority of cross-pair predictions failed, with a median accuracy of approximately 50% (corresponding to random predictions) [47]. Taken together, these findings demonstrated

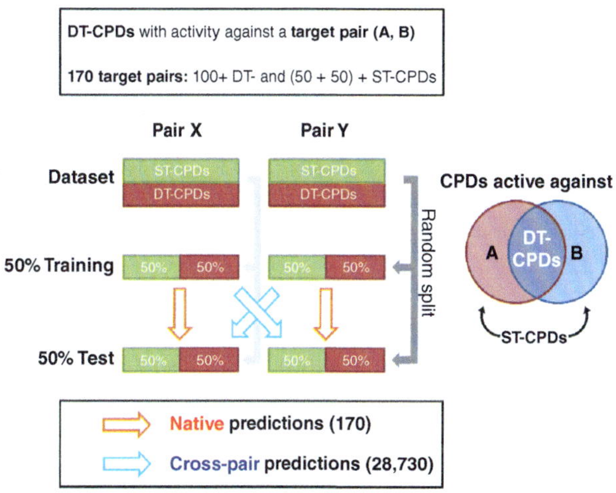

Figure 1.4 Target-pair-based compound test system for diagnostic ML investigating potential structural signatures of DT-CPDs. Adapted from Feldmann and Bajorath [47].

that structural features distinguishing between DT- and corresponding ST-CPDs existed and depended on target pairs. Hence, distinguishing features were not generalizable across target pairs.

Subsequently, structural features determining accurate predictions were identified for exemplary pairs of unrelated targets [48] using the SHapley Additive exPlanations (SHAP) approximation of the Shapley value concept from game theory [49] that was adapted as an explainable artificial intelligence (XAI) approach for rationalizing ML predictions [49, 50]. Features determining accurate predictions were characteristic of DT-CPDs and absent in ST-CPDs. In DT-CPDs, these structural features formed coherent and well-defined substructures [48]. These findings showed that explainable ML can be used to identify structural motifs implicated in MT activity.

1.8 Conclusion

In this introductory chapter, a brief account of the origins and history of the polypharmacology concept is provided and the Janus-headed nature of polypharmacology is emphasized. Despite potential pitfalls, polypharmacology has evolved into a viable therapeutic strategy for the treatment of multifactorial diseases that is here to stay and will further expand. MT activity of small molecules is at the root of polypharmacology and, in addition to its relevance for drug discovery, of considerable interest from a basic research perspective, as also discussed. Despite increasing insights into different facets of polypharmacology, a number of open questions remain to be addressed. For example, the magnitude of polypharmacology among current drugs still needs to be quantified, beyond expectations and case-by-case studies, which might require Herculean efforts (or novel experimental concepts). Furthermore, serendipity continues to play a major role in polypharmacology-oriented drug discovery, leaving much room for the development of further advanced drug design strategies. These and other questions related to polypharmacology provide ample opportunities for attractive future research.

References

1 Regenmortel, M.H.V. (2004). Reductionism and complexity in molecular biology: scientists now have the tools to unravel biological complexity and overcome the limitations of reductionism. *EMBO Rep.* 5: 1016–1020.

2 Morphy, R., Ka, C., and Rankovic, Z. (2004). From magic bullets to designed multiple ligands. *Drug Discov. Today* 9: 641–651.

3 Verlinde, C.L. and Hol, W.G. (1994). Structure-based drug design: progress, results and challenges. *Structure* 2: 577–587.

4 Anderson, A.C. (2003). The process of structure-based drug design. *Chem. Biol.* 10: 787–797.

5 Ideker, T., Galitski, T., and Hood, L. (2001). A new approach to decoding life: systems biology. *Annu. Rev. Genomics Hum. Genet.* 2: 343–372.

6 Butcher, E.C., Berg, E.L., and Kunkel, E.J. (2004). Systems biology in drug discovery. *Nat. Biotechnol.* 22: 1253–1259.

7 Hopkins, A.L. (2007). Network pharmacology. *Nat. Biotechnol.* 25: 1110–1111.

8 Ainsworth, C. (2011). Networking for new drugs. *Nat. Med.* 17: 1166–1169.

9 Knight, Z.A., Lin, H., and Shokat, K.M. (2010). Targeting the cancer kinome through polypharmacology. *Nat. Rev. Cancer* 10: 130–137.

10 Tschöp, M.H., Finan, B., Clemmensen, C. et al. (2016). Unimolecular polypharmacy for treatment of diabetes and obesity. *Cell Metab.* 24: 51–62.

11 Benek, O., Korabecny, J., and Soukup, O. (2020). A perspective on multi-target drugs for Alzheimer's disease. *Trends Pharmacol. Sci.* 41: 434–445.

12 Weston-Green, K. (2022). Antipsychotic drug development: from historical evidence to fresh perspectives. *Front. Psychiatry* 13: 903156.

13 Löscher, W. (2021). Single-target versus multi-target drugs versus combinations of drugs with multiple targets: preclinical and clinical evidence for the treatment or prevention of epilepsy. *Front. Pharmacol.* 12: 730257.

14 Hopkins, A.L., Mason, J.S., and Overington, J.P. (2006). Can we rationally design promiscuous drugs? *Curr. Opin. Struct. Biol.* 16: 127–136.

15 Peters, J.U. (2013). Polypharmacology – foe or friend? *J. Med. Chem.* 56: 8955–8971.

16 Anighoro, A., Bajorath, J., and Rastelli, G. (2014). Polypharmacology: challenges and opportunities in drug discovery. *J. Med. Chem.* 57: 7874–7887.

17 Bolognesi, M.L. and Cavalli, A. (2016). Multitarget drug discovery and polypharmacology. *ChemMedChem* 11: 1190–1192.

18 Campillos, M., Kuhn, M., Gavin, A.C. et al. (2008). Drug target identification using side-effect similarity. *Science* 321: 263–266.

19 Berger, S.I. and Iyengar, R. (2011). Role of systems pharmacology in understanding drug adverse events. *Wiley Interdiscip. Rev. Syst. Biol. Med.* 3: 129–135.

20 Pushpakom, S., Iorio, F., Eyers, P.A. et al. (2019). Drug repurposing: progress, challenges and recommendations. *Nat. Rev. Drug Discov.* 18: 41–58.

21 Cohen, P., Cross, D., and Jänne, P.A. (2021). Kinase drug discovery 20 years after imatinib: progress and future directions. *Nat. Rev. Drug Discov.* 20: 551–569.

22 Ryszkiewicz, P., Malinowska, B., and Schlicker, E. (2023). Polypharmacology: promises and new drugs in 2022. *Pharmacol. Rep.* 75: 755–770.

23 Proschak, E., Stark, H., and Merk, D. (2018). Polypharmacology by design: a medicinal chemist's perspective on multitargeting compounds. *J. Med. Chem.* 62: 420–444.

24 Li, X., Li, X., Liu, F. et al. (2021). Rational multitargeted drug design strategy from the perspective of a medicinal chemist. *J. Med. Chem.* 64: 10581–10605.

25 Morphy, R. and Rankovic, Z. (2005). Designed multiple ligands. An emerging drug discovery paradigm. *J. Med. Chem.* 48: 6523–6543.

26 Moser, D., Wisniewska, J.M., Hahn, S. et al. (2012). Dual-target virtual screening by pharmacophore elucidation and molecular shape filtering. *ACS Med. Chem. Lett.* 3: 155–158.

27 Rastelli, G. and Pinzi, L. (2015). Computational polypharmacology comes of age. *Front. Pharmacol.* 6: 157.

28 Feldmann, C. and Bajorath, J. (2022). Advances in computational polypharmacology. *Mol. Inform.* 41: 2200190.

29 Berman, H.M., Westbrook, J., Feng, Z. et al. (2000). The protein data bank. *Nucleic Acids Res.* 28: 235–242.

30 Gilberg, E., Stumpfe, D., and Bajorath, J. (2018). X-ray-structure-based identification of compounds with activity against targets from different families and generation of templates for multitarget ligand design. *ACS Omega* 3: 106–111.

31 Gilberg, E., Gütschow, M., and Bajorath, J. (2019). Promiscuous ligands from experimental structures, binding conformations, and protein family dependent interaction hotspots. *ACS Omega* 4: 1729–1737.

32 Gaulton, A., Hersey, A., Nowotka, M. et al. (2017). The ChEMBL database in 2017. *Nucleic Acids Res.* 45: D945–D954.

33 Baell, J.B. and Holloway, G.A. (2010). New substructure filters for removal of pan assay interference compounds (PAINS) from screening libraries and for their exclusion in bioassays. *J. Med. Chem.* 53: 2719–2740.

34 Bruns, R.F. and Watson, I.A. (2012). Rules for identifying potentially reactive or promiscuous compounds. *J. Med. Chem.* 55: 9763–9772.

35 Irwin, J.J., Duan, D., Torosyan, H. et al. (2015). An aggregation advisor for ligand discovery. *J. Med. Chem.* 58: 7076–7087.

36 Hu, Y. and Bajorath, J. (2017). Entering the 'big data' era in medicinal chemistry: molecular promiscuity analysis revisited. *Future Sci. OA* 3: FSO179.

37 Wang, Y., Bryant, S.H., Cheng, T. et al. (2017). PubChem BioAssay: 2017 update. *Nucleic Acids Res.* 45: D955–D963.

38 Feldmann, C., Miljković, F., Yonchev, D., and Bajorath, J. (2019). Identifying promiscuous compounds with activity against different target classes. *Molecules* 24: 4185.

39 Wang, Y. and Ma, H. (2015). Protein kinase profiling assays: a technology review. *Drug Discov. Today Technol.* 18: 1–8.

40 Mestres, J., Gregori-Puigjané, E., Valverde, S., and Solé, R.V. (2009). The topology of drug–target interaction networks: implicit dependence on drug properties and target families. *Mol. BioSyst.* 5: 1051–1057.

41 Law, V., Knox, C., Djoumbou, Y. et al. (2014). DrugBank 4.0: shedding new light on drug metabolism. *Nucleic Acids Res.* 42: D1091–D1097.

42 Hu, Y. and Bajorath, J. (2014). Monitoring drug promiscuity over time. *F1000Research* 3: 218.

43 Han, L., Wang, Y., and Bryant, S.H. (2009). A survey of across-target bioactivity results of small molecules in PubChem. *Bioinformatics* 25: 2251–2255.

44 Hu, Y. and Bajorath, J. (2013). High-resolution view of compound promiscuity. *F1000Research* 2: 144.

45 Hu, Y., Jasial, S., and Bajorath, J. (2015). Promiscuity progression of bioactive compounds over time. *F1000Research* 4: 118.

46 Jasial, S., Hu, Y., and Bajorath, J. (2016). Determining the degree of promiscuity of extensively assayed compounds. *PLoS ONE* 11: e0153873.

47 Feldmann, C. and Bajorath, J. (2021). Machine learning reveals that structural features distinguishing promiscuous and non-promiscuous compounds depend on target combinations. *Sci. Rep.* 11: 7863.

48 Feldmann, C., Philipps, M., and Bajorath, J. (2021). Explainable machine learning predictions of dual-target compounds reveal characteristic structural features. *Sci. Rep.* 11: 21594.

49 Lundberg, S.M. and Lee, S. (2017). A unified approach to interpreting model predictions. *Adv. Neural Inf. Proces. Syst.* 30: 4766–4775.

50 Chen, H., Covert, I.C., Lundberg, S.M., and Lee, S.I. (2023). Algorithms to estimate Shapley value feature attributions. *Nat. Mach. Intell.* 5: 590–601.

Part A

Polypharmacology as a Safety Concern in Drug Discovery

2

The Safety Relevance and Interpretation of Compound Off-target Interactions

Eric A.G. Blomme, Jonathon R. Green, Prathap Kumar S. Mahalingaiah, Terry R. Van Vleet, and Andy Vo

2.1 Introduction

Polypharmacology generally refers to the activity of pharmaceutical agents on multiple targets or disease pathways. It can represent an intended or unintended design of a molecule. In this chapter, the focus is mostly on the unintended aspect of polypharmacology and on small molecules. Even after careful optimization, most small molecules bind to multiple targets in addition to their intended pharmacological target. Hence, small-molecule drugs have been reported to bind to approximately 6–12 unintended targets, largely based on data obtained from high-throughput screens (i.e., radiometric-based biochemical assay against various receptors, ion channels, and transporters or kinase panels) [1, 2]. In many cases, this additional non-intended pharmacology is the basis of adverse effects [3]. These off-targets are usually lower-affinity targets in comparison to the intended pharmacological target, and consequently, the interactions may not always manifest themselves as clinical effects. However, in preclinical toxicity studies, higher doses and exposures are used to characterize the toxicity profile of molecules in a dose-response manner, and in these situations, interactions with these off-targets can become meaningful and drive the observed toxic effects. Likewise, in humans, differences in sensitivity or exposure are observed due to variability of individual pharmacokinetics (PK, e.g., genetic polymorphisms in the targets or drug-metabolizing enzymes, disease state, comedications, dietary components), misuse, or accidental exposure [4].

Differences in potency against the intended pharmacological target versus the unintended targets (also referred to as selectivity) dictate the biological and toxicological relevance of these lower-affinity targets when evaluated in the context of *in vivo* exposures (systemic plasma exposures, but also tissue exposures for drug transporter substrates) [4]. Given the complexity driving systemic and tissue concentrations, as well as the variability observed in human populations, accurately assessing the potential safety liabilities of off-target interactions of drug candidates can prove to be quite challenging and is far from a perfect science.

In some cases, unintended polypharmacology has resulted in a beneficial outcome (i.e., improved efficacy) [5]. For example, a recent study showed that off-target toxicity is a common and often the primary mechanism of action for cancer drugs [6]. In that study, using CRISPR-Cas9 mutagenesis of several cancer-relevant targets (e.g., HDAC6, MAPK14/p38α, PAK4, PBK, and PIM1) in breast, colorectal, lung, and melanoma cell lines, the authors showed that the efficacy of various drugs was unaffected by the genetic loss of their putative target, indicating that efficacy was essentially driven by off-target interactions [6].

According to a survey conducted by the Innovation and Quality (IQ) Consortium, attrition of molecules in the preclinical phase is mainly due to toxicity in rodent and non-rodent models, some of which may not be relevant to humans [7]. Off-target interactions are an important cause of this preclinical toxicity-related attrition. However, assessing those off-target effects, especially for those not relevant to humans, can be daunting for several reasons. These factors include differences in affinity across the same target of different species, scarcity of screening assays for targets from preclinical species, lack of sufficiently robust computational tools for predicting interactions with targets from preclinical species (largely due to the scarcity of screening data using targets from preclinical species), and lack of accurate tissue expression information for protein targets in commonly used animal models and strains. Consequently, off-target safety screens are mostly focused on human protein targets. However, to better understand interspecies differences, screening against key preclinical targets may also be necessary. This information would be

Polypharmacology: Strategies for Multi-Target Drug Discovery, First Edition. Edited by Jens-Uwe Peters.

helpful for putting preclinical toxicity findings, or a lack thereof, into perspective. Recently, a preclinical species body atlas of gene expression (RNAseq) was published and released [8]. In this effort, constitutive gene expression data for approximately 50 tissues in CD1 mice, Sprague Dawley rats, Beagle dogs, and Cynomolgus monkeys was generated to aid in assessing tissue/species expression of drug targets and in contextualizing the results from molecular pharmacology assays [8].

2.2 Assessing Off-Target Interactions of Small Molecules

The most common method used in pharmaceutical research and development (R&D) for predicting the off-target liability of small molecules is by screening compounds in *in vitro* panels of ligand binding assays [9]. Two complementary screening strategies are recommended: (i) testing against pharmacological targets that are similar to the target of interest and (ii) testing against a panel of off-targets associated with adverse side effects (also sometimes referred to as anti-targets) [10–12]. It is common to include in the second category assays central nervous system (CNS) targets that have been associated with drug abuse potential (i.e., opioid, cannabinoid, *N*-methyl-D-aspartate, ion channel complexes [e.g., calcium, potassium, and chloride], and transporters [e.g., dopamine, serotonin, and gamma-aminobutyric acid GABA]), as described in the Food and Drug Administration (FDA) guidance on the assessment of abuse potential (FDA CDER Guidance; https://www.fda.gov/media/116739/download). In addition, specific secondary molecular pharmacology screens represent an important aspect of a weight-of-evidence carcinogenicity risk assessment (WOE-CRA), as outlined in ICH S1B (https://www.ema.europa.eu/en/documents/scientific-guideline/ich-s1br1-guideline-testing-carcinogenicity-pharmaceuticals-step-5_en.pdf). While there is yet no consensus on what these WOE-CAR panels should include, the objective is to assess the binding potential of the parent molecule and its metabolites against targets that may inform on the carcinogenic potential of the molecules (e.g., nuclear and hormone receptors).

Although testing compounds against as many off-target assays as possible could be beneficial, this approach is not practical and would be resource-prohibitive. Hence, these secondary pharmacology panels need to be designed to incorporate the most relevant anti-targets. For example, anti-targets that are infrequent hits or with less concerning biology provide limited value. Conversely, other targets are frequently hit (i.e., promiscuous targets), and they may represent a useful indication of the general promiscuity of the agent being tested. For example, Rao et al reported *in vitro* screening results of 15 compounds, corresponding to *in silico* off-target predictions. The *in vitro* assay panel showed that of the 15 compounds, 5 interacted with Nav1.2, 4 with Cav1.5 receptors, and three with 5HT (2a and 2c) receptors [13]. This reflects the prevalence of off-target interactions by small molecules and suggests that some targets are particularly more likely to engage than others.

Many pharmaceutical companies and contract research organizations (CROs) use kinase assay panels to sample kinome interaction diversity [14]. There are several types of assay platforms available for kinase profiling, each with their advantages and limitations [15]. In contrast to other first-tier secondary pharmacology assays, kinome screening platforms are more comprehensive (often covering all or most of the entire kinome), can be functional (i.e., activity-based or enzymatic), and use a dose-response approach. However, many of these assays still lack the appropriate cellular context to fully appreciate the consequences of target engagement since they utilize isolated proteins or truncated protein domains without all relevant binding partners [11].

In vitro binding assays have some common limitations. The most commonly used formats rarely measure the actual "active" concentration of the available compound in the test well. Hence, compounds with poor solubility may show inaccurately low potency because they are out of solution. Prodrugs that only become active once metabolized may not show activity, and therefore, the parent molecule (i.e., the active drug species) should also be assessed. Likewise, it is difficult to predict the off-target activity in these assays for allosteric modulators [16]. In some assay formats with fluorescent output, fluorescent test compounds naturally may cause interference [17]. Compounds with stability issues, such as light sensitivity, may also yield incorrect results if not properly protected from light to prevent degradation during shipment, formulation, storage, and testing. Furthermore, some mechanisms of toxicity are still undefined, while others are species-specific. Therefore, usage of cells from the appropriate animal species can be critical. For example, it would be a poor choice to use rodent-derived cardiomyocytes for predicting QT interval prolongation in humans given that rodents have minimal K_v11.1 (hERG) current and that the ionic mechanism of repolarization differs between rodents and humans [18]). Finally, as already stated above, most screening panels are conducted in cells or targets of human origin, which may not reflect the binding in preclinical species and hence may not inform on preclinical toxicity. Nonetheless, few screens based on proteins from preclinical species are currently available and can be used to confirm binding in preclinical species [19]. Likewise, exploring differences in expression between species using a body atlas can help contextualize binding results. Altogether, this additional context could help identify preclinical toxicities of no relevance to humans.

2.3 Interpretation of Data from Secondary Pharmacology Assays

When interpreting ligand binding data, several aspects should be considered. Importantly, secondary pharmacology screens are typically assays measuring binding affinity to an anti-target (and not functional activity). Compounds may bind to the anti-target but may be either activating or inhibiting the target or even have no functional activity. Additionally, these binding assays are conducted using high concentrations (typically 10 μM), and these concentrations may never be achieved *in vivo* to result in functional activity. Therefore, follow-up testing in a functional assay of compounds binding at prespecified levels (e.g., 50% at 10 μM) is highly recommended.

Another variable to consider is protein binding. For molecules with substantial plasma protein binding, *in vitro* IC50s can be lower than *in vivo* IC50s because *in vitro* assays generally have little to no plasma serum. Hence, a better correlation may be obtained using free concentration values [20]. Finally, for CNS-related anti-targets, the degree of blood-brain barrier penetration needs to be understood, as some binding data may not be relevant for non-CNS-penetrant compounds [21].

When interpreting the safety significance of data generated from secondary pharmacology screens, several key factors should be contemplated. These include (i) the significance of any safety signal associated with the target, (ii) the likelihood of detection in preclinical models, (iii) the current ability to monitor the safety effects clinically, and (iv) the potential reversibility of these effects. Thus, when targets in screening panels are tiered for concern based on these factors, four categories emerge (Figure 2.1). Category 1 includes targets associated with severe potential safety effects that are unlikely to be detected preclinically, cannot be monitored clinically, and are not reversible. Strong consideration for termination is warranted for compounds significantly binding these targets. Category 2 includes targets associated with a significant safety signal (but with a potential therapeutic index (TI) based on tissue distribution) that is monitorable and possibly reversible. Further experimentation complemented by strong modeling is warranted in this situation before advancing this type of compounds. Category 3 encompasses targets with a manageable safety liability (but severity may depend on the TI) that is monitorable and reversible. This data should be considered when developing the clinical protocol. Finally, category 4 consists of those targets with a low safety concern potential that is unlikely to be development-limiting given the therapeutic indication in spite of being undesirable. While this data may not support the discontinuation of a compound, it is still useful as it may suggest an overall lower probability of technical success, may impact the WOE-based carcinogenicity risk assessment, or may result in additional testing, such as for drug abuse potential.

Indeed, secondary pharmacology screens are an important component in the assessment of the abuse potential

Category 1: significant safety signal and strong evidence for development discontinuation; severe effects noted; effects unlikely to be successfully monitored or are not reversible.

Category 2: significant safety signal, but severity will depend on therapeutic index and tissue distribution; not considered a developmental halting signal on its own, but rather to be considered in context to TI; effects may be monitorable; some effects are not reversible; ability to transverse the blood brain barrier is important to signal interpretation for relevant targets.

Category 3: moderate safety signal, but severity will depend on therapeutic index and tissue distribution; not considered a developmental halting signal on its own, but rather to be considered in context to TI; effects may be monitorable and reversible; ability to transverse the blood brain barrier is important to signal interpretation for relevant targets.

Category 4: signal to be aware of, but unlikely to result in development discontinuation; instead, effects can be used for explanation of phenotypic findings; effects are likely monitorable and reversible and non-life threatening.

Figure 2.1 Definitions of categories for the interpretation of secondary pharmacology profiles. Phenotypic effects for all tiers will depend on whether the test article is an agonist versus antagonist to the noted receptor and may depend on permeability to the relevant compartment such as central nervous system.

of an experimental agent. Hence, several abuse liability targets (e.g., opioid and dopamine receptors) are typically included in molecular pharmacology panels and used as a first-tier assessment of the abuse potential of compounds. This first-tier assessment will be complemented further by an understanding of CNS exposure at the clinical dose (level of CNS penetration), as well as clinical data evaluating the incidence of drug abuse adverse events from pharmacovigilance data. Based on the evaluation of this data, additional preclinical *in vivo* testing (drug discrimination, self-administration, and drug-dependence studies) may be warranted for an appropriate assessment of the abuse potential of small-molecule agents (https://www.fda.gov/media/116739/download).

A general promiscuity evaluation can also be a useful outcome of screening panels, irrespective of the targets engaged. High numbers of off-target interactions indicate the potential for engaging additional unscreened targets. An index of promiscuity could be represented by the ratio of confirmed interactions divided by the number of screened targets as an overall measure of the selectivity of a small molecule. Small diverse panels of targets may be used for this purpose at early discovery stages to generate an initial assessment of general promiscuity for chemical matter prior to larger off-target screens. Such an approach may be most beneficial for traditionally promiscuous targets, such as kinases [11]. The main use of a general promiscuity assessment is to guide chemists and prioritize more appealing early compounds to advance to more costly/resource-intensive screens. An additional consideration for a bioprofiling screen is to include many of the category 1 targets, especially from diverse families, to avoid promiscuity and get an early evaluation of the targets of most concern at the earliest stages of drug discovery and development. Although overall promiscuity within certain categories of targets can be seen as detrimental, overlapping at others may also unearth beneficial polypharmacology. Figure 2.2 details the current landscape of FDA-approved kinase inhibitors, the number of kinases for which they are approved to inhibit, and the families to which those kinases belong. As shown, most of these inhibitors are approved to inhibit multiple kinases, and in several instances, approvals span multiple kinase families. Figure 2.3 outlines the bioactivity landscape across approximately 4000 compounds and 47 targets from Roche's in-house data collection [22].

The impact of small-molecule polypharmacology on fundamental cellular processes such as cell division and electron transport chain (ETC) function during off-target screens should be noted. Microtubules or mitochondrial ETC proteins have been established as targets of promiscuous small molecules and may be excluded using a variety of assays. However, when these activities are part of a small-molecule-intended polypharmacology profile, interpretation and action may become more challenging [9]. In the case of ETC modulators, inhibitors of complexes I and V do not exhibit acute cytotoxic effects when tested in many primary cell culture systems [23, 24], suggesting that cell viability alone may not be sufficient to detect all mitochondrial toxicants. Yet, many ETC inhibitors result in significant toxic liabilities [25]. While mitochondrial toxicants generally can be detected in preclinical toxicology species as changes in mitochondrially rich tissues (e.g., heart, liver, kidney, brain, gastrointestinal tract), these toxicities are often delayed and may not be apparent in shorter-term preclinical toxicity studies, unless sensitive approaches such as toxicogenomics are used [26, 27]. Hence, routine assessment of small molecules in advanced mitochondrial cell health assays (such as the Seahorse Cell Mito Stress Assay) is warranted, especially in chemical space previously associated with mitochondrial effects [23, 24].

Even though off-target proteomic interaction of drugs is commonly associated with toxicities or adverse events, recent reports showing ubiquitous transcriptome interaction of FDA-approved protein targeted drugs suggest that some adverse effects may be related to changes in RNA function [28, 29]. Hence, characterization of a drug's interactions with the human transcriptome may add further value to understanding and de-risking of off-target-mediated toxicities. However, demonstrating first that off-target engagement with cellular RNAs can cause meaningful biological perturbations leading to specific toxicity/adverse changes will be important before adopting this type of screen widely in early drug development.

2.4 Off-Target Interactions of Biologics: Polyreactivity and Polyspecificity

Biologics, exemplified by therapeutic monoclonal antibodies (mAbs), are known for their exquisite specificity compared to small molecules. The initial belief was that mAbs or their derivatives (e.g., antibody fusions and fragments, bispecific, trifunctional antibodies) bind specifically to their intended targets, typically a single target for the traditional mAbs or two to three targets for bispecific or trispecific mAbs or derivatives. However, it is now well recognized that off-target interactions and poor molecular characteristics are also a cause of development termination

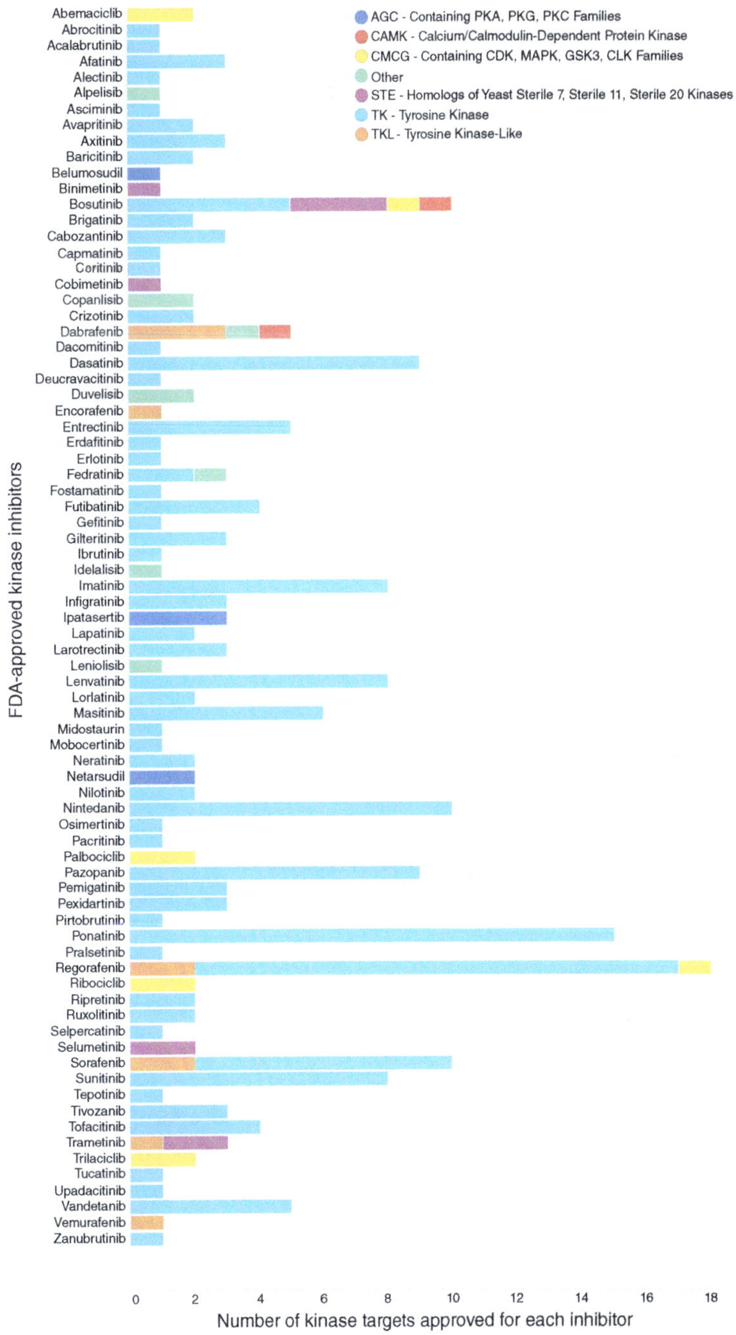

Figure 2.2 FDA-approved kinase inhibitors and their approved targets. There are currently 78 FDA-approved kinase inhibitors, which are approved to collectively target 68 unique kinases (data not shown) across seven kinase families. Adapted from PKIDB.com. In many instances, inhibitors are approved to target kinases in multiple families; for example, bosutinib is approved to treat ten targets across four kinase families. In terms of magnitude, the majority of these FDA approvals (59) belong to the tyrosine kinase family. The remaining six families are each linked to seven, or fewer, approved drugs.

for mAbs and derivatives [30]. Multiple mechanisms drive what is typically called polyreactivity and polyspecificity.

For natural antibodies, polyreactivity denotes the ability of antibodies to bind a variety of antigens and results from a more conformationally flexible antigen binding pocket [31]. It is now known that there are a large number of natural antibodies that can bind to a variety of totally unrelated self and foreign antigens (i.e., polyreactive antibodies). In other words, polyreactivity is a conserved feature of the immune system.

For therapeutic mAbs, the term "polyreactivity" is mostly used to refer to the suboptimal molecular characteristics that lead to poor PK, less-than-ideal potency, low bioavailability, or unacceptable immunogenicity liabilities. This polyreactivity, also sometimes referred to as mAb "stickiness," is mostly driven by poor physicochemical properties (e.g., excessive and imbalanced positive charge or hydrophobicity in the variable regions), leading to low-affinity extracellular matrix and/or membrane interactions [30]. This is an important developability issue and,

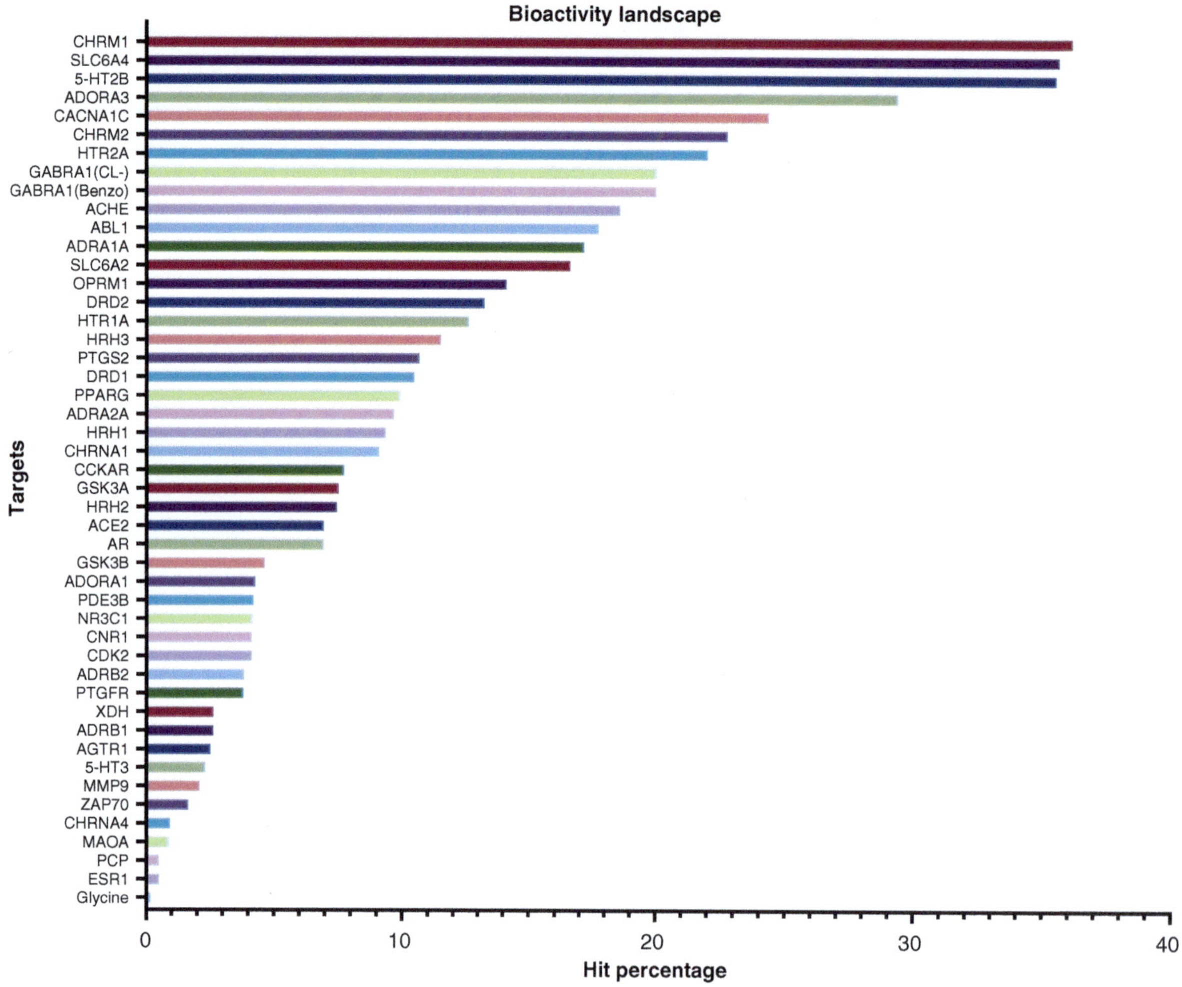

Figure 2.3 Target bioactivity landscape. Bioactivity profiling for 47 targets with at least 500 compounds screened ranked from most frequently hit. Data from https://www.ncbi.nlm.nih.gov/pmc/articles/PMC9077900/.

as such, has been the topic of extensive research efforts leading to substantial improvement in the molecular design of mAbs [30, 32].

Conversely, polyspecificity is the preferred term to denote the off-target binding of therapeutic mAbs at meaningful affinity levels to unintended host proteins. The contribution of polyspecificity to safety-related attrition and development success is not fully understood, because it is not until recently that rapid, reasonably priced, and robust screening selectivity panels have become available to adequately assess the off-target interactions of mAbs. Currently, two primary methods have been developed and are routinely used to identify polyspecificity against the human membrane proteome: array-type platforms using protein expressed on human cell membranes (e.g., Retrogenix platform) and flow-cytometry-based methods using cell- or virus-like particles transfected to express human membrane proteins [30]. Like for small molecules, binding hits in these screens need to be confirmed in an orthogonal assay, and the potential risk of the confirmed off-target interaction needs to be assessed.

2.5 Case Study Examples

2.5.1 COX-2 Inhibition

Cyclooxygenase 2 (COX-2) is an enzyme routinely included in secondary pharmacology profiling panels. This is mainly based on the cardiovascular safety liability (myocardial infarction) in humans associated with few approved COX-2 inhibitors particularly with prolonged and high-dose usage [33, 34]. Even though not all COX-2 selective inhibitors are associated with adverse cardiovascular effects, off-target interaction with this target is considered undesirable for the progression of new molecules in early development. There are multiple potential mechanisms that have been proposed for COX-2-inhibition-mediated cardiovascular risk, including effects on vasal tone and blood pressure, increased platelet aggregation, and activation of vascular endothelial cells [35, 36].

To further confirm and understand the potential implications of off-target COX-2 inhibition at the early preclinical development stage, a recent study tested four different approved/withdrawn COX-2 inhibitors (rofecoxib, celecoxib, etodolac, and meloxicam) with different ranges of COX-2 selectivity. The study consisted of multiple *in vitro* and *in vivo* preclinical models including endpoints covering all potential mechanisms associated with an increased risk of adverse cardiovascular findings in humans [37]. Results of this study suggested that all COX-2 inhibitors tested have no direct effects on platelet function (*in vitro* platelet aggregation assay), endothelial activation (*in vitro* assay for activation markers), or *in vivo* cardiovascular parameters (dog cardiovascular study evaluating mean arterial pressure, heart rate, and left ventricular contractility). Only rofecoxib showed a constriction response in isolated intact femoral artery endothelial cell tissue bath experiments when vessels were not denuded, suggesting a role for an endothelial-cell-mediated response. Overall, these results highlighted a disconnect between COX-2 selective inhibition and cardiovascular risk in preclinical models. This work also suggested that adverse cardiovascular effects related to COX-2 inhibitors are not correlated with their selectivity or potency for COX-2 inhibition and are also not predicted by available preclinical models used routinely to predict cardiovascular risk for new compounds in drug discovery and development. Considering this data and other factors which likely influence heterogeneity in the human response to COX-2 inhibitors and the increased COX-2 expression in inflammatory and infectious conditions [38, 39], it is reasonable to speculate that inhibition of COX-2 is not the only factor contributing to the increased incidence of cardiovascular adverse findings reported for COX-2 inhibitor drugs.

Altogether, this illustrates the importance of regularly leveraging new data to revisit the composition of selectivity screening panels and tiering categories based on confidence in the level of concern and, when appropriate, to use orthogonal confirmatory assays for decision-making in early discovery and development.

2.5.2 Acetylcholinesterase/Neuropathy Target Esterase

Acetylcholinesterase (ACHase) is an enzyme commonly included in off-target screening panels [12]. This is due largely to the reported acute effects of ACHase inhibition on the cardiovascular and respiratory systems [40]. ACHase inhibition results in the well-known cholinergic syndrome, from the increase in acetylcholine (due to a lack of degradation by ACHase) and the overstimulation of muscarinic receptors [40]. Symptomatically, this manifests as bradycardia, wheezing, diaphoresis, miosis, diarrhea, and salivation. Fortunately, this syndrome is detectable in preclinical models; however, earlier detection can avoid the progression of compounds in *in vivo* studies, and these human enzyme-based screens can confirm the likelihood of preclinical to clinical translation.

Chronic asymptomatic inhibition of ACHase used to be associated with organophosphate-induced delayed polyneuropathy (OPIDP), as well as chronic organophosphate-induced neuropsychiatric disorder (COPIND). OPIDN is characterized histologically by distal degeneration of long axons in the spinal cord and peripheral nerves. COPIND is likely related to CNS, rather than to peripheral nervous system (PNS) effects. Unfortunately, these adverse conditions are difficult to detect in standard shorter-term preclinical safety studies conducted prior to clinical trials, can be significant clinically, and are typically not reversible.

More recently, these neuropathies have been determined to be due to the inhibition of a second esterase, neuropathy target esterase (NTE). NTE inhibitors often also inhibit ACHase, which is why ACHase was originally thought to be the initiating molecular event associated with OPIDN. The highest concentrations of NTE are in the endoplasmic reticulum (ER) membranes from neurons and lymphocytes. In adults, NTE is expressed throughout the brain, particularly in the cortex, in the Purkinje cells of the cerebellum, and in the hippocampus. Because of this, NTE tests are now recommended by both the Environmental Protection Agency (EPA) and the Organization for Economic Cooperation and Development (OECD). No such guidance has been issued to date by the FDA. Interestingly, not all NTE inhibitors cause OPIDP/COPIND. Organophosphates that "age" (i.e., irreversibly bind to the NTE active site) are associated with these syndromes, but some carbamate insecticides (which do not "age" on the enzyme) have also been associated with OPIDP. Off-target screening of ACHase inhibitors, with sufficient margins to support progression, may therefore benefit from NTE inhibition screening to rule out the potential for chronic effects.

2.5.3 VEGF Inhibition

Vascular endothelial growth factor (VEGF) inhibitors, as a consequence of their mechanism of action, can produce many unwanted cardiovascular and renal effects, which often hinder their intended beneficial impact and length of use. Aimed as an effective treatment to deter tumor angiogenesis, this class of drugs (and its downstream counterparts) has been shown to affect blood pressure, coronary vasculature, cardiomyocytes, cardiac function, induction of arrhythmias such as bradycardia and atrial fibrillation, thrombosis, and the renal system [41–43].

Of these effects, hypertension is particularly well known and documented. Studies suggest that 30–80% of patients taking VEGF inhibitors will experience a rapid dose-dependent increase in blood pressure and reach a state of hypertension (systolic > 140 mm Hg or diastolic > 90 mm Hg). Due to exclusion criteria, the real-world incidence may even be much higher. For many of these patients/subjects, increases in blood pressure often reached grade 3 hypertension (systolic > 160 mm Hg or diastolic >100 mm Hg) or higher [42, 44–46]. For the most part, these increases are reversible upon discontinuation and are thought to be caused by a multitude of changes in vascular tone, ranging from decreased nitric oxide production along with increases in endothelin, to potential loss of microvasculature, changes in vascular stiffness, increases in pulmonary vascular resistance, and decreased cardiac output [41, 46]. Pre-existing conditions, age, and potential latent effects from prior kinase inhibitor use may also contribute to causality and magnitude, but nonetheless, changes in blood pressure related to the use of VEGF inhibitors need to be routinely and cautiously monitored [47].

Hypertension, along with potential cardiovascular and renal liabilities, is considered a class effect with VEGF inhibitors. Thus, during drug development, VEGF inhibition may become an important consideration when identified in a selectivity screen, as unintended, off-target inhibition of VEGF may lead to safety liabilities.

2.6 Physicochemical Properties

There is strong evidence that the physicochemical properties of small molecules impact their probability of success. In particular, drug candidates in the past two decades have tended to have higher molecular weight (MW) and be more lipophilic, a trend coined by Hann as "molecular obesity" [48]. These evolving physicochemical properties have been associated with several challenges, such as difficulty to formulate, lower intestinal absorption, promiscuity, and toxicity [49]. For example, numerous novel small molecules are not compliant with Lipinski's "Rule of 5" (i.e., beyond rule of 5/BRO5 chemical space), indicating that achieving acceptable oral bioavailability will be challenging and that new principles are warranted to successfully optimize BRO5 molecules [50]. In addition, beyond their impact on oral bioavailability, several physicochemical attributes have been correlated with a higher probability of being linked to safety issues. Specifically, the Pfizer 3/75 rule was derived from a large retrospective evaluation that showed that the total polar surface area (TPSA, <75 $Å^2$) and calculated logP (ClogP > 3) are strongly correlated with *in vivo* preclinical toxicity [51]. This increased propensity to induce toxicity *in vivo* could be linked to a higher level of promiscuity in a 48-assay radiometric-based biochemical assay against various receptors, ion channels, and transporters panel, as defined by over 50% binding at 10 μM in three or more of the 48 assays. An analysis similar to that of Pfizer's was conducted at AbbVie, leading to comparable conclusions [9]. This data suggests that nonpolar, lipophilic compounds are associated with more adverse outcomes due to off-target effects, an observation consistent with that of others [52, 53]. This is not surprising, as most protein binding sites are hydrophobic and hence lipophilic interactions are major contributors to ligand affinity. Therefore, highly lipophilic molecules tend to be more promiscuous [54].

Data on the influence of MW on promiscuity is conflicting. On the one hand, data from Pfizer's high-throughput screen demonstrated an inverse relationship between MW and promiscuity [55]. On the other hand, safety screening data from Novartis suggest that high-MW molecules are more promiscuous [52]. High MW is usually associated with a higher degree of molecular complexity (i.e., larger molecules are usually more complex), which may result in more specific pharmacophoric features and higher selectivity. Conversely, high MW is also associated with higher lipophilicity and consequently potentially higher promiscuity. Hence, given the contradictory evidence, it is best to assume that MW has limited influence on promiscuity [54].

The ionic state of molecules is another important parameter to consider for promiscuity. In general, molecules that are ionized at physiological pH are less promiscuous due to lower hydrogen bonding and other polar interactions [54]. Bases are in general more promiscuous than acids, neutral compounds, zwitterions, and uncharged bases [56]. This higher promiscuity of basic molecules is considered related to the positive charge at the basic center often observed with multiple pharmacophores (e.g., aminergic and non-aminergic GPCRs, ion channels). This partly explains why in selectivity panels aminergic GPCRs like 5-HT2b are frequent hits [54, 57].

It is important to bear in mind that other molecular properties (e.g., structural features, pKa) and parameters (e.g., potency at anti-targets, tissue distribution, and PK) also need to be considered when designing molecules for intended targets that will have sufficient selectivity. Hence, another report from four companies (Astra-Zeneca, Eli Lilly, Glaxo-Smith-Kline, and Pfizer) that partnered to evaluate a dataset of oral development-stage molecules (i.e., molecules already selected based on stringent criteria for toxicity properties, including *in vivo* data) found that ClogP and TPSA did not influence *in vivo* toxicological outcomes, suggesting that high-MW lipophilic compounds can retain sufficient selectivity to become safe drugs [58]. Interestingly, candidates failing for safety reasons in phase I were significantly more lipophilic than those advancing to phase II (mean c Log P = 3.1) [56].

2.7 *In Silico* Methods to Predict Off-Target Interactions

In addition to *in vitro* screening panels, computational approaches have been proposed to identify potential off-target interactions and/or associated safety outcomes. Many of these methods or tools are summarized in Table 2.1.

The majority of computational approaches to predict target interactions and outcomes rely on utilizing molecular descriptors and/or fingerprints that characterize a compound's molecular structure [73]. These molecular descriptors contain data regarding both physical and chemical attributes of a molecule. These features can then be used to apply quantitative structure–activity relationship (QSAR) and chemical similarity methodologies to cluster similar compounds and identify features that may be associated with a specific target interaction [13, 74]. However, these methods are often limited and may yield poor predictivity for prospective compounds due to differences in chemical space and potential activity cliffs (i.e., lack of data and differences in activity despite small changes in chemical structure) [75]. Beyond this, other approaches have also been explored that utilize protein-structure-based approaches or genomic reference datasets. These approaches have been shown to work but are also limited due to the lack of readily available protein structures and/or drug-treated omics datasets.

More recently, artificial intelligence (AI) and machine-learning (ML) methodology have further enhanced computational methods to better leverage complex molecular descriptors to profile drug interactions and polypharmacology. The off-target safety assessment workflow as described by Rao et al. utilizes an ensemble method incorporating 2D and 3D structural prediction models to predict potential interactions across 7000 targets. The method was applied and validated against 15 AbbVie internal compounds with known off-target interactions supported by *in vitro* experimental data and was found to capture 56.8% of confirmed off-target interactions in addition to primary pharmacology [13]. Naga et al. recently developed an open-source ML framework, Off-TargetP, to predict interactions for Roche's internal 50 off-target panels spanning across approximately 4000 compounds [22]. A comparative analysis was performed between five methods including deep learning and automated ML (AutoML) approaches. Methods involving neural networks often yielded the best performance but were significantly more time-consuming relative to AutoML methods due to required subject-matter expertise in both biology and data science to fine-tune hyperparameters. Most of the neural network models for each target achieved a favorable outcome where the balanced accuracy exceeded 70% for 28 targets, ranged between 60% and 70% for 12 targets and dropped below 60% for 7 targets [22]. Target space and model accuracy is currently limited by interaction data but will continue to improve as these assays become higher-throughput and data become more readily available.

2.8 Predicting Antibody Specificity

In contrast to small molecules, there are still no consistent approaches in the industry on how (and when in the development cascade) to screen for mAb specificity. As already mentioned, there is a paucity of data to fully gauge the impact that polyspecificity has on the probability of technical success of mAbs. However, the lack of mAb specificity can result in unexpected issues, most notably accelerated clearance (and hence poor PK and short half-life) and safety-related events. Hence, recent reports suggest that screening for polyspecificity before development activities (and the associated substantial investments) initiate can be beneficial. This is illustrated by the following examples of safety issues uncovered in the clinic or during nonclinical toxicity studies.

Blockers of the program cell death protein 1 (PD-1) have become a central component of many oncology treatments given their long-term survival benefits. Several anti-PD-1 mAbs have been advanced to the clinic and even approved as therapeutics, and for the most part, their safety profile has been consistent clinically [76]. Surprisingly, one anti-PD-1, camrelizumab, induced capillary hemangiomas in patients [77]. This is in stark contrast to the numerous other anti-PD-1 mAbs evaluated in clinical trials, suggesting a specificity issue. Further investigations using array-based human proteome screens (Retrogenix platform) showed that this was due to an unexpected binding to and agonism of the pro-angiogenic receptor VEFGR2 [78]. Likewise, several cases of thrombocytopenia in cynomolgus monkeys following mAb dosing have been reported in the last few years. Some of these cases have been further investigated and shown to be driven by

Table 2.1 In silico tools and references to perform off-target predictions and/or assess safety.

Name	Description	Reference
AutoDock	A suite of automated docking tools used to predict protein–ligand docking based on protein-structure-based methods	[59]
BERT	Text classifier to identify and extract drug–target interactions from PubMed	[60]
ChemProt3.0	QSAR-based target prediction for 850 proteins	[61]
CSNAP (Chemical Similarity Network Analysis Pull-down)	Compound target identification based on network similarity graphs	[62]
DINES (drug–target interaction network inference engine based on supervised analysis)	Predicted drug–target interactions from drug profiles (e.g., chemical structures, drug side effects, amino acid sequences, and protein domains)	[63]
FLAME	Open-source software for integrating bioactivity data and building target and safety prediction models	[64]
LINCS	MOA prediction by querying a database of reference compounds with large-scale genomic datasets	[65]
Off-Target Safety Assessment (OTSA)	Predicts off-target interactions for over 4000 proteins using an ensemble method that integrates chemical similarity and machine-learning methods	[13]
Off-TargetP	Workflows (including AutoML methods) for predicting Roche's off-target panel consisting of 50 targets	[22]
SABPred	Tool to predict developability and various potential outcomes for antibodies	[66]
SuperPred 3.0	Target prediction using logistic regression machine-learning models across 646 targets	[67]
SwissTargetPrediction	Utilizes 2D and 3D similarity to predict 3068 target interactions across mice, rats, and humans	[68]
TargetHunter	Prediction of targets by querying target profiles using the Targets Associated with its Most Similar Counterparts (TAMOSIC) algorithm	[69]
Therapeutic Antibody Profiler (TAP)	Web application that utilizes antibody input sequences to assess the risk of developability	[70]
Therapeutic Antibody Developability Analysis (TA-DA)	Method to assess the risk of developability by profiling antibody descriptors	[71]
TransMHC-II	MHC-II binding prediction model built using both protein language and image classifier models to evaluate immunogenicity	[72]

off-target binding. For example, an anti-epidermal growth factor receptor (EGFR) mAb was shown to cause acute thrombocytopenia in cynomolgus monkeys, but not in rats, and to bind and activate cynomolgus monkey platelets *in vitro*. This was shown to be related to binding to the integrin αIIbβ3, an important platelet surface receptor [79]. Similarly, Loberg et al. reported a severe dose-dependent toxicity in cynomolgus monkeys (characterized by severe clinical signs; thrombotic changes in the heart, lungs, and brain; acute and chronic thrombocytopenia) following dosing with ABT-736, an anti-amyloid β mAb developed for the treatment of Alzheimer's disease [80]. Additional investigations demonstrated that ABT-736 binds with high affinity to monkey and human plasma protein platelet factor 4 (PF-4), a cytokine present in platelet granules and promoting coagulation. This off-target binding resulted in a clinical presentation similar to a syndrome known as heparin-induced thrombocytopenia, where anti-heparin autoantibodies bind PF-4, resulting in thrombocytopenia and increased thrombotic events [80].

Unfortunately, the majority of computational tools do not yet support off-target prediction for larger molecules such as biologics and degradomers. In recent years, these types of molecules are becoming increasingly important, and the need to assess bioactivity of these compounds will be essential. Currently, the most popular method is utilizing Retrogenix Cell Microarray Technology to detect the interactions of biologics for over 6200 human plasma membrane and secreted proteins [81]. Building computational models for predicting large molecule interactions remains a significant challenge, with key hurdles being the novelty of this drug space, limited availability of binding data, and high cost of generating such datasets. Despite these challenges, significant progress has been made in developing computational methods that leverage protein informatics to predict post-translational modifications, developability, and immunogenicity. Prediction of N-linked glycosylation, asparagine deamidation, and methionine oxidation risks can play a crucial role in identifying potential liabilities that can affect protein structure, antigen binding, aggregation, and *in vivo* half-life, ultimately leading to improved drug development strategies for antibodies. [82–84]. Tools such as the Therapeutic Antibody Profiler (TAP) and Therapeutic Antibody Developability Analysis (TA-DA) can be used to assess antibody developability that may be associated to liabilities such as polyspecificity and immunogenicity. These tools largely use antibody structural properties (e.g., hydrophobic patches) based on the entire structure or specific CDR regions for model development [70, 71]. Most recently, new methods such as TransMHCII are utilizing the emergence of AlphaFold2 and protein language models to evaluate immunogenicity risk assessment by predicting antibody major histocompatibility complex class II (MHC-II) binding affinity of peptides [72].

Altogether, current evidence indicates that in contrast to initial beliefs, therapeutic mAbs can be associated with relevant off-target interactions. Consequently, screening to assess their specificity at early stages is warranted using both experimental and computational approaches.

References

1 Metz, J.T. and Hajduk, P.J. (2010). Rational approaches to targeted polypharmacology: creating and navigating protein-ligand interaction networks. *Curr. Opin. Chem. Biol.* 14: 498–504.

2 Peon, A., Naulaerts, S., and Ballester, P.J. (2017). Predicting the reliability of drug-target interaction predictions with maximum coverage of target space. *Sci. Rep.* 7: 3820.

3 Van Vleet, T.R., Liguori, M.J., Lynch, J.J. 3rd, et al. (2019). Screening strategies and methods for better off-target liability prediction and identification of small-molecule pharmaceuticals. *SLAS Discov.* 24: 1–24.

4 Eaton, D.L. and Klaassen, C.D. (2001). Principles of toxicology. In: *Casarett and Doull's toxicology: The Basic Science of Poisons*, (ed. Curtis D. Klaassen) 11–34. New York: McGraw-Hill.

5 Janga, S.C. and Tzakos, A. (2009). Structure and organization of drug-target networks: insights from genomic approaches for drug discovery. *Mol. BioSyst.* 5: 1536–1548.

6 Lin, A., Giuliano, C.J., Palladino, A. et al. (2019). Off-target toxicity is a common mechanism of action of cancer drugs undergoing clinical trials. *Sci. Transl. Med.* 11: eaaw8412.

7 Ralston, S. (2017). Pre-development attrition of pharmaceuticals: how to identify the bad actors early. *Toxicologist* 156: 312.

8 Krause, C., Suwada, K., Blomme, E.A.G. et al. (2022). Preclinical species gene expression database: development and meta-analysis. *Front. Genet.* 13: 1078050.

9 Blomme, E.A. and Will, Y. (2016). Toxicology strategies for drug discovery: present and future. *Chem. Res. Toxicol.* 29: 473–504.

10 Bowes, J., Brown, A.J., Hamon, J. et al. (2012). Reducing safety-related drug attrition: the use of in vitro pharmacological profiling. *Nat. Rev. Drug Discov.* 11: 909–922.

11 Green, J.R., Mahalingaiah, P.K.S., Gopalakrishnan, S.M. et al. (2023). Off-target pharmacological activity at various kinases: potential functional and pathological side effects. *J. Pharmacol. Toxicol. Methods* 123: 107468.

12 Lynch, J.J. 3rd, Van Vleet, T.R., Mittelstadt, S.W., and Blomme, E.A.G. (2017). Potential functional and pathological side effects related to off-target pharmacological activity. *J. Pharmacol. Toxicol. Methods* 87: 108–126.

13 Rao, M.S., Gupta, R., Liguori, M.J. et al. (2019). Novel computational approach to predict off-target interactions for small molecules. *Front. Big Data* 2: 25.

14 Goedken, E.R., Argiriadi, M.A., Banach, D.L. et al. (2015). Tricyclic covalent inhibitors selectively target Jak3 through an active site thiol. *J. Biol. Chem.* 290: 4573–4589.

15 Wang, Y. and Ma, H. (2015). Protein kinase profiling assays: a technology review. *Drug Discov. Today Technol.* 18: 1–8.

16 Bibevski, S., Zhou, Y., McIntosh, J.M. et al. (2000). Functional nicotinic acetylcholine receptors that mediate ganglionic transmission in cardiac parasympathetic neurons. *J. Neurosci.* 20: 5076–5082.

17 Itoh, Y., Suzuki, M., Matsui, T. et al. (2016). False HDAC inhibition by aurone compound. *Chem. Pharm. Bull.* 64: 1124–1128.

18 Gintant, G., Sager, P.T., and Stockbridge, N. (2016). Evolution of strategies to improve preclinical cardiac safety testing. *Nat. Rev. Drug Discov.* 15: 457–471.

19 Mandlekar, S.V., Rose, A.V., Cornelius, G. et al. (2007). Development of an in vivo rat screen model to predict pharmacokinetic interactions of CYP3A4 substrates. *Xenobiotica* 37: 923–942.

20 Gintant, G. (2011). An evaluation of hERG current assay performance: translating preclinical safety studies to clinical QT prolongation. *Pharmacol. Ther.* 129: 109–119.

21 Wager, T.T., Liras, J.L., Mente, S., and Trapa, P. (2012). Strategies to minimize CNS toxicity: in vitro high-throughput assays and computational modeling. *Expert Opin. Drug Metab. Toxicol.* 8: 531–542.

22 Naga, D., Muster, W., Musvasva, E., and Ecker, G.F. (2022). Off-targetP ML: an open source machine learning framework for off-target panel safety assessment of small molecules. *J. Cheminform.* 14: 27.

23 Chan, R. and Benet, L.Z. (2018). Evaluation of the relevance of DILI predictive hypotheses in early drug development: review of in vitro methodologies vs. BDDCS classification. *Toxicol. Res.* 7: 358–370.

24 Wills, L.P. (2017). The use of high-throughput screening techniques to evaluate mitochondrial toxicity. *Toxicology* 391: 34–41.

25 Low Wang, C.C., Galinkin, J.L., and Hiatt, W.R. (2015). Toxicity of a novel therapeutic agent targeting mitochondrial complex I. *Clin. Pharmacol. Ther.* 98: 551–559.

26 Dykens, J.A. and Will, Y. (2007). The significance of mitochondrial toxicity testing in drug development. *Drug Discov. Today* 12: 777–785.

27 Gu, Y.G., Weitzberg, M., Clark, R.F. et al. (2007). N-3-[2-(4-alkoxyphenoxy)thiazol-5-yl]-1-methylprop-2-ynylcarboxy derivatives as acetyl-coA carboxylase inhibitors-improvement of cardiovascular and neurological liabilities via structural modifications. *J. Med. Chem.* 50: 1078–1082.

28 Disney, M.D. (2019). Targeting RNA with small molecules to capture opportunities at the intersection of chemistry, biology, and medicine. *J. Am. Chem. Soc.* 141: 6776–6790.

29 Fang, L., Velema, W.A., Lee, Y. et al. (2023). Pervasive transcriptome interactions of protein-targeted drugs. *Nat. Chem.* 15: 1374–1383.

30 Cunningham, O., Scott, M., Zhou, Z.S., and Finlay, W.J.J. (2021). Polyreactivity and polyspecificity in therapeutic antibody development: risk factors for failure in preclinical and clinical development campaigns. *MAbs* 13: 1999195.

31 Notkins, A.L. (2004). Polyreactivity of antibody molecules. *Trends Immunol.* 25: 174–179.

32 Kohli, N., Jain, N., Geddie, M.L. et al. (2015). A novel screening method to assess developability of antibody-like molecules. *MAbs* 7: 752–758.

33 Shi, S. and Klotz, U. (2008). Clinical use and pharmacological properties of selective COX-2 inhibitors. *Eur. J. Clin. Pharmacol.* 64: 233–252.

34 Topol, E.J. (2004). Failing the public health--rofecoxib, Merck, and the FDA. *N. Engl. J. Med.* 351: 1707–1709.

35 Hamberg, M., Svensson, J., and Samuelsson, B. (1975). Thromboxanes: a new group of biologically active compounds derived from prostaglandin endoperoxides. *Proc. Natl. Acad. Sci. USA* 72: 2994–2998.

36 McAdam, B.F., Catella-Lawson, F., Mardini, I.A. et al. (1999). Systemic biosynthesis of prostacyclin by cyclooxygenase (COX)-2: the human pharmacology of a selective inhibitor of COX-2. *Proc. Natl. Acad. Sci. USA* 96: 272–277.

37 Koshman, Y.E., Bielinski, A.L., Bird, B.M. et al. (2023). Disconnect between COX-2 selective inhibition and cardiovascular risk in preclinical models. *J. Pharmacol. Toxicol. Methods* 120: 107251.

38 Akarasereenont, P., Techatrisak, K., Chotewuttakorn, S., and Thaworn, A. (1999). The induction of cyclooxygenase-2 in IL-1beta-treated endothelial cells is inhibited by prostaglandin E2 through cAMP. *Mediat. Inflamm.* 8: 287–294.

39 Syeda, F., Grosjean, J., Houliston, R.A. et al. (2006). Cyclooxygenase-2 induction and prostacyclin release by protease-activated receptors in endothelial cells require cooperation between mitogen-activated protein kinase and NF-kappaB pathways. *J. Biol. Chem.* 281: 11792–11804.

40 Ecobichon, D.J. (2001). Toxic effects of pesticides. In: *Casarett and Doull's Toxicology. The Basic Science of Poisons*, (ed. Curtis D. Klaassen) 763–810. New York: McGraw-Hill.

41 Dobbin, S.J.H., Petrie, M.C., Myles, R.C. et al. (2021). Cardiotoxic effects of angiogenesis inhibitors. *Clin. Sci.* 135: 71–100.

42 Pandey, A.K., Singhi, E.K., Arroyo, J.P. et al. (2018). Mechanisms of VEGF (vascular endothelial growth factor) inhibitor-associated hypertension and vascular disease. *Hypertension* 71: e1–e8.

43 Yang, Y.; Kowalkowski, K.; Ciurlionis, R.; Buck, W. R.; Glaser, K. B.; Albert, D. H. and Blomme, E. A. G. (2021), Identification of VEGF signaling inhibition-induced glomerular injury in rats through site-specific urinary biomarkers. *Int. J. Mol. Sci.* 22:12629.

44 Agarwal, M., Thareja, N., Benjamin, M. et al. (2018). Tyrosine kinase inhibitor-induced hypertension. *Curr. Oncol. Rep.* 20: 65.

45 Grossman, E. and Messerli, F.H. (2012). Drug-induced hypertension: an unappreciated cause of secondary hypertension. *Am. J. Med.* 125: 14–22.

46 Maki-Petaja, K.M., McGeoch, A., Yang, L.L. et al. (2021). Mechanisms underlying vascular endothelial growth factor receptor inhibition-induced hypertension: the HYPAZ trial. *Hypertension* 77: 1591–1599.

47 Touyz, R.M., Herrmann, S.M.S., and Herrmann, J. (2018). Vascular toxicities with VEGF inhibitor therapies-focus on hypertension and arterial thrombotic events. *J. Am. Soc. Hypertens.* 12: 409–425.

48 Hann, M.M. (2011). Molecular obesity, potency and other addictions in drug discovery. *Royal Soc. Chem.* 2: 6.

49 Meanwell, N.A. (2011). Improving drug candidates by design: a focus on physicochemical properties as a means of improving compound disposition and safety. *Chem. Res. Toxicol.* 24: 1420–1456.

50 DeGoey, D.A., Chen, H.J., Cox, P.B., and Wendt, M.D. (2018). Beyond the rule of 5: lessons learned from AbbVie's drugs and compound collection. *J. Med. Chem.* 61: 2636–2651.

51 Hughes, J.D., Blagg, J., Price, D.A. et al. (2008). Physiochemical drug properties associated with in vivo toxicological outcomes. *Bioorg. Med. Chem. Lett.* 18: 4872–4875.

52 Azzaoui, K., Hamon, J., Faller, B. et al. (2007). Modeling promiscuity based on in vitro safety pharmacology profiling data. *ChemMedChem* 2: 874–880.

53 Peters, J.U., Schnider, P., Mattei, P., and Kansy, M. (2009). Pharmacological promiscuity: dependence on compound properties and target specificity in a set of recent Roche compounds. *ChemMedChem* 4: 680–686.

54 Kabir, A. and Muth, A. (2022). Polypharmacology: the science of multi-targeting molecules. *Pharmacol. Res.* 176: 106055.

55 Hopkins, A.L., Mason, J.S., and Overington, J.P. (2006). Can we rationally design promiscuous drugs? *Curr. Opin. Struct. Biol.* 16: 127–136.

56 Leeson, P.D. and Springthorpe, B. (2007). The influence of drug-like concepts on decision-making in medicinal chemistry. *Nat. Rev. Drug Discov.* 6: 881–890.

57 Whitebread, S., Hamon, J., Bojanic, D., and Urban, L. (2005). Keynote review: in vitro safety pharmacology profiling: an essential tool for successful drug development. *Drug Discov. Today* 10: 1421–1433.

58 Waring, M.J., Arrowsmith, J., Leach, A.R. et al. (2015). An analysis of the attrition of drug candidates from four major pharmaceutical companies. *Nat. Rev. Drug Discov.* 14: 475–486.

59 Goodsell, D.S., Sanner, M.F., Olson, A.J., and Forli, S. (2021). The AutoDock suite at 30. *Protein Sci.* 30: 31–43.

60 Aldahdooh, J., Vaha-Koskela, M., Tang, J., and Tanoli, Z. (2022). Using BERT to identify drug-target interactions from whole PubMed. *BMC Bioinformatics* 23: 245.

61 Kringelum, J., Kjaerulff, S.K., Brunak, S. et al. (2016). ChemProt-3.0: a global chemical biology diseases mapping. *Database* 2016: bav123.

62 Lo, Y.C., Senese, S., Li, C.M. et al. (2015). Large-scale chemical similarity networks for target profiling of compounds identified in cell-based chemical screens. *PLoS Comput. Biol.* 11: e1004153.

63 Yamanishi, Y., Kotera, M., Moriya, Y. et al. (2014). DINIES: drug-target interaction network inference engine based on supervised analysis. *Nucleic Acids Res.* 42: W39–W45.

64 Pastor, M., Gomez-Tamayo, J.C., and Sanz, F. (2021). Flame: an open source framework for model development, hosting, and usage in production environments. *J. Cheminform.* 13: 31.

65 Iwata, M. and Yamanishi, Y. (2019). The use of large-scale chemically-induced transcriptome data acquired from LINCS to study small molecules. *Methods Mol. Biol.* 1888: 189–203.

66 Dunbar, J., Krawczyk, K., Leem, J. et al. (2016). SAbPred: a structure-based antibody prediction server. *Nucleic Acids Res.* 44: W474–W478.

67 Gallo, K., Goede, A., Preissner, R., and Gohlke, B.O. (2022). SuperPred 3.0: drug classification and target prediction-a machine learning approach. *Nucleic Acids Res.* 50: W726–W731.

68 Daina, A., Michielin, O., and Zoete, V. (2019). SwissTargetPrediction: updated data and new features for efficient prediction of protein targets of small molecules. *Nucleic Acids Res.* 47: W357–W364.

69 Wang, L., Ma, C., Wipf, P. et al. (2013). TargetHunter: an in silico target identification tool for predicting therapeutic potential of small organic molecules based on chemogenomic database. *AAPS J.* 15: 395–406.

70 Raybould, M.I.J. and Deane, C.M. (2022). The therapeutic antibody profiler for computational developability assessment. *Methods Mol. Biol.* 2313: 115–125.

71 Negron, C., Fang, J., McPherson, M.J. et al. (2022). Separating clinical antibodies from repertoire antibodies, a path to in silico developability assessment. *MAbs* 14: 2080628.

72 Yu, X., Negron, C., Huang, L., and Veldman, G. (2023). TransMHCII: a novel MHC-II binding prediction model built using a protein language model and an image classifier. *Antib. Ther.* 6: 137–146.

73 Grisoni, F., Consonni, V., and Todeschini, R. (2018). Impact of molecular descriptors on computational models. *Methods Mol. Biol.* 1825: 171–209.

74 Concu, R., Cordeiro, M., Perez-Perez, M., and Fdez-Riverola, F. (2023). MOZART, a QSAR multi-target web-based tool to predict multiple drug-enzyme interactions. *Molecules* 28: 1182.

75 Dablander, M., Hanser, T., Lambiotte, R., and Morris, G.M. (2023). Exploring QSAR models for activity-cliff prediction. *J. Cheminform.* 15: 47.

76 Alturki, N.A. (2023). Review of the immune checkpoint inhibitors in the context of cancer treatment. *J. Clin. Med.* 12: 4301.

77 Teng, Y., Guo, R., Sun, J. et al. (2019). Reactive capillary hemangiomas induced by camrelizumab (SHR-1210), an anti-PD-1 agent. *Acta Oncol.* 58: 388–389.

78 Finlay, W.J.J., Coleman, J.E., Edwards, J.S., and Johnson, K.S. (2019). Anti-PD1 'SHR-1210' aberrantly targets pro-angiogenic receptors and this polyspecificity can be ablated by paratope refinement. *MAbs* 11: 26–44.

79 Zhang, Y., Sun, J., Tan, M. et al. (2018). Species-specific involvement of integrin alphaIIbbeta3 in a monoclonal antibody CH12 triggers off-target thrombocytopenia in cynomolgus monkeys. *Mol. Ther.* 26: 1457–1470.

80 Loberg, L.I., Chhaya, M., Ibraghimov, A. et al. (2021). Off-target binding of an anti-amyloid beta monoclonal antibody to platelet factor 4 causes acute and chronic toxicity in cynomolgus monkeys. *MAbs* 13: 1887628.

81 Freeth, J. and Soden, J. (2020). New advances in cell microarray technology to expand applications in target deconvolution and off-target screening. *SLAS Discov.* 25: 223–230.

82 Pakhrin, S.C., Aoki-Kinoshita, K.F., Caragea, D., and Kc, D.B. (2021). DeepNGlyPred: a deep neural network-based approach for human N-linked glycosylation site prediction. *Molecules* 26: 7314.

83 Tavella, D., Ouellette, D.R., Garofalo, R. et al. (2022). A novel method for in silico assessment of Methionine oxidation risk in monoclonal antibodies: Improvement over the 2-shell model. *PLoS ONE* 17: e0279689.

84 Yan, Q., Huang, M., Lewis, M.J., and Hu, P. (2018). Structure based prediction of asparagine deamidation propensity in monoclonal antibodies. *MAbs* 10: 901–912.

3

Off-target Activity and Adverse Drug Reactions

A Practical Perspective

Dimitar Yonchev

3.1 Personal Perspective

Pharmacology was always one of the most interesting subjects during my undergraduate studies in pharmacy. My colleagues and I were taught in detail pathophysiology and the mechanisms of action of drugs alongside long lists of their side effects. While great care was taken to elaborate on drugs' pharmacodynamic (PD) effects, mechanistic understanding of similar interactions with undesirable targets potentially causing adverse drug reactions (ADRs) often remained elusive or undiscussed. Understandably, the primary focus of the learning was to recognize clinical manifestations of side effects in patients and prevent potential drug–drug interactions in everyday practice.

Later, during my work in a community pharmacy, I was often confronted by patients receiving their prescriptions with questions about ADRs and their causes. Despite my very recent textbook knowledge of pharmacology, I remember sometimes struggling to pinpoint the exact mechanistic reason for an ADR, at least not without additional research. A few years later, while writing my PhD thesis, I was surprised by the apparent lack of publicly available data on drug–target–ADR associations, which seemed to me like a gap in knowledge about an important healthcare topic. It was only until I began working in drug discovery R&D that I started grasping the true complexity of *in vitro–in vivo* translation. From a data science point of view, this challenge seemed to be further exacerbated by the limited quality and quantity of relevant data for systematic analysis. Thus, the opportunity to share knowledge and data with the wider scientific community made working on the topic even more exciting and interesting.

3.2 Introduction

In the evolving landscape of drug discovery and development, the competing objectives of maximizing therapeutic efficacy while minimizing ADRs remain paramount [1, 2]. In the context of target-based drug discovery, a drug's PD effect is achieved by specific desired interactions with one or more biological target(s), i.e., its primary ("on-target") activity [3]. In contrast, the concept of secondary ("off-target") activity refers to the ability of a drug to interact with biological targets other than its intended primary site(s) of action [4]. While, in general, promiscuous interactions can be leveraged for therapeutic benefit following the paradigm of polypharmacology [5] and drug repurposing [6], they are often the culprits behind ADRs, posing significant challenges to drug development and patient safety. Prominent examples of polypharmacology-driven drug repurposing include thalidomide for the treatment of leprosy and cancer (including its application as proteolysis-targeting chimeras (PROTACs) in oncology) [7], promiscuous multi-kinase inhibitors such as imatinib for treating different forms of cancer [8], and antipsychotic drugs such as clozapine for the treatment of schizophrenia [9].

However, the successful exploitation of off-target effects remains much more limited compared to the negative consequences of serious ADRs, which may lead to longer preclinical development, attrition in clinical trials, and market withdrawals. Important examples include the antihistaminic terfenadine and the 5-hydroxytryptamine 4 (5-HT4) receptor agonist cisapride, both causing a blockage of the human Ether-à-go-go-Related Gene (hERG) potassium channel, which leads to QT interval prolongation and cardiac arrhythmia known as torsades de pointes [10]. While the off-target–ADR relationships in these cases are mechanistically characterized, ADRs can also be the consequence of a drug's on-target activity, when the downstream effects are not well understood or effects in unintended tissues/organs are not accounted for. Such examples are the painkiller rofecoxib, the type 2 diabetes drug troglitazone, and the repurposed drug sildenafil. Rofecoxib acts through the selective inhibition of the cyclooxygenase 2

Polypharmacology: Strategies for Multi-Target Drug Discovery, First Edition. Edited by Jens-Uwe Peters.

(COX-2) enzyme, which can lead to an imbalance in the prostaglandin synthesis pathway and increase the risk of cardiovascular events [10]. Troglitazone's primary PD mechanism as a peroxisome proliferator-activated receptor γ (PPARγ) agonist can cause severe hepatotoxicity [10]. Sildenafil is a phosphodiesterase 5 (PDE5) enzyme inhibitor that was initially developed to treat angina pectoris and later repurposed for the treatment of erectile dysfunction, pulmonary arterial hypertension, and some forms of cancer [11, 12].

From a pharmacokinetic (PK) perspective, enzymes and transporters such as the cytochrome P450 (CYP450) enzyme and the P-glycoprotein (P-gp) efflux pump play an important role in drug metabolism and disposition, which may cause undesirable off-target effects of metabolites and drug–drug interactions [13, 14]. These and many other examples highlight the complexity in the origins of ADRs, and multiple approaches have been proposed to predict ADRs by utilizing genetics [15–17], genomics/proteomics [18–20], chemical structure [21–23], pharmacovigilance [24], and multimodal data [25]. Despite these efforts, most methods rarely provide practical translational insights and guidance, i.e., how can ADRs be avoided through preclinical compound optimization from *in vitro* and *in vivo* to humans. *In vitro* safety pharmacology, also referred to as "secondary pharmacology" as it deals with off-target drug effects, attempts to address these challenges and has become a regulatory requirement alongside *in vivo* safety assessments upon submission of investigational new drug applications to health authorities [26, 27].

With the increasing availability of biochemical and functional assays developed internally in pharmaceutical companies and offered by contract research organizations, compounds have been routinely screened in a high-throughput manner against off-target panels of various sizes and compositions with the goal of early de-risking preclinical and clinical development programs [28]. This has led to the accumulation of protein–ligand data collected over the years during hit-to-lead and lead optimization campaigns, which can be used to develop machine-learning models for quantitative structure–activity relationship (QSAR) predictions of compound off-target profiles [29–31]. Which off-targets constitute a screening assay panel is determined by combining statistical methods and the weight of evidence for target–ADR associations [32]. Despite significant overlap in the composition of *in vitro* off-target panels across different pharmaceutical companies, there are still considerable variations, highlighting the lack of full consensus on which off-target activities are related to (pre)clinical ADRs [10]. Since prospective prediction of the strength of such relationships is limited and mostly of qualitative nature, testing of already marketed drugs becomes necessary to illuminate off-target–ADR associations. Unfortunately, such data is rarely disclosed, and the already publicly available data often comes from heterogeneous sources, which limits large-scale quantitative assessment.

The general scope of this chapter is to provide a practical perspective on associations between drugs' *in vitro* target activity and ADRs. First, the challenges as well as necessary steps in data collection and processing for large-scale quantitative analysis will be emphasized. Next, the clinical relevance of *in vitro* target activities will be assessed, and target–ADR associations will be systematically evaluated using different statistical methods enabling the identification of the strongest relationships, including the most representative activity measures. Finally, the utility of this approach to uncover previously unknown target–ADR relationships and propose mechanisms for known ADRs will be demonstrated. The work is based on a recent publication, in which my co-authors and I have reported a systematic analysis of target–ADR relationships based on a large secondary pharmacology database of approximately 150,000 *in vitro* off-target activity measurements of marketed drugs, which was assembled at Novartis and released to the public [33].

3.3 Secondary Pharmacology and Adverse Drug Reactions

3.3.1 Important Considerations for Data Analysis

Postulating relationships between drug off-targets and ADRs represents a complex task because it requires establishing a statistically strong link between observations taking place at the opposite ends of the drug development process (from lead optimization up to post-market monitoring) and spanning over several translational steps. Hence, a systematic quantitative analysis of such relationships necessitates the collection of chemical substance data, *in vitro* assay data from compound profiling experiments, PK data on human drug exposure, and clinical as well as pharmacovigilance data on ADRs. Such data are is heterogeneous and, in each case, requires multiple harmonization steps.

For chemical substance data, unambiguous compound representations such asInternational Chemical Identifier (InChI) keys [34] were used to summarize information related to chirality, different salt forms, prodrugs, or metabolites of a compound and ensure the integration of data from various sources. In the case of *in vitro* assay data, normalized concentration–response curves were used to determine compound AC50 values, i.e., the concentration at which a drug exhibits 50% of its biological activity.

To enable comparison across different assay formats, "AC50" was used as a substitute for other measurement types such as IC50 or EC50, while the original modes of action were recorded. AC50 values across multiple assay runs (replicates) were summarized with a geometrical mean and censored data (e.g., "AC50 > 30 μM" measurements) were included only after careful consideration. Changing protocols for the same off-target assay may affect measurement results, and therefore, concordance analysis was carried out to determine which assay readouts can be pooled together. When integrating activity data from multiple sources such as ChEMBL [35], DrugCentral [36], or licensed databases, consistency among species, targets, assay modes of action, activity measure types, and units was ensured using identifiers from UniProt [37], Entrez [38], and BioAssay Ontology [39]. For the PK data, obtaining consistent human drug exposure measurements such as the maximal plasma concentration (Cmax) and plasma protein binding (PPB) is complicated by the heterogeneity in doses, patient populations, or drug metabolites reported in the different data sources. This was, at least partially, overcome by selecting a representative value for each drug, based on correlation analysis with already validated data. Hence, the final PK dataset was assembled by combining a high-quality manually curated dataset of drug Cmax values, values from the licensed Pharmapendium database (www.pharmapendium.com), and additional values reported in key reviews [40]. In the case of ADR data, a consistent vocabulary of medical terms from the Medical Dictionary for Regulatory Activities (MedDRAa) [41] was utilized, and only ADRs with a likelihood ratio > 5 were selected from the FDA Reporting System (FAERS) and the Side Effect Resource (SIDER) [42] databases. Finally, the different compound annotations (*in vitro* activity, *in vivo* exposure, and ADRs) were brought together to characterize target–ADR relationships. Such relationships can be derived only when a compound's pharmacologically active form exhibits an effect in an off-target assay, has sufficiently high human exposure, and is unambiguously associated with an ADR. In addition, associations can be postulated with various degrees of certainty using different methods for statistical significance. If one or more of the above conditions or predefined statistical thresholds are not met, target–ADR relationships are deemed unlikely or potentially not relevant. Importantly, conclusions in either direction can be drawn only when there are sufficient test frequency and data coverage, i.e., having enough active and inactive compounds in an assay as well as ADR-positive and ADR-negative drugs.

Following these methods, the secondary pharmacology database (SPD) was assembled, constituting the largest publicly available off-target activity resource at the time of its release in 2023. It contains measurements for 1958 marketed drugs and 200 safety pharmacology assays summarized into 168 distinct assay groups, which overall amounted to 121 097 unique (not replicated) drug–target interactions. The median test frequency for drugs was 66 (20–161), and each assay had a median of 794 drugs (30–1942) tested. Overall, 95% of all and 36% of the active (AC50 < 1 μM) readouts did not overlap with data from ChEMBL – the largest resource for curated medicinal chemistry literature data. For the subset of overlapping drug–target pairs, only moderate concordance of AC50 values was observed, owing to the heterogeneity of assay types and methods. Noticeably, AC50 differences between SPD and ChEMBL were much more pronounced for less potent compounds (>10 uM in SPD but <10 μM in ChEMBL) than for highly active compounds (<0.1 μM in both SPD and ChEMBL), confirming the publication bias toward more potent molecules.

3.3.2 Assessing the Clinical Relevance of Drug–Target Associations

Whether a drug's off-target activity results in an ADR depends on various factors, such as target expression and the drug's effective concentration at the target site [43]. While the former is governed by physiological aspects such as tissue type, genetic polymorphisms, epigenetic regulation, hormonal levels, and patient's age and sex, the exposure of a drug can be controlled by its dose and dosing interval. A commonly used parameter to assess the physiological relevance of a drug's *in vitro* measured target activity is the safety margin expressed by the ratio between AC50 and free plasma concentration at the highest approved therapeutic dose [2]. Accordingly, small safety margins manifest in physiologically relevant effects and mitigation of ADRs can be achieved by balancing between low on-target and high off-target margins [44]. Within the subset of drugs in the SPD, for which accurate PK data was available, margin values were categorized based on whether the drug's activity was measured against its reported on-target(s), against its published off-target(s), or against previously undescribed and potentially novel off-target(s). The value distributions are shown in Figure 3.1.

The on-target category yielded 429 interactions, of which 28% and 12% exceeded free Cmax margins of 10 and 50, respectively. To better understand the on-target margin distributions, values were grouped by the drugs' mode of action (MoA), which revealed the lowest median margin for the β2 adrenergic receptor ADRB2 (0.04; nine on-target drugs) and the highest for the dopamine transporter SLC6A3 (253.87; six on-target drugs). These findings highlight that not every drug may confer its therapeutic benefits

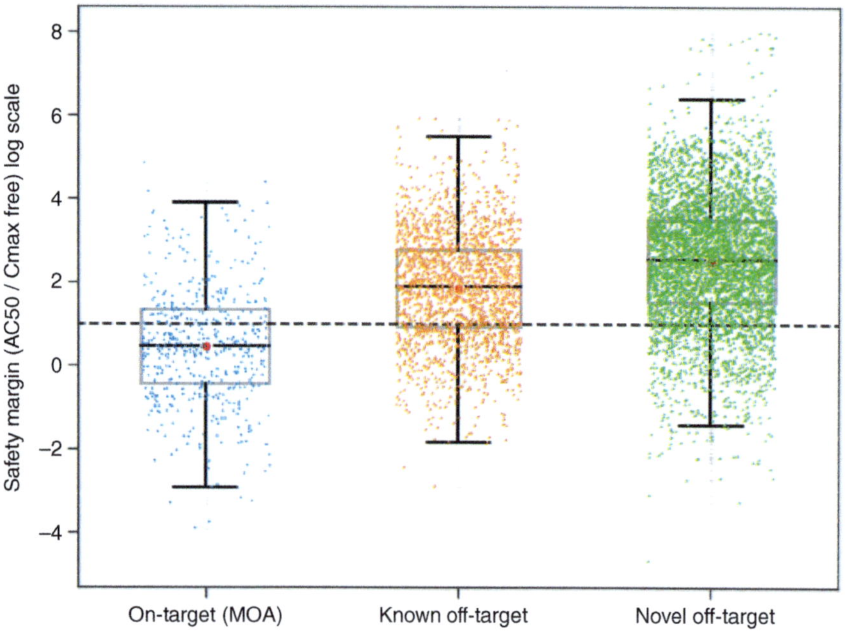

Figure 3.1 Safety margin distributions of marketed drugs. Shown are boxplots of the safety margins (AC50/free Cmax; log10 scale) for marketed drugs tested in the SPD. Each point corresponds to an individual drug's margin and is categorized according to the target it was measured for: on-target (MoA) activity (blue, left), known (literature-reported) off-target activity (orange, middle), or novel (unreported in the literature) off-target activity (green, right). The horizontal dashed line marks the safety margin of 10 used as a threshold to determine the physiological relevance of a drug–target activity.

via its reported primary mechanism(s), but instead its efficacy can be often driven by polypharmacological effects.

Investigation of the remaining interactions revealed 517 published and 684 previously unreported potentially physiological off-target activities (safety margin ≤10). Ranking drugs according to their promiscuity (percent assays with AC50 readouts <10 μM for any target) demonstrated an over-proportional increase in physiologically relevant off-target activities, especially for unpublished ones. For example, four physiologically relevant on-target and 23 such off-target activities (five previously unreported) were detected for the antidepressant nefazodone (35% promiscuity). Notable outliers from this trend were the kinase inhibitor sunitinib with only a single physiologically relevant margin (on-target) despite 51% target promiscuity and the antibiotic cefepime with 0% target promiscuity yet six physiologically relevant off-target safety margins. In both cases, the results are largely driven by the free Cmax of these drugs (low for sunitinib and high for cefepime). Further examples of selected novel off-target activities are summarized in Table 3.1 together with publications providing context to the findings.

3.3.3 Systematic Evaluation of Target–ADR Relationships

Drug activities measured in the SPD were further used to examine the strength of target–ADR associations reported in the literature. A set of 60 targets accounting for 1–34 MedDRA-coded ADRs from the SIDER and FAERS databases was correlated with three types of drug activity measures: (i) unadjusted AC50 values as reported in the SPD, (ii) the total safety margin defined by the ratio of AC50 to total Cmax, and (iii) the free safety margin defined by the ratio of AC50 to free Cmax. Of the 719 published target–ADR pairs, 240 (33%) were confirmed as statistically significant via Kruskal–Wallis (KW) test [45], with p-value $< 10^{-3}$. These included ADRs associated with serotonin receptors (e.g., HTR1A, 100% validated), adrenergic receptors (e.g., ADRA1A, 100% validated), and muscarinic receptors (e.g., CHRM1, 88% validated). Furthermore, multivariate statistical analysis using Lasso logistic regression [46] was used to better understand the lack of statistical significance in many reported target–ADR relationships. The analysis indicated a higher likelihood for significant relationships with an increasing number of highly active drugs (AC50 < 1 μM) in an assay and for ADRs related to the nervous system and psychiatric disorders. To expand the analysis beyond literature-reported data, all possible target–ADR relationships with sufficient data coverage in the SPD were systematically examined. Juxtaposition of drug activities for 124 targets to 2647 MedDRA terms for ADRs yielded 562,744 relationships where each pair had the three types of activity measures described above. Statistical significance based on a combination of thresholds (ROC AUC ≥ 0.7 and KW p-value $< 10^{-6}$) was observed for 1992 associations, which were then summarized to 1432 pairs for further analysis by removing redundant MedDRA terms.

Validating target–ADR relationships identified by univariate analysis using thresholds relies on pooling together both drug on- and off-target activity measurements. One drawback of this approach is that drugs with increasing target promiscuity can cause ADRs via different targets, which may confound univariate analysis results. To overcome

Table 3.1 Selected novel *in vitro* off-target activities of marketed drugs.

Drug	MoA	On-target (s)	Indications	Assay	AC50 (μM)	Free Cmax margin	Promiscuity (%)	Literature (PMID)
Citalopram	Antagonist	SLC6A4	Depressive disorder, generalized anxiety disorder, major depressive disorder	HRH2 binding	0.350	1.093	13.5	12119661 20199498 31121087
Glipizide	Antagonist	ABCC8, KCNJ11	Diabetes mellitus type 2	CNR1 inhibition	0.019	0.556	1.0	32297119
Losartan	Antagonist	AGTR1	Diabetic renal disease, hypertension with left ventricular hypertrophy, hypertensive disorder, hypertensive urgency	CNR1 inhibition	0.001	0.116	3.9	32297119
Rotigotine	Agonist	DRD2	Parkinson's disease, restless legs	ADRA1A agonist	0.003	9.262	24.3	18704368
Zolpidem	Antagonist	GABRA1, GABRA2, GABRA3, GABRB2, GABRB3, GABRG2	Initial insomnia, insomnia	CHRM1 inhibition	0.210	3.373	2.9	31379728

	NR3C1			PGR			Adrenal cortical hyperfunction
	AC50 weight 1	Total Cmax weight 2	Free Cmax weight 3	AC50 weight 4	Total Cmax weight 5	Free Cmax weight 6	
Drug 1							No
Drug 2							No
Drug 3							Yes
Drug 4							No
Drug 5							No
Drug 6							Yes
Drug 7							No
Drug 8							No
...							
Drug N							

(a)

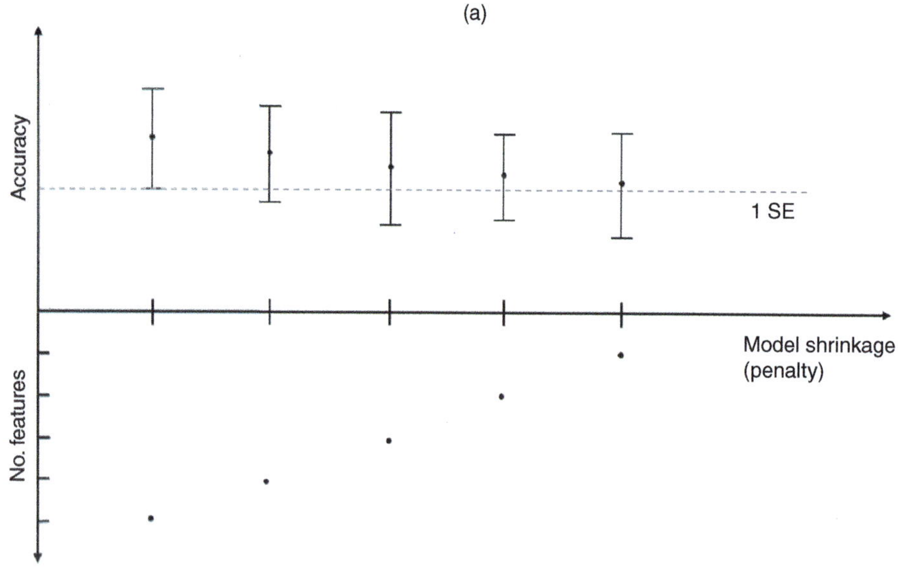

(b)

Figure 3.2 Principle of multivariate statistical modeling. (a) Shown is a schematic drawing of a dataset for multivariate statistical modeling using Lasso logistic regression. Each row represents a single drug labeled according to whether it causes the ADR "adrenal cortical hyperfunction" (positive, red; negative, green). Each column represents the drug's activity measured as either unadjusted AC50 value, total margin, or free margin against two different targets (NR3C1 and PGR). Each of these represents a feature in the dataset and is given a weight coefficient. In this example, the single most important activity measurement remaining in the final reduced dataset is highlighted with darker colors. The exact measurement values and drug names are omitted for clarity. (b) Exemplary illustration of the principle for multivariate statistical modeling using Lasso logistic regression. A classification model for the ADR is run multiple times while increasing the penalization parameter (*x*-axis), which results in shrinkage of the feature coefficients toward 0, leading to gradual feature elimination. Once the model accuracy (upper *y*-axis) decreases below one standard error of the best-performing model run (1 SE, horizontal dashed line), the procedure is interrupted, and the number of remaining features (lower y-axis) is recorded. In this example, a single feature is retained in the final dataset.

this, multivariate Lasso-penalized logistic regression was employed to model the probability of occurrence of an ADR in the presence of significantly associated targets identified above. Unadjusted AC50 and total and free margin values per drug–target pair were tabulated for all drugs, positive or negative, for a given ADR. For each ADR, this resulted in a $n \times m$ matrix where n is the number of instances (positive + negative drugs) and m is the number of features (number of drug targets × three activity measures). This represented the initial dataset for building a classification model that discriminates ADR-positive from ADR-negative compounds (Figure 3.2a). Each ADR model was then repeatedly trained, while gradually increasing the Lasso shrinkage parameter in each subsequent run, resulting in the penalization of the feature weight coefficients and elimination of less important (and hence potentially redundant) features across different runs (Figure 3.2b). The penalization procedure for each model was discontinued once the prediction accuracy decreased by one standard error below the best-performing model run. Thus, the final feature set for each ADR model was reduced only to the most significant target activities (one or more), which are likely to give rise to the ADR. Each experiment was run multiple times to ensure that important features are consistently selected across different trials. Since the main objective of these experiments was explainability, i.e., feature extraction, and not optimizing model performance, classification accuracy was used only for orientation across model trials. It is important to note that while this linear method is intuitive and was used for simplicity and complementarity to univariate analysis, it has limitations and does not preclude employing other approaches for feature extraction and model explanation such as SHapley Additive exPlanation (SHAP) values [47], random forest Gini impurity coefficients [48], or principal component analysis (PCA) [49].

Applying the method to a combined set of unpublished target–ADR pairs meeting strict significance criteria (ROC AUC ≥ 0.7 and KW $p < 10^{-6}$) and published associations satisfying a less stringent criterion (KW $p < 10^{-3}$, assuming higher confidence in literature-reported findings) led to the discarding of 801 target–ADR associations initially labeled as significant by univariate analysis using cutoffs. As a result, 442 target–ADR pairs were significant in both uni- and multivariate analysis, and further 189 pairs, where data for multivariate analysis was not sufficient, were significant only through univariate thresholds. Most of the examined ADRs were related to modulation of the glucocorticoid receptor (NR3C1), histamine H1 receptor (HRH1), serotonin transporter (SLC6A4), dopamine D3 (DRD3), adrenergic alpha-1A (ADRA1A), and dopamine D1 (DRD1) receptors.

For example, the target relationships of the ADR "adrenal cortical hyperfunction" were modeled using two separate datasets from SIDER and FAERS containing 18 versus 469 and 17 versus 588 ADR-positive and ADR-negative drugs, respectively, with 11 overlapping positives. In both cases, of all six features in the dataset (three activity measures for NR3C1 and PGR each), only the unadjusted AC50 for NR3C1 was retained in the final model, rendering it the single most discriminative activity measure for this ADR. Another notable example is the ADR "extrapyramidal disorder," for which the number of target activities was reduced from 28 to 3 (unadjusted AC50 values for SLC6A4, HRH1, and DRD3), of which one association (DRD3) was absent from the literature reviews.

An interesting observation was made with respect to the different target activity measures used in these analyses. Unadjusted AC50 values alone were the predominant activity measure accounting for statistical significance in univariate analysis for both published and unpublished target–ADR relationships. In contrast, relationships confirmed via multivariate analysis showed a higher proportion of margin-based activity measures (total and free margin). Preference for margin-based values was also independently observed during univariate analysis for associations with an increasing proportion of drugs with on-target activity, i.e., with confirmed MoA. These findings highlight an important point for ADR determination: target–ADR associations supported by safety margins are likely to be more plausible and representative indicators of ADR risk compared to simple unadjusted AC50 values as it has been suggested also in several other important contributions to the field [10, 26, 50].

3.4 A Practical Perspective

The analysis described above yielded 631 significant target–ADR relationships, 482 (76%) of which were previously not reported in the resources used for the study. This provided the opportunity to detect potentially unknown clinically relevant associations by corroborating qualitative observations with quantitative evidence from the statistical analysis described above. ADRs with at least 20% of associated drugs having off-target activities (unadjusted AC50 and/or margin-based) were investigated by examining primary literature and revealed several previously unreported relationships summarized together with supporting literature in Table 3.2. Notable examples include associations such as adrenergic receptor ADRA2C inhibition for auditory hallucinations/paranoia, dopaminergic DRD2/DRD3 binding for mammary/menstruation-related ADRs, and progesterone receptor (PGR) agonism for hyperpigmentation.

Furthermore, analysis results were used to propose putative mechanisms for already known ADRs based on overlapping significant associations with off-target activities from SPD. The majority of newly explained relationships included cardiac and respiratory ADRs, most notably in connection with OPRK1 drug activity. OPRK1, CHRM2, and ADRA1A activities for multiple drugs were also related to various motor disorders. Other interesting examples include accommodation disorders reported for the GABA-A receptor agonist zolpidem [51] and 5-HT1A receptor agonist eletriptan [52], which could be linked to muscarinic receptor binding (CHRM1 and CHRM2, respectively) as well as psychotropic and endocrinal ADRs reported for the selective serotonin reuptake inhibitor citalopram [53], which could be potentially attributed to dopamine receptor DRD1 and histamine receptor HRH2 binding, respectively. These and other selected examples are provided in Table 3.3 and are discussed in detail in the original publication [33].

Table 3.2 Novel target–ADR relationships with support in biomedical literature.

Assay name	MedDRA name	KW *p*-value*	ROC AUC	Significant measures	Significant sources	Literature (PMID)
ADRA1A Binding (NIBR assay)	Eosinophilia	3.6E−08	0.85	Total (not free) margin	FAERS	32194050
	Tardive dyskinesia	1.9E−23	0.87	AC50, free margin	SIDER	3015270 19565482
ADRA2C inhibition	Hallucination, auditory	1.1E−11	0.86	AC50, total (not free) margin	FAERS	9526020 34930904 11317218
	Paranoia	1.2E−09	0.82	AC50, free margin	FAERS	9526020 34930904 11317218
AR binding (agonist)	Erythema nodosum	2.6E−10	0.70	AC50	SIDER	27075133
	Ovarian and fallopian tube cysts and neoplasms	2.4E−09	0.70	AC50	SIDER	30791431
	Retinal embolism and thrombosis	6.5E−07	0.74	AC50	SIDER	29040227
	Ovarian neoplasms malignant 55(excluding germ cell)	5.9E−12	0.82	AC50	SIDER	30791431
CHRM1 inhibition	Micturition urgency	4.0E−09	0.73	AC50	SIDER	35117285
	Dental caries	3.9E−07	0.73	AC50	SIDER	31289718 12974516
CHRM2 binding	Ileus paralytic	2.7E−10	0.76	AC50, free margin	SIDER	14607264
DRD2 binding	Breast enlargement	2.4E−08	0.72	AC50	SIDER	16169407 19896213 24490071
	Menstruation irregular	2.4E−08	0.71	AC50	SIDER	24490071
	Salivary hypersecretion	7.9E−11	0.71	AC50	SIDER	16421461
DRD3 binding	Tardive dyskinesia	4.3E−31	0.90	AC50, free margin	FAERS, SIDER	10379516 11490179 16513329
	Extrapyramidal disorder	5.1E−33	0.87	AC50, free margin	FAERS, SIDER	19506579
	Gambling disorder	2.8E−09	0.83	AC50	SIDER	26192187
	Hyperprolactinemia	1.6E−17	0.86	AC50, free margin	SIDER	16169407 33854317
	Blood prolactin increased	4.2E−14	0.88	AC50, free margin	SIDER	33854317
HRH1 binding	Hypomania	5.0E−17	0.88	AC50, free margin	SIDER	34572558
HTR2B antagonist	Lethargy	7.4E−07	0.71	AC50	SIDER	30666218
	Sleep disorders NEC	1.9E−08	0.72	AC50	SIDER	30666218
OPRK1 binding	Salivary hypersecretion	2.8E−08	0.74	AC50	SIDER	Entrez Gene
PGR agonist	Hyperpigmentation disorders	7.8E−13	0.71	Free margin	SIDER	17477341
SLC6A4 binding	Generalized tonic-clonic seizure	1.5E−08	0.76	Free margin	FAERS	31849820
	Galactorrhea	2.0E−21	0.76	AC50, free margin	SIDER	14997175
	Blood prolactin increased	2.5E−07	0.78	AC50, free margin	SIDER	14997175

*E denotes 10, e.g., 2.5E−07 = 2.5×10^{-7}.

Table 3.3 Newly explained ADR mechanisms based on drug off-target activities.

Drug (on-target)	Off-target	AC50 (μM)	Explained ADRs
Amiodarone (KCNH2)	HRH1	2.4	Akinesia, bradykinesia, decreased activity, disorientation, extrapyramidal disorder, hypokinesia, inappropriate antidiuretic hormone secretion
Amlodipine (CACNA1C)	ADRB2	1.8	Sinus bradycardia
	DRD2	2.9	Parkinsonism
	OPRK1	1.2	Cardiorespiratory arrest, drug abuse
Citalopram (SLC6A4)	DRD1	3.9	Akathisia, catatonia, choreoathetosis, dyskinesia, ejaculation disorder, extrapyramidal disorder, libido increased, movement disorder, orgasmic disorders and disturbances, parkinsonism, priapism, psychotic disorder, serotonin syndrome, substance-related and addictive disorders, tardive dyskinesia, torticollis, trismus
	HRH2	0.35	Hyperprolactinemia
Donepezil (ACHE)	ADRA1A	4.4	Drooling
	CHRM2	1.1	Extrapyramidal disorder
Eletriptan (HTR1B, D, F)	CHRM2	4.2	Accommodation disorder
Fluoxetine (SLC6A4)	OPRK1	1.5	Cardiorespiratory arrest, coordination abnormal, delusion, drug abuse
Lurasidone (DRD2, HTR2A)	ADRA1A	0.075	Blood prolactin increased, cogwheel rigidity, drooling, muscle rigidity, oculogyric crisis, sexual dysfunction, torticollis
	CHRM2	3.5	Dyskinesia
	DRD1	0.13	Dyskinesia, sedation, substance-related and addictive disorders
Pregabalin (CACNA2D1, 2)	ADRA1A	3.7	Autonomic nervous system imbalance, blood prolactin increased, dizziness postural, miosis, orgasmic disorders and disturbances, restlessness, sudden death, torticollis, trismus, urinary incontinence
Quetiapine (DRD2, HTR2A)	OPRK1	1.6	Cardiorespiratory arrest, choreiform movements, coordination abnormal, delusion, drug withdrawal syndrome
	OPRM1	1.8	Drug dependence, respiratory arrest
Ranitidine (HRH2)	CHRM2	2.6	Dyskinesia
Sibutramine (SLC6A2, 3, 4)	HTR2A	1	Torticollis
	OPRK1	4.8	Delusion
Tramadol (OPRM1)	CHRM2	1.5	Dyskinesia, parkinsonism
Trimipramine (SLC6A2, 3, 4)	CHRM1	0.31	Accommodation disorder, mydriasis
	CHRM2	0.16	Accommodation disorder, extrapyramidal disorder, mydriasis, parotid gland enlargement
Valbenazine (SLC18A2)	HTR1A	1.4	Tardive dyskinesia
Zolpidem (GABRA1)	CHRM1	0.21	Accommodation disorder, mydriasis

In conclusion, perhaps the most important aspect and ultimate goal of such work should be to share the actual data and analysis workflows (code, files etc.) with the public. In a fragmented and heterogeneous safety pharmacology and ADR data landscape, large high-quality resources such as the one described in this chapter are invaluable sources of knowledge to support progress in the field and can serve as inspiration for others to develop better and safer drugs. This way, one can only hope that the future generations of medical doctors, pharmacists, scientists, and patients will have a more illuminated understanding of ADRs, their causes, and how they could be mitigated.

Acknowledgments

The author would like to thank Dr. Jeffrey Sutherland, Dr. Laszlo Urban, Dr. Christian Feldmann, and Dr. Helen-Louise Lightfoot for their reviews and helpful suggestions.

References

1 Hughes, J., Rees, S., Kalindjian, S., and Philpott, K. (2011). Principles of early drug discovery. *Br. J. Pharmacol.* 162: 1239–1249.

2 Muller, P.Y. and Milton, M.N. (2012). The determination and interpretation of the therapeutic index in drug development. *Nat. Rev. Drug Discov.* 11: 751–761.

3 Croston, G.E. (2017). The utility of target-based discovery. *Expert Opin. Drug Discov.* 12: 427–429.

4 Urban, L., Whitebread, S., Hamon, J., Mikhailov, D., and Azzaoui, K. (2012). Screening for Safety-Relevant Off-Target Activities. In: *Polypharmacology in Drug Discovery* (ed. J.-U. Peters), 15–46. Hoboken: Wiley.

5 Peters, J.-U. (2013). Polypharmacology – foe or friend? *J. Med. Chem.* 56: 8955–8971.

6 Oprea, T.I. and Mestres, J. (2012). Drug repurposing: far beyond new targets for old drugs. *AAPS J.* 14: 759–763.

7 Ito, T. and Handa, H. (2020). Molecular mechanisms of thalidomide and its derivatives. *Proc. Japan Acad., Ser. B* 96: 189–203.

8 Islam, S., Wang, S., Bowden, N. et al. (2022). Repurposing existing therapeutics, its importance in oncology drug development: kinases as a potential target. *Br. J. Clin. Pharmacol.* 88: 64–74.

9 Anighoro, A., Bajorath, J., and Rastelli, G. (2014). Polypharmacology: challenges and opportunities in drug discovery. *J. Med. Chem.* 57: 7874–7887.

10 Bowes, J., Brown, A.J., Hamon, J. et al. (2012). Reducing safety-related drug attrition: the use of *in vitro* pharmacological profiling. *Nat. Rev. Drug Discov.* 11: 909–922. https://doi.org/10.1038/nrd3845.

11 Pushpakom, S. (2022). Introduction and historical overview of drug repurposing opportunities. In: *Drug Repurposing* (ed. D. Cavalla), 1–13. London: The Royal Society of Chemistry https://doi.org/10.1039/9781839163401-00001.

12 Cruz-Burgos, M., Losada-Garcia, A., Cruz-Hernandez, C.D. et al. (2021). New approaches in oncology for repositioning drugs: the case of PDE5 inhibitor sildenafil. *Front. Oncol.* 11.

13 Manikandan, P. and Nagini, S. (2018). Cytochrome P450 structure, function and clinical significance: a review. *Curr. Drug Targets* 19: 38–54.

14 Zolk, O. and Fromm, M.F. (2011). Transporter-mediated drug uptake and efflux: important determinants of adverse drug reactions. *Clin. Pharmacol. Ther.* 89: 798–805.

15 Carss, K.J., Deaton, A.M., Del Rio-Espinola, A. et al. (2023). Using human genetics to improve safety assessment of therapeutics. *Nat. Rev. Drug Discov.* 22: 145–162.

16 Duffy, Á., Verbanck, M., Dobbyn, A. et al. (2020). Tissue-specific genetic features inform prediction of drug side effects in clinical trials. *Sci. Adv.* 6: eabb6242.

17 Nguyen, P.A., Born, D.A., Deaton, A.M. et al. (2019). Phenotypes associated with genes encoding drug targets are predictive of clinical trial side effects. *Nat. Commun.* 10: 1579.

18 Ge, F. and He, Q.-Y. (2009). Genomic and proteomic approaches for predicting toxicity and adverse drug reactions. *Expert Opin. Drug Metab. Toxicol.* 5: 29–37.

19 Cui, Y. and Paules, R.S. (2010). Use of transcriptomics in understanding mechanisms of drug-induced toxicity. *Pharmacogenomics* 11: 573–585.

20 Takarabe, M., Kotera, M., Nishimura, Y. et al. (2012). Drug target prediction using adverse event report systems: a pharmacogenomic approach. *Bioinformatics* 28: i611–i618.

21 Bender, A., Scheiber, J., Glick, M. et al. (2007). Analysis of pharmacology data and the prediction of adverse drug reactions and off-target effects from chemical structure. *ChemMedChem* 2: 861–873.

22 Dearden, J.C. (2003). In silico prediction of drug toxicity. *J. Comput. Aided Mol. Des.* 17: 119–127.

23 Yamanishi, Y., Pauwels, E., and Kotera, M. (2012). Drug side-effect prediction based on the integration of chemical and biological spaces. *J. Chem. Inf. Model.* 52: 3284–3292.

24 Schotland, P., Racz, R., Jackson, D.B. et al. (2021). Target adverse event profiles for predictive safety in the postmarket setting. *Clin. Pharmacol. Ther.* 109: 1232–1243.

25 Krix, S., DeLong, L.N., Madan, S. et al. (2023). MultiGML: multimodal graph machine learning for prediction of adverse drug events. *Heliyon* 9: e19441.

26 Whitebread, S., Dumotier, B., Armstrong, D. et al. (2016). Secondary pharmacology: screening and interpretation of off-target activities – focus on translation. *Drug Discov. Today* 21: 1232–1242. https://doi.org/10.1016/j.drudis.2016.04.021.

27 Papoian, T., Chiu, H.J., Elayan, I. et al. (2015). Secondary pharmacology data to assess potential off-target activity of new drugs: a regulatory perspective. *Nat. Rev. Drug Discov.* 14: 294–294.

28 Whitebread, S., Bojanic, D., Urban, L., and Hamon, J. (2005). Preclinical safety pharmacology is designed to flag adverse clinical effects at a low cost and can prevent high attrition rates during clinical trials in vitro safety pharmacology profiling: an essential tool for successful drug development. www.drugdiscoverytoday.com (accessed 12 September 2024).

29 Lounkine, E., Keiser, M.J., Whitebread, S. et al. (2012). Large-scale prediction and testing of drug activity on side-effect targets. *Nature* 486: 361–367.

30 Rao, M.S., Gupta, R., Liguori, M.J. et al. (2019). Novel computational approach to predict off-target interactions for small molecules. *Front. Big Data* 2: 25.

31 Heyndrickx, W., Mervin, L., Morawietz, T. et al. (2023). Melloddy: cross-pharma federated learning at unprecedented scale unlocks benefits in QSAR without compromising proprietary information. *J. Chem. Inf. Model.* 64: 2331–2344. https://doi.org/10.1021/acs.jcim.3c00799.

32 Bendels, S., Bissantz, C., Fasching, B. et al. (2019). Safety screening in early drug discovery: an optimized assay panel. *J. Pharmacol. Toxicol. Methods* 99: 106609.

33 Sutherland, J.J., Yonchev, D., Fekete, A., and Urban, L. (2023). A preclinical secondary pharmacology resource illuminates target-adverse drug reaction associations of marketed drugs. *Nat. Commun.* 14: 4323.

34 Heller, S.R., McNaught, A., Pletnev, I. et al. (2015). InChI, the IUPAC international chemical identifier. *J. Cheminform.* 7: 23.

35 Zdrazil, B., Felix, E., Hunter, F. et al. (2024). The ChEMBL database in 2023: a drug discovery platform spanning multiple bioactivity data types and time periods. *Nucleic Acids Res.* 52: D1180–D1192.

36 Avram, S., Bologa, C.G., Holmes, J. et al. (2021). DrugCentral 2021 supports drug discovery and repositioning. *Nucleic Acids Res.* 49: D1160–D1169.

37 UniProt Consortium (2019). UniProt: a worldwide hub of protein knowledge. *Nucleic Acids Res.* 47: D506–D515.

38 Agarwala, R., Barrett, T., Benson, D.A. et al. (2018). Database resources of the National Center for Biotechnology Information. *Nucleic Acids Res.* 46: D8–D13.

39 Visser, U., Abeyruwan, S., Vempati, U. et al. (2011). BioAssay Ontology (BAO): a semantic description of bioassays and high-throughput screening results. *BMC Bioinform.* 12: 257.

40 Smit, I.A., Afzal, A.M., Allen, C.H. et al. (2021). Systematic analysis of protein targets associated with adverse events of drugs from clinical trials and postmarketing reports. *Chem. Res. Toxicol.* 34: 365–384.

41 Jaasu, N.M., Kamaraj, R., and Seetharaman, R. (2018). MedDRA (Medical dictionary for regulatory activities). *Res. J. Pharm. Technol.* 11: 4751.

42 Kuhn, M., Letunic, I., Jensen, L.J., and Bork, P. (2016). The SIDER database of drugs and side effects. *Nucleic Acids Res.* 44: D1075–D1079.

43 Summerfield, S.G., Yates, J.W.T., and Fairman, D.A. (2022). Free drug theory – no longer just a hypothesis? *Pharm. Res.* 39: 213–222.

44 Hamon, J., Whitebread, S., Techer-Etienne, V. et al. (2009). *In vitro* safety pharmacology profiling: what else beyond hERG? *Future Med. Chem.* 1: 645–665.

45 McKight, P.E. and Najab, J. (2010). Kruskal-Wallis test. In: *The Corsini Encyclopedia of Psychology*, vol. 1, 1–10. Wiley https://doi.org/10.1002/9780470479216.corpsy0491.

46 Tibshirani, R. (1996). Regression shrinkage and selection via the lasso. *J. R. Stat. Soc. Ser. B Stat. Methodol.* 58: 267–288.

47 Lundberg, S. and Lee, S.-I. (2017). A unified approach to interpreting model predictions. arXiv:1705.07874.

48 Nembrini, S., König, I.R., and Wright, M.N. (2018). The revival of the Gini importance? *Bioinformatics* 34: 3711–3718.

49 Bro, R. and Smilde, A.K. (2014). Principal component analysis. *Anal. Methods* 6: 2812–2831.

50 Shah, F., Leung, L., Barton, H.A. et al. (2015). Setting clinical exposure levels of concern for Drug-Induced Liver Injury (DILI) using mechanistic *in vitro* assays. *Toxicol. Sci.* 147: 500–514.

51 Greenblatt, D.J. and Roth, T. (2012). Zolpidem for insomnia. *Expert. Opin. Pharmacother.* 13: 879–893.

52 Diener, H.-C. (2005). Eletriptan in migraine. *Expert. Rev. Neurother.* 5: 43–53.

53 Pollock, B.G. (2001). Citalopram: a comprehensive review. *Expert. Opin. Pharmacother.* 2: 681–698.

4

Off-Target Screening Strategies

Sonia Roberts and Helen L. Lightfoot

4.1 Introduction

When we are considering off-target(s) for a compound of interest, we tend to focus on the non-desired targets of the compound: the targets your compound binds to, modulates, and consequently produces an adverse drug reaction (ADR). One must consider the context here, as an ADR off-target for one drug may not be an ADR for another drug; patient population, indication, target tissues, and drug kinetics, for example, all can play a role in determining if an off-target will produce an adverse effect in preclinical studies or patients. Compounds that are promiscuous often hit many off-targets and produce ADRs. Off-targets can also be beneficial and have positive implications on the biodistribution and activity of compounds. Each therapeutic can be categorized according to its structural type (small molecule, large molecule, and RNA molecule) or modality (small molecule, protein replacement, antibody, oligonucleotide therapy, or cell and gene therapy). Each of these demonstrates certain characteristics and is associated with a higher or lower propensity for adverse off-target interactions [1]. Small-molecule drugs are considered to have a high risk of off-target interactions due to their ability to interact with multiple molecular targets, which are relatively accessible due to the drug's chemical properties and also possible diverse drug delivery approaches that can be utilized. For larger molecules and more complex therapeutics, selectivity for their desired target is achieved by specialized drug delivery and their intrinsically more focused target-type interactions. Regardless of the increased challenges associated with selectivity, small-molecule therapeutics still have many benefits for patients (e.g., oral delivery, accessible globally, and reduced cost), and they currently dominate the market of therapeutic drugs. In this chapter, we will discuss safety-related off-target evaluations for protein-targeting small molecules as well as new modalities (summarized in Table 4.1). We will focus on basic insight into the rationale for off-target safety profiling as well as the processes that are presently applied or could be applied in future for those modalities that are still in infancy. As part of this, we will also discuss early mitigations of key ADR-related off-target effects using structure–activity relationship (SAR) analysis developed from fingerprints of chemical structures together with data from binding experiments to specific off-targets. We will however not consider here how we use pharmacokinetic data and *in vitro* ADR off-target profiling to predict clinical ADRs of drug candidates as this will be covered by another chapter in this book.

4.2 Small Molecules

The design of drug-like compounds started with what is typically termed "small molecules." These are low-molecular-weight compounds (up to MW 1500 g/mole) that are designed to have good pharmacokinetic properties, for example, through the application of Lipinski's Rule of 5 [2]. Consequently, they have good oral bioavailability by achieving adequate permeability, solubility, and gastrointestinal uptake, balanced with appropriate renal and biliary clearance. The starting point for identifying a small molecule lead frequently comes from high-throughput screening, i.e., testing a wide range of chemically different molecules for affinity or activity at the desired target. Typically, these molecules are starting points for medicinal chemists to use their skills by refining the chemistry building blocks to enhance potency and selectivity. Quantitative structure–activity relationships (QSARs) define the direction in which the chemistry goes to create a potentially new drug. In more recent times, machine learning and artificial intelligence is beginning to be used to influence decision-making on which compounds should be synthesized, saving time and resources. The quality of machine-learning models is driven by the data available and may be improved by collaborations between industry

Polypharmacology: Strategies for Multi-Target Drug Discovery, First Edition. Edited by Jens-Uwe Peters.

Table 4.1 Overlap across the modalities discussed in this chapter in terms of their targets, delivery, and off-/on-target considerations. ADME: chemical absorption, distribution, metabolism, and excretion. ASO: antisense oligonucleotide. QSAR: quantitative structure–activity relationships.

Modality	Molecular target			Target localization			Delivery	Off-target potential	Selectivity and Safety approaches
	DNA	RNA	Protein	Extra-cellular	Plasma membrane	Intra-cellular			
Small molecule PROTAC RIBOTAC	●	●	●	●	●	●	All	High	QSAR ADME profiling Broad screening Targeted drug delivery Mechanism-driven safety assays, Target engagement approaches
ASO		●		●	●	●	Parenteral	Low	Off-target prediction Broad screening Targeted drug delivery Mechanism-driven safety assays
Antibody			●	●	●		Parenteral	Low	Protein engineering Immunology sensitivity screens
Cell therapy				●			In vitro modified Parenteral	Low	Immunology sensitivity screens

and academia where sensitive intellectual property can be protected [3]. The search for high-affinity, selective, and bioavailable small-molecule compounds is a quest that all drug hunters make. There is no exact path, and the outcome may also be different. The route you take will be driven by experience, expectations, and resources. During the selection of a clinical candidate, many small-molecule compounds are synthesized, and in addition to measuring affinity at the desired target, representative molecules from each chemical class are routinely screened for selectivity. It is not feasible to assess the selectivity of all compounds against all proteins. With this in mind, secondary pharmacology has a focus on those off-targets that are of high concern and clinical relevance [4–7]. Secondary pharmacology data can be important to the drug development process at all stages of development. Many experimental approaches can be used to address the secondary pharmacology of small molecules [8]. The general process of secondary pharmacology data generation and evaluation is summarized in Figure 4.1. Radioligand binding is still the mainstay for evaluating the selectivity profile of small molecules. However, there are drawbacks to this approach as firstly it tends to be highly sensitive and may lead to false positives, and secondly, it is important to understand the pharmacology of any interaction to understand its risk; hence, functional assays will be required. The cost of functional assays can be similar to radioligand binding; however, it may be necessary to test in different modes to understand if the off-target effect will be of clinical concern. Some companies now use only functional assays for secondary pharmacology screening. If you can identify a functional panel that fits your budget, turnaround times, and chemical diversity within your portfolio, then this is a desirable position. However, for many companies, there is a strategic balance that allows rapid primary screening of many compounds and screening in functional assays. Multiple assay formats are available, with the largest vendors (Eurofins, Wuxi) providing a wide range of targets and assay options. For cost-effective screening, both companies offer several panels for small-molecule secondary pharmacology screening, which use either or both biochemical and functional assay formats. A cross-pharma survey across 14 large and 4 medium-sized companies identified that the secondary pharmacology targets reported by Bowes et al. [5] and Lynch et al. [6] are still forming the mainstay of our off-target profiling strategy, with an average of 49 targets profiled in a primary panel and 99 in a secondary panel [9]. Selecting the most appropriate targets and assay formats is somewhat challenging, and different approaches are taken across companies. To select the secondary pharmacology screening panel used at Roche, the Bioprint® dataset from Eurofins was analyzed using multiple pairwise comparisons to identify targets with a similar binary fingerprint [4]. Statistical ranking based on the cumulative hit rates

Secondary pharmacology assay panel

- ***Diverse panel*** of receptors, transporters, ion channels, and enzymes with safety-relevant rationale
- ***Patient- or tissue-relevant off-target sites*** should also be considered where exposures may be high or sustained during treatment e.g., GI tract, CNS, vascular, etc.

Methods

- ***Assay format:*** binding or functional
- ***End-points***: radioligand, fluorescence, luminescence, Ca^{2+} flux, second messenger, etc. and the criteria for a positive flag ("hit") e.g. 50% inhibition of binding
- ***Analysis methods:*** e.g., nonlinear regression (variable slope or 3 parameters), normalised or non-normalised data

Results

- ***Raw data:*** Tabular and graphical format, ideally electronically transferable for additional analysis or review
- ***Summary data:*** Mean effect, number of replicates, variability (via a standard deviation or individual values)
- ***Key endpoints:*** e.g., IC_{50}, EC_{50} maximum effect, Hill slope, inhibition constant (Ki)

Data interpretation

- ***Relevance of "hits"*** in the assay panel and potential safety impact
- ***Selectivity margin*** for desired target versus off target
- ***Biological impact and correlation*** with in vivo findings
- ***Potency*** of off-target with respect to local tissue/plasma concentrations
- ***Monitoring*** for potential off-target effect in clinic

Figure 4.1 Summary of the key aspects in established occupancy-driven secondary pharmacology evaluations for small molecules, PROTACs, and RIBOTACs.

(>50% inhibition of binding at 10 μM) identified that 95% of hits were covered by 47 assays. With three additions, based on known clinical pharmacology, this forms the Roche screening panel. This panel consists solely of radioligand or enzyme assays and requires follow-up functional testing to establish the pharmacological profile of any "hits." [4] Testing at a single, high concentration (typically 10 μM), which is ideally approximately 100-fold greater than the anticipated free efficacious plasma concentration, enables an overview of the promiscuity and safety of molecules during the lead selection phase. A combination of functional follow-up assays and larger assay panels is typically used as the compound progress to the stage of clinical candidate selection. The assay format selected may depend on the purpose, pharmacology, availability, and cost of testing. The caveat of radioligand binding is its tendency to detect more positives compared to functional assays; however, this is often assay specific and translation to clinical risk will require functional assessment. In addition, single-point data is not adequate to evaluate the affinity (Kd) of a compound, and additional testing will be needed to achieve this level of data accuracy. Jenkinson et al. reviewed the practical aspects of secondary pharmacology and covered many of the elements relevant to small molecules [8].

4.3 Proteolysis-Targeting Chimeras (PROTACs)

Proteolysis-targeting chimeras (PROTACs) are heterobifunctional organic compounds that drive target protein degradation [10, 11] (Figure 4.2a). A PROTAC consists of two small-molecule ligands joined together by a linker. One ligand binds a protein target of interest (POI), and the other ligand binds an E3 ubiquitin ligase and is referred to as the warhead. PROTAC-mediated co-recruitment of the POI and the E3 ligase results in ubiquitylation and degradation of the POI by the ubiquitin–proteasome system. PROTACs have brought a wave of advantages to target-mediated drug discovery, including the ability to catalytically target non-functional protein pockets bringing an increase in possible selective drug targets and lower doses [14, 15]. There are of course still limitations with this approach as discussed in Gao et al. [14], for example limitations in POI selection and on the ligases that can be applied. There have been at least nineteen biotechnology investments from large pharmaceutical companies in targeted protein degradation approaches over the past five years, which include both investments into platforms and specific degrading compounds, most of which are in the discovery phase, giving an idea of the status of this field. In addition, as of early 2023, there were fifteen clinical PROTACs (two in phase II (Figure 4.2a), 10 in phase I, and 3 at Investigational New Drug (IND) stage), with all except one (which is developed for autoimmune disease) being developed for the treatment of various cancers. The majority of these PROTACs are surprisingly administered orally and when reported use the cereblon (CRBN) ligase warhead. The ability to orally administer these large non-drug-like compounds resides in the extensive work carried out to formulate the parameters required to enhance their oral bioavailability, such as their optimal polar surface area (PSA) and molecular weight [16].

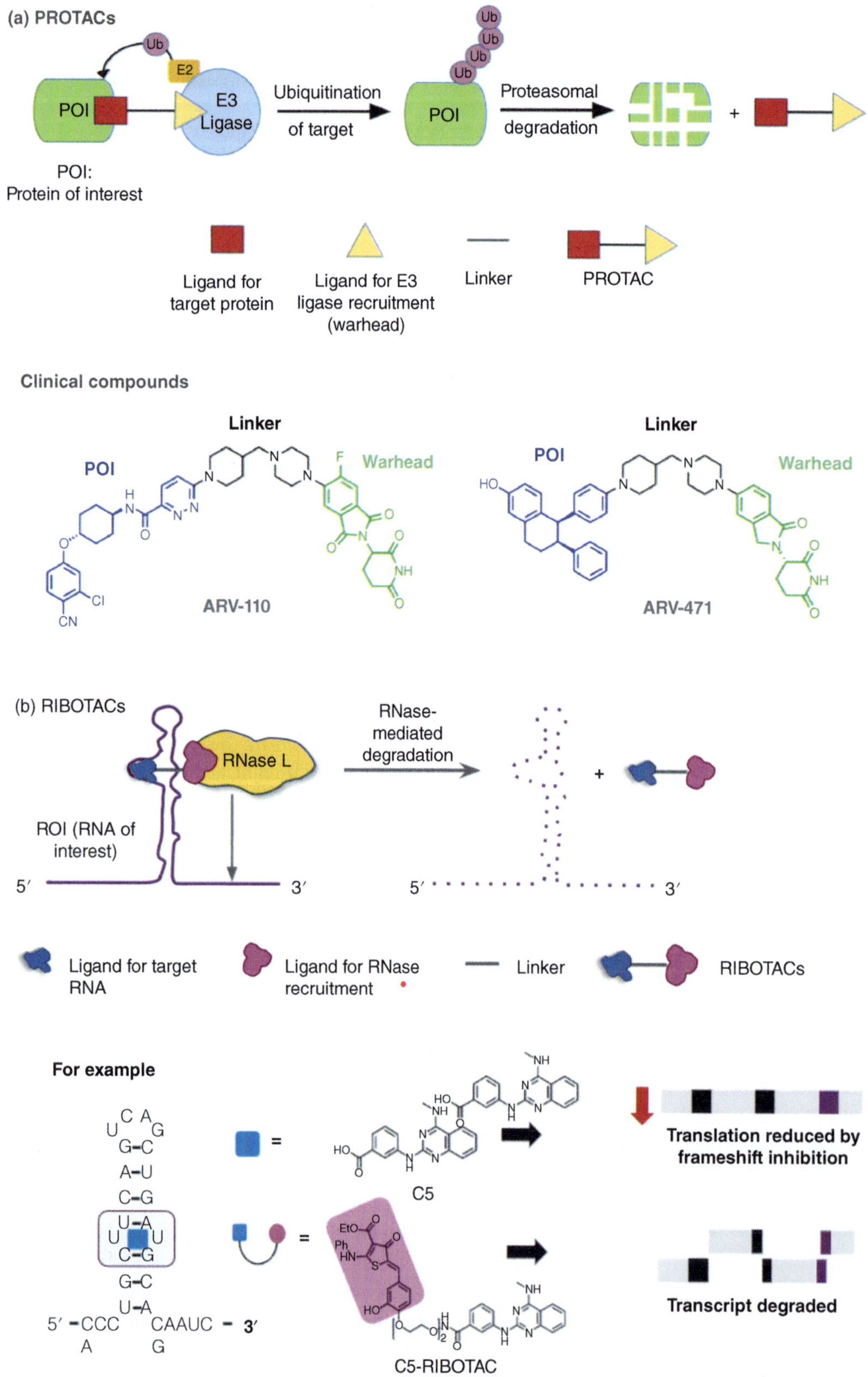

Figure 4.2 The mechanism of action and examples of (a) PROTACs. Adapted from Békés et al. [12]. Ub: Ubiquitination. E2: Ubiquitin-conjugating enzyme. (b) RIBOTACs. Adapted from Haniff et al. [13].

PROTACs are identified in a similar manner as discussed above for occupancy-mediated compounds (small molecules as described above) and have challenges with off-target selectivity that is both similar to and different from that observed for small molecules. Firstly, PROTACs can degrade proteins other than the intended target (bystander degradation) [17]. Bystander degradation is where a protein that is not directly bound by the PROTAC becomes ubiquitinated and consequently degraded. Such bystander degradation can be due to interactions with neo-substrates, where such neo-substrates become degraded as a consequence of ubiquitination by the E3 ligase. For example, prevalent ligase ligands used in PROTACs are the clinical immuno-modulatory drugs (IMiDs) thalidomide, pomalidomide, and lenalidomide, which bind to the E3 ligase CRBN. These IMiDs recruit neo-substrates (for example, SALL4, IKZF1, IKZF3, and p63) to CRBN, which induces neo-substrate degradation. CRBN-mediated neo-substrate degradation has been linked to numerous clinical toxicities, including reproductive toxicity (SALL4 and p63), hematotoxicity, and immunotoxicity (IKZF1, IKZF3) [17]. Other adverse effects have also been reported for IMiDs; however, it is not clear if these toxicities are associated with neo-substrate degradation [17]. Hence, more focused efforts are needed to determine whether the E3 ligase ligands of PROTACs cause specific substrate degradation and assess the consequent safety liabilities. Finally, as for most drug compounds, off-target engagement can also result from binary interactions of the compound with off-target proteins as discussed above for small molecules (Figure 4.1).

A cross-pharma survey identified the main perceived issues and sources of toxicity for PROTACs, demonstrating and confirming that the focus here is on binary interactions, and neo-substrate or off-target protein degradation, as well as identifying the disruption of the proteasome function/accumulation of native substrates as an additional issue [18]. Ideally, PROTAC safety evaluations should consider both degradation-independent and degradation-dependent effects, with the former driven by target occupancy and the latter by target degradation, as well as effects on the proteasome and its substrates. Degradation-dependent selectivity and effects on the proteasome can be monitored via global proteomics, approaches that have been broadly applied across industry and academia in cells, *in vivo*, and in the clinic. When assessing the risks of degradation-mediated selectivity challenges of PROTACs, it can be crucial to consider the levels and stoichiometry of the E3 ligase and off-target protein, as well as the turnover of the off-target protein, all of which can dictate an off-target safety liability. Off-target proteins degraded by PROTACs need to also be assessed in terms of their safety risks upon reducing their levels in biologically relevant tissues. For established targets, this can be straightforward and based on preclinical and clinical insights, with the risk placed in context with tissue selectively and the efficacious exposures. However, for novel targets outside of standard off-target safety screening panels, this would require hypothesis-driven clinical translations from *in vitro* and *in vivo* data. Degradation-independent selectivity assessments are at this stage similar to that observed for other small-molecule compounds, with a high weight of evidence of safety targets screened against representative series members and extended panels screened against only lead and candidate compounds. Special considerations for such assays need to be taken into account such as solubility of these large molecules and a potential "hook effect" (bell-shaped concentration response due to saturation of binding sites). An alternative approach would be the application of affinity-based chemical proteomics. Chemical proteomics profiling would identify the binding partners of the PROTAC of interest, as well as the specific E3 ligase and POI ligands (along with their respective proteomics dissociation constants [KDs]). Such information would also allow the assessment of safety risks derived from the degradation profile linked to specific binary binding events. On a promising note, in the past five years, PROTACs have passed through several clinical trials, as mentioned mainly for oncology indications and targeting a variety of targets, with clinical benefit observed in patients; however, no PROTAC-mediated safety implications have been noted to date.

Although here we have focused on PROTACS, other targeted protein degradation approaches are emerging for example lysosome-targeting chimeras (LYTACs), antibody-based PROTACs (AbTAC), and autophagy-targeting chimeras (AUTACs) [19]. However, compared to PROTACs, these are in their infancy, and it can be expected that novel and overlapping off-target safety concerns will need to be evaluated for each.

4.4 Small Molecules Targeting RNA (smRNA)

Numerous studies have shown small molecules can bind to RNA and in many cases consequently affect cellular RNA function, with several compounds showing highly selective behavior for their RNA targets of interest [13, 20–25]. It is therefore highly likely that protein-targeting compounds have altered RNA biology via interacting with RNA molecules/targets and consequently triggered pharmacology. In the drug discovery world, this has largely been overlooked, which could in part be due to the historical assumption that RNA-binding compounds intercalate with RNA and display clear genotoxic side effects. Key examples for smRNAs come from work from pharmaceutical companies and from academia. Firstly, we have the discovery of an smRNA that targets a flavin mononucleotide (FMN) Riboswitch [26], followed by the development of splicing modulators that target the survival of motor neuron 2 (SMN2) protein for the treatment of spinal muscular atrophy (SMA) [27, 28]. All of these were identified by initially carrying out targeted phenotypic screening approaches. In addition, among many other academics, the Disney lab (https://disney.scripps.ufl.edu/) have identified numerous selective smRNA compounds modulating a variety of targets, some of which resemble present-day drugs. For example, the team found that drugs targeting topoisomerases and kinases can also bind and potentially modulate RNA, in this case microRNAs (miRNAs), with the potential of their effects on the miRNA miR-21 contributing to the clinical mode of action (MoA) of these compounds [29]. Newer research has shown a variety of chemotypes can bind RNA, many of which are helpfully stored in several public databases (SMMRNA, NALDB, and R-BIND). So far, RNA-targeting-specific chemotypes are yet to be identified; however, examples of physical properties enriched in RNA-binding compounds have been noted and include higher total charge, fraction sp3, logD, and more flat, rod-like structures, with smRNA–RNA interactions mostly driven through pi-stacking and hydrogen bonding [30]. At this stage, designing RNA-specific compounds, which are selective for RNA as opposed to other macromolecules, is still not possible; hence, we should still expect a degree of promiscuity across RNA- and protein-binding pharmacophores with compounds that target either class. On this note, compounds have been identified from affinity screening which, based on the data assessed, can selectively bind to RNA only [31]. In regard to predicting off-targets of smRNA, a combined two-dimensional combinatorial screen (2DCS) and cheminformatics platform (Inforna) used to characterize RNA hairpin secondary structure motif–compound pairs could help understand RNA off-target of such compounds by identifying these motifs throughout the transcriptome [32] and assessing selectivity *in silico* at an early stage. Another approach to this was recently published where the selectivity of an RNA-binding motif of an smRNA was predicted by looking for this motif throughout the published cellular RNAs where we have a strong idea of their structures, such as the ribosomal RNA and pre-miRNAs [33]. In the future, predictive approaches could also provide insights into the risks of protein-targeting drugs modulating RNA off-targets.

Identification of protein off-targets for RNA- and protein-targeting compounds would be identical and is discussed above and described in Figure 4.1. This would involve off-target prediction models, off-target screening panels, and possibly OMICS and target engagement approaches, for example. For the identification of chemistry-driven RNA off-targets, again the backbone of the approaches used would be the same as for protein off-target identification strategies; for example, target engagement, in silico predictions, OMICS, and panel screening can again be applied. Here, such approaches are presently being established. Practically, RNA off-targets can be assessed through pull-downs or cleavage assays in cells or tissues using key compounds of interest. Numerous probes have been reported in the literature that covalently attach to or cleave RNA target(s) upon interaction of an attached smRNA with the RNA target(s), with detection of the RNA target via 5′-RNA-linker-mediated rapid amplification of cDNA ends (5′-RLM-RACE) or transcriptomics [34–38]. Further evaluations of smRNA interactions with RNA off-targets can be assessed using target site mapping such as Chem-CLIP-Map-Seq, PEARL-seq, or SHAPE, helping to identify key functional RNA motifs/structures involved [38–40]. This information could then be used for the development of a high weight of evidence *in vitro* off-target screening panel or used in off-target predictions as discussed above. In general, the application of tool compounds and other gene modulation approaches in *in vitro* and *in vivo* systems will help validate a novel RNA off-target role in the pathology of interest and help understand translational information. Of note, at this stage, most – if not all – RNA motifs identified will require *in vitro* and *in vivo* investigations, and validations likely in the absence of known tool compounds. In addition, when considering the development and use of *in vitro* off-target panel screening for RNA off-targets, one would need to assess if the RNA structural motifs identified are amenable for panel screening based on their ability to be functionally recapitulated *in vitro*. The design of such panels would also be somewhat different to that for protein off-targets and might consist of a variety of outputs such as conformational alterations, direct binding interactions, and modulation of functional ligand binding sites. Yet translational applications across *in vitro*, preclinical, and clinical evaluation would potentially follow suit as standard. Similar to PROTACs, for smRNA, we can also consider mechanistic toxicity, such as the ability of splice-altering compounds to affect the general splicing of RNAs; the ability of compounds that target ribosomal RNA to effect the translation of RNAs in general; or the ability of compounds that target bacterial riboswitches to affect the gut microbiota in humans [41].

An extension of smRNA is RIBOnuclease-TArgeting Chimeras (RIBOTACs). A RIBOTAC is a compound that consists of a molecule which interacts with an RNA-degrading enzyme (RNase L) combined with an smRNA through a linker (Figure 4.2b). This bifunctional compound functions in a similar manner to PROTACS, with several proof-of-concept studies *in vitro* and in animals [13, 42, 43]. Safety concerns to be considered here include RNase L-mediated toxicity, off-target RNA and protein assessments, and non-standard safety screen profiling due to their unusual physicochemical properties. RNase L is ubiquitous and has been shown to be proinflammatory through its role in the activation of NLR family pyrin domain containing 3 (NLRP3), and its deficiency leads to a compromised antiviral response [44]. To date, when studied in cells and in animals, RIBOTACs have not shown any effects on the immune system [42, 45]. Cleavage of off-target RNA by RIBOTACs should be seen by transcriptomics. Transcriptomics would not, however, consider their ability to bind but not cleave RNA or protein targets. This could be assessed by similar approaches to that used for smRNA, with the caveat of additional challenges of these large molecules, e.g., reduced solubility in *in vitro* assays and a potential "hook effect" (bell-shaped concentration–response curves due to saturation of binding sites) as shown for PROTACs. Newer developed RIBOTAC equivalents, referred to as proximity-induced nucleic acid degraders (PINADs), may bypass several challenges with present-day RIBOTACs, particularly those focused on poor physicochemical properties and the requirement for RNAse L [46]. PINADs have already shown activity *in vitro*, in cellulo, and *in vivo* [46]; however, these compounds are still in the early stages at present.

4.5 Antisense Oligonucleotides

Antisense oligonucleotides (ASOs) are compounds that tend to consist of consecutive nucleic acid bases with a variety of modifications in the backbone and bases. They bind to (hybridization) RNA and alter gene expression by modulating the function of the target RNA, leading to altered splicing and altered target RNA and/or protein levels. Although these molecules are chemical in nature, their off-target evaluations are very different to that of small molecules. Their off-target evaluations can cover two main areas, hybridization-dependent and hybridization-independent, and are covered in several good reviews, for example [47, 48]. In contrast to toxicities that result from the desired target-mediated interactions (or exaggerated pharmacology) [49], hybridization-dependent off-targets result from oligonucleotide drugs binding to other targets in the same manner as that of the desired target, although how they modulate the off-targets may differ from that of the primary target (splicing versus translational inhibition, for example) [50]. Hybridization-dependent off-target effects can be initially assessed by predicting which part of the genome or transcriptome will bind your antisense oligonucleotide compound via direct hybridization. This initial analysis can then be followed with experimental validation of key off-targets [50], a full risk assessment in the context of the patient, the effect on the desired target, the amount of compound reaching the tissue of interest, the washout time of the compound, and the indication if deemed a risk. Preclinical evaluations, similar to other modalities, can further help in the risk assessment for off-targets. However, if sequence species differences exist, one could consider the use of a surrogate molecule against the desired target in the preclinical model of interest [48]. Hybridization-independent off-targets can be both dependent and independent of sequence. Sequence-independent effects tend to be driven by the concentration of the drug in the blood plasma (plasma Cmax), for example transient prolongation of coagulation time and activation of the alternative complement system, which can have effects on blood pressure and result in vascular inflammation; however, plasma concentrations rarely exceed the threshold for these effects, and *in vitro* assays can assess this risk early [51–54]. Sequence-dependent effects tend not to be dependent on the Cmax and include proinflammatory effects and accumulation-driven toxicity in high-exposure organs (such as liver and kidney). Thankfully, predictive *in vitro* models and design strategies for the latter sequence-dependent effects have been developed further reducing such ASO toxicities [55–57]. There are only a few formal guidelines for therapeutic oligos, so it is still a little unclear what the expectations are from health authorities. Best practices are usually defined by several white papers published by cross-pharma groups and the Japanese regulators [51, 58–62], and key reviews [47]. The species used in *in vivo* toxicology assessment can be crucial here based on the fact that most single-stranded phosphorothioate backbone ASOs cause rat-specific issues such as chronic progressive nephropathy, which can potentially interfere with longer-term toxicology studies [47].

4.6 Large Molecules

The structural complexity of large molecules, which can include antibodies, antibody–drug conjugates, antibody-like compounds (ScFvs, Fabs), and whole chimeric antigen receptor (CAR) cell therapies, have steric hindrance limitations for access and binding to macromolecules, thus reducing their promiscuity. Large molecules therefore interact more specifically with their binding targets, with additional stability limitations also contributing to their low risk of systemic toxicity. However, class-related adverse effects may occur; for example, engineered T-cell therapy is associated with immune-related adverse events and neurotoxicity in oncology patients [63]. CD19 CAR T-cell therapy is reportedly associated with 60% of patients experiencing neurotoxic events [64], which may be associated with on-target, off-tumor, and off-target toxicity. Cell microarray technology can be used to screen for off-target interactions and demonstrate on-target specificity. The Retrogenix Cell Microarray Technology is frequently used to identify binding by the single-chain fragment variable (scFv) of biologic therapeutics or whole CAR T cells to cell surface receptors and secreted proteins by using >4000 full-length proteins that are overexpressed in human cells [65]. This platform was used to support the clinical submission for Kymriah, a CD19 CAR T-cell therapeutic for the treatment of B-cell acute lymphoblastic leukemia [65]. Beyond this binding assay, safety profiling for large molecules is routinely performed in animal models that are similar to humans, e.g., non-human primates or humanized mouse models. More recent progress in new approach methodologies, e.g., three-dimensional organoid models, now attempt to more closely parallel the tumor microenvironment by combining tumor and immune cell to test T-cell efficacy and immunogenicity [66]. These models offer great potential but currently remain exploratory with more work required to achieve regulatory acceptance.

4.7 Regulatory Aspects

The International Council on Harmonization (ICH), which highlights the regulatory guidance provided for human pharmaceuticals, was founded in 1990, leading to a framework of standards which have significantly harmonized the data used to evaluate the quality, safety, and efficacy of new medicinal products. ICH is a global organization that connects regulators, industry, and other stakeholders. The guidelines it produces influence the data required to support clinical evaluations, e.g., IND applications (US), Common Drug Review (Canada), and Investigator's Brochures (Europe). Regulatory ICH guidance on secondary pharmacology is included in S7A, i.e., Safety Pharmacology Studies for Human Pharmaceuticals, where adverse effects related to the target site and off-target site (secondary pharmacology) are considered under the umbrella term of safety pharmacology. The objectives of safety pharmacology studies are identified as:

1. To identify undesirable pharmacodynamic properties of a substance that may have relevance to its human safety;
2. To evaluate adverse pharmacodynamic and/or pathophysiological effects of a substance observed in toxicology and/or clinical studies;
3. To investigate the mechanism of the adverse pharmacodynamic effects observed and/or suspected.

Secondary pharmacology data is expected to be provided in the clinical submission package, i.e., an IND application in the United States and the Clinical Trial Application (CTA) used in Europe. Being part of the drug development process, it can be used to provide information on the selectivity profile of the molecule, the design of safety studies, and the relevance for human safety.

Regulatory bodies, such as the Food and Drug Administration (FDA) in the United States and European Medicines Agency (EMA) in Europe, do also closely monitor the secondary pharmacology of drugs before granting approval for public use. Regulatory secondary pharmacology data for small molecules is considered to be covered by radioligand ligand binding or enzyme assay data, which suggests a potential for adverse effects, although this might not meet all needs. No further detail is provided on which off-targets are considered critical or advised. Therefore the extent, assay format, and timing of secondary pharmacology data submitted to the regulatory agencies vary. Papoian et al. discussed the relevance of secondary pharmacology data to regulatory reviewers, while also providing examples where follow-up actions would provide an increased understanding of effects that may impact patient safety [67]. As discussed earlier, since the publication of 44 targets with clear flags for human safety by a pharma consortium in 2012 [5], most regulatory secondary pharmacology screening datasets include these targets [9]. Yet additional testing should be considered based on scientific knowledge of the target, dosing route, local tissue exposures, metabolite profile, and excretion route, with generally broader panels assessed in the later phases of drug discovery. Regulators may also request profiling in additional assays associated with patient risk, e.g., for CNS therapeutics targets associated with suicidality ideation and behavior (5-HT1A antagonist, 5-HT2A agonist, 5-HTT antagonist, CaN antagonist, CB1 antagonist, D1 antagonist, NET antagonist, nACh agonist, VMAT2 antagonist) [68]. Such requests may be made at any stage of the drug development process, including the clinical phase for adverse effect exploration. Novel early screening approaches involving *in silico* assessments and promiscuity panels are yet to be officially applied in regulatory documentations.

4.8 Future Outlook

Secondary pharmacology evaluation of pipeline small molecules is historical and mainstream, with processes in place within most pharmaceutical companies and many biotechs. Here, factors such as translatability of *in vitro* assessments of key targets (functional and binding data) across preclinical species, the use of cost-effective modular screening panels allowing a combination of different targets to be screened, and finally translational insights from *in vitro* target specific screens to target-related human adverse effects are example areas that would benefit from future resources. On the contrary, assessments of protein off-target evaluations for newer modalities such as PROTACs and nucleic-acid-targeting oligonucleotides are much more in their infancy, with a focus on specific mechanistic understandings, *in silico* predictions, and/or OMICS approaches. Potentially, in time, high weight of evidence of off-target screening might also be applied here for these newer modalities; however, to date, these compounds are not commonly screened in such panels due to factors such as their physicochemical properties and/or a lack of a good rationale. In the case of oligonucleotides, bound proteins that could direct adverse safety properties of these compounds have been identified (as well as general pharmacokinetics and pharmacodynamics); however, question marks still exist on whether these nucleic acids are really ever naked in the cell and/or whether these interactions are non-functional or artificial. Other aspects of secondary pharmacology that have not been discussed here include the use of drug combinations and organoid systems and how off-target evaluations and mitigations should be applied in these cases; hopefully, we will hear more about these aspects in the near future.

Acknowledgments

We would like to thank Dimitar Yonchev for providing helpful comments and proofreading this manuscript.

Addendum

Since the compilation of this chapter, Brennan et al [69] reported on a cross-company survey which explored the current state of secondary pharmacology screening. This includes recommendations to extend assessments beyond current safety-associated panels by the addition of further safety-associated targets, many of which are kinases.

References

1 Tambuyzer, E., Vandendriessche, B., Austin, C.P. et al. (2020). Therapies for rare diseases: therapeutic modalities, progress and challenges ahead. *Nat. Rev. Drug Discov.* 19 (2): 93–111.

2 Lipinski, C.A. (2004). Lead- and drug-like compounds: the rule-of-five revolution. *Drug Discov. Today Technol.* 1 (4): 337–341.

3 Volkamer, A., Riniker, S., Nittinger, E. et al. (2023). Machine learning for small molecule drug discovery in academia and industry. *Artif. Intell. Life Sci.* 3: 100056.

4 Bendels, S., Bissantz, C., Fasching, B. et al. (2019). Safety screening in early drug discovery: an optimized assay panel. *J. Pharmacol. Toxicol. Methods* 99: 106609.

5 Bowes, J., Brown, A.J., Hamon, J. et al. (2012). Reducing safety-related drug attrition: the use of in vitro pharmacological profiling. *Nat. Rev. Drug Discov.* 11 (12): 909–922.

6 Lynch, J.J., Vleet, T.R.V., Mittelstadt, S.W., and Blomme, E.A.G. (2017). Potential functional and pathological side effects related to off-target pharmacological activity. *J. Pharmacol. Toxicol. Methods* 87: 108–126.

7 Nguyen, P.A., Born, D.A., Deaton, A.M. et al. (2019). Phenotypes associated with genes encoding drug targets are predictive of clinical trial side effects. *Nat. Commun.* 10 (1): 1579.

8 Jenkinson, S., Schmidt, F., Ribeiro, L.R. et al. (2020). A practical guide to secondary pharmacology in drug discovery. *J. Pharmacol. Toxicol. Methods* 105: 106869.

9 Valentin, J.-P., Guillon, J.-M., Jenkinson, S. et al. (2018). In vitro secondary pharmacological profiling: An IQ-DruSafe industry survey on current practices. *J. Pharmacol. Toxicol. Methods* 93: 7–14.

10 Li, K. and Crews, C.M. (2022). PROTACs: past, present and future. *Chem. Soc. Rev.* 51 (12): 5214–5236.

11 Pettersson, M. and Crews, C.M. (2019). PROteolysis TArgeting Chimeras (PROTACs) – past, present and future. *Drug Discov. Today Technol.* 31: 15–27.

12 Békés, M., Langley, D.R., and Crews, C.M. (2022). PROTAC targeted protein degraders: the past is prologue. *Nat. Rev. Drug Discov.* 21 (3): 181–200.

13 Haniff, H.S., Tong, Y., Liu, X. et al. (2020). Targeting the SARS-CoV-2 RNA genome with small molecule binders and ribonuclease targeting chimera (RIBOTAC) degraders. *ACS Central Sci.* 6 (10): 1713–1721.

14 Gao, H., Sun, X., and Rao, Y. (2020). PROTAC technology: opportunities and challenges. *ACS Med. Chem. Lett.* 11 (3): 237–240.

15 Guenette, R.G., Yang, S.W., Min, J. et al. (2022). Target and tissue selectivity of PROTAC degraders. *Chem. Soc. Rev.* 51 (14): 5740–5756.

16 Cecchini, C., Pannilunghi, S., Tardy, S., and Scapozza, L. (2021). From conception to development: investigating PROTACs features for improved cell permeability and successful protein degradation. *Front. Chem.* 9: 672267.

17 Moreau, K., Coen, M., Zhang, A.X. et al. (2020). Proteolysis-targeting chimeras in drug development: a safety perspective. *Br. J. Pharmacol.* 177 (8): 1709–1718.

18 Hemkens, M., Stamp, K., Loberg, L.I. et al. (2023). Industry perspective on the nonclinical safety assessment of heterobifunctional degraders. *Drug Discov. Today* 28 (8): 103643.

19 Zhao, L., Zhao, J., Zhong, K. et al. (2022). Targeted protein degradation: mechanisms, strategies and application. *Signal Transduct. Target. Ther.* 7 (1): 113.

20 Abulwerdi, F.A., Xu, W., Ageeli, A.A. et al. (2019). Selective small-molecule targeting of a triple helix encoded by the long noncoding RNA, MALAT1. *ACS Chem. Biol.* 14 (2): 223–235.

21 Haniff, H.S., Knerr, L., Chen, J.L. et al. (2020). Target-directed approaches for screening small molecules against RNA targets. *SLAS Discov.* 25 (8): 869–894.

22 Haniff, H.S., Liu, X., Tong, Y. et al. (2022). A structure-specific small molecule inhibits a miRNA-200 family member precursor and reverses a type 2 diabetes phenotype. *Cell Chem. Biol.* 29 (2): 300–311.e10.

23 Rizvi, N.F., Howe, J.A., Nahvi, A. et al. (2018). Discovery of selective RNA-binding small molecules by affinity-selection mass spectrometry. *ACS Chem. Biol.* 13 (3): 820–831.

24 Sivaramakrishnan, M., McCarthy, K.D., Campagne, S. et al. (2017). Binding to SMN2 pre-mRNA-protein complex elicits specificity for small molecule splicing modifiers. *Nat. Commun.* 8 (1): 1476.

25 Zhang, P., Park, H.-J., Zhang, J. et al. (2020). Translation of the intrinsically disordered protein α-synuclein is inhibited by a small molecule targeting its structured mRNA. *Proc. Natl. Acad. Sci. USA* 117 (3): 1457–1467.

26 Howe, J.A., Wang, H., Fischmann, T.O. et al. (2015). Selective small-molecule inhibition of an RNA structural element. *Nature* 526: 672–677.

27 Naryshkin, N.A.; Weetall, M.; Dakka, A.; Narasimhan, J.; Zhao, X.; Feng, Z.; Ling, K.K.Y.; Karp, G.M.; Qi, H.; Woll, M.G.; Chen, G.; Zhang, N.; Gabbeta, V.; Vazirani, P.; , Bhattacharyya, A.; Furia, B.; Risher, N.; Sheedy, J.; Kong, R.; Ma, J.; Turpoff, A.; Lee, C.-S.; Zhang, X.; Moon, Y.-C.; Trifillis, P.; Welch, E.M.; Colacino, J.M.; Babiak, J.; Almstead, N.G.; Peltz, S.W.; Eng, L.A.; Chen, K.S.; Mull, J.L.; Lynes, M.S.; Rubin, L.L.; Fontoura, P.; Santarelli, L.; Haehnke, D.; McCarthy, K.D.; Schmucki, R.; Ebeling, M.; Sivaramakrishnan, M,; Ko, C.-P.; Paushkin, S.V.; Ratni, H.; Gerlach, I.; Ghosh, A.; Metzger, F.; (2014), Motor neuron disease. SMN2 splicing modifiers improve motor function and longevity in mice with spinal muscular atrophy. *Science* 345, 688–693.

28 Palacino, J., Swalley, S.E., Song, C. et al. (2015). SMN2 splice modulators enhance U1–pre-mRNA association and rescue SMA mice. *Nat. Chem. Biol.* 11 (7): 511–517.

29 Velagapudi, S.P., Costales, M.G., Vummidi, B.R. et al. (2018). Approved anti-cancer drugs target oncogenic non-coding RNAs. *Cell Chem. Biol.* 25 (9): 1086–1094.e7.

30 Padroni, G., Patwardhan, N.N., Schapira, M., and Hargrove, A.E. (2020). Systematic analysis of the interactions driving small molecule–RNA recognition. *RSC Med. Chem.* 11 (7): 802–813.

31 Rizvi, N.F.; Maria, J.P., Jr; Nahvi, A.; Klappenbach, J.; Klein, D.J.; Curran, P.J.; Richards, M.P.; Chamberlin, C.; Saradjian, P.; Burchard, J.; Aguilar, R.; Lee, J.T.; Dandliker, P.J.; Smith, G.F.; Kutchukian, P.; Nickbarg, E.B. (2020), Targeting RNA with small molecules: identification of selective, RNA-binding small molecules occupying drug-like chemical space. *SLAS Discov.* 25, 4, 384–396.

32 Disney, M.D., Winkelsas, A.M., Velagapudi, S.P. et al. (2016). Inforna 2.0: a platform for the sequence-based design of small molecules targeting structured RNAs. *ACS Chem. Biol.* 11 (6): 1720–1728.

33 Tong, Y., Lee, Y., Liu, X. et al. (2023). Programming inactive RNA-binding small molecules into bioactive degraders. *Nature* 618 (7963): 169–179.

34 Benhamou, R.I., Angelbello, A.J., Andrews, R.J. et al. (2020). Structure-specific cleavage of an RNA repeat expansion with a dimeric small molecule is advantageous over sequence-specific recognition by an oligonucleotide. *ACS Chem. Biol.* 15 (2): 485–493.

35 Disney, M.D., Velagapudi, S.P., Li, Y. et al. (2019). Identifying and validating small molecules interacting with RNA (SMIRNAs). *Methods Enzymol.* 623 (2019): 45–66.

36 Lasham, A., Herbert, M., ‘t Wallant, N.C. et al. (2010). A rapid and sensitive method to detect siRNA-mediated mRNA cleavage in vivo using 5′ RACE and a molecular beacon probe. *Nucleic Acids Res.* 38 (3): e19–e19.

37 Suresh, B.M., Li, W., Zhang, P. et al. (2020). A general fragment-based approach to identify and optimize bioactive ligands targeting RNA. *Proc. Natl. Acad. Sci.* 117 (52): 33197–33203.

38 Mukherjee, H., Blain, J.C., Vandivier, L.E. et al. (2020). PEARL-seq: a photoaffinity platform for the analysis of small molecule-RNA interactions. *ACS Chem. Biol.* 15 (9): 2374–2381.

39 Velagapudi, S.P., Li, Y., and Disney, M.D. (2019). A cross-linking approach to map small molecule-RNA binding sites in cells. *Bioorg. Med. Chem. Lett.* 29 (12): 1532–1536.

40 Mailler, E., Paillart, J., Marquet, R. et al. (2019). The evolution of RNA structural probing methods: from gels to next-generation sequencing. *Wiley Interdiscip. Rev.: RNA* 10 (2): e1518.

41 Zhang, Y.-J., Li, S., Gan, R.-Y. et al. (2015). Impacts of gut bacteria on human health and diseases. *Int. J. Mol. Sci.* 16 (4): 7493–7519.

42 Costales, M.G., Aikawa, H., Li, Y. et al. (2020). Small-molecule targeted recruitment of a nuclease to cleave an oncogenic RNA in a mouse model of metastatic cancer. *Proc. Natl. Acad. Sci.* 117 (5): 2406–2411.

43 Disney, M.D., Suresh, B.M., Benhamou, R.I., and Childs-Disney, J.L. (2020). Progress toward the development of the small molecule equivalent of small interfering RNA. *Curr. Opin. Chem. Biol.* 56: 63–71.

44 Zhao, C. and Zhao, W. (2020). NLRP3 inflammasome – a key player in antiviral responses. *Front. Immunol.* 11: 211.

45 Zhang, P., Liu, X., Abegg, D. et al. (2021). Reprogramming of protein-targeted small-molecule medicines to RNA by ribonuclease recruitment. *J. Am. Chem. Soc.* 143 (33): 13044–13055.

46 Mikutis, S., Rebelo, M., Yankova, E. et al. (2023). Proximity-induced nucleic acid degrader (PINAD) approach to targeted RNA degradation using small molecules. *ACS Central Sci.* 9 (5): 892–904.

47 Frazier, K.S. (2015). Antisense oligonucleotide therapies. *Toxicol. Pathol.* 43 (1): 78–89.

48 Goyenvalle, A., Jimenez-Mallebrera, C., van Roon, W. et al. (2023). Considerations in the preclinical assessment of the safety of antisense oligonucleotides. *Nucleic Acid Ther.* 33 (1): 1–16.

49 Kornbrust, D., Cavagnaro, J., Levin, A. et al. (2013). Oligo safety working group exaggerated pharmacology subcommittee consensus document. *Nucleic Acid Ther.* 23 (1): 21–28.

50 Lindow, M., Vornlocher, H.-P., Riley, D. et al. (2012). Assessing unintended hybridization-induced biological effects of oligonucleotides. *Nat. Biotechnol.* 30 (10): 920–923.

51 Henry, S.P., Seguin, R., Cavagnaro, J. et al. (2016). Considerations for the characterization and interpretation of results related to alternative complement activation in monkeys associated with oligonucleotide-based therapeutics. *Nucleic Acid Ther.* 26 (4): 210–215.

52 Shen, L., Engelhardt, J.A., Hung, G. et al. (2016). Effects of repeated complement activation associated with chronic treatment of cynomolgus monkeys with 2′-o-methoxyethyl modified antisense oligonucleotide. *Nucleic Acid Ther.* 26 (4): 236–249.

53 Crooke, S.T., Baker, B.F., Kwoh, T.J. et al. (2016). Integrated safety assessment of 2′-o-methoxyethyl chimeric antisense oligonucleotides in nonhuman primates and healthy human volunteers. *Mol. Ther.* 24 (10): 1771–1782.

54 Yu, R.Z., Kim, T.-W., Hong, A. et al. (2007). Cross-species pharmacokinetic comparison from mouse to man of a second-generation antisense oligonucleotide, ISIS 301012, targeting human apolipoprotein B-100. *Drug Metab. Dispos.* 35 (3): 460–468.

55 Dieckmann, A., Hagedorn, P.H., Burki, Y. et al. (2018). A sensitive in vitro approach to assess the hybridization-dependent toxic potential of high affinity Gapmer oligonucleotides. *Mol. Ther. Nucleic. Acids* 10: 45–54.

56 Moisan, A., Gubler, M., Zhang, J.D. et al. (2017). Inhibition of EGF uptake by nephrotoxic antisense drugs in vitro and implications for preclinical safety profiling. *Mol. Ther. Nucleic. Acids* 6: 89–105.

57 Paz, S., Hsiao, J., Cauntay, P. et al. (2017). The distinct and cooperative roles of toll-like receptor 9 and receptor for advanced glycation end products in modulating in vivo inflammatory responses to select CpG and Non-CpG oligonucleotides. *Nucleic Acid Ther.* 27 (5): 272–284.

58 Berman, C.L., Barros, S.A., Galloway, S.M. et al. (2016). OSWG recommendations for genotoxicity testing of novel oligonucleotide-based therapeutics. *Nucleic Acid Ther.* 26 (2): 73–85.

59 Henry, S.P., Arfvidsson, C., Arrington, J. et al. (2022). Assessment of the immunogenicity potential for oligonucleotide-based drugs. *Nucleic Acid Ther.* 32 (5): 369–377.

60 Berman, C.L., Cannon, K., Cui, Y. et al. (2014). Recommendations for safety pharmacology evaluations of oligonucleotide-based therapeutics. *Nucleic Acid Ther.* 24 (4): 291–301.

61 Hirabayashi, Y., Maki, K., Kinoshita, K. et al. (2021). Considerations of the Japanese Research Working Group for the ICH S6 & related issues regarding nonclinical safety assessments of oligonucleotide therapeutics: comparison with those of biopharmaceuticals. *Nucleic Acid Ther.* 31 (2): 114–125.

62 Tessier, Y., Achanzar, W., Mihalcik, L. et al. (2021). Outcomes of the European Federation of Pharmaceutical Industries and Associations oligonucleotide working group survey on nonclinical practices and regulatory expectations for therapeutic oligonucleotide safety assessment. *Nucleic Acid Ther.* 31 (1): 7–20.

63 Sterner, R.C. and Sterner, R.M. (2021). CAR-T cell therapy: current limitations and potential strategies. *Blood Cancer J.* 11 (4): 69.

64 Donnadieu, E., Luu, M., Alb, M. et al. (2022). Time to evolve: predicting engineered T cell-associated toxicity with next-generation models. *J. Immunother. Cancer* 10 (5): e003486.

65 Freeth, J. and Soden, J. (2020). New advances in cell microarray technology to expand applications in target deconvolution and off-target screening. *SLAS Discov.* 25 (2): 223–230.

66 Mu, P., Zhou, S., Lv, T. et al. (2023). Newly developed 3D in vitro models to study tumor–immune interaction. *J. Exp. Clin. Cancer Res.* 42 (1): 81.

67 Papoian, T., Chiu, H.-J., Elayan, I. et al. (2015). Secondary pharmacology data to assess potential off-target activity of new drugs: a regulatory perspective. *Nat. Rev. Drug Discov.* 14 (4): 294–294.

68 Muller, P.Y., Dambach, D., Gemzik, B. et al. (2015). Integrated risk assessment of suicidal ideation and behavior in drug development. *Drug Discov. Today* 20 (9): 1135–1142.

69 Brennan, R.J., Jenkinson, S., Brown, A. et al. (2024). The state of the art in secondary pharmacology and its impact on the safety of new medicines. *Nat Rev Drug Discov* 23: 525–545. https://doi.org/10.1038/s41573-024-00942-3.

5

Molecular Properties and Structural Motifs Related to Pharmacological Promiscuity

Jens-Uwe Peters

5.1 Introduction

The discovery of the first β-blocker, propranolol, in 1967 provided a novel treatment option for hypertension and other cardiovascular disorders [1]. Similarly, the introduction of the β-agonist, isoprenaline, in 1947 improved the treatment of asthma. However, these drugs came with side effects. The adrenergic β-receptor is expressed in both the heart and the lungs. Propranolol led to dangerous bronchoconstriction in asthmatic patients, due to its antagonism of β-receptors in the lungs. Isoprenaline was dangerous when accidentally overdosed, probably because of its agonizing effect on the β-receptors in the heart [2]. In 1967, it was recognized that two types of β-receptors exist. The β_1 receptor is located mainly in the heart, and the β_2 receptor in the lungs. This gave the impetus for the discovery of a second generation of β-blockers, such as atenolol (1976). Atenolol is selective for the β_1 receptor and is therefore safe to use in asthmatic patients. Similarly, β_2-selective agonists were introduced as safe bronchodilators for the treatment of asthma, such as salbutamol in 1967. These are the earliest examples of improving drug safety by improving target selectivity.

Selective serotonin reuptake inhibitor (SSRI) antidepressants are another example of improving drug safety by improving target selectivity. Pharmacologically promiscuous tricyclic antidepressants are lethal in overdose due to their off-target effects on cardiac ion channels (e.g., imipramine, 1957). This is an important drawback in depressed patients, because of potential suicide attempts. After the discovery of serotonin reuptake inhibition as the principal mode of action, SSRIs were introduced and are today regarded as exceptionally safe drugs for the treatment of depression (e.g., fluoxetine, 1987) [3].

Selectivity as a concept in drug discovery evolved further in the 1990s. Advances in molecular biology allowed the setup of "safety panels" – a collection of 40–180 (or more) safety-relevant targets for the screening of drug candidates. It became the goal of many drug discovery programs to achieve high activity at a single therapeutic target and high selectivity over unintended targets. More recently, smaller (and therefore cheaper) panels were proposed for the screening of early compounds [4, 5].

In the years 1996–2001, an unusual number of drugs were withdrawn from the market due to adverse effects. Some of these adverse effects were traced to unintended off-target activity, such as blockade of the cardiac hERG channel (astemizole, terfenadine, grepafloxacin) or agonism of the 5-HT_{2B} receptor (fenfluramine) [6–8]. Other off-target activities were linked to adverse effects as well. For instance, orthostatic hypotension, which may lead to fainting, was linked to α_1 adrenergic receptor antagonism [9], whereas weight gain was attributed to histamine H_1 receptor antagonism [10]. These findings accentuated the importance of selectivity over "anti-targets," i.e., targets that cause adverse effects [11].

From 2006 onwards, it was recognized that some molecular properties are related to pharmacological promiscuity, i.e., the tendency to "hit" multiple targets. Similarly, some structural motifs were found to be predictive for promiscuity across large target families, such as aminergic G-protein-coupled receptors (GPCRs) or kinases, and small target families, such as metalloproteases or carbonic anhydrases. The following sections summarize these studies.

5.2 Basicity and Protonation State

The protonation state of a compound depends on its basicity: compounds with a [b]pK_a of 7 or higher will be protonated at near-neutral pH. Such compounds are generally more promiscuous than neutrals, acids, and zwitterions in typical safety screens [12–14].

An analysis of the BioPrint® dataset showed that the promiscuity of basic compounds increases sharply at a

Polypharmacology: Strategies for Multi-Target Drug Discovery, First Edition. Edited by Jens-Uwe Peters.

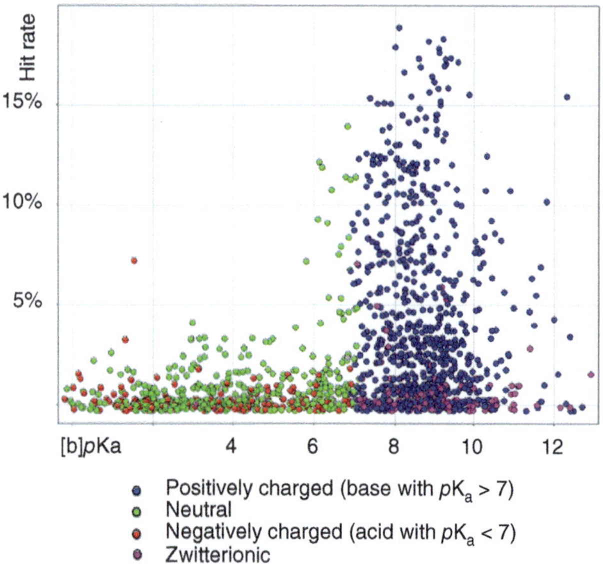

Figure 5.1 Analysis of compounds with a basic center in the BioPrint® dataset: Compounds with a basic center and a [b]pK_a > 6 are (partly) protonated under assay conditions and are more promiscuous than compounds with a [b]pK_a < 6. Zwitterions have a reduced promiscuity, similar to unprotonated compounds. The hit rate of a compound is the percentage of targets across the whole BioPrint® panel (excluding CYP 450 enzymes), which are "hit" by this compound, i.e., which bind the compound with a pIC_{50} > 6 in a radioligand displacement assay. For instance, the topmost datapoint in the diagram represents a compound with sub-micromolar affinity for approximately 18% of the BioPrint® targets.

[b]pK_a of 6 (Figure 5.1). This suggests that the protonation state determines promiscuity, rather than merely the presence of a basic center. For a basic compound with a [b]pK_a of 6, 10% of all molecules will be protonated at neutral pH, according to the Henderson–Hasselbalch equation. This fraction of protonated compounds is large enough to elicit a pharmacological response. In contrast, basic compounds with a [b]pK_a < 6 are not significantly protonated under assay conditions, and their promiscuity potential is comparable to neutrals and acids. (The BioPrint® dataset is a nearly complete matrix of approximately 2400 drugs and drug-like compounds, screened against 141 safety-relevant targets [15].)

The promiscuity of basic compounds with a [b]pK_a > 6 stems from interactions with a limited set of targets, such as aminergic GPCRs, cationic ion channels, opioid receptors, and biogenic amine transporters [13, 14, 16]. These targets are known to bind positively charged ligands. Many of these targets are safety-relevant. Table 5.1 lists some such targets, which are frequently hit by basic compounds and may cause adverse effects. These targets accept a wide range of compounds, so that >20% of all protonable compounds

Table 5.1 Targets binding >20% of all basic compounds in the in the BioPrint® dataset with sub-micromolar affinity. "Basic" means here the presence of a basic center with a [b]pK_a > 6. "Hit rate" means the percentage of targets across the whole BioPrint® panel (excluding CYP 450 enzymes), which bind the compound with a pIC_{50} > 6 in a radioligand displacement assay.

Receptor	Hit rate of basic compounds	Associated adverse effects [17]
5-HT_{2B}	36%	Valvulopathy, pulmonary hypertension (agonists); lethargy, sleep disorders (antagonist)
5-HT_{2A}	30%	Cognitive impairment, hallucination (agonists)
α_{1b}	27%	Orthostatic hypotension (antagonists)
D_{2S}	26%	Confusion, emesis (agonists); orthostatic hypotension (antagonists)
α_{2b}	23%	Sedation (agonists)
5-HT_{2c}	21%	Anxiety (agonists); weight gain (antagonists)
α_{2c}	21%	Auditory hallucination, paranoia
D_3	21%	Tardive dyskinesia, extrapyramidal disorder (antagonists)

in the BioPrint® dataset bind to them [13]. This percentage is given as the "hit rate" (for definition, see table legend).

Often, the activity of a compound at one of these targets is predictive for activities at other similar targets. For instance, a compound binding to the serotonergic 5-HT_{2B} receptor will bind with >50% probability to numerous other 5-HT receptor subtypes, as well as to the D_1–D_3, α_1, α_2, and H_1 receptors ("binding" here means having sub-micromolar affinity). Affinity for one of the muscarinic receptors usually predicts affinity for all five of them [18]. Screening at a few targets may therefore be sufficient to evaluate a compound's potential for general promiscuity in a cost-efficient way [13, 16, 19].

Because of the high promiscuity potential of compounds with a [b]pK_a > 6, such compounds should be screened early in the drug discovery process, e.g., at the hit-to-lead stage. This early screening can be done cost-efficiently at a limited set of predictive targets, e.g., the six targets in the header line of Figure 5.2. For more sophisticated panels comprising a larger set of targets, including non-aminergic targets with lower hit rates but high safety relevance, see [4, 5].

5.2.1 Mitigating the Promiscuity of Basic Compounds

An obvious strategy to mitigate the promiscuity of basic compounds is the omission or replacement of the basic

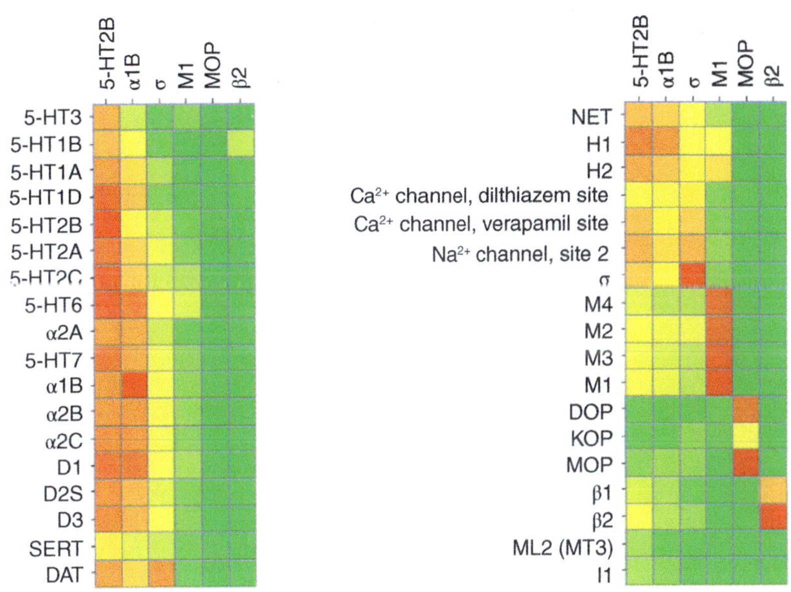

Figure 5.2 A small panel of representative, frequently hit targets may be useful for the early identification of promiscuous compounds. The heat maps show the sensitivity of six selected assays (header line) for each of the most frequently hit targets (columns left of heat maps). The six selected receptors in the header line are predictive for the majority of the frequently hit receptors. Heat map colors code the sensitivity of assays A (six columns) for assays B (36 rows) (number of compounds hitting A and B/number of compounds hitting B; pure red = 100%, pure yellow = 50%, pure green = 0). The targets are grouped according to the number of shared ligands: each assay was described by a vector of pIC_{50} values of all the compounds measured against the assay. The assays were clustered using hierarchical clustering (complete linkage, similarity metric: Pearson's correlation coefficient). Reproduced with permission from [13].

center. If this is not possible, the following strategies may be tried:

5.2.1.1 Reduction of Basicity below a [b]pK_a of 6

Figure 5.1 suggests that the reduction of basicity below a [b]pK_a of 6 reduces the potential for promiscuity. A matched pairs analysis of AstraZeneca compounds [16] corroborates this notion: all pairs with a > 0.5 log unit reduction in basicity show also a reduction in promiscuity in a panel of 130 enzymes, ion channels, and receptors. Some of these molecular pairs are given in Figure 5.3. Excellent reviews on the basicity of amines and the modulation of amine [b]pK_a may inspire the design of improved compounds with reduced basicities [20, 21].

5.2.1.2 Creation of a Zwitterion

Another strategy is the creation of a zwitterion, e.g., by introducing a COOH function into a basic compound. The promiscuity of zwitterions is comparable to neutrals and acids (Figure 5.1). A case in point is the antihistamines terfenadine and fexofenadine (Figure 5.4). Terfenadine hits five off-targets in the BioPrint® panel with sub-micromolar affinity, whereas fexofenadine hits none. Terfenadine was withdrawn from the market because it caused cardiac arrhythmia in some patients, likely due to a blockade of the hERG channel. It was replaced by fexofenadine without this liability, which is today an important treatment for allergies.

5.2.1.3 Reduction of the Number of Aromatic Rings to 1 or 0

Basic compounds with two or more aromatic rings are more liable to promiscuity than compounds with just one or no

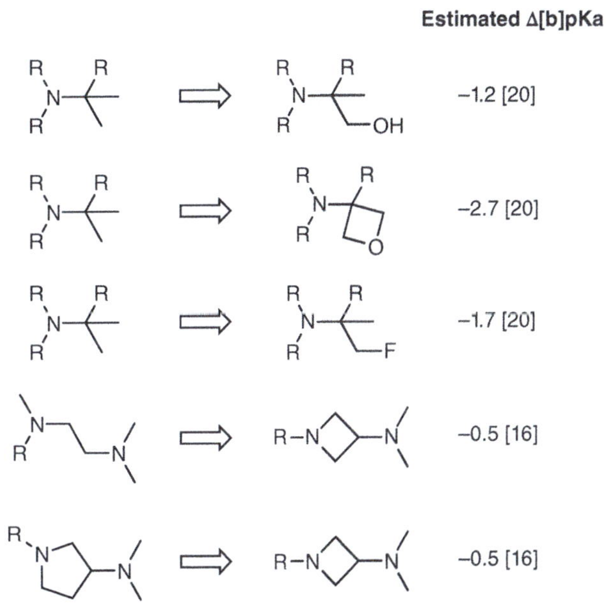

Figure 5.3 Matched molecular pairs leading to a reduction in general promiscuity in a panel of 130 enzymes, ion channels, and receptors. Brown et al. [16]/American Chemical Society.

aromatic ring. Whereas basic compounds with two or three aromatic rings had average hit rates of 6% in the BioPrint® dataset, basic compounds with just one aromatic ring had an average hit rate of 2% and basic compounds with no aromatic ring had an average hit rate of 0.6% and were comparable to neutral compounds with no aromatic ring [13].

5.2.1.4 Replacement of a Tertiary or Secondary Amine by a Primary Amine

A Pfizer study showed that primary amines are less promiscuous than secondary and tertiary amines: on average, in an

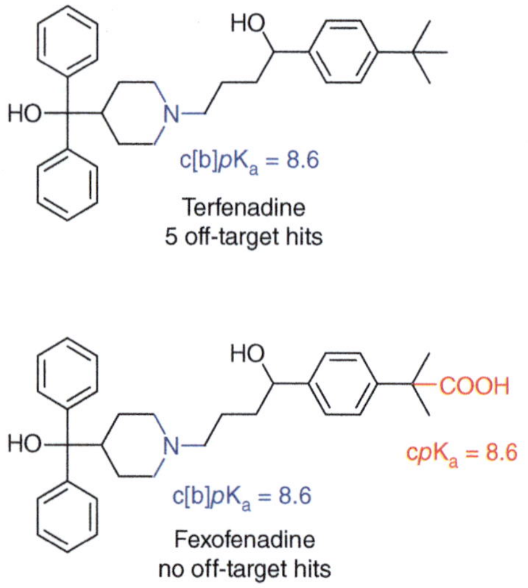

Figure 5.4 The zwitterionic fexofenadine hits no off-targets in the BioPrint® panel, whereas the positively charged terfenadine hits five off-targets with an $IC_{50} < 1\,\mu M$. (Both compounds also "hit" additionally their therapeutic target, the histamine H_1 receptor). Reproduced with permission of Peters et al. [13]/Elsevier.

assay panel of 100 targets, secondary and tertiary amines hit approximately 10 targets, whereas primary amines hit only approximately 4 targets. For comparison, neutrals hit on average 1.8 targets. (In this study, >70% single point activity at 10 μM was considered active.) Primary amines had also a higher median toxicity threshold (>6 μM) than secondary and tertiary amines (~1 μM) [22].

5.2.2 Should We Avoid Basic Amines?

In recent years, scientists have tended to avoid basic amines as drug candidates. For instance, an analysis of Pfizer compounds proceeding to toxicology studies noted a decline in basic compounds. In follow-up conversations, concerns regarding pharmacological promiscuity were brought up as a reason to avoid basic amines [22].

Nevertheless, for each decade since 1950, basic compounds were over-represented in approved drugs, as compared to neutrals and acids [23]. This may be due to the favorable physicochemical and pharmacokinetic properties of amines [24]. Amines tend to be better soluble than neutrals and tend to have a lower plasma protein binding than acids. Moreover, amines with a [b]pK_a around seven usually permeate biological membranes well, possibly due to their equilibrium between protonation states [25]. Finally, a range of targets seems to require basic amines. For instance, all DPP-IV inhibitors, and all aminergic GPCR, NOP, and opioid receptor agonists, are basic amines.

It may also be noted that some basic drugs may have a binding affinity for aminergic off-targets, which do not translate into adverse effects. For instance, SSRIs are not truly selective, but bind to several aminergic targets, such as the 5-HT_{2a} and 5-HT_{2c} receptors (fluoxetine), the α_{1b} receptor (sertraline), or the M_1 receptor (paroxetine). Nevertheless, these drugs are regarded as safe. Also, some frequently hit targets have no clear links to adverse effects. For instance, the serotonergic receptors 5-$HT_{4\text{-}7}$ attract high hit rates but are not associated with adverse effects [26]. The serotonergic 5-HT_{5a} receptor is a case in point: ligands of this receptor do not produce adverse behavioral or physiological responses [27], and knock-out animals show a very mild phenotype [28].

More generally, a study of AstraZeneca clinical candidates showed that polypharmacological compounds had no increased propensity for safety findings in preclinical or phase I studies, which would have led to their termination [29].

For these reasons, compounds with a basic center should not be generally avoided, but should be screened early in a drug discovery project against a small set of amine-preferring, yet diverse, targets (e.g., the targets in the header line of Figure 5.2). Basic compounds also have an increased propensity for hERG blockade, CYP2D6 inhibition, and phospholipidosis [30, 31]. These mechanisms can potentially cause adverse effects and should be tested early in a drug discovery project as well.

5.3 Lipophilicity

Lipophilic compounds tend to be more promiscuous in almost all published studies [13, 14, 16, 19, 32–34]. An illustrative example is β-blockers, which have similar structures but span a wide range of lipophilicity (Figure 5.4). Less lipophilic β-blockers with a clogP < 2 are less promiscuous and do not hit more than five off-targets in the BioPrint® panel, whereas all lipophilic β-blockers with a clogP > 2.5 are promiscuous and hit between 7 and 20 off-targets.

Also, a matched-pairs analysis of AstraZeneca compounds [16] shows that most pairs with a substantial reduction in lipophilicity also show a reduction in promiscuity in their panel of 130 targets. A few examples are shown in Figure 5.6.

The influence of some other parameters is predictable from their relationship with lipophilicity: compounds with a lower polar surface area tend to be more lipophilic and tend to be more promiscuous [35]. The same is true for compounds with fewer hydrogen bond donors or acceptors and fewer aromatic rings [12, 13, 35, 36].

However, to substantially reduce the promiscuity potential within a compound series, a reduction in lipophilicity by several orders of magnitude may be necessary. This

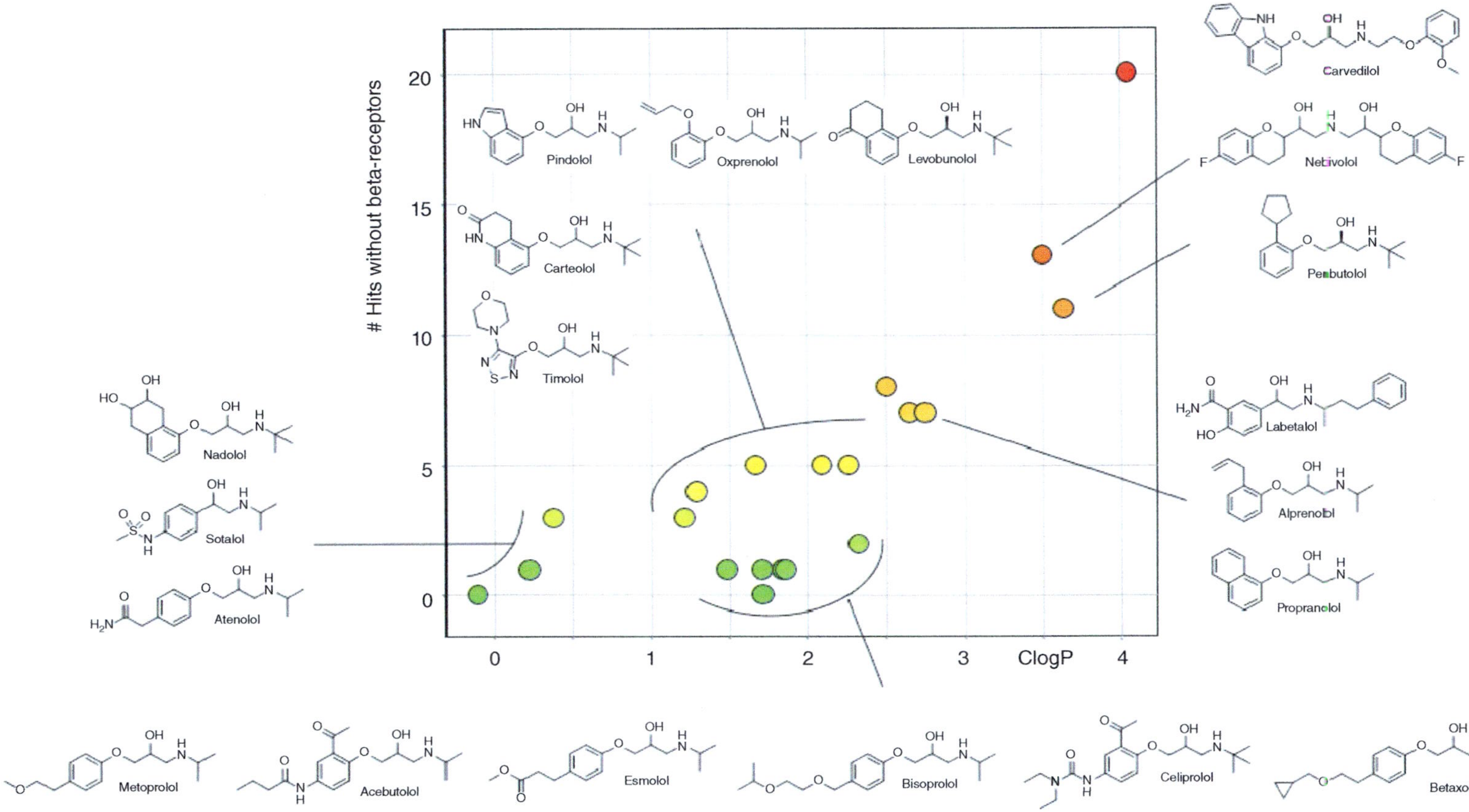

Figure 5.5 Dependence of off-target hits on the calculated lipophilicity of marketed β-blockers. Lipophilic β-blockers tend to be more promiscuous. "Off-target hits" denotes the number of sub-micromolar hits in a safety screening panel of 129 targets (BioPrint® dataset excluding β-adrenergic receptors and CYP enzymes).

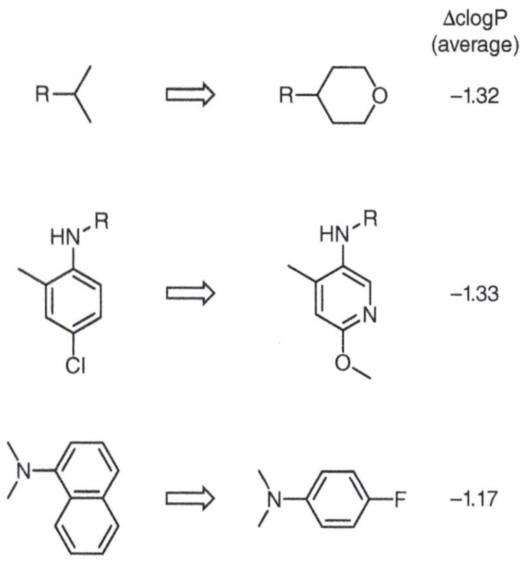

Figure 5.6 Matched molecular pairs leading to a reduction in general promiscuity in a panel of 130 enzymes, ion channels, and receptors. Brown et al. [16]/American Chemical Society.

will be difficult to achieve in most series. For inspiration, an excellent analysis of matched pairs and lipophilicity can be found at [37]. An instance of successful promiscuity optimization by reduction in lipophilicity has been published [38].

5.4 Molecular Weight

A link between molecular weight (MW) and promiscuity is often assumed. However, results from numerous studies are contradictory: MW and promiscuity correlate in Roche HTS data [39] and in AstraZeneca and Novartis safety panel data [16, 40], but correlate inversely in Pfizer HTS data [32] and show no correlation in Takeda project compounds [19]. The BioPrint® data shows a positive correlation for neutral compounds and a more complex relationship with basic compounds, with the maximum promiscuity in the drug-like MW range of 300–500 [12, 13]. Ligands in the Protein Data Bank show no correlation [41]. In analyses of high-potency promiscuity, a positive correlation was found among ChEMBL compounds [42], but not among Roche project compounds [14]. Altogether, MW does not seem to be a useful predictor of promiscuity, at least not for small molecules in the MW range of 200–600.

5.5 Other Parameters

Compounds tend to be less promiscuous if they are structurally complex, of low flexibility, or decorated with many side chains [41, 43–47]. Such compounds require a high shape complementarity from a protein's binding site, which may be the reason for their improved selectivity [48, 49].

5.6 Structural Motifs

5.6.1 Aminergic Motifs

In drug discovery, basic and non-basic compounds have an increased propensity for promiscuity across aminergic targets if they originate from a drug discovery project with an aminergic therapeutic target [14]. On average, a drug with activity for an aminergic target binds to six other GPCRs [50, 51].

Some structural motifs are particularly prone to promiscuity across aminergic targets, for instance the "tricyclic motif" in drugs such as chlorpromazine, clozapine, and imipramine; or the "ergoline motif" in drugs such as pergolide [13]. Compounds with such structural motifs often have a rich polypharmacology across aminergic targets (Table 5.2). The structural basis for this polypharmacology has been reviewed [57–59].

5.6.2 Kinase Motifs

It is well known that kinase inhibitors are often active across many kinases. Sometimes, polypharmacology across several kinases may be desired. In many other cases, selective kinase inhibitors have been sought. Optimization toward selectivity is one of the challenges in kinase inhibitor research.

Polypharmacological kinase inhibitors frequently contain kinase hinge-binding motifs. Some such compounds show high promiscuity across the kinome [60]. The "2-0" rule helps identify compounds with an increased likelihood of kinase activity by counting typical structural fragments: more than two heteroaromatic nitrogens (N or NH) or any (more than 0) aromatic NH substituents and nitriles. Much more predictive is the presence of a bisarylaniline motif, in particular, of a "heteroaryl-NH-aryl" motif as defined in Figure 5.7; such compounds often inhibit a variety of kinases [61]. Other motifs have been reviewed as well [62]. However, these motifs alone do not necessarily lead to kinase activity and promiscuity, because additional interactions are required [63].

5.6.3 Other Motifs

Several functional motifs can lead to promiscuity across smaller target families [64]. For instance, a large structural variety of drugs with a primary sulfonamide bind across the 15 human carbonic anhydrases [65], and

Table 5.2 The tricyclic motif (red) in drugs such as promethazine, chlorpromazine, clozapine, and imipramine, and the ergoline motif (blue) in drugs such as pergolide often lead to a rich polypharmacology across aminergic targets. The activity at the receptors is given in K_i [nM] [52–56].

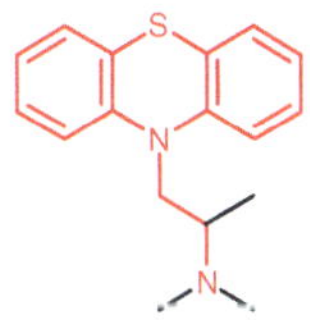

Promethazine
antihistamine
(First generation)

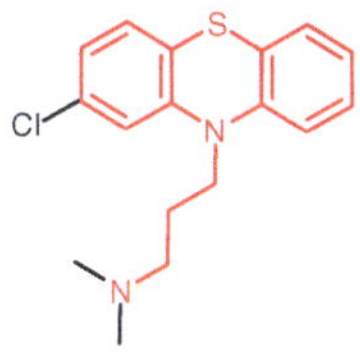

Chlorpromazine
antipsychotic

Clozapine
antipsychotic
(atypical/Second generation)

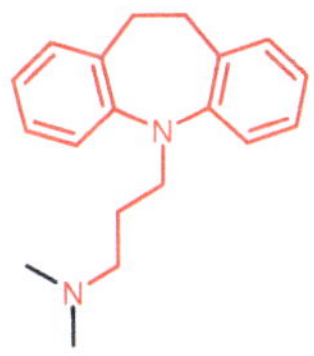

Imipramine
antidepressant

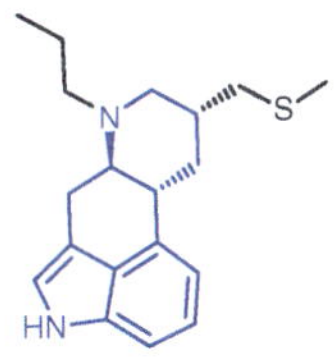

Pergolide

Target	Promethazine	Chlorpromazine	Clozapine	Imipramine	Pergolide
α_{1A} Receptor	32	0.28	1.62	32	1047
α_{1B} Receptor	21	0.81	7		692
α_{2A} Receptor	256	184	37	3100	50
α_{2B} Receptor	24	28	26.5		32
α_{2C} Receptor	353	46	6		68
5-HT_{1A} Receptor	1484	3115	123.7	5800	1.9
5-HT_{1B} Receptor		1489	519		282
5-HT_{1D} Receptor		452	1356		13
5-HT_{1E} Receptor		344			
5-HT_{2A} Receptor	19	2.75	5.35	80	8.3
5-HT_{2B} Receptor	43		8.37		7.1
5-HT_{2C} Receptor	6.48	25	9.44	120	295
5-HT_3 Receptor		776	241	970	
5-HT_{5A} Receptor		118	3857		
5-HT_6 Receptor	1128	19.5	13.49	190	30
5-HT_7 Receptor		21	17.95	1000	1
D_1 Receptor	1372	114.8	266.25	>10000	339
D_2 Receptor	260	7.244	157	620	30
D_3 Receptor	190	6.9	269.08	387	5.5
D_4 Receptor		32.36	26.36		59
H_1 Receptor	0.33	4.25	1.13	7.6	1698
H_2 Receptor	1146	174	153	550	
M_1 Receptor	3.32	47	6.17	42	>10000
M_2 Receptor	12	433	36.67	88	
M_3 Receptor	4.15	47	19.25	60	
M_4 Receptor	1.06	151	15.33	112	
M_5 Receptor	3.31		15.5	83	
NET	4203	2443	3168	20	
SERT	2130	7590	1624	1.3	

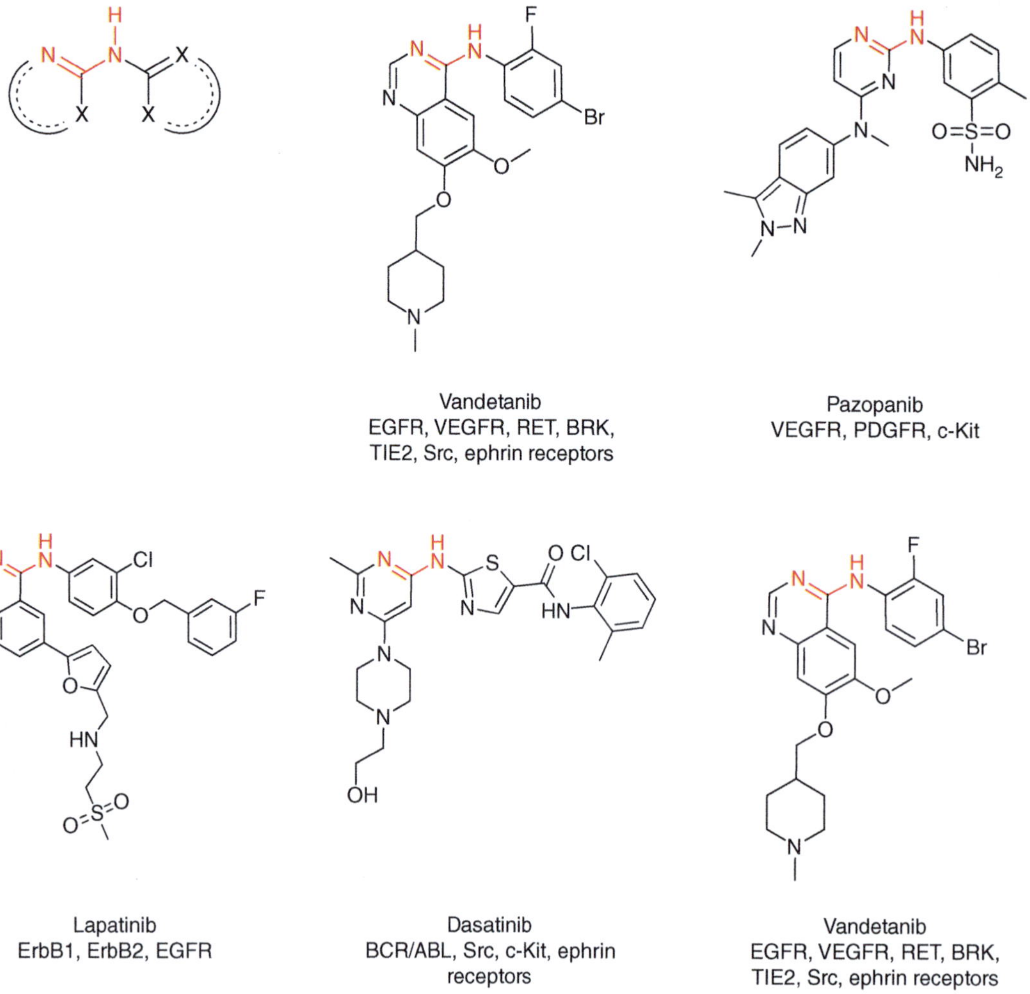

Figure 5.7 The "heteroaryl-NH-aryl" kinase hinge-binding motif [61] and examples.

many hydroxamic acids bind across the 23 human matrix metalloproteases [66].

5.7 Conclusion

Numerous studies investigated the correlations of pharmacological promiscuity with molecular properties, especially lipophilicity, ionization state/pK_a, and MW.

Most studies emphasize the relationship of lipophilicity with pharmacological promiscuity: more lipophilic compounds tend to be more promiscuous. However, this insight is often not useful for the optimization of compounds toward high selectivity. To achieve a significant reduction in the promiscuity potential of a lipophilic compound or series, lipophilicity needs to be reduced by several orders of magnitude. Such a reduction will often be hard to achieve. However, this insight emphasizes the importance of low lipophilicity as a key parameter for lead optimization and for the selection of hit and lead series [12, 67–74].

The practically most important predictor for promiscuity is ionization state/basicity. Compounds with a basic center and a [b]pK_a > 6 are (partly) protonated and positively charged. Such compounds have a high propensity to bind to aminergic GPCRs, amine transporters, opioid receptors, and cation channels. This propensity is further increased if such a compound is lipophilic, contains two aromatic rings in the vicinity of the basic center, or originates from an aminergic project. To assess a compound's pharmacological promiscuity, it will usually be screened against a safety panel of 40–180 targets. Such safety panels contain typically approximately 15–25% of targets, which prefer positively charged ligands and otherwise accept a wide variety of structures. Therefore, many positively charged compounds (i.e., bases with a [b]pK_a > 6) have high hit rates in such safety panels. The connection of basicity with promiscuity could be interpreted as an artifact, due to a bias of safety panels toward "aminergic" targets. However, this "aminergic" activity is also the dominant source of pharmacological promiscuity in clinically used drugs and is predictive of adverse effects [75–77].

Some aminergic targets attract a high hit rate from positively charged compounds (bases with a [b]$pK_a > 6$). Several studies mention the serotonergic 5-HT_{2B} receptor and related aminergic GPCRs, to which 20–30% of randomly chosen, positively charged compounds bind with sub-micromolar affinity. However, this does not necessarily translate to an undesired functional effect or to an adverse effect *in vivo*, and follow-up studies will often de-risk such "aminergic" compounds [78, 79].

Some structural motifs can lead to a pronounced promiscuity across aminergic targets. For instance, many older antihistamines, antipsychotics, and antidepressants contain a "tricyclic motif." All drugs with such a motif have a rich polypharmacology and bind to many targets with nanomolar affinity.

Thus, compounds with a basic center and a [b]$pK_a > 6$ have an increased potential to be pharmacologically promiscuous in safety screens. Such compounds should be tested early in the drug discovery process, e.g., in the hit-to-lead stage, against a small panel of predictive targets. This will help to recognize pharmacological promiscuity early and to prioritize compound series for further optimization.

References

1 Black, J. (1989). Drugs from emasculated hormones: the principle of syntopic antagonism. *Science* 245 (4917): 486–493.

2 Sneader, W. (2005). *Drug Discovery – A History*, 188–195. New York: Wiley.

3 Wong, D.T., Perry, K.W., and Bymaster, F.P. (2005). Case history: the discovery of fluoxetine hydrochloride (Prozac). *Nat. Rev. Drug Disc.* 4 (9): 764–774.

4 Bendels, S., Bissantz, C., Fasching, B. et al. (2019). Safety screening in early drug discovery: an optimized assay panel. *J. Pharmacol. Toxicol. Methods* 99: 106609.

5 Bowes, J., Brown, A.J., Hamon, J. et al. (2012). Reducing safety-related drug attrition: the use of in vitro pharmacological profiling. *Nat. Rev. Drug Discov.* 11 (12): 909–922.

6 Vaz, R.J. and Klabunde, T. (2008). *Antitargets: Prediction and Prevention of Drug Side Effects*, 3–11. Weinheim: Wiley-VCH.

7 Smith, D.A. and Schmid, E.F. (2006). Drug withdrawals and the lessons within. *Curr. Opin. Drug Discov. Devel.* 9 (1): 38–46.

8 Bender, A.M., Parr, L.C., Livingston, W.B. et al. (2023). 2B determined: the future of the serotonin receptor 2B in drug discovery. *J. Med. Chem.* 66 (16): 11027–11039.

9 Michelsen, J.W. and Meyer, J.M. (2007). Cardiovascular effects of antipsychotics. *Expert. Rev. Neurother.* 7 (7): 829–839.

10 Kroeze, W.K., Hufeisen, S.J., Popadak, B.A. et al. (2003). H1-histamine receptor affinity predicts short-term weight gain for typical and atypical antipsychotic drugs. *Neuropsychopharmacology* 28 (3): 519–526.

11 Lynch III, J.J., Van Vleet, T.R., Mittelstadt, S.W., and Blomme, E.A.G. (2017). Potential functional and pathological side effects related to off-target pharmacological activity. *J. Pharmacol. Toxicol. Methods* 87: 108–126.

12 Leeson, P.D. and Springthorpe, B. (2007). The influence of drug-like concepts on decision-making in medicinal chemistry. *Nat. Rev. Drug Discov.* 6 (11): 881–890.

13 Peters, J.-U., Hert, J., Bissantz, C. et al. (2012). Can we discover pharmacological promiscuity early in the drug discovery process? *Drug Discov. Today* 17 (7–8): 325–335.

14 Peters, J.-U., Schnider, P., Mattei, P., and Kansy, M. (2009). Pharmacological promiscuity: dependence on compound properties and target specificity in a set of recent Roche compounds. *ChemMedChem* 4 (4): 680–686.

15 Krejsa, C.M., Horvath, D., Rogalski, S.L. et al. (2003). Predicting ADME properties and side-effects: the BioPrint approach. *Curr. Opin. Drug Discov. Devel.* 6 (4): 470–480.

16 Brown, D.G., Smith, G.F., and Wobst, H.J. (2020). Promiscuity of in vitro secondary pharmacology assays and implications for lead optimization strategies. *J. Med. Chem.* 63: 6251–6275.

17 Sutherland, J.J., Yonchev, D., Fekete, A., and Urban, L. (2023). A preclinical secondary pharmacology resource illuminates target-adverse drug reaction associations of marketed drugs. *Nat. Comm.* 14: 4323–4340.

18 Myslivecek, J. (2022). Multitargeting nature of muscarinic orthosteric agonists and antagonists. *Front. Physiol.* 13: 974160.

19 Sameshima, T., Yukawa, T., Hirozane, Y. et al. (2020). Small-scale panel comprising diverse gene family targets to evaluate compound promiscuity. *Chem. Res. Toxicol.* 33 (1): 154–161.

20 Morgenthaler, M., Schweizer, E., Hoffmann-Röder, A. et al. (2007). Predicting and tuning physicochemical properties in lead optimization: amine basicities. *ChemMedChem* 2: 1100–1115.

21 Martin, R.E., Plancq, B., Gavelle, O. et al. (2007). Remote modulation of amine basicity by a phenylsulfone and a phenylthio group. *ChemMedChem* 2: 285–287.

22 Lee, E.C.Y., Steeno, G., Wassermann, A.M. et al. (2017). Amine promiscuity and toxicology analysis. *Bioorg. Med. Chem. Lett.* 27 (3): 653–657.

23 Leeson, P.D., St-Gallay, S.A., and Wenlock, M.C. (2011). Impact of ion class and time on oral drug molecular properties. *MedChemComm* 2 (2): 91–105.

24 Charifson, P.S. and Walters, W.P. (2014). Acidic and basic drugs in medicinal chemistry: a perspective. *J. Med. Chem.* 57 (23): 9701–9717.

25 Manallack, D.T., Prankerd, R.J., Yuriev, E. et al. (2013). The significance of acid/base properties in drug discovery. *Chem. Soc. Rev.* 42 (2): 485–496.

26 Urban, L., Whitebread, S., Hamon, J. et al. (2012). Screening for safety-relevant off-target activities. In: *Polypharmacology in Drug Discovery* (ed. J.-U. Peters), 15–46. Hoboken: Wiley.

27 Volk, B., Nagy, B.J., Vas, S. et al. (2010). Medicinal chemistry of 5-HT5A receptor ligands: a receptor subtype with unique therapeutical potential. *Curr. Top. Med. Chem.* 10 (5): 554–578.

28 Grailhe, R., Waeber, C., Dulawa, S.C. et al. (1999). Increased exploratory activity and altered response to LSD in mice lacking the 5-HT5A receptor. *Neuron* 22 (3): 581–591.

29 Muthas, D., Boyer, S., and Hasselgren, C. (2013). A critical assessment of modeling safety-related drug attrition. *MedChemComm* 4 (7): 1058–1065.

30 Bell, I.M., Bilodeau, M.T., and Lagrutta, A.A. (2012). Activity at cardiovascular ion channels: a key issue for drug discovery. In: *Polypharmacology in Drug Discovery*, vol. 2012 (ed. J.-U. Peters), 83–109. Wiley.

31 Fischer, H., Atzpodien, E.-A., Csato, M. et al. (2012, 2012). In silico assay for assessing phospholipidosis potential of small druglike molecules: training, validation, and refinement using several data sets. *J. Med. Chem.* 55 (1): 126–139.

32 Hopkins, A.L., Mason, J.S., and Overington, J.P. (2006). Can we rationally design promiscuous drugs? *Curr. Opin. Struct. Biol.* 16 (1): 127–136.

33 Hansson, M., Pemberton, J., Engkvist, O. et al. (2014). On the relationship between molecular hit rates in high-throughput screening and molecular descriptors. *J. Biomol. Screen.* 19 (5): 727–737.

34 Tarcsay, A. and Keseru, G.M. (2013). Contributions of molecular properties to drug promiscuity. *J. Med. Chem.* 56: 1789–1795.

35 Hughes, J.D., Blagg, J., Price, D.A. et al. (2008). Physiochemical drug properties associated with in vivo toxicological outcomes. *Bioorg. Med. Chem. Lett.* 18 (17): 4872–4875.

36 Young, R.J., Green, D.V., Luscombe, C.N., and Hill, A.P. (2011). Getting physical in drug discovery II: the impact of chromatographic hydrophobicity measurements and aromaticity. *Drug Discov. Today* 16 (17–18): 822–830.

37 Landry, M.L. and Crawford, J.J. (2020). LogD contributions of substituents commonly used in medicinal chemistry. *ACS Med. Chem. Lett.* 11 (1): 72–76.

38 Whitlock, G.A., Fish, P.V., Fray, M.J. et al. (2008). Pyridyl-phenyl ether monoamine reuptake inhibitors: impact of lipophilicity on dual SNRI pharmacology and off-target promiscuity. *Bioorg. Med. Chem. Lett.* 18 (9): 2896–2899.

39 Peters, J.-U. (2012). Pharmacological promiscuity and molecular properties. In: *Polypharmacology in Drug Discovery* (ed. J.-U. Peters), 47–62. Hoboken: Wiley.

40 Azzaoui, K., Hamon, J., Faller, B. et al. (2007). Modeling promiscuity based on in vitro safety pharmacology profiling data. *ChemMedChem* 2 (6): 874–880.

41 Haupt, V.J., Daminelli, S., and Schroeder, M. (2013). Drug promiscuity in PDB: protein binding site similarity is key. *PLoS ONE* 8 (6): e65894.

42 Gleeson, M.P., Hersey, A., Montanari, D., and Overington, J. (2011). Probing the links between in vitro potency, ADMET and physicochemical parameters. *Nat. Rev. Drug Discov.* 10 (3): 197–208.

43 Sturm, N., Desaphy, J., Quinn, R.J. et al. (2012). Structural insights into the molecular basis of the ligand promiscuity. *J. Chem. Inf. Model.* 52 (9): 2410–2421.

44 Lovering, F. (2013). Escape from Flatland 2: complexity and promiscuity. *MedChemComm* 4 (3): 515–519.

45 Yang, Y., Chen, H., Nilsson, I. et al. (2010). Investigation of the relationship between topology and selectivity for drug-like molecules. *J. Med. Chem.* 53 (21): 7709–7714.

46 Monteleone, S., Fuchs, J.E., and Liedl, K.R. (2017). Molecular connectivity predefines polypharmacology: aliphatic rings, chirality, and sp3 centers enhance target selectivity. *Front. Pharmacol.* 8: 552/1–552/9.

47 He, M.W., Lee, P.S., and Sweeney, Z.K. (2015). Promiscuity and the conformational rearrangement of drug-like molecules: insight from the protein data bank. *ChemMedChem* 10 (2): 238–244.

48 Hann, M.M., Leach, A.R., and Harper, G. (2001). Molecular complexity and its impact on the probability of finding leads for drug discovery. *J. Chem. Inf. Comput. Sci.* 41 (3): 856–864.

49 Leach, A.R. and Hann, M.M. (2011). Molecular complexity and fragment-based drug discovery: ten years on. *Curr. Opin. Chem. Biol.* 15 (4): 489–496.

50 Mestres, J., Gregori-Puigjane, E., Valverde, S., and Sole, R.V. (2009). The topology of drug-target interaction

networks: implicit dependence on drug properties and target families. *Mol. BioSyst.* 5 (9): 1051.
51 Jalencas, X. and Mestres, J. (2013). On the origins of drug polypharmacology. *MedChemComm* 4 (1): 80–87.
52 https://en.wikipedia.org/wiki/Promethazine (accessed 3 October 2023).
53 https://en.wikipedia.org/wiki/Chlorpromazine (accessed 3 October 2023).
54 https://en.wikipedia.org/wiki/Clozapine (accessed 3 October 2023).
55 https://en.wikipedia.org/wiki/Imipramine (accessed 3 October 2023).
56 https://en.wikipedia.org/wiki/Pergolide (accessed 3 October 2023).
57 Sexton, P.M. and Christopoulos, A. (2018). To bind or not to bind: unraveling GPCR polypharmacology. *Cell* 172: 636–638.
58 Peng, Y., McCorvy, J.D., Harpsoee, K. et al. (2018). 5-HT2C receptor structures reveal the structural basis of GPCR polypharmacology. *Cell* 172 (4): 719–730.e14.
59 Kampen, S., Duy Vo, D., Zhang, X. et al. (2021). Structure-guided design of G-protein-coupled receptor polypharmacology. *Angew. Chem. Int. Ed.* 60 (33): 18022–18030.
60 Kirchner, S. (2012). Kinases as antitargets in genotoxicity. In: *Polypharmacology in Drug Discovery* (ed. J.-U. Peters), 63–81. Hoboken: Wiley.
61 Aronov, A.M., McClain, B., Stuver Moody, C., and Murcko, M.A. (2008). Kinase-likeness and kinase-privileged fragments: toward virtual polypharmacology. *J. Med. Chem.* 51 (5): 1214–1222.
62 Attwood, M.M., Fabbro, D., Sokolov, A.V. et al. (2021). Trends in kinase drug discovery: targets, indications and inhibitor design. *Nat. Rev. Drug Discov.* 20 (11): 839–861.
63 Cohen, P., Cross, D., and Janne, P.A. (2021). Kinase drug discovery 20 years after imatinib. *Nat. Rev. Drug Discov.* 20 (7): 551–569.
64 Ehrt, C., Brinkjost, T., and Koch, O. (2019). Binding site characterization – similarity, promiscuity, and druggability. *MedChemComm* 10 (7): 1145–1159.
65 Supuran, C.T. (2012). Carbonic anhydrases: off targets, add-on activities, or emerging novel targets? In: *Polypharmacology in Drug Discovery* (ed. J.-U. Peters), 459–491. Hoboken: Wiley.
66 Yadav, R.K., Gupta, S.P., Sharma, P.K., and Patil, V.M. (2011). Recent advances in studies on hydroxamates as matrix metalloproteinase inhibitors: a review. *Curr. Med. Chem.* 18 (11): 1704–1722.
67 Price, D.A., Blagg, J., Jones, L. et al. (2009). Physicochemical drug properties associated with in vivo toxicological outcomes: a review. *Expert Opin. Drug Metab.Toxicol.* 5 (8): 921–931.
68 van de Waterbeemd, H., Smith, D.A., and Jones, B.C. (2001). Lipophilicity in PK design: methyl, ethyl, futile. *J. Comput. Aided Mol. Des.* 15 (3): 273–286.
69 Arnott, J.A. and Planey, S.L. (2012). The influence of lipophilicity in drug discovery and design. *Expert Opin. Drug Discov.* 7 (10): 863–875.
70 Yukawa, T. and Naven, R. (2020). Utility of physicochemical properties for the prediction of toxicological outcomes: Takeda perspective. *ACS Med. Chem. Lett.* 11 (2): 203–209.
71 Bayliss, M.K., Butler, J., Feldman, P.L. et al. (2016). Quality guidelines for oral drug candidates: dose, solubility and lipophilicity. *Drug Discov. Today* 21 (10): 1719–1727.
72 Miller, R.R., Madeira, M., Wood, H.B. et al. (2020). Integrating the impact of lipophilicity on potency and pharmacokinetic parameters enables the use of diverse chemical space during small molecule drug optimization. *J. Med. Chem.* 63 (21): 12156–12170.
73 Bunally, S.B., Luscombe, C.N., and Young, R.J. (2019). Using physicochemical measurements to influence better compound design. *SLAS Discov.* 24 (8): 791–801.
74 Smith, D.A. (2016). Physicochemistry and the off-target effects of drug molecules. *Drug Discov. Toxicol.* 55 67.
75 Yildirim, M.A., Goh, K.-I., Cusick, M.E. et al. (2007). Drug-target network. *Nat. Biotechnol.* 25 (10): 1119–1126.
76 Gregori-Puigjane, E. and Mestres, J. (2008). A ligand-based approach to mining the chemogenomic space of drugs. *Comb. Chem. High Throughput Screen.* 11 (8): 669–676.
77 Brianso, F., Carrascosa, M.C., Oprea, T.I., and Mestres, J. (2011). Cross-pharmacology analysis of G protein-coupled receptors. *Curr. Top. Med. Chem.* 11 (15): 1956–1963.
78 Whitebread, S., Dumotier, B., Armstrong, D. et al. (2016). Secondary pharmacology: screening and interpretation of off-target activities – focus on translation. *Drug Discov. Today* 21 (8): 1232–1242.
79 Jenkinson, S., Schmidt, F., Rosenbrier Ribeiro, L. et al. (2020). A practical guide to secondary pharmacology in drug discovery. *J. Pharmacol. Toxicol. Methods* 105: 106869.

6

Kinase Liabilities in Early Drug Discovery

Stephan Kirchner

6.1 Introduction

Mainly based on advances in the molecular understanding of cellular signaling pathways in the development of cancer, protein kinases have been recognized as potential drug targets since more than 40 years. Today, protein kinase inhibitors have evolved as one of the most important drug classes for the treatment of cancer, and the number of small-molecule kinase inhibitors (SMKIs) in non-oncology indications such as inflammatory, transplantation, or autoimmune disorders is increasing constantly [1–3].

However, the development of new SMKIs remained challenging. Many of the kinase inhibitors that have been approved or are under investigation show multi-kinase inhibition. Targeting the highly conserved adenosine triphosphate (ATP)-binding pocket often impedes to achieve sufficient selectivity and potency to compete with the high ATP concentrations present in human cells. Non-translating efficacy from noncellular to cellular and from *in vitro* to *in vivo* conditions and off-target toxicity issues for ATP-competitive inhibitors are a major reason for high attrition rates [4].

6.2 Protein Kinases and Inhibitor Binding Sites

Protein kinases regulate most aspects of cellular life by protein phosphorylation. Most of them act in networks of kinases and signaling cascades and are modulated by autophosphorylation and phosphorylation by other kinases. They catalyze the transfer of the terminal phosphate of ATP to substrates that usually contain a serine, threonine, or tyrosine residue (serine/threonine and tyrosine kinases). The common secondary structure consists of 12 subdomains that fold into a bilobed catalytic core structure. The ATP-binding site is located in a deep cleft between the lobes. For ATP binding, the adenine ring forms hydrogen bonds with the hinge region – which connects the amino-terminal and carboxy-terminal domains of the kinase – whereas the ribose and triphosphate groups of ATP bind in a hydrophilic channel extending to the substrate binding site with conserved residues that are essential for catalytic reaction (Figure 6.1 shows human cyclin-dependent kinase 2 (Cdk2) complexed with ATP).

Key to the regulation of kinase activity is a highly conserved activation loop with DFG and APE motifs (referring to the single-letter amino acid code) at the start and end of the loop with varying lengths. The conformation of the activation loop can largely vary depending on the phosphorylation of one or more residues located in the activation segment at the center. Catalytically competent conformations are usually phosphorylated, and inactive conformations of the activation loop block the substrate-binding site. Other kinase domains link to subcellular localization of the protein or other signaling modules.

Most SMKIs target the ATP-binding site. They typically mimic the hydrogen bonding by the adenine ring of ATP and build one to three hydrogen bonds with the hinge region of the target kinase [1, 7–9].

The **type I inhibitors** target the ATP-binding site of the kinase in its active conformation with a phosphorylated activation loop and are catalytically active for ATP phosphate transfer. Perhaps simply for historical reasons, they constitute the majority of ATP-competitive inhibitors as many kinase inhibitors were originally discovered by functional assays based on the active kinase conformation. Type 1 inhibitors typically consist of heterocyclic ring systems that can occupy the adenine region of the ATP-binding site and serve as a scaffold for additional side chains that occupy the adjacent hydrophobic regions I and II. Functional interaction with the ribose or the phosphate-binding region of ATP could serve as a basis for selectivity of kinase inhibition; however, despite rare exceptions, the majority of type 1 kinase inhibitors do not exploit these sites (Figure 6.2a).

Polypharmacology: Strategies for Multi-Target Drug Discovery, First Edition. Edited by Jens-Uwe Peters.

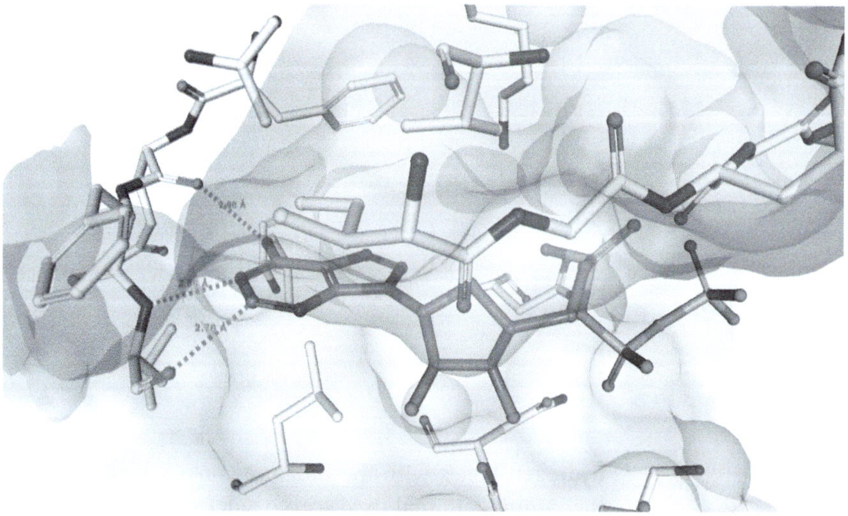

Figure 6.1 X-ray crystal structure of ATP binding to CDK2 (PDB identifier 1FIN [ID 1FIN]; the image was created using NGL viewer [5] and Research Collaboratory for Structural Bioinformatics Protein Data Bank [RCSB PDB]) [6]. Monomeric Cdk2 is inactive as active site residues involved in ATP orientation are disrupted and access of the protein substrate to ATP is prevented.

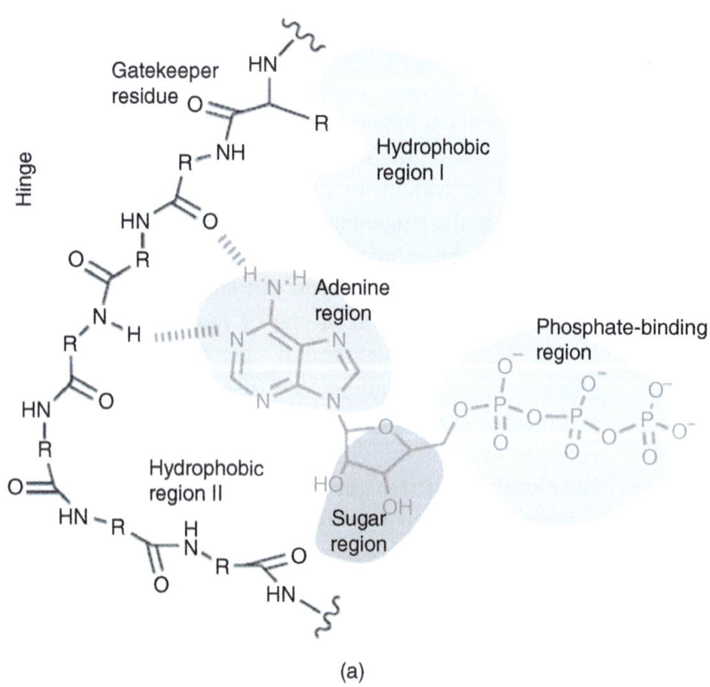

Figure 6.2a Subregions of the conserved ATP-binding site. Liu and Gray [8].

In contrast, **type 2 kinase inhibitors** target the inactive conformation of kinases. They recognize an additional hydrophobic binding pocket immediately adjacent to the region occupied by ATP and referred to as the "allosteric site." This pocket is created by a shift of phenylalanine from the DFG motif and forms a unique activation loop rearrangement (called DFG-out conformation) that is characteristic of kinases in an inactive conformation. Type 2 kinase inhibitors share a common pharmacophore that exploits a conserved set of hydrogen bonds in the "allosteric" site (Figure 6.2b).

The ability to induce dramatic conformational changes is not unique to type II kinase inhibitors and demonstrates that the kinase active site is highly mobile and can remodel to accommodate a variety of inhibitors [8].

Allosteric kinase inhibitors bind outside the ATP-binding pocket and modulate kinase activity in an allosteric manner. They differ in the activation mechanisms that they prevent, and ligands binding at allosteric sites actually regulate the kinase activity either positively or negatively. Depending on the location of their binding site, they are subdivided into **type 3 inhibitors** binding close to the ATP-binding site in a pocket induced by an αC-out conformation (in some cases by an inactive conformation of the A-loop) and **type 4 inhibitors,** which bind to remote allosteric sites located distantly from

Figure 6.2b Allosteric ATP-binding site. Liu and Gray [8].

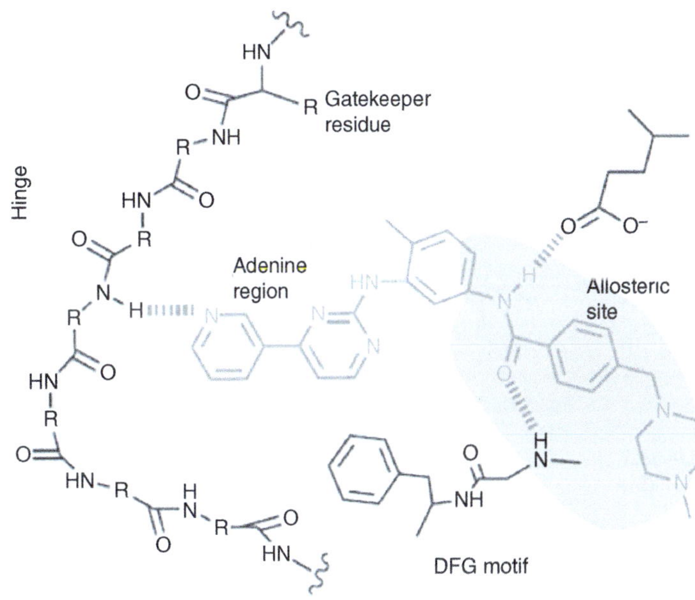

the ATP-binding site [1]. As allosteric kinase inhibitors exploit binding sites and regulatory mechanisms outside the highly conserved kinase domains, they tend to show higher selectivity because they target unique features of a specific kinase. Remarkably, allosteric type 3 inhibitors targeting focal adhesion kinase (FAK), tropomyosin receptor kinase A (TRKA), PAK1, and insulin-like growth factor 1 receptor (IGF1R) kinases by binding to the canonical back pocket and stabilizing a DFG-out conformation were reported to show isoform selectivity against closely related family members [1, 10–13].

Covalent binding inhibitors form irreversible, covalent bonds to the active site of kinases, mainly cysteine residues that are accessible for covalent modification by the Michael addition reaction with the inhibitor. Approximately 200 different kinases across the kinome have a cysteine located in the vicinity of the ATP pocket that could be targeted by irreversible inhibitors [14, 15]. Despite this large number of kinases that could be addressed for inhibition by this approach, selectivity could be proposed to be lower than for allosteric inhibitors if inhibition of ATP-binding targets directly the ATP-binding site.

More recently, some additional strategies were applied to kinase inhibitors with the aim to improve potency, selectivity, and pharmacokinetic properties. Macrocyclization was successfully used for the anaplastic lymphoma kinase (ALK) inhibitor lorlatinib to lock the bioactive conformation of its non-cyclized template crizotinib, thereby increasing efficacy substantially [16–18].

Bifunctional inhibitors combining type 1 inhibitors by chemical linkers were used to bring inhibitors with different binding sites together, for example, ATP mimetic inhibitors with Src homology 2 (SH2) ligands or substrate-competitive inhibitors. Molecular glues inducing targeted protein degradation by inducing the association of ubiquitin ligases with their target and proteolysis-targeting chimeras (PROTACs) or degraders also belong to this class [1].

Although the application of these strategies is often associated with increased selectivity of kinase inhibitors, their large molecular sizes make clinical development very challenging.

Binding site and binding mode are essential determinants for the selectivity of kinase inhibition and thereby for potential side effects. Recognition of ATP and catalyzation of the phosphotransfer are the common features of all kinases (as well as for other nucleotide-dependent enzymes), and the ATP-binding pocket is highly conserved. Type 1 inhibitors tend therefore to have a higher potential for cross-reactivities to other kinases and unwanted side effects than type 2 inhibitors. The amino acids around the "allosteric site" targeted by type 2 inhibitors are less conserved than the adenine and hydrophobic regions of the ATP-binding pocket, and there is considerable more conformational variability in the inactive state of kinases allowing for more specific inhibition. Allosteric inhibitors provide even more opportunities to push for selectivity and avoid unwanted cross-reactivities as might also be reflected by an increasing number of inhibitors of this type. In contrast, the covalent inhibitors share not only the potential for inhibiting additional unwanted kinases, but they might also lead to toxicities by the modification of

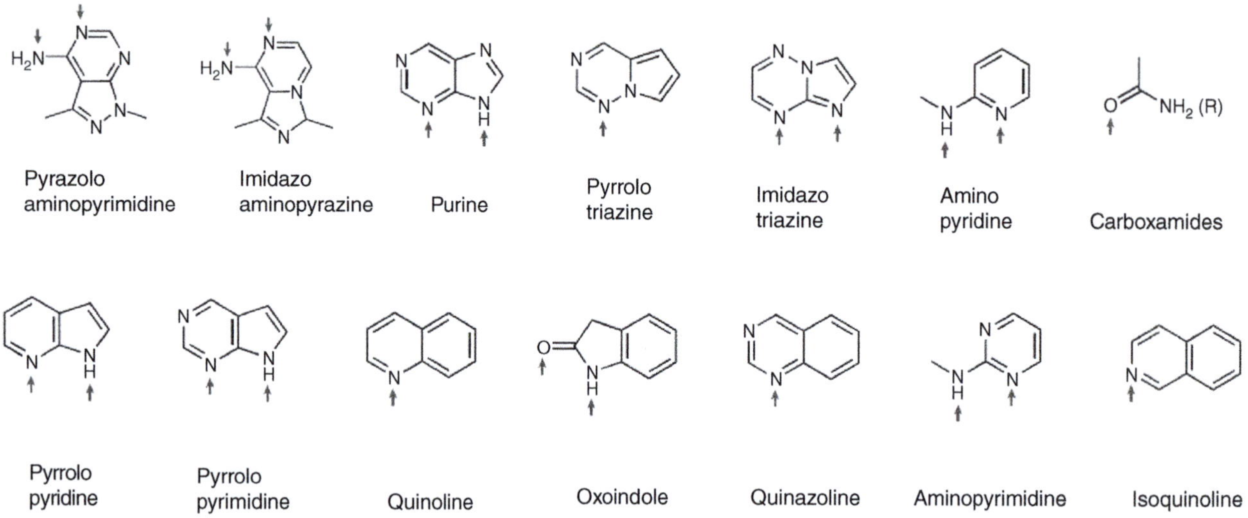

Figure 6.3 Major hinge-binding motifs used to target the adenine region of the purine-binding site in approved kinase inhibitors. Hydrogen bond donors and acceptors are indicated by arrows. Attwood et al. [1]/Springer Nature.

unrelated targets such as the formation of deoxyribonucleic acid (DNA) adducts, ultimately leading to mutations.

Over the last decades, SMKIs evolved as a key drug class. Initially, targeting kinases has proven to be an effective strategy to treat cancer. However, due to the important mechanistic impact of kinases to control many essential signaling pathways and cellular functions beyond cell growth and apoptosis, the range of indications expanded constantly. There are currently 87 SMKIs approved by regulatory agencies including indications such as rheumatoid arthritis and others [1–3, 19].

As most SMKIs target the structurally and functionally conserved ATP-binding site, there is a high potential for unwanted toxicities (Figure 6.3). Major side effects commonly observed in preclinical discovery and clinical development include cardiotoxicity and deregulated cell cycle signaling eventually leading to bone marrow and/or genetic toxicity.

It is important to note that the toxicological potential of SMKIs may arise from unselective off-target kinase inhibition or exaggerated pharmacological effects by the targeted kinase. Other ATP-binding enzyme classes such as polymerases, motor proteins, and topoisomerases might be considered as well for off-target binding and undesirable side effects.

6.3 Kinase-regulated Cardiac Functions and Potential Consequences of Inhibition

Many pathways involved in essential functions of the heart are regulated by kinases, and many SMKIs used for anticancer therapies have been associated with cardiotoxicity in the clinic. Adverse cardiac events often observed include QT prolongation, hypertension, reduced left ventricular ejection fraction (LVEF), congestive heart failure (CHF), acute coronary syndromes (ACS), and myocardial infarction (MI) [20–22].

Focusing on the impact of kinases on physiological or pathological signaling pathways in the heart, this likely might relate to disrupted cardiovascular energy homeostasis. Predominantly regulated by protein kinases that maintain a balance of ongoing energy generation, 5′-AMP-activated protein kinase (AMPK) acts as the master sensor of energy status in cardiomyocytes and is essential to avoid CHF [22]. Calcium/calmodulin-dependent protein kinase II is a key player of calcium homeostasis [23], and any disturbance to calcium regulation can lead to dysfunctional cardiac conduction and result in pathologic hypertrophy [24]. The phosphoinositide 3-kinase (PI3K)–AKT pathway plays a crucial role in the regulation of cardiomyocyte survival by acting on a number of targets and preventing apoptosis [25]. Specific kinases of concern for potential cardiac side effects were reviewed by Force and Kolaja and are summarized in Table 6.1 [22].

6.4 Core Kinases Driving the Cell Division Cycle and Consequences of Interference

Cell division is a tightly coordinated sequence of events to ensure duplication and equal segregation of the entire cellular content, resulting in genetically identical daughter cells. The replication of DNA takes place during interphase

Table 6.1 Important kinases in the heart and vasculature based on genetically modified or kinase inhibitor-treated mice.

Kinase	(Putative) roles of kinase in heart/vasculature
RAF1/BRAF	Antiapoptotic; preserves LV function under stress. KO condition: LV dysfunction and HF in the absence of additional stress; DNTG condition: reduced hypertrophy but LV dysfunction due to cell death
PI3K (p110α)	Physiological heart growth; cardiomyocyte survival
PI3K (p110γ)	Regulates contractility and pathological hypertrophy
PDK1	Cardiomyocyte survival and β adrenergic responsiveness
AKT 1, 2, or 3	Regulators of cardiomyocyte survival, growth, and metabolism
mTOR	mTORC1 regulates protein synthesis; inhibition leads to energy preservation under stress; mTORC2 regulates AKT activation
AMPK	Sensor of energy stress; inhibits mTORC1, preserving energy stores. KO of AMPKα2 increased hypertrophy and LV dysfunction after TAC
GSK3α/β	Together with AMPK, inhibits mTORC1; deletion of GSKβ protective in post-MI remodeling; deletion of GSK3α leads to HF in the setting of stress
PDGFRs	β-Isoform is crucial in angiogenesis and heart's response to PO
VEGFRs	Crucial in angiogenesis and the heart's response to PO; antihypertensive effects
EGFR (ERBB1)	Helps to maintain LV function in the setting of chronic catecholamine stimulation; mediates pro-survival signaling
ERBB2, ERBB4	Cardiomyocyte survival and homeostasis; maintenance of LV function
KIT	Promotes CSC and immature cardiomyocyte differentiation; promotes homing to sites of MI, promoting repair
ABL/ARG	Maintains ER homeostasis. LV dysfunction is seen in rodents treated with imatinib
JAK2	JAK2 and STAT3 protective in many pathological settings
FAK	Antihypertrophic and antifibrotic in heart
DMPK	Myotonic dystrophy type 1 is caused by excess repeats of the 3′ UTR region of DMPK
LTK	Activation of LTK results in cardiac hypertrophy and cardiomyocyte degeneration
ROCK	Pro-fibrotic and pro-apoptotic in the setting of PO
LKB1	Activates AMPK which is pro-angiogenic in the heart
LDB3, ZASP, and/or Cypher	Mutations in ZASP are associated with skeletal muscle myopathies
ERK1/2	Generally promotes survival and may modulate physiological (but not pathological) hypertrophy
PKCα	Adverse effects on the heart in the setting of PO
cGMP-dependent PK	One of the four nodal kinases in HF; activated by PDE5 inhibitors; inhibits apoptosis, hypertrophy, and β-adrenergic responses
PIM Kinase	Pro-survival; activated by AKT; regulated at the level of gene expression
CAMKII	Nodal kinase in HF; pro-hypertrophic; promotes decompensation in the setting of PO
GRK2 and/or GRK5	Downregulates β-adrenergic signaling through recruitment of β-arrestin
ASK1	Promotes pathological hypertrophy and remodeling; pro-apoptotic
PTEN phosphatase	Antihypertrophic; impairs cardiomyocyte survival with stress and antagonizes PI3K signaling

Annotation: AMPK, 5′ AMP-activated protein kinase; ARG, Abl-related gene; ASK1, apoptosis signal-regulating kinase 1; CAMKII, Ca2+/calmodulin-dependent protein kinase II; DMPK, dystrophia myotonica protein kinase; EGFR, epidermal growth factor receptor (also known as ERBB1); ERBB2, human epidermal growth factor receptor 2; ERBB4, human epidermal growth factor receptor 4; ERK2, extracellular signal-regulated kinase 2 (also known as MAP kinase 1); FAK, focal adhesion kinase; GSK3, glycogen synthase kinase 3; JAK2, Janus kinase 2; LDB3, LIM domain binding 3; LKB1, serine/threonine kinase 11 (also known as STK11); LTK, leukocyte receptor tyrosine kinase; mTOR C1, mammalian target of rapamycin complex 1; PDE5, phosphodiesterase 5; PDGFR, platelet-derived growth factor receptor; PDK1, phosphoinositide-dependent protein kinase 1; PI3K, phosphoinositide 3-kinase; PKC, protein kinase C; PTEN, phosphatase and tensin homolog; ROCK, RHO-associated kinase (RHO-kinase); VEGFR, vascular endothelial growth factor receptor; CSCs, cardiac stem cells; DNTG, dominant-negative transgenic; HF, heart failure; KO, knockout; LV, left ventricular; MI, myocardial infarction; PO, pressure overload; TAC, trans-aortic constriction.

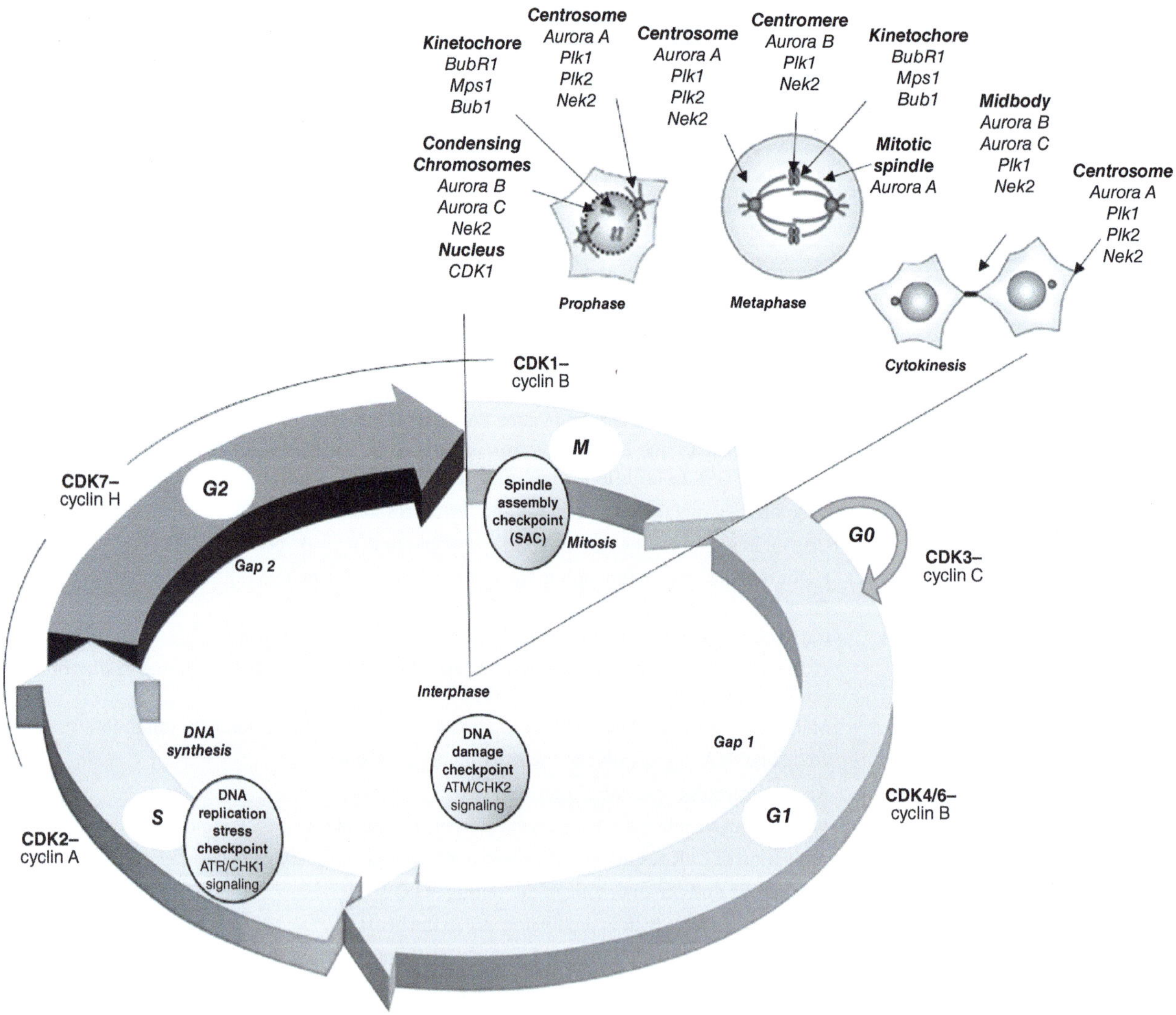

Figure 6.4 Major cell cycle kinases regulating mammalian cell cycle progression. DNA damage, replication stress, and chromosome/spindle attachments are monitored by specific checkpoint control mechanisms. The dynamic localization of major M-phase kinases during mitosis depends on specific functions of the particular kinases.

and is separated into a distinct phase called the "S phase." In the M phase (mitosis), following condensation into chromosomes, the duplicated DNA is then distributed to the newly formed daughter cells. Cytokinesis completes the cell cycle at the end of mitosis by the separation of the cellular contents, and cells can either exit the cell cycle or enter into a new cell division (Figure 6.4).

The CDK family plays a central role in multiple signaling pathways controlling transcription and driving the progression through the different stages of the cell division cycle (Cdc). Their activity is mainly regulated by cyclins and proteins structurally defined by the presence of the so-called cyclin box and transcribed in a cell cycle–dependent manner. The activities of core cell cycle CDKs such as Cdk4, Cdk6, Cdk2, and Cdk1 are additionally under the control of multiple transcription factors (TFs) or transient association with other regulatory elements such as the retinoblastoma (Rb) protein or inhibitory proteins like Cip1, Kip1/2, or Ink4A-D (Table 6.2).

In association with cyclin H and Mat1, Cdk7 functions as a CDK-activating kinase (CAK) complex by directly phosphorylating several of the CDKs mentioned above. Furthermore, this complex can also be found associated with the DNA-dependent helicase Xpd, thereby regulating the coordination and progression of mitosis [27, 28].

Cell cycle–dependent changes in CDK activity orchestrate cell cycle entry, progression, and completion. The interdependence of CDK activities and cyclin accumulation during different phases by cell cycle–regulated transcription and inhibition of protein degradation ensures

Table 6.2 CDK family members involved in cell cycle regulation.

Kinase	Main activating cyclin	Other interacting proteins	Substrates	Function
CDK1	A1, A2, B1, B2 (E, B3)	Cks	Actopaxin, adenomatous polyposis coli, amphiphysin 1, APC, BARD1, BRCA2, caldesmon, Cdc7, Cdc20, Cdc25A, Cdc25 C, Cdh1, Cdk7, C/EBPb, CKII, dynein, dystrophin, EF-1, Eg5, EGFR, FANCG, Fos, GFAP, GM130, GRASP65, histone H1, hHR6A, HMG I(Y), IFAP300, KRC, lamins A, B, and C, lamin B receptor, Lats1, MAP1B, MAP4, Marcks, MCM2, MCM4, MKLP1, Myb, NBP60, neurofilament H, NFI, Nir2, Npm, NPC, nucleolin, Nucks, numatrin, Orc1, p18, p47, p53, p54NRB, PAP, plectin, PP1-I2, pRb, R2, Rab4, Rap1GAP, RCC1, RIIa, S6K1, Sam68, separase, Ski, survivin, mSTI1, Tau, vimentin, thymidine kinase	G2-M
CDK2	A1, A2, E1, E2 (D1, D2, B1, B3)		BARD1, B-Myb, BRCA1, BRCA2, CBP/p300, Cdc6, Cdc7, Cdk7, Cdt1, C/EBPb, DP1, hHR6A, HIRA, Ku70, Marcks, MCM2, MCM4, MyoD, NPAT, Npm, p107, p21Cip1, p27Kip1, p53, pRb, R2, RPA, Smad3, thymidine kinase	G1-S
CDK3	E1, E2, A1, A2, C	E2F/DP	Cables1	G0-G1-S
CDK4	D1, D2, D3	MyoD	Cdt1, Marcks, p107, p130, pRb, Smad3	G1-S
CDK6	D1, D2, D3		p107, p130, pRb	G1-S
CDK7	H	Mat1	Cdk1–Cdk6, p53, RARg, RNA pol II	Cdk-activating kinase, transcription

Annotation: A domain 1, A domain containing protein 1; BRCA1, breast cancer 1; CAK, Cdk-activating kinase; CBP/p300, CREB-binding protein/p300; Cdc, cell division cycle; C/EBPb, CCAAT/enhancer-binding protein b; Cks, Cdc28-dependent kinase subunit; CKII, casein kinase II; DP1, E2F dimerization partner 1; EF-1, elongation factor 1; EGFR, epidermal growth factor receptor; FANCG, Fanconi anemia, complementation group G; GAP, GTPase-activating protein; GFAP, glial fibrillary acidic protein; GM130, Golgi matrix protein 130; GRASP65, Golgi reassembly stacking protein 1, 65kDa; hHR6A, human homolog of Rad6; HIRA, HIR histone cell cycle regulation defective homolog A; HMG, high-mobility group; IFAP300, intermediate filament-associated protein, 300 kDa; KRC, kB-binding and recognition component; MAP, microtubule-associated protein; Marcks, myristoylated alanine-rich C-kinase substrate; Mat1, Menage a trois 1, a targeting subunit of cyclin-dependent kinase-activating kinase and general transcription factor IIH kinase; MCM, minichromosome maintenance; MKLP1, mitotic kinase-like protein 1; mSTI1, murine stress-inducible protein 1; NBP60, NLS-binding protein 60 kDa; NF-I, nuclear factor I; NPAT, nuclear protein, ataxia telangiectasia; Npm, nucleophosmin (B23, NO38); NPC, nuclear pore complex; PAP, phosphatidic acid phosphatase; PP, protein phosphates; pRb, retinoblastoma protein; R2, ribonucleotide reductase 2; RanBPM, Ran-binding protein M; RNPS1, RNA-binding protein S1; RARg, retinoic acid receptor-g; RCC1, regulator of chromosome condensation 1; RIIa, type II cAMP-dependent protein kinase; RNA pol II, RNA polymerase II; S6K1, S6 kinase; TRAP, TNF receptor-associated protein. Abbreviations: APC, anaphase-promoting complex; BARD1, BRCA1-associated RING [26].

both sequential and unidirectional cell cycle progression [29, 30].

The accumulation of cyclin D–CDK4/6 in the prereplicative G1 phase is required for cell cycle entry and prevents it from entering the nondividing G0 resting phase. Initially, cyclin D–CDK 4/6 activates the Rb protein and triggers E2F-dependent transcription, resulting in the accumulation of cyclin E and cyclin A. Following the inactivation of Rb protein by cyclin E–CDK2 activity, cells pass a decision stage (also called restriction point) and irreversibly undergo cell division. Via a positive feedback loop, cyclin E–CDK2 activity further activates E2F-dependent transcription, leading to increased cyclin E–CDK2 activity and cyclin A–CDK2 complexes. Through the inactivation of anaphase-promoting complex (APC)/CCDH1 activity, this then results in the accumulation of cyclin A–CDK2 activity and initiates the S phase. Following the successful completion of DNA replication, the accumulation of cyclin A/B–CDK1 complex in the post-replicative G2 phase then initiates a second decision stage. Cyclin A/B–CDK1 activity prepares for mitotic entry [31, 32] and drives nuclear envelope breakdown, building the mitotic spindle [33] and condensation, alignment, and segregation of replicated chromosomes during mitosis [27, 34, 35].

As for interphase pathways, reversible protein phosphorylation is a crucial regulatory element of mitotic signaling

[29, 30, 36]. A small number of kinases orchestrates almost all central events from entry into mitosis to cytokinesis in distinct spatial and temporal contexts. Activation of cyclin B–Cdk1 is considered as the master guardian synchronizing essential steps during mitosis [31], and multiple targets are phosphorylated by CDK1 activity [37, 38]. Activation of M-phase (mitotic) kinases polo-like kinase 1 (Plk1), Aurora A, and Aurora B is in the center of regulating mitotic processes. Major M-phase kinases involved are listed in Table 6.3.

The main role of PLK1 is in the progression of mitosis, with an established regulatory function in mitotic entry, maturation of the centrosome, spindle assembly, and cytokinesis [39–42]. It localizes to centrosomes, kinetochores, and the midbody depending on its multiple functions in different mitotic phases (Figure 6.4). Aurora kinases in complex with their cofactors Bora, centrosomal protein of 192 kDa (CEP192), inner centromere protein (INCENP), and targeting protein for Xklp2 (TPX2) (multifunctional scaffold proteins) and the Aurora-PLK1 signaling cascades organized by Bora, CEP192, and INCENP control all pathways of spindle assembly and centrosome cycle and additional other crucial aspects of mitosis [43–74].

Aurora serine/threonine kinases are essential to maintain chromosome integrity in mammalian cells. Aurora A is a key player in the regulation of the centrosome cycle and prepares for bipolar spindle formation by centrosome separation. It is also necessary for maintaining the stability of the spindle (presumably by phosphorylation of motor proteins like the kinesin-related protein Eg5 and proteins involved in astral microtubule nucleation) [47, 48].

Table 6.3 Major M-phase-specific and checkpoint kinases.

Kinase	Roles of kinase in cell cycle regulation
Aurora A (also known as Aurora 2)	Functions during prophase of mitosis and is required for correct duplication and separation of the centrosomes (the microtubule organizing centers in eukaryotic cells). Interaction with Aurora A is mediated by major cofactor CEP192. Activated by Bora and TPX2
Aurora B (also known as Aurora 1)	Controls SAC and is required to promote correct kinetochore–microtubule attachment. Recruitment of Mps1, inhibition of recruitment of PP1, SKA complex, astrin: SKAP. Binds to other CPC subunits as regulatory protein INCENP
Aurora C	Germ cell-specific activity
PLK1	Multiple functions in mitosis. Early trigger for G2/M transition, supports functional maturation of the centrosome in late G2/early prophase and establishment of the bipolar spindle (phosphorylation of cdc25C). Phosphorylates and activates components of the anaphase-promoting complex (APC). Activation of PLK1 depends on the coordinated action of Aurora A and Bora
PLK2	Essential for mitotic centriole replication.
Bub1	Kinetochore recruitment of BubR1:Bub3 and Cdc20. Binds to Bub3, Cdc20, P-MELT sequences
BubR1	Component of MCC. Binds to Bub3, Mad2, Cdc20, Bub1:Bub3 complex
Mps1	Phosphorylation of MELT repeats of Knl1. Binds to Ndc80 for kinetochore recruitment. Directly activating SAC
Nek 2, Nek 6, Nek 7, Nek 9	Nek 2 promotes centrosome splitting at G2/M and chromosome segregation and regulates chromosome segregation. Catalytic activity is inhibited following irradiation, preventing centrosome separation.
	Nek 6 and Nek 7 are required for mitotic spindle formation and cytokinesis, promote centrosome separation, and are activated downstream of Nek 9
ATM	DNA damage checkpoint kinase, recruited and activated by DNA double-strand breaks, phosphorylates multiple downstream effectors including p53, CHK2, and H2AX, master regulator of the DNA damage checkpoint leading to cell cycle arrest, DNA repair, or apoptosis
CHK2	Downstream effector kinase of ATM. Involved in DNA repair, cell cycle arrest, or apoptosis in response to DNA damage, phosphorylates multiple targets including cdc25 phosphatases responsible for dephosphorylating and activating CDKs
ATR	DNA replication stress checkpoint kinase, activated in response to single-strand breaks, phosphorylates key downstream effector CHK1, thereby initiating a signal transduction cascade that culminates in cell cycle arrest
CHK1	Downstream effector kinase of ATR, coordinates DNA damage and cell cycle checkpoint response, targets WEE1 and PLK1 kinases, and inhibits cdk1 by phosphorylating WEE1
WEE1	Downstream effector kinase of CHK1

Annotation: ATM, ataxia telangiectasia mutated serine/threonine kinase; ATR, ataxia telangiectasia and Rad3-related protein serine/threonine kinase; Aurora, aurora kinase; Bub 1, budding uninhibited by benzimidazoles 1 kinase; BubR1, Bub1-related kinase or MAD3/Bub1b; CHK1, checkpoint kinase 1; CHK2, checkpoint kinase 2; Mps1, monopolar spindle protein kinase; Nek, never in mitosis A-related kinase; PLK, polo-like kinase; WEE1, nuclear serine/threonine kinase of 96 kDa.

Localization of Aurora A is restricted to the centrosomes and spindle poles during mitosis. As part of the chromosome passenger complex (CPC), the localization of Aurora B depends on the stage of the cell cycle (Figure 6.4). It interacts with INCENP and stays with the kinetochore until the metaphase–anaphase transition before relocalization to the midbody [45]. Aurora B is also required for the regulation of kinetochore–microtubule interactions, couples chromosome alignment with anaphase, and regulates chromosome condensation by phosphorylating the centromere histone variant centromere protein A (CENP-A) [43].

In addition to Plk1, Aurora A and Aurora B multiple never in mitosis A (NIMA)-related kinase (NEK) family members act in networks to regulate specific events of mitosis, including centrosome separation, spindle assembly, and cytokinesis. Nek kinases are defined by their homology with the *Aspergillus nidulans* mitotic kinase NIMA. Nek2 localizes to centrosomes during interphase and regulates centrosome separation during mitosis by interaction and phosphorylation of several centrosomal proteins including centrosomal Nek2-associated protein 1 (cNap-1), rootletin, and b-catenin [49–53]. Additionally, Nek2 has been shown in association with condensed chromatin, the midbody, and the kinetochores in dividing cells underlying a role for chromosome segregation [54] and involvement in centriole function has been linked with Plk1 interaction. Nek6 and Nek7 were found to act downstream of Nek9 signaling and regulate the mitotic spindle and cytokinesis [55]. Phosphorylation by CDK1 and PLK1 during mitosis activates Nek9 and ultimately leads to Nek6/7-dependent phosphorylation of centrosomal Eg5, which is required for centrosome separation [56].

Chromosome segregation and mitotic exit are irreversibly initiated by the activation of the APC/cyclosome (APC/C) (CCDC20) and the degradation of cyclins, resulting in the inactivation of CDK activity [43]. The newly formed daughter cells might either exit from further division and stay in the nondividing G0 phase or reenter a new cell cycle by the accumulation of cyclin D–CDK4/6 activity.

Astrin, a spindle-associated protein; Bora, protein aurora borealis; Cdc20, cell division cycle protein 20 homolog; Cep192, centrosomal protein of 192 kDa; INCENP, inner centromere protein; Mad2, mitotic arrest deficient 2 spindle checkpoint protein; CPC, chromosome passenger complex; H2AX, H2A histone family member; Knl1, kinetochore scaffold 1 protein; MCC, mitotic checkpoint complex; MELT, Met–Glu–Leu–Thr motifs; p-MELT, phosphorylated MELT sequences; Ndc80, kinetochore protein homolog; p53, tumor suppressor protein P53; PP1, SAC-silencing phosphatase; TPX2, targeting protein for Xklp2; APC, anaphase-promoting complex; SAC, spindle assembly checkpoint (also known as mitotic or metaphase checkpoint); SKA, spindle and kinetochore-associated complex; SKAP, phosphorylated spindle and kinetochore-associated complex.

6.5 Cell Cycle Checkpoints Controlling Cell Division

Conserved checkpoint control mechanisms operate during cell division to ensure genomic integrity [57]. Transient phosphorylation by specific checkpoint kinases (CHKs) constitutes a central regulatory element of the underlying signaling to monitor DNA damage, replication stress, and mitotic spindle assembly. Activation of a checkpoint and induction of cell cycle arrest (essentially by inhibiting cyclin E/A–CDK2, cyclin D–CDK4/6, and APC/C activities) allows to either reenter the cell cycle upon successful repair/correction of damage or eliminate excessively damaged cells by apoptosis and mitotic catastrophe mechanisms.

Following damage caused by exogenous agents, including radiation, reactive oxygen species (ROS), or, e.g., DNA-reactive chemotherapeutic, the DNA damage checkpoint can be activated by double-strand DNA breaks (DSBs). Cell cycle arrest is triggered at either G1/S phases or G2/M phases of the cell cycle and is mediated by checkpoint protein kinases ataxia telangiectasia mutated (ATM) and CHK2 (Figure 6.4). ATM phosphorylates multiple substrates in addition to CHK2 to activate the DNA repair response, but most important with regard to cell cycle regulation are the TF p53 and the CDK-activating phosphatases Cdc25A, Cdc25B, and Cdc25C [58–61, 75]. Activation of the ATM/CHK2 cascade ultimately results in p53 stabilization and upregulation of antiproliferative genes including CDK inhibitor p21 [62–67]. p53 is mandatory for G1/S arrest but likely contributes to G2/M arrest as well despite some redundancy with the DNA replication checkpoint.

In contrast to the DNA damage checkpoint, the DNA replication stress checkpoint operates exclusively in the S phase. The accumulation of single-stranded DNA, arising from impaired DNA replication forks, triggers activation of the checkpoint protein kinase ataxia telangiectasia and Rad3-related (ATR), which is a member of the phosphoinositide 3-kinase-related kinase (PIKK) family as ATM, and its effector kinase CHK1. Downstream regulation of the ATR/CHK1 signaling cascade involves CHK1-dependent inactivation of phosphatase CDC25 and activation of kinase nuclear serine/threonine kinase of 96 kDa (WEE1) to inhibit cyclin A/B–CDK1/2 activity and allow for completion of DNA replication before mitotic entry [68–70].

The spindle assembly checkpoint (SAC) is activated in mitosis and essentially coordinates the spindle attachment of chromosomes with cell cycle progression to prevent mitotic exit in the presence of unattached or improperly attached chromosomes [71].

As for the DNA damage and DNA replication stress checkpoints, reversible protein phosphorylation by kinases is a crucial regulator of SAC signaling and its downstream effects [71–73]. The SAC operates as a multifunctional protein complex, which is recruited upon phosphorylation by Aurora B and CDK1 kinases to unattached or incorrectly attached kinetochores (a multiprotein complex at chromosome centromeres and responsible for attachment to the spindle microtubules) [71, 74, 76]. In response to the incomplete attachment of identical sister chromatids to the mitotic spindle, the SAC triggers the formation and activation of the downstream mitotic checkpoint complex (MCC) to stabilize and inhibit the APC/C [77–80], a ubiquitin ligase initiating chromosome segregation and mitotic exit by polyubiquitination of cyclin B and securin [81, 82].

Aurora B–dependent recruitment of Mps1 kinase to kinetochores [83, 84] is considered crucial to the structural and temporal regulation of SAC signaling and APC/C regulation at kinetochores [84, 85]. Phosphorylation of the Knl1 phosphodomain at several Met–Glu–Leu–Thr (MELT) motifs creates binding sites for the hierarchical recruitment of additional SAC proteins, including Bub3, Bub1, BubR1 (also known as Bub1b), Mad1, Mad2, and Cdc20 [86–88]. These proteins constitute important subunits in the assembly and function of the MCC, and their interaction with the APC is required for the correct timing of mitosis [89–98]. BubR1 contributes to the inhibition of the APC/C by preventing the binding of Cdc20 [89].

The SAC signaling cascade effectively inhibits chromosome segregation and mitotic exit until each chromosome is correctly attached to the mitotic spindle in bipolar orientation. Then, the onset of anaphase is irreversible initiated by the inactivation of SAC activity and the involvement of Mps1 by feedback regulation [99, 100]. Finally, the destruction of the MCC liberates cell division cycle 20 (CDC20) to operate as a co-activator of APC/C, resulting in mitotic exit [101].

The impairment of kinase-regulated cell division and checkpoint signaling could result in genetic damage caused by accumulation of direct or replication stress-induced DNA damage and reduced mitotic fidelity. Formation of micronuclei, multinucleation, macronuclei (indicating polyploidy), and aneuploid cells (chromosome numbers deviate from multiples of the haploid number of chromosomes) is resulting from chromosomal dispersal, chromosome breaks, centrosome damage, and cytokinesis failure [26, 102, 103].

Exit from the cell cycle prevents the survival of cells with high levels of DNA damage, mitotic defects, and checkpoint errors that do not allow for repair [26, 102, 103]. Cell cycle exit in interphase depends on the DNA damage checkpoint and could be either reversible, through initiating quiescence, or nonreversible, by triggering senescence or apoptosis [104]. Cell death during mitosis in response to exhaustive spindle assembly and chromosome segregation defects is caused by a process called mitotic catastrophe and occurs by apoptosis, necrosis, or autophagic cell death. Aneuploid, polyploid, and multipolar cells are most susceptible to mitotic catastrophe [105]. It is associated with G2/M arrest by prolonged activation of cyclin B1 [106], but the exact mechanistic role of the SAC for mitotic catastrophe is not yet clear. The inhibition of SAC proteins, as well as prolonged SAC activation by depletion of essential centrosomal proteins such as Aurora kinase A, Ninein, or tumor overexpression gene (TOG), was shown to promote or trigger mitotic catastrophe and subsequent cell death. Remarkably, the efficiency of cell death is modulated by active SAC kinase BubR1 [43, 103, 107, 108]. Overexpression or inhibition of chromosome passenger kinase Aurora B is also associated with chromosome segregation and cytokinesis errors or polyploidization and genomic instability [109] and promotes mitotic catastrophe as does inhibition of PLK1, either by alone or by co-inhibition with Aurora kinases. Additionally, mitotic catastrophe might also result from exogenous and endogenous sources in interphase causing premature entry into mitosis by high levels of DNA replicative stress and dysfunctional DNA damage checkpoint control [106, 110, 111].

Removal of defective and genomically unstable cells caused by disturbed cell cycle and checkpoint regulation is of highly antiproliferative nature and includes exit from cell cycle by apoptosis, necrosis, mitotic catastrophe, and senescence/quiescence mechanisms. Failure to prevent the proliferation of defective mitotic cells with genome damage leads to the accumulation of persistent genetic instability [103], which is considered a hallmark in tumor cells.

As discussed above, reversible protein phosphorylation is a crucial regulator of cell division and checkpoint control (major kinases involved are listed in Tables 6.2 and 6.3). Therefore, disturbed kinase regulation (by either inhibition or expression changes) plays a key role in antiproliferative activity and genomic instability under various stress conditions. Based on underlying mechanisms, rapidly dividing cells and highly proliferative tissues are a preferential target for on- and off-target toxicities of SMKIs. Despite genetic instability, bone marrow suppression is among the most commonly observed side effects of SMKIs. Loss of hematopoietic precursors can result in deficiencies of their corresponding immune (leucocytes) and blood cells (thrombocytes and erythrocytes), increasing the risk of affecting the immune system, tissue hypoxia, overwhelming infection, malignant neoplasms, or hemorrhage.

Different thresholds of kinase activities are required to execute particular cellular functions of a kinase, and the molecular profile of cell division perturbations will determine biological consequences and the physiologic fate of cells [112]. The decision whether cell division defects will drive a cell to an irreversible antiproliferative fate of death or lead to persistent genome instability and unrestrained growth of defective cells will not only depend on levels of inhibition and/or expression changes in involved kinases. The relative order of kinase activity thresholds is likely impacted by cell line-dependent differences including expression, the cellular environment (pH and ion concentrations), and activities of phosphatases and other regulator proteins. The pharmacology of a specific SMKI has to be considered additionally as the biological effects of complete loss of a kinase function and partial kinase activity loss of function for extended times might differ substantially and also contribute to organ- or species-dependent differences in response to kinase inhibition.

6.6 Selectivity Profiling of Kinase Inhibition

It is likely that off-target effects of SMKIs affect both efficacy and toxicity profiles. Efficacy often does not translate from cell-free activity to *in vitro* and *in vivo activity*, and off-target kinase-associated safety issues still constitute a major cause for failure in kinase drug development programs (but also in non-kinase programs) [4]. Aiming for selectivity is considered essential to reduce the risk of unwanted side effects and is part of standard screening cascades for kinase profiling in early drug discovery phases.

Determining kinase inhibition and selectivity profiles is quite challenging for inhibitors targeting the conserved ATP-binding site and competing with cellular ATP. Michaelis–Menten constants for ATP (specific for each kinase) are typically much lower than ATP levels in mammalian cells (2–8 mM) [113, 114]. The Cheng–Prusoff equation can be used to estimate the effective inhibitor potency (IC_{50}) of a kinase inhibitor in a cellular environment or assay from an IC_{50} value in a functional biochemical assay using purified kinases and their catalytic domains to phosphorylate a substrate (Figure 6.5).

The IC_{50} of ATP-competitive kinase inhibitors depends on the ATP concentration used in an assay and on the

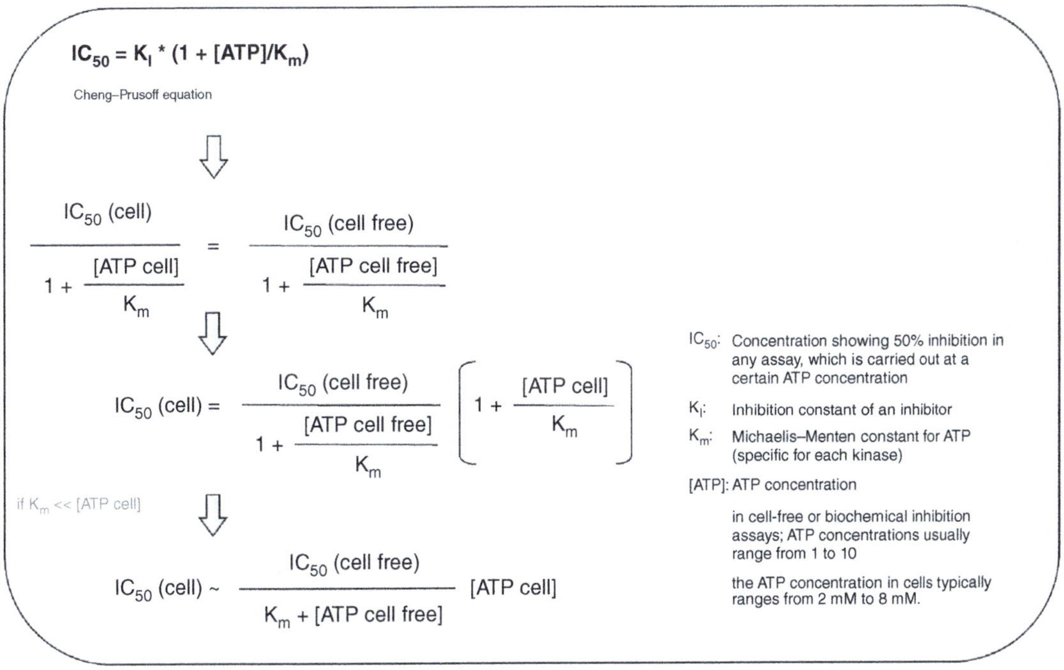

Figure 6.5 Most kinase inhibitors are reversible and ATP-competitive; therefore, the IC_{50} depends on the intrinsic affinity of the inhibitor (the dissociation constant, K_i) as well as the competition from ATP under the specific assay conditions (the [ATP] concentration and the K_M). The relation of these variables to each other is described by the Cheng–Prusoff equation. Cheng and Prusoff [115]/with the permission of Elsevier.

affinity of the target kinase for ATP, which is expressed by the Michaelis–Menten constant for ATP (K_m) [116]. Depending on the K_m of the target kinase, inhibitors often show less cellular activity (are less active in cellular assays) compared to cell-free activity (activity in cell-free assays). The selectivity profile of kinase inhibitors depends on the ATP concentrations and on the K_m values of the involved off-target kinases. An inhibitor that has similar K_i values against multiple kinases (cell-free environment) will show more potent cellular inhibition for kinases that have a higher K_m.

This means high physiological concentrations of ATP as in human cells (2–8 mM) might reduce the effective potency of ATP-competitive kinase inhibitors on the target kinase. On the other side, toxic concentrations as usually achieved for *in vitro* assessment of side effects might associate with low ATP concentrations. This could increase the effective potency of ATP-competitive kinase inhibitors for off-target kinases and reduce selectivity. Therefore, concurrent measurement of ATP concentrations in toxicity assays might add to the interpretation of *in vitro* toxicity assessment of kinase inhibitors [117].

In addition, cell-free binding and activity assays suffer from a lack of reflection of the complex cellular regulation of kinase activities and do not consider the impact of redundancies in cellular signaling cascades. The activity of kinases is highly dependent and modulated by the intracellular environment including pH, ion concentrations, substrate levels, (time- and tissue-dependent) expression and post-translational modifications, cellular localization, additional regulator proteins, and activity of phosphatases [118, 119]. To reliably link inhibitor concentrations at the site of action to pharmacological effects and phenotypic readouts, as well as to prevent unwanted adverse side effects, efficient methods accurately determining target engagement across the kinome in living cells are needed [114, 120].

Over the last years, chemoproteomic approaches using cell-permeable broad-spectrum kinase inhibitor probes turned out to be a promising tool to investigate kinase selectivity in living cells [121]. XO44, a sulfonyl fluoride–based pan-kinase inhibitor, was the first probe used to covalently label multiple intracellular kinases by binding to a conserved lysine in the ATP pocket with high efficiency and competing with high intracellular concentrations of ATP [121]. XO44 labeled up to 133 endogenous kinases in human Jurkat cells including the key cell cycle kinases CDK1, CDK2, CDK4, CDK6, and CDK7, Aurora A and Aurora B, and Nek 2, Nek7, and Nek9 and CHKs CHK1, CHK2, and WEE1 [121]. Modified and optimized scaffolds of this probe were also successfully applied to other cell lines and *in vivo* investigations in mice [122, 123]. Further development of new probes expanding the kinome coverage across different cell lines and physiological conditions could further increase the space to investigate inhibition and selectivity profiles translating into adverse side effects. In a long-term perspective, this might allow for the identification of signature profiles based on more relevant quantitative inhibition and selectivity data in living cells, thereby improving prediction and reducing major off-target kinase liabilities such as cardiotoxicity, bone marrow toxicity, and genetic instability caused by disruption of kinase signaling.

References

1 Attwood, M.M., Fabbro, D., Sokolov, A.V. et al. (2021). Trends in kinase drug discovery: targets, indications and inhibitor design. *Nat. Rev. Drug Discov.* 20 (11): 839–861.

2 Roskoski, R. Jr., (2019). Properties of FDA-approved small molecule protein kinase inhibitors. *Pharmacol. Res.* 144: 19–50.

3 Wu, P., Nielsen, T.E., and Clausen, M.H. (2015). FDA-approved small-molecule kinase inhibitors. *Trends Pharmacol. Sci.* 36: 422–439.

4 Cohen, P. and Alessi, D.R. (2013). Kinase drug discovery–what's next in the field? *ACS Chem. Biol.* 8: 96–104.

5 Rose, A.S., Bradley, A.R., Valasatava, Y. et al. (2018). NGL viewer: web-based molecular graphics for large complexes. *Bioinformatics* 34: 3755–3758.

6 Jeffrey, P.D., Russo, A.A., Polyak, K. et al. (1995). Mechanism of CDK activation revealed by the structure of a cyclinA-CDK2 complex. *Nature* 376: 313–320.

7 Huang, D., Zhou, T., Lafleur, K. et al. (2010). Kinase selectivity potential for inhibitors targeting the ATP binding site: a network analysis. *Bioinformatics* 26 (2): 198–204.

8 Liu, Y. and Gray, N.S. (2006). Rational design of inhibitors that bind to inactive kinase conformations. *Nat. Chem. Biol.* 2: 358–364.

9 Zhang, J., Yang, P.L., and Gray, N.S. (2009). Targeting cancer with small molecule kinase inhibitors. *Nat. Rev. Cancer* 9 (1): 28–39.

10 Tomita, N., Hayashi, Y., and Suzuki, S. (2013). Structure-based discovery of cellular-active allosteric

inhibitors of FAK. *Bioorg. Med. Chem. Lett.* 23: 1779–1785.

11 Bagal, S.K., Omoto, K., and Blakemore, D.C. (2019). Discovery of allosteric, potent, subtype selective, and peripherally restricted TrkA kinase inhibitors. *J. Med. Chem.* 62: 247–265.

12 Karpov, A.S., Amiri, P., Bellamacina, C. et al. (2015). Optimization of a dibenzodiazepine hit to a potent and selective allosteric PAK1 inhibitor. *ACS Med. Chem. Lett.* 6: 776–781.

13 Heinrich, T., Grädler, U., Böttcher, H. et al. (2010). Allosteric IGF-1R inhibitors. *ACS Med. Chem. Lett.* 1: 199–203.

14 Cohen, M.S., Zhang, C., Shokat, K.M., and Taunton, J. (2005). Structural bioinformatics-based design of selective, irreversible kinase inhibitors. *Science* 308 (5726): 1318–1321.

15 Kwak, E.L., Sordella, R., Bell, D.W. et al. (2005). Irreversible inhibitors of the EGF receptor may circumvent acquired resistance to gefitinib. *Proc. Natl. Acad. Sci. USA* 102 (21): 7665–7670.

16 Vyse, S. and Huang, P.H. (2019). Targeting EGFR exon 20 insertion mutations in non-small cell lung cancer. *Signal Transduct. Target. Ther.* 4: 1–10.

17 Bauer, T.M., Shaw, A.T., Johnson, M.L. et al. (2020). Brain penetration of lorlatinib: cumulative incidences of CNS and Non-CNS progression with lorlatinib in patients with previously treated ALK-positive non-small-cell lung cancer. *Target. Oncol.* 15: 55–65.

18 Zou, H.Y., Li, Q., Engstrom, L.D. et al. (2015). PF-06463922 is a potent and selective next-generation ROS1/ALK inhibitor capable of blocking crizotinib-resistant ROS1 mutations. *Proc. Natl. Acad. Sci. USA* 112: 3493–3498.

19 Traynor, K. (2012). FDA approves tofacitinib for rheumatoid arthritis. *Am. J. Health Syst. Pharm.* 69: 2120.

20 Force, T., Krause, D.S., and Van Etten, R.A. (2007). Molecular mechanisms of cardiotoxicity of tyrosine kinase inhibition. *Nat. Rev. Cancer* 7: 332–344.

21 Kerkelä, R., Grazette, L., Yacobi, R. et al. (2006). Cardiotoxicity of the cancer therapeutic agent imatinib mesylate. *Nat. Med.* 12: 908–916.

22 Force, T. and Kolaja, K.L. (2011). Cardiotoxicity of kinase inhibitors: the prediction and translation of preclinical models to clinical outcomes. *Nat. Rev. Drug Discov.* 10 (2): 111–126.

23 Zhang, T., Kohlhaas, M., Backs, J. et al. (2007). CaMKIIδ isoforms differentially affect calcium handling but similarly regulate HDAC/MEF2 transcriptional responses. *J. Biol. Chem.* 282: 35078–35087.

24 Barry, S.P., Davidson, S.M., and Townsend, P.A. (2008). Molecular regulation of cardiac hypertrophy. *Int. J. Biochem. Cell Biol.* 40: 2023–2039.

25 Matsui, T., Tao, J., del Monte, F. et al. (2008). Akt activation preserves cardiac function and prevents injury after transient cardiac ischemia in vivo. *Circulation* 104: 330–335.

26 Marcos Malumbres and Mariano Barbacid (2005). Mammalian cyclin-dependent kinases. *Trends Biochem. Sci.* 30 (11): 630–641.

27 Gavet, O. and Pines, J. (2010). Activation of cyclin B1-Cdk1 synchronizes events in the nucleus and the cytoplasm at mitosis. *J. Cell Biol.* 189: 247–259.

28 Zachos, G., Rainey, M.D., and Gillespie, D.A.F. (2003). Chk1-deficient tumour cells are viable but exhibit multiple checkpoint and survival defects. *EMBO J.* 22: 713–723.

29 Fisher, R.P. (2012). The CDK network: linking cycles of cell division and gene expression. *Genes Cancer* 3: 731–738.

30 Simmons Kovacs, L.A., Orlando, D.A., and Haase, S.B. (2008). Transcription networks and cyclin/CDKs: the yin and yang of cell cycle oscillators. *Cell Cycle* 7: 2626–2629.

31 Gavet, O. and Pines, J. (2010). Progressive activation of cyclinB1-Cdk1 coordinates entry to mitosis. *Dev. Cell* 18: 533–543.

32 Taubenberger, A.V., Baum, B., and Matthews, H.K. (2020). The mechanics of mitotic cell rounding. *Front. Cell Dev. Biol.* 8: 687.

33 Agircan, F.G., Schiebel, E., and Mardin, B.R. (2014). Separate to operate: control of centrosome positioning and separation. *Philos. Trans. R. Soc. Lond. Ser. B Biol. Sci.* 369 (1650): 20130461.

34 Santos, S.D.M., Wollman, R., Meyer, T., and Ferrell, J.E. (2012). Spatial positive feedback at the onset of mitosis. *Cell* 149: 1500–1513.

35 Pines, J. and Hunter, T. (1991). Human cyclins A and B1 are differentially located in the cell and undergo cell cycle-dependent nuclear transport. *J. Cell Biol.* 115: 1–17.

36 Morgan, D.O. (2007). *The Cell Cycle: Principles of Control*. London: New Science Press.

37 Dephoure, N., Zhou, C., Villén, J. et al. (2008). A quantitative atlas of mitotic phosphorylation. *Proc. Natl. Acad. Sci. USA* 105: 10762–10767.

38 Blethrow, J.D., Glavy, J.S., Morgan, D.O., and Shokat, K.M. (2008). Covalent capture of kinase-specific phosphopeptides reveals Cdk1-cyclin B substrates. *Proc. Natl. Acad. Sci. USA* 105: 1442–1447.

39 Chiappa, M., Petrella, S., Damia, G. et al. (2022). Present and future perspective on PLK1 inhibition in cancer treatment. *Front. Oncol.* 12: 903016.

40 Barr, F.A., Sillje, H.H., and Nigg, E.A. (2004). Polo-like kinases and the orchestration of cell division. *Nat. Rev. Mol. Cell Biol.* 5: 429–440.

41 Golan, A., Yudkovsky, Y., and Hershko, A. (2002). The cyclin-ubiquitin ligase activity of cyclosome/APC is jointly activated by protein kinases Cdk1-cyclin B and Plk. *J. Biol. Chem.* 277 (18): 15552–15557.

42 do Carmo Avides, M., Tavares, A., and Glover, D.M. (2001). Polo kinase and Asp are needed to promote the mitotic organizing activity of centrosomes. *Nat. Cell Biol.* 3: 421–424.

43 Musacchio, A. (2015). The molecular biology of spindle assembly checkpoint signaling dynamics. *Curr. Biol.* 25: R1002–R1018.

44 Tavernier, N., Frank, S.F., and Pintard, L. (2021). Aurora A kinase activation: different means to different ends. *J. Cell Biol.* 220 (9): e202106128.

45 Adams, R.R., Wheatley, S.P., Gouldsworthy, A.M. et al. (2000). INCENP binds the Aurora-related kinase AIRK2 and is required to target it to chromosomes, the central spindle and cleavage furrow. *Curr. Biol.* 10 (17): 1075–1078.

46 Girdler, F., Gascoigne, K.E., Eyers, P.A. et al. (2006). Validating Aurora B as an anti-cancer drug target. *J. Cell Sci.* 119: 3664–3675.

47 Giet, R., Uzbekov, R., Cubizolles, F. et al. (1999). The Xenopus laevis aurora-related protein kinase pEg2 associates with and phosphorylates the kinesin-related protein XlEg5. *J. Biol. Chem.* 274 (21): 15005–15013.

48 Giet, R., McLean, D., Descamps, S. et al. (2002). Drosophila Aurora A kinase is required to localize D-TACC to centrosomes and to regulate astral microtubules. *J. Cell Biol.* 156 (3): 437–451.

49 Faragher, A.J. and Fry, A.M. (2003). Nek2A kinase stimulates centrosome disjunction and is required for formation of bipolar mitotic spindles. *Mol. Biol. Cell* 14: 2876–2889.

50 Fry, A.M., Meraldi, P., and Nigg, E.A. (1998). A centrosomal function for the human Nek2 protein kinase, a member of the NIMA family of cell cycle regulators. *EMBO J.* 17: 470–481.

51 Fry, A.M., Mayor, T., Meraldi, P. et al. (1998). C-Nap1, a novel centrosomal coiled-coil protein and candidate substrate of the cell cycle-regulated protein kinase Nek2. *J. Cell Biol.* 141: 1563–1574.

52 Bahe, S., Stierhof, Y.D., Wilkinson, C.J. et al. (2005). Rootletin forms centriole-associated filaments and functions in centrosome cohesion. *J. Cell Biol.* 171: 27–33.

53 Bahmanyar, S., Kaplan, D.D., Deluca, J.G. et al. (2008). beta-Catenin is a Nek2 substrate involved in centrosome separation. *Genes Dev.* 22: 91–105.

54 Lou, Y., Yao, J., Zereshki, A. et al. (2004). NEK2A interacts with MAD1 and possibly functions as a novel integrator of the spindle checkpoint signaling. *J. Biol. Chem.* 279: 20049–20057.

55 O'Regan, L. and Fry, A.M. (2009). The Nek6 and Nek7 protein kinases are required for robust mitotic spindle formation and cytokinesis. *Mol. Cell Biol.* 29: 3975–3990.

56 Bertran, M.T., Sdelci, S., Regue, L. et al. (2011). Nek9 is a Plk1- activated kinase that controls early centrosome separation through Nek6/7 and Eg5. *EMBO J.* 30: 2634–2647.

57 Sancar, A., Lindsey-Boltz, L.A., Unsal-Kacmaz, K., and Linn, S. (2004). Molecular mechanisms of mammalian DNA repair and the DNA damage checkpoints. *Annu. Rev. Biochem.* 73: 39–85.

58 Bensimon, A., Schmidt, A., Ziv, Y. et al. (2010). ATM-dependent and -independent dynamics of the nuclear phosphoproteome after DNA damage. *Sci. Signal.* 3: rs3.

59 Shieh, S.Y., Ahn, J., Tamai, K. et al. (2000). The human homologs of checkpoint kinases Chk1 and Cds1 (Chk2) phosphorylate p53 at multiple DNA damage-inducible sites. *Genes Dev.* 14: 289–300.

60 Hirao, A., Kong, Y.Y., Matsuoka, S. et al. (2000). DNA damage-induced activation of p53 by the checkpoint kinase Chk2. *Science* 287: 1824–1827.

61 Chehab, N.H., Malikzay, A., Appel, M., and Halazonetis, T.D. (2000). Chk2/hCds1 functions as a DNA damage checkpoint in G(1) by stabilizing p53. *Genes Dev.* 14: 278–288.

62 Canman, C.E., Lim, D.S., Cimprich, K.A. et al. (1998). Activation of the ATM kinase by ionizing radiation and phosphorylation of p53. *Science* 281: 1677–1679.

63 Stommel, J.M. and Wahl, G.M. (2004). Accelerated MDM2 auto-degradation induced by DNA-damage kinases is required for p53 activation. *EMBO J.* 23: 1547–1556.

64 Tibbetts, R.S., Brumbaugh, K.M., Williams, J.M. et al. (1999). A role for ATR in the DNA damage-induced phosphorylation of p53. *Genes Dev.* 13: 152–157.

65 Dulic, V., Kaufmann, W.K., Wilson, S.J. et al. (1994). p53-dependent inhibition of cyclin-dependent kinase activities in human fibroblasts during radiation-induced G1 arrest. *Cell* 76: 1013–1023.

66 Little, J.B., Nagasawa, H., Keng, P.C. et al. (1995). Absence of radiation-induced G1 arrest in two closely related human lymphoblast cell lines that differ in p53 status. *J. Biol. Chem.* 270: 11033–11036.

67 Kastan, M.B., Zhan, Q., El-Deiry, W.S. et al. (1992). A mammalian cell cycle checkpoint pathway utilizing p53 and GADD45 is defective in ataxia-telangiectasia. *Cell* 71: 587–597.

68 Abraham, R.T. (2001). Cell cycle checkpoint signaling through the ATM and ATR kinases. *Genes Dev.* 15: 2177–2196.

69 Jackson, S.P. and Bartek, J. (2009). The DNA-damage response in human biology and disease. *Nature* 461: 1071–1078.

70 McGowan, C.H. and Russell, P. (2004). The DNA damage response: sensing and signaling. *Curr. Opin. Cell Biol.* 16: 629–633.

71 Foley, E.A. and Kapoor, T.M. (2013). Microtubule attachment and spindle assembly checkpoint signalling at the kinetochore. *Nat. Rev. Mol. Cell Biol.* 14: 25–37.

72 Lara-Gonzalez, P., Westhorpe, F.G., and Taylor, S.S. (2012). The spindle assembly checkpoint. *Curr. Biol.* 22: R966–R980.

73 London, N. and Biggins, S. (2014). Signalling dynamics in the spindle checkpoint response. *Nat. Rev. Mol. Cell Biol.* 15: 736–747.

74 Hayward, D., Alfonso-Pérez, T., and Gruneberg, U. (2019). Orchestration of the spindle assembly checkpoint by CDK1-cyclin B1. *FEBS Lett.* 593: 2889–2290.

75 Matsuoka, S., Ballif, B.A., Smogorzewska, A. et al. (2007). ATM and ATR substrate analysis reveals extensive protein networks responsive to DNA damage. *Science* 316: 1160–1166.

76 Grieco, D. and Serpico, A.F. (2020). Recent advances in understanding the role of Cdk1 in the spindle assembly checkpoint. *F1000Research* 9: F1000 Faculty Rev-57.

77 Yamaguchi, M., VanderLinden, R., Weissmann, F. et al. (2016). Cryo-EM of mitotic checkpoint complex-bound APC/C reveals reciprocal and conformational regulation of ubiquitin ligation. *Mol. Cell* 63: 593–607.

78 Alfieri, C., Chang, L., Zhang, Z. et al. (2016). Molecular basis of APC/C regulation by the spindle assembly checkpoint. *Nature* 536: 431–436.

79 Sudakin, V., Chan, G.K.T., and Yen, T.J. (2001). Checkpoint inhibition of the APC/C in HeLa cells is mediated by a complex of BUBR1, BUB3, CDC20, and MAD2. *J. Cell Biol.* 154 (925): 936.

80 Liu, S.-T. and Zhang, H. (2016). The mitotic checkpoint complex (MCC): looking back and forth after 15 years. *AIMS Mol. Sci.* 3: 597–634.

81 Primorac, I. and Musacchio, A. (2013). Panta rhei: the APC/C at steady state. *J. Cell Biol.* 201: 177–189.

82 Chang, L. and Barford, D. (2014). Insights into the anaphase-promoting complex: a molecular machine that regulates mitosis. *Curr. Opin. Struct. Biol.* 29: 1–9.

83 Rieder, C.L., Cole, R.W., Khodjakov, A., and Sluder, G. (1995). The checkpoint delaying anaphase in response to chromosome monoorientation is mediated by an inhibitory signal produced by unattached kinetochores. *J. Cell Biol.* 130: 941–948.

84 Santaguida, S., Tighe, A., D'Alise, A.M. et al. (2010). Dissecting the role of MPS1 in chromosome biorientation and the spindle checkpoint through the small molecule inhibitor reversine. *J. Cell Biol.* 190: 73–87.

85 Vigneron, S., Prieto, S., Bernis, C. et al. (2004). Kinetochore localization of spindle checkpoint proteins: who controls whom? *Mol. Biol. Cell* 15: 4584–4596.

86 Shepperd, L.A., Meadows, J.C., Sochaj, A.M. et al. (2012). Phosphodependent recruitment of Bub1 and Bub3 to Spc7/KNL1 by Mph1 kinase maintains the spindle checkpoint. *Curr. Biol.* 22: 891–899.

87 London, N., Ceto, S., Ranish, J.A., and Biggins, S. (2012). Phosphoregulation of Spc105 by Mps1 and PP1 regulates Bub1 localization to kinetochores. *Curr. Biol.* 22: 900–906.

88 Yamagishi, Y., Yang, C.H., Tanno, Y., and Watanabe, Y. (2012). MPS1/Mph1 phosphorylates the kinetochore protein KNL1/Spc7 to recruit SAC components. *Nat. Cell Biol.* 14: 746–752.

89 Overlack, K., Primorac, I., Vleugel, M. et al. (2015). A molecular basis for the differential roles of Bub1 and BubR1 in the spindle assembly checkpoint. *elife* 4: e05269.

90 Krenn, V., Overlack, K., Primorac, I. et al. (2014). KI motifs of human Knl1 enhance assembly of comprehensive spindle checkpoint complexes around MELT repeats. *Curr. Biol.* 24: 29–39.

91 Vleugel, M., Tromer, E., Omerzu, M. et al. (2013). Arrayed BUB recruitment modules in the kinetochore scaffold KNL1 promote accurate chromosome segregation. *J. Cell Biol.* 203: 943–955.

92 Vleugel, M., Omerzu, M., Groenewold, V. et al. (2015). Sequential multisite phospho-regulation of KNL1-BUB3 interfaces at mitotic kinetochores. *Mol. Cell* 57: 824–835.

93 Moyle, M.W., Kim, T., Hattersley, N. et al. (2014). A Bub1-Mad1 interaction targets the Mad1-Mad2 complex to unattached kinetochores to initiate the spindle checkpoint. *J. Cell Biol.* 204: 647–657.

94 Primorac, I., Weir, J.R., Chiroli, E. et al. (2013). Bub3 reads phosphorylated MELT repeats to promote spindle assembly checkpoint signaling. *elife* 2: e01030.

95 Lischetti, T., Zhang, G., Sedgwick, G.G. et al. (2014). The internal Cdc20 binding site in BubR1 facilitates both spindle assembly checkpoint signalling and silencing. *Nat. Commun.* 5: 5563.

96 Zhang, G., Lischetti, T., and Nilsson, J. (2014). A minimal number of MELT repeats supports all the functions of KNL1 in chromosome segregation. *J. Cell Sci.* 127: 871–884.

97 Diaz-Martinez, L.A., Tian, W., Li, B. et al. (2015). The Cdc20-binding Phe box of the spindle checkpoint protein BubR1 maintains the mitotic checkpoint complex during mitosis. *J. Biol. Chem.* 290: 2431–2443.

98 Di Fiore, B., Davey, N.E., Hagting, A. et al. (2015). The ABBA motif binds APC/C activators and is shared by APC/C substrates and regulators. *Dev. Cell* 32: 358–372.

99 Leng, M., Chan, D.W., Luo, H. et al. (2006). MPS1-dependent mitotic BLM phosphorylation is important for chromosome stability. *Proc. Natl. Acad. Sci. USA* 103: 11485–11490.

100 Palframan, W.J., Meehl, J.B., Jaspersen, S.L. et al. (2006). Anaphase inactivation of the spindle checkpoint. *Science* 313: 680–684.

101 Kapanidou, M., Curtis, N.L., and Bolanos-Garcia, V.M. (2017). Cdc20 at the crossroads between chromosome segregation and mitotic exit. *Trends Biochem. Sci.* 42: 193–205.

102 Castedo, M., Perfettini, J.-L., Roumier, T. et al. (2004). Cell death by mitotic catastrophe: a molecular definition. *Oncogene* 23 (16): 2825–2837.

103 Vitale, I., Galluzzi, L., Castedo, M., and Kroemer, G. (2011). Mitotic catastrophe a mechanism for avoiding genomic instability. *Nat. Rev. Mol. Cell Biol.* 12 (6): 385–392.

104 Matthews, H.K., Bertoli, C., and de Bruin, R.A.M. (2022). Cell cycle control in cancer. *Nat. Rev. Mol. Cell Biol.* 23: 74–88.

105 Denisenko, T.V., Sorokina, I.V., Gogvadze, V., and Zhivotovsky, B. (2016). Mitotic catastrophe and cancer drug resistance: a link that must to be broken. *Drug Resist. Updat.* 24: 1–12.

106 Chan, Y.W., Chen, Y., and Poon, R.Y.C. (2009). Generation of an indestructible cyclin B1 by caspase-6-dependent cleavage during mitotic catastrophe. *Oncogene* 28 (2): 170–183.

107 Kimura, M., Yoshioka, T., Saio, M. et al. (2013). Mitotic catastrophe and cell death induced by depletion of centrosomal proteins. *Cell Death Dis.* 4 (4): e603–e603.

108 Nakayama, Y. and Inoue, T. (2016). Antiproliferative fate of the tetraploid formed after mitotic slippage and its promotion; a novel target for cancer therapy based on microtubule poisons. *Molecules* 21 (5): 663.

109 Dar, A.A., Gof, L.W., Majid, S. et al. (2010). Aurora kinase inhibitors-rising stars in cancer therapeutics? *Mol. Cancer Ther.* 9 (2): 268–278.

110 Sazonova, E.V., Petrichuk, S.V., Kopeina, G.S. et al. (2021). A link between mitotic defects and mitotic catastrophe: detection and cell fate. *Biol. Direct* 16: 25.

111 Mc Gee, M.M. (2015). Targeting the mitotic catastrophe signaling pathway in cancer. *Mediat. Inflamm.* 2015: 146282.

112 Lera, R.F. and Burkard, M.E. (2012). High mitotic activity of polo-like kinase 1 is required for chromosome segregation and genomic integrity in human epithelial cells. *J. Biol. Chem.* 287 (51): 42812–42825.

113 Greiner, J.V. and Glonek, T. (2021). Intracellular ATP concentration and implication for cellular evolution. *Biology* 10: 1166.

114 Vasta, J.D., Corona, C.R., Wilkinson, J. et al. (2018). Quantitative, wide-spectrum kinase profiling in live cells for assessing the effect of cellular ATP on target engagement. *Cell Chem. Biol.* 25: 206–214.

115 Cheng, Y. and Prusoff, W.H. (1973). Relationship between the inhibition constant (K1) and the concentration of inhibitor which causes 50 per cent inhibition (I50) of an enzymatic reaction. *Biochem. Pharmacol.* 22: 3099–3108.

116 Knight, Z.A. and Shokat, K.M. (2005). Features of selective kinase inhibitors. *Chem. Biol.* 12: 621–637.

117 Ziegler, R., Häusermann, F., Kirchner, S., and Polonchuk, L. (2021). Cardiac safety of kinase inhibitors – improving understanding and prediction of liabilities in drug discovery using human stem cell-derived models. *Front. Cardiovasc. Med., Sec. Cardio-Oncol.* 8: 639824.

118 Cann, M.L., McDonald, I.M., East, M.P. et al. (2017). Measuring kinase activity – a global challenge. *J. Cell. Biochem.* 118: 3595–3606.

119 Wodicka, L.M., Ciceri, P., Davis, M.I. et al. (2010). Activation state-dependent binding of small molecule kinase inhibitors: structural insights from biochemistry. *Chem. Biol.* 17: 1241–1249.

120 Savitski, M.M., Reinhard, F.B.M., Franken, H. et al. (2014). Tracking cancer drugs in living cells by thermal profiling of the proteome. *Science* 346: 1255784.

121 Zhao, Q., Ouyang, X., Wan, X. et al. (2017). Broad-spectrum kinase profiling in live cells with lysine-targeted sulfonyl fluoride probes. *J. Am. Chem. Soc.* 139: 680–685.

122 Yang, T., Cuesta, A., Wan, X. et al. (2022). Reversible lysine-targeted probes reveal residence time-based kinase selectivity. *Nat. Chem. Biol.* 1–8.

123 Guiley, K.Z., Stevenson, J.W., Lou, K. et al. (2019). p27 allosterically activates cyclin-dependent kinase 4 and antagonizes palbociclib inhibition. *Science* 366 (6471): eaaw2106.

7

Activity at Cardiovascular Ion Channels

A Key Issue for Drug Discovery

Ian M. Bell and Armando A. Lagrutta

7.1 Introduction

Voltage-gated ion channels that conduct sodium, calcium, and potassium ions are critically important to the normal function of the heart [1]. Consequently, pharmacological activity at cardiovascular ion channels is a key safety issue for many drug discovery programs. Within the scope of nonclinical safety assessment, the discipline of safety pharmacology is concerned with adverse pharmacodynamic effects of a drug in the therapeutic or supra-therapeutic range [2]. Evaluation of cardiovascular risk in nonclinical models is one of the major or "core" safety pharmacology areas that both the pharmaceutical industry and regulatory agencies recognize as critical for the overall success of drug discovery and development efforts. This is underscored by the inclusion of safety pharmacology guidelines among the globally harmonized recommendations provided under the umbrella of the ICH (International Conference on Harmonization of Technical Requirements for Registration of Pharmaceuticals for Human Use) and their subsequent adoption by regulatory agencies in the United States, the European Union, and Japan [3]. The ICH guideline S7A, on Safety Pharmacology Studies for Human Pharmaceuticals, adopted by regulatory agencies in 2000, recommends a core battery of *in vivo* observations, investigating the cardiovascular, respiratory, and nervous systems, preferably in non-anesthetized animals chronically instrumented for telemetry measurements. Core cardiovascular studies mentioned in the document include measurements of blood pressure, heart rate, and the electrocardiogram (ECG). S7A also mentions the possible need of follow-up studies within each of the core systems, and supplemental studies on other systems [4]. The ICH S7B guideline on Safety Pharmacology Studies for Assessing the Potential for Delayed Ventricular Repolarization (QT interval prolongation) by Human Pharmaceuticals, approved for wide implementation in 2005, emphasizes *in vivo* studies of long QT prolongation in animal models and *in vitro* studies of I_{Kr}/hERG current, and refers to an "integrated risk assessment" in general terms [5]. Recent revisions to the S7B guideline and to the related efficacy guideline E14, on The Clinical Evaluation of QT/QT_C Interval Prolongation and Proarrhythmic Potential for Nonantiarrhythmic Drugs, written as Questions and Answers, guide investigators through the process of requesting waivers for the conduct of clinical thorough-QT studies [6]. As defined in the E14 guideline, the "thorough QT/QTc study" is intended to determine whether the drug has a threshold pharmacologic effect on cardiac repolarization, as detected by QT/QTc prolongation. The threshold level of regulatory concern is around 5 ms, as evidenced by an upper bound of the 95% confidence interval around the mean effect on QTc of 10 ms. The study is typically carried out in healthy volunteers (as opposed to individuals at increased risk of arrhythmias) and is used to determine whether or not the effect of a drug on the QT/QTc interval in target patient populations should be studied intensively during later stages of drug development. It is not intended to identify drugs as being proarrhythmic [7]. Some of the underlying considerations in the recent Q&A's relate to predicting proarrhythmic risk, rather than determining I_{Kr}/hERG current inhibition or prolongation of the QT interval of the ECG. This concept was intensely explored in an industry-regulatory consortium named the Comprehensive In Vitro Proarrhythmia Assay Project [8].

The heightened safety concerns around QT prolongation and proarrhythmic risk by drugs for *all* therapeutic indications were initially specific for antiarrhythmic drug programs, with widely publicized instances of mortality in the clinic, implicating cardiovascular ion channels. In the late 1980s, results from clinical trials of sodium channel blockers as antiarrhythmic agents revealed increased mortality when compared to placebo, despite significant suppression of premature ventricular contractions [9]. Then, in the early 1990s, reports of case studies began to link the risk of Torsades de Points (TdP), a ventricular

Polypharmacology: Strategies for Multi-Target Drug Discovery, First Edition. Edited by Jens-Uwe Peters.

polymorphic arrhythmia, and ensuing ventricular fibrillation and death, to drug action on a cardiac potassium ion channel not long before considered a reasonable candidate for antiarrhythmic therapy [10, 11]. These TdP findings implicated a diverse set of drugs for various therapeutic indications. Four major drugs were removed from the market in the 1990s (terfenadine, astemizole, grepafloxacin, and cisapride), availability restrictions were imposed on several others, and a general desire to improve preclinical determination of this risk for new investigational drugs began to gain momentum. The reader is referred to more detailed reviews on this background [12–15].

A separate, converging line of research has been the study of inherited arrhythmogenic disorders implicating particular ion channel genes. The elucidation of monogenic inherited disorders linked to arrhythmias has provided insights into arrhythmogenic mechanisms, and helped identify a pool of susceptible ion channel candidates and ion channel-mediated mechanisms of arrhythmia. For instance, rare ion channelopathies involving the cardiac sodium channel gene have been linked to inherited arrhythmias, such as Brugada syndrome, sick sinus syndrome, and long QT syndrome. Similarly, mutations in the genes encoding subunits of the rapid and slow delayed rectifier cardiac potassium currents, I_{Kr} and I_{Ks}, and the inward rectifier cardiac potassium current I_{K1}, have been linked to inherited long QT syndrome [16–21]. Looking at other physiological processes where cardiovascular ion channels play a role, for example, genetic studies of blood pressure regulation, has pointed to other signal transduction pathways implicating ion channels as effector molecules. Monogenic disorders of renal salt and volume regulation include polymorphisms in epithelial sodium channel genes [22].

The cardiac action potential (AP), a representation of the changes in electrical potential differences across cardiac cell membranes as a function of time, is an ion channel-driven process. Figure 7.1a is a diagrammatic representation of the AP and its ECG correlate. Major contributors to the cardiac AP are depolarizing sodium current (shown on the upstroke of the AP), calcium current contributing to the plateau phase of the AP (not shown), and repolarizing potassium currents (shown on the downstroke of the AP). Other important contributors are the inward rectifier potassium current and other ion channels, exchangers, and pumps that help set the resting membrane potential and the degree of refractoriness to depolarization in various cardiac tissues. The ECG measures voltage changes at the body surface related to the cardiac AP changes. A signature waveform with discrete deflections, designated P, QRS, T, and sometimes U, corresponds to the self-propagating wave of depolarizations and

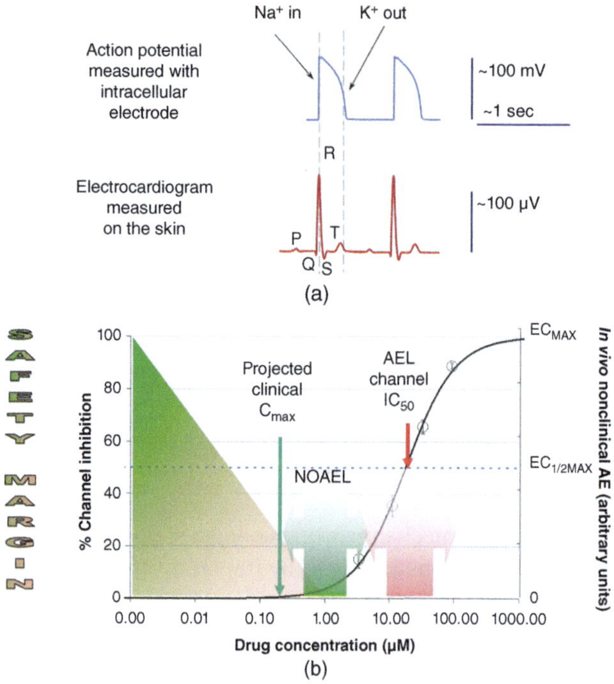

Figure 7.1 (a) Diagram of cardiac ion channel activity indicators. The cardiac AP measures membrane potential changes evoked by flux of ions across ion channels in cells, tissues, or the whole heart. The contribution of sodium channels and delayed rectifier potassium channels are shown. The ECG measures voltage changes related to the cardiac AP changes. Signature waveform features is shown. (b) Diagram of efficacious and cardiac adverse effects as function of drug concentration.

repolarizations from the sinoatrial node, the pacemaker region of the heart, through the atria, the atrioventricular node, and finally extending to the ventricles. Notably, the PR interval corresponds to atrial depolarization and wave propagation through the cardiac conducting system; the QRS interval, to the coordinated depolarization of both ventricles; and the QT interval, to ventricular repolarization activity. The ECG has become an invaluable diagnostic tool of physiological and pathological activity. For example, abnormal elevation of the ST segment, abnormal lengthening of the QRS peak, and abnormal prolongation of the QT interval are useful clinical markers of arrhythmias.

Drug-acquired adverse effects in the heart can be summarized as follows: inhibitors of sodium currents can reduce conduction by inhibiting the depolarization of cardiac myocytes; inhibitors of calcium currents can lead to vasodilation and suppression of cardiac contractility, and they can also affect the conduction of the sinoatrial and atrioventricular nodes; finally, inhibitors of potassium currents can delay cardiac repolarization. Insights into drug-acquired channelopathies began to be revealed with the use of pharmacological tools originally designed as antiarrhythmics.

A prime example is the pharmacological dissection of I_{Kr} and I_{Ks}, fast and slow components of delayed rectifier potassium current in guinea pig ventricular myocytes, using the class III antiarrhythmic agent E-4031 [23]. That initial effort led to an understanding of the complementary and somewhat overlapping properties of I_{Kr} and I_{Ks}, thought to form the basis for a ventricular repolarization reserve. Evidence linking arrhythmia risk factors with a reduction in this reserve has been presented [24–26]. At the cellular level, inhibition of I_{Kr} or I_{Ks} prolongs AP duration (APD), a situation that can degenerate in afterdepolarizations, and particularly early afterdepolarizations (EADs), during the plateau phase of the cardiac AP, when the cardiac cell membrane is vulnerable to minor alterations in the balance of inward and outward currents [27]. In this context, abnormal calcium influx could contribute to EADs. In general, ion channel structure-function analysis has facilitated better understanding of drug-acquired ion channel disorders (*vide infra*).

Ion channels are also implicated in the maintenance of vascular tone. The role of L-type calcium channels, calcium-activated potassium channels, voltage-dependent potassium channels, and transient receptor potential channels, among others, in vascular smooth muscle, as well as the role of epithelial sodium channels in the kidney, have been studied and reviewed [22, 28–31].

Monitoring activity on an expanded panel of cardiac ion channels, to include as many of the players implicated by the genetic evidence mentioned above, has been facilitated by higher throughput screens (*vide infra*). Counter-screening activity of anti-targets is prioritized, not only according to the promiscuity of the anti-target, but also according to potential clinical consequences. Two examples illustrating this prioritization philosophy are hERG-implicating QT prolongation and calcium-channel-implicating hypotension. On the one hand, there is an imperfect connection between TdP and QT prolongation and between QT prolongation and hERG channel block [12, 15, 27, 32–35], but hERG channel activity has been identified as the most common contributor to drug-acquired QT prolongation, and QT prolongation is correlated with rare, though potentially catastrophic adverse effects. The promiscuity shown by the hERG channel towards many chemical series continues to merit special attention in frontline screens and lead-optimization chemistry efforts. On the other hand, calcium channel inhibitors have a long history of use as antihypertensive agents, and calcium channel activity correlated to nonclinical hypotension may be a clinically manageable risk.

The goal of a safety-minded drug discovery and nonclinical development strategy is to minimize clinical risk. A helpful general principle is the concept of provisional safety margins based on the *in vitro* potency on the cardiac ion channel and the projected therapeutic concentration. Figure 7.1b is a diagram extending the principle of provisional safety margins, originally applied to hERG potency and QT prolongation risk, to any cardiac adverse effect implicating a cardiovascular ion channel. The diagram plots on a semilogarithmic scale the "effective," unbound drug concentrations that produce efficacious and cardiac adverse effects. The provisional safety margin is the separation between the IC_{50} value measured at the ion channel and the projected "effective" therapeutic concentration. For simplicity, only adverse effects are plotted as a function of "effective" drug concentration, and the *in vitro* cardiac ion channel effect is perfectly superimposed on the adverse effect (AE) measured *in vivo*. The *in vivo* dose just below a dose producing an AE is commonly known as the "no observed adverse effect level" (NOAEL). The "effective" concentration underlying this NOAEL is another critical parameter that can be compared with the "effective" therapeutic concentration. In the diagram, the safety margin is illustrated as a triangle that comes to a point, or dissipates, when the "effective" therapeutic concentration approaches or surpasses critical threshold points in the concentration response curves for nonclinical adverse effects (e.g., ion channel IC_{20}). A comprehensive review of clinical and nonclinical data indicated that a 30-fold provisional safety margin is a prudent starting point for comparisons between the hERG channel IC_{50} value and the "effective" therapeutic concentration; that is, the unbound fraction of the therapeutic concentration [36]. Several recent analyses validate the merits of the principle but also point to complexity and caveats on the optimal safety margin [33–35].

7.2 Screening Methods

In this section, we review technologies used to counter-screen potentially unsafe, adverse activity on cardiovascular ion channels. The use of heterologously expressed ion channel subunits in stably transfected mammalian cell lines [37] is now commonplace for screening applications. Reconstitution of the native phenotype by inclusion of all the appropriate ion channel subunits is the ideal, but practical solutions can be reached expressing only the major ion channel subunits, especially with limited understanding of the native channel composition.

Because a large number of chemical series interact with the hERG channel, intense effort has been placed on designing and optimizing assays to detect hERG channel activity in drug discovery and development. Non-functional *in vitro* assays, such as ligand displacement binding (for example, dofetilide, astemizole, and MK-499) provide high throughput and correlate reasonably well with the functional characterization of hERG channel inhibition obtained by conventional patch clamp studies [38, 39]. Commonly mentioned caveats for this type of assay are the inability to detect inhibition that does not compete with the binding of the specific ligand and the inability to distinguish inhibition from activation. In contrast, conventional patch clamp directly measures the current (ion flux/unit time) passing through hERG or other ion channels, and provides high information content but relatively low throughput. This technique can provide the most complete and reliable data about the interaction between a drug and the various conformational states of a channel (such as open, closed, and inactivated states), desirable mechanistic insights into a biophysically complex molecule like hERG. Patch clamp measures the activity of ion channels directly and in real time. Increased throughput has been made possible with automated patch-clamp electrophysiology, and these screens have been incorporated into discovery and preclinical development strategies. In these platforms, the traditional single micropipette electrode has been replaced with a planar substrate with an array of microapertures. Several instruments utilizing this technology are commercially available. Parallel recording options vary from 4 to 8 cells (Nanion, Patchliner®), 48 cells (Sophion, QPatch), up to 384 cells per experiment (Molecular Devices, IonWorks®; Nanion, SyncroPatch® 384). Recording quality on the Nanion instruments and QPatch is comparable to that observed in conventional, manual patch clamp electrophysiology. Also, asynchronous operation and integrated fluidics allow for application of multiple compounds or concentrations and efficient investigation of dose–response relations with these instruments. Thus, throughput in these high-quality parallel cell recording systems can be from 2000 up to 20,000 data points per day, compared to less than 100 for a conventional patch-clamp setup, making them suitable for support of lead optimization efforts [40–45]. The more recent addition of well-controlled and fast solution exchange, temperature control, and the availability of the current clamp mode, required to analyze standard cell lines and excitable cells such as stem cell-derived cardiomyocytes in a more physiologically relevant environment, have increased the value that these instruments can bring to drug discovery programs. Functional assays bridging the gap between automated patch clamp and binding assays can be useful, with a clear understanding of the specific readout and its limitations as it relates to the ion channel function. Examples include rubidium flux measured by mass spectroscopy or thallium flux measured by fluorescence as surrogates of potassium current and polarization measured by fluorescence as a surrogate of voltage-dependent ion current [46–49].

Test systems heterologously overexpressing cloned human ion channel subunits answer questions concerning direct interactions between drug and ion channel, acute or delayed [37]. These test systems cannot be expected to reconstitute native signal transduction pathways and possible indirect drug effects on the ion channels. Isolated cardiac myocytes or multicellular cardiac preparations from animal model systems aim to fill this gap, but attention should be given to the specific signal transduction pathways being examined to ensure test system suitability. When used with *in vivo* studies on the same species, these preparations provide mechanistic insights into nonclinical results. Technically, fully differentiated myocyte preparations are difficult to adapt to the automated patch clamp systems described above, limiting their use to conventional experiments measuring ion channel currents or APs. More recently, human-induced pluripotent stem cells (hiPSCs) yielding cardiomyocyte syncytia have been utilized in plate-based assays to measure either field potentials equivalent to ECGs or readouts related to cardiac contractility [50]. To date, in the published literature, the electrical activity of cardiomyocytes from these sources can be at best described as immature, with syncytia composed of a gradient of nodal-like, atrial-like, and ventricular-like cells that beat spontaneously. These hiPSC cardiomyocyte models have been used in conjunction with a biophysical "*in silico*" model of the human ventricular cardiomyocyte aiming at predicting arrhythmogenicity based on the integrated activity of cardiac ion channels [6, 8]. The practical utility of these models to predict drug arrhythmogenic potential has been documented and recently reviewed within the CiPA consortium and is contemplated in the recent ICH E14/S7B Q&A's [6, 8, 50].

7.3 Structural Insights into the Interaction Between Drugs and CV Ion Channels

Pharmacological approaches to characterize the inhibition of voltage-gated ion channels have been complemented by structural data, historically obtained in surrogates (e.g., homologous bacterial channels, truncated constructs of the channel of interest) with improved molecular stability, facilitating successful expression, purification, and structural analysis, either by X-ray diffraction of crystals, or more recently, by cryogenic electron microscopy (cryo-EM) technology [51–57]. These insights, combined with functional studies based on site-directed mutagenesis, provide a better understanding of how compounds bind to these ion channels. Voltage-gated ion channel families share common pore-forming structural features [58, 59]. The pore-forming subunit of sodium and calcium channels is comprised of four homologous domains (DI-DIV), each one containing six transmembrane α-helices (S1–S6). Figure 7.2 shows an illustrated representation of the cryo-EM reconstruction of a well-structured rat Nav1.5 α-subunit core transmembrane construct at 3.2–3.5 Å resolution (rNav1.5c). This core α-subunit construct, with deletions in DI-DII (L462-G658), DII-DIII (S1069-P1189), and the C-terminus (L1898-V2019), was selected for its improved homogeneity and expression and high voltage-activated Na^+ conductance.

Voltage-gated potassium channels consist of tetramers of α-subunits, each one with S1–S6 domains analogous to the domains within sodium and calcium channels. The selectivity filter and inner pore of these channels are formed by a pore module (S5, P loop, and S6). While state dependence of ion channel inhibition and site of action vary, a common mechanism is binding to the pore accessed intracellularly upon channel opening. Compounds inhibiting multiple ion channels with micromolar or submicromolar potencies are common and include local anesthetics and many antiarrhythmic drugs.

The biophysical and structural properties of the hERG channel have attracted particular attention. Peculiar biophysical properties include inward rectification due to a fast, voltage-dependent inactivation process at depolarized potentials and large "tail" currents at repolarized potentials, as the channel quickly recovers from inactivation and deactivates more slowly [60]. Like other voltage-dependent potassium channels, the hERG α-subunit encodes six transmembrane domains and a voltage sensor. Figure 7.3 shows an illustrated representation of the cryo-EM reconstruction of a well-structured hERG construct at 3.8 Å resolution ($hERG_T$). In this construct, the hERG gene was modified to remove two segments of the hERG channel (amino acid residues 141–350 and 871–1005), which suppressed molecular fluctuation and increased molecular stability, and led to successful expression, purification, and structural analysis [54, 55].

Unique structural features of the hERG channel include the sequence encoding the pore region, with a Gly-Phe-Gly signature sequence (GFG) instead of the prototypical potassium channel Gly-Tyr-Gly sequence (GYG) in the selectivity filter (Figure 7.4), and a number of aromatic residues lining the pore, thought to contribute to its promiscuous drug binding [54, 55, 61].

Also, the hERG channel has an Ile-Phe-Gly (IFG) instead of a Pro-X-Pro motif in the S6 or inner helix lining the pore, with F656 contributing to the hERG-specific drug binding pocket [54, 55]. All these features appear to play a role in the drug promiscuity of hERG [61, 62] and in observations, such as drug trapping [63]. In addition, the linker between the fifth transmembrane domain and the pore region is unusually long (43 amino acids) and rich in charged residues shown to be involved in the fast inactivation process [64]. Sequences encoding the intracellular domains

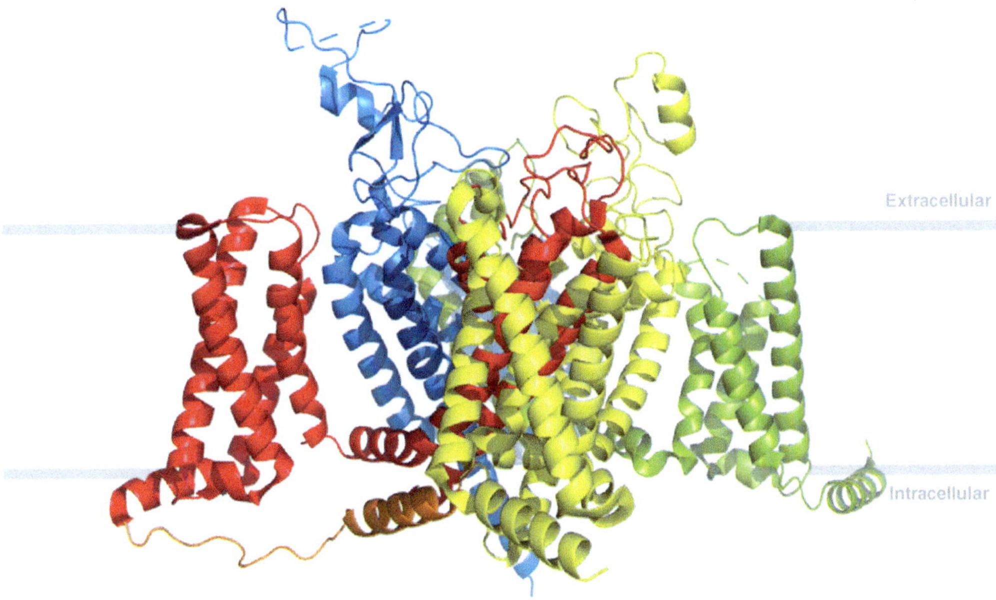

(a)

Figure 7.2 Cryo-EM structure of $rNa_v1.5_c$. Jiang et al. 57/with permission of Elsevier. (a) Side view of ribbon representation of the $rNa_v1.5_c$ showing the overall structure of the channel. *DI* is shown in blue; *DII* is shown in green; *DIII* is shown in yellow; *DIV* is shown in red; and the *DIII*-*DIV* linker is shown in orange. (b) Top view of the $rNa_v1.5_c$ channel using the same color scheme as in (a).

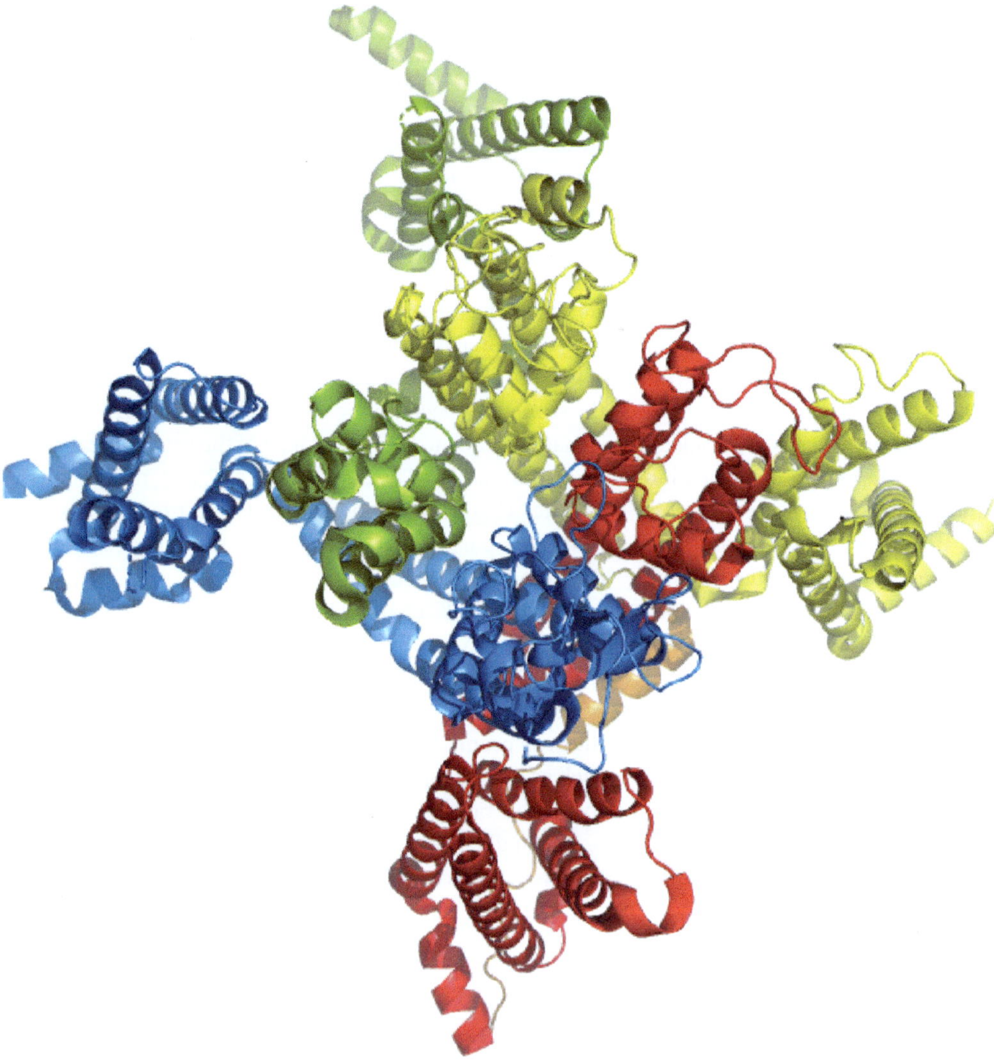

(b)

Figure 7.2 *(Continued)*

are also unique in hERG. The extended C-terminal region contains a cAMP binding domain, and the N-terminus contains a 135-amino acid sequence, crystallized and defined structurally as a Per Arnt Sim (PAS) domain, implicated in signal transduction processes [65]. Native hERG1 channels comprise at least two subunits, hERG1a and hERG1b. The hERG1a subunit contains the (PAS) domain that regulates channel gating through interactions with the C-terminal cyclic nucleotide-binding homology domain and the cytoplasmic S4–S5 linker. The hERG1b subunit has a much shorter and unique N-terminus that lacks a PAS domain. The importance of the PAS domain and heteromeric arrangement in channel biophysical properties and pharmacology have been recently described [66]. Drug interactions with the hERG channel have been elucidated at the molecular level by probing the potency of these compounds against hERG channels with a battery of single amino acid substitutions in domains hypothesized to line the pore, such as the S6 helix and the pore helix. This strategy has identified determinants of ion channel affinity that appear to be specific to a particular chemical series, or common to drugs across several chemical series, and it has led to the conclusion that aromatic residues Y652 and F656 in the S6 helix of hERG contribute to the binding affinity of a range of diverse hERG blockers [54, 58, 67–71]. Most other voltage-gated potassium channels have aliphatic residues (DI or DIV) at analogous positions [62], suggesting that Y652 and F656 may be important contributors to the promiscuity of hERG.

However, mutation of F656 to aliphatic amino acids did not lead to a dramatic decrease in cisapride binding,

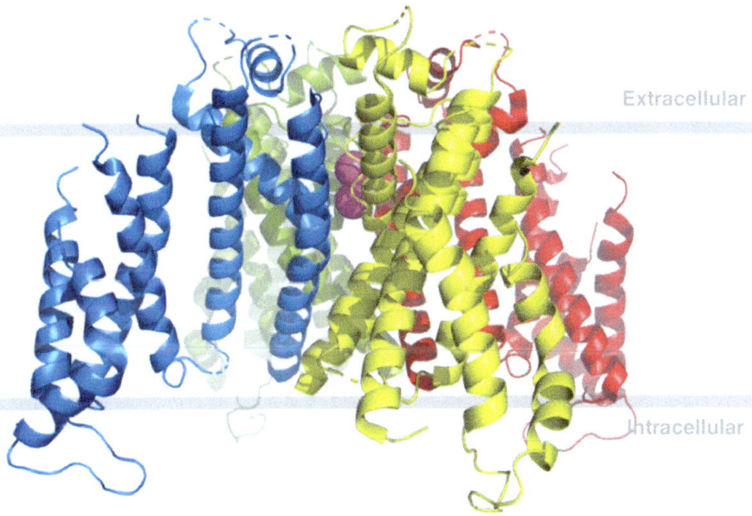

(a)

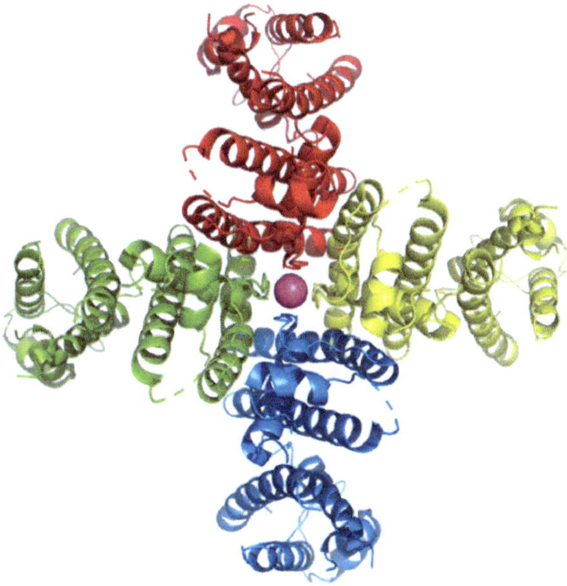

(b)

Figure 7.3 Cryo-EM structure of $hERG_T$ core region. Asai et al. [55]/with permission of Elsevier. (a) Side view of ribbon representation of the $hERG_T$ core region and voltage sensor domains with each subunit shown in different color (blue, yellow, red, and green) and the potassium ions in the pore shown in magenta. (b) Top view of the $hERG_T$ core region and voltage sensor domains using the same color scheme as in (a).

suggesting primarily lipophilic contributions to binding rather than a dependence on π-π stacking for this hERG inhibitor [72]. Consistent with this dependence on lipophilicity, inhibition of hERG by MK-499, cisapride, and terfenadine correlated well with the two-dimensional approximation of the van der Waals hydrophobic surface area for the side chain at position 656 [72]. There is a greater requirement for an aromatic residue for Y652 which has been attributed to π-cationic interactions with molecules containing basic functionality. But it has been argued that the losses of potency observed when Y652 is replaced by aliphatic residues are not as great as would be expected for the loss of a π-cation interaction [73]. Neutral drugs that bind to hERG are presumed to depend on π-stacking or lipophilic contacts with Y652. The studies that point to π-stacking and hydrophobic interactions between aromatic residues lining the large cavity of the hERG channel pore and aromatic residues suitably positioned in a drug molecule create a picture that is consistent with the binding of high affinity drugs, like methanesulfonanilides, and some low affinity drugs, like vesnarinone, to hERG channels. The different contributions of the F656 and Y652 residues across the four α-subunits have been explored using site-directed mutagenesis of tandem dimers in the context of cisapride binding [74]. Alanine replacement of Y652 on adjacent subunits reduced binding of cisapride, suggesting that adjoining Y652 residues are required for binding. Mutation of a single F656 on adjacent

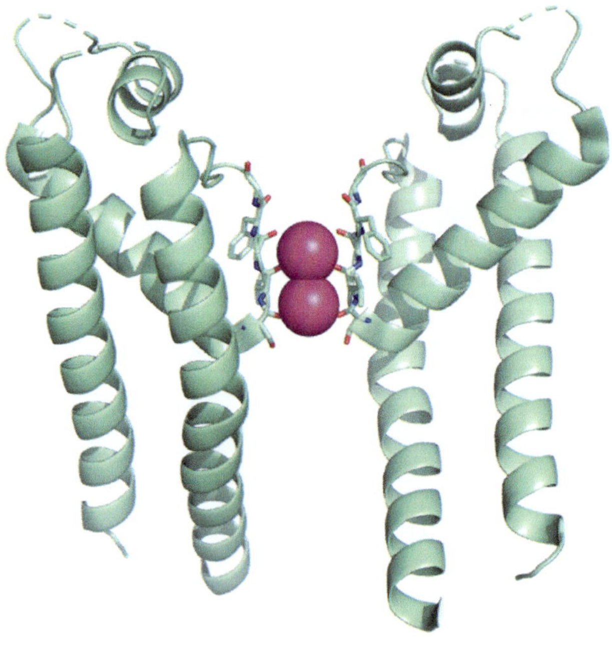

Figure 7.4 Selectivity filter of $hERG_T$ core region. Asai et al. [55]/with permission of Elsevier. Side view of ribbon representation of the $hERG_T$ core region with the residues that comprise the selectivity filter shown as sticks and the potassium ions shown in magenta. For clarity, only two opposing subunits are shown.

subunits did not have a dramatic effect on binding. In general, tandem dimer data is indicative of additivity of cisapride interactions with Y and F on the same subunit and cooperativity with these residues on the adjacent subunit.

Drugs differ with respect to their interactions with other residues, particularly those closer to the selectivity filter. Residues T623, S624, and V625 near the pore helix seem to be important for the binding of most open channel blockers, and subtle differences have been identified in the interactions of drugs, such as bepridil and the methanesulfonanilides, with other residues in these critical pore lining domains, for example with G648 and V659 flanking the S6 helix [54, 55, 58, 69, 70, 75].

Prototypical small molecule blockers of voltage-gated sodium and calcium channels block the inner pore upon channel opening in voltage dependent manner by direct, intracellular access [29, 76]. Local anesthetics, among other sodium channel inhibitors have been shown to bind with greater affinity to the open or inactivated state than the resting state of channels. Since sodium channels are formed from a single α-subunit containing four homologous transmembrane domains, the inner pore regions lack the fourfold symmetry of voltage-gated potassium channels. Site-directed mutagenesis has identified molecular determinants for drug blockade across a range of sodium channels. Local anesthetics and antiarrhythmics have been shown to interact with specific sites in domains I, III, and IV of voltage-gated Na channels. Transmembrane S6 segments are crucial for functional inhibition of a range of Nav subtypes, and a number of pore-lining S6 residues, typically lipophilic amino acids, are involved in inhibitor binding [77]. Recent cryo-EM studies of flecainide bound to rNav1.5c indicate interaction of flecainide determinants with the side chain of Phe1762 in the DIV S6 segment, a binding pose for flecainide that is consistent with previous structure-function studies of Nav1.5 and Nav1.2 [57, 76–78].

Voltage-gated calcium channels, similar to sodium channels, present an asymmetric pore domain defined by four homologous S5 and S6 domains [59]. The L-type calcium channel pore is defined by pore-lining inner helices of S6, outer helices S5, and the P-loops from the four motifs of the α1 subunit. Primarily hydrophobic residues have been shown to be significant contributors to binding. Three major classes of L-type antagonists have been investigated: dihydropyridines, phenylalkylamines, and benzothiazepines, and the three classes have been shown to bind to different but overlapping sites along the S6 segment [79, 80]. Alanine scanning and dihydropyridine binding showed a range of DIII S6 and DIV S6 amino acids that most significantly impacted binding. Alanine scanning and phenylalkylamine binding showed a range of partially overlapping hydrophobic amino acids in DIV S6 [81]. In the dihydropyridine structures, a phenyl ring has been shown to be critical to potency. A common feature in phenylalkylamines is the presence of two methoxy-substituted aryl rings flanking an alkylamine chain. High-resolution cryo-EM structures for Cav1.2 channels remain elusive. Recently, cryo-EM structures of Cav1.1 in complex with nifedipine (a dihydropyridine), diltiazem (a benzothiazepine), and verapamil (a phenylalkylamine), at resolutions of 2.9, 3.0, and 2.7Å, respectively, have informed prior mutagenesis data. While diltiazem and verapamil traverse the central cavity of the pore domain, directly blocking ion permeation, nifedipine occupies a fenestration site at the interface of repeats III and IV [56].

Along with X-ray crystal structures and extensive mutagenesis experiments, recent cryo-EM structures of cardiovascular ion channels with bound inhibitors have provided useful insights concerning key interactions [55–58], as well as alternative access routes by some ligands into the deep pore, such as fenestrations [56, 57]. Despite great progress, the ideal of proactively avoiding ion channel activity by molecular design based on this body of knowledge, particularly for hERG, has not been fully realized at this time.

Many efforts have been undertaken to predict interactions through modeling and to impact molecule design through the development of *in silico* models for predicting hERG channel inhibition [82, 83]. Useful insights into some common features among hERG inhibitors have emerged from a growing list of ligand-based SAR studies [84]. Common properties of inhibitors, such as lipophilicity and basicity, have been noted as frequent drivers of hERG binding affinity. The inner pore domains of ion channels offer a highly lipophilic environment for the binding of drugs. The large lipophilic surfaces in the inner pore are expected to drive binding events primarily through entropic contributions. The combination of newer computational approaches, such as novel artificial intelligence methods [83], with recent advances in understanding of the structure of the hERG channel [54, 55] should enhance the ability to predict and avoid hERG inhibition during the design of new compounds in drug discovery programs.

It has been commonly noted that hERG is a particularly promiscuous target, though data on the frequency of finding actives has been lacking in the literature. Internal data from Merck & Co., Inc., on compounds screened in a high-throughput binding assay has shown that ~45% of compounds tested had $IC_{50} < 10\ \mu M$ (of about 400,000 compounds tested, unpublished data). Approximately 10% have $IC_{50} < 1\ \mu M$, which contrasts with Nav1.5 and Cav1.2 both showing around 1% of compounds with $IC_{50} < 1\ \mu M$ in FLIPR assay formats (unpublished data). While it is possible that this dataset could be biased because compounds may be chosen for the assay specifically because hERG is anticipated to be an issue, this dataset is sufficiently large to suggest the extent of the problem. Analyses of a large set of AstraZeneca's compounds tested through a functional hERG assay (Ionworks whole-cell electrophysiology) also demonstrate the high frequency of hERG actives in medicinal chemistry program molecules [85]. The analysis provided a breakdown that emphasized the importance of lipophilicy and basicity as factors that drive high affinity binding to the channel. Molecules with basic functionality and a calculated logD = 3 were shown to have a 76% probability of having a hERG $IC_{50} \leq 10\ \mu M$. Neutral compounds with the same lipophilicity had a 24% probability of a hERG $IC_{50} \leq 10\ \mu M$, and acidic compounds had an 8% probability of a hERG $IC_{50} \leq 10\ \mu M$. For less lipophilic basic molecules (calculated logD = 2), the probability of a hERG $IC_{50} \leq 10\ \mu M$ dropped from 76% to 50%. These data illustrate the particular challenge of the promiscuity of the hERG channel during lead optimization. The data are also consistent with what has been demonstrated to be important structural elements of the channel, with basicity and lipophilic functionalities being particularly suited to interact with features of the hERG channel.

7.4 Medicinal Chemistry Approaches

While drug development programs can struggle with a number of undesirable cardiovascular ion channel activities, more attention by far has been paid to the hERG potassium channel. It is a truism that this channel is highly promiscuous, and the structural data, revealing a large, lipophilic pore lined with aromatic amino acids, are consistent with this observation (*vide supra*). Therefore, it is not surprising that more compounds on more programs appear to possess significant affinity for hERG compared with other ion channels. Moreover, the consequences of I_{Kr} blockade are of such gravity that this activity has received a great deal of attention in the early phase of drug discovery projects. Consequently, this overview of medicinal chemistry approaches to tackling cardiovascular ion channel issues will focus primarily on activity at the hERG channel.

The high capacity and robust nature of radioligand binding assays for the hERG channel, such as those that utilize [^{3}H]-dofetilide [86] or [^{35}S]-MK-499 [38] as the radiotracer, has made them a common counterscreen in many large pharmaceutical companies. With such an assay, it is relatively inexpensive and technically feasible to screen essentially all active compounds from a program of interest [87]. The ready availability of such hERG binding data has been a huge asset for medicinal chemistry programs trying to address this undesirable activity. Additionally, when hundreds of thousands of compounds are screened in this way, very large datasets are generated, and these may be mined for structure-activity relationships and trends that can guide future efforts to ameliorate the activity in question.

One approach to such data mining is the "molecular transformations" methodology described by Sheridan et al. [88], which applies a similar strategy to one employed routinely by medicinal chemists: comparing pairs of similar compounds to discern structure-activity trends. The methodology allows this basic medicinal chemistry skill to be applied to large datasets. Using molecular descriptors, pairs of similar compounds in the dataset are identified, and the "transformation" – the structural difference between the two molecules – is defined. The compound pairs may be placed into groups that share this same transformation, and any activity trend can be readily examined to see if the transformation increased or decreased the activity of interest on average. For example, the transformations can be sorted to identify those structural modifications that correlated with the largest average activity changes. This kind of analysis is, of course, limited to the compounds that have been made and tested in a given assay. The dataset is inherently limited in terms of the available pairs of close analogues that have been assayed and also potentially biased by the relative representation of different structural classes and chemical modifications. Nonetheless, this methodology is an empirical method for finding discrete modifications that can have significant effects on activity across a range of structural classes.

Applying this transformation methodology to the dataset generated with the [^{35}S]-MK-499 hERG binding assay run at our company identified a number of structural changes that correlated with reduced affinity for the ion channel. In this analysis of approximately 200,000 data points, molecular transformations were selected for which there were at least 50 different examples across at least two structural classes of compounds, and for which there was a significant reduction in affinity for the hERG channel (average IC_{50} increased ≥ threefold). A pair of compounds illustrating a representative transformation is shown in Scheme 7.1.

In this example, the conversion of the *N*-alkylpiperazine **1** to the *N*-acylpiperazine **2** led to a 760-fold reduction in affinity for the hERG channel [89]. The modification responsible for this change in potency is from amine to amide, and it amounts to the removal of a basic center and the addition of a polar carbonyl group, both of which could reduce affinity for the potassium channel. While most transformations do not lead to such a dramatic change in potency, other representative results from this molecular transformations analysis are shown in Scheme 7.2. Examples include: removal of a basic amine; conversion of a carboxylic ester to the corresponding acid; removal of a lipophilic substituent from an aryl ring; incorporation of an alcohol or other polar moiety; incorporation of a nitrogen atom into an aryl ring; and removal of a phenyl ring. In principle, the transformations shown in Scheme 7.2 represent an unbiased analysis of the available data and illustrate structural changes that successfully reduced affinity for the hERG channel across a range of structures and compound classes. This approach identifies general modifications that work across different structural classes, and these appear to work, at least in part, because they lower lipophilicity and/or remove basic functionality. Other modifications may also be beneficial (see Table 7.1). but they are often less general in their application.

The outline below summarizes strategies that have been shown to be effective for reduction of affinity for the hERG potassium channel. In addition to the approaches uncovered by the molecular transformations methodology, there are other strategies that, by their very nature, are not readily observed with this technique. One approach is to modify the overall topology of the molecule, for example, by constraining a linear molecule to give a

1
hERG IC_{50} = 68 nM

2
hERG IC_{50} = 52,000 nM

Scheme 7.1 A representative transformation analysis.

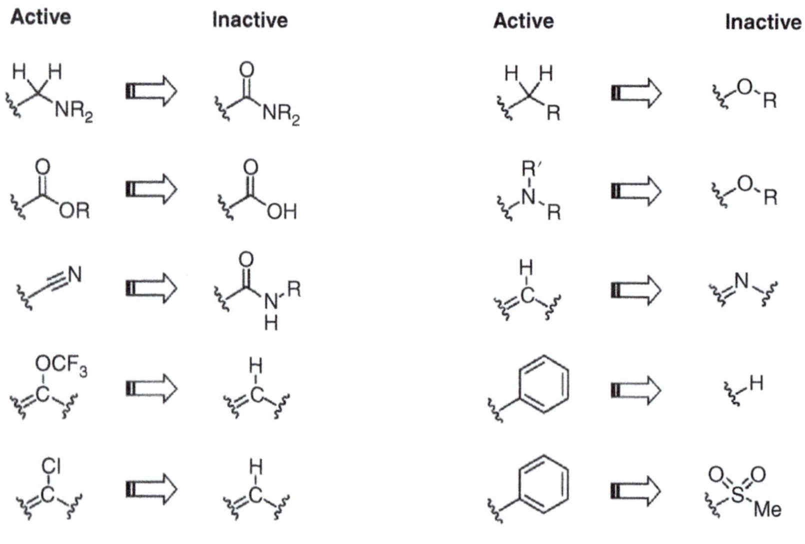

Scheme 7.2 Structural changes that reduce hERG affinity.

Table 7.1 Strategies for reducing affinity for the hERG channel.

- Reduce basicity
  - Lower pK_a of, or delete, basic functionality
- Reduce lipophilicity
  - Add polar groups, for example, alcohol or ether
  - Remove hydrophobic groups
- Reduce π-stacking interactions
  - Remove or modify aryl rings
- Subtle modification of substituents, stereochemistry, or overall topology
  - It's a protein-ligand interaction!

macrocycle. Good examples of this strategy were observed for imidazole-containing farnesyltransferase inhibitors. Perhaps more strikingly, subtle modifications of substituents or stereochemistry can, in some cases, have a dramatic effect on binding to the hERG channel. Such observations underline the key point that this binding event is between a small molecule and a protein.

Taken as a whole, the strategies described in Table 7.1 are similar to those previously described in other reviews [84]. Some of these modifications, in particular many of those identified by the molecular transformations methodology, essentially work by reducing lipophilicity. Ideally, selectivity for the desired target is enhanced, and the compound relies to a greater extent upon specific interactions with the target protein instead of on general hydrophobicity or, to put this in more formal terms, the lipophilic ligand efficiency is increased [90]. Because small molecules can bind to hydrophobic pockets directly below the selectivity filter in the hERG channel [55], these strategies may be particularly effective ways to reduce affinity for this ion channel but, in principle, they could also be applied to other selectivity issues.

To investigate this possibility, we applied similar transformation methodology to data from an in-house L-type calcium channel binding assay which uses [^{3}H]-diltiazem as the radioligand. The results of this analysis were generally similar to the results obtained with the hERG channel as represented in Scheme 7.2, and can be broadly classified as modifications that increase polarity or reduce basicity of the test compounds. These observations are consistent with structural and mutagenesis data, which indicate that the channel has affinity for ligands with lipophilic and aromatic moieties (*vide supra*). Interestingly, when similar methodology was applied to data from another common counterscreen – a pregnane X receptor (PXR) activation assay using a luciferase reporter gene, similar results to those shown in Scheme 7.2 were not obtained and, in fact, we identified very few transformations that satisfied our criteria (average IC_{50} or EC_{50} increased ≥ threefold and at least 2 structural classes of compounds).

While it is certainly possible that these very different results obtained for the calcium channel and pregnane X receptor reflect some intrinsic bias in the structures contained in the datasets, both datasets were of a similar size (approximately 50,000 data points) and both contained data for compounds from many drug discovery programs. Thus, our observations suggest that the general strategies deduced from these analyses, namely reducing lipophilicity and basicity, are particularly effective for ion channel targets like these calcium and potassium channels and may be less effective for ameliorating other off-target activities like the nuclear hormone receptor PXR. A possible rationalization for this is that the large, hydrophobic cavities of some ion channels are highly promiscuous, but ligand affinity for them can be reduced by elimination of basic centers (which can be key recognition elements) or by reduction of lipophilicity in the ligand. For other receptors, such as PXR, the binding site for small molecules may be better defined and discrete modifications, which may be quite specific to a particular series of ligands, may often be more effective than a general lowering of logP.

No discussion of general strategies for addressing affinity for ion channels would be complete without a reminder that these approaches are not guaranteed to work. Indeed, in some cases they may have exactly the opposite effect and increase affinity for the ion channel. There are numerous published examples that describe how minor modifications of substituents or stereochemistry can have a dramatic effect on binding to the hERG channel. Such observations underline the key point that this binding event is between a small molecule and a protein. While it has been noted that systematic modification of molecular properties, such as pK_a and logP, can be a useful strategy for reducing affinity for ion channels, it is important to remember that a protein-ligand interaction is responsible for such activity.

To further illustrate the general strategies we outline for addressing undesirable affinity for the hERG channel, some representative examples from the literature are shown in Schemes 7.3 and 7.4. These examples represent a small fraction of the publications describing efforts to address off-target activity at the hERG channel, but they nonetheless serve to illustrate the concepts involved with some real-world examples.

Scheme 7.3a presents some examples in which hERG affinity was ameliorated by removing basic functionality. Compound **3** in Scheme 7.3a was an early compound from the NR2B subtype selective NMDA receptor antagonist program. In this case, the hERG affinity appeared to be associated with the basic 4-aminopyridine functionality in **3** (CpK_a = 8.21), and replacement with the significantly less-basic 2-aminopyrimidine in **4** (CpK_a = 4.03)

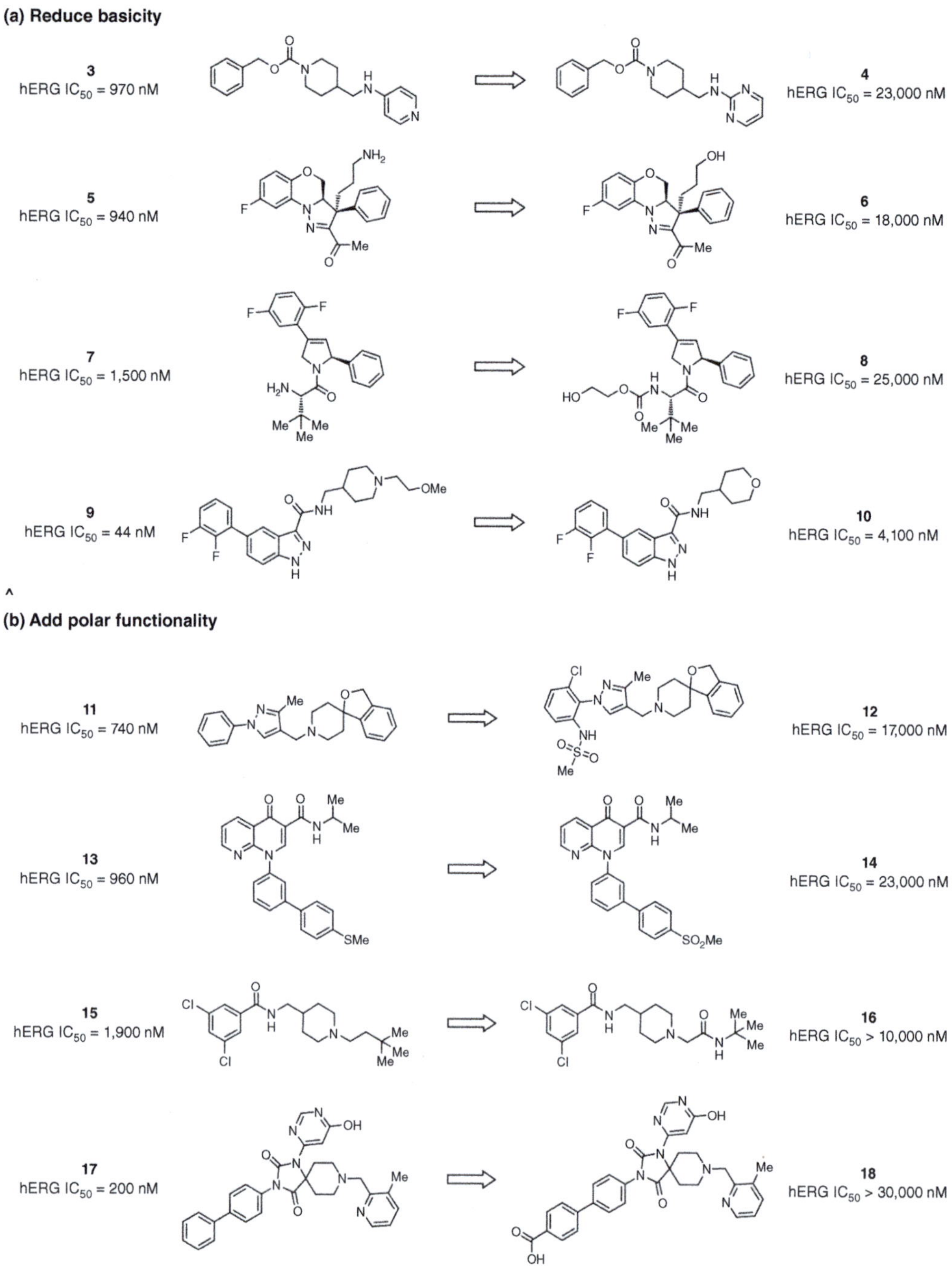

Scheme 7.3 hERG inhibition data for selected compounds.

significantly reduced the affinity [91]. A very clear example of removing basic functionality is shown in the transformation of the KSP inhibitor **5** to the analog **6**, in which a primary amine is replaced with an alcohol to afford a 20-fold reduction in hERG affinity [92]. Compound **7**, another KSP inhibitor, also contained a problematic basic amine, and in this case, conversion to the carbamate **8** proved to be a successful strategy [93]. The GSK-3β inhibitor **9** was determined to be a potent inhibitor of the hERG channel, and it was hypothesized that the basic piperidine ring contributed to this activity. Indeed, replacement of the piperidine with a tetrahydropyran ring, to give compound **10**, led to a 90-fold reduction in the hERG IC$_{50}$ value [94].

Judicious incorporation of polar functionality can lead to attenuation of affinity for the lipophilic hERG channel while maintaining activity at the target of interest. Some representative examples of the effects of polarity on hERG affinity are shown in Scheme 7.3b. The addition of a methanesulfonamide group to the terminal phenyl ring in

(a) Remove or modify aryl rings

19
hERG IC_{50} = 530 nM

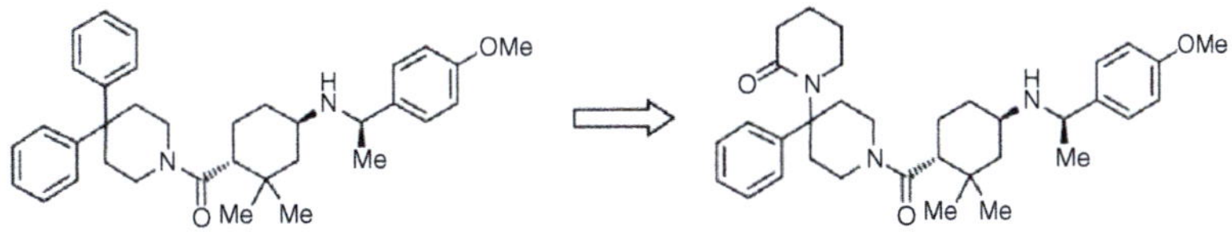

20
hERG IC_{50} > 10,000 nM

21
hERG IC_{50} = 250 nM

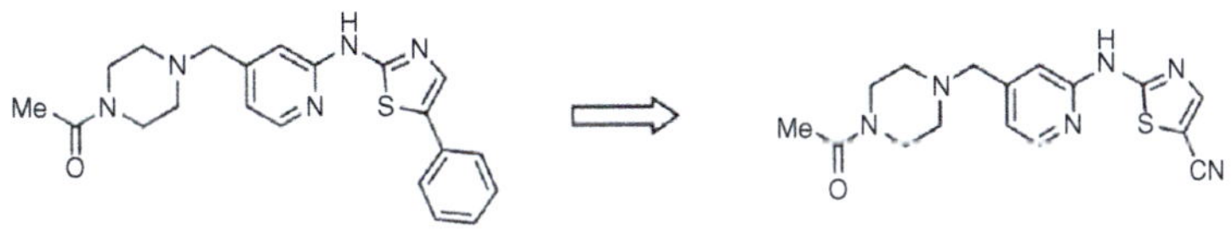

22
hERG IC_{50} = 32,000 nM

23
hERG IC_{50} = 590 nM

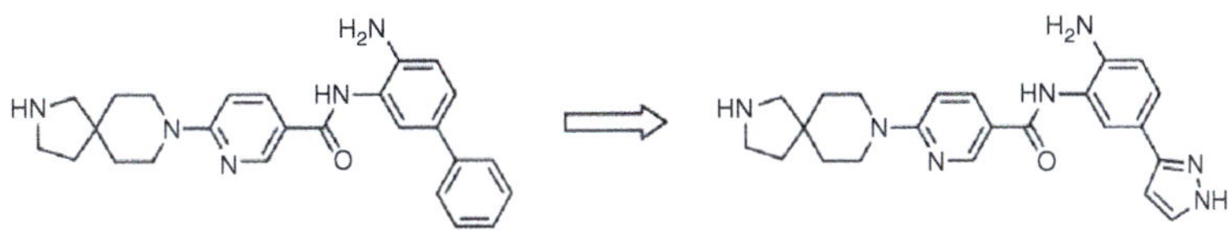

24
hERG IC_{50} > 10,000 nM

25
hERG IC_{50} = 1,600 nM

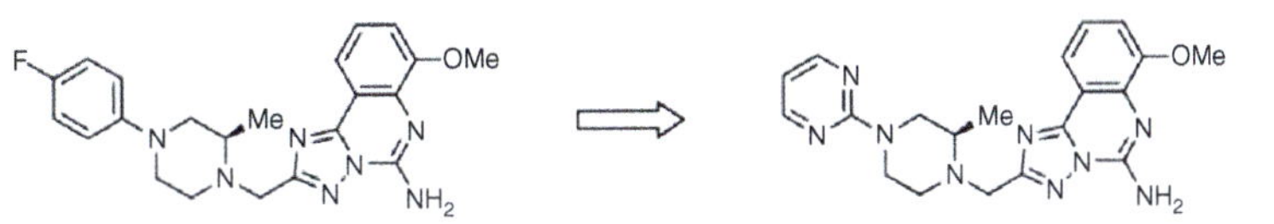

26
hERG IC_{50} > 60,000 nM

27
hERG IC_{50} = 290 nM

28
hERG IC_{50} > 30,000 nM

(b) Modify topology, substitution, or stereochemistry

29
hERG IC_{50} = 100 nM

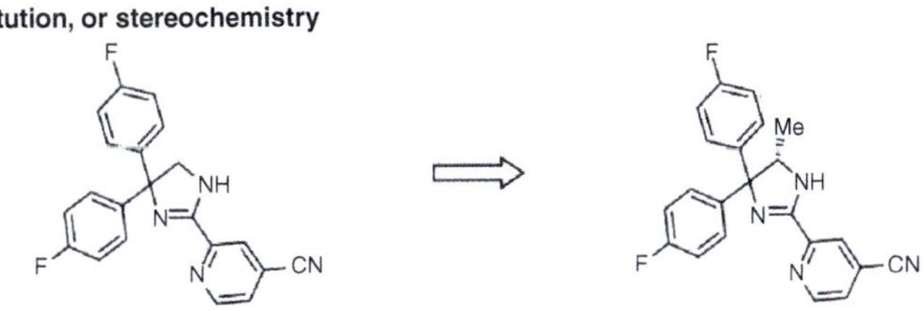

30
hERG IC_{50} = 1700 nM

31
hERG IC_{50} = 350 nM

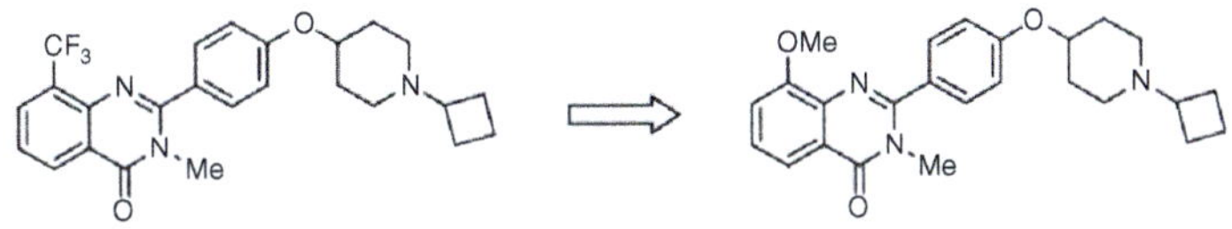

32
hERG IC_{50} > 10,000 nM

33
hERG IC_{50} = 1900 nM

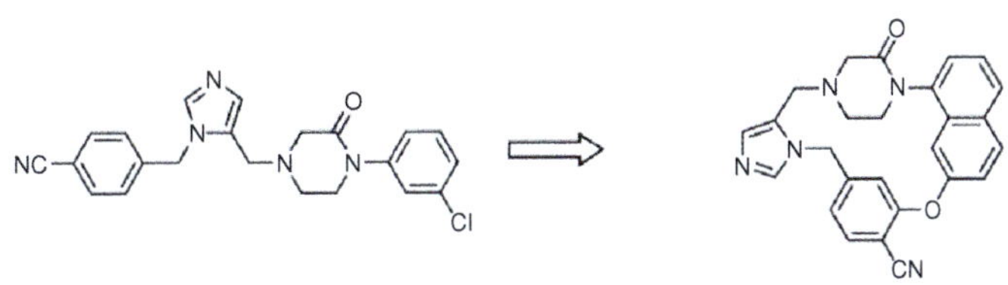

34
hERG IC_{50} = 9700 nM

35
hERG IC_{50} = 80 nM

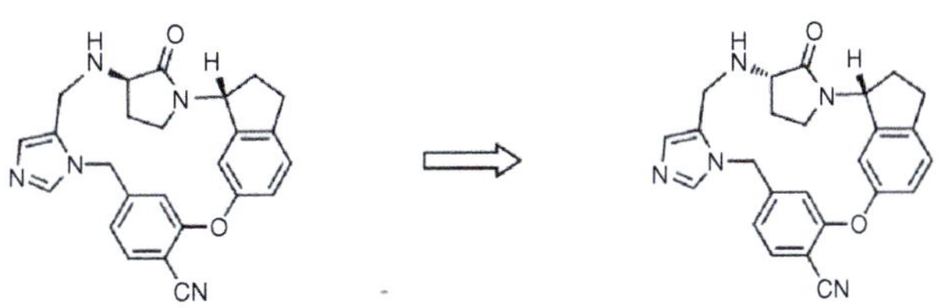

36
hERG IC_{50} = 4,700 nM

Scheme 7.4 hERG inhibition data for selected compounds.

the ORL1 receptor antagonist **11** apparently led to a significant lowering of hERG activity [95]. Similarly, conversion of the lipophilic thioether in the PDE4 inhibitor **13** to the corresponding sulfone led to compound **14**, which had comparable activity against PDE4 but 20-fold lower affinity for the hERG channel [96]. In a series of T-type calcium channel antagonists, it was discovered that replacement of a methylene group in **15** by an amide moiety in **16** provided a compound with a hERG IC_{50} >10 μM [97]. One of the most reliable ways to reduce affinity for hERG is to incorporate an acidic functional group. An example of this strategy is the addition of a carboxylic acid moiety to the prolyl hydroxylase inhibitor **17** to afford compound **18**, which had significantly attenuated hERG affinity [98].

The modification or removal of aryl rings is a widely used approach to reducing affinity for the hERG potassium channel. The success rate may be related to a reduction of π-π stacking interactions between the aromatic residues that line the channel cavity and the aryl rings in the ligand, or it may also be related to a reduction in the lipophilicity of the small molecule, which usually accompanies such modifications. Either way, this strategy has been very successful for addressing hERG issues, and a few specific examples are shown in Scheme 7.4a. The neuropeptide Y Y1 receptor antagonist **19** contained a 4,4-diphenylpiperidine moiety, and it was found that replacement of one of the phenyl rings by a heterocycle, such as the piperidinone in **20**, afforded a significant reduction in affinity for hERG [99]. An even more dramatic example is shown by the KDR kinase inhibitors **21** and **22**: substitution of a cyano group in place of the terminal phenyl ring reduced hERG binding by more than two orders of magnitude [100]. The modification of a terminal phenyl ring often seems to provide a beneficial effect, and three additional examples were observed for the HDAC inhibitor **23** [101], the adenosine A_{2A} receptor antagonist **25** [102], and the SHP2 inhibitor **27** [103]. In all three cases, replacement of the phenyl ring with a heterocycle led to a substantial amelioration of hERG activity.

The final strategy we describe for improvement of hERG activity is the one that should serve as a reminder that we are dealing with a protein-ligand interaction: subtle modification of substituents or stereochemistry or modification of the overall topology of the molecule. By their very nature, these kinds of structure-activity effects tend to be case-specific and difficult to predict, but that does not mean that they should not be considered. Some examples are given in Scheme 7.4b. Affinity for the hERG potassium channel was significantly reduced by the addition of a methyl substituent to the neuropeptide Y Y5 receptor antagonist **29** [104]. Replacement of a trifluoromethyl moiety in the quinazolinone-based H_3 receptor inverse agonist **31** with the methoxy group in compound **32** also had a marked effect on hERG activity [105]. In a series of imidazole-containing farnesyltransferase inhibitors, it was discovered that macrocyclization had a beneficial effect, as illustrated by the conversion of the linear molecule **33** into the macrocycle **34** [106]. However, some related macrocyclic farnesyltransferase inhibitors had high affinity for the hERG channel, such as the 3-aminopyrrolidinone **35** [107]. Interestingly, a simple inversion of the 3-aminopyrrolidinone stereocenter produced a 60-fold reduction in hERG affinity.

As the compounds shown in Schemes 7.3 and 7.4 illustrate, the hERG potassium channel is able to bind a wide variety of structures with high affinity. This promiscuity increases the probability that a given drug discovery program will have to address this particular issue. We have outlined some strategies herein which have been found to ameliorate activity at hERG and other ion channels for a range of compound classes. While they are not guaranteed to succeed, these approaches offer a starting point for addressing undesirable cardiovascular ion channel activity.

7.5 Conclusion

The process of minimizing off-target effects demands significant time and resources during drug development. For many years, the potential for unwanted cardiovascular effects has been ascertained utilizing *in vivo* cardiovascular models. Over time, an understanding has emerged about the relationship of specific targets to cardiovascular effects, with ion channels being a particularly important class of targets. Among the many ion channels that can affect cardiovascular parameters, none has become as pervasive a problem as hERG. The remarkable promiscuity of the hERG channel in terms of binding to small molecules has made it particularly imperative to develop strategies for understanding and designing out activity against the channel. Regulatory guidance has gradually developed around assessing hERG activity risks in clinical candidates. Recent years have seen the development of a greater understanding of the role of hERG in QTc prolongation and the discovery of high-throughput methods to measure the inhibition of hERG *in vitro*. There have been many insights into the mechanism of inhibition of ion channels, including hERG. These advances have been achieved through detailed biophysical characterizations, the determination of the structures of multiple ion channels using X-ray crystallography or cryoelectron microscopy, detailed mutagenesis efforts, computational modeling, and through investigating inhibitor SAR relationships. In this latter category, transformation analysis has been highlighted as one particularly useful way to decipher key molecular sub-structures that contribute to increases in hERG activity. The continued advancement of our understanding in all of these areas will allow the mitigation of off-target activity risk earlier in the drug discovery process. The ultimate goal of avoiding hERG and other off-targets during the design of molecules will lead to the greatest benefit to the efficient development of safe and effective therapeutics.

References

1 Roden, D.M., Balser, J.R., George, A.L. Jr., and Anderson, M.E. (2002). Cardiac ion channels. *Annu. Rev. Physiol.* 64: 431–475.

2 Pugsley, M.K., Authier, S., and Curtis, M.J. (2008). Principles of safety pharmacology. *Br. J. Pharmacol.* 154: 1382–1399.

3 Anonymous (2000). International conference on harmonisation of technical requirements for registration of pharmaceuticals for human use (ICH). http://www.ich.org (accessed 14 September 2024).

4 Anonymous (2000). Guidance for industry S7A. Safety pharmacology studies for human pharmaceuticals. http://www.ich.org (accessed 14 September 2024).

5 Anonymous (2005). Guidance for industry S7B. Nonclinical evaluation of the potential for delayed ventricular repolarization (QT interval prolongation) by human pharmaceuticals. http://www.ich.org (accessed 14 September 2024).

6 Anonymous (2022). Clinical and nonclinical evaluation of QT/QTc interval prolongation and proarrhythmic potential questions and answers (E14/S7B Q&As). http://www.ich.org (accessed 14 September 2024).

7 Anonymous (2005). The clinical evaluation of QT/QTC interval prolongation and proarrhythmic potential for nonantiarrhythmic drugs (E14). http://www.ich.org (accessed 14 September 2024).

8 Strauss, D.G., Gintant, G., Li, Z. et al. (2019). Comprehensive in vitro proarrhythmia assay (CiPA) update from a cardiac safety research consortium/health and environmental sciences institute/FDA meeting. *Ther. Innov. Regul. Sci.* 53: 519–525.

9 Pratt, C.M. and Moyé, L.A. (1995). The cardiac arrhythmia suppression trial: casting suppression in a different light. *Circulation* 91: 245–247. https://doi.org/10.1161/01.CIR.91.1.245.

10 Honig, P.K., Wortham, D.C., Zamani, K. et al. (1993). Terfenadine-ketoconazole interaction - pharmacokinetic and electrocardiographic consequences. *JAMA* 269: 1513–1518.

11 Vorperian, V.R., Zhou, Z.F., Mohammad, S. et al. (1996). Torsade de pointes with an antihistamine metabolite: potassium channel blockade with desmethylastemizole. *J. Am. Coll. Cardiol.* 28: 1556–1561.

12 Farkas, A.S. and Nattel, S. (2010). Minimizing repolarization-related proarrhythmic risk in drug development and clinical practice. *Drugs* 70: 573–603.

13 Kannankeril, P.J. and Roden, D.M. (2007). Drug-induced long QT and torsade de pointes: recent advances. *Curr. Opin. Cardiol.* 22: 39–43.

14 Lagrutta, A.A. and Salata, J.J. (2006). Ion channel safety issues in drug development. In: *Voltage-Gated Ion Channels as Drug Targets* (ed. D.J. Triggle, M. Gopalakrishnan, D. Rampe, and W. Zheng), 444–465. Weinheim: Wiley-VCH Verlag.

15 Lagrutta, A.A., Trepakova, E.S., and Salata, J.J. (2008). The hERG channel and risk of drug-acquired cardiac arrhythmia: an overview. *Curr. Top. Med. Chem.* 8: 1102–1112.

16 Curran, M.E., Splawski, I., Timothy, K.W. et al. (1995). A molecular-basis for cardiac-arrhythmia - hERG mutations cause long Qt syndrome. *Cell* 80: 795–803.

17 Lehnart, S.E., Ackerman, M.J., Benson, W. et al. (2007). Inherited arrhythmias - a national heart, lung, and blood institute and office of rare diseases workshop consensus report about the diagnosis, phenotyping, molecular mechanisms, and therapeutic approaches for primary cardiomyopathies of gene mutations affecting ion channel function. *Circulation* 116: 2325–2345.

18 Roberts, R. (2006). Genomics and cardiac arrhythmias. *J. Am. Coll. Cardiol.* 47: 9–21.

19 Splawski, I., Shen, J.X., Timothy, K.W. et al. (2000). Spectrum of mutations in long-QT syndrome genes KVLQT1, HERG, SCN5A, KCNE1, and KCNE2. *Circulation* 102: 1178–1185.

20 Wang, Q., Curran, M.E., Splawski, I. et al. (1996). Positional cloning of a novel potassium channel gene: KVLQT1 mutations cause cardiac arrhythmias. *Nat. Genet.* 12: 17–23.

21 Wang, Q., Shen, J.X., Splawski, I. et al. (1995). Scn5a mutations associated with an inherited cardiac-arrhythmia, long Qt syndrome. *Cell* 80: 805–811.

22 Rosskopf, D., Schurks, M., Rimmbach, C., and Schafers, R. (2007). Genetics of arterial hypertension and hypotension. *Naunyn Schmiedeberg's Arch. Pharmacol.* 374: 429–469.

23 Sanguinetti, M.C. and Jurkiewicz, N.K. (1990). 2 components of cardiac delayed rectifier k+ current - differential sensitivity to block by class-iii antiarrhythmic agents. *J. Gen. Physiol.* 96: 195–215.

24 Roden, D.M. (2006). Long QT syndrome: reduced repolarization reserve and the genetic link. *J. Intern. Med.* 259: 59–69.

25 Roden, D.M. (2008). Repolarization reserve - a moving target. *Circulation* 118: 981–982.

26 Roden, D.M. (2008). Cellular basis of drug-induced torsades de pointes. *Br. J. Pharmacol.* 154: 1502–1507.

27 Anderson, M.E. (2006). QT interval prolongation and arrhythmia: an unbreakable connection? *J. Intern. Med.* 259: 81–90.

28 Cauvin, C., Loutzenhiser, R., and Vanbreemen, C. (1983). Mechanisms of calcium antagonist-induced vasodilation. *Annu. Rev. Pharmacol. Toxicol.* 23: 373–396.

29 Triggle, D.J. (2007). Calcium channel antagonists: clinical uses – past, present and future. *Biochem. Pharmacol.* 74: 1–9.

30 Inoue, R., Jensen, L.J., Shi, J. et al. (2006). Transient receptor potential channels in cardiovascular function and disease. *Circ. Res.* 99: 119–131.

31 Sonkusare, S., Palade, P.T., Marsh, J.D. et al. (2006). Vascular calcium channels and high blood pressure: pathophysiology and therapeutic implications. *Vasc. Pharmacol.* 44: 131–142.

32 Lee, N., Authier, S., Pugsley, M.K., and Curtis, M.J. (2010). The continuing evolution of torsades de pointes liability testing methods: Is there an end in sight? *Toxicol. Appl. Pharmacol.* 243: 146–153.

33 Pollard, C.E., Abi Gerges, N., Bridgland-Taylor, M.H. et al. (2010). An introduction to QT interval prolongation and non-clinical approaches to assessing and reducing risk. *Br. J. Pharmacol.* 159: 12–21.

34 Valentin, J.P., Pollard, C., Lainee, P., and Hammond, T. (2010). Value of non-clinical cardiac repolarization assays in supporting the discovery and development of safer medicines. *Br. J. Pharmacol.* 159: 25–33.

35 Leishman, D.J., Abernathy, M.M., and Wang, E.B. (2020). Revisiting the hERG safety margin after 20 years of routine hERG screening. *J. Pharmacol. Toxicol. Methods* 105: 106900. https://doi.org/10.1016/j.vascn.2020.106900.

36 Redfern, W.S., Carlsson, L., Davis, A.S. et al. (2003). Relationships between preclinical cardiac electrophysiology, clinical QT interval prolongation and torsade de pointes for a broad range of drugs: evidence for a provisional safety margin in drug development. *Cardiovasc. Res.* 58: 32–45.

37 Clare, J.J. (2006). Functional expression of ion channels in mammalian systems. In: *Expression and Analysis of Recombinant Ion Channels* (ed. J.J. Clare and D.J. Trezise), 79–110. Weinheim: Wiley-VCH Verlag.

38 Raab, C.E., Butcher, J.W., Connolly, T.M. et al. (2006). Synthesis of the first sulfur-35-labeled hERG radioligand. *Bioorg. Med. Chem. Lett.* 16: 1692–1695.

39 Wang, J.X., Della Penna, K., Wang, H. et al. (2003). Functional and pharmacological properties of canine ERG potassium channels. *Am. J. Physiol.-Heart Circul. Physiol.* 284: H256–H267.

40 Tao, H.M., Ana, D.S., Guia, A. et al. (2004). Automated tight seal electrophysiology for assessing the potential hERG liability of pharmaceutical compounds. *Assay Drug Dev. Technol.* 2: 497–506.

41 Guo, L. and Guthrie, H. (2005). Automated electrophysiology in the preclinical evaluation of drugs for potential QT prolongation. *J. Pharmacol. Toxicol. Methods* 52: 123–135.

42 Ghetti, A., Guia, A., and Xu, J. (2007). Automated voltage-clamp technique. In: *Methods in Molecular Biology* (ed. P. Molnar and J.J. Hickman), 59–69. Totawa, NJ: Humana Press.

43 Mathes, C., Friis, S., Finley, M., and Liu, Y. (2009). QPatch: the missing link between hts and ion channel drug discovery. *Comb. Chem. High Throughput Screen.* 12: 78–95.

44 Stoelzle, S., Obergrussberger, A., Brüggemann, A. et al. (2011). State-of-the-art automated patch clamp devices: heat activation, action potentials, and high throughput in ion channel screening. *Front. Pharmacol.* 2: 76. https://doi.org/10.3389/fphar.2011.00076.

45 Farre, C., Haythornthwaite, A., Haarmann, C. et al. (2009). Port-a-patch and patchliner: high fidelity electrophysiology for secondary screening and safety pharmacology. *Comb. Chem. High Throughput Screen.* 12: 24–37.

46 Dunlop, J., Bowlby, M., Peri, R. et al. (2008). Ion channel screening. *Comb. Chem. High Throughput Screen.* 11: 514–522.

47 Huang, C.J., Harootunian, A., Maher, M.P. et al. (2006). Characterization of voltage-gated sodium-channel blockers by electrical stimulation and fluorescence detection of membrane potential. *Nat. Biotechnal.* 24: 439–446.

48 Titus, S.A., Beacham, D., Shahane, S.A. et al. (2009). A new homogeneous high-throughput screening assay for profiling compound activity on the human ether-a-go-go-related gene channel. *Anal. Biochem.* 394: 30–38.

49 Zhang, J.H., Wilson, G.F., Soerens, A.G. et al. (2009). Functional cardiomyocytes derived from human induced pluripotent stem cells. *Circ. Res.* 104: E30–E41.

50 Gintant, G., Pfeiffer Kaushik, E., Feaster, T. et al. (2020). Repolarization studies using human stem cell-derived cardiomyocytes: validation studies and best practice recommendations. *Regul. Toxicol. Pharmacol.* 117: 104756. https://doi.org/10.1016/j.yrtph.2020.104756.

51 Doyle, D.A., Morais Cabral, J., Pfuetzner, R.A. et al. (1998). The structure of the potassium channel: molecular basis of K+ conduction and selectivity. *Science* 280: 69–77.

52 Jiang, Y.X., Lee, A., Chen, J.Y. et al. (2002). The open pore conformation of potassium channels. *Nature* 417: 523–526.

53 Long, S.B., Campbell, E.B., and Mackinnon, R. (2005). Crystal structure of a mammalian voltage-dependent shaker family K+ channel. *Science* 309: 897–903.

54 Wang, W. and MacKinnon, R. (2017). Cryo-EM structure of the open human ether-a-gogo-related K+ channel hERG. *Cell* 169: 422–430. https://doi.org/10.1016/j.cell.2017.03.048.

55 Asai, T., Adachi, N., Moriya, T. et al. (2021). Cryo-EM structure of K+-bound hERG channel complexed with the blocker astemizole. *Structure* 29: 203–212. https://doi.org/10.1016/j.str.2020.12.007.

56 Zhao, Y., Huang, G., Wu, J. et al. (2019). Molecular basis for ligand of a mammalian voltage-gated Ca^{2+} channel. *Cell* 177: 1495–1506.e12. https://doi.org/10.1016/j.cell.2019.04.043.

57 Jiang, D., Shi, H., Tonggu, L. et al. (2020). Structure of the cardiac sodium channel. *Cell* 180: 122–134. https://doi.org/10.1016/j.cell.2019.11.041.

58 Brewer, K.R., Kuenze, G., Vanoye, C.G. et al. (2020). Structures illuminate cardiac ion channel functions in health and in Long QT syndrome. *Front. Pharmacol.* 11: 00550. https://doi.org/10.3389/fphar.2020.00550.

59 Yu, F.H., Yarov-Yarovoy, V., Gutman, G.A., and Catterall, W.A. (2005). Overview of molecular relationships in the voltage-gated ion channel superfamily. *Pharmacol. Rev.* 57: 387–395.

60 Vandenberg, J.I., Torres, A.M., Campbell, T.J., and Kuchel, P.W. (2004). The HERG K+ channel: progress in understanding the molecular basis of its unusual gating kinetics. *Eur. Biophys. J. Biophys. Lett.* 33: 89–97.

61 Sanguinetti, M.C. and Mitcheson, J.S. (2005). Predicting drug-hERG channel interactions that cause acquired long QT syndrome. *Trends Pharmacol. Sci.* 26: 119–124.

62 Mitcheson, J.S. (2008). hERG potassium channels and the structural basis of drug-induced arrhythmias. *Chem. Res. Toxicol.* 21: 1005–1010.

63 Mitcheson, J.S., Chen, J., and Sanguinetti, M.C. (2000). Trapping of a methanesulfonanilide by closure of the HERG potassium channel activation gate. *J. Gen. Physiol.* 115: 229–240.

64 Clarke, C.E., Hill, A.P., Zhao, J. et al. (2006). Effect of S5P alpha-helix charge mutants on inactivation of hERG K+ channels. *J. Physiol.* 573: 291–304.

65 MacKinnon, R. (2004). Potassium channels and the atomic basis of selective ion conduction (Nobel lecture). *Angew. Chem.-Int. Ed.* 43: 4265–4277.

66 Ukachukwu, C.U., Jimenez-Vazquez, E.N., Jain, A., and Jones, D.K. (2024). hERG1 channel subunit composition mediates proton inhibition of rapid delayed rectifier potassium current (IKr) in cardiomyocytes derived from hiPSCs. *J. Biol. Chem.* 299: 102778. https://doi.org/10.1016/j.jbc.2022.102778.

67 Kamiya, K., Niwa, R., Mitcheson, J.S., and Sanguinetti, M.C. (2006). Molecular determinants of HERG channel block. *Mol. Pharmacol.* 69: 1709–1716.

68 Witchel, H.J., Dempsey, C.E., Sessions, R.B. et al. (2004). The low-potency, voltage-dependent HERG blocker propafenone--molecular determinants and drug trapping. *Mol. Pharmacol.* 66: 1201–1212.

69 Mitcheson, J.S., Chen, J., Lin, M. et al. (2000). A structural basis for drug-induced long QT syndrome. *Proc. Natl. Acad. Sci. U.S.A* 97: 12329–12333.

70 Perry, M., Stansfeld, P.J., Leaney, J. et al. (2006). Drug binding interactions in the inner cavity of HERG channels: molecular insights from structure-activity relationships of clofilium and ibutilide analogs. *Mol. Pharmacol.* 69: 509–519.

71 Sanchez-Chapula, J.A., Navarro-Polanco, R.A., Culberson, C. et al. (2002). Molecular determinants of voltage-dependent human ether-a-go-go related gene (HERG) K+- channel block. *J. Biol. Chem.* 277: 23587–23595.

72 Fernandez, D., Ghanta, A., Kauffman, G.W., and Sanguinetti, M.C. (2004). Physicochemical Features of the hERG channel drug binding site. *J. Biol. Chem.* 279: 10120–10127.

73 Farid, R., Day, T., Friesner, R.A., and Pearlstein, R.A. (2006). New insights about HERG blockade obtained from protein modeling, potential energy mapping, and docking studies. *Biorg. Med. Chem.* 14: 3160–3173.

74 Myokai, T., Ryu, S., Shimizu, H., and Oiki, S. (2008). Topological mapping of the asymmetric drug binding to the human ether-Ã -go-go-related gene product (HERG) potassium channel by use of tandem dimers. *Mol. Pharmacol.* 73: 1643–1651.

75 Perry, M., de Groot, M.J., Helliwell, R. et al. (2004). Structural determinants of HERG channel block by clofilium and ibutilide. *Mol. Pharmacol.* 66: 240–249.

76 Ragsdale, D.S., McPhee, J.C., Scheuer, T., and Catterall, W.A. (1996). Common molecular determinants of local anesthetic, antiarrhythmic, and anticonvulsant block of voltage-gated Na+ channels. *Proc. Natl. Acad. Sci. U.S.A* 93: 9270–9275.

77 Catterall, W.A. (2000). From ionic currents to molecular mechanisms: the structure and function of voltage-gated sodium channels. *Neuron* 26: 13–25.

78 Liu, H., Atkins, J., and Kass, R.S. (2004). Common molecular determinants of flecainide and lidocaine block of heart Na^+ channels: evidence from experiments with neutral and quaternary flecainide analogues. *J. Gen. Physiol.* 121: 199–214.

79 Hockerman, G.H., Peterson, B.Z., Johnson, A. et al. (1997). Molecular determinants of drug binding and action on L-type calcium channels. *Annu. Rev. Pharmacol. Toxicol.* 37: 361–396.

80 Mitterdorfer, J., Grabner, M., Kraus, R.L. et al. (1998). Molecular basis of drug interaction with L-Type Ca^{2+} channels. *J. Bioenerg. Biomembr.* 30: 319–334.

81 Dilmac, N., Hilliard, N., and Hockerman, G.H. (2004). Molecular determinants of frequency dependence and Ca^{2+} potentiation of verapamil block in the pore region of Cav1.2. *Mol. Pharmacol.* 66: 1236–1247.

82 Villoutreix, B.O. and Taboureau, O. (2015). Computational investigations of hERG channel blockers: new insights and current predictive models. *Adv. Drug Deliv. Rev.* 86: 72–82.

83 Siramshetty, V.B., Nguyen, D.T., Martinez, N.J. et al. (2020). Critical assessment of artificial intelligence methods for prediction of hERG channel inhibition in the "big data" era. *J. Chem. Inf. Model.* 60: 6007–6019.

84 Jamieson, C., Moir, E.M., Rankovic, Z., and Wishart, G. (2006). Medicinal chemistry of hERG optimizations: highlights and hang-ups. *J. Med. Chem.* 49: 5029–5046.

85 Waring, M.J. and Johnstone, C. (2007). A quantitative assessment of hERG liability as a function of lipophilicity. *Bioorg. Med. Chem. Lett.* 17: 1759–1764.

86 Finlayson, K., Turnbull, L., January, C.T. et al. (2001). $[^3H]$Dofetilide binding to HERG transfected membranes: a potential high throughput preclinical screen. *Eur. J. Pharmacol.* 430: 147–148.

87 Priest, B.T., Bell, I.M., and Garcia, M.L. (2008). Role of hERG potassium channel assays in drug development. *Channels* 2: 87–93.

88 Sheridan, R.P., Hunt, P., and Culberson, J.C. (2006). Molecular transformations as a way of finding and exploiting consistent local QSAR. *J. Chem. Inf. Model.* 46: 180–192.

89 Evans, B.E. and Rodzinak, K.J. Unpublished results. Merck Research Laboratories.

90 Leeson, P.D. and Springthorpe, B. (2007). The influence of drug-like concepts on decision-making in medicinal chemistry. *Nat. Rev. Drug Discov.* 6: 881–890.

91 Liverton, N.J., Bednar, R.A., Bednar, B. et al. (2007). Identification and characterization of 4-methylbenzyl 4-[(pyrimidin-2-ylamino)methyl]piperidine-1-carboxylate, an orally bioavailable, brain penetrant NR2B selective N-methyl-D-aspartate receptor antagonist. *J. Med. Chem.* 50: 807–819.

92 Garbaccio, R.M., Tasber, E.S., Neilson, L.A. et al. (2007). Kinesin spindle protein (KSP) inhibitors. Part 7: design and synthesis of 3,3-disubstituted dihydropyrazolobenzoxazines as potent inhibitors of the mitotic kinesin KSP. *Bioorg. Med. Chem. Lett.* 17: 5671–5676.

93 Fraley, M.E., Garbaccio, R.M., Arrington, K.L. et al. (2006). Kinesin spindle protein (KSP) inhibitors. Part 2: the design, synthesis, and characterization of 2,4-diaryl-2,5-dihydropyrrole inhibitors of the mitotic kinesin KSP. *Bioorg. Med. Chem. Lett.* 16: 1775–1779.

94 Prati, F., Buonfiglio, R., Furlotti, G. et al. (2020). Optimization of indazole-based GSK-3 inhibitors with mitigated hERG issue and in vivo activity in a mood disorder model. *ACS Med. Chem. Lett.* 11: 825–831.

95 Yoshizumi, T., Miyazoe, H., Ito, H. et al. (2008). Design, synthesis, and structure-activity relationship study of a novel class of ORL1 receptor antagonists based on N-biarylmethyl spiropiperidine. *Bioorg. Med. Chem. Lett.* 18: 3778–3782.

96 Guay, D., Boulet, L., Friesen, R.W. et al. (2008). Optimization and structure-activity relationship of a series of 1-phenyl-1,8-naphthyridin-4-one-3-carboxamides: Identification of MK-0873, a potent and effective PDE4 inhibitor. *Bioorg. Med. Chem. Lett.* 18: 5554–5558.

97 Yang, Z.Q., Barrow, J.C., Shipe, W.D. et al. (2008). Discovery of 1,4-substituted piperidines as potent and selective inhibitors of T-type calcium channels. *J. Med. Chem.* 51: 6471–6477.

98 Vachal, P., Miao, S., Pierce, J.M. et al. (2012). 1,3,8-Triazaspiro[4.5]decane-2,4-diones as efficacious pan-inhibitors of hypoxia-inducible factor prolyl hydroxylase 1-3 (HIF PHD1-3) for the treatment of anemia. *J. Med. Chem.* 55: 2945–2959.

99 Cho, K., Ando, M., Kobayashi, K. et al. (2009). Design, synthesis and evaluation of a novel cyclohexanamine class of neuropeptide Y Y1 receptor antagonists. *Bioorg. Med. Chem. Lett.* 19: 4781–4785.

100 Bilodeau, M.T., Balitza, A.E., Koester, T.J. et al. (2004). Potent N-(1,3-thiazol-2-yl)pyridin-2-amine vascular endothelial growth factor receptor tyrosine kinase inhibitors with excellent pharmacokinetics and low

affinity for the hERG ion channel. *J. Med. Chem.* 47: 6363–6372.

101 Methot, J.L., Hamblett, C.L., Mampreian, D.M. et al. (2008). SAR profiles of spirocyclic nicotinamide derived selective HDAC1/HDAC2 inhibitors (SHI-1:2). *Bioorg. Med. Chem. Lett.* 18: 6104–6109.

102 Deng, Q., Lim, Y.-H., Anand, R. et al. (2015). Use of molecular modeling aided design to dial out hERG liability in adenosine A_{2A} receptor antagonists. *Bioorg. Med. Chem. Lett.* 25: 2958–2962.

103 Bagdanoff, J.T., Chen, Z., Acker, M. et al. (2019). Optimization of fused bicyclic allosteric SHP2 inhibitors. *J. Med. Chem.* 62: 1781–1792.

104 Sato, N., Ando, M., Ishikawa, S. et al. (2009). Discovery of tetrasubstituted imidazolines as potent and selective neuropeptide Y Y5 receptor antagonists: reduced human ether-a-go-go related gene potassium channel binding affinity and potent antiobesity effect. *J. Med. Chem.* 52: 3385–3396.

105 Mizutani, T., Nagase, T., Ito, S. et al. (2008). Development of novel 2-[4-(aminoalkoxy)phenyl]-4(3H)-quinazolinone derivatives as potent and selective histamine H-3 receptor inverse agonists. *Bioorg. Med. Chem. Lett.* 18: 6041–6045.

106 Dinsmore, C.J., Bogusky, M.J., Culberson, J.C. et al. (2001). Conformational restriction of flexible ligands guided by the transferred NOE experiment: potent macrocyclic inhibitors of farnesyltransferase. *J. Am. Chem. Soc.* 123: 2107–2108.

107 Bell, I.M., Gallicchio, S.N., Abrams, M. et al. (2002). 3-Aminopyrrolidinone farnesyltransferase inhibitors: design of macrocyclic compounds with improved pharmacokinetics and excellent cell potency. *J. Med. Chem.* 45: 2388–2409.

Part B

Polypharmacology as an Opportunity in Different Disease Areas

8

Toward Mechanism-based Therapies and Network Pharmacology

Cristian Nogales, Zina Piper, Zeinab Mamdouh, and Mayra Pacheco Pachado

8.1 A Crisis in the Pharmaceutical Industry

The field of drug discovery is grappling with a crisis, often encapsulated by the concept of Eroom's law, i.e., Moore's law spelled backward. Eroom's law highlights the trend in pharmaceutical research and development (R&D) that, despite increased investment and advancements in technology and research, the cost of developing new drugs has risen exponentially, while the number of new drugs approved per billion dollars spent has decreased [1]. The causes of this crisis are several, including high regulatory requirements and the "better than the Beatles" problem, where new drugs must compete with existing drugs widely used in clinics [2]. In addition, current pharmacological treatments are also highly ineffective. This is reported in many disease phenotypes by high numbers needed to treat (NNT), which refers to the large number of patients who need to be treated for one to benefit from the drug [3]. NNTs have also been increasing, suggesting that newer drugs often offer marginal improvements over existing therapies [4]. This trend is particularly noticeable in areas such as psychiatry and chronic conditions, where the heterogeneity of patient responses and the complex nature of the diseases make it challenging to develop highly effective treatments. For instance, in the field of antidepressants, while these drugs are effective for some, the overall benefit compared to placebo can be modest for the majority of patients [5]. The healthcare system faces thus a double challenge of (i) managing increasing drug development costs while (ii) addressing the need for more precise treatments that benefit a larger number of patients.

Additionally, the shift toward targeting more complex and less understood diseases, such as Alzheimer's disease, has led to higher failure rates in drug development programs [6, 7]. The issue is that we do not understand diseases mechanistically and we mostly target disease symptoms. This symptomatic approach to disease treatment is partly due to the way that diseases are often defined and classified. Traditionally, diseases have been grouped under broad umbrella disease terms based on the organ affected – e.g., heart failure – or the most prominent disease symptoms – e.g., hypertension – rather than on their underlying molecular mechanisms. For instance, cancer is classified primarily by the organ of origin, yet the genetic and molecular characteristics significantly vary within a single organ-specific cancer type [8]. The current organ- and symptom-based classification systems have led to a "one-size-fits-all" approach to treatment, which is ineffective due to the heterogeneous nature of complex diseases. In addition, when we try to understand the disease mechanisms, the focus has traditionally been on identifying single proteins as pharmacological targets, i.e., the "one-disease, one-target, one-drug" dogma. However, this approach is limited, particularly in the context of complex diseases, where the underlying mechanism is most certainly a complex network [9]. Moreover, drugs are also promiscuous, meaning they can bind multiple, sometimes unknown, proteins, which can lead to off-target effects that can be both beneficial and detrimental [10, 11]. Altogether, these issues highlight the need for a profound paradigm shift from single targets to disease networks in order to obtain a more holistic, but also more precise, view of drugs and their targets.

8.2 Disease Modules as Targets for Precision Medicine

Systems medicine is an interdisciplinary field that recognizes that human health and disease are the result of intricate interactions within biological systems and between these systems and the environment. Systems medicine builds upon the principles of systems biol-

Polypharmacology: Strategies for Multi-Target Drug Discovery, First Edition. Edited by Jens-Uwe Peters.

ogy, which emerged in the late 1990s as a response to the growing complexity of biological data and the need for integrative approaches to understand biological systems [12]. The application of systems biology to medicine led to the development of systems medicine, which emphasizes a holistic view of disease mechanisms. This approach differs from traditional reductionist approaches, which focus on individual components such as single gene or protein to explain diseases. The field of systems medicine computationally integrates different data types (e.g., genomics, transcriptomics, and proteomics) to develop models that predict the underlying mechanisms of diseases [13]. Network medicine expands and complements systems medicine by focusing on the complex networks of molecular and cellular interactions that underlie diseases. A disease is thus viewed as the outcome of several biological components interacting with each other forming an intricate pathological network, rather than a single-gene product [14]. A fundamental breakthrough came from the development of the first network of all human diseases, i.e., the diseasome [15, 16], where diseases were connected by shared risk genes. In this network, disease clusters appeared with diseases linked by common risk genes, hinting at a common causal mechanism. This comprehensive disease network was further expanded by other multiscale disease networks that explored the relationship of diseases to drugs, shared symptoms, or comorbidities [17]. The resulting disease clusters contained disease phenotypes affecting different organs, reinforcing the idea that current organ- and symptom-based disease classifications are outdated. These phenotypes are thus just disease symptoms of an underlying common causal molecular mechanism, which once elucidated will become the new disease definitions, the endotypes.

In network medicine, the causal mechanisms of diseases are known as disease modules. Disease modules are an observation in network medicine, where components associated with a disease, e.g., genes or proteins, segregate in the same neighborhood of the human interactome creating functional units [17, 18]. Disease modules can be identified using disease-specific data to infer the disease module directly. Alternatively, they can be constructed de novo using already existing molecular interaction networks such as protein–protein interaction (PPI) networks. The DREAM challenge showed that these methods are well-suited for finding disease modules [19]. However, not all disease module identification methods are the same, and new methods are constantly being developed to better capture the complex interplay of cell signaling. We can broadly group them into different categories, as exemplified by the DREAM challenge, e.g., (i) kernel clustering, (ii) modularity optimization, (iii) random-walk-based, (iv) local methods, (v) ensemble methods, (vi) hybrid methods. These methods use different approaches; for instance, random-walk-based methods use random walks or diffusion processes as their basis. On the other hand, local methods mostly use expansive approaches where a set of highly clinically relevant disease seed genes is used as the starting point to identify the disease module.

Generating a curated list of seeds is a crucial step for disease module identification, and it is usually achieved by combining various sources and databases. Predominantly, proteins encoded by their respective genes are preferred due to the comprehensive molecular PPI data available. Thus, the general strategy in seed-based local methods is to map the list of protein seeds to an existing protein interactome that includes all experimentally validated proton pump inhibitors (PPIs). This step provides a first indication of how the seeds are interacting with one another. However, to be able to see the full scope of the underlying network, the seeds have to be expanded *in silico* to include additional functionally related genes [20]. This can be achieved by utilizing graph theories such as the "guilt-by-association" principle, which represents the idea that interacting proteins, and therefore the genes encoding them, likely partake in similar cellular functions and influence associated traits [21]. Additionally, other types of data explaining proteins' relationships can be integrated, including transcriptional and gene co-expression data. However, the more data included, the higher the likelihood of introducing noise and unnecessary complexities. It is thus important to ensure that only biologically meaningful proteins and disease-specific interactions remain in the final disease subnetwork or disease module [22]. This evidence is also supported by recent studies highlighting the spatial constraints of cells, characterized by highly organized environments known as signalosomes [23]. Disease module identification methods deal with this issue in different ways. For instance, a cutoff value can be selected to correct and remove highly connected proteins in the interactome that are poorly connected in the extracted disease network or disease module, since they most likely do not play a significant role in the causal disease mechanism. Additionally, community detection algorithms such as the Louvain method can be leveraged to identify topological clusters and uncover the final disease modules [24].

Identifying all underlying mechanisms of a disease or disease modules is known as disease endotyping. In this process, diseases segregate into several mechanistic endotypes, represented by one or a combination of causal disease module(s) [25]. Disease modules would thus be the causal mechanisms leading to complex diseases where multiple key components of a network are affected. Those networks are also further heavily influenced by

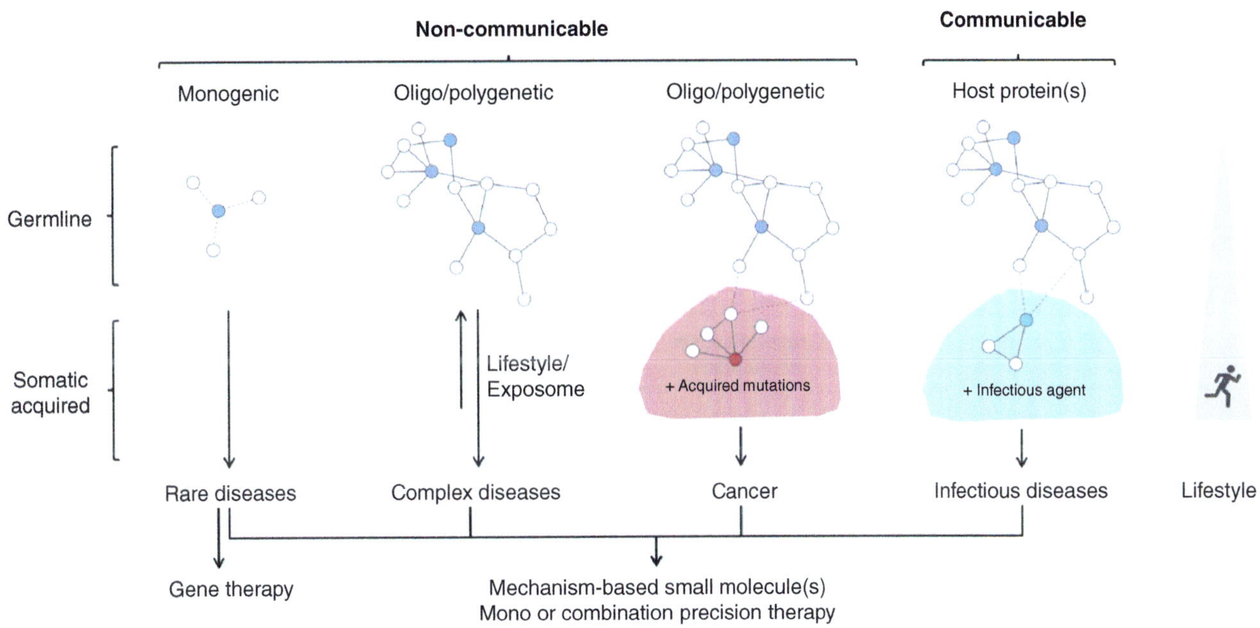

Figure 8.1 Mechanism-based disease types. With the exception of rare monogenic diseases, the majority of diseases are caused by complex networks or disease modules, characterized by germline and/or acquired mutations, which are significantly impacted by the lifestyle or exposome of an individual. Additionally, diseases can be classified as communicable diseases – including rare diseases, complex diseases, and cancer – and noncommunicable or infectious diseases. All of these disease types would benefit from a network pharmacology therapy targeting the underlying disease mechanism or disease module.

an individual's lifestyle or exposome. Nevertheless, if we attempt to classify diseases mechanistically, we can identify other disease types (Figure 8.1). For instance, rare or autoimmune diseases are usually caused by a single gene or protein instead of a whole network. They can be treated mechanistically with small molecules that directly target the affected gene or protein. It is only in these cases where the current therapeutic dogma of "one-disease, one-target, one-drug" appears to be valid. However, given the current advances and ongoing research in gene therapy, in those cases where the disease is caused by a single affected gene, gene therapy might be a better and curative option. On the other hand, cancer is also most likely caused by a disease network or disease modules where acquired mutations disrupt the normal healthy signaling of the network, leading to the development of the tumor. Additionally, in communicable diseases, it is the infectious agent who interrupts the physiological signaling of the host. These disease types caused by disease modules or networks can benefit significantly from a direct mechanism-based pharmacological approach with small molecule(s), i.e., a precision network pharmacology approach. Disease modules thus become the new targets for precision medicine, around which to develop (i) diagnostic strategies for patient identification and stratification and (ii) mechanism-based treatments that correct the affected disease network. Importantly, these newly identified disease modules need to be translated into signaling modules or pathways to comprehend how the signaling cascade functions in healthy and diseased individuals. During this process, expert knowledge is incorporated and biological evidence is curated to deliver a biologically meaningful signaling pathway from the disease module.

8.3 Mechanism-based Therapies and Network Pharmacology

Once diseases are endotyped, we can develop mechanistic pharmacological therapies centered around the underlying mechanism or disease module. Network pharmacology is a novel approach, which highlights a paradigm shift in drug discovery and development, moving away from the traditional "one-disease, one-target, one-drug" model toward a more holistic approach. It is important to differentiate mechanism-based network pharmacology from a similar concept stemming from the principles of traditional Chinese herbal medicine, which also often uses combinations of herbs. The key distinction lies in the methodology and scientific rigor; network pharmacology relies on quantitative data and computational

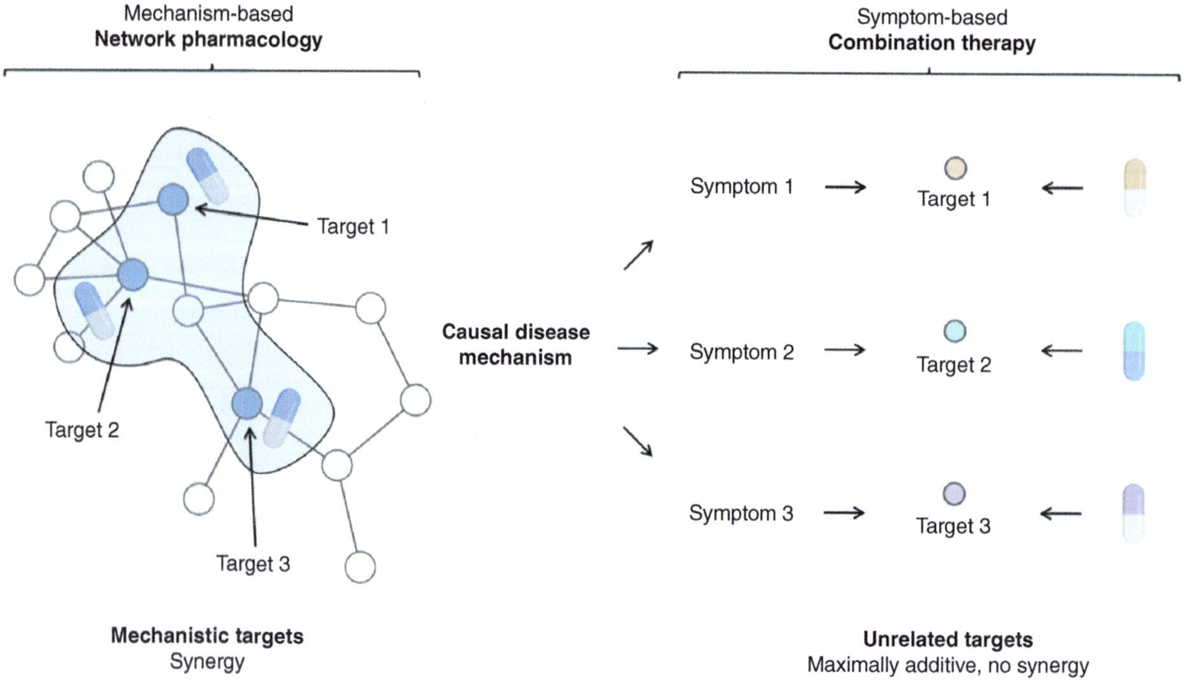

Figure 8.2 Mechanism-based network pharmacology. Network pharmacology treatments target the underlying disease module with two or more drugs directed at key components of the network, resulting in a highly synergistic effect. Alternatively, combination therapy is symptom-based, meaning that the combined drugs target the disease symptoms and thus targets unrelated to the causal disease mechanism. As a result, combination therapy treatments show no synergy and the combined drugs will be additive at best.

models to predict interactions and outcomes, whereas traditional Chinese medicine is rather based on empirical experience and historical use [26]. Mechanism-based network pharmacology therapies use two or more drugs acting mechanistically on the same causal disease module, targeting key components of the disease network in a highly synergistic manner [27]. In comparison with more traditional combination therapy, which combines drugs based on their known effects on the disease symptoms while ignoring the disease cause, network pharmacology is a precision medicine approach that directly targets the underlying mechanism of the disease (Figure 8.2). Because we target disease modules, one of the key advantages of network pharmacology is the potential for synergistic drug combinations; i.e., the combined effect of the drugs is greater than the sum of their individual effects. This synergy can lead to more effective treatments at lower doses, reducing the risks of drug toxicity and side effects [9, 28]. Mechanism-based network pharmacology holds thus great promise for drug development and new therapies, with the potential to enhance efficacy, reduce side effects, and uncover synergistic drug combinations that directly target the root of the disease.

The potential for network pharmacology is even greater because once we know the disease module and thus the disease cause, we can repurpose drugs that target this disease module from other indications giving them new therapeutic uses. Drug repurposing offers thus several advantages over traditional drug development, such as reduced development costs, shorter development to reach clinical use, and a higher success rate, since the safety profile, side effects, and pharmacokinetic properties of the drugs are already well-known. In the last few years, drug repurposing has gained significant attention as a rapid response to emerging health crises, such as the COVID-19 pandemic, where existing drugs were screened for antiviral effects against severe acute respiratory syndrome coronavirus 2 (SARS-CoV-2) [29–31]. Additionally, drug repurposing is significantly relevant in the development of treatments for rare and neglected diseases that lack funding due to the anticipated low investment return for traditional drug discovery. Computational methods play a crucial role in drug repurposing by identifying potential drug–disease and drug–target associations and predicting drug efficacy. For instance, cheminformatics combines elements of chemistry and computer science to solve chemical problems through data analysis and the development of computational models. A substantial area of cheminformatics is the study of molecular interactions between drugs and their targets through the analysis of binding pockets, i.e., specific regions on a target protein where ligands, such as drugs, can form stable interactions. The comprehensive collection of binding sites in proteins is known as the "pocketome" and provides crucial information for drug discovery and molecular modeling. Understanding the structure and properties of these pockets is essential for the design of new drugs and the improvement of existing ones [32–34]. All in all, mechanism-based network pharmacology therapies have high potential by directly targeting the underlying disease mechanism or disease module. Not only would the patient benefit from such a therapy by restoring the healthy signaling, but they might also avoid developing future comorbidities associated with their disease module. Additionally, drug repurposing presents itself as a way to shorten the drug development process, significantly accelerating the availability of treatments for newly discovered disease mechanisms.

8.4 Implementing Mechanism-based Therapies

Incorporating signaling disease modules as the foundational element in defining diseases (**Figure 8.1**) would provide great benefits to the treatment of patients. Mechanism-based therapies – including network pharmacology – offer a more nuanced and effective approach to treatment. These therapeutic strategies are not only conceptual but are also being implemented in clinical settings, where patients are being stratified not only on their symptoms but also according to their underlying disease mechanism. This approach represents a significant shift toward precision medicine and offers potential great outcomes for the treatment of diseases.

8.4.1 Cancer

The standard-of-care (SOC) regimens for cancer have traditionally consisted of surgery, radiotherapy, and/or chemotherapy. After single-agent therapies, combinational or sequential regimens using cytotoxic drugs became the standard practice to maximize therapeutic responses to short-lived remissions [35, 36]. Thus, each tumor benefits solely from the drugs to which it is most sensitive from the provided combinations [37]. It was not until the late 1990s when rituximab, an anti-CD20 monoclonal antibody, was approved for CD20-positive B-cell non-Hodgkin's lymphoma revolutionizing anticancer therapies and ushering in a new era of targeted anticancer agents [38]. Targeted cancer therapies proved therapeutically beneficial for many cancer types, including lung, pancreatic, colorectal, breast, lymphoma, melanoma, and leukemia. However, acquired resistance resulting from the numerous genetic and epigenetic changes occurring in cancer cells still leads to relapse despite a positive initial response [39, 40].

Cancer exhibits vast heterogeneity, spanning from one tumor type to another, from patient to patient with the same tumor type (between-tumor heterogeneity), and even from cell to cell within the same tumor (within-tumor heterogeneity) [17]. In typical solid tumors like breast cancer, or pancreatic adenocarcinoma, approximately 33–66 genes tend to exhibit major somatic mutations. However, the number of mutations can vary significantly across different tumor types, with some displaying considerably more or fewer mutations than the average [41]. The shift toward genomic profiling, as opposed to relying solely on anatomic or histologic staging, has enhanced our comprehension of the common genetic characteristics across various types of human cancers, progressively unlocking remarkable prospects for precision oncology and targeted cancer therapies [42, 43]. Clinical trials are increasingly directed by biomarkers with the aim of predicting which patients will respond to specific drugs [35]. However, these clinical trials focus on targeting single genetic alterations or specific pathways, which in reality are simply mind maps of all possible signaling events around a particular signaling principle. The entirety of those canonical pathways does not constitute experimentally validated coherent protein–protein interaction networks or functional disease modules. Validated disease modules, however, frequently include combinations of two or more canonical pathways. Therefore, relying on classical pathways can be misleading when defining a disease module.

Cancer-relevant disease modules created by network medicine approaches take into consideration PPIs from the interactome allowing for opportunities for developing guidelines to treat cancer patients based on their genetic aberrations. Targeting the perturbed modules by applying the principles of network pharmacology, and combining drugs aimed at key points of the affected modules, would enable the development of new potential curative therapies [27]. Additionally, drug repurposing opens the potential use of a wide range of approved drugs, shortening the period to get these new combinations approved for treating patients [44]. Developing new network pharmacology drug treatments together with a better understanding of the mechanisms of cancer resistance would ultimately lead to precision medicine, thus treating cancer patients based on the molecular events underlying their malignancies.

8.4.2 Inborn Errors of Metabolism

Although we have previously described that rare diseases are mostly monogenic, this is not always entirely the case and there are examples that could still benefit from a network pharmacology approach. For instance, inborn errors of metabolism (IEM) are a large clinically heterogeneous group of rare inherited metabolic disorders affecting approximately 1 in 2500 births [45]. They are caused by different single-gene defects in enzymes responsible for the synthesis or catabolism of proteins, carbohydrates, or fats, resulting in a blockage in the corresponding metabolic pathway. This causes the accumulation of substances that are toxic or disrupt normal functions as well as deficiencies of substances essential for normal functioning. As a result, basically all biochemical pathways are impacted, presenting a variety of multiorgan symptoms at any age (from prenatal period to adulthood) with severity ranging from benign to severe or even life-threatening conditions [46]. With the recent development of omics data, genomic and metabolomic screenings have become more and more common in patients, helping the understanding of the diseases [47]. Nevertheless, patient stratification and endotyping remain a challenge in IEMs, due to the diversity of these diseases and the limited number of patients [48, 49]. Classification attempts have been made; for instance, the Study of Inborn Errors of Metabolism (SSIEM) group initially categorized approximately 600 diseases into 15 distinct categories based on the overall disease group or on the molecular pathways involved alongside known pathomechanisms in 2011 [50]. The International Classification of Inherited Metabolic Disorders (ICIMD) advisory group, which evolved from the SSIEM group, further expanded this classification, now categorizing 1450 disorders into 24 categories [51]. Diagnostics are achieved by genetic sequencing and metabolite analyses, including newborn screening. First-line treatment options include nutritional therapies alongside (co)enzyme replacement treatment to improve residual enzyme activity and prevent substrate accumulation and supplement-deficient metabolites. In addition, the removal of toxic substrates and cell and organ transplantation are common practices and gene therapy is being investigated [52]. Despite adequate and early-start dietary and crisis management, severe long-term consequences affecting multiple organs, i.e., cardiomyopathy or kidney failure, may still occur in later adulthood, questioning the alleged effective disease management (i.e., no metabolic crises) of these treatments long term [53, 54].

Even though IEMs are caused by single-gene mutations, the paradigm of monogenic inheritance ("one gene, one disease") has been questioned due to the heterogeneity of the phenotypes resulting from different mutations in the same gene. This suggests that perturbations might extend through the connections of the fundamental cellular networks surrounding the affected gene, just as in complex or common diseases [49, 50, 55]. Furthermore, a clear genotype–phenotype relationship could not be detected in IEMs, and it has been observed that multiple modifying factors significantly influence the phenotypes [56–58]. Generally, IEMs have hardly been studied from a network perspective; however, one study generated an IEM-specific interactome (IEMi) and compared PPIs among the seemingly distinct disease categories of the SSIEM [55]. IEMs appear to be highly interconnected; i.e., more than 50% of all connections are shared between disease groups 1, 2, 3, 4, 10, and 11, and particularly, disease group 1 takes a central role within this network. Furthermore, 168 genes were grouped together in the largest connected component (LCC) comprising 14 disease groups out of the 15 SSIEM classes. This could potentially explain not only the high heterogeneity within a disease group but also the shared phenotypes and genes between disease groups.

Network approaches have thus immense potential in IEMs by identifying mechanistic disease modules and concomitant biomarkers, thus allowing for patient stratification. Advanced available algorithms enable the comprehensive integration and interpretation of multiscale omics and sequencing datasets allowing for pathogenic variant networks [59]. By combining IEMs based on shared interactions, it might be possible to finally overcome the "rare disease, rare data" problem. This in turn will contribute to elucidating pathomechanisms and allowing identification of potential drug targets and mechanism-based pharmacological therapies [50].

8.4.3 Human Immunodeficiency Virus (HIV)

Communicable diseases can also greatly benefit from mechanistic approaches that directly target the viral proteins, the host proteins, or both. For instance, the management and prevention of HIV have improved considerably with the access to mechanistic therapies that directly target the virus replication machinery. Understanding the viral life cycle has been instrumental for developing targeted therapies that inhibit specific stages of HIV replication, thereby managing the disease and improving patient outcomes. In the area of HIV prevention, pre-exposure prophylaxis (PrEP) involves the use of antiretroviral drugs by individuals who are at high risk of HIV infection but are currently HIV-negative. This drug regime includes a combination of two antiretroviral drugs, tenofovir disoproxil fumarate (TDF) and emtricitabine

(FTC), which are effective in blocking key steps in the HIV replication process within the host cells. These drugs are nucleoside reverse transcriptase inhibitors (NRTIs) that target the viral reverse transcriptase enzyme, which is key for the conversion of viral ribonucleic acid (RNA) into deoxyribonucleic acid (DNA), stopping the viral replication cycle. By blocking this step, PrEP effectively prevents the establishment of a productive infection, reducing the risk of acquiring HIV. Different clinical trials have shown that PrEP reduces significantly the infection rates among users who adhere to the regimen [60, 61], leading to its approval by the U.S. Food and Drug Administration for HIV prevention in 2012. PrEP can currently be prescribed and taken in the form of pills or shots significantly stagnating the spread of HIV.

For individuals living with HIV, therapies are aimed at suppressing the viral load to undetectable levels in order to prevent disease progression and reduce the risk of transmission. This goal is achieved with antiretroviral therapies, which typically involve a combination of drugs targeting different stages of the HIV life cycle. These include NRTIs, non-NRTIs, protease inhibitors (PIs), integrase strand transfer inhibitors (INSTIs), and entry inhibitors. This mechanistic multidrug approach not only inhibits viral replication but also reduces the likelihood of resistance development and has transformed HIV from a fatal disease to a manageable chronic condition [62]. Nevertheless, ongoing R&D in this field continues to refine these strategies, promising a future where HIV can be effectively controlled and ultimately eradicated.

8.4.4 ROS/cGMP Cluster of Cardiovascular, Respiratory, Metabolic, and Neurological Disease Phenotypes

The network of human diseases, i.e., the diseasome, showed that disease phenotypes affecting different organs might have a common underlying disease mechanism. For instance, in this network, a heterogeneous multiorgan disease cluster appeared, including cardiovascular, respiratory, metabolic, and neurological disease phenotypes. These disease phenotypes are connected by risk genes related to reactive oxygen species (ROS) formation and cyclic guanosine monophosphate (cGMP) (ROCG) signaling in one joined signaling network [63]. Using the associated disease genes as seeds in a local disease module identification approach, a ROCG disease module emerged, containing some, but not all, elements of ROS and cGMP signaling [22]. This result reinforces the idea that disease modules are composed of some elements of different canonical pathways and that we cannot merely rely on classical pathways to treat diseases. The ROCG disease module was further validated with two top-performing algorithms of the DREAM challenge [19] with all three *in silico* methods pointing at the nicotinamide adenine dinucleotide phosphate (NADPH) oxidase 5 (NOX5) as a direct neighbor of endothelial nitric oxide cGMP signaling. These findings further highlight the need for a new approach to signaling and pharmacology centered around disease modules. *In vivo* and *in vitro* studies have already demonstrated the potential of a network pharmacology approach targeting the ROCG disease module [64, 65]. The approach is also currently being tested in clinical trials on diseases of the ROCG cluster such as ischemic stroke and heart failure with preserved ejection fraction. One of such trials is the "Networked Drug REpurposing for Mechanism-based neuroPrOtection in Ischaemic STROKE (REPO-STROKE IIa)" trial (NCT05762146). This is also a drug repurposing trial where registered compounds – (i) riociguat, a soluble guanylate cyclase (sGC) stimulator approved for pulmonary artery hypertension; (ii) propylthiouracil, marketed for the treatment of hyperthyroidism; and (iii) perphenazine, an antiemetic and antipsychotic drug – are being used for their ability to target key components of the ROCG module in a network pharmacology drug combination.

8.5 Summary and Conclusions

The emergence of network medicine has opened new avenues for drug discovery and pharmacology. By moving away from the traditional one-disease, one-target, one-drug approach, and identifying mechanistic disease endotypes, new mechanism-based treatments can be developed that directly tackle the underlying disease cause. This shift is crucial to address the current drug development crisis in the pharmaceutical industry, currently struggling with high costs and low success rates. In addition, complex diseases are better understood as intricate networks of molecular interactions better targeted through a network pharmacology regime. Network pharmacology has the potential to enhance the precision and efficacy of treatments by targeting disease modules that are composed of elements of canonical pathways. Embracing the complexity of diseases and leveraging the power of network-based approaches will be key to overcoming the current crisis in drug discovery and development, paving the way for a new era in pharmacology and significantly enhancing patient outcomes.

References

1 Scannell, J.W., Blanckley, A., Boldon, H., and Warrington, B. (2012). Diagnosing the decline in pharmaceutical R&D efficiency. *Nat. Rev. Drug Discov.* 11 (3): 191–200.

2 Munos, B. (2009). Lessons from 60 years of pharmaceutical innovation. *Nat. Rev. Drug Discov.* 12: 959–968.

3 Schork, N.J. (2015). Personalized medicine: time for one-person trials. *Nature* 520 (7549): 609–611.

4 Leucht, S., Hierl, S., Kissling, W. et al. (2012). Putting the efficacy of psychiatric and general medicine medication into perspective: review of meta-analyses. *Br. J. Psychiatry* 200 (2): 97–106.

5 Kirsch, I., Deacon, B.J., Huedo-Medina, T.B. et al. (2008). Initial severity and antidepressant benefits: a meta-analysis of data submitted to the food and drug administration. *PLoS Med.* 5 (2): e45.

6 Pammolli, F., Magazzini, L., and Riccaboni, M. (2011). The productivity crisis in pharmaceutical R&D. *Nat. Rev. Drug Discov.* 10 (6): 428–438.

7 Cummings, J., Feldman, H.H., and Scheltens, P. (2019). The "rights" of precision drug development for Alzheimer's disease. *Alzheimers Res. Ther.* 11 (1): 1–14.

8 Weinberg, R.A. (2014). Coming full circle-from endless complexity to simplicity and back again. *Cell* 157 (1): 267–271.

9 Hopkins, A.L. (2008). Network pharmacology: the next paradigm in drug discovery. *Nat. Chem. Biol.* 4 (11): 682–690.

10 Keiser, M.J., Setola, V., Irwin, J.J. et al. (2009). Predicting new molecular targets for known drugs. *Nature* 462 (7270): 175–181.

11 Reddy, A.S. and Zhang, S. (2013). Polypharmacology: drug discovery for the future. *Expert. Rev. Clin. Pharmacol.* 6 (1): 41–47.

12 Kitano, H. (2002). Systems biology: a brief overview. *Science* 295 (5560): 1662–1664.

13 Auffray, C., Chen, Z., and Hood, L. (2009). Systems medicine: the future of medical genomics and healthcare. *Genome Med.* 1 (1): 1–11.

14 Barabási, A.L., Gulbahce, N., and Loscalzo, J. (2011). Network medicine: a network-based approach to human disease. *Nat. Rev. Genet.* 12 (1): 56–68.

15 Goh, K.I., Cusick, M.E., Valle, D. et al. (2007). The human disease network. *Proc. Natl. Acad. Sci. U.S.A* 104 (21): 8685–8690.

16 Goh, K.I. and Choi, I.G. (2012). Exploring the human diseasome: the human disease network. *Brief. Funct. Genomics* 11 (6): 533–542.

17 Menche, J., Sharma, A., Kitsak, M. et al. (2015). Uncovering disease-disease relationships through the incomplete interactome. *Science* 347 (6224): 1257601. https://science.sciencemag.org/content/347/6224/1257601.short.

18 Loscalzo, J. and Barabasi, A.L. (2011). Systems biology and the future of medicine. *Wiley Interdiscip. Rev. Syst. Biol. Med.* 3 (6): 619–627.

19 Choobdar, S., Ahsen, M.E., Crawford, J. et al. (2019). Assessment of network module identification across complex diseases. *Nat. Methods* 16 (9): 843–852.

20 Sadegh, S., Skelton, J., Anastasi, E. et al. (2021). Network medicine for disease module identification and drug repurposing with the NeDRex platform. *Nat. Commun.* 12 (1): 6848.

21 Vidal, M., Cusick, M.E., and Barabási, A.L. (2011). Interactome networks and human disease. *Cell* 144 (6): 986–998.

22 Elbatreek MH, Sadegh S, Anastasi E, et al. (2020), NOX5-induced uncoupling of endothelial NO synthase is a causal mechanism and theragnostic target of an age-related hypertension endotype. *PLoS Biol.* 18; e3000885. https://doi.org/10.1371/journal.pbio.3000885

23 Viana, M.P., Chen, J., Knijnenburg, T.A. et al. (2023). Integrated intracellular organization and its variations in human iPS cells. *Nature* 613 (7943): 345–354.

24 Traag, V.A., Waltman, L., and van Eck, N.J. From Louvain to Leiden: guaranteeing well-connected communities. *Sci. Rep.* 9 (1): 5233.

25 Vallance, P. (2010). An audience with Patrick Vallance. *Nat. Rev. Drug Discov.* 9 (11): 834.

26 Li, S. and Zhang, B. (2013). Traditional Chinese medicine network pharmacology: theory, methodology and application. *Chin. J. Nat. Med.* 11 (2): 110–120.

27 Nogales, C., Mamdouh, Z.M., List, M. et al. (2022). Network pharmacology: curing causal mechanisms instead of treating symptoms. *Trends Pharmacol. Sci.* 43 (2): 136–150.

28 Zimmermann, G.R., Lehár, J., and Keith, C.T. (2007). Multi-target therapeutics: when the whole is greater than the sum of the parts. *Drug Discov. Today* 12 (1–2): 34–42.

29 Pushpakom, S., Iorio, F., Eyers, P.A. et al. (2019). Drug repurposing: progress, challenges and recommendations. *Nat. Rev. Drug Discov.* 18 (1): 41–58.

30 Harrison, C. (2020). Coronavirus puts drug repurposing on the fast track. *Nat. Biotechnol.* 38 (4): 379–381.

31 Dotolo, S., Marabotti, A., Facchiano, A., and Tagliaferri, R. (2020). A review on drug repurposing applicable to COVID-19. *Brief. Bioinform.* 22 (2): 726–741.

32 Hajduk, P.J. and Greer, J. (2007). A decade of fragment-based drug design: strategic advances and lessons learned. *Nat. Rev. Drug Discov.* 6 (3): 211–219.

33 Krasowski, A., Muthas, D., Sarkar, A. et al. (2011). DrugPred: a structure-based approach to predict protein druggability developed using an extensive nonredundant data set. *J. Chem. Inf. Model.* 51 (11): 2829–2842.

34 Kufareva, I., Ilatovskiy, A.V., and Abagyan, R. (2011). Pocketome: an encyclopedia of small-molecule binding sites in 4D. *Nucleic Acids Res.* 40 (D1): D535–D540.

35 Palmer, A.C. and Sorger, P.K. (2017). Combination cancer therapy can confer benefit via patient-to-patient variability without drug additivity or synergy. *Cell* 171 (7): 1678–91.e13.

36 Schirrmacher, V. (2019). From chemotherapy to biological therapy: a review of novel concepts to reduce the side effects of systemic cancer treatment. *Int. J. Oncol.* 54 (2): 407–419.

37 DeVita, V.T. Jr., and Chu, E. (2008). A history of cancer chemotherapy. *Cancer Res.* 68 (21): 8643–8653.

38 Gonzalez-Fierro, A. and Dueñas-González, A. (2021). Drug repurposing for cancer therapy, easier said than done. *Semin. Cancer Biol.* 68: 123–131.

39 Seebacher, N.A., Stacy, A.E., Porter, G.M., and Merlot, A.M. (2019). Clinical development of targeted and immune based anti-cancer therapies. *J. Exp. Clin. Cancer Res.* 38 (1): 156.

40 Rebucci, M. and Michiels, C. (2013). Molecular aspects of cancer cell resistance to chemotherapy. *Biochem. Pharmacol.* 85 (9): 1219–1226.

41 Vogelstein, B., Papadopoulos, N., Velculescu, V.E. et al. (2013). Cancer genome landscapes. *Science* 339 (6127): 1546–1558.

42 Chen, Y., McGee, J., Chen, X. et al. (2014). Identification of druggable cancer driver genes amplified across TCGA datasets. *PLoS ONE* 9 (5): e98293.

43 Kandoth, C., McLellan, M.D., Vandin, F. et al. (2013). Mutational landscape and significance across 12 major cancer types. *Nature* 502 (7471): 333–339.

44 Chartier, M., Morency, L.P., Zylber, M.I., and Najmanovich, R.J. (2017). Large-scale detection of drug off-targets: hypotheses for drug repurposing and understanding side-effects. *BMC Pharmacol. Toxicol.* 18 (1): 18.

45 Jeanmonod, R., Asuka, E., and Jeanmonod, D. (2023). Inborn errors of metabolism. In: *StatPearls*. Treasure Island (FL): StatPearls Publishing.

46 Ezgu, F. (2016). Inborn errors of metabolism. *Adv. Clin. Chem.* 73: 195–250.

47 Arnold, G.L. (2018). Inborn errors of metabolism in the 21st century: past to present. *Ann. Transl. Med.* 6 (24): 467.

48 Lanpher, B., Brunetti-Pierri, N., and Lee, B. (2006). Inborn errors of metabolism: the flux from Mendelian to complex diseases. *Nat. Rev. Genet.* 7 (6): 449–460.

49 Argmann, C.A., Houten, S.M., Zhu, J., and Schadt, E.E. (2016). A next generation multiscale view of inborn errors of metabolism. *Cell Metab.* 23 (1): 13–26.

50 Zschocke, J. (2014). SSIEM classification of inborn errors of metabolism. In: *Physician's Guide to the Diagnosis, Treatment, and Follow-Up of Inherited Metabolic Diseases* (ed. N. Blau, M. Duran, K.M. Gibson, and C.D. Vici), 817–830. Berlin, Heidelberg: Springer.

51 Ferreira, C.R., Rahman, S., Keller, M., and Zschocke, J. (2021). ICIMD advisory group. An international classification of inherited metabolic disorders (ICIMD). *J. Inherit. Metab. Dis.* 44 (1): 164–177.

52 Thomas, J.A., Van Hove, J.L.K., Larson, A.A., and Baker, P.R. II, Inborn errors of metabolism. In: *Current Diagnosis & Treatment: Pediatrics* (ed. M. Bunik, W.W. Hay, M.J. Levin, and M.J. Abzug). McGraw Hill Medical.

53 Tuncel, A.T., Boy, N., Morath, M.A. et al. (2018). Organic acidurias in adults: late complications and management. *J. Inherit. Metab. Dis.* 41 (5): 765–776.

54 Deodato, F., Boenzi, S., Santorelli, F.M., and Dionisi-Vici, C. (2006). Methylmalonic and propionic aciduria. *Am. J. Med. Genet. C Semin. Med. Genet.* 142C (2): 104–112.

55 Woidy, M., Muntau, A.C., and Gersting, S.W. (2018). Inborn errors of metabolism and the human interactome: a systems medicine approach. *J. Inherit. Metab. Dis.* 41 (3): 285–296.

56 Scriver, C.R. and Waters, P.J. (1999). Monogenic traits are not simple: lessons from phenylketonuria. *Trends Genet.* 15 (7): 267–272.

57 Dipple, K.M. and McCabe, E.R. (2000). Modifier genes convert "simple" Mendelian disorders to complex traits. *Mol. Genet. Metab.* 71 (1–2): 43–50.

58 Schadt, E.E. (2009). Molecular networks as sensors and drivers of common human diseases. *Nature* 461 (7261): 218–223.

59 Oubounyt, M., Adlung, L., Patroni, F. et al. (2023). Inference of differential key regulatory networks and mechanistic drug repurposing candidates from scRNA-seq data with SCANet. *Bioinformatics* 39 (11): btad644. https://doi.org/10.1093/bioinformatics/btad644.

60 Grant, R.M., Lama, J.R., Anderson, P.L. et al. (2010). Preexposure chemoprophylaxis for HIV prevention in men who have sex with men. *N. Engl. J. Med.* 363 (27): 2587–2599.

61 Baeten, J.M., Donnell, D., Ndase, P. et al. (2012). Antiretroviral prophylaxis for HIV prevention in heterosexual men and women. *N. Engl. J. Med.* 367 (5): 399–410.

62 Cohen, M.S., Chen, Y.Q., McCauley, M. et al. (2011). Prevention of HIV-1 infection with early antiretroviral therapy. *N. Engl. J. Med.* 365 (6): 493–505.

63 Langhauser, F., Casas, A.I., Dao, V.T.V. et al. (2018). A diseasome cluster-based drug repurposing of soluble guanylate cyclase activators from smooth muscle relaxation to direct neuroprotection. *NPJ Syst. Biol. Appl.* 4: 8.

64 Casas, A.I., Hassan, A.A., Larsen, S.J. et al. (2019). From single drug targets to synergistic network pharmacology in ischemic stroke. *Proc. Natl. Acad. Sci. U.S.A* 116 (14): 7129–7136.

65 Casas, A.I., Kleikers, P.W., Geuss, E. et al. (2019). Calcium-dependent blood-brain barrier breakdown by NOX5 limits postreperfusion benefit in stroke. *J. Clin. Invest.* 130: 1772–1778.

9

Advancements in Rational Multi-Targeted Drug Discovery

Improving the Efficacy-Safety Balance of Small Molecule Cancer Therapeutics

Balaguru Ravikumar, Anna Cichońska, Navriti Sahni, Tero Aittokallio, and Rayees Rahman

9.1 Introduction

While it is widely recognized that complex polypharmacological interactions contribute to both therapeutic effects and adverse side effects, there is still a lack of systematic approaches for improving this efficacy-safety balance through integrated experimental-computational strategies. This is particularly challenging in the field of oncology, where the intricate nature of cancer pathophysiology often necessitates the use of polypharmacology. A comprehensive survey of approved small molecule drugs for oncology reveals that these drugs often have multiple potent targets, making them polypharmacological in nature [1].

The ability to modulate multiple targets simultaneously is advantageous for achieving desirable therapeutic outcomes, especially in the context of complex diseases such as cancer, where single-target approaches often fail due to compensatory resistance mechanisms. However, such multi-targeted drugs (MTDs) also have promiscuous interactions with potent off-targets, which can lead to toxic adverse events (AEs), illustrating the double-edged sword of polypharmacology. These adverse effects are often identified as black box warnings when the drugs are introduced to the market, highlighting the need for improved predictive AE assessment early in the drug development process. Furthermore, approximately 90% drug candidates that pass phase I trial fail in phases II–III. This is a substantial attrition rate that poses immense costs to the industry. Upon a closer look, 70% of the failure rate can be attributed to lack of clinical efficacy of the drug and toxicity observed in phases II and III [2]. To address this prominent issue, we must leverage the power of cutting-edge computational approaches alongside experimental methods. It is essential to design computational and experimental strategies that enable the rational design of selective MTDs with both efficacy and safety in mind. Figure 9.1 provides a schematic illustration and an overview of different approaches to MTD design. Only through such integrated strategies we can aim to reduce the high attrition rates in drug development and bring more effective and safer therapies to patients with improved success rates.

In this chapter, we provide an overview of the latest progress in the preclinical anti-cancer small molecule drug discovery, with a focus on optimizing the balance between selectivity and toxicity. We begin by examining the underlying complexity of cancer and its treatment strategies, the principles of polypharmacology, and summarizing its role in drug resistance and efficacy, as well as the need for differentiating between therapeutic targets and anti-targets. Following this, we review recent advancements in both computational and experimental techniques that assist in improving the balance between drug efficacy and safety. Finally, we discuss the challenges and limitations of the current approaches and offer insights into future directions.

9.2 Cancer and the Existing Treatment Strategies

9.2.1 Overview of Cancer and its Complexity

Cancer is a multi-faceted group of clonal diseases that originate in a single progenitor cell with DNA damage. This damage may result from genetic predisposition or through various non-genetic causes, such as exposure to carcinogens over time, inflammation, continuous stress, or mitotic errors during replication [3, 4]. Carcinogenesis can also occur due to tumor predisposition syndromes (TPSs) and cancer syndromes, and individuals who inherit these specific germline mutations (syndromes) experience an increased risk of tumor development, with cancer arising more rapidly [5]. If the cell proliferates before repair, the damage becomes irreversible, and the error is

Polypharmacology: Strategies for Multi-Target Drug Discovery, First Edition. Edited by Jens-Uwe Peters.

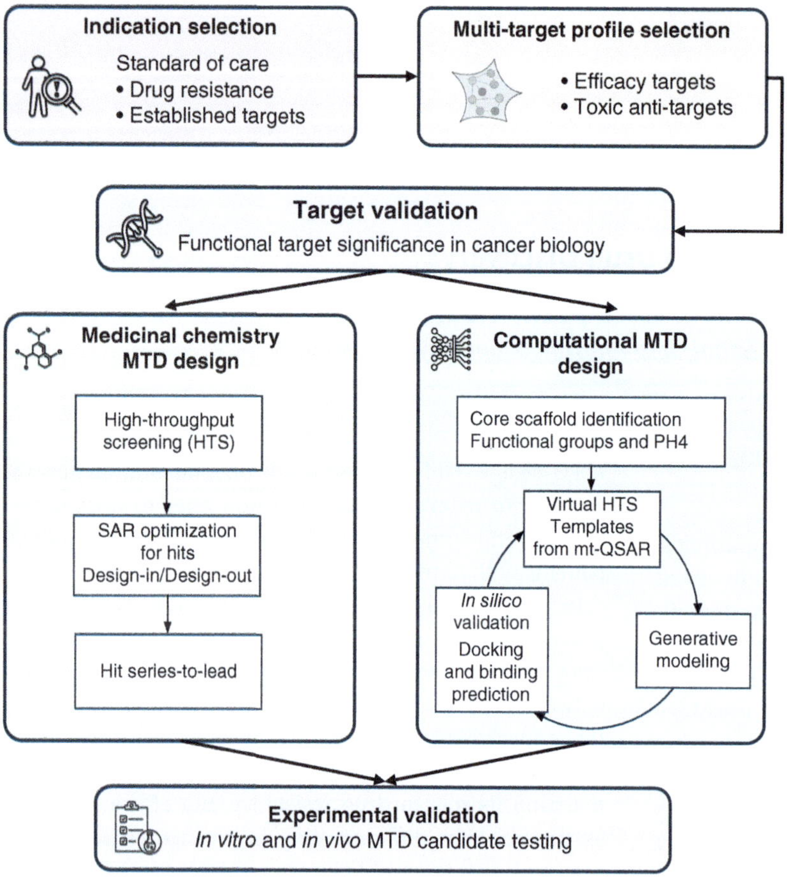

Figure 9.1 Flowchart of the integrated approach to multi-targeted drug (MTD) design. The process begins with indication selection, where potential drug resistance and targets of the standard of care are considered. This informs the multi-target profile selection that focuses on targets for enhanced drug efficacy and avoids anti-targets responsible for toxic side effects. The selected target profile is validated by assessing its functional significance in the specific cancer biology context. An MTD can be designed using either medicinal chemistry or a computational strategy. Medicinal chemistry for MTD design involves high-throughput screening (HTS), followed by hit optimization and the development of a hit-to-lead series. On the other hand, computational MTD design starts with the identification of the core scaffold among existing inhibitors, considering functional groups and pharmacophores (PH4) toward the desired target profile. Virtual HTS, based on the identified templates, is followed by generative modeling and *in silico* validation through docking and binding prediction using machine learning, carried out iteratively. Finally, experimental validation through *in vitro* and *in vivo* MTD candidate testing is performed.

propagated to all subsequent lineages, leading to neoplasia. Without intervention, these neoplastic cells can progress through tumorigenesis, gradually accumulating a series of alterations and acquiring the hallmarks of cancer [6, 7]. Such inherited DNA damage causes genetic and epigenetic alterations that may give rise to oncogenes with a dominant gain of function, tumor suppressor genes with a recessive loss of function, or mutations in DNA repair genes, all of which cause uncontrolled proliferation of the cells [8]. Through iterative rounds of clonal expansion and clonal selection, a subset of malignant clones may become capable of metastasis [9]. In addition to enabling metastasis, such clonal populations often harbor distinct genetic and phenotypic features, driving both intra- and inter-tumor heterogeneity [10]. These heterogenous tumor cells exist in a complex ecosystem and interact with diverse non-neoplastic stromal cells, collectively known as the tumor microenvironment (TME) [11]. The complexity of the cancer pathophysiology, diversity of cancer types, the genetic makeup of tumors, their heterogeneity, and interactions with the TME pose significant challenges to all aspects of cancer biology, ranging from diagnosis and staging to treatment.

9.2.2 Historical Perspective on Cancer Drug Discovery

Until the 1900s, radical excision of the tumor was the only therapeutic option available for cancer treatment. This approach was often ineffective for patients with advanced tumor pathology or when surgery was unsuccessful. However, a significant milestone in treatment intervention occurred in the mid-1900s with the discovery of cytotoxic antibiotics and the subsequent use of chemotherapy to treat various hematological malignancies [12]. In the early 1980s, advancements in the fields of molecular and cell biology led to a more refined understanding of the molecular mechanisms underlying tumorigenesis. This knowledge contributed to the development of targeted therapies that specifically intervene with altered oncogenes or tumor suppressors driving the disease progression in specific cancers [13]. Such targeted therapies encompass all interventions that use either small molecules or macromolecules to modulate the activity of specific cellular proteins that are involved in the initiation, promotion, conversion, or progression phases of tumorigenesis [14]. A comprehensive review discussing the evolution of cancer therapeutics is detailed elsewhere [15]. In this chapter, we specifically focus on small molecule targeted therapies and their role in the current landscape of cancer treatment.

9.2.3 Drug Resistance in Targeted Therapies

As of 2022, 48% of FDA-approved cancer therapies are targeted drugs, with 31% being small molecule kinase inhibitors (KIs) [1]. Though targeted therapies have revolutionized cancer treatment strategies, they still face significant challenges, the most prominent of which is drug resistance. Resistance often emerges through several mechanisms, including genetic mutations, epigenetic alterations, and drug-induced changes in TME. These adaptations not only hinder the efficacy of targeted drugs but also pose challenges in predicting their therapeutic outcomes. Almost all tumors develop intrinsic or acquired resistance to targeted drugs after their clinical use in patients, and this has led to the development of second- and eventually third- or fourth-generation anti-cancer drugs to overcome the resistance mechanisms [16]. For example, crizotinib, a multi-kinase ALK inhibitor, was approved for metastatic ALK-positive non-small cell lung cancer (NSCLC) in 2011. However, after approximately 12 months of clinical use, most patients acquired resistance mutations (L1196M, G1269A), leading to relapse. This led to the development of ceritinib, alectinib, and brigatinib KIs, all of which were potent specifically towards crizotinib-resistant ALK mutations, and brigatinib was approved as the first-line therapy in 2017 [17, 18]. As with crizotinib, exposure to these second-generation ALK inhibitors led to the development of resistance mutations (G1202R, V1180L), which was later addressed by the development of lorlatinib, a third-generation ALK inhibitor approved in 2018 [19].

Furthermore, therapeutic agents that are initially intended to target a single protein often exhibit unforeseen side effects and dose-limiting toxicity, which can impact their efficacy. Many of these undesirable AEs can be attributed to drug's promiscuity arising from unintended off-target interactions. Often, the unmanageable toxicity resulting from on-target or off-target interactions of the drug candidates contribute to the high failure rate (70%) observed in phase II and III trials [2]. At the same time, it is worth noting that the mechanism of action (MoA) of many drug candidates currently undergoing clinical trials can also be attributed to off-target interactions [20]. A comprehensive review summarizing the myriad of factors underlying drug resistance mechanisms in cancer is detailed elsewhere [21].

9.2.4 Rational Combination Therapies

Rational combination of targeted therapies is one strategy currently adopted in clinical practice to combat drug resistance [22]. Compared to using a single agent, administering two or more anti-cancer agents with different MoA has been found to elicit a synergistic or otherwise improved clinical response [23]. These agents also have different mechanisms of resistance and distinct toxicity profiles. The rationale of selecting such combinations requires understanding the resistance mechanism in the current treatment strategy. For instance, when treated with vemurafenib, intrinsic resistance was observed in $BRAF^{V660E}$ driven Metastatic ColoRectal Cancer (MCRC). Upon investigation, it was found that BRAF inhibition causes feedback activation through EGFR, impacting the clinical efficacy of vemurafenib in $BRAF^{V600E}$ driven CRC [24]. This led to the use of encorafenib, a $BRAF^{V600E}$ inhibitor, and cetuximab, a potent EGFR inhibitor, in combination as the standard of care in $BRAF^{V600E}$-resistant MCRC [25].

In cases where resistance is acquired through mutations either in the upstream or downstream of the targeted pathway, as in the case of reactivation of MAPK pathway when treated with BRAF inhibitor, a combination of BRAF and MEK inhibitors has evolved as the standard of care in BRAF-mutant melanoma [26]. Preclinical studies have also explored the idea of using drug cocktails, a combination of three to four drugs, to be used concurrently to engage multiple targets (multi-node inhibition), and this approach was shown to be efficacious in EGFR inhibitor-resistant NSCLC tumors. As of 2023, the FDA has approved a total of 186 combination therapies. These include targeted KIs, immunotherapies, monoclonal antibodies, cytotoxic agents, and chemotherapeutic drugs [27].

9.2.5 Rational Targeted Polypharmacology

Similar to combination therapies, rational polypharmacological drug design is another approach used to mitigate drug resistance. "Polypharmacology" refers to the promiscuous nature of many drugs that inhibit multiple targets with varying degrees of efficacy and potency [28, 29]. It is important to note that these co-inhibited targets can come from distinct protein families or from different sub-families. As briefly discussed in Section 9.2.3, the promiscuous target profile, which includes both intended on-targets and unintended off-targets, plays a key role in dictating the efficacy and safety properties of many of the drugs. Pharmacotoxicology and pharmacogenomics studies have revealed that the observed adverse drug reactions (ADRs) or drug-induced AEs stem from these on- and off-target interactions [30, 31]. Interestingly, off-target interactions can also drive the MoA of many clinical drug candidates [20]. While most approved anti-cancer KIs are inherently multi-targeting, a rational design approach for targeted polypharmacology involves creating therapeutic agents that specifically interact with carefully selected on-targets driving tumorigenesis or drug resistance, while avoiding anti-targets whose inhibition could lead to significant toxicities in patients [32].

Rational design of MTDs has gained traction recently. The application of this approach in clinical practice is exemplified by a notable case study: the recent approval of momelotinib for the treatment of myelofibrosis (MF). MF is a myeloproliferative neoplasm characterized by the activation of the JAK-STAT signaling pathway. MF leads to disrupted bone marrow fibrosis, impairing the production of normal blood cells, and also increases the risk of evolving into acute myeloid leukemia (AML) [33]. Momelotinib is a multi-kinase inhibitor designed to inhibit the ACVR1-SMAD and JAK-STAT pathways. It primarily targets JAK1 and JAK2 kinases, which drive the progression of MF, but whose inhibition is also known to exacerbate myelosuppression in patients (as seen in other JAK2 inhibitors such as ruxolitinib). In addition to JAK kinases, momelotinib additionally has potent interactions towards ACVR1 (ALK2) protein kinase, which results in decreased hepcidin and restored erythropoiesis in MF, alleviating on-target myelosuppression. In 2023, momelotinib received approval for use in MF patients with anemia, irrespective of prior first-line treatment, including the administration of approved JAK inhibitor ruxolitinib [34].

A review of ClincialTrails.gov reveals a notable disparity: there are currently many more combination therapies under exploration compared to MTDs [22]. This difference can largely be attributed to the fact that the rational design of MTDs is typically defined and initiated during the early or preclinical stages of drug development. In contrast, combination therapies often utilize existing, approved anti-cancer drugs. Nevertheless, a single drug that modulates multiple targets presents clear advantages over combination therapies. For instance, the pharmacokinetic profile of a single agent is more predictable than that of drug combinations. Furthermore, combination therapy frequently leads to both acute and delayed treatment-related toxicity, which may stem from the drugs themselves or from unforeseen drug-drug interactions. Another challenge with combination therapies is the complication in designing population-specific dosages and therapeutic regimens due to the complex interplay of multiple drugs. A single-agent, MTD may reduce the burden of patients by simplifying treatment protocols and potentially lowering the risk of AEs compared to the multi-drug regimens required in combination therapies. Additionally, the development of multi-targeted therapies often presents fewer logistical challenges compared to combination drugs, as they do not require complex agreements between multiple pharmaceutical or biotech companies when combining drugs from different sources [28].

9.2.6 Protein Kinases and Small Molecule KIs

Protein kinases are regulatory enzymes important to cell growth, proliferation, and differentiation. Working in conjunction with phosphatases, they phosphorylate and dephosphorylate enzymes, to transduce signals through various signaling cascades. Aberrations of kinases in such regulatory pathways and their association in proliferation diseases, such as cancer, were explored in 1990s which eventually led to the concept of "signal transduction therapy" [35, 36]. The feasibility of this strategy was investigated using various naturally derived molecules such as quercetin, genistein, lavendustin, and erbstatin that target Receptor Tyrosine Kinases (RTK, TK). Staurosporine, is one such naturally derived tool compound that is currently used as a broad-spectrum kinase binder in biochemical and cellular assays. However, since these natural molecules often exhibit low potency and poor selectivity, they are often used as tool compounds to design more potent and selective "tyrphostins" (TYRosine PHOSphorylation INhibitors), as demonstrated by Yaish et al. for growth factor receptors [37]. The subsequent application of this approach has revolutionized anti-cancer drug discovery, establishing the human kinome as viable druggable targets for various cancer treatment strategies, and leading to the design and development of novel KIs [16, 38].

Roskoski classified KIs into six different classes (types I–VI) by analyzing the structures of protein-ligand complexes of 21 approved KIs. The classification was based on the conformation of the kinase, the characteristics of ligand binding, and the features of the binding pocket [39]. The human protein kinome consists of ~530 kinases, out of which 448 have an accompanying assay to measure their activity. There are 312 kinases with at least one crystal structure, and 129 kinases have a probe molecule for activity and lead investigations [40]. Adenosine triphosphate (ATP) acts as the cofactor for most kinases, and compared to the substrate-binding domain, the ATP-binding fold in kinases is more conserved and structurally well-defined [41]. As a result, the majority of designed KIs mimic ATP and are primarily ATP-competitive type-I inhibitors. This high degree of conservation across kinases in their binding pocket residues [41, 42] is the primary factor that influences their promiscuity. This is reiterated when reviewing both the panel of approved KIs and the numerous preclinical compounds from high-throughput kinase screening

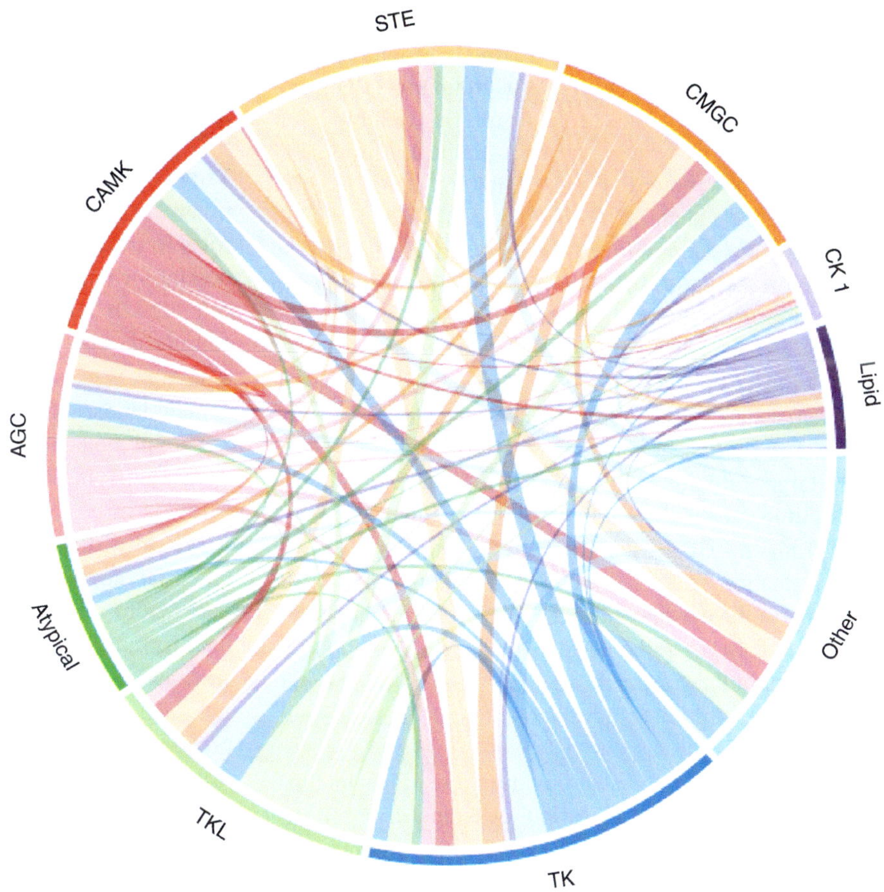

Figure 9.2 A polypharmacological interaction map across kinase subfamilies is represented as a chord diagram and depicted using a kinome-wide screening study of 72 inhibitors across 446 kinases. Adapted from Davis et al. [43]. Each chord represents a distinct kinase subfamily, the area of which shows the total number of compounds that were found to be active (≤100 nM) for the given subfamily. The links and their corresponding size indicate the number of compounds that were found to be active between subfamilies.

studies [43, 44]. They suggest that KIs are inherently multi-targeting, and whether this promiscuity is beneficial or detrimental depends on the disease type and the specific target profile associated with each KI (Figure 9.2).

9.2.7 Review of FDA-Approved Multi-Targeted Inhibitors

This section highlights a subset of FDA-approved inhibitors used here to underscore the benefits of MTDs. Cabozantinib, like crizotinib, is a KI with a multi-target activity profile. Its primary targets include c-Met, VEGFR-1/2/3, RET, c-Kit, TRK2, TYRO3, and TIE2, and it has been approved for the treatment of metastatic renal cell carcinoma (MRCC) and thyroid cancer [45]. Similarly, the first-generation FLT3 inhibitors for AML, such as sunitinib, tandutinib, ponatininb, midostaurin, and lestaurtinib, are all MTDs that inhibit various RTKs, including PDGFRs, c-KIT, VEGFRs, RAF, ABL, LYN, and JAKs. However, due to AEs observed during clinical trials, all except midostaurin were later discontinued [46]. However, lestaurtinib, which targets JAK2, FLT3, and TRKA, is currently being investigated for a wide range of cancers such as Hodgkin lymphoma, MF, and prostate cancer [47]. Cediranib, a pan-VEGFR inhibitor currently in phase II study for glioblastoma, is a potent multi-kinase inhibitor targeting VEGFRs, PDGFRs, and c-Kit [48]. Entrectinib, a recently approved TRK inhibitor for solid tumors, is an MTD targeting TRKs, ROS1, and ALK [49]. Lapatinib, approved for HER2-positive breast cancer, is an MTD that targets HER2, ErB2, and EGFR. In addition to these type-I binders, sorafenib, a type-II pan-Raf inhibitor with inhibitory activity across various RTKs, including VEGFRs, PDGFR, FLT1, and RET, was approved in 2005 for hepatocellular carcinoma, RCC, and thyroid cancer [50]. Although contributing to efficacy and reducing the risk of drug resistance, the typically wide target profile observed for most of the KIs highlighted here was discovered serendipitously [28, 29, 51].

There has been a significant effort recently to design "hybrid compounds," where two or more pharmacophores are rationally embedded in a single molecule to enhance the multi-target potential of an inhibitor. The most notable examples, especially in anti-cancer drug treatments are kinase/histone deacetylase (HDAC) hybrid inhibitors [52]. HDACs are a family of ubiquitous proteins that play an essential role in tumorigenesis; silencing them has been shown to affect the cell cycle, cell growth, differentiation, and apoptosis. Currently, four HDAC inhibitors (HDACis) have been approved by the FDA for the treatment of peripheral T-cell lymphoma and multiple myeloma. However, while potent, these inhibitors cause significant side effects such as thrombocytopenia, neutropenia, diarrhea, nausea, vomiting, fatigue, and cardiotoxicity [53]. Hybrid kinase/HDAC inhibitors were designed as a solution to overcome these limitations, wherein the hybrids are formed by merging the pharmacophores of HDACis, usually hydroxamic or benzamide, with a KI fragment using a linker fragment. A notable example is CUDC-101, a first-in-class hybrid HDAC, EGFR/HER2 multi-target inhibitor that combines the pharmacophores of erlotinib, an approved EGFR inhibitor, with hydroxamic from the HDACis. CUDC-101 has entered phase I dose escalation study and is used in combination with cisplatin for the treatment of head and neck squamous cell cancer [54]. In a similar fashion to erlotinib, other KIs, such as lapatinib, pazopanib, etc., have been used in hybrid kinase/HDAC inhibitor development and are reviewed elsewhere [52].

Conversely, the high discontinuation rate of KIs in clinical trials is largely due to various AEs and drug-related toxicity, mediated mostly by their off-target interactions [20]. Notably, a positive correlation was observed between the toxicity and the number of kinases a KI inhibits [55]. Both on-target and off-target toxicities have been linked to ADRs such as hypertension, skin reactions, and proteinuria. However, pharmacovigilance studies identified cardiac dysfunction as a significant AE associated with the majority of KIs. Multi-targeting KIs interact with several signaling pathways, and those that cause cardiovascular AEs include targets such as EGFR, ErbB2, HER2, PI3K, and AMPK [56]. For example, sunitinib, a small molecule multi-kinase inhibitor approved for RCC and GIST, targets PDGFR-α/β, VEGFR-1/2/3, RET, c-Kit, CSF1R, and FLT3 among others. It was also investigated for a treatment of AML, however, it was later discontinued due to its cardiotoxicity, mediated by its off-target inhibition of AMPK [57]. Other MTDs like imatinib, dasatinib, ponatinib, and nilotinib, which inhibit ABL, PDGFR, SRC, KIT, BRAF, DDR1, DDR2, are prone to causing ADRs such as QT prolongation, pulmonary hypertension, and vascular dysfunction [58]. A comprehensive survey of ADRs mediated by KIs in various cancer therapies is summarized in a recent review [59]. However, it is crucial to note the complex relationship between a drug's clinical efficacy, its selectivity, and the adverse effect toxicity it causes. These are essential factors that need to be balanced when designing MTDs.

9.3 Safety and Efficacy: A Double-Edged Sword

9.3.1 Approaches for Target Selection for MTDs

Target selection in precision oncology is a critical process involving the identification of specific genes that play roles in cancer progression, survival, and metastasis. It is crucial for targeted therapies to selectively inhibit cancer cell growth and proliferation without harming normal cells [60]. Thus, developing targeted therapies requires a comprehensive understanding of the molecular and genetic landscapes of both cancerous and normal tissues. For small molecules, target selection also depends on their druggability and the potential for therapeutic intervention [61, 62].

Typically, the first step in target selection is identifying potential key drivers of cancer through genomic, proteomic, or metabolomic assays. Additionally, retrospective bioinformatics analysis of published large-scale datasets, such as The Cancer Genome Atlas (TCGA) and The Cancer Dependency Map (DepMap), can suggest potential protein targets for drug discovery using multi-omics data from patient samples and cell lines, respectively [63, 64]. Following the initial identification of a potential target, the next step is target validation, which confirms the functional significance of the potential target in cancer biology. Target validation may involve several experimental assays, such as genetic knockouts or site-directed mutagenesis studies, that rigorously demonstrate the functional impact of the target on cancer cell viability, proliferation, and metastasis [60, 65]. These experiments must also show limited phenotypic change in normal tissue types for selectivity. While the process for discovering and validating specific, single targets for cancer development follows a typical experimental timeline, dissecting the relationships across multiple targets, simultaneously, is an emerging paradigm that lacks an established validation process.

Selecting targets for multi-targeted therapies requires a nuanced understanding of cancer biology, particularly across various networks of signaling pathways that contribute to tumorigenesis. One approach for revealing target synergy is to evaluate the relationship between proteins within a single cancer signaling pathway. For instance, protein kinases RAF and MEK, both part of the MAPK kinase pathway, show that inhibition of one kinase can lead to compensatory upregulation of the other kinase in certain cellular contexts [66, 67]. Thus, MEK and RAF inhibitors are often combined to cover this form of resistance signaling that can occur from the usage of either a standalone MEK or RAF inhibitor.

Alternatively, multiple targets can be chosen across several parallel pathways to improve efficacy and avoid resistance to single pathway inhibition. For instance, a novel combination strategy of bromodomain and extra-terminal (BET) inhibitors with FLT3 protein KIs is being tested for FLT3-mutant AML [68]. Here, BET inhibitors primarily modulate gene expression, while FLT3 inhibitors disrupt the PI3K-AKT, STAT5, and Ras-MAPK signaling pathways [69, 70]. This combination has been shown preclinically to be synergistically lethal in AML cells driven by FLT3 mutations [71]. Importantly, many small molecule KIs are known to engage bromodomain proteins with high potency, opening the door for MTDs across diverse protein families [72].

Target synergy can also arise from engaging diverse cell types. For instance, anti-PD/PD-L1 antibodies are often combined with targeted therapies. This combination both sensitizes the immune system to cancer cells (via PD/PDL1 blockade) and concurrently acts cytotoxically on cancer cells (via the targeted therapy) [73]. For example, in MRCC, pembrolizumab is used in conjunction with axitinib to improve patient outcomes compared to sunitinib monotherapy [74]. While there are not yet any approved small molecule immunotherapies, emerging targets such as the protein kinase HPK1 and diacylglycerol kinases (DGKs), both of which modulate the immune response, could be potential synergistic co-targets with targeted therapies [75, 76].

Another approach for discovering novel target synergies involves identifying synthetic lethal pairs. Synthetic lethality occurs when mutations or alterations in two distinct genes simultaneously lead to cancer cell death, whereas neither mutation alone does, thereby leading to selective inhibition of cancer cells while sparing the wild-type cells [77–79]. MTD design that engages with pairs of synthetically lethal targets can lead to improved efficacy, as observed with the targeting of Wee1 and PKMYT1 in ovarian cancer [80].

Although there is significant potential in identifying synergistic co-targets for MTD design, confirming target synergy in terms of enhanced efficacy presents several significant challenges, specifically in terms of experimental validation and clinical translation. One of the primary challenges involves the complexity of accurately modeling the interactions between potential synergistic targets *in vivo*. Traditional, single-target validation benefits from straightforward, well-established models [81]. However, multi-targeted approaches require more complex models that can accurately reflect the dynamic interplay between multiple pathways and the broader cellular environment. Another significant challenge is the variability of synthetic lethality across different genetic backgrounds and cancer subtypes. Certain synthetically lethal interactions may not be universally applicable across all cancer cells due to differences in genetic makeup [77]. Finally, the risk of toxicity increases with the number of targets engaged [55]. For instance, synthetically lethal pairs should be tested across various healthy cell types to identify potential risks to non-cancerous cells due to unintended interactions within essential cellular pathways.

9.3.2 Anti-Targets for MTD Design

Unlike in traditional drug discovery, where high selectivity for the principal target is forefront in small molecule optimization, MTD discovery often results in drugs with broader activity profiles. This increased promiscuity is believed to enhance efficacy, albeit potentially at the expense of increased toxicity due to undesirable interactions with off-targets. It is therefore essential to identify and consider potential anti-targets early in drug optimization to model safety and to minimize AEs. For instance, the hERG channel (human Ether-a-go-go Related Gene) is a critical anti-target in drug discovery due to its association with cardiotoxicity when engaged by small molecules [82]. Similarly, inhibition of the CYP450 enzyme, such as CYP3A4, can lead to severe toxicity through drug-drug interactions, underscoring the need for vigilant monitoring [83].

The comprehensive identification of anti-targets that contribute to toxicity remains an evolving field. Secondary pharmacology assays, such as Eurofin's BioPrint and PreclinicalGPS are designed to biochemically profile therapeutic candidates against a broad range of targets to identify potential off-target liabilities. A notable advancement was the development of a secondary pharmacology database by Novartis, which assessed 1958 drugs across 200 secondary pharmacology assays, illuminating key targets linked to ADRs, as cataloged in SIDER and the FDA Adverse Event Reporting System (FAERS) [84, 85].

Although these resources and assays are not infallible predictors of toxicity, they play a crucial role in the early identification and optimization of key anti-targets during MTD development. This approach can lead to drugs with wider selectivity windows between their intended effects and specific anti-targets, potentially offering a better therapeutic index than single-target alternatives. A prime example is the FLT3 inhibitor gilteritinib, which, despite its promiscuous binding profile, demonstrates a significant selectivity for FLT3 over KIT [86]. There is evidence that off-target interactions with KIT may drive myelosuppression seen with other FLT3 inhibitors [87]. In clinical trials compared to quizartinib, a more selective FLT3 inhibitor but with potent KIT activity, the incidence of neutropenia was higher in patients than with gilteritinib (34% versus 14% respectively) [86, 88]. Also importantly, quizartinib was approved with a black box warning for risk of QT prolongation possibly due to off-target interactions with cardiac delayed rectifier potassium channels (IKs), whereas gilteritinib was approved with no such warning [89].

Overall, careful consideration of anti-targets during MTD discovery can result in drugs that maintain a balance between efficacy and safety, potentially offering improved therapeutic indices over drugs that are designed to bind to a single target.

9.4 Rational Design of MTDs

9.4.1 Rational Design of MTDs Through Medicinal Chemistry

The evolution of medicinal chemistry strategies from designing and optimizing compounds for single targets to multi-targets reflects a significant paradigm shift in drug discovery and development. Initially, the focus was to rationally design drugs that target a single biological target which was believed to be critical for disease progression. This approach was based on the "lock-and-key" model, where the aim was to design drugs ("keys") that fit precisely into a biological target ("lock") and modulate its activity. The goal was to achieve high selectivity, minimize off-target effects, and thereby reduce side effects. HTS and computational modeling were key tools used in identifying and optimizing these selective compounds. They were essential to both identify hits and sequentially perform structure-activity relationship (SAR) studies on these hits to improve potency, selectivity, pharmacokinetic properties, absorption, distribution, metabolism, excretion, and toxicity (ADMET) properties, and safety profile, resulting in a lead compound [90].

Creating MTDs through the lens of medicinal chemistry involves a deliberate and sophisticated approach aimed at simultaneously addressing multiple pathological pathways, a strategy particularly beneficial in the realm of KIs [91]. The rational design of such drugs necessitates a multi-faceted approach, integrating computational modeling, structural biology, and an in-depth understanding of the disease's molecular underpinnings. For instance, the development process may involve the use of crystallography to determine the three-dimensional (3D) structures of target kinases, alongside molecular docking studies to predict how potential inhibitors might interact with multiple targets at the molecular level. This is complemented by SAR analyses to refine the compound's ability to engage effectively with these targets, optimizing therapeutic efficacy while minimizing off-target effects [92].

Integrating the concepts of merging functional groups, scaffolds, linking pharmacophores, fusing pharmacophores, and scaffold hopping into the rational design of MTDs through medicinal chemistry enhances the specificity and efficacy of these compounds, as shown in Figure 9.3. The use of merged functional groups and

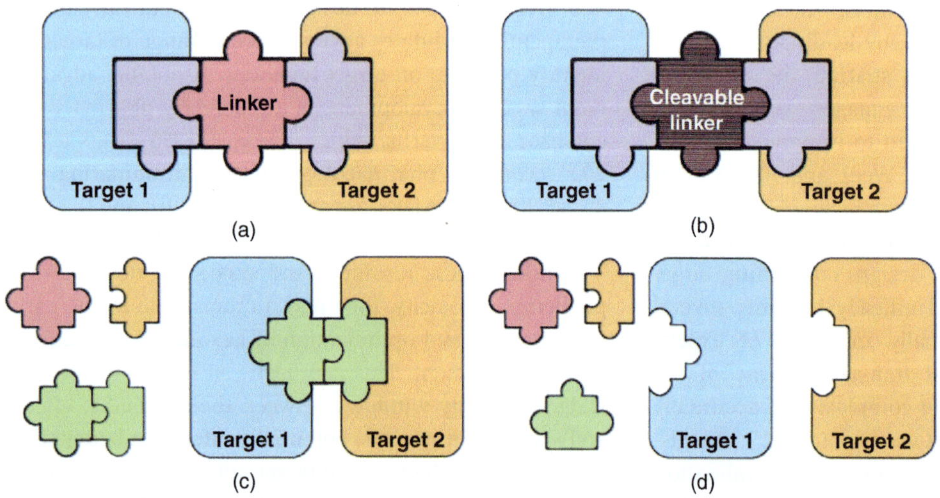

Figure 9.3 A broad classification of various medicinal chemistry strategies adopted for multi-targeted drug design. Distinct functional groups, pharmacophores, and linkers are represented as puzzle pieces that bind to targets 1 and 2. (a) Linked pharmacophore design, where two distinct pharmacophores are linked using a linker fragment. (b) Linked pharmacophore design where the linkers are cleavable or biodegradable. (c) Fused pharmacophore design, where diverse pharmacophores or functional groups are fused to a single molecule. (d) Merged pharmacophore design, where pharmacophores or functional groups are fused to a single molecule that usually has a lower molecular weight and better ADME properties than (a)–(c).

scaffolds is a cornerstone in designing MTDs, allowing chemists to create molecules with enhanced binding affinities and selectivity for multiple targets. This approach involves the integration of functional groups or fragments that are known to interact favorably with different active sites across various targets. For instance, incorporating heterocyclic scaffolds, such as pyrimidines or pyridines, which are common in KIs like sorafenib, into a single molecule can provide the structural basis for multi-target activity [93]. These scaffolds serve as the foundation upon which additional functional groups can be appended to fine-tune the interaction with each target [28]. Linking pharmacophores with a chemical linker represents another approach, enabling the simultaneous targeting of distinct binding sites within or across proteins. Linkers can be designed to be flexible, to accommodate movements within the binding site, or rigid, to constrain the pharmacophores in an orientation favorable for binding to the on-targets. Among the numerous instances reported, there is one example of dual inhibitor targeting JAK and HDAC obtained by conjugation of zinc binding group of vorinostat (HDAC inhibitor) and vital pharmacophore of ruxolitinib (JAK inhibitor) [94]. Recent advancements have introduced biodegradable linkers that enhance the selectivity and efficacy of these hybrid molecules, addressing limitations related to bioavailability [95]. ZRX1 is an example of compound combining capecitabine (a prodrug of 5-fluorouracil, 5 FU that interferes with DNA synthesis) and FD105 (an EGFR inhibitor) via a cleavable linker, designed to release 5-FU and FD105 post-metabolism for synergistic effects [96, 97].

Another strategy for MTD design involves fusing pharmacophores, that is, integrating two pharmacophoric elements into a single, cohesive molecular entity. This strategy creates a new compound that retains the ability to interact with multiple targets through a single, unified structure, rather than separate entities joined by a linker. Fused pharmacophores can lead to molecules with improved pharmacokinetic properties, such as increased oral bioavailability and stability, while still targeting multiple disease-relevant kinases. For instance, Mahboobi et al. demonstrated pharmacophore fusion by integrating the zinc binding group of an HDAC inhibitor and the core pharmacophore of imatinib within a single synthetic small molecule, enhancing the potency of imatinib [96, 98]. Scaffold hopping is another approach to designing new compounds with desired activity against specific kinase targets by replacing the core structure (scaffold) while retaining its pharmacophoric elements. Wang and colleagues developed a dual PARP/PI3K inhibitor targeting BRCA-proficient cancer by combining piperazine and pyrimidine, then improved its PI3Kα and PARP-1 inhibition by replacing pyrimidine with 1,3,5-triazine through scaffold hopping [99]. Scaffold hopping involves identifying a lead compound with desired activity against a specific kinase target, and then systematically modifying its core scaffold to generate new molecules that can inhibit additional kinase targets [100].

The application of these design concepts in medicinal chemistry has been instrumental in the development of MTDs that offer improved therapeutic outcomes. The continuous evolution of drug design strategies, supported by technological advancements in drug discovery platforms, highlights the dynamic nature of this field. However, the field faces a significant hurdle in advancing SAR analysis, necessitating novel tools for simultaneous evaluation and design towards targeting multiple biological pathways [28]. As research progresses, the integration of merging functional groups, scaffold hybridization, pharmacophore linking, and fusion is expected to aid in the design of more effective and safer multi-targeted therapies.

9.4.2 Computational Methods to Design MTDs

A majority of multi-targeting KIs highlighted in this chapter were based on a serendipitous discovery. In the preclinical stage of drug discovery, hits that inhibited the desired multi-target profile are identified mostly by chance, during the HTS process. These hits are later optimized as MTDs by maintaining or improving their *in vitro* potency and *in vivo* efficacy [101]. For example, the dual p38α/MAPK and phosphodiesterase 4 (PDE-4) inhibitor CBS-3595 was identified during the hit profiling stage and was later optimized to a potential lead by balancing its efficacy and potency properties [102]. Most of this optimization occurs during the lead generation phase, which follows one of two strategies. The first is the design-in approach, where the objective is to enhance the compound's affinity towards a set of targets while retaining its affinity to the existing target. Second is the design-out approach, where the optimization is carried out on a molecule that engages with multiple targets concurrently, and the objective is to diminish its affinity towards undesirable off-targets, thereby improving its selectivity [103]. In both cases, a systematic method to quantitatively estimate the multi-target structure-activity relationship (mt-QSAR) is essential for optimization [104]. QSAR-Co-X is one such toolkit that uses the Box–Jenkins approach to support mt-QSAR modeling. Its effectiveness as a chemometric model was demonstrated in finding preclinical inhibitors targeting PI3K isoforms [105, 106]. Compared to conventional computer-aided drug discovery

(CADD) tools, which focus on single targets, computational approaches that cater to the needs of MTDs are very limited.

A pivotal computational work that revealed the extent of polypharmacology in known inhibitors, diseases, and target spaces was the global polypharmacological interaction analysis conducted by Paloini et al. [107]. This analysis later paved way for the concept of network polypharmacology, giving rise to several approaches that modeled the relationship between protein families and compound promiscuity [108]. A statistical model that uses compound promiscuity as a measure to better understand target similarities is the similarity ensemble approach. Originally employed to analyze the polypharmacological maps of G protein-coupled receptors (GPCRs), this model was later used to predict ADRs of 653 drugs by identifying their unforeseen off-target interactions across 73 proteins [109]. This approach was subsequently applied to understand the chemogenomic landscape of protein kinases and used as a drug repurposing model to identify co-targets of known KIs [110]. A comprehensive review of the tools, databases, along with ligand, target, and network-centric models for rational polypharmacology is detailed elsewhere [111].

Recent advancements in computational methods that predict the 3D protein structures, such as AlphaFold2 and ColabFold, are crucial in designing MTDs [112, 113]. Apart from co-crystallized structures, the availability of these predicted structures enables computational chemists to conduct tandem molecular docking studies across multiple targets. Moreover, the availability of these protein structures aids in the designing of pharmacophore models. These models encompass the protein-ligand binding characteristic stemming from the seed compounds and the 3D structure of the targets, as a way to virtually screen libraries and validate designs [114]. For instance, the Gaussian Ensemble Screening approach is one such model that encodes the ligand's promiscuity feature as a polypharmacological fingerprint which, in turn, is used to virtually screen and identify compounds with a similar polypharmacological profile [115]. In an effort to eliminate candidates with poor ADMET properties due to key off-target interactions, such pharmacophore models have been designed for specific anti-targets, such as hERG, cytochrome P450, and P-glycoprotein, and are used for virtual screening [116].

Most of the established computational approaches for designing MTDs are fragment-based methods which, as highlighted in Section 9.4.1, involve fusing, merging, linking of fragments, and scaffold hoping. Automating and scaling such fragment analysis strategies will be of vital importance, as it reduces the time and effort invested by medicinal chemists when design MTDs. For instance, an automated method demonstrating the evolution of known inhibitor to accommodate a polypharmacological or selectivity profile (design-in/design-out) through multi-objective optimization and Bayesian modeling was developed by Besnard et al. [117]. In this study, 75% of designed ligands had the intended multi-target profile while maintaining cellular target engagement. In some cases, integrating computational approaches with an experimental/synthesizing framework has been crucial in identifying MTDs. A *de novo* iterative fragment growing strategy that starts by identifying multi-targeting seeds (common substructure fragments shared among targets of interest) was used to iteratively design, synthesize, and test dual inhibition of cyclooxygenase (Cox) and leukotriene A hydrolase (LTA_4H) [118]. The upcoming section details the use of artificial intelligence (AI)-based generative and binding prediction models to iteratively design MTDs.

9.4.3 AI Models and Their Role in MTDs

The field of drug design is being notably transformed by the advent of AI. Generative modeling leverages deep learning (DL) algorithms to design virtual libraries of compounds with desired properties. These models are trained on large datasets of chemical structures, represented as SMILES strings or molecular graphs, to generate novel molecules with therapeutic potential. The most popular architectures include variational autoencoders (VAEs), adversarial autoencoders (AAEs), generative adversarial networks (GANs), recurrent neural networks (RNNs), and generative pre-trained transformers (GPTs) [119].

VAEs encode molecular structures into a low-dimensional latent space [120]. New molecules are generated by sampling from the latent space distribution and decoding the samples back into molecular representations. AAEs are similar to VAEs but enable more controlled generation of molecular structures by imposing additional constraints on the latent space to better capture the distribution of real molecular data. This is achieved by incorporating a discriminator module alongside the encoder and decoder. The discriminator's role is to differentiate between the latent space's distribution and a predefined prior distribution, ensuring a more refined control over the generated molecular structures. GANs, on the other hand, directly generate new data without an explicit encoding-decoding process. Instead, GANs consist of a generator network that creates molecules, and a discriminator network that evaluates them, iteratively improving the quality of generated molecules in a competitive framework.

RNNs, especially those enhanced with memory cells like long short-term memory units, alongside GPTs exploit the sequential nature of molecular SMILES strings. Each character or element in the SMILES notation, representing an atom or structural motif, is considered a token. Both RNNs and GPTs generate molecules token by token based on a learned chemical grammar. RNNs process each token sequentially, predicting the next token based on previously processed information. In contrast, GPTs, with their transformer architecture, predict the next token by considering the entire sequence context, not just the preceding sequence, offering a more sophisticated approach to capturing complex patterns in the data and enabling more nuanced, contextually aware predictions. Additionally, graph-based generative models leveraging graph convolutional networks and attention mechanisms have gained traction for their ability to represent molecules in a more structurally intuitive manner, directly generating atoms and bonds in a graph.

The integration of these generative models into the early stages of the drug discovery process can significantly accelerate the identification of novel, high-quality small molecules. However, accounting for a specific multi-target profile remains a challenge, with only a few approaches so far attempting to address it. Innovations in this domain include the application of deep generative models to preferentially generate molecules with multiple targets. Blaschke and Bajorath's pioneering work demonstrated the feasibility of steering a publicly available REINVENT model [121], based on RNN with reinforcement learning (RL), towards designing MTDs while reducing the likelihood of producing compounds with single or no active targets. This was achieved through fine-tuning the REINVENT model via transfer learning [122]. In another study, Tan et al. developed an RNN generative model coupled with a multitask deep neural network (MTDNN) discriminative model to design compounds modulating a spectrum of neurotransmitter receptor targets simultaneously. Six analogs of the generated high-scoring compound were experimentally evaluated, and one exhibited not only the desired polypharmacological profile but also demonstrated an antipsychotic effect in animal models [123]. Towards achieving the efficacy-safety balance, DrugEx2.0 serves as an exemplary multi-objective algorithm that utilizes Pareto optimization to design compounds with high activity towards two adenosine receptor protein targets while minimizing affinity for the hERG channel, binding to which causes cardiac toxicity. The algorithm interplays an RNN-based generative model with machine learning activity predictors within an RL framework [124].

In recent years, the use of machine learning for predicting compound-protein interactions has become a highly explored field, with many modern methods capable of simultaneously predicting activities across multiple protein targets and anti-targets. A diverse range of learning algorithms has been proposed for this task, including random forest, kernel methods, gradient boosting machines, and various DL approaches, utilizing, for example, convolutional neural networks or graph neural networks (GNNs) [125, 126]. Compound-protein interaction prediction can be considered a classification or regression task. In classification, the model is trained to distinguish between active and inactive compounds on a given protein target of interest, whereas in regression, the model aims to predict a quantitative value measuring the compound's affinity toward the target, for example, dissociation constant K_d, inhibition constant K_i, or half maximal inhibitory concentration IC_{50}.

Traditional *in silico* approaches to compound-protein activity prediction aim to relate the structural properties of chemical molecules to their activity profiles, known as QSAR models [127–130]. However, most recent models exploit the descriptors of both compounds and protein targets, referred to as proteochemometric models, under the assumption that similar compounds are likely to interact with similar proteins [125, 131, 132]. Various compound and protein descriptors are utilized, including SMILES, molecular fingerprints, and graphs for compounds. Proteins are typically represented by amino acid sequences or, increasingly, by 3D structure-based descriptors or embeddings from pretrained large language models. For example, the recent ConPLex deep learning model [133] represents compounds using Morgan fingerprints, while protein features are derived from a pretrained ProtBert language model [134]. The ConPLex model then calculates compound-protein binding affinity by projecting these features into a shared embedding space and performing a dot product, simultaneously learning to distinguish true drugs from synthetically generated decoys in a contrastive learning framework. Another recent DL model, KDBNet, utilizes GNNs to learn structure representations from the 3D coordinates of compounds and protein binding pockets. The learned representations are then used to predict the binding affinity, including a measure of the prediction's uncertainty [135]. In fact, quantifying the confidence or the uncertainty of the models and their predictions for compound-target activities is increasingly performed using methods such as conformal prediction [136]. This is important when prioritizing which *in silico* designed molecules to explore further and eventually synthesize for experimental studies. Another recent trend is the integration of a variety of protein and compound descriptors in a single model [137]. Such strategy leverages the unique strengths of different descriptor types towards developing more accurate approaches for proteome-wide compound activity prediction that can aid MTD design.

9.5 Perspective, Limitations, and Challenges

In this chapter, we have reviewed the state-of-the-art of rational MTD discovery, with a focus on both computational and experimental approaches to optimize the efficacy-safety balance toward clinical applications. Several examples described in the previous sections demonstrate the enormous potential of MTDs in anticancer drug discovery; however, there also remain critical challenges with this polypharmacological approach, especially how to strike a balance between the improved efficacy and potential toxicity of multi-targeted co-inhibition both in the early development stages and in future clinical applications of the MTDs. Exciting DL models, such as AlphaFold for predicting protein structures based solely on the amino acid sequences [112], and ProGen for generating protein sequences with a predictable function [138], are enabling new systematic approaches for improved protein and drug design across wide protein families. However, these learning algorithms require large amounts of high-quality cellular response data before one can associate protein sequences and structures to experimentally measured drug binding affinities and other phenotypic responses. Such target activity data are still relatively sparse for many protein families and are scattered in many in-house and public data resources. While there are ongoing standardization efforts, such as those implemented in ChEMBL [139] and DrugTargetCommons [140], manual curation and annotation of quantitative drug-target bioactivity data poses still a major bottleneck for scaling up the DL-based drug discovery efforts. Phenotypic data combined with ML models are expected to improve the future efficiency of target-based drug discovery approaches and distinguish between on- and off-target mechanisms [141].

Another challenge in rational MTD discovery is how to best define the efficacy targets to inhibit and the anti-targets to avoid? Identification of maximally selective compound-target pairs is effectively a bi-objective optimization problem in which one tries to simultaneously optimize the balance between the compound's potency against the target(s) of interest (disease target(s)) and the compound's selectivity against other off-targets [142]. The challenge comes from the fact that not only the disease targets, but also the anti-targets may be highly context-specific. Hence, careful computational and experimental investigation of the optimal drug-target space is required for each application case. In anticancer applications, an additional challenge is how to selectively inhibit the mutant protein variants which often drive the disease development or resistance to monotherapies. At the same time, it is crucial to avoid inhibiting the wild-type versions of these proteins, as this often leads to co-inhibition of non-cancerous cells leading to side effects. There are still rather limited amounts of training data available for distinguishing between mutant and wild-type activities, either from cell-based target activity profiling or phenotypic screening, which are required for ML-based approaches. While single-cell sequencing data can help for designing selective co-inhibition patterns that maximize the synergy-efficacy-toxicity balance [143, 144], ideally patient-level information will be needed when going towards clinical applications. However, most preclinical researchers have limited access to real-word efficacy and toxicity data, even if these would be highly useful for first validation and no-go decisions for specific cases where either physician-reported or patient-reported AE information is available.

Despite these limitations and challenges, we anticipate that emerging research will provide novel solutions for rational MTD discovery in the coming years, which will enable the identification of effective and safe treatments for cancers and other complex indications.

References

1 Scott, E.C., Baines, A.C., Gong, Y. et al. (2023). Trends in the approval of cancer therapies by the FDA in the twenty-first century. *Nat. Rev. Drug Discov.* 22 (8): 625–640.

2 Sun, D., Gao, W., Hu, H., and Zhou, S. (2022). Why 90% of clinical drug development fails and how to improve it? *Acta Pharm. Sin. B* 12 (7): 3049–3062.

3 Levine, M.S. and Holland, A.J. (2018). The impact of mitotic errors on cell proliferation and tumorigenesis. *Genes Dev.* 32 (9–10): 620–638.

4 Wang, Q. (2016). Cancer predisposition genes: molecular mechanisms and clinical impact on personalized cancer care: examples of Lynch and HBOC syndromes. *Acta Pharmacol. Sin.* 37 (2): 143–149.

5 Carbone, M., Arron, S.T., Beutler, B. et al. (2020). Tumour predisposition and cancer syndromes as models to study gene–environment interactions. *Nat. Rev. Cancer* 20 (9): 533–549.

6 Hanahan, D. and Weinberg, R.A. (2011). Hallmarks of cancer: the next generation. *Cell* 144 (5): 646–674.

7 Martincorena, I. and Campbell, P.J. (2015). Somatic mutation in cancer and normal cells. *Science* 349 (6255): 1483–1489.
8 Liu, G., Zhan, X., Dong, C., and Liu, L. (2017). Genomics alterations of metastatic and primary tissues across 15 cancer types. *Sci. Rep.* 7 (1): 13262.
9 Sethi, N. and Kang, Y. (2011). Unravelling the complexity of metastasis – molecular understanding and targeted therapies. *Nat. Rev. Cancer* 11 (10): 735–748.
10 Greaves, M. and Maley, C.C. (2012). Clonal evolution in cancer. *Nature* 481 (7381): 306–313.
11 Mayer, S., Milo, T., Isaacson, A. et al. (2023). The tumor microenvironment shows a hierarchy of cell-cell interactions dominated by fibroblasts. *Nat. Commun.* 14 (1): 5810.
12 Bolzán, A.D. and Bianchi, M.S. (2018). DNA and chromosome damage induced by bleomycin in mammalian cells: an update. *Mutat. Res./Rev. Mutat. Res.* 775: 51–62.
13 Hartmann, J., Haap, M., Kopp, H.-G., and Lipp, H.-P. (2009). Tyrosine kinase inhibitors – a review on pharmacology, metabolism and side effects. *CDM* 10 (5): 470–481.
14 Tsimberidou, A.-M. (2015). Targeted therapy in cancer. *Cancer Chemother. Pharmacol.* 76 (6): 1113–1132.
15 Falzone, L., Salomone, S., and Libra, M. (2018). Evolution of cancer pharmacological treatments at the turn of the third millennium. *Front. Pharmacol.* 9: 1300.
16 Zhong, L., Li, Y., Xiong, L. et al. (2021). Small molecules in targeted cancer therapy: advances, challenges, and future perspectives. *Sig. Transduct Target Ther.* 6 (1): 201.
17 Markham, A. (2017). Brigatinib: first global approval. *Drugs* 77 (10): 1131–1135.
18 Friboulet, L., Li, N., Katayama, R. et al. (2014). The ALK inhibitor ceritinib overcomes crizotinib resistance in non–small cell lung cancer. *Cancer Discov.* 4 (6): 662–673.
19 Basit, S., Ashraf, Z., Lee, K., and Latif, M. (2017). First macrocyclic 3rd -generation ALK inhibitor for treatment of ALK/ROS1 cancer: clinical and designing strategy update of lorlatinib. *Eur. J. Med. Chem.* 134: 348–356.
20 Lin, A., Giuliano, C.J., Palladino, A. et al. (2019). Off-target toxicity is a common mechanism of action of cancer drugs undergoing clinical trials. *Sci. Transl. Med.* 11 (509): eaaw8412.
21 Vasan, N., Baselga, J., and Hyman, D.M. (2019). A view on drug resistance in cancer. *Nature* 575 (7782): 299–309.
22 Jin, H., Wang, L., and Bernards, R. (2022). Rational combinations of targeted cancer therapies: background, advances and challenges. *Nat. Rev. Drug Discov.* 1–22.
23 Morphy, J.R. and Harris, C.J. (ed.) (2012). *Designing Multi-Target Drugs*. London: Royal Society of Chemistry.
24 Prahallad, A., Sun, C., Huang, S. et al. (2012). Unresponsiveness of colon cancer to BRAF(V600E) inhibition through feedback activation of EGFR. *Nature* 483 (7387): 100–103.
25 Tabernero, J., Grothey, A., Van Cutsem, E. et al. (2021). Encorafenib plus Cetuximab as a new standard of care for previously treated *BRAF* V600E–mutant metastatic colorectal cancer: updated survival results and subgroup analyses from the BEACON study. *J. Clin. Oncol.* 39 (4): 273–284.
26 Flaherty, K.T., Infante, J.R., Daud, A. et al. (2012). Combined BRAF and MEK inhibition in melanoma with BRAF V600 mutations. *N. Engl. J. Med.* 367 (18): 1694–1703.
27 Fudio, S., Sellers, A., Pérez Ramos, L. et al. (2022). Anti-cancer drug combinations approved by US FDA from 2011 to 2021: main design features of clinical trials and role of pharmacokinetics. *Cancer Chemother. Pharmacol.* 90 (4): 285–299.
28 Anighoro, A., Bajorath, J., and Rastelli, G. (2014). Polypharmacology: challenges and opportunities in drug discovery: miniperspective. *J. Med. Chem.* 57 (19): 7874–7887.
29 Peters, J.-U. (2013). Polypharmacology – Foe or Friend? *J. Med. Chem.* 56 (22): 8955–8971.
30 Garon, S.L., Pavlos, R.K., White, K.D. et al. (2017). Pharmacogenomics of off-target adverse drug reactions. *Brit. J. Clinical Pharma.* 83 (9): 1896–1911.
31 Rudmann, D.G. (2013). On-target and Off-target-based toxicologic effects. *Toxicol. Pathol.* 41 (2): 310–314.
32 Knight, Z.A., Lin, H., and Shokat, K.M. (2010). Targeting the cancer kinome through polypharmacology. *Nat. Rev. Cancer* 10 (2): 130–137.
33 Schieber, M., Crispino, J.D., and Stein, B. (2019). Myelofibrosis in 2019: moving beyond JAK2 inhibition. *Blood Cancer J.* 9 (9): 74.
34 Azhar, M., Kincaid, Z., Kesarwani, M. et al. (2022). Momelotinib is a highly potent inhibitor of FLT3-mutant AML. *Blood Adv.* 6 (4): 1186–1192.
35 Levitzki, A. (1994). Signal-transduction therapy: a novel approach to disease management. *Eur. J. Biochem.* 226 (1): 1–13.
36 Levitzki, A. (2013). Tyrosine kinase inhibitors: views of selectivity, sensitivity, and clinical performance. *Annu. Rev. Pharmacol. Toxicol.* 53 (1): 161–185.

37 Yaish, P., Gazit, A., Gilon, C., and Levitzki, A. (1988). Blocking of EGF-dependent cell proliferation by EGF receptor kinase inhibitors. *Science* 242 (4880): 933–935.

38 Wilson, L.J., Linley, A., Hammond, D.E. et al. (2018). New perspectives, opportunities, and challenges in exploring the human protein kinome. *Cancer Res.* 78 (1): 15–29.

39 Roskoski, R. (2024). Properties of FDA-approved small molecule protein kinase inhibitors: a 2024 update. *Pharmacol. Res.* 200: 107059.

40 Anderson, B., Rosston, P., Ong, H.W. et al. (2023). How many kinases are druggable? A review of our current understanding. *Biochem. J.* 480 (16): 1331–1363.

41 Huang, D., Zhou, T., Lafleur, K. et al. (2010). Kinase selectivity potential for inhibitors targeting the ATP binding site: a network analysis. *Bioinformatics* 26 (2): 198–204.

42 Van Linden, O.P.J., Kooistra, A.J., Leurs, R. et al. (2014). KLIFS: a knowledge-based structural database to navigate kinase–ligand interaction space. *J. Med. Chem.* 57 (2): 249–277.

43 Davis, M.I., Hunt, J.P., Herrgard, S. et al. (2011). Comprehensive analysis of kinase inhibitor selectivity. *Nat. Biotechnol.* 29 (11): 1046–1051.

44 Elkins, J.M., Fedele, V., Szklarz, M. et al. (2016). Comprehensive characterization of the published kinase inhibitor set. *Nat. Biotechnol.* 34 (1): 95–103.

45 Choueiri, T.K., Escudier, B., Powles, T. et al. (2016). Cabozantinib versus everolimus in advanced renal cell carcinoma (METEOR): final results from a randomised, open-label, phase 3 trial. *Lancet Oncol.* 17 (7): 917–927.

46 Daver, N., Schlenk, R.F., Russell, N.H., and Levis, M.J. (2019). Targeting FLT3 mutations in AML: review of current knowledge and evidence. *Leukemia* 33 (2): 299–312.

47 Hexner, E.O., Serdikoff, C., Jan, M. et al. (2008). Lestaurtinib (CEP701) is a JAK2 inhibitor that suppresses JAK2/STAT5 signaling and the proliferation of primary erythroid cells from patients with myeloproliferative disorders. *Blood* 111 (12): 5663–5671.

48 Batchelor, T.T., Duda, D.G., Di Tomaso, E. et al. (2010). Phase II study of cediranib, an oral pan-vascular endothelial growth factor receptor tyrosine kinase inhibitor, in patients with recurrent glioblastoma. *JCO* 28 (17): 2817–2823.

49 Doebele, R.C., Drilon, A., Paz-Ares, L. et al. (2020). Entrectinib in patients with advanced or metastatic NTRK fusion-positive solid tumours: integrated analysis of three phase 1–2 trials. *Lancet Oncol.* 21 (2): 271–282.

50 Di Costanzo, G.G., De Stefano, G., Tortora, R. et al. (2015). Sorafenib off-target effects predict outcomes in patients treated for hepatocellular carcinoma. *Future Oncol.* 11 (6): 943–951.

51 Zheng, W., Zhao, Y., Luo, Q. et al. (2017). Multi-targeted anticancer agents. *Curr. Top. Med. Chem.* 17: 28.

52 Luan, Y., Li, J., Bernatchez, J.A., and Li, R. (2019). Kinase and histone deacetylase hybrid inhibitors for cancer therapy. *J. Med. Chem.* 62 (7): 3171–3183.

53 Fu, R., Sun, Y., Sheng, W., and Liao, D. (2017). Designing multi-targeted agents: an emerging anticancer drug discovery paradigm. *Eur. J. Med. Chem.* 136: 195–211.

54 Galloway, T.J., Wirth, L.J., Colevas, A.D. et al. (2015). A phase i study of CUDC-101, a multitarget inhibitor of HDACs, EGFR, and HER2, in combination with chemoradiation in patients with head and neck squamous cell carcinoma. *Clin. Cancer Res.* 21 (7): 1566–1573.

55 Jabbour, E., Deininger, M., and Hochhaus, A. (2011). Management of adverse events associated with tyrosine kinase inhibitors in the treatment of chronic myeloid leukemia. *Leukemia* 25 (2): 201–210.

56 Eschenhagen, T. (2018). Exaggerated cardiotoxicity of sunitinib in stressed 3-dimensional heart muscles. *JACC Basic Trans. Sci.* 3 (2): 277–279.

57 Kerkela, R., Woulfe, K.C., Durand, J. et al. (2009). Sunitinib-induced cardiotoxicity is mediated by off-target inhibition of amp-activated protein kinase. *Clin. Transl. Sci.* 2 (1): 15–25.

58 Ancker, O., Wehland, M., Bauer, J. et al. (2017). The adverse effect of hypertension in the treatment of thyroid cancer with multi-kinase inhibitors. *Int. J. Mol. Sci.* 18 (3): 625.

59 Shyam Sunder, S., Sharma, U.C., and Pokharel, S. (2023). Adverse effects of tyrosine kinase inhibitors in cancer therapy: pathophysiology, mechanisms and clinical management. *Sig.: Transduct. Target Ther.* 8 (1): 262.

60 Manchado, E., Huang, C.-H., Tasdemir, N. et al. (2016). A pipeline for drug target identification and validation. *Cold Spring Harb. Symp. Quant. Biol.* 81: 257–267.

61 Hahn, W.C., Bader, J.S., Braun, T.P. et al. (2021). An expanded universe of cancer targets. *Cell* 184 (5): 1142–1155.

62 Radoux, C.J., Vianello, F., McGreig, J. et al. (2022). The druggable genome: twenty years later. *Front. Bioinform.* 2.

63 Weinstein, J.N., Collisson, E.A., Mills, G.B. et al. (2013). The cancer genome atlas pan-cancer analysis project. *Nat. Genet.* 45 (10): 1113–1120.

64 Tsherniak, A., Vazquez, F., Montgomery, P.G. et al. (2017). Defining a cancer dependency map. *Cell* 170 (3): 564–576.e16.

65 Thomas, A.M., Serafini, M., Grant, E.K. et al. (2023). Mutate and conjugate: a method to enable rapid in-cell target validation. *ACS Chem. Biol.* 18 (11): 2405–2417.

66 Subbiah, V., Baik, C., and Kirkwood, J.M. (2020). Clinical development of BRAF plus MEK inhibitor combinations. *Trends Cancer.* 6 (9): 797–810.

67 Kun, E., Tsang, Y.T.M., Ng, C.W. et al. (2021). MEK inhibitor resistance mechanisms and recent developments in combination trials. *Cancer Treat. Rev.* 92: 102137.

68 Lee, L., Hizukuri, Y., Severson, P. et al. (2019). A novel combination regimen of BET and FLT3 inhibition for FLT3-ITD acute myeloid leukemia. *Blood* 134 (Supplement_1): 1373.

69 Donati, B., Lorenzini, E., and Ciarrocchi, A. (2018). BRD4 and cancer: going beyond transcriptional regulation. *Mol. Cancer* 17 (1): 164.

70 Takahashi, S. (2011). Downstream molecular pathways of FLT3 in the pathogenesis of acute myeloid leukemia: biology and therapeutic implications. *J. Hematol. Oncol.* 4: 13.

71 Fiskus, W., Sharma, S., Qi, J. et al. (2014). BET protein antagonist JQ1 is synergistically lethal with FLT3 tyrosine kinase inhibitor (TKI) and overcomes resistance to FLT3-TKI in AML cells expressing FLT-ITD. *Mol. Cancer Therap.* 13 (10): 2315–2327.

72 Schwalm, M.P. and Knapp, S. (2022). BET bromodomain inhibitors. *Curr. Opin. Chem. Biol.* 68: 102148.

73 Yi, M., Zheng, X., Niu, M. et al. (2022). Combination strategies with PD-1/PD-L1 blockade: current advances and future directions. *Mol. Cancer* 21 (1): 28.

74 Rini, B.I., Plimack, E.R., Stus, V. et al. (2019). Pembrolizumab plus axitinib versus sunitinib for advanced renal-cell carcinoma. *N. Engl. J. Med.* 380 (12): 1116–1127.

75 Sawasdikosol, S. and Burakoff, S. (2020). A perspective on HPK1 as a novel immuno-oncology drug target. *elife* 9: e55122.

76 Riese, M.J., Moon, E.K., Johnson, B.D., and Albelda, S.M. (2016). Diacylglycerol kinases (DGKs): novel targets for improving T cell activity in cancer. *Front. Cell Dev. Biol.* 4, e912.

77 Akimov, Y. and Aittokallio, T. (2021). Re-defining synthetic lethality by phenotypic profiling for precision oncology. *Cell Chem. Biol.* 28 (3): 246–256.

78 Kaelin, W.G. (2005). The concept of synthetic lethality in the context of anticancer therapy. *Nat. Rev. Cancer* 5 (9): 689–698.

79 Ashworth, A., Lord, C.J., and Reis-Filho, J.S. (2011). Genetic interactions in cancer progression and treatment. *Cell* 145 (1): 30–38.

80 Benada, J., Bulanova, D., Azzoni, V. et al. (2023). Synthetic lethal interaction between WEE1 and PKMYT1 is a target for multiple low-dose treatment of high-grade serous ovarian carcinoma, *NAR. Cancer* 5, 3: zcad029.

81 Chen, D., An, X., Ouyang, X. et al. (2019, 2019). In vivo pharmacology models for cancer target research. In: *Target Identification and Validation in Drug Discovery: Methods and Protocols* (ed. J. Moll and S. Carotta), 183–211. New York, NY: Springer.

82 Priest, B., Bell, I.M., and Garcia, M. (2008). Role of hERG potassium channel assays in drug development. *Channels* 2 (2): 87–93.

83 Guengerich, F.P. (2021). A history of the roles of cytochrome P450 enzymes in the toxicity of drugs. *Toxicol Res.* 37 (1): 1–23.

84 Sutherland, J.J., Yonchev, D., Fekete, A., and Urban, L. (2023). A preclinical secondary pharmacology resource illuminates target-adverse drug reaction associations of marketed drugs. *Nat. Commun.* 14 (1): 4323.

85 Kuhn, M., Letunic, I., Jensen, L.J., and Bork, P. (2016). The SIDER database of drugs and side effects. *Nucleic Acids Res.* 44 (D1): D1075–D1079.

86 Lee, L.Y., Hernandez, D., Rajkhowa, T. et al. (2017). Preclinical studies of gilteritinib, a next-generation FLT3 inhibitor. *Blood* 129 (2): 257–260.

87 Warkentin, A.A., Lopez, M.S., Lasater, E.A. et al. Overcoming myelosuppression due to synthetic lethal toxicity for FLT3-targeted acute myeloid leukemia therapy. *elife* 3: e03445.

88 Perl, A.E., Lu, Q., Fan, A. et al. (2020). Clinical outcomes following treatment with gilteritinib or quizartinib in patients with relapsed/refractory FLT3-ITD+ acute myeloid leukemia. *Blood* 136 (Supplement 1): 42–43.

89 Kang, D., Ludwig, E., Jaworowicz, D. et al. (2021). Concentration–QTc analysis of quizartinib in patients with relapsed/refractory acute myeloid leukemia. *Cancer Chemother. Pharmacol.* 87 (4): 513–523.

90 Wermuth, C.G., Aldous, D., Raboisson, P., and Rognan, D. (2015, 2015). *The Practice of Medicinal Chemistry*, 4the. Academic Press.

91 Morphy, R. and Rankovic, Z. (2009). Designing multiple ligands – medicinal chemistry strategies and challenges. *Curr. Pharm. Des.* 15 (6): 587–600.

92 Cavalli, A., Bolognesi, M.L., Minarini, A. et al. (2008). Multi-target-directed ligands to combat neurodegenerative diseases. *J. Med. Chem.* 51 (3): 347–372.

93 Wilhelm, S., Carter, C., Lynch, M. et al. (2006). Discovery and development of sorafenib: a multikinase inhibitor for treating cancer. *Nat. Rev. Drug Discov.* 5 (10): 835–844.

94 Yao, L., Mustafa, N., Tan, E.C. et al. (2017). Design and synthesis of ligand efficient dual inhibitors of Janus Kinase (JAK) and Histone Deacetylase (HDAC) based on ruxolitinib and vorinostat. *J. Med. Chem.* 60 (20): 8336–8357.

95 Soltan, O.M., Shoman, M.E., Abdel-Aziz, S.A. et al. (2021). Molecular hybrids: a five-year survey on structures of multiple targeted hybrids of protein kinase inhibitors for cancer therapy. *Eur. J. Med. Chem.* 225: 113768.

96 Li, X., Li, X., Liu, F. et al. (2021). Rational multi-targeted drug design strategy from the perspective of a medicinal chemist. *J. Med. Chem.* 64 (15): 10581–10605.

97 Katsoulas, A., Rachid, Z., McNamee, J.P. et al. (2008). Combi-targeting concept: an optimized single-molecule dual-targeting model for the treatment of chronic myelogenous leukemia. *Mol. Cancer Therap.* 7 (5): 1033–1043.

98 Mahboobi, S., Dove, S., Sellmer, A. et al. (2009). Design of chimeric histone deacetylase- and tyrosine kinase-inhibitors: a series of imatinib hybrides as potent inhibitors of wild-type and mutant BCR-ABL, PDGF-Rbeta, and histone deacetylases. *J. Med. Chem.* 52 (8): 2265–2279.

99 Wang, J., Li, H., He, G. et al. (2020). Discovery of novel dual poly(ADP-ribose)polymerase (PARP) and Phosphoinositide 3-kinase (PI3K) Inhibitors as a promising strategy for cancer therapy. *J. Med. Chem.* 63: 122–163.

100 Hu, Y., Stumpfe, D., and Bajorath, J. (2016). Recent advances in scaffold hopping. *J. Med. Chem.* 60 (4): 1238–1246.

101 Proschak, E., Stark, H., and Merk, D. (2019). Polypharmacology by design: a medicinal chemist's perspective on multitargeting compounds. *J. Med. Chem.* 62 (2): 420–444.

102 Albrecht, W., Unger, A., Bauer, S.M., and Laufer, S.A. (2017). Discovery of *N* -{4-[5-(4-fluorophenyl)-3-methyl-2-methylsulfanyl-3 *H* -imidazol-4-yl]-pyridin-2-yl}-acetamide (CBS-3595), a dual p38α MAPK/PDE-4 inhibitor with activity against TNFα-related diseases. *J. Med. Chem.* 60 (13): 5290–5305.

103 Morphy, R. and Rankovic, Z. (2005). Designed multiple ligands. An emerging drug discovery paradigm. *J. Med. Chem.* 48 (21): 6523–6543.

104 Speck-Planche, A. (2019). Multicellular target QSAR model for simultaneous prediction and design of anti-pancreatic cancer agents. *ACS Omega* 4 (2): 3122–3132.

105 Halder, A.K. and Dias Soeiro Cordeiro, M.N. (2021). QSAR-Co-X: an open source toolkit for multitarget QSAR modelling. *J. Cheminform.* 13 (1): 29.

106 Halder, A.K. and Cordeiro, M.N.D.S. (2019). Development of multi-target chemometric models for the inhibition of class I PI3K enzyme isoforms: a case study using QSAR-Co Tool. *IJMS* 20 (17): 4191.

107 Paolini, G.V., Shapland, R.H.B., Van Hoorn, W.P. et al. (2006). Global mapping of pharmacological space. *Nat. Biotechnol.* 24 (7): 805–815.

108 Hopkins, A.L. (2008). Network pharmacology: the next paradigm in drug discovery. *Nat. Chem. Biol.* 4 (11): 682–690.

109 Lounkine, E., Keiser, M.J., Whitebread, S. et al. (2012). Large-scale prediction and testing of drug activity on side-effect targets. *Nature* 486 (7403): 361–367.

110 Ravikumar, B., Timonen, S., Alam, Z. et al. (2019). Chemogenomic analysis of the druggable kinome and its application to repositioning and lead identification studies. *Cell Chem. Biol.* 26 (11): 1608–1622.e6.

111 Ravikumar, B. and Aittokallio, T. (2018). Improving the efficacy-safety balance of polypharmacology in multi-target drug discovery. *Expert Opin. Drug Discov.* 13 (2): 179–192.

112 Jumper, J., Evans, R., Pritzel, A. et al. (2021). Highly accurate protein structure prediction with AlphaFold. *Nature* 596 (7873): 583–589.

113 Mirdita, M., Schütze, K., Moriwaki, Y. et al. (2022). ColabFold: making protein folding accessible to all. *Nat. Methods* 19 (6): 679–682.

114 Wang, X., Pan, C., Gong, J. et al. (2016). Enhancing the enrichment of pharmacophore-based target prediction for the polypharmacological profiles of drugs. *J. Chem. Inf. Model.* 56 (6): 1175–1183.

115 Pérez-Nueno, V.I., Karaboga, A.S., Souchet, M., and Ritchie, D.W. (2014). GES polypharmacology fingerprints: a novel approach for drug repositioning. *J. Chem. Inf. Model.* 54 (3): 720–734.

116 Thai, K.-M., Ngo, T.-D., Tran, T.-D., and Le, M.-T. (2013). Pharmacophore modeling for antitargets. *Curr. Top. Med. Chem.* 13 (9): 1002–1014.

117 Besnard, J., Ruda, G.F., Setola, V. et al. (2012). Automated design of ligands to polypharmacological profiles. *Nature* 492 (7428): 215–220.

118 Shang, E., Yuan, Y., Chen, X. et al. (2014). *De Novo* design of multitarget ligands with an iterative fragment-growing strategy. *J. Chem. Inf. Model.* 54 (4): 1235–1241.

119 Bilodeau, C., Jin, W., Jaakkola, T. et al. (2022). Generative models for molecular discovery: recent

advances and challenges. *WIREs Computat. Mol. Sci.* 12 (5): e1608.

120 Ochiai, T., Inukai, T., Akiyama, M. et al. (2023). Variational autoencoder-based chemical latent space for large molecular structures with 3D complexity. *Commun. Chem.* 6 (1): 1–14.

121 Blaschke, T., Arús-Pous, J., Chen, H. et al. (2020). Reinvent 2.0: an AI tool for de novo drug design. *J. Chem. Inf. Model.* 60 (12): 5918–5922.

122 Blaschke, T. and Bajorath, J. (2022). Fine-tuning of a generative neural network for designing multi-target compounds. *J. Comput. Aided Mol. Des.* 36 (5): 363–371.

123 Tan, X., Jiang, X., He, Y. et al. (2020). Automated design and optimization of multitarget schizophrenia drug candidates by deep learning. *Eur. J. Med. Chem.* 204: 112572.

124 Liu, X., Ye, K., Van Vlijmen, H.W.T. et al. (2021). DrugEx v2: de novo design of drug molecules by Pareto-based multi-objective reinforcement learning in polypharmacology. *J. Cheminform.* 13 (1): 85.

125 Rifaioglu, A.S., Atas, H., Martin, M.J. et al. (2019). Recent applications of deep learning and machine intelligence on in silico drug discovery: methods, tools and databases. *Brief. Bioinform.* 20 (5): 1878–1912.

126 Cichońska, A., Ravikumar, B., Allaway, R.J. et al. (2021). Crowdsourced mapping of unexplored target space of kinase inhibitors. *Nat. Commun.* 12 (1): 3307.

127 Brown, A.C. and Fraser, T.R. (1868). On the connection between chemical constitution and physiological action; with special reference to the physiological action of the salts of the ammonium bases derived from strychnia, Brucia, Thebaia, Codeia, Morphia, and Nicotia. *J. Anat. Physiol.* 2 (2): 224–242.

128 Tropsha, A. (2010). Best practices for QSAR model development, validation, and exploitation. *Mol. Inform.* 29 (6–7): 476–488.

129 Cherkasov, A., Muratov, E.N., Fourches, D. et al. (2014). QSAR modeling: where have you been? Where are you going to? *J. Med. Chem.* 57 (12): 4977–5010.

130 Tropsha, A., Isayev, O., Varnek, A. et al. (2023). Integrating QSAR modelling and deep learning in drug discovery: the emergence of deep QSAR. *Nat. Rev. Drug Discov.* 1–15.

131 Cortés-Ciriano, I., Ain, Q.U., Subramanian, V. et al. (2015). Polypharmacology modelling using proteochemometrics (PCM): recent methodological developments, applications to target families, and future prospects. *Med. Chem. Commun.* 6 (1): 24–50.

132 Bongers, B.J., Ijzerman, A.P., and Van Westen, G.J.P. (2019). Proteochemometrics – recent developments in bioactivity and selectivity modeling. *Drug Discov. Today Technol.* 32–33: 89–98.

133 Singh, R., Sledzieski, S., Bryson, B. et al. (2023). Contrastive learning in protein language space predicts interactions between drugs and protein targets. *Proc. Natl. Acad. Sci.* 120 (24): e2220778120.

134 Elnaggar, A., Heinzinger, M., Dallago, C. et al. (2022). ProtTrans: toward understanding the language of life through self-supervised learning. *IEEE Trans. Pattern Anal. Mach. Intell.* 44 (10): 7112–7127.

135 Luo, Y., Liu, Y., and Peng, J. (2023). Calibrated geometric deep learning improves kinase–drug binding predictions. *Nat. Mach. Intell.* 1–12.

136 Alvarsson, J., McShane, S.A., Norinder, U., and Spjuth, O. (2021). Predicting with confidence: using conformal prediction in drug discovery. *J. Pharm. Sci.* 110 (1): 42–49.

137 Zhu, Y., Zhao, L., Wen, N. et al. (2023). DataDTA: a multi-feature and dual-interaction aggregation framework for drug-target binding affinity prediction. *Bioinformatics* 39 (9): btad560.

138 Madani, A., Krause, B., Greene, E.R. et al. (2023). Large language models generate functional protein sequences across diverse families. *Nat. Biotechnol.* 41 (8): 1099–1106.

139 Zdrazil, B., Felix, E., Hunter, F. et al. (2024). The ChEMBL database in 2023: a drug discovery platform spanning multiple bioactivity data types and time periods. *Nucleic Acids Res.* 52 (D1): D1180–D1192.

140 Tang, J., Tanoli, Z.-R., Ravikumar, B. et al. (2018). Drug target commons: a community effort to build a consensus knowledge base for drug-target interactions. *Cell Chem. Biol.* 25 (2): 224–229.e2.

141 Sadri, A. (2023). Is target-based drug discovery efficient? Discovery and “off-target” mechanisms of all drugs. *J. Med. Chem.* 66 (18): 12651–12677.

142 Wang, T., Pulkkinen, O.I., and Aittokallio, T. (2022). Target-specific compound selectivity for multi-target drug discovery and repurposing. *Front. Pharmacol.* 13: 1003480.

143 Ianevski, A., Lahtela, J., Javarappa, K.K. et al. (2021). Patient-tailored design for selective co-inhibition of leukemic cell subpopulations. *Sci. Adv.* 7 (8): eabe4038.

144 Sheridan, C. (2024). Can single-cell biology realize the promise of precision medicine? *Nat. Biotechnol.* 42 (2): 159–162.

10

Polypharmacology

A Return to the Origins of Antibacterial Discovery

Lynn L. Silver

10.1 Introduction

Most registered antibiotics (including drugs derived from both natural and synthetic compounds) are not directed toward protein targets. They almost always target more than one protein, multiple sites on a protein, or the products of multiple genes. In the last case, this can be a structure that is synthesized by a pathway of macromolecular synthesis (such as peptidoglycan or one of its intermediates) or the multiple copies of identical genes, such as ribosomal ribonucleic acid (rRNA), where hitting more than one gene product is necessary to cause an inhibitory effect [1]. Two major systemic antibiotic classes targeting proteins are β-lactams and fluoroquinolones (FQs), which hit multiple penicillin-binding proteins (PBPs) and the bacterial type II topoisomerases, deoxyribonucleic acid (DNA) gyrase, and topoisomerase IV, respectively.

Most classes of registered antibiotics were discovered via phenotypic whole-cell screening, with subsequent identification of their molecular targets. The last classes of registered antibiotics were discovered in the early 1980s [2]. Over time, resistance arose to all these classes, but it did not occur overnight, as can happen with pre-existing single chromosomal mutations. The success of these classes in monotherapy can be correlated with their low frequencies of single-step resistance, which, as I have posited [1], was due to their multitargeting. No antibiotic selecting for overnight resistance in the clinic would be developable for monotherapy. While multitargeting is insufficient to avoid rapid resistance development, it may be necessary.

My chapter in the previous edition of this book was entitled "Polypharmacology as an Emerging Trend in Antibacterial Discovery." I've come to call this iteration "Polypharmacology, a Return to the Origins of Antibacterial Discovery." Our initial antibiotic armamentarium exploited polypharmacology serendipitously. A purposeful reemphasis of it could be beneficial.

10.2 The Failure of Single-target-based Discovery of Antibiotics

By the 1980s, the previously productive search for new, developable antibiotics began to wane. With public recognition of increasing antimicrobial resistance (AMR), pressure to develop novel entities increased as well. The older, phenotypic screening methods of discovery being unproductive, the industry switched direction. The advent of DNA sequencing of bacteria in 1995 [3] led to the focus of antibiotic discovery on screening for or synthesizing inhibitors of the genomically identified and enticing single-gene targets shown to be essential for growth and survival. Much of Big Pharma directed its antibiotic discovery programs to this targeted approach for at least fifteen years. Funders like the National Institutes of Health (NIH) favored this strategy as well. Indeed, many *in vitro* inhibitors of bacterial enzymes were discovered, but only a subset inhibited bacterial growth. This was due mainly to their inability to reach the intracellular molecules they targeted. Nevertheless, numerous antibacterial enzyme inhibitors were disclosed – many targeting their proposed binding sites (i.e., validated as hitting the proposed target). Where measured, most were subject to single mutations leading to decreased susceptibility (raising minimum inhibitory concentration [MIC] at least 16-fold) [4]. With multitargeted agents, stepwise resistance can arise through multiple endogenous (in the pathogen) mutations, sometimes in the target, in other functions affecting compound accumulation at the target site, or in regulating expression. Resistance can result where the permeation of a periplasmic- or cytoplasmic-targeted agent requires specific uptake pathways or efflux avoidance. This is especially true in Gram-negative bacteria with their duplex membranes. Such loss of uptake or enhancement of efflux [5] can lead to resistance via single mutations. On the other hand, single gene products (the canonical

Polypharmacology: Strategies for Multi-Target Drug Discovery, First Edition. Edited by Jens-Uwe Peters.

drug target) can have a low likelihood of mutating to a resistant state if such mutations lead to reduced fitness of the organism. Yet, as Dan Andersson posits, compensatory mutations can often arise that allow survival of isolates with the mutated function [6]. Most AMR, however, involves the import of genes from environmental sources by HGT. HGT occurs over years, while endogenous chromosomal mutational resistance can pre-exist in pathogens before drug selection and can appear overnight once challenged.

Thus, the strategy of pursuing unexploited targets to fight AMR by avoiding cross-resistance with standard-of-care (SoC) agents had largely overlooked the requirement that new entities themselves must avoid rapid resistance development. Furthermore, a lack of understanding of possible rules of bacterial permeability hampered the success of the *in vitro*, biochemical target-based approach. Indeed, this antibacterial discovery strategy has been considered a failure [2, 7–12]. Often reported as being due to financial and marketing considerations, much of Big Pharma left the antibacterial discovery area; however, the extended period of low productivity was a serious contributing factor.

New discovery scenarios have been explored, including biosynthetic prospecting (exploiting the genome sequencing of antibiotic-producing organisms to find and optimize novel, often unexpressed, natural products), the pursuit of antivirulence functions, and a return to phenotypic screening. While proponents of target-based discovery have argued that primary screening for whole-cell (kill-the-bug) activity made the ultimate assignment of the molecular targets difficult, new technologies have been developed to better identify them. Purposeful multitargeting has been part of these efforts, as discussed below. Multitargeting aims not only to inactivate more than one target molecule but also to avoid the possibility of a single mutational event leading to significant resistance. It may take the form of compounds having multiple binding sites on a single-target molecule that a single mutational change cannot alleviate. Peptides could make such contacts – and they are being increasingly explored.

Since the publication of my previous chapter [13], a variety of authors discussing antibacterial discovery have focused on or included in a broader review the area of polypharmacology or multitargeting in antibacterials [14–23]. This chapter will cover some intentional efforts toward multitargeting and output from more serendipitous screens that revealed polypharmacological compounds of interest.

10.3 Attempts at Purposeful Multitargeting

10.3.1 Mur Ligases and a Caveat

The peptidoglycan synthesis pathway has been recognized as a target of interest since the finding that penicillin and beta-lactams inhibit cell wall synthesis. In the genomic age of choosing enzymatic targets, the cytoplasmically located members of the Mur pathway, which produces the pentapeptide intermediate for incorporation into the peptidoglycan complex, have been a focus of discovery. There have been efforts toward finding inhibitors, *in vitro*, of single enzymes of the pathway, as reviewed in [24–26]. While a variety of inhibitors of the Mur ligases, MurC–MurF, has been found, only a few have antibacterial activity – and in only one case (MurC inhibition in permeable *Pseudomonas aeruginosa* and *Escherichia coli* [27]) has that activity been shown to be due solely to inhibition of the targeted enzyme. It should also be noted that MurF inhibitors (diarylquinolines) have been produced that do show antibacterial activity in permeable *E. coli*, at least in part related to proven inhibition of peptidoglycan synthesis [28]. Some Mur ligase inhibitors with antibacterial activity have been shown, instead, to cause membrane damage [29].

As noted in the above reviews, efforts at finding single inhibitors led to the discovery that certain chemical platforms yielded inhibitors of multiple enzymes. This is not surprising in that the mechanism for amino acid addition by the sequential Mur ligase enzymes is more or less conserved [30]. Programs aimed at multitargeting – to reduce resistance or gain potency – have been undertaken [24, 31]. As with most other attempts, none of these had antibacterial activity demonstrated to be due to the inhibition of peptidoglycan synthesis. It is likely that the inhibitors are unable to enter the bacterial cytoplasm – as the pathway intermediates are rich in phosphates and nucleotides. For example, inhibitors designed as transition state analogs are unlikely to penetrate the cytoplasmic membrane without specific permeases. Fosfomycin, a MurA inhibitor that is a useful antibiotic, takes advantage of such permeases for its uptake. There is further speculation about other reasons for the lack of validated antibacterial activity of Mur ligase inhibitors [30].

In enzyme-based antibacterial discovery, where inhibitors are evaluated by *in vitro* biochemical assays, such as the search for Mur ligase inhibitors, it is common that antibacterial activity seen is not due to inhibition of the *in vitro* identified target, but is due to nonspecific mechanisms (such as membrane perturbation). Killing bacteria by nonspecific mechanisms is very common, and evidence should be sought to exclude such mechanisms. To claim that target-based discovery has produced an antibacterial compound requires validation to formally tie the antibacterial activity to *in vitro* target inhibition. This is necessary to support chemical design and optimization and to reduce the likelihood of off-target toxicity.

One validation method is to select for resistant mutations and show them to be resident in the desired target gene, but this is not always straightforward. Nontarget-based mutations, such as decreased permeability or increased efflux may greatly outnumber target mutations. In cases of possible polypharmacology, single mutations may not lead to increased MICs. This may be a desirable attribute, but it is more likely a sign that nonspecific inhibition or multitargeting is occurring. Other phenotypic methods can be used [2, 32], including measurement of macromolecular synthesis [33] and various omics approaches. While it is certainly not necessary for an antibiotic to have a single mechanism of action (the opposite is desirable), any contemplation of this discovery area requires attention to true target validation.

10.3.2 Protein Synthesis Inhibition by Ribosomal RNA Binders

As noted above, the many protein synthesis inhibiting antibiotics that bind to ribosomal RNA (rRNA) display a sort of polypharmacology in that almost all pathogens have multiple copies of the rRNA genes (save *Mycobacterium tuberculosis* with one copy and *Helicobacter pylori* with two copies) and mutations in single copies are recessive. Resistance to these antibiotics usually requires horizontally transmitted mechanisms, modifying either the antibiotics or their binding sites on the ribosome. Lincosamides (among other classes) bind to the peptidyl transferase center (PTC) on the 60S ribosomal subunit. Various rRNA methylases (such as *cfr* and *erm* gene products [34]) and ribosome-binding proteins (such as LsaA [35]) block the binding of the registered lincosamide, clindamycin (CLI) (Figure 10.1a), and other antibiotics in the PTC. Synthetic lincosamides were made by a collaboration between the Polikanov group at the University of Chicago and the Myers group at Harvard, with the aim of adding additional moieties capable of restoring rRNA binding to such resistant ribosomes. One such compound, iboxamycin (IBX) (Figure 10.1b), made by combining an oxepanoproline unit

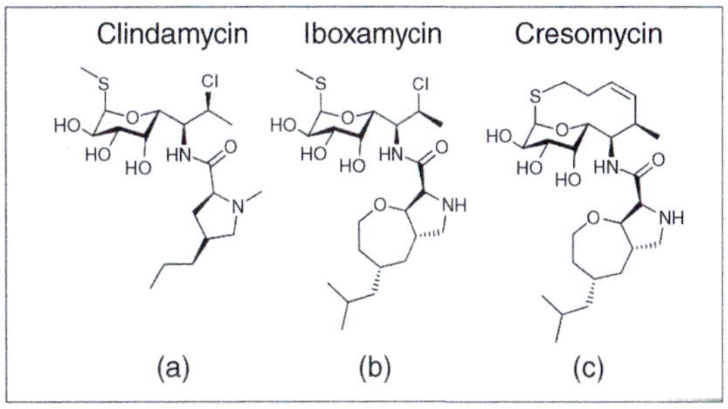

Figure 10.1 Structures of lincosamides of increasing potency and activity against resistant ribosomes. (a) Clindamycin (b) iboxamycin, and (c) cresomycin.

with the aminooctose moiety of CLI, had broad spectrum activity and was highly active against CLI-resistant strains including those encoding Cfr, Erm, and LsaA [36]. Structural studies indicate that the alterations afford tighter binding that overcomes these resistance mechanisms. IBX is efficacious in treating murine thigh infections by *Staphylococcus aureus*, *Streptococcus pyogenes*, *Acinetobacter baumannii*, and *E. coli*. The Gram-negative activity of IBX is not seen with CLI. The same academic groups continued this work, arriving at a further improved product, cresomycin (CRM) (Figure 10.1c), a synthetic bridged macrobicyclic compound [37]. This was achieved by optimizing better ribosome binding by preorganization and conformational restriction. Preorganization is a design concept [38, 39] based on the modeling of the host (receptor) and guest (inhibitor) to maximize the complementarity of binding components. Structural constraint, in this case, was achieved through strategic cyclization. Interestingly, none of the atoms within the macrobicyclic moiety of CRM contact the ribosome. The constrained structure influences the conformation of the molecule such that interactions between other parts of the "guest" and the ribosome "host" are enhanced. MICs of CLI, IBX, and CRM against a collection of clinical isolates, including a variety of resistant strains, showed increasing potency in the order CLI < IBX < CRM [37]. For example, CLI, IBX, and CRM have MICs on strain *S. aureus* AR-0693 *ermA* of >32, 4, and 0.125, respectively.

10.3.3 tRNA Synthetases as Targets

Mupirocin, a narrow-spectrum natural product topical agent, inhibits bacterial isoleucyl transfer RNA (tRNA) synthetase (*IleS*) and is used as a topical agent for nasal decolonization of *S. aureus* with high-dosage

(2%) ointment. Mutations in the *ileS* gene give rise to low-level resistance, while high-level resistance is due to the plasmid-borne *mupA* gene encoding a modified isoleucine tRNA synthetase [40]. There have been a variety of programs aimed at the discovery of inhibitors of tRNA synthetases [41–43]. A synthetic oxaborole inhibitor of leucyl tRNA synthetase, GSK2251052, was a clinical candidate, but it failed in phase II trials due to rapid high-level resistance development [44]. This was possibly the first broadly acknowledged case where a single-targeted agent failed in the clinic for that reason. Interestingly, a related compound, GSK3036656 (ganfeborole), is being developed for use against *M. tuberculosis* [45], where therapy is always by combinations of drugs, thus retarding resistance development. Several groups, acting on the idea of multi-targeting to prevent or retard rapid resistance, attempted to find inhibitors of more than one bacterial tRNA synthetase [20, 42]. Thus far, this approach has not proved fruitful. Often, expanding the spectrum of inhibition of these enzymes leads to unwanted activity against mammalian tRNA synthetases [42, 46].

10.3.4 New Topoisomerase Inhibitors

The bacterial type II topoisomerases have two pairs of identical subunits: the α subunits that interact with DNA, catalyzing strand breakage, covalent self-attachment to DNA, and rejoining of the topologically altered (supercoiled or decatenated) DNA; and the β-subunits that bind adenosine triphosphate (ATP) to transduce energy required for strand passage to form negative supercoils (gyrase) or decatenate daughter chromosomes (topoisomerase IV). In *E. coli*, the genes for these proteins are referred to as *gyrA* and *gyrB* for DNA gyrase and *parC* and *parE* for the α- and β-subunits of topoisomerase IV. This pair of targets is the prime example of antibacterial polypharmacology in resistance reduction: two essential enzymes with very similar active site motifs that can be inhibited by a single pharmacophore.

The FQs, such as ciprofloxacin, inhibit both enzymes. In *E. coli*, mutations are seen first in DNA gyrase (*gyrA*), usually at S83L, as it is the more sensitive target. Secondary mutations leading to increased efflux often arise. Further MIC increase can occur with mutations in *parC* (Topo IV), primarily at S80I. Residues *gyrA* S83 and *parC* S80 each coordinate a water molecule that participates in a bridge with Mg^{++} chelated by the FQ, mediating the interaction of drug and enzyme [47]. If dual inhibitors are well-balanced, inhibiting both targets at similar concentrations, then the window between the MIC of the more sensitive target and the lesser will be small, making the likelihood of stepwise mutation low. See Drlica [48] for an excellent explanation of the mutant selection window (MSW) and the role of dual targeting in resistance avoidance. Later FQs, such as moxifloxacin, had more balanced activities against the two enzymes [49]. The sequence where FQ resistance maps – in both enzymes – is called the quinolone resistance-determining region (QRDR) and is located on the α-subunit. Unlike the primarily HGT-encoded resistance that other commonly used antibiotics have selected, most FQ resistance arises through chromosomal mutations, as noted above. Due to their widespread use over the last 37 years, resistance to the FQs has increased.

As introduced in the previous edition [13], increased FQ resistance has engendered a significant effort to find dual inhibitors of the bacterial type II topoisomerases that are not cross-resistant with the FQs, so-called novel bacterial topoisomerase inhibitors (NBTIs). Recently, two different novel NBTIs have shown positive results in phase III clinical trials and are approaching registration.

10.3.4.1 Gepotidacin

Gepotidacin (Figure 10.2a), a triazaacenaphthylene, is under development by GlaxoSmithKline (GSK) for

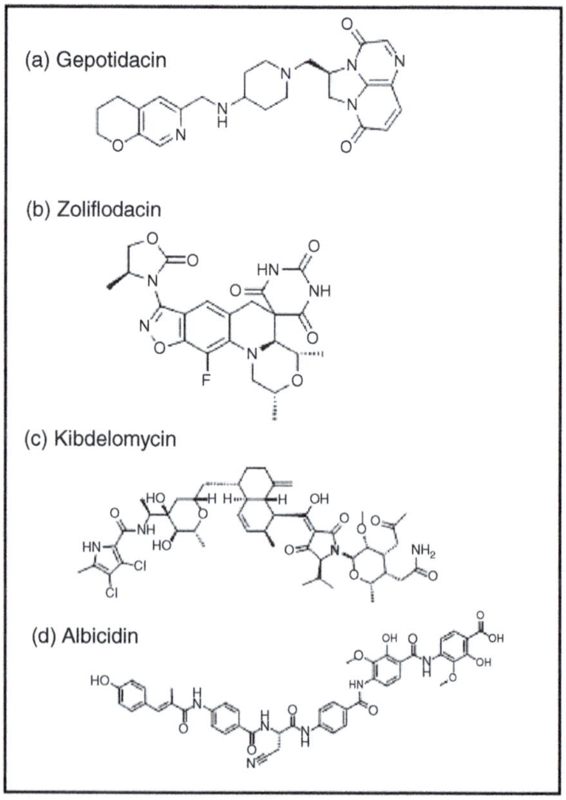

Figure 10.2 Structures of novel DNA gyrase/topoisomerase IV inhibitors. (a) Gepotidacin, (b) zoliflodacin (c) kibdelomycin, and (d) albicidin.

oral treatment of uncomplicated urinary tract infections (UUTIs), and recent phase III trials indicate success in the treatment of urogenital gonorrhea [50]. It is a well-balanced inhibitor of *E. coli* gyrase and topoisomerase IV [51]. This is important because it underlines the need for both targets to be inhibited at similar intracellular concentrations to benefit from the resistance reduction aspects of polypharmacology. In *E. coli*, it appears that gepotidacin directly contacts residues *gyrA* D82 and *parC* D79. Note that these residues are directly adjacent to the critical FQ resistance sites, yet no appreciable cross-resistance has been seen thus far. Single mutations of neither D82N nor D79N in an engineered *E. coli* affected the wild-type gepotidacin MIC more than twofold, while the double-mutant MIC increased >128-fold [52].

10.3.4.2 Zoliflodacin

Zoliflodacin (ZOL) (Figure 10.2b), a spiropyrimidinetrione, was discovered at AstraZeneca and is now being developed by Global Antibiotic Research & Development Partnership (GARDP) as an oral treatment for *Neisseria gonorrhoeae* including drug-resistant strains [53]. There appears to be no cross-resistance between ciprofloxacin and ZOL (in that ciprofloxacin-resistant clinical isolates never exposed to ZOL are susceptible to ZOL). The frequency of resistance (FoR) to ZOL is low: 1.3×10^{-11} to $\leq2.5\times10^{-13}$ [54], but FoRs appear to vary with growth and selection conditions. Mutations from a single-step selection and those raised in a second round of ZOL selection [55] are in GyrB. They raise the MIC of ZOL but lower the MIC of ciprofloxacin. As second-round mutants yielded further mutations in a limited region in GyrB [55], and not the β-subunit of topoisomerase II (ParE), DNA gyrase appears to be the primary target of ZOL. It may be that polypharmacology is not responsible for the low FoR here; it may be due to highly avid binding by ZOL.

The natural products discussed below are dual topoisomerase inhibitors that have been in early preclinical studies.

10.3.4.3 Kibdelomycin

Kibdelomycin (Figure 10.2c) is a natural product shown to be an inhibitor of the β-subunits of DNA gyrase and topoisomerase IV that demonstrates a low FoR [56]. It has a mostly Gram-positive spectrum but does show activity against some clinical strains of *A. baumannii* [57]. It shows no cross-resistance with novobiocin or coumermycin (gyrase B inhibitors that bind in the ATP site [58]) nor FQs, but has a unique binding mode [59]. Kibdelomycin is U-shaped, with two arms binding to sites on the B subunits: one arm inserts in the ATP-binding site, while the other makes hydrophobic and polar contacts on the enzyme's surface. To reach clinical candidacy, kibdelomycin requires further optimization.

10.3.4.4 Albicidin

A *Xanthomonas albilineans* polyketide–peptide product, albicidin (Figure 10.2d), disclosed in 1985 [60] and found to inhibit DNA replication, has recently been shown to be an inhibitor of DNA gyrase and topoisomerase IV [61, 62]. Albicidin exhibits not only binding to the two topoisomerases but also dual binding to each of the enzymes. It has good activity against Gram-negative pathogens, due to its uptake via the Tsx transporter of nucleotides [63]. Although albicidin is multitargeted, resistance arises through mutations in the *tsx* transporter gene and also via overexpression of a small chromosomal protein, GyrI [64], which competes with topoisomerases for binding to the antibiotic [65].

10.4 Cell Surface Targets and Macrocyclic Peptides (MCPs)

10.4.1 Relevant Permeability Properties of Bacteria

Phospholipid-containing cytoplasmic membranes, such as those bounding Gram-positive bacteria, prevent the entry of hydrophilic and polar molecules, such as vancomycin, into cells but admit (by diffusion) relatively neutral or lipophilic compounds, such as rifampicin, generally following Lipinski's rule of 5 [66]. Gram-negative bacteria have a duplex membrane, the cytoplasmic or inner membrane (IM), having entry properties like that of Gram-positives. The lipopolysaccharide (LPS)-containing outer membrane (OM) presents a barrier to the entry of large and hydrophobic molecules into the cell. Polar molecules less than 600 MW can generally enter the periplasm via porins that transit the OM. However, as with Gram-positives, such molecules will be barred from diffusing through the IM. Additionally, efflux pumps can remove a variety of compounds from the periplasm and the cytoplasm, depositing them extracellularly. Thus, Gram-positives are relatively welcoming to sufficiently lipophilic antibiotics, while the Gram-negative cytoplasm is well protected by the orthogonal sieving properties of the dual membranes and the action of efflux pumps. There are, however, exceptions to these general permeability concepts. Facilitated diffusion portals can allow the passage of specific solutes across the OM and even into the cytoplasm. In all species, certain hydrophilic and polar compounds may cross the IM using specific permeases, energy-requiring processes, or the involvement of proton-motive force (PMF). Discussions of bacterial permeability in relation to antibiotic drug discovery may be found in references [67–69].

10.4.2 Cell Surface Targets

The overall barriers to easy penetration of antibiotics into cells, especially into Gram-negatives, have been an obstacle to the development of novel antibiotics. While there has been recent progress toward formulating physicochemical rules for endowing compounds with characteristics compatible with bacterial entry (to engage with cytoplasmic targets) [70–72], the benefits of exploiting cell surface targets have not been overlooked. Many of our important antibacterial agents have cell surface (extracytoplasmic) targets, such as glycopeptides and polymyxins (as noted below), and β-lactams. These and other natural products evolved to interact with essential cell surface structures that function as antibacterial targets, including peptidoglycan, its lipid-linked intermediates, membranes themselves, and both periplasmic and integral membrane proteins responsible for synthesizing or transporting essential cell components.

10.4.3 Polypharmacology and the Benefits of MCPs

The often-large natural products that interact with cell surfaces may bind to multiple receptors, including sites on the same or adjacent molecules, a facet of polypharmacology. These multiple interactions can afford binding affinity as well as reduce resistance potential.

Indeed, there are reports of such cell-surface-directed, multitargeting antibiotics. Many of these are macrocyclic peptides (MCPs), which are non-ribosomal peptides modified to cyclize either wholly or in part. MCPs are more conformationally constrained than linear peptides and may take up forms that bind to specific and relatively large and complex expanses of target molecules. These structured peptides have been made synthetically, often in large numbers, and screened directly for activity. Synthetic methods have also been used to diversify and optimize the original structures. MCPs are relevant here in a discussion of polypharmacology since their size and complexity often endow them with multiple target-binding sites that don't yield to resistance due to single mutations in target molecules. On the other hand, in the discussion of the relationship of polypharmacology to reduction of target-based resistance, it is important to differentiate those inhibitory molecules that bind to multiple sites on one or more proteins or polynucleotides (where mutations leading to reduced binding can occur within the coding sequence of the target), from inhibitors binding to nonprotein (or non-nucleoside) molecules whose structure cannot be altered by single base changes (or mutational events). In the latter case, binding to multiple sites could well improve efficacy by greatly increasing binding affinity or adding new inhibitory mechanisms.

10.4.4 Classic MCPs Targeting Cell Surfaces

Many commonly used antibiotics are MCPs, often with appended structural groups, such as lipopeptides like polymyxin and glycopeptides like vancomycin.

10.4.4.1 Polymyxin

Polymyxin is a macrocyclic lipopeptide that remains in use despite its toxicity because it retains activity against many multidrug-resistant (MDR) Gram-negative bacterial species, especially Acinetobacter. Its cyclic cationic portion binds to the lipid A portion of LPS of the OM, while its fatty acid tail penetrates and disrupts the cytoplasmic membrane. Thus, the molecular targets of polymyxin are not proteins but the products of multistep, multigene pathways. Resistance to polymyxin is usually due to decreasing the negative charge of lipid A by adding cationic groups, such as phosphoethanolamine (PEA), at appropriate sites [73]. Such changes may be the result of chromosomal mutations leading to the expression of endogenous stress response systems (including those producing PEA transferase [PEt]) or to plasmid-borne *mcr* genes encoding PEts. Mcr resistance has been rising rapidly and globally. Several recent programs have been aimed at the reduction of polymyxin toxicity, yielding three such agents in clinical trials: SPR206, QPEX9003, and MRX-8 [74]. Even if toxicity is reduced, the increase in polymyxin resistance will remain a constant problem. Inhibitors of PEts may provide a path forward [75].

10.4.4.2 Vancomycin

Vancomycin is a natural product antibiotic active against Gram-positives that gained wide use with the advent of methicillin-resistant *S. aureus* (MRSA). It binds to a surface component, the terminal D-Ala-D-Ala moiety of uncross-linked Lipid II (undecaprenyl-PPi-N-acetylmuramoyl-[N-acetylglucosamine]-L-alanyl-γ-D-glutamyl-L-lysyl-D-alanyl-D-alanine), which is accessible on the surface of Gram-positive species. This binding prevents the transglycosylation and transpeptidation of nascent peptidoglycan [76]. While stepwise mutations may give rise to intermediate levels of resistance [77] (such as vancomycin intermediate *S. aureus* [VISA] strains), the major vancomycin resistance determinants, mostly seen in Enterococci [78], have been horizontally transmitted from the producing microorganisms [79].

10.4.5 New Gram-positive MCP Agents

10.4.5.1 Teixobactin

Teixobactin (Figure 10.3a) was discovered in a phenotypic screen of the products of specially isolated and grown "unculturable" bacteria [81]. It is a potent inhibitor of Gram-positive pathogens including MRSA, Enterococci, *Clostridioides difficile*, and *Bacillus anthracis*, demonstrating in vivo efficacy. Interestingly, it was originally touted as being able to "kill pathogens without detectable resistance." This was based on *in vitro* resistance selection experiments and on the fact that the producing organism is a Gram-negative bacterium (*Eleftheria terrae*) that is not itself sensitive to teixobactin due to impermeability; thus, it would likely not have evolved a self-protective mechanism that could be transmitted to pathogens (as is the case with horizontally transmitted vancomycin resistance). Further investigation showed that teixobactin binds to the PPi-N-acetylmuramic acid (PPi-MurNAc) portion of the Gram-positive surface molecules, Lipid II and Lipid III. These lipids are the precursors of peptidoglycan and wall teichoic acid, respectively. The structure of Lipid II is essentially immutable, in that it is not directly encoded by genes – hence, endogenous resistance selection is indeed unlikely [82]. Recent work uncovered the complex mode of action of teixobactin – as illustrated in Figure 10.3d. The cyclic cationic C-terminal binds to PPi-MurNAc, while the linear tail interacts with membrane lipid. Upon landing on the membrane surface, teixobactin monomers form head-to-tail oligomeric β-sheets that slowly elongate into lateral fibrillar sheets forming a supramolecular structure that blocks peptidoglycan synthesis and causes membrane defects [80]. Cellular lysis of *S. aureus* by teixobactin is mediated by autolysin AtlA.

10.4.5.2 Clovibactin

Clovibactin (Figure 10.3b) is a natural product MCP produced by an *E. terrae* isolate, closely related to the producer of teixobactin [83]. It has broad-spectrum Gram-positive bactericidal activity and binds to the PPi moiety of Lipid I, Lipid II, Lipid III, and the C55-carrier lipid, undecaprenyl pyrophosphate. It does not bind to PPi in solution, only on membranes. While its overall interaction with membranes is similar to that of teixobactin, in that it first forms oligomers and then fibrils, it appears that recognition of PPi occurs after oligomers form. Furthermore, the supermolecular structures are dissimilar, owing to differences in compound structure and specific binding moiety. Peptidoglycan synthesis is blocked by the inhibition of the many lipid-PPi intermediate-involved steps; lysis does not require AtlA.

10.4.5.3 Cilagicin

Biosynthetic prospecting led to the discovery of a cyclic lipopeptide, cilagicin [84] (Figure 10.3c), with broad-spectrum Gram-positive activity and dual targeting of C-55-Pi and C-55-PPi. The authors were unable to isolate single-step resistant mutants and thus attempted serial passage selection of cilagicin in parallel with the C-55-Pi binder, amphomycin, and the C-55-PPi binder, bacitracin, as controls. Both amphomycin and bacitracin rapidly selected mutants to 256-fold MIC and eightfold MIC, respectively, while the cilagicin MICs rose at most twofold after 25 days of passage. No autoresistance functions were found in the producing organism, *Paenibacillus mucilaginosus*, a Gram-variable bacterium that appears to have an OM, which might protect it from self-toxicity. The effect of cilagicin is consistent with inhibition of peptidoglycan synthesis, as would be expected from C-55-Pi and C-55-PPi binders, and it does not lead to cytotoxicity in human embryonic kidney 293 (HEK-293) cells nor membrane depolarization or cell lysis of *S. aureus*. It should, however, be noted that the lack of genetic proof does not eliminate the possibility of additional mechanisms of action.

10.4.6 Gram-negative Inhibitors Targeting Cell Surfaces, Including OMPTAs

LPS makes up most of the surface of Gram-negative bacteria while few essential proteins are exposed. As shown in Figure 10.4e, two major protein complexes present are the β-barrel–assembly machinery (BAM) complex and Lpt system, which act as transporters of OM proteins (OMPs) and LPS to their OM locations, respectively. The inventors of murepavadin (see below) at Polyphor Ltd. called inhibitors of these surface proteins OMP targeting antibiotics (OMPTAs) [86]. It has been suggested that such exploitation of OM targets may provide a solution to the dearth of novel anti-Gram-negative agents in the pipeline [87].

10.4.6.1 Darobactin

Darobactin (Figure 10.4a), an MCP discovered by screening a small set of isolates of the Gram-negative, nematode-symbiont genera Photorhabdus and Xenorhabdus [88], is active against Gram-negatives. It selected for single-resistant mutations at a frequency of 8×10^{-9} at 4× MIC on solid medium but not at higher compound concentrations. Serial passage in liquid yielded three mutations all in BamA. An engineered *E. coli* strain containing these three mutations was resistant to darobactin but had low virulence, as did a single mutant. Cardiolipin (a membrane lipid) enhances the interaction of darobactin and BamA, likely interacting with the compound. The authors note that impaired virulence of resistant mutants and some

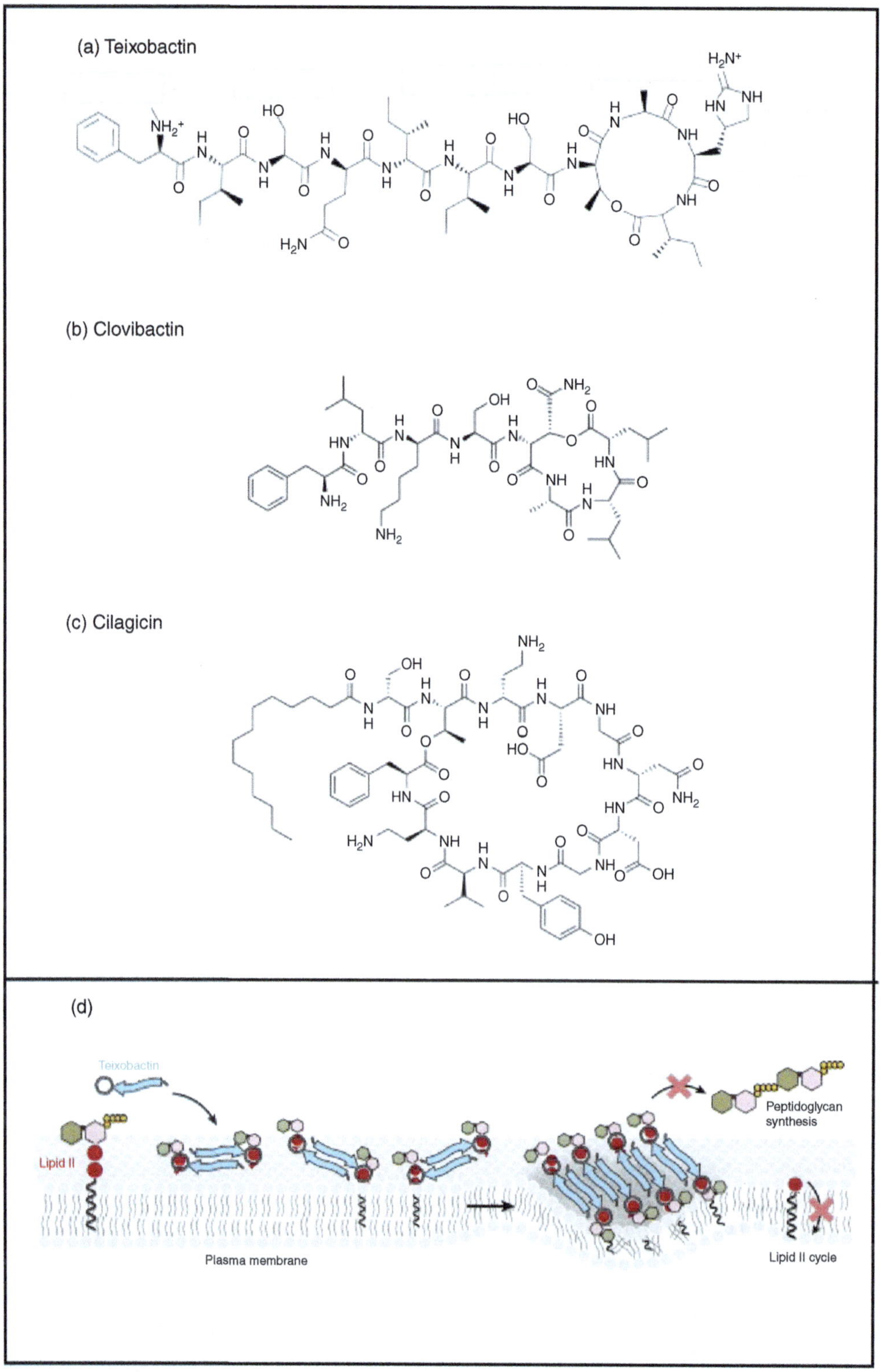

Figure 10.3 Structures of macrocyclic peptides (MCPs) that target cell surface structures. (a) Teixobactin, (b) clovibactin, (c) cilagicin, and (d) dual mechanism of action of teixobactin. Figure 10.3d is from Shukla et al. [80]/Springer Nature/CC BY 4.0.

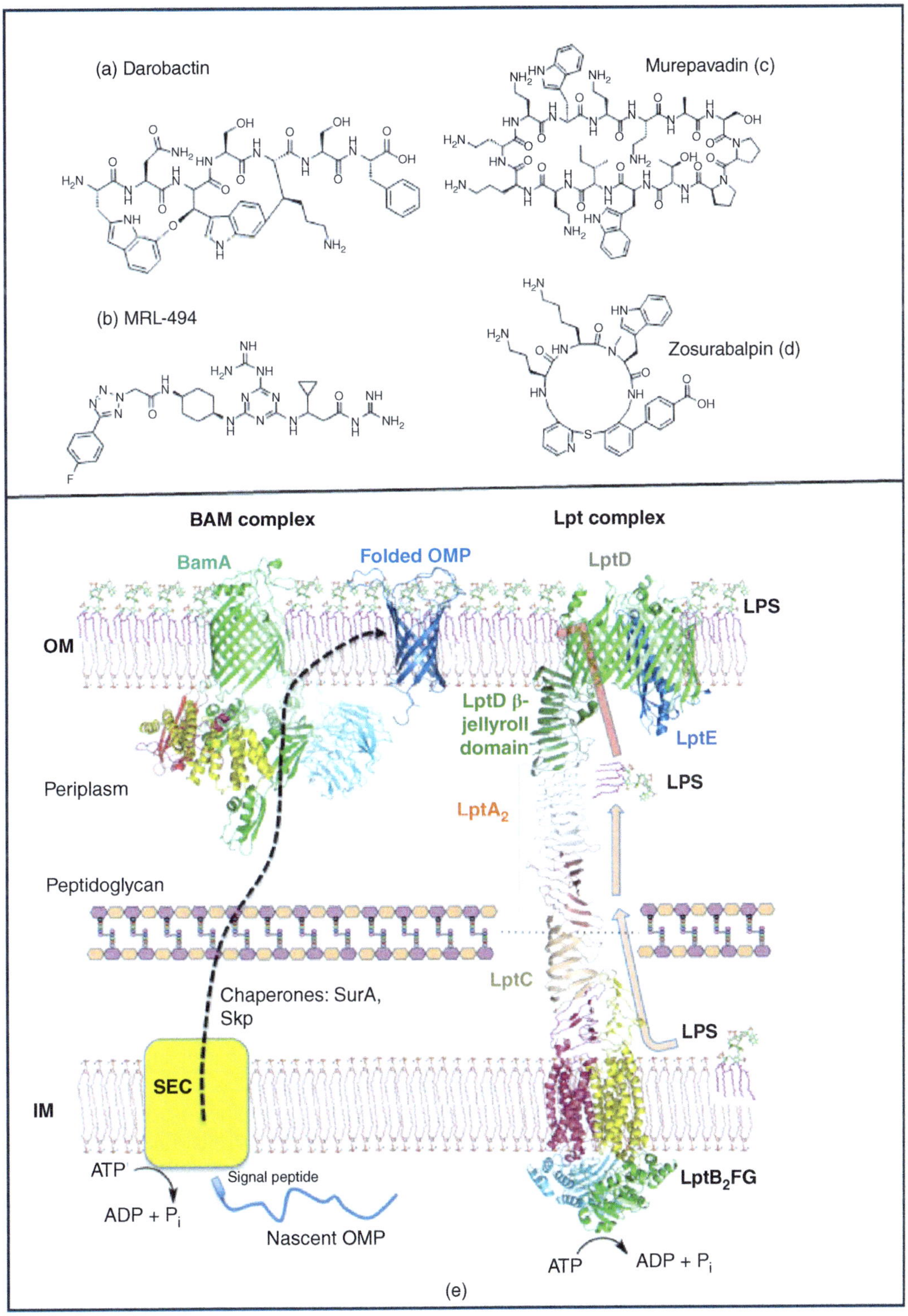

Figure 10.4 Inhibitors of BAM and Lpt complexes of the Gram-negative outer membrane (OM). Compounds (a) darobactin, and (b), MRL-494, inhibit BamA. Compounds (c) murepavadin and (d) zosurabalpin, inhibit the Lpt complex. E. Bam and Lpt complexes span the outer and inner (cytoplasmic) membranes. Figure 10.4e is from [85]/Frontiers Media/CC BY.40.

participation of cardiolipin in binding to BamA protein may limit resistance development [89].

10.4.6.2 MRL-494

Though not macrocyclic, MRL-494 (Figure 10.4b) [86] is relevant here. It was obtained in a screen for small-molecule disrupters of OM at sub-MIC that had MICs against *E. coli* unaffected by efflux [90]. MRL-494 inhibits the assembly of OMPs by the Bam complex, specifically by interacting with BamA, the only surface-exposed member of the complex. Interestingly, it was found to disrupt the membranes of Gram-positives but not the IM of Gram-negatives, most likely because it is unable to cross the OM.

10.4.6.3 Murepavadin

Murepavadin (POL7080; Figure 10.4c) [86] is a synthetic tethered (hence macrocyclic) peptide that was found to have potent selective activity against *P. aeruginosa*. It binds to *P. aeruginosa* LptD inhibiting the installation of LPS in the OM. Selectivity is due to the presence of a ~120 residue insert in the N-terminal region of *P. aeruginosa* LptD but not in that of other Gram-negatives [91]. Murepavadin has an extremely low occurrence of resistance, although a specific 18 residue duplication within the *P. aeruginosa*-specific insert in LptD did lead to resistance while retaining transport activity. After initial successful phase I trials (by Roche), murepavadin was returned to Polyphor, where it reached phase III trials for intravenous (i.v.) treatment of hospital-acquired pneumonia (HAP)/ventilator-associated pneumonia (VAP) driven by *P. aeruginosa*. Thereafter, however, the development of the i.v. formulation was curtailed due to nephrotoxicity. Development of a nasal formulation for *P. aeruginosa* in cystic fibrosis (CF) and non-CF bronchiectasis continued, initially under the aegis of Spexis, with apparent good phase I results. Basilea has recently acquired the Spexis antibacterial pipeline and has yet to announce its plans for murepavadin.

10.4.6.4 Zosurabalpin

Zosurabalpin (Figure 10.4d) is a synthetic MCP, discovered in a screen of 44,000 tethered peptides for activity against *A. baumannii* [92]. Zosurabalpin targets the IM $LptB_2FGC$ LPS-transporter complex (Figure 10.4d), with resistance mutations occurring in LptF and LptG. Detailed mechanistic studies [93] indicate that zosurabalpin binds to and traps LPS within the LptF–LptG section of the transporter, which is located on the periplasmic side of the IM. As zosurabalpin resistance frequencies are not low ($\leq 1 \times 10^{-8}$), it would seem that further optimization is necessary. Nevertheless, zosurabalpin has recently been disclosed to be in clinical trials [94].

10.5 Conclusions

Polypharmacology of antibacterial agents can include multiple binding sites on one or more target molecules. This leads to tighter binding, possible synergy between inhibitory functions, and, in many cases, reduction in resistance potential over inhibitors with single binding sites. Many of our existing antibiotics have these properties. While much of the past two decades of antibacterial discovery has favored single-targeted agents, an effort that has had little success, many recently reported agents have returned, often serendipitously, to the polypharmacological paradigm. In addition to synthetic approaches to multiple binders, compounds that are themselves natural products or are designed based on them are often large and rather complex, such as MCPs with various molecular appendages. The size and physicochemical nature of these may preclude their entry into cells, and thus, their binding sites are on the cell surface. This can avoid the problem of endowing new agents with the ability to enter cells and avoid resistance due to reduced permeability/increased efflux. Multiple targeting can also reduce target-based endogenous mutational resistance. While it may seem circular reasoning to relate the properties of MCPs (large size, constrained structure, exclusion from the cytoplasm, and multitargeting of surface components), it should be recognized that the original natural products evolved to have these attributes. The origins of antibiotic discovery centered on natural products. Acknowledging and incorporating the beneficial qualities, especially polypharmacology, that natural products bring to antibacterial drug discovery can overcome or avoid some of the rate-limiting steps of antibacterial discovery, including bacterial entry and rapid resistance development [2].

References

1 Silver, L.L. (2007). Multi-targeting by monotherapeutic antibacterials. *Nat. Rev. Drug Discov.* 6 (1): 41–55.

2 Silver, L.L. (2011). Challenges of antibacterial discovery. *Clin. Microbiol. Rev.* 24 (1): 71–109.

3 Fleischmann, R.D., Adams, M.D., White, O. et al. (1995). Whole-genome random sequencing and assembly of haemophilus influenzae rd. *Science* 269 (5223): 496–512.

4 Silver, L.L. (2017). The antibiotic future. In: *Topics in Medicinal Chemistry* (ed. J.F. Fisher, M. Miller, and S. Mobashery), 1–37. Berlin, Heidelberg: Springer.

5 Silva, K.P.T., Sundar, G., and Khare, A. (2023). Efflux pump gene amplifications bypass necessity of multiple target mutations for resistance against dual-targeting antibiotic. *Nat. Commun.* 14 (1): 3402.

6 Andersson, D.I. (2006). The biological cost of mutational antibiotic resistance: any practical conclusions? *Curr. Opin. Microbiol.* 9: 461–465.

7 Brötz-Oesterhelt, H. and Sass, P. (2010). Postgenomic strategies in antibacterial drug discovery. *Future Microbiol.* 5 (10): 1553–1579.

8 Sutterlin, H.A., Malinverni, J.C., Lee, S.H. et al. (2017). Antibacterial new target discovery: sentinel examples, strategies, and surveying success. In: *Antibacterials*; Topics in Medicinal Chemistry (ed. J.F. Fisher, S. Mobashery, and M.J. Miller), 1–29. Berlin, Heidelberg: Springer.

9 Dick, T. and Young, D. (2011). How antibacterials really work: impact on drug discovery. *Future Microbiol.* 6 (6): 603–604.

10 Projan, S.J. (2003). Why is big Pharma getting out of antibacterial drug discovery? *Curr. Opin. Microbiol.* 6 (5): 427–430.

11 Brown, E.D. and Wright, G.D. (2016). Antibacterial drug discovery in the resistance era. *Nature* 529 (7586): 336–343.

12 Chopra, I. (2012). The 2012 Garrod lecture: discovery of antibacterial drugs in the 21st century. *J. Antimicrob. Chemother.* 68: 496–505.

13 Silver, L.L. (2012). Polypharmacology as an emerging trend in antibacterial discovery. In: *Polypharmacology in Drug Discovery* (ed. J. Peters), 167–202. Hoboken, NJ: Wiley.

14 Gray, D.A. and Wenzel, M. (2020). Multitarget approaches against multiresistant superbugs. *ACS Infect. Dis.* 6 (6): 1346–1365.

15 Feng, J., Zheng, Y., Ma, W. et al. (2023). Multitarget antibacterial drugs: an effective strategy to combat bacterial resistance. *Pharmacol. Therap.* 252: 108550.

16 Wetzel, C., Lonneman, M., and Wu, C. (2021). Polypharmacological drug actions of recently FDA approved antibiotics. *Eur. J. Med. Chem.* 209: 112931.

17 Privalsky, T.M., Soohoo, A.M., Wang, J. et al. (2021). Prospects for antibacterial discovery and development. *J. Am. Chem. Soc.* 143 (50): 21127–21142.

18 Bremner, J. (2021). *Single Molecule Non-cleavable Multiply Active Antibacterials. Multiple Action-Based Design Approaches to Antibacterials*, 51–119. Singapore: Springer.

19 Abrusán, G. and Marsh, J.A. (2019). Ligands and receptors with broad binding capabilities have common structural characteristics: an antibiotic design perspective. *J. Med. Chem.* 62: 9357–9374.

20 Randall, C.P., Rasina, D., Jirgensons, A., and O'Neill, A.J. (2016). Targeting multiple aminoacyl-tRNA synthetases overcomes the resistance liabilities associated with antibacterial inhibitors acting on a single such enzyme. *Antimicrob. Agents Chemother.* 60 (10): 6359–6361.

21 Feng, X., Zhu, W., Schurig-Briccio, L.A. et al. (2015). Antiinfectives targeting enzymes and the proton motive force. *Proc. Natl. Acad. Sci.* 112 (51): E7073–E7082.

22 Oldfield, E. and Feng, X. (2014). Resistance-resistant antibiotics. *Trends Pharmacol. Sci.* 35 (12): 664–674.

23 East, S.P. and Silver, L.L. (2013). Multitarget ligands in antibacterial research: progress and opportunities. *Expert Opin. Drug Discov.* 8 (2): 143–156.

24 Hrast, M., Sosic, I., Sink, R., and Gobec, S. (2014). Inhibitors of the peptidoglycan biosynthesis enzymes MurA-F. *Bioorg. Chem.* 55: 2–15.

25 Hervin, V., Roy, V., and Agrofoglio, L.A. (2023). Antibiotics and antibiotic resistance-mur ligases as an antibacterial target. *Molecules* 28 (24).

26 Silver, L.L. (2013). Viable screening targets related to the bacterial cell wall. *Ann. N. Y. Acad. Sci.* 1277 (1): 29–53.

27 Shahul Hameed, P., Manjrekar, P., Chinnapattu, M. et al. (2014). Pyrazolopyrimidines establish MurC as a vulnerable target in pseudomonas aeruginosa and *Escherichia coli*. *ACS Chem. Biol.* 9 (10): 2274–2282.

28 Baum, E.Z., Crespo-Carbone, S.M., Foleno, B.D. et al. (2009). MurF inhibitors with antibacterial activity: effect on muropeptide levels. *Antimicrob. Agents Chemother.* 53 (8): 3240–3247.

29 Hrast, M., Anderluh, M., Knez, D. et al. (2014). Design, synthesis and evaluation of second generation MurF inhibitors based on a cyanothiophene scaffold. *Eur. J. Med. Chem.* 73: 83–96.

30 Kouidmi, I., Levesque, R.C., and Paradis-Bleau, C. (2014). The biology of Mur ligases as an antibacterial target. *Mol. Microbiol.* 94: 242–253.

31 Hrast, M., Frlan, R., Knez, D. et al. (2021). Mur ligases inhibitors with azastilbene scaffold: expanding the structure–activity relationship. *Bioorg. Med. Chem. Lett.* 40: 127966.

32 O'Neill, A.J. and Chopra, I. (2004). Preclinical evaluation of novel antibacterial agents by microbiological and molecular techniques. *Expert Opin. Investig. Drugs* 13 (8): 1045–1063.

33 Cunningham, F.X., Lafond, T.P., and Gantt, E. (2000). Evidence of a role for LytB in the nonmevalonate pathway of isoprenoid biosynthesis. *J. Bacteriol.* 182 (20): 5841–5848.

34 Jeremia, L., Deprez, B., Dey, D. et al. (2023). Ribosome-targeting antibiotics and resistance via ribosomal RNA methylation. *RSC Med. Chem.* 14 (4): 624–643.

35 Murina, V., Kasari, M., Hauryliuk, V., and Atkinson, G.C. (2018). Antibiotic resistance ABCF proteins reset the peptidyl transferase centre of the ribosome to counter translational arrest. *Nucleic Acids Res.* 46 (7): 3753–3763.

36 Mitcheltree, M.J., Pisipati, A., Syroegin, E.A. et al. (2021). A synthetic antibiotic class overcoming bacterial multidrug resistance. *Nature* 599 (7885): 507–512.

37 Wu, K.J.Y., Tresco, B.I.C., Ramkissoon, A. et al. (2024). An antibiotic preorganized for ribosomal binding overcomes antimicrobial resistance. *Science* 383 (6684): 721–726.

38 Cram, D.J. (1986). Preorganization – from solvents to spherands. *Angew. Chem. Int. Ed. Engl.* 25 (12): 1039–1057.

39 Goto, K. (2017). The concept of preorganization. In: *Comprehensive Supramolecular Chemistry II* (ed. J.L. Atwood), 61–71. Oxford: Elsevier.

40 Park, S.Y., Kim, S.M., and Park, S.D. (2012). The prevalence, genotype and antimicrobial susceptibility of high- and low-level mupirocin resistant methicillin-resistant *Staphylococcus aureus*. *Ann. Dermatol.* 24 (1): 32–38.

41 Gallant, P., Finn, J., Keith, D., and Wendler, P. (2000). The identification of quality antibacterial drug discovery targets: a case study with aminoacyl-tRNA synthetases. *Emerg. Therap. Targ.* 4 (1): 1–9.

42 Hurdle, J.G., O'Neill, A.J., and Chopra, I. (2005). Prospects for aminoacyl-tRNA synthetase inhibitors as new antimicrobial agents. *Antimicrob. Agents Chemother.* 49 (12): 4821–4833.

43 Kim, S., Lee, S.W., Choi, E.C., and Choi, S.Y. (2003). Aminoacyl-tRNA synthetases and their inhibitors as a novel family of antibiotics. *Appl. Microbiol. Biotechnol.* 61 (4): 278–288.

44 O'Dwyer, K., Spivak, A., Ingraham, K. et al. (2015). Bacterial resistance to leucyl-tRNA synthetase inhibitor GSK2251052 develops during treatment of complicated urinary tract infections. *Antimicrob. Agents Chemother.* 59: 289–298.

45 Diacon, A.H., Barry, C.E., Carlton, A. et al. (2024). A first-in-class leucyl-tRNA synthetase inhibitor, ganfeborole, for rifampicin-susceptible tuberculosis: a phase 2a open-label, randomized trial. *Nat. Med.* 30: 896–904.

46 Ochsner, U.A. and Jarvis, T.C. (2013). Aminoacyl-tRNA synthetase inhibitors. In: *Antibiotics: Targets, Mechanisms and Resistance* (ed. C. Gualerzi, L. Brandi, A. Fabbretti, and C. Pon), 387–410. Weinheim, Germany: Wiley-VCH.

47 Aldred, K.J., Breland, E.J., Vlčková, V. et al. (2014). Role of the water–metal ion bridge in mediating interactions between quinolones and *Escherichia coli* topoisomerase IV. *Biochemistry* 53 (34): 5558–5567.

48 Drlica, K. (2003). The mutant selection window and antimicrobial resistance. *J. Antimicrob. Chemother.* 52 (1): 11–17.

49 Schedletzky, H., Wiedemann, B., and Heisig, P. (1999). The effect of moxifloxacin on its target topoisomerases from *Escherichia coli* and *Staphylococcus aureus*. *J. Antimicrob. Chemother.* 43 (suppl_2): 31–37.

50 Mullard, A. (2024). GSK's first-in-class antibiotic secures another phase III win, approaches regulatory run. *Nat. Rev. Drug Discov.* 23: 238.

51 Oviatt, A.A., Gibson, E.G., Huang, J. et al. (2024). Interactions between *Gepotidacin* and *Escherichia coli* gyrase and topoisomerase IV: genetic and biochemical evidence for well-balanced dual-targeting. *ACS Infect. Dis.* 10 (4): 1137–1151.

52 Szili, P., Draskovits, G., Révész, T. et al. (2019). Rapid evolution of reduced susceptibility against a balanced dual-targeting antibiotic through stepping-stone mutations. *Antimicrob. Agents Chemother.* 63 (9): e00207–e00219.

53 Bradford, P.A., Miller, A.A., O'Donnell, J., and Mueller, J.P. (2020). Zoliflodacin: an oral spiropyrimidinetrione antibiotic for the treatment of neisseria gonorrheae, including multi-drug-resistant isolates. *ACS Infect. Dis.* 6 (6): 1332–1345.

54 Foerster, S., Drusano, G., Golparian, D. et al. (2019). In vitro antimicrobial combination testing of and evolution of resistance to the first-in-class spiropyrimidinetrione zoliflodacin combined with six therapeutically relevant antimicrobials for Neisseria gonorrhoeae. *J. Antimicrob. Chemother.* 74: 3521–3529.

55 Alm, R.A., Lahiri, S.D., Kutschke, A. et al. (2015). Characterization of the novel DNA gyrase inhibitor AZD0914: low resistance potential and lack of cross-resistance in Neisseria gonorrhoeae. *Antimicrob. Agents Chemother.* 59 (3): 1478–1486.

56 Phillips, J.W., Goetz, M.A., Smith, S.K. et al. (2011). Discovery of kibdelomycin, a potent new class of bacterial type II topoisomerase inhibitor by chemical-genetic profiling in *Staphylococcus aureus*. *Chem. Biol.* 18 (8): 955–965.

57 Singh, S.B., Dayananth, P., Balibar, C.J. et al. (2015). Kibdelomycin is a bactericidal broad-spectrum aerobic antibacterial agent. *Antimicrob. Agents Chemother.* 59 (6): 3474–3481.

58 Maxwell, A. and Lawson, D.M. (2003). The ATP-binding site of type II topoisomerases as a target for antibacterial drugs. *Curr. Top. Med. Chem.* 3 (3): 283–303.

59 Lu, J., Patel, S., Sharma, N. et al. (2014). Structures of kibdelomycin bound to *Staphylococcus aureus* GyrB and ParE showed a novel u-shaped binding mode. *ACS Chem. Biol.* 9: 2023–2031.

60 Birch, R.G. and Patel, S.S. (1985). Preliminary characterization of an antibiotic produced by Xanthomonas albilineans which inhibits DNA synthesis in *Escherichia coli. Microbiology* 131 (5): 1069–1075.

61 Michalczyk, E., Hommernick, K., Behroz, I. et al. (2023). Molecular mechanism of topoisomerase poisoning by the peptide antibiotic albicidin. *Nat. Catal.* 6 (1): 52–67.

62 Hashimi, S.M. (2019). Albicidin, a potent DNA gyrase inhibitor with clinical potential. *J. Antibiot.* 72 (11): 785–792.

63 Birch, R.G., Pemberton, J.M., and Basnayake, W.V. (1990). Stable albicidin resistance in *Escherichia coli* involves an altered outer-membrane nucleoside uptake system. *J. Gen. Microbiol.* 136 (1): 51–58.

64 Krishnamoorthy, G., Leus, I.V., Weeks, J.W. et al. (2017). Synergy between active efflux and outer membrane diffusion defines rules of antibiotic permeation into gram-negative bacteria. *MBio* 8 (5): 1 16.

65 Saathoff, M., Kosol, S., Semmler, T. et al. (2023). Gene amplifications cause high-level resistance against albicidin in gram-negative bacteria. *PLoS Biol.* 21 (8): e3002186.

66 Lipinski, C. (2004). Lead- and drug-like compounds: the rule-of-five revolution. *Drug Discov. Today* 1 (4): 337–341.

67 Nichols, W.W. (2012). Permeability of bacteria to antibacterial agents. In: *Antibiotic Discovery and Development* (ed. T.J. Dougherty and M.J. Pucci), 849–879. New York: Springer.

68 Basarab, G.S., Eakin, A.E., Nichols, W.W., and Rao, Y. (2024). Chapter 32 – Design of antibacterial agents. In: *Molecular Medical Microbiology*, (Third Edition)e (ed. Y.-W. Tang, M.Y. Hindiyeh, D. Liu, et al.), 615–631. Academic Press.

69 Silver, L.L. (2016). A gestalt approach to gram-negative entry. *Bioorg. Med. Chem.* 24 (24): 6379–6389.

70 Richter, M.F., Drown, B.S., Riley, A.P. et al. (2017). Predictive compound accumulation rules yield a broad-spectrum antibiotic. *Nature* 545: 299–304.

71 Benedetto Tiz, D., Kikelj, D., and Zidar, N. (2018). Overcoming problems of poor drug penetration into bacteria: challenges and strategies for medicinal chemists. *Expert Opin. Drug Discov.* 13 (6): 497–507.

72 Cain, B.N. and Hergenrother, P.J. (2024). Using permeation guidelines to design new antibiotics-a passage into pseudomonas aeruginosa. *Clin. Transl. Med.* 14 (3): e1600.

73 Shahzad, S., Willcox, M.D.P., and Rayamajhee, B. (2023). A review of resistance to polymyxins and evolving mobile colistin resistance gene (mcr) among pathogens of clinical significance. *Antibiotics* 12 (11).

74 Slingerland, C.J. and Martin, N.I. (2024). Recent advances in the development of polymyxin antibiotics: 2010–2023. *ACS Infect. Dis.* 10 (4): 1056–1079.

75 Samantha, A. and Vrielink, A. (2020). Lipid A phosphoethanolamine transferase: regulation, structure and immune response. *J. Mol. Biol.* 432 (18): 5184–5196.

76 Walsh, C. (1999). Deconstructing vancomycin. *Science* 284 (5413): 442–443.

77 Schaaff, F., Reipert, A., and Bierbaum, G. (2002). An elevated mutation frequency favors development of vancomycin resistance in *Staphylococcus aureus. Antimicrob. Agents Chemother.* 46 (11): 3540–3548.

78 Murray, B.E. (1997). Vancomycin-resistant enterococci. *Am. J. Med.* 102 (3): 284–293.

79 Marshall, C.G., Lessard, I.A.D., Park, I.-S., and Wright, G.D. (1998). Glycopeptide antibiotic resistance genes in glycopeptide-producing organisms. *Antimicrob. Agents Chemother.* 42 (9): 2215–2220.

80 Shukla, R., Lavore, F., Maity, S. et al. (2022). Teixobactin kills bacteria by a two-pronged attack on the cell envelope. *Nature* 608 (7922): 390–396.

81 Ling, L.L., Schneider, T., Peoples, A.J. et al. (2015). A new antibiotic kills pathogens without detectable resistance. *Nature* 517: 455–459.

82 Hughes, D., Lawrence, W., Peel, J. et al. (2024). A resistance-evading antibiotic for treating anthrax. *Preprint* (Version 1). Available at Research Square 2024.

83 Shukla, R., Peoples, A.J., Ludwig, K.C. et al. (2023). An antibiotic from an uncultured bacterium binds to an immutable target. *Cell* 186 (19): 4059–73.e27.

84 Wang, Z., Koirala, B., Hernandez, Y. et al. (2022). Bioinformatic prospecting and synthesis of a bifunctional lipopeptide antibiotic that evades resistance. *Science* 376 (6596): 991–996.

85 Robinson, J.A. (2019). Folded synthetic peptides and other molecules targeting outer membrane protein complexes in Gram-negative bacteria. *Front. Chem.* 7: 45.

86 Luther, A., Moehle, K., Chevalier, E. et al. (2017). Protein epitope mimetic macrocycles as biopharmaceuticals. *Curr. Opin. Chem. Biol.* 38: 45–51.

87 Walker, S.S. and Black, T.A. (2021). Are outer-membrane targets the solution for MDR Gram-negative bacteria? *Drug Discov. Today* 26 (9): 2152–2158.

88 Imai, Y., Meyer, K.J., Iinishi, A. et al. (2019). A new antibiotic selectively kills Gram-negative pathogens. *Nature* 576: 459–464.

89 Kaur, H., Jakob, R.P., Marzinek, J.K. et al. (2021). The antibiotic darobactin mimics a β-strand to inhibit outer membrane insertase. *Nature* 593 (7857): 125–129.

90 Hart, E.M., Mitchell, A.M., Konovalova, A. et al. (2019). A small-molecule inhibitor of BamA impervious to efflux and the outer membrane permeability barrier. *Proc. Natl. Acad. Sci.* 116: 21748–21757.

91 Srinivas, N., Jetter, P., Ueberbacher, B.J. et al. (2010). Peptidomimetic antibiotics target outer-membrane biogenesis in *Pseudomonas aeruginosa*. *Science* 327 (5968): 1010–1013.

92 Zampaloni, C., Mattei, P., Bleicher, K. et al. (2024). A novel antibiotic class targeting the lipopolysaccharide transporter. *Nature* 625 (7995): 566–571.

93 Pahil, K.S., Gilman, M.S.A., Baidin, V. et al. (2024). A new antibiotic traps lipopolysaccharide in its intermembrane transporter. *Nature* 625 (7995): 572–577.

94 Guenther, A., Millar, L., Messer, A. et al. (2023). 2126. Safety, tolerability, and pharmacokinetics (PK) in healthy participants following single dose administration of zosurabalpin, a novel pathogen-specific antibiotic for the treatment of serious acinetobacter infections. *Open Forum Infect. Dis.* 10: ofad500.1749.

11

Multi-Specific Binding Strategy

A Novel Multi-Targeted Antiviral Modality

Yang Zhou, Shujing Xu, Dang Ding, Kai Tang, Xinyong Liu, Meehyein Kim, and Peng Zhan

Viral infections pose significant threats not only to human public health but also to socio-economic development. The novel coronavirus infection caused by Severe Acute Respiratory Syndrome Coronavirus 2 (SARS-CoV-2) has claimed the lives of millions of people, negatively impacting public health security and global economic growth [1, 2]. Besides, infectious disease burdens triggered by other viruses, such as human immunodeficiency virus (HIV), influenza virus, hepatitis B virus (HBV), and Ebola virus, still highlight urgent need for therapeutic resolution [3–6]. With the emergence of drug-resistant strains of these viruses, the world is facing significant challenges in combating potential epidemic or pandemic viral infections [7–9]. Consequently, more creative breakthrough strategies are required for providing the prevention and treatment solutions of current as well as future viral infectious diseases.

Since the 1970s, drug design has progressively evolved into a "one drug, one target" paradigm, leading to the discovery of hundreds of drugs. The classical approach to drug development usually entails initial identification of a specific target, followed by screening of drug candidates and structure-guided designing of the compounds in a way to improve their target-specificity and physical properties, including solubility, membrane permeability, and metabolic stability. Before clinical trials, it should be sufficiently proved that a drug candidate exerts its stimulatory or inhibitory functions by directly binding to the target and subsequently exerts therapeutic effects in an animal model via systemic administration without organ failure.

In contrast, the multi-specific binding strategy builds upon the foundation of classical drug design rationale while adding an additional layer of functionality. One approach involves recruiting the target to a cell-specific effector, allowing this effector to modulate the target. An example illustrating this concept is the proteolysis targeting chimera (PROTAC). Another approach involves confining the drug's action to a specific site, primarily to minimize adverse drug reactions. This is exemplified by antibody-drug conjugates (ADCs) [10].

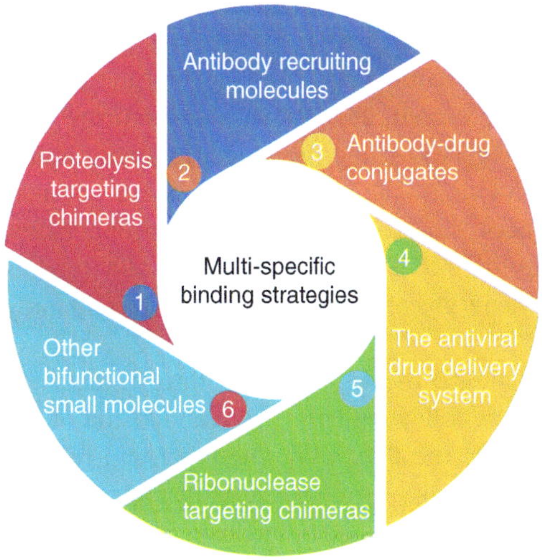

Figure 11.1 Outline of multi-specific binding strategies in the antiviral field.

In recent years, the rapid development of multi-specific drugs has become a prominent research focus [11, 12]. This article provides a comprehensive overview of various multi-specific binding strategies in antiviral drug development from the perspective of medicinal chemistry. These strategies include PROTACs, antibody recruiting molecules (ARMs), ADCs, antiviral drug delivery aystems, and ribonuclease targeting chimeras (RIBOTACs), as illustrated in Figure 11.1. The objective is to offer guidance for current and future endeavors in antiviral drug development.

Polypharmacology: Strategies for Multi-Target Drug Discovery, First Edition. Edited by Jens-Uwe Peters.

11.1 Proteolysis Targeting Chimera (PROTAC)

The concept of PROTAC was originally introduced by Professor Craig Crews and collaborators in 2001 [13]. It harnesses the eukaryotic protein degradation machineries, such as the natural ubiquitin-proteasome system (UPS), that reduce protein levels rather than inhibit their functionality, thereby achieving therapeutic goals for various diseases. The PROTAC molecule is a bifunctional entity comprising three essential components: a ligand targeting the protein of interest (POI), another ligand engaging with an E3 ubiquitin ligase, and a linker bridging the two ligands (Figure 11.2) [14]. E3 ubiquitin ligases in the PROTAC technology encompass several types, with cereblon (CRBN), von Hippel-Lindau (VHL), tripartite motif-containing protein 25 (TRIM25), and murine double minute 2 (MDM2) being the most prevalent options [15, 16]. Additionally, the nature of the linker substantially influences the structural rigidity, hydrophobicity, and solubility of PROTAC molecules, with alkyl chains and polyethylene glycol (PEG) chains being commonly employed [17].

PROTAC facilitates the degradation of the POI mediated by UPS, with the following basic process: one end of the PROTAC binds to the POI, while the other end recruits the E3 ubiquitin ligase, forming a ternary complex. Subsequently, the E3 ubiquitin ligase mediates the transfer of ubiquitin from the E1 ubiquitin-activating enzyme and the E2 ubiquitin-conjugating enzyme onto the POI, thereby resulting in multi-ubiquitination of the POI, which is then recognized and degraded by the cellular proteasome [18, 19]. Concomitantly, the PROTAC molecule is released to engage in the degradation of the next POI. Unlike the traditional small molecules that necessitate relatively high concentrations within the body to occupy the active site of the target, PROTAC operates in an "event-driven" manner, requiring catalytic quantities to be effective, thus circumventing off-target toxicities associated with drug administration [20, 21]. Furthermore, PROTAC addresses the challenges of targeting "undruggable proteins" that are difficult for small molecules to access, such as scaffold proteins, transcription factor modulators, or epigenetic-related proteins in cells [22]. Moreover, drug resistance is an inevitable concern during the clinical application of conventional small molecular inhibitors, whereas the advantage of PROTAC's weak binding to active sites gradually becomes evident in overcoming drug resistance [23, 24].

In recent years, the PROTAC strategy has progressively found applicability in the field of antiviral drug development.

In 2014, Montrose et al. [25] reported a study on the application of PROTAC technology in reducing HBV X protein levels. In their study, a novel cell-penetrating PROTAC (**1**, Figure 11.3) demonstrated significant effectiveness. This compound comprises three components: an oligomerization domain at the N-terminus of the X protein, an instability domain at the C-terminus of the X protein, and a fused polyarginine cell-penetrating peptide (CPP) (Figure 11.3). The oligomerization domain counteracts the pro-apoptotic function of the X protein, CPP enhances the cell-penetrating capability of the PROTAC, and the instability domain designates the X protein for proteasomal degradation. The research indicated that compound **1**

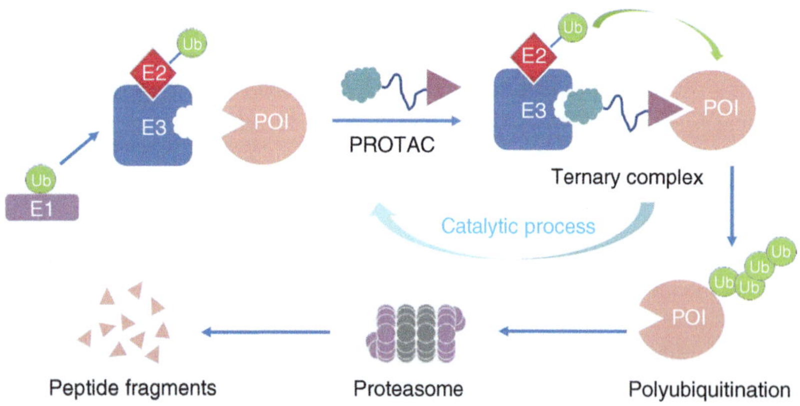

Figure 11.2 A mechanistic overview of PROTAC-mediated protein degradation. PROTAC: proteolysis targeting chimera; E3: E3 ubiquitin ligase; E1: E1 ubiquitin-activating enzyme; E2: E2 ubiquitin-conjugating enzyme; POI: Protein of interest; Ub: Ubiquitin.

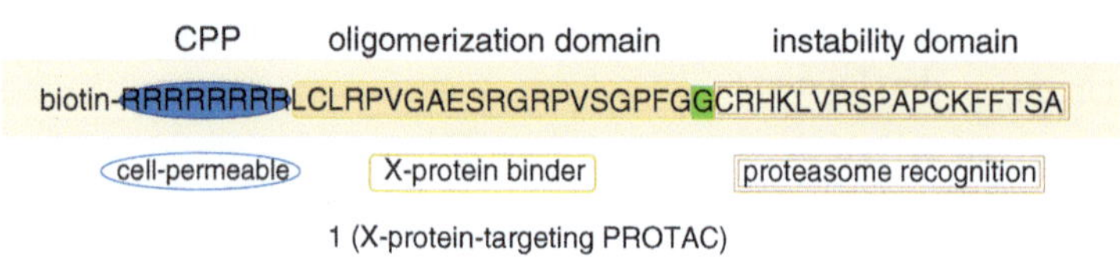

Figure 11.3 Amino acid sequence of the HBV X-protein-targeting PROTAC. CPP: polyarginine cell-penetrating peptide.

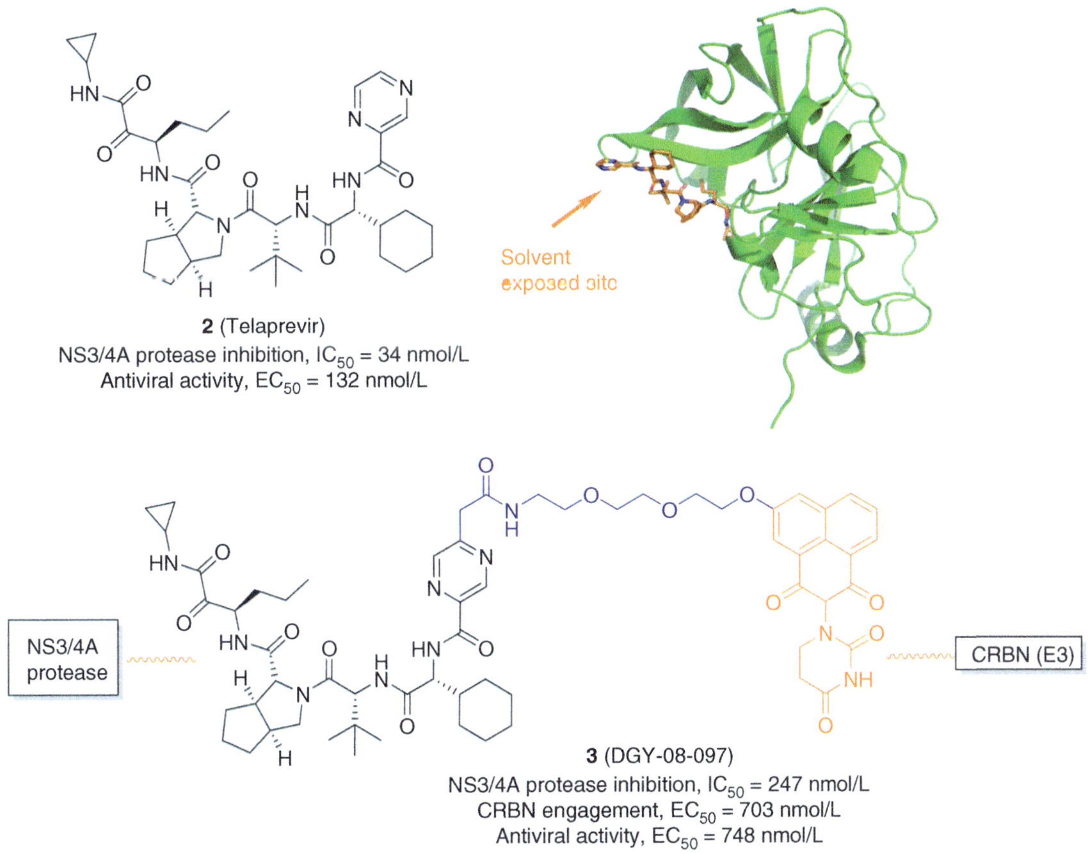

Figure 11.4 Chemical structures of **2**, **3** and co-crystal structure of **2** bound to the HCV NS3/4A protease complex (PDB code: 3SV6); CRBN: cereblon.

could degrade both the full-length and C-terminally truncated forms of the X protein in HepG2 liver cancer cells. However, further investigations are still required to explore the potential of this peptide-based PROTAC in inhibiting HBV replication and treating chronic hepatitis.

Telaprevir (**2**, Figure 11.4) is the first reversible covalent inhibitor that directly targets the NS3/4A protease of hepatitis C virus (HCV). However, it suffers from low target affinity and is prone to the development of drug resistance [26, 27]. To address these limitations, in 2019, de Wispelaere et al. [28] leveraged the crystal structure of telaprevir in complex with the HCV NS3/4A protease (PDB code: 3SV6) and strategically linked a CRBN ligand to the pyrimidine ring of telaprevir within the solvent-accessible pocket, resulting in the design of the PROTAC molecule DGY-08-097 (EC_{50} = 748 nmol/L, **3**, Figure 11.4) for targeted viral protein degradation. Protein immunoblotting assays and quantitative proteomic studies demonstrated that compound **3** reduces intracellular NS3 protein levels in a concentration-dependent manner, verifying target protein degradation-mediated antiviral actions. Moreover, **3** retains activity against two variants (HCV-NS3-V55A, IC_{50} = 288 nmol/L; HCV-NS3-A156S, IC_{50} = 1561 nmol/L), partially addressing telaprevir's resistance issue. This study validates the feasibility of the small-molecule PROTAC technology as an effective strategy for suppressing viral replication.

In 2022, Li et al. [29] utilized oleanolic acid (OA, **4**, Figure 11.5) as a ligand for targeting influenza viral hemagglutinin (HA) and designed two classes of novel pentacyclic triterpenoid-based PROTAC molecules by employing CRBN (compounds C1-C3, **5**, Figure 11.5) and VHL (compounds V1-V6, **6**, Figure 11.5) E3 ligases. This study marked the first application of PROTAC technology in the field of anti-influenza research. Mechanistic validation experiments revealed that V3 exhibited concentration-dependent degradation of HA (EC_{50} = 1.44 μmol/L), with the degradation process reliant on proteasomes. High-resolution mass spectrometry analysis of photo-crosslinked complexes further demonstrated that **4** exhibited a preference for binding to non-active sites, while V3 exhibited a novel mode of antiviral action through degradation, rather than inhibiting its function. This work represents a significant

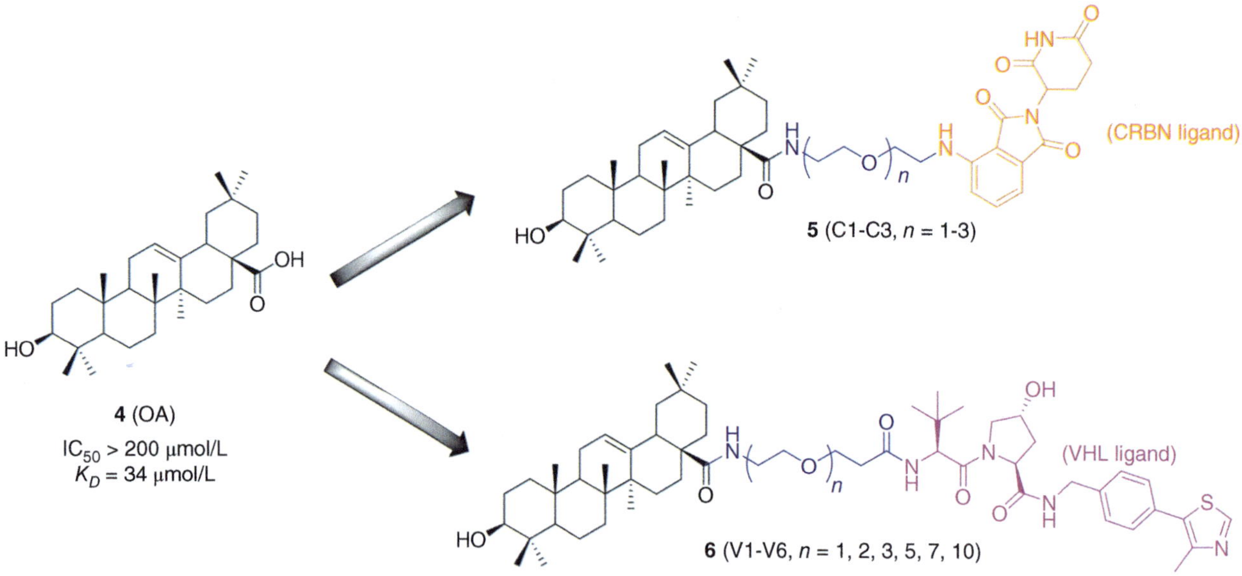

Figure 11.5 Chemical structures of **4–6**. OA: oleanolic acid; VHL: von Hippel–Lindau.

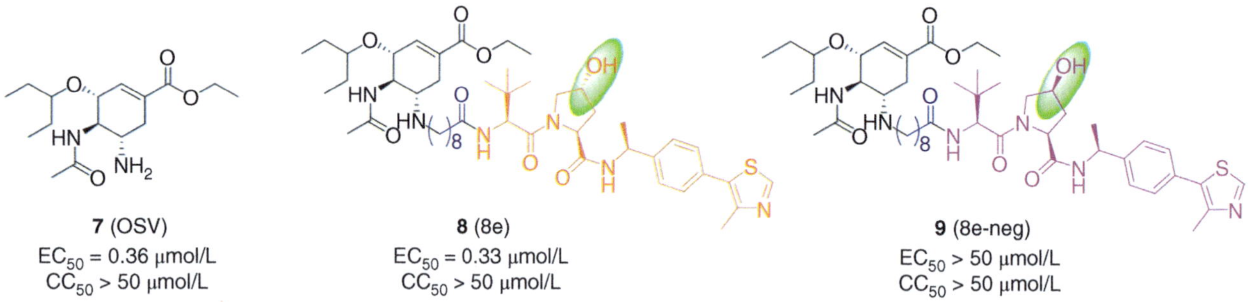

Figure 11.6 Chemical structures of **7–9**.

milestone in the application of this technique within the realm of influenza therapeutics.

In 2022, Xu et al. [30] utilized the existing anti-influenza drug oseltamivir (OSV, **7**, Figure 11.6), a neuraminidase (NA) inhibitor, as a basis to design the PROTAC compound **8e** (**8**, Figure 11.6). In MDCK cells, this compound exhibited favorable activity and low cytotoxicity against the wild-type influenza virus strain H1N1 ($EC_{50} = 0.33$ μmol/L, $CC_{50} > 50$ μmol/L). Protein immunoblotting assays revealed that **8** significantly reduced the expression level of NA, while the negative control with different chirality, compound 8e-neg (**9**, Figure 11.6), failed to exhibit such an effect. This suggests that the antiviral activity of **8** might be attributed to the degradation of NA. Further mechanistic validation experiments indicated that **8** achieves NA degradation in a concentration-dependent manner through the ubiquitin-proteasome pathway. Additionally, it effectively countered the OSV-resistant H1N1 strain (H1N1 H274Y), further confirming the potential of the PROTAC strategy in addressing antiviral drug resistance issues.

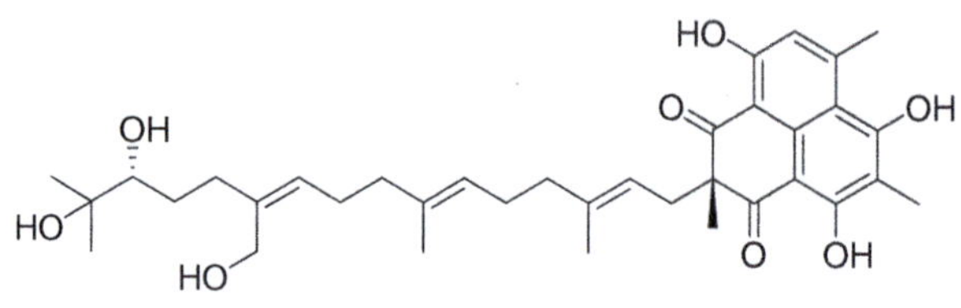

10 (APL-16-5)
$EC_{50} = 0.28$ μmol/L
$CC_{50} > 100$ μmol/L

Figure 11.7 Chemical structure of **10.**

In 2022, Zhao et al. [31] reported a microorganism-derived compound named APL-16-5 ($EC_{50} = 0.28$ μmol/L, $CC_{50} > 100$ μmol/L,**10**, Figure 11.7) with antiviral activity against influenza A virus (IAV). Mechanistic experiments revealed that compound **10** interacted with the E3 ubiquitin ligase TRIM25 and the IAV polymerase PA subunit, leading to ubiquitination of the PA protein and subsequent degradation through a proteasome-dependent mechanism.

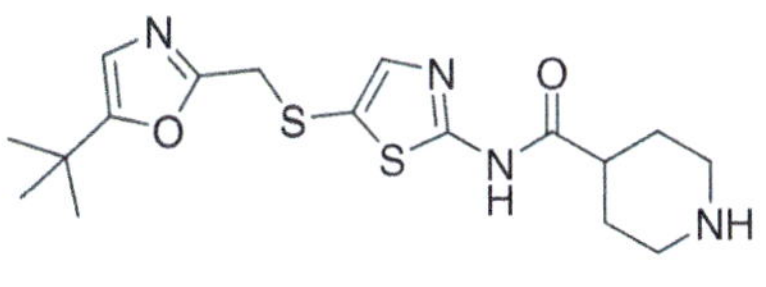

11 (SNS032)
$EC_{50} = 0.105 \pm 0.004$ µmol/L
$CC_{50} = 0.240 \pm 0.022$ µmol/L

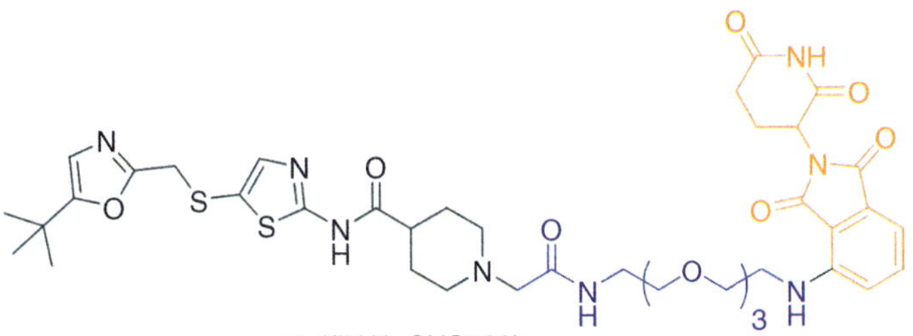

12 (THAL-SNS032)
$EC_{50} = 0.025 \pm 0.001$ µmol/L
$CC_{50} = 0.125 \pm 0.014$ µmol/L

Figure 11.8 Chemical structures of **11** and **12**.

The cyclin-dependent kinase (CDK) inhibitor SNS032 ($EC_{50} = 0.105 \pm 0.004$ µmol/L, **11**, Figure 11.8) exhibits significant activity against human cytomegalovirus (HCMV) [32, 33]. In 2021, Hahn et al. [33] employed **11** as a lead compound to design a PROTAC molecule, named THAL-SNS032, which revealed potent anti-HCMV activity ($EC_{50} = 0.025$ µmol/L, $CC_{50} = 0.125$ µmol/L,**12**, Figure 11.8). Compound **12** is derived from **11** by linking it with the CRBN ligand thalidomide through a linker. Protein immunoblotting experiments demonstrated that under conditions of low cytotoxicity, **12** induced concentration-dependent degradation of CDK and displayed broad-spectrum antiviral activity (murine human cytomegalovirus, MCMV, $EC_{50} = 0.21 \pm 0.09$ µmol/L, $CC_{50} = 1.00 \pm 0.22$ µmol/L; SARS-CoV-2, $EC_{50} = 0.11 \pm 0.02$ µmol/L, $CC_{50} = 64.9 \pm 16.2$ µmol/L).

3CL^{pro} (3C-like protease), also known as main protease (M^{pro}), is an essential non-structural protein in coronaviruses that plays a crucial role in the replication and transcription of viral progeny. In 2022, Panlong Pharmaceuticals designed two classes of PROTAC molecules targeting the 3CL^{pro} candidate drug PLC-01 (**13**, Figure 11.9) using a novel mechanism of protein degradation, marking the first report in the field of coronaviruses [34, 35]. Researchers employed different linkers to conjugate **13** with CRBN ligands (thalidomide) and VHL ligands, resulting in six series of PROTAC molecules. Fluorescence resonance experiments and protein immunoblot assays demonstrated that certain PROTAC molecules (i.e., **14–23**, Figure 11.9) exhibited both inhibitory activity against 3CL^{pro} (IC_{50} = 50-100 nmol/L) and effective degradation activity (DC_{50} = 50-100 nmol/L).

Recently, a new class of antiviral small molecules that induce the degradation of SARS-CoV-2 M^{Pro} has been reported [36]. The research team started with previously developed M^{Pro} inhibitors MPI8 (**24**, Figure 11.10) and MPI29 (**25**, Figure 11.10), designed and synthesized a new class of small-molecule antivirals. Among them, MPD2 (**26**, Figure 11.10) was demonstrated to effectively reduce M^{Pro} protein levels in 293T cells ($IC_{50} = 41$ nmol/L, $DC_{50} = 296$ nmol/L) in a time-dependent, CRBN-mediated, and proteasome-driven mechanism. Furthermore, MPD2 showed significant efficacy in diminishing the levels of M^{Pro} protein in SARS-CoV-2-infected A549-ACE2 cells, concurrently demonstrating potent anti-SARS-CoV-2 activity ($EC_{50} = 492$ nmol/L). This proof-of-concept study highlights the potential of PROTAC-mediated M^{Pro} targeted protein degradation as an innovative and promising approach in COVID-19 drug discovery.

Enterovirus A 71 (EV71) is a neurotropic virus that may lead to acute flaccid paralysis, encephalitis, cardiopulmonary failure, or even death. No vaccine and defensive drug controlling EV71 is currently available, novel and efficient antiviral drug is therefore urgently needed. 3D^{pol} (RNA-dependent RNA polymerase (RdRp)) has been an important target for anti-EV71 drug development [37]. In 2023, Zhou et al. designed two classes of PROTAC compounds for the broad-spectrum antiviral drug Remdesivir (**27**, Figure 11.11) and its intermediate GS-441524 (**28**, Figure 11.11) [38]. Among their derivatives, compounds **29-34** (Figure 11.11) displayed significantly superior antiviral activity compared to the parent molecule **28** and were comparable to **27**. On the other hand, compounds **29-34** were capable of ubiquitin-dependent degradation of the key enzyme 3D^{pol} required for EV71 replication through an *in vitro* pathway. This enhancement in activity for such compounds over the active metabolite **28** of **27** significantly improved their effectiveness, reaching a level similar to that of **27**.

The *nef* gene of lentiviruses in primates, including HIV-1, HIV-2, and simian immunodeficiency virus (SIV), encodes a unique membrane-associated protein that plays a crucial role in virus replication. Recently, Nef-directed PROTACs effectively prevented HIV-1 multiplication in donor PBMCs and restored Nef-mediated MHC-I and CD4 downregulation in T cells. All HIV-1 Nef functions are expected to be reversed by targeted degradation of Nef, which may also aid in the restoration of adaptive immune responses against HIV-1 reservoir cells *in vivo* [39].

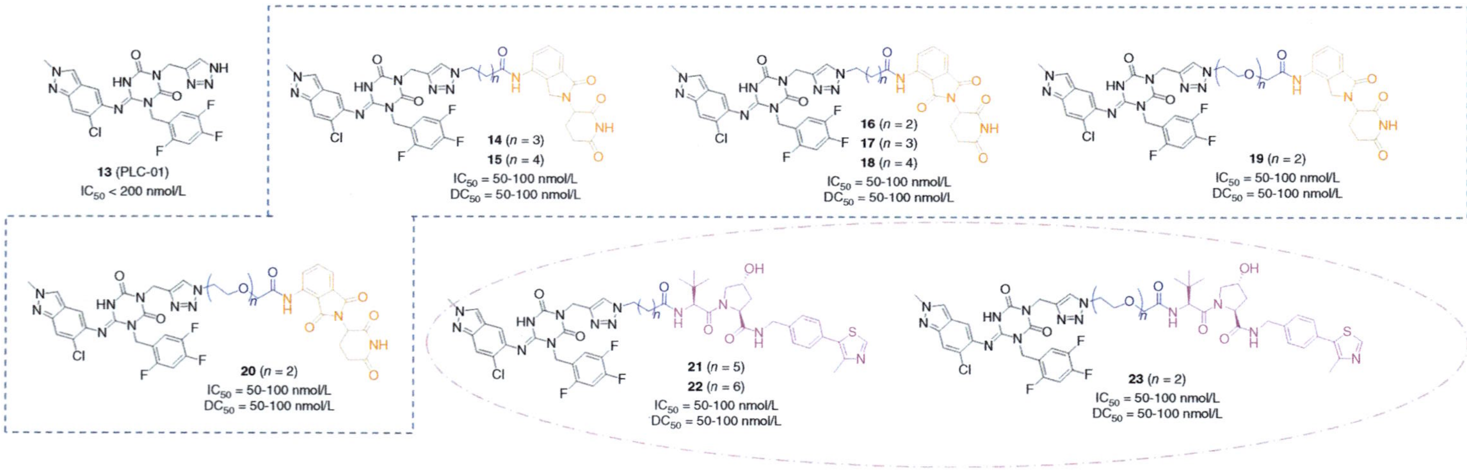

Figure 11.9 Chemical structures of **13–23**.

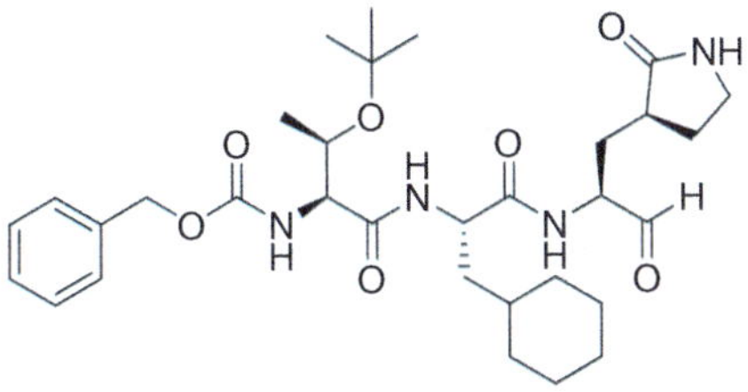

24 (MPI8)

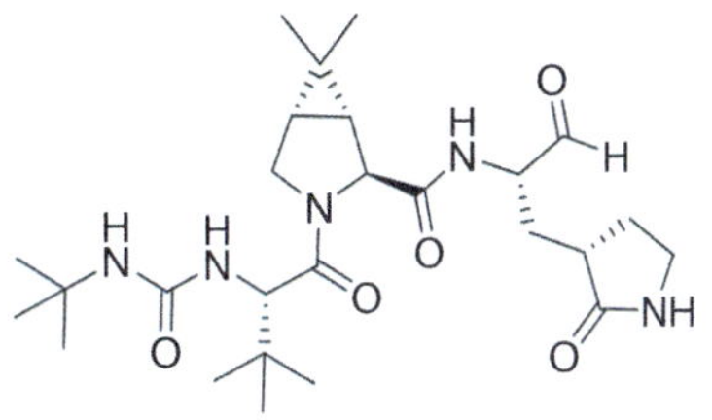

25 (MPI29)

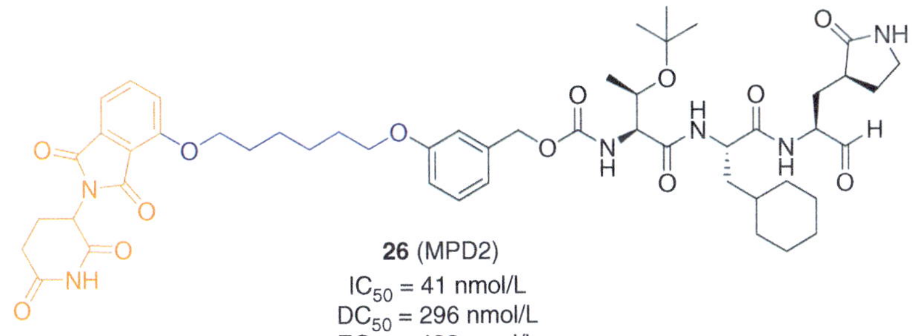

26 (MPD2)
IC_{50} = 41 nmol/L
DC_{50} = 296 nmol/L
EC_{50} = 492 nmol/L
CC_{50} = 120 μmol/L

Figure 11.10 Chemical structures of **24–26.**

11.2 Antibody Recruiting Molecules

The specific recognition and binding between antibodies (Ab) and antigens (Ag) form the foundation of biomedical therapies. Based on their origin, antibodies can be categorized into exogenous antibodies and endogenous antibodies. Endogenous antibodies are naturally generated in response to non-self antigens encountered in daily life and are present in the human bloodstream, including IgM, IgG, and other immunoglobulins [40]. Notable examples of endogenous antibodies found in human serum include antibodies against dinitrophenol (DNP), *L*-rhamnose (Rha), *α*-Gal, and aldolase [41]. Recent research has indicated that recruiting endogenous antibodies to the surface of target cells can activate the immune system to recognize infected cells and induce their destruction through immune mechanisms mediated by the Fc region [42]. ARMs have emerged as a novel class of molecules based on this concept.

ARMs are a type of bispecific small molecules composed of three domains: the antibody-binding terminus (ABT), the target-binding terminus (TBT), and the linker connecting the two [43]. ARMs facilitate the formation of a ternary complex between endogenous antibodies and pathogenic factors (such as proteins, viruses, or infected cells). This complex then recruits additional antibodies or various immune effector cells, ultimately inducing the clearance of target cells through local or systemic immune responses [44]. Firstly, in complement-dependent cytotoxicity (CDC), the Fc region of the endogenous antibody binds to complement C1q, leading to the formation of the membrane attack complex (MAC) on the surface of the target cell. This triggers cell lysis by causing an influx of ions and disrupting the cell membrane, eliminating target cells such as pathogen-infected cells and abnormally proliferating cancer cells [45]. Secondly, the Fc region of endogenous antibodies can interact with the Fcγ receptor (FcγR) on immune cells, linking the humoral and cellular immune responses [46]. Thirdly, phagocytic cells can eliminate target cells through antibody-dependent cellular phagocytosis (ADCP) [47]. Fourthly, in antibody-dependent cell-mediated cytotoxicity (ADCC), the Fc segment binds to Fc receptors (such as FcγRIIIa/CD16) on effector cells, mediating the killing of target cells (tumor cells, infected cells, pathogenic immune cells, etc.) [48]. In recent years, antibody-recruiting-based strategies have been progressively applied in the field of antiviral therapy.

HIV infection relies on the fusion between the viral envelope (Env) and the host cell membrane, a process mediated by the Env glycoproteins gp120 and gp41. Gp120 interacts with the cell surface receptors CD4 and co-receptors CCR5 or CXCR4. The transmembrane subunit gp41 forms a six-helix bundle (six-helix bundle, 6HB) by assembling the C-terminal heptad repeat region (CHR) and N-terminal heptad repeat region (NHR). This assembly facilitates membrane fusion between

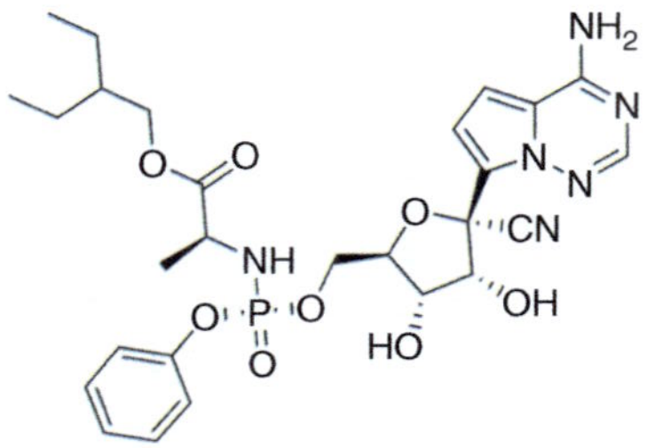

27 (Remdesivir)
EC_{50} = 1.03 μmol/L
CC_{50} > 50 μmol/L

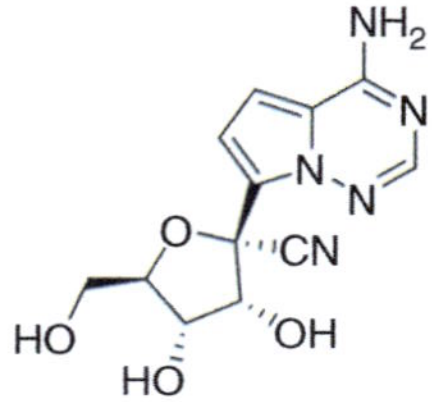

28 (GS-441524)
EC_{50} = 8.58 μmol/L
CC_{50} > 50 μmol/L

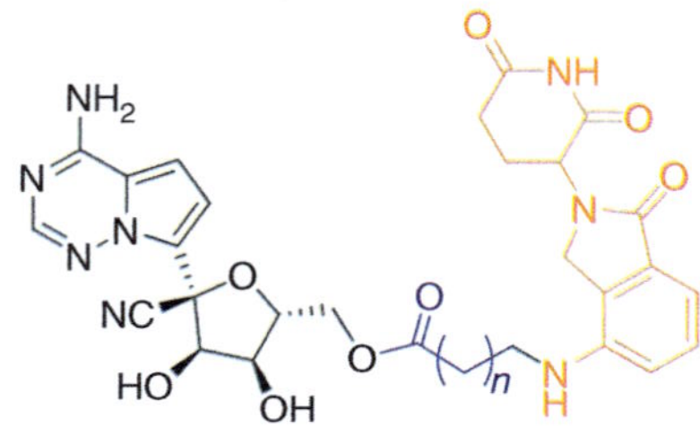

29 (*n* = 8)
EC_{50} = 2.35 μmol/L
CC_{50} > 100 μmol/L

30 (*n* = 9)
EC_{50} = 1.32 μmol/L
CC_{50} > 100 μmol/L

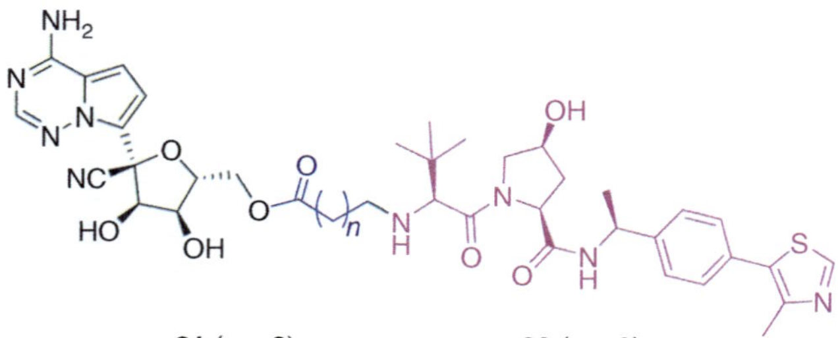

31 (*n* = 8)
EC_{50} = 1.35 μmol/L
CC_{50} > 100 μmol/L

32 (*n* = 9)
EC_{50} = 1.32 μmol/L
CC_{50} > 100 μmol/L

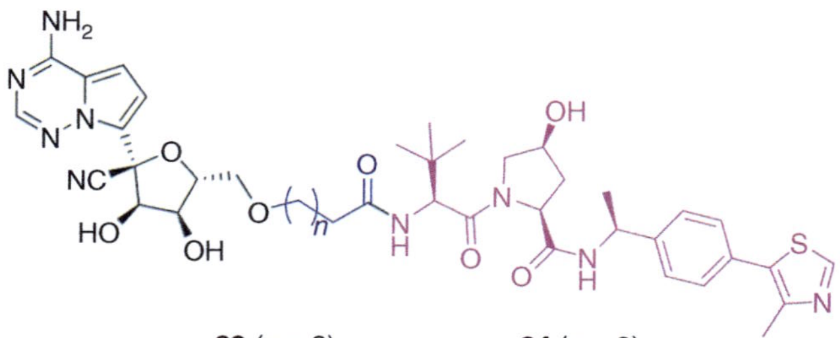

33 (*n* = 8)
EC_{50} = 4.45 μmol/L
CC_{50} > 50 μmol/L

34 (*n* = 9)
EC_{50} = 2.01 μmol/L
CC_{50} > 50 μmol/L

Figure 11.11 Chemical structures of **27**–**34**.

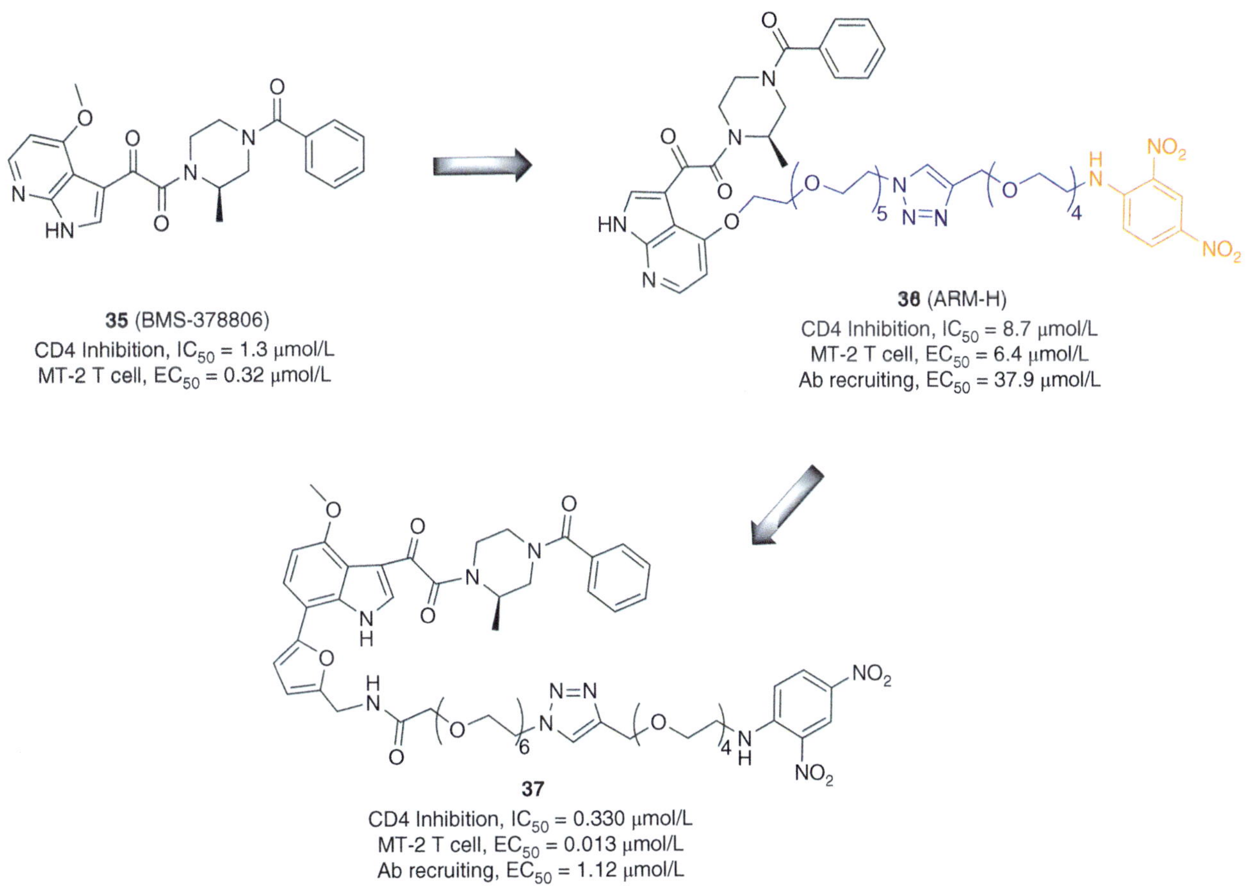

Figure 11.12 Chemical structures of **35**–**37**.

the virus and host cell, and inhibiting this fusion process effectively curtails HIV replication [49]. In 2009, Parker et al. [50] utilized the membrane fusion inhibitor BMS-378806 as a lead compound (**35**, Figure 11.12) and employed an antibody-recruiting strategy to design an antibody-recruiting molecule (ARM-H) for HIV treatment (**36**, Figure 11.12). Compound **36** was obtained by linking **35** to DNP through a linker. Enzyme-linked immunosorbent assay (ELISA) demonstrated that **35** bound to the viral glycoprotein gp120, thereby inhibiting its interaction with CD4 (**36**, IC_{50} = 8.7 μmol/L; **35**, IC_{50} = 1.3 μmol/L). The researchers further evaluated their inhibitory effects on HIV-1 entry into human MT-2 T cells (**36**, EC_{50} = 6.4 μmol/L; **35**, EC_{50} = 0.32 μmol/L), and **36** did not exhibit cytotoxicity. Furthermore, **36** could recruit anti-DNP antibodies in a concentration-dependent manner, forming a ternary complex comprising gp120, **36**, and anti-DNP antibodies, thereby activating the immune system's CDC to induce death of infected cells.

In 2014, the same research team [51] combined medicinal chemistry, biochemistry, cell-based assays, and computational chemistry to develop an optimized ARM-H analog (**37**, Figure 11.12). This analog displayed enhanced potency in inhibiting virus replication (MT-2 T cells) compared to **36**, with a higher affinity for gp120 and improved antibody recruitment ability (MT-2 T cell EC_{50} = 12 μmol/L, CD4 inhibition IC_{50} = 19.9 μmol/L, Ab recruiting EC_{50} = 37.9 μmol/L). In summary, compounds **37** and **37** are distinctive in that they represent a multi-specific binding strategy based on antibody recruitment. Through two synergistic mechanisms, they target different stages of the virus life cycle by inhibiting viral entry while also targeting cells expressing Env glycoprotein for immune recognition and clearance.

Influenza viral surface glycoprotein NA inherently possesses antigenic properties, and these can be enhanced through effective semi-antigen modification to elicit an immune response for clearing infected cells. In 2020, Liu et al. [52] designed a conjugate of the NA inhibitor zanamivir (A/Wisconsin/629-D00015/2009 H1N1 pdm09, K_D = 0.3 nmol/L; A/Netherlands/22/2003 H3N2, K_D = 1.0 nmol/L; B/Florida/4/2006 Yamagata Lineage, K_D = 2.4 nmol/L; B/Brisbane/60/2008 Victoria Lineage, K_D = 7.2 nmol/L; ZNV, **38**, Figure 11.13) with the highly immunogenic semi-antigen DNP to create the zanamivir-DNP conjugate (zan-DNP, **39**, Figure 11.13).

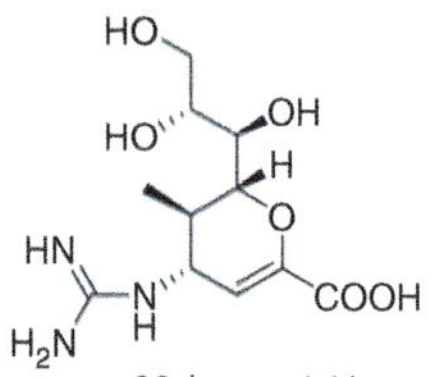

38 (zanamivir)
A/Wisconsin/629-D00015/2009 H1N1 pdm09, K_D = 0.3 nmol/L
A/Netherlands/22/2003 H3N2, K_D = 1.0 nmol/L
B/Florida/4/2006 Yamagata Lineage, K_D = 2.4 nmol/L
B/Brisbane/60/2008 Victoria Lineage, K_D = 7.2 nmol/L

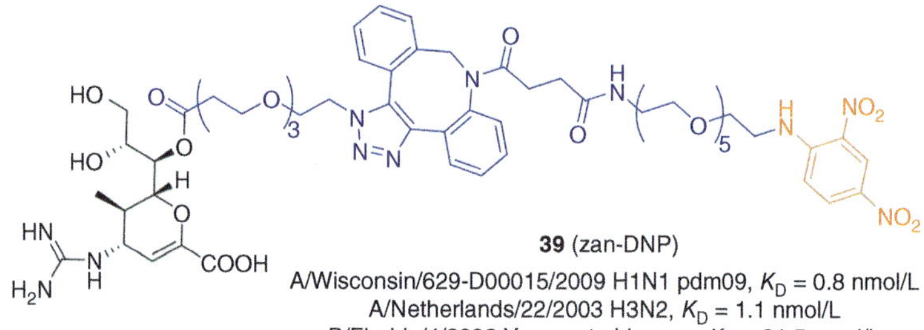

39 (zan-DNP)
A/Wisconsin/629-D00015/2009 H1N1 pdm09, K_D = 0.8 nmol/L
A/Netherlands/22/2003 H3N2, K_D = 1.1 nmol/L
B/Florida/4/2006 Yamagata Lineage, K_D = 21.5 nmol/L
B/Brisbane/60/2008 Victoria Lineage, K_D = 58.4 nmol/L

Figure 11.13 Chemical structures of **38** and **39**.

Surface plasmon resonance experiments and confocal microscopy images demonstrated that **39** exhibited strong binding affinity to NA (A/Wisconsin/629-D00015/2009 H1N1 pdm09, K_D = 0.8 nmol/L; A/Netherlands/22/2003 H3N2, K_D = 1.1 nmol/L; B/Florida/4/2006 Yamagata Lineage, K_D = 21.5 nmol/L; B/Brisbane/60/2008 Victoria Lineage, K_D = 58.4 nmol/L). Furthermore, it could recruit anti-DNP antibodies in a concentration-dependent manner. This conjugate effectively inhibited the NAs of both influenza A and B viruses and recruited anti-DNP antibodies to infected cells, thereby triggering the immune system's CDC and ADCC against the virus-infected cells. Additionally, a single dose of **39** administered to mice infected with viruses at 100 times the median lethal dose (MLD50) could eliminate late-stage infections caused by representative strains of influenza A and B viruses (e.g., A/Puerto Rico/8/1934 H1N1, A/Puerto Rico/8/1934 Reassortant X-31 H3N2, A/California/07/200 H1N1 pdm09, B/Florida/4/2006). The treatment remained effective even for mice severely infected for up to 3 days, suggesting that this multi-specific drug might successfully address influenza virus infections that are challenging to treat with current therapies.

11.3 Antibody-Drug Conjugates (ADCs)

ADCs represent a new generation of targeted chemoimmunotherapy that utilizes macromolecules, such as antibodies (Ab), as delivery vehicles for small-molecule cytotoxic drugs, in which the two parts are connected with linker molecules. Originally, the tripartite ADCs deployed the antigen-specific targeting capability of Ab to recognize biomarkers distinctively or prominently expressed on the target cell surface. This allows efficient delivery of cytotoxic drugs to the interior of the targeted cells via endocytosis, leading to their apoptosis, as well as programming the immune repertoire to summon effector T cells, natural killer cells, or macrophages, eventually eliminating cancer cells [53]. Unlike ARMs which are also bispecific small molecules but designed to bind to endogenous antibodies via non-covalent interaction, ADCs employ covalent binding between antibodies and cytotoxic agents to induce cell killing. The unique composition of ADCs results in distinct pharmacokinetics, pharmacodynamics, and resistance mechanisms compared to other types of drugs. In recent years, the strategy of antibody-drug conjugation has gradually been applied in the field of antiviral drug development.

The co-receptor CCR5 plays a pivotal role in the fusion process between the lentiviruses and their susceptible cells. In 2010, Gavrilyuk et al. [54] designed and synthesized the 38C2–Aplaviroc conjugate (**41**, Figure 11.14) by linking the CCR5 inhibitor Aplaviroc (**40**, Figure 11.14) to the aldolase antibody 38C2 through a linker. Flow cytometry experiments and neutralization assays demonstrated that **41** could bind to the co-receptor CCR5, effectively inhibiting entry of HIV, SIV, and simian/human immunodeficiency virus (SHIV) into target cells (HIV-$1_{JR\text{-}FL}$ IC_{50} = 0.57 nmol/L; SIVmac239 IC_{50} = 0.1 nmol/L; $SHIV_{162P3}$ IC_{50} = 0.89 nmol/L). In addition, through stability assay of the **41** in human serum, researchers found that the conjugate possessed a chemical stability that would allow for extended pharmacokinetics. This study highlights the potential application of the antibody-drug conjugation strategy by presenting the chemical programming of Ab with an antiviral agent.

In 2013, Sato et al. [55] reported a study involving the construction of ADCs based on the HIV membrane fusion inhibitors BMS-378806 (**35**, Figure 11.12) and BMS-488043 (**42**, Figure 11.15). The researchers synthesized N-acyl-β-lactam derivatives **43** (Figure 11.15) and **44** (IC_{50} = 67.5 nmol/L, Figure 11.14) of the two membrane fusion inhibitors **35** and **42**. These compounds were further conjugated with mAb 38C2 to produce two conjugates, namely ADC **45** (Figure 11.15) and **46** (IC_{50} = 128 nmol/L, Figure 11.15). While **43** and **45** showed no significant activity, **44** and **46** were able to effectively inhibit HIV infection. This indicates that **42** possesses suitable conjugation sites for constructing ADCs, which could potentially facilitate the development of novel bispecific antibodies.

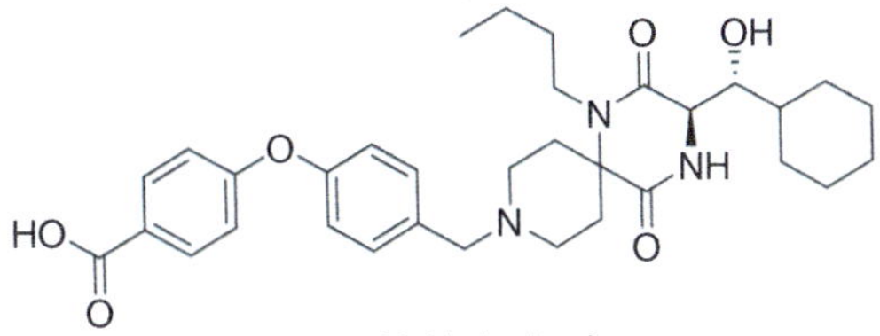

40 (Aplaviroc)

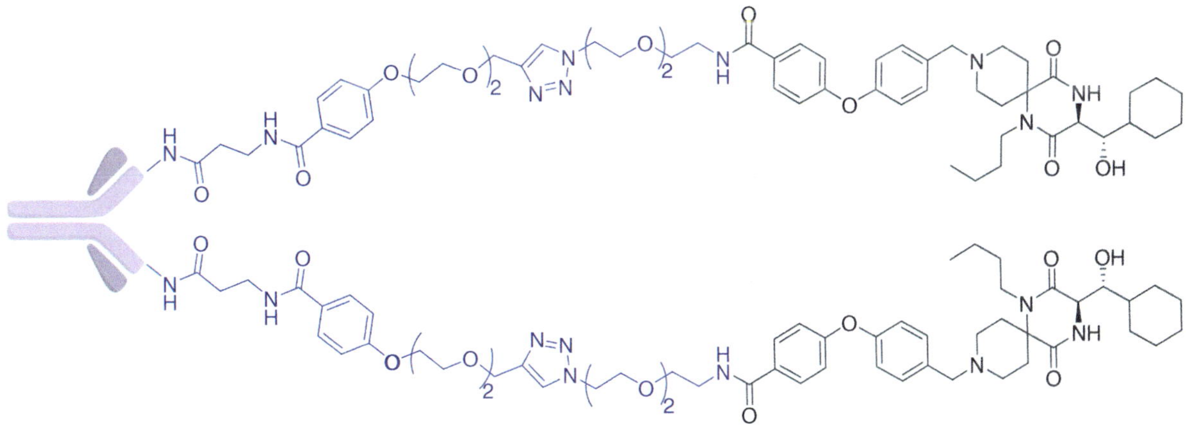

41 (38C2-Aplaviroc)
HIV-1_{JR-FL} IC_{50} = 0.57 nmol/L
SIV_{mac239} IC_{50} = 0.1 nmol/L
$SHIV_{162P3}$ IC_{50} = 0.89 nmol/L

Figure 11.14 Chemical structures of **40** and **41**. mAb 38C2: aldolase antibody 38C2.

In 2014, Asano et al. [56] identified effective conjugation sites on the CCR5 inhibitor Maraviroc (HIV-1_{JR-FL} IC_{50} = 1.6 ± 0.25 nmol/L, **47**, Figure 11.16) and designed ADCs by connecting it to mAb 38C2 through a linker. They developed Maraviroc ADCs **48** (HIV-1_{JR-FL} IC_{50} = 19 ± 2.6 nmol/L, Figure 11.16) and **49** (HIV-1_{JR-FL} IC_{50} = 7.7 ± 0.50 nmol/L, Figure 11.16). Notably, the activity of the triazole-derived **49** from **47** was not significantly compromised. Thus, this study identified feasible binding sites on **47** and successfully employed the multi-specific binding strategy of antibody-drug conjugation to exert antiviral effects.

In addition, two novel types of ADCs developed by Regeneron Pharmaceuticals and Cidara Therapeutics companies also hold promising potential for the development of antiviral drugs against influenza virus [57, 58]. Overall, ADCs have significant prospects in protecting diverse human viral infections.

11.4 Antiviral Drug Delivery Systems

11.4.1 Nucleic Acid Delivery Molecules

With the emergence of more transmissible variants of SARS-CoV-2, the development of antiviral drugs capable of rapidly adapting to the virus's continuous mutations has become increasingly urgent [59]. SARS-CoV-2 is a positive-sense single-stranded RNA virus that enters host cells through the human angiotensin-converting enzyme 2 (hACE2) receptor mainly. The viral genomic RNA is replicated through subgenomic mRNAs, enabling the synthesis of viral proteins. Research has shown that small interfering RNA (siRNA) can target both of these RNA molecules simultaneously, achieving dual gene silencing [60]. As an oligonucleotide drug, siRNAs can be directly linked to peptides, antibodies, or other small molecules to facilitate their membrane permeability or navigation to target cells and tissues by altering their *in vivo* distribution or metabolic dynamics [61]. In 2022, Traube et al. reported the design and synthesis of chemically stable siRNA targeting SARS-CoV-2 [62]. The researchers integrated an alkyne moiety at the 3′ end of siRNA, enabling further chemical modification through Cu(I)-catalyzed click chemistry. Following this strategy, siRNA was conjugated with a hACE2 receptor ligand to create siRNA-modified entities, specifically siRNA-hACE2-binding peptide conjugates (**50**, Figure 11.17a), thereby achieving targeted delivery. The study demonstrated that the receptor-binding peptide-conjugated siRNA could reduce virus replication and virus-induced cell apoptosis in 3D bronchial epithelial

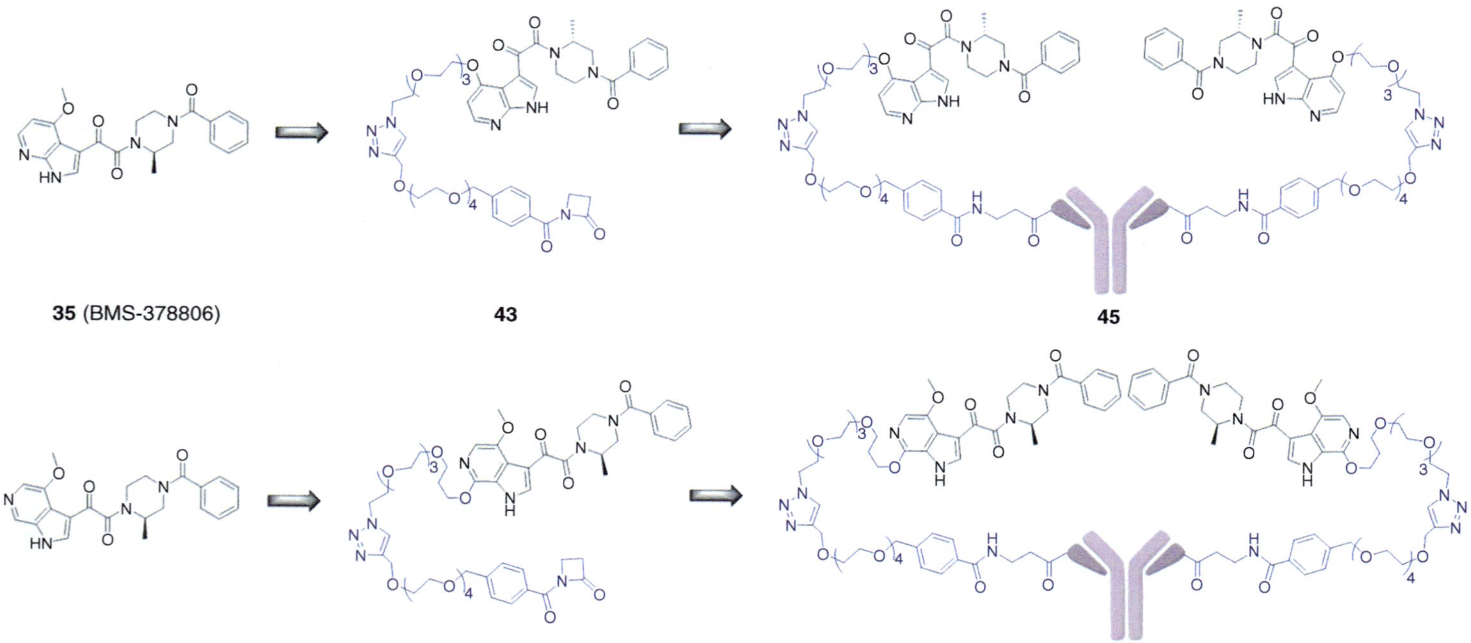

35 (BMS-378806) **43** **45**

42 (BMS-488043) **44** IC_{50} = 67.5 nmol/L **46** IC_{50} = 128 nmol/L

Figure 11.15 Chemical structures of **35** and **42–46**.

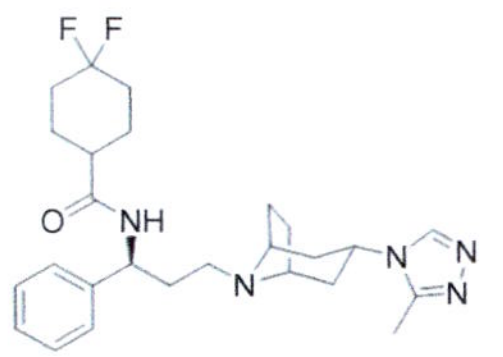

47 (Maraviroc)
HIV-$1_{JR\text{-}FL}$ IC_{50} = 1.6 ± 0.25 nmol/L

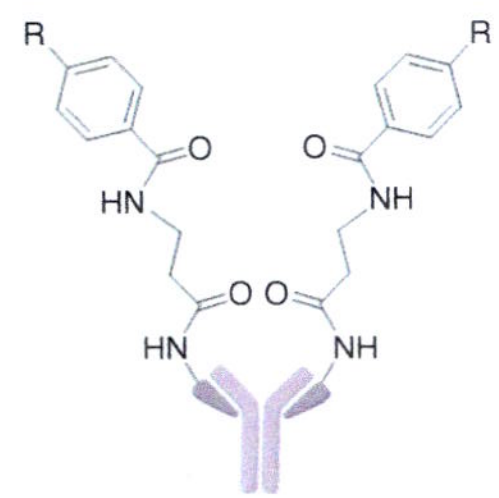

48 (R = X)
HIV-$1_{JR\text{-}FL}$ IC_{50} = 19 ± 2.6 nmol/L
49 (R = Y)
HIV-$1_{JR\text{-}FL}$ IC_{50} = 7.7 ± 0.50 nmol/L

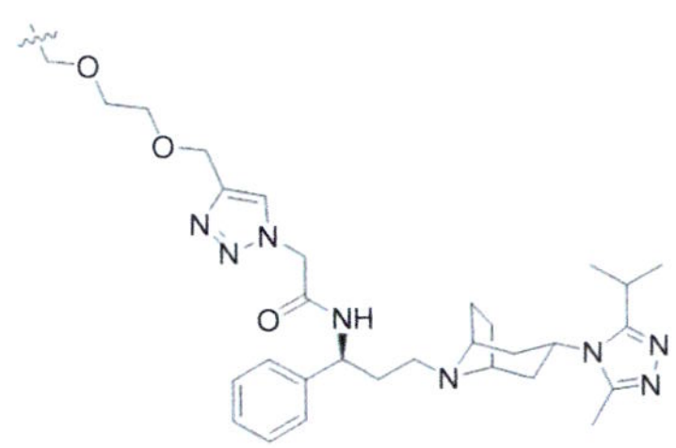

X

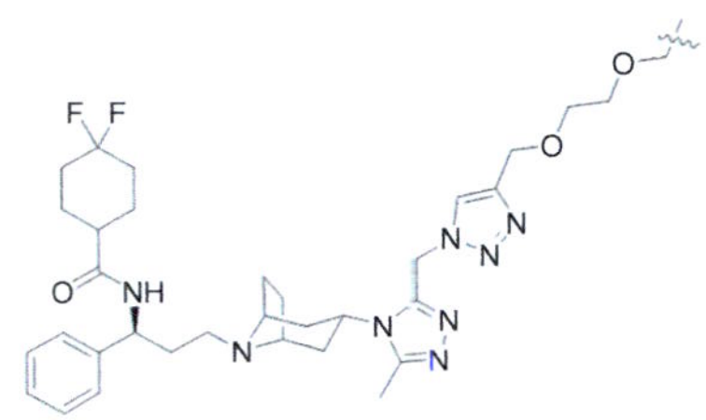

Y

Figure 11.16 Chemical structures of **47–49**.

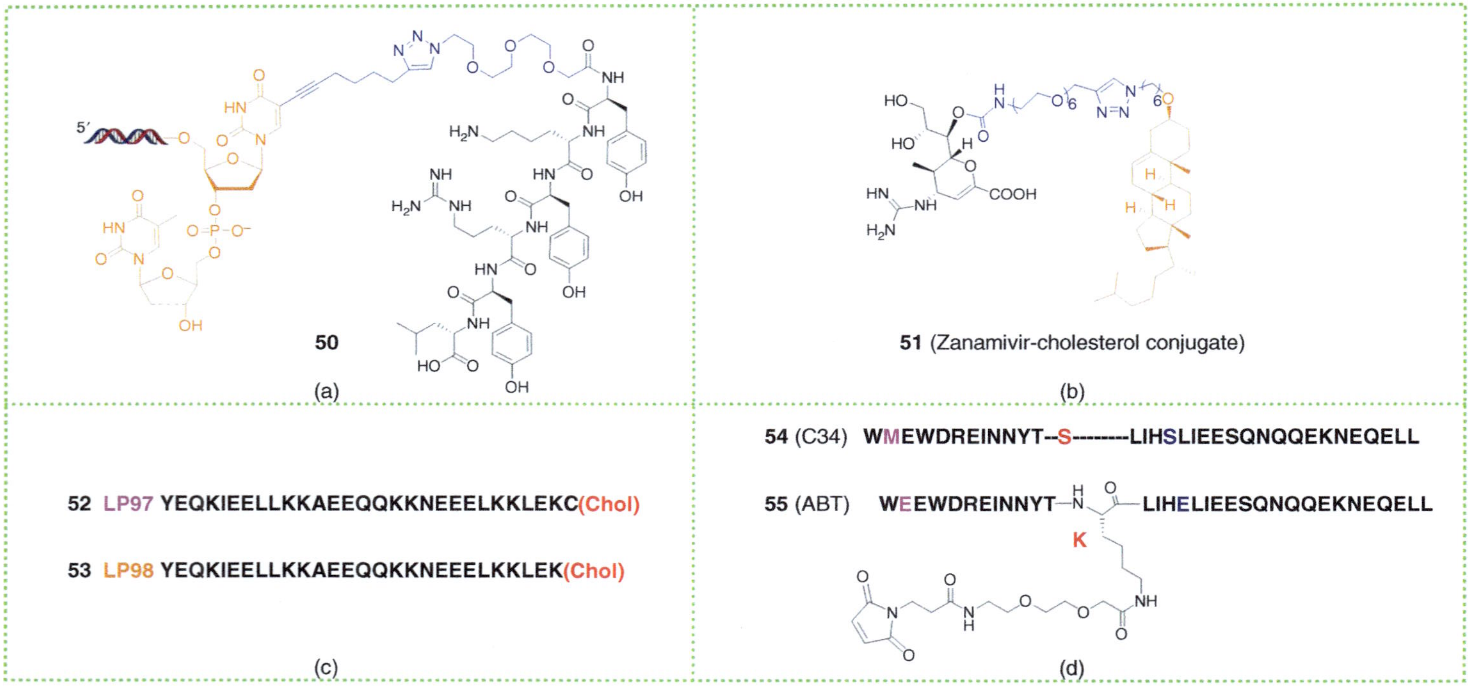

Figure 11.17 Chemical structures of **50** (a), **51** (b), **52** and **53** (c), **54** and **55** (d).

cultures. In summary, the approach of coupling siRNA to receptor ligands through click chemistry facilitates specific siRNA delivery, thereby exerting antiviral effects more selectively in virus-infected cells.

11.4.2 Cholesterol Conjugated Molecules

Cholesterol is abundant in the membranes of eukaryotic cells. Cholesterol conjugation can spontaneously insert modified nucleic acids and peptides into the lipid bilayer and subsequently promote their cellular uptake [63]. This membrane-targeting strategy is particularly advantageous for enhancing the antiviral efficacy of NA inhibitors, as they inhibit the NA activity on the surface of influenza virus-infected cells, thereby blocking the release and spread of progeny viruses [64]. In 2021, Lv et al. [65] discovered a cholesterol-conjugated zanamivir (ZNV) derivative (**51**, Figure 11.17b) as a long-acting NA inhibitor with potent activity against influenza virus (A/California/07/2009 H1N1). This compound conferred 100% survival to mice after single or multiple dosing against wild-type or mutant H1N1 influenza viruses carrying the OSV-resistant H275Y mutation. Compared to free ZNV ($t_{1/2}$ = 0.3 h in rats), compound **51** exhibited significantly improved antiviral activity and plasma half-life ($t_{1/2}$ = 7.6 h in rats). A single dose of **51** provided protection to mice from lethal attacks of wild-type or H275Y mutant H1N1 influenza viruses. Mechanistic studies confirmed that the conjugated molecule targeted the cell membrane and entered host cells, thereby inhibiting NA function as well as the assembly of progeny virus particles. Therefore, this study demonstrates that cholesterol conjugation may be an effective strategy for enhancing the pharmacokinetics and therapeutic efficacy of small-molecule drugs when targeting cell surface-localized viral proteins.

Enfuvirtide (T20), also known as T20, is a peptide derivative that targets the C-terminal heptad repeat (CHR) region. It is currently the only HIV fusion inhibitor approved by the FDA and is used in combination therapy for HIV-1 [66]. However, its antiviral activity is relatively low, and some mutant strains have developed resistance to it, highlighting the need for the development of new generations of HIV fusion-inhibitory lipopeptides. In 2022, Xue et al. [67] used T20 as a lead compound and employed a cholesterol conjugation strategy to develop the lipopeptide inhibitors LP-97 (**52**) and LP-98 (**53**, Figure 11.17c). These inhibitors exhibited highly efficient and sustained antiviral activity both *in vitro* and *in vivo*, with IC_{50} values ranging from 0.28 to 7.31 pmol/L. Experimental results showed that a single low-dose treatment with **53** significantly reduced viral load and maintained long-term virus suppression. Additionally, pretreatment with **53** completely blocked infection by the virus in the rectum and vagina. In summary, cholesterol conjugation holds enormous potential as an effective HIV treatment or prevention strategy.

11.4.3 Human Serum Albumin Drug Loading

Human serum albumin (HSA) is the most abundant protein in the blood serum, with concentrations ranging from 30 to 50 g/L. It serves as a natural transporter for various molecules within the human body [68]. Being a major protein in human blood, HSA possesses several advantages, such as stability in the body, safety, low immunogenicity, biodegradability, ease of purification, and good water solubility. The non-covalent binding of small-molecule drugs to HSA can provide protection against enzymatic degradation and renal clearance, thereby extending their half-life in a physiological condition [69]. Consequently, HSA serves as an ideal drug carrier for targeted delivery and for enhancing the pharmacokinetic properties of drugs.

The membrane fusion inhibitor T20, used for HIV treatment, is a peptide drug with a short half-life in the body ($t_{1/2}$ = 3.46-4.35 h), necessitating frequent injections [70]. Therefore, the development of long-acting peptide-based anti-HIV drugs using new strategies holds significant importance. Albuvirtide (ABT, **55**, Figure 11.17d) was designed based on the C34 sequence template (**54**, Figure 11.17d), with 3-maleimidopropionic acid (MPA) modification. In 2018, ABT was approved in China as a novel anti HIV peptide-based drug. The 13th residue, serine (S), was replaced with lysine (K), allowing for single MPA modification at that position. The other two non-target binding residues were substituted with glutamic acid to enhance solubility, stability, and antiviral activity. Results showed that **55** irreversibly binds to HSA, extending its half-life. **55** can form a stable helical structure with the target sequence, effectively blocking cell-cell membrane fusion mediated by 6-HB (EC_{50} = 0.82 μmol/L) and HIV-1 Env (EC_{50} = 1.27 nmol/L). Importantly, **55** can inhibit the entry of various HIV-1 subtypes and mutants into cells, including T20-resistant strains [70]. Furthermore, a Phase III clinical trial (TALENT study) indicated that the injectable long-acting HIV-1 drug **55**, used in combination with ritonavir-boosted lopinavir (LPV/r), is both safe and effective [71]. Therefore, the discovery of **55** highlights the potential of HSA as a drug carrier for the development of long-acting antiviral drugs.

11.5 Ribonuclease Targeting Chimeras

Currently, research has revealed the association of various diseases with molecules like mRNA, miRNA, and long

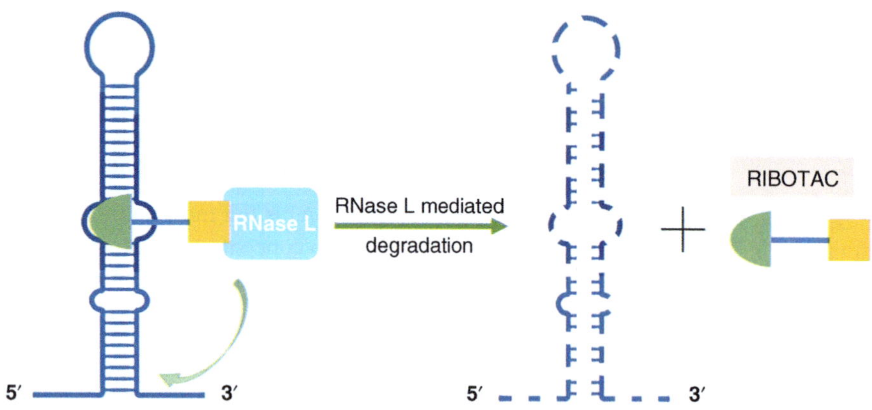

Figure 11.18 A mechanistic overview of RIBOTAC-mediated RNA degradation. RNase L: ribonuclease L; RIBOTAC: ribonuclease targeting chimera.

non-coding RNAs [72]. Researchers commonly employ antisense oligonucleotides (ASOs) and siRNA to target and degrade RNA molecules. Although several drugs of this class, such as inotersen, nusinersen, and patisiran, are already in clinical use, they still face challenges like poor cellular uptake, limited tissue-specific delivery (except for liver and kidney), inefficient endosomal escape, and potential off-target toxicities [73, 74]. Therefore, the development of novel technologies that can target RNA without the limitations of oligonucleotide therapies is of utmost importance for the future advancement of RNA-targeted therapeutics.

Ribonuclease L (RNase L) is an enzyme that degrades invading RNA within cells and is a component of the innate immune response. It exists as an inactive monomer expressed at low levels in all cells but becomes activated by dimerization during viral infections, exhibiting intrinsic substrate specificity [75]. Leveraging the biological function of RNase L, researchers have developed the RIBOTAC technology for targeted RNA degradation [76]. RIBOTAC molecules consist of three components: a small molecule that binds to RNA, a small molecule that recruits RNase L, and a linker connecting the two (as shown in Figure 11.18). The function of RIBOTAC is to recruit RNase L to a target transcript, leading to selective cleavage and degradation [77]. Immunologically, interferon responses can also be triggered by small RNA products cleaved by RNase L with unusual 5′-OH and 2′,3′-cyclic phosphate that are bound to upstream viral RNA sensors, RIG-I and MDA5.

RIBOTAC, as a novel molecule, holds potential for targeting different types of RNA and offers advantages compared to ASO and siRNA [78]: (i) Improved pharmacokinetic properties: Being a small molecule, RIBOTAC is more likely to enter tissues compared to ASO and siRNA. (ii) Catalytic activity: Once RIBOTAC binds and induces degradation of one RNA, it can bind to another RNA. This enables even low concentrations of RIBOTAC to achieve degradation of multiple copies of the target RNA.

In 2020, based on compound C5 (K_D = 11 nmol/L, **56**, Figure 11.19a), Haniff et al. [79] designed a biologically active small molecule C5-RIBOTAC (**57**, Figure 11.19a) targeting the functional structure of the SARS-CoV-2 RNA genome. Structural analysis of the SARS-CoV-2 genome revealed the presence of structurally conserved sites within the frameshifting element (FSE), which controls the translation of critical viral replication proteins (pp1a and pp1ab polyproteins). It was found that **56** could bind and stabilize FSE, significantly inhibiting the frameshifting ability of SARS-CoV-2 FSE.

Using **56** as an RNA-binding small molecule, researchers connected it to a molecule recruiting RNase L through a linker to create the RNA degrader **57.** Experimental results demonstrated that the multi-specific strategy based on RIBOTAC enhanced the antiviral activity of **56** by at least 10-fold (according to comprehensive assessments using different criteria). Mechanistic validation showed that **57** exerts its effect by inducing the degradation of the entire SARS-CoV-2 RNA genome, with this degradation process relying on RNase L. Overall, this study suggests that RIBOTAC could represent a new avenue for developing viral RNA genome degraders.

G-quadruplexes (G4) are crucial components of the SARS-CoV-2 RNA genome, and the β coronaviral pseudoknot plays a pivotal role in the replication process of SARS-CoV-2 [77]. Given the essential roles of these two oligonucleotides, in 2023, Mikutis et al. [80] utilized PDS (**58**, K_D = 26 nmol/L, Figure 11.19b) targeting G4 and MTDB (**59**, K_D = 56.1 μmol/L, Figure 11.19b) targeting the betacoronaviral pseudoknot as lead compounds. They designed two RIBOTAC molecules, **60** (K_D = 137 μmol/L, Figure 11.19b) and **61** (K_D = 16.6 μmol/L, Figure 11.19b), by connecting the RNA degradation moiety imidazole with a PEG linker.

Research indicated that **60** could not only bind and degrade G4 *in vitro* but also reduce virus replication in a concentration-dependent manner (IC_{50} = 1 μmol/L).

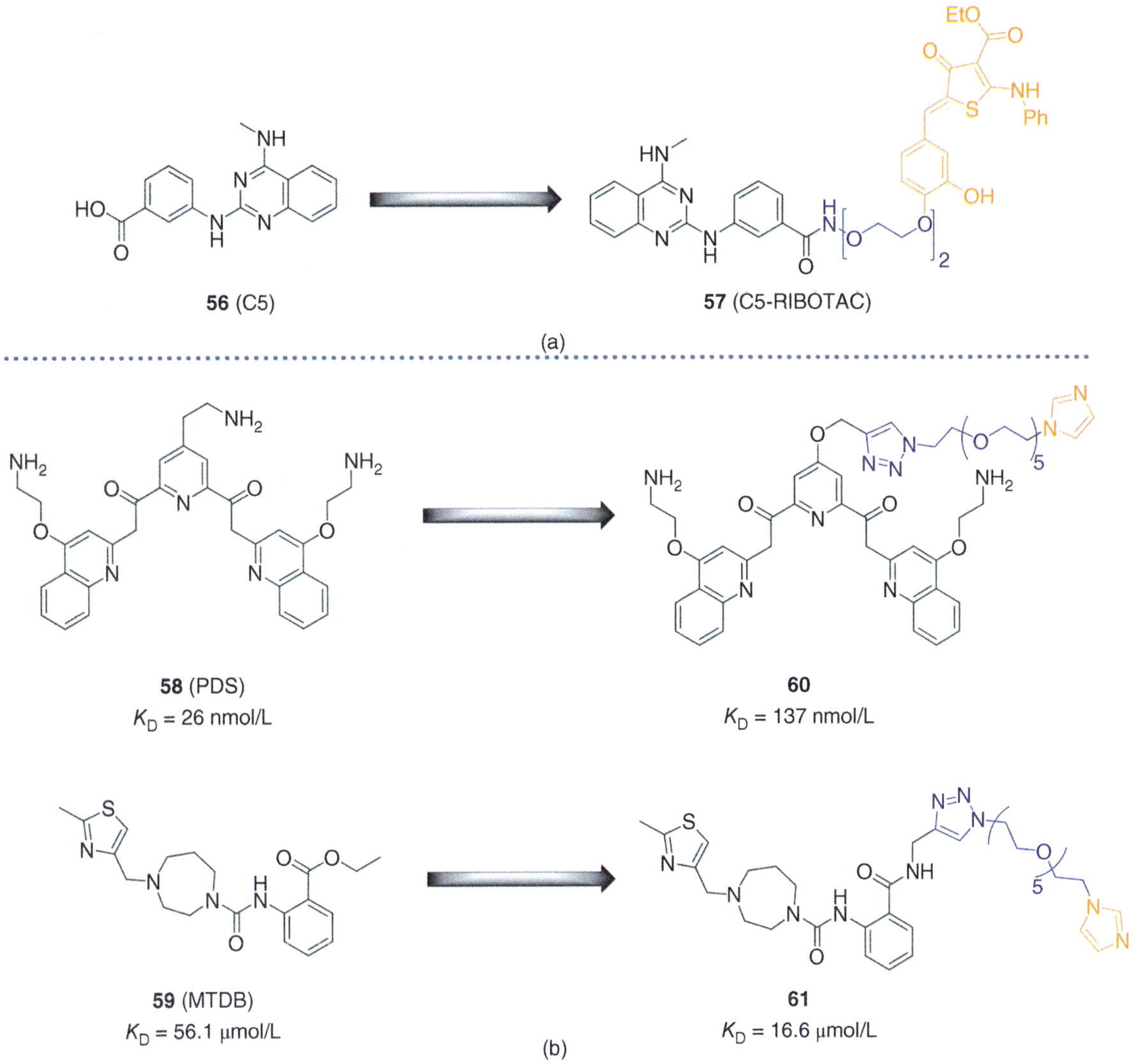

Figure 11.19 Chemical structures of **56** and **57** (**a**), **58–61** (**b**).

Similarly, **61** could bind and degrade its target RNA *in vitro* and exhibited significant antiviral activity at low micromolar concentrations. Furthermore, **61** was able to inhibit SARS-CoV-2 replication in mice. This study demonstrated that RIBOTAC is both tolerable and effective *in vivo*, holding potential for treating viral infections.

11.6 Other Bifunctional Small Molecules

In addition to the aforementioned examples of antiviral drug research based on multi-specific binding, there are several other novel dual-functional small-molecule compounds with antiviral activity.

In 2019, Rashad et al. [81] hypothesized that covalent conjugation of gp120 protein antagonist and host cell coreceptor (CoR) protein antagonist could enhance anti-HIV activity. They designed a dual-functional synergistic molecule, LJC240-L4-UM15 (Bal.01, IC_{50} = 0.359 nmol/L, **62**, Figure 11.20a), by covalently linking the CCR5 small molecule antagonist LJC240 and the gp120 peptide triazole antagonist UM15 to form a dual-functional chimeric entity. Compared to the non-covalent mixture of the two inhibitors, compound **62** retained the specific targeting functions of individual components. It exhibited potent efficacy in inhibiting cell infection by different pseudoviruses, with IC_{50} values as low as subnanomolar levels (Bal.01, IC_{50} = 0.359 ± 0.05 nmol/L; JRFL, IC_{50} = 4 ± 0.6 nmol/L; YU2, IC_{50} = 4.8 ± 0.008 nmol/L), greatly surpassing the potency of the non-covalent mixture of individual components (JRFL, IC_{50} = 34 ± 3.6 nmol/L). This work demonstrated that the dual-functional chimeric

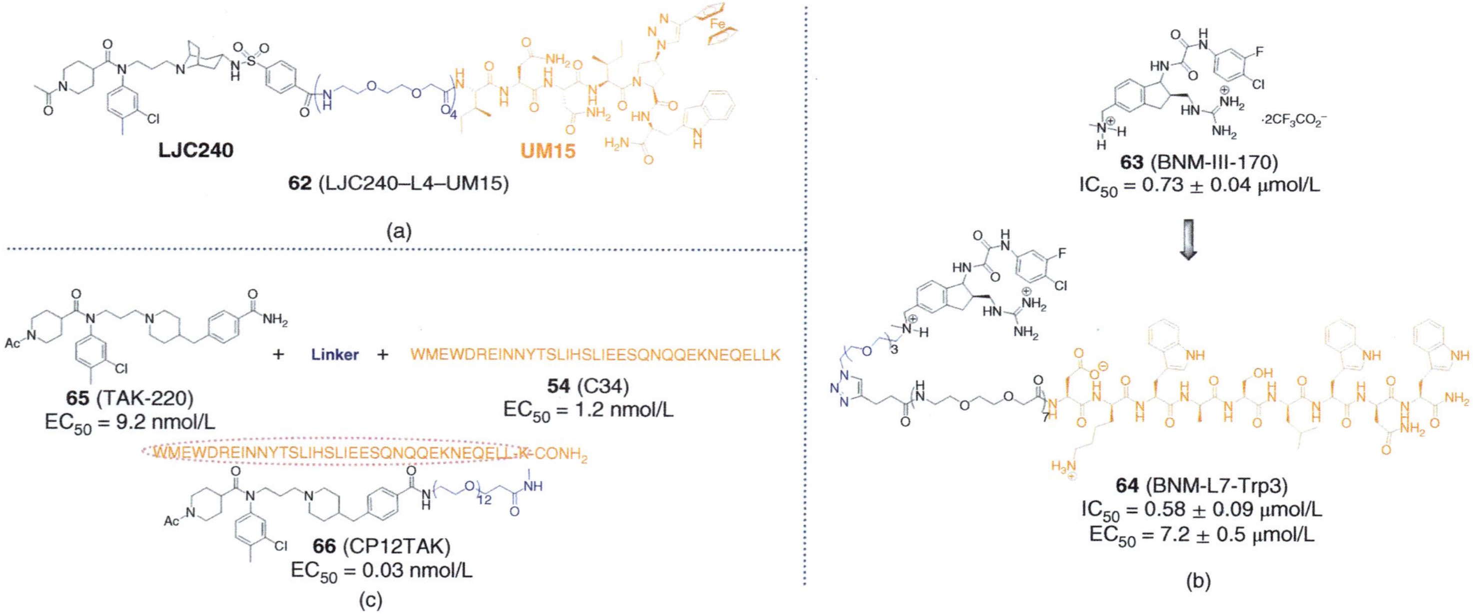

Figure 11.20 Chemical structures of **62** (a); **63**, **64** (b); **54**, **65**, **66** (c).

entity can engage in and synergistically inhibit the process of host cell and viral membrane fusion.

Compound BNM-III-170 (Bal.01, $IC_{50} = 0.73 \pm 0.04$ μmol/L, **63**, Figure 11.20b) is a small molecule CD4 mimic that can bind to the gp120 subunit within the HIV-1 Env glycoprotein. In 2021, Gaffney et al. [82] conjugated **63** with a peptide containing three tryptophan residues (Trp3) that can bind to the gp41 subunit. They designed and synthesized a class of dual-functional HIV-1 inhibitors, BNM-L7-Trp3 (Bal.01, $IC_{50} = 0.58 \pm 0.09$ μmol/L, $EC_{50} = 7.2 \pm 0.5$ μmol/L, **64**, Figure 11.20b), which can simultaneously bind to both gp120 and gp41. ELISA assays demonstrated that using **63** alone effectively inhibited virus-cell infection, while using Trp3 alone did not exhibit an inhibitory effect on virus infection. Only the covalently linked compound **64** was able to induce virus cleavage. Molecular simulations revealed that **61** has a propensity to embed into the viral membrane structure and synergize with Trp3 to disrupt the viral membrane structure. This might be a potential reason for the non-specific viral cleavage effect produced by the dual-functional inhibitor. Furthermore, the researchers found that the dual-functional inhibitor could effectively clear cells infected with HIV-1, suggesting its potential therapeutic effect for AIDS treatment.

In 2021, Wang et al. [83] reported a dual-functional HIV-1 entry inhibitor composed of the pyridine-4-carboxamide CCR5 antagonist TAK-220 ($EC_{50} = 9.2$ nmol/L, **65**, Figure 11.20c) and the fusion inhibitory peptide C34 targeting the gp41 NHR ($EC_{50} = 1.2$ nmol/L, **54**, Figure 11.20c). The synthesized chimeric molecule was constructed by linking the two pharmacophores with a PEG chain. One of the chimeras, CP12TAK ($EC_{50} = 0.03$ nmol/L, **66**, Figure 11.20c), exhibited exceptionally potent antiviral activity, being 40-fold and 306-fold more active than **54** and **65**, respectively. In addition to R5-tropic viruses, **66** also strongly inhibited the infection of X4-tropic HIV-1 strains. These findings suggest that **66**, as a novel mechanism-based anti-HIV-1 agent, is worth further development. Moreover, these results indicate that the multi-specific binding strategy holds promise for extending to the treatment of other Env virus infections.

11.7 Summary and Outlook

The emergence of chronic infectious diseases such as HIV/AIDS and hepatitis B, as well as highly pathogenic acute viruses like Ebola, avian influenza, and SARS-CoV-2, has underscored the necessity of developing highly efficient and low-toxicity antiviral drugs using novel strategies. The multi-specific binding approach has proven to be an effective tool against these pathogenic viruses, and this article summarizes its applications in the field of antiviral research in recent years. Strategies such as PROTAC, ARM, ADC, antiviral drug delivery, and RIBOTAC that utilize multi-specific binding ability have been employed to discover new drug entities, enhance antiviral potency, and address issues of drug resistance [84].

However, these strategies also come with certain limitations. PROTACs must traverse the cell membrane to achieve intracellular protein degradation, but high molecular weight often correlates with limited water solubility and cell permeability, resulting in poor bioavailability. Additionally, RIBOTACs also face challenges related to their typically large molecular size, potentially affecting drug-like properties. Furthermore, the scarcity of endogenous antibody species presents a significant hurdle in designing ARM drugs, and the complex production process or lack of amino acid sequence-specific manipulation of ADC drugs poses a major barrier to their development. In nucleic acid delivery strategies, the intrinsic metabolic instability and off-target gene silencing effects of oligonucleotides hinder their progression into clinical studies. Cholesterol-conjugated molecules exhibit poor oral bioavailability and the low loading efficiency. Limited applicability of drug carriers like HSA remains critical issue that needs to be addressed in achieving effective targeted drug delivery.

Metalloenzymes like HIV integrase, HIV RNase H, and influenza endonuclease are also significant antiviral targets, often hindered by poor membrane permeability and weak cellular activity [85]. Moreover, HIV RNase H possesses shallow ligand-binding pockets, rendering it a challenging "undruggable" target. These limitations could potentially be addressed using targeted protein degradation strategies. For ARM strategies, diversifying ARM design can be achieved by targeting different connecting sites on lead compounds to attach endogenous antibodies. Furthermore, while each HSA molecule can only accommodate a limited number of drug molecules, the drug-loading capacity can be enhanced by encapsulating drugs within HSA nanoparticles [69]. This approach allows controlled release at the target site, potentially improving therapeutic efficacy. The combination of various multi-specific binding strategies holds promise in addressing issues related to poor drug-like properties. For instance, the integration of PROTACs and RIBOTACs with HSA drug delivery systems could enhance drug pharmacokinetics. In conclusion, innovative multi-specific binding strategies offer a potential solution to overcome the challenges associated with antiviral drug development, particularly for targets such as metalloenzymes and "undruggable" sites. By combining different approaches and utilizing advanced drug delivery systems, researchers aim to enhance the efficacy and bioavailability of antiviral agents in the fight against infectious diseases.

Considering the high variability of viruses and the complexity of their pathogenic mechanisms, DNA-encoded chemistry technology [86], genome editing technology [87], nucleic acid aptamer technology [88–91], and ligands based on protein self-assembly [92] are expected to be applied, providing insights for the development of highly effective antiviral drugs. Additionally, the advancement of structural biology with a focus on antiviral drug targets and computational chemistry (such as predicting ternary complex formation) is poised to significantly facilitate the design of PROTAC-based antiviral molecules [93–95]. Furthermore, natural products, characterized by intricate structures and diverse pharmacological activities, offer a rich resource for discovering novel antiviral drugs. For instance, the Tat protein plays a critical role in HIV-1 transcription. The natural product Triptolide can function as a Tat degrader *via* the proteasomal pathway, leading to the reduction of Tat protein levels and demonstrating inhibitory activity against HIV-1 replication at picomolar to nanomolar levels [96].

Integrating medicinal chemistry with bioinformatics and artificial intelligence technologies enables high-throughput phenotypic screening and reverse pharmacophore matching virtual screening of approved drugs, facilitating the exploration of new indications. Furthermore, the establishment of rapid drug screening systems and robust assessment frameworks for safety and efficacy will expedite the development of antiviral drugs.

With the rapid advancement of multi-specific binding strategies, we can expect the reporting of antiviral drugs with higher activity and improved drug-like properties. This will contribute to the development of novel antiviral drug molecules, addressing existing viral infections and potential outbreaks of other viruses threatening human life.

References

1 Tang, D., Comish, P., and Kang, R. (2020). The hallmarks of COVID-19 disease. *PLoS Pathog.* 16: e1008536.

2 Hu, B., Guo, H., Zhou, P. et al. (2022). Characteristics of SARS-CoV-2 and COVID-19. *Nat. Rev. Microbiol.* 20: 315.

3 Gottlieb, M.S. (1981). Pneumocystis pneumonia--Los Angels. *MMWR Morb. Mortal Wkly. Rep.* 30: 250–252.

4 Hutchinson, E.C. (2018). Influenza virus. *Trends Microbiol.* 26: 809–810.

5 Shih, C., Yang, C.C., Choijilsuren, G. et al. (2018). Hepatitis B virus. *Trends Microbiol.* 26: 386–387.

6 Jacob, S. T.; Crozier, I.; Fischer, W.A., 2nd, (2020), et al. Ebola virus disease, *Nat. Rev. Dis. Primers* 6, 13.

7 Wang, P., Nair, M.S., Liu, L. et al. (2021). Antibody resistance of SARS-CoV-2 variants B.1.351 and B.1.1.7. *Nature* 593: 130–135.

8 Blassel, L., Zhukova, A., Villabona-Arenas, C.J. et al. (2021). Drug resistance mutations in HIV: new bioinformatics approaches and challenges. *Curr. Opin. Virol.* 51: 56–64.

9 Papatheodoridis, G.V. and Deutsch, M. (2008). Resistance issues in treating chronic hepatitis B. *Future Microbiol.* 3: 525–538.

10 Deshaies, R.J. (2020). Multispecific drugs herald a new era of biopharmaceutical innovation. *Nature* 580: 329–338.

11 Wang, Y. and Yang, S. (2020). Multispecific drugs: the fourth wave of biopharmaceutical innovation. *Signal Transduct. Target. Ther.* 5: 86.

12 Zhong, X. and D'Antona, A.M. (2021). Recent advances in the molecular design and applications of multispecific biotherapeutics. *Antibodies* 10: 13.

13 Sakamoto, K.M., Kim, K.B., Kumagai, A. et al. (2001). PROTACs: chimeric molecules that target proteins to the Skp1-Cullin-F box complex for ubiquitination and degradation. *Proc. Natl. Acad. Sci. USA* 98: 8554–8559.

14 Li, X., Pu, W., Zheng, Q. et al. (2022). Proteolysis-targeting chimeras (PROTACs) in cancer therapy. *Mol. Cancer* 21: 99.

15 Kannt, A. and Dikic, I. (2021). Expanding the arsenal of E3 ubiquitin ligases for proximity-induced protein degradation. *Cell. Chem. Biol.* 28: 1014–1031.

16 Jevtic, P., Haakonsen, D.L., and Rape, M. (2021). An E3 ligase guide to the galaxy of small-molecule-induced protein degradation. *Cell. Chem. Biol.* 28: 1000–1013.

17 Chen, Y., Tandon, I., Heelan, W. et al. (2022). Proteolysis-targeting chimera (PROTAC) delivery system advancing protein degraders towards clinical translation. *Chem. Soc. Rev.* 51: 5330–5350.

18 Schapira, M., Calabrese, M.F., Bullock, A.N. et al. (2019). Targeted protein degradation: expanding the toolbox. *Nat. Rev. Drug Discov.* 18: 949–963.

19 Paiva, S.L. and Crews, C.M. (2019). Targeted protein degradation: elements of PROTAC design. *Curr. Opin. Chem. Biol.* 50: 111–119.

20 Gabizon, R., Shraga, A., Gehrtz, P. et al. (2020). Efficient targeted degradation *via* reversible and irreversible covalent PROTACs. *J. Am. Chem. Soc.* 142: 11734–11742.

21 Li, K. and Crews, C.M. (2022). PROTACs: past, present and future. *Chem. Soc. Rev.* 51: 5214–5236.

22 Dang, C.V., Reddy, E.P., Shokat, K.M. et al. (2017). Drugging the "undruggable" cancer targets. *Nat. Rev. Cancer* 17: 502–508.

23 Sun, Y., Zhao, X., Ding, N. et al. (2018). PROTAC-induced BTK degradation as a novel therapy for mutated BTK C481S induced ibrutinib-resistant B-cell malignancies. *Cell Res.* 28: 779–781.

24 Buhimschi, A.D., Armstrong, H.A., Toure, M. et al. (2018). Targeting the C481S ibrutinib-resistance mutation in bruton's tyrosine kinase using PROTAC-mediated degradation. *Biochemistry* 57: 3564–3575.

25 Montrose, K. and Krissansen, G.W. (2014). Design of a PROTAC that antagonizes and destroys the cancer-forming X-protein of the hepatitis B virus. *Biochem. Biophys. Res. Commun.* 453: 735–740.

26 Poveda, E. and García, F. (2013). Resistencia a telaprevir [telaprevir resistance]. *Enferm. Infecc. Microbiol. Clin.* 31 (Suppl 3): 26–32.

27 Muir, A.J. (2011). Telaprevir for the treatment of chronic hepatitis C infection. *Expert Rev. Anti-Infect. Ther.* 9: 1105–1114.

28 de Wispelaere, M., Du, G., Donovan, K.A. et al. (2019). Small molecule degraders of the hepatitis C virus protease reduce susceptibility to resistance mutations. *Nat. Commun.* 10: 3468.

29 Li, H., Wang, S., Ma, W. et al. (2022). Discovery of pentacyclic triterpenoid PROTACs as a class of effective hemagglutinin protein degraders. *J. Med. Chem.* 65: 7154–7169.

30 Xu, Z., Liu, X., Ma, X. et al. (2022). Discovery of oseltamivir-based novel PROTACs as degraders targeting neuraminidase to combat H1N1 influenza virus. *Cell Insight* 1: 100030.

31 Zhao, J., Wang, J., Pang, X. et al. (2022). An anti-influenza A virus microbial metabolite acts by degrading viral endonuclease PA. *Nat. Commun.* 13: 2079.

32 Jiang, L., Wen, C., Zhou, H. et al. (2022). Cyclin-dependent kinase 7/9 inhibitor SNS-032 induces apoptosis in diffuse large B-cell lymphoma cells. *Cancer Biol. Ther.* 23: 319–327.

33 Hahn, F., Hamilton, S.T., Wangen, C. et al. (2021). Development of a PROTAC-based targeting strategy provides a mechanistically unique mode of anti-cytomegalovirus activity. *Int. J. Mol. Sci.* 22: 12858.

34 Liang, C.Y., Xin, L., Tian, L. et al. (2022). PROTACs based on VHL ligand targeting coronavirus 3CL protease and preparation method and application thereof. US 11518759B1[P].

35 Liang, C.Y., Xin, L., Tian, L. et al. (2022). PROTACs targeting coronavirus 3CL protease and preparation method and application thereof. US 11530195B1[P].

36 Alugubelli, Y.R., Xiao, J., Khatua, K. et al. (2023), Discovery of first-in-class PROTAC degraders of SARS-CoV-2 main protease. *Preprint*. bioRxiv 2023.09.29.560163. https://doi.org/10.1101/2023.09.29.560163.

37 Li, Y., Yu, J., Qi, X. et al. (2019). Monoclonal antibody against EV71 3Dpol inhibits the polymerase activity of RdRp and virus replication. *BMC Immunol.* 20: 6.

38 Zhou, H.B., Wu, S.W., Ma, X. Y. (2023). A PROTAC compound of remdesivir or its intermediate and its preparation method and anti-EV71 application. CN 115819423A[P].

39 Emert-Sedlak, L.A., Tice, C.M., Shi, H. et al. (2023), PROTAC-mediated degradation of HIV-1 Nef efficiently restores cell-surface CD4 and MHC-I expression and blocks HIV-1 replication. *Preprint*. bioRxiv; 2023.08.14.553289.

40 Wang, X., Chen, H., Chiodo, F. et al. (2019). Detection of human IgM and IgG antibodies by means of galactofuranose-coated and rhamnose-coated gold nanoparticles. *Matters* 5: e201908000004.

41 Jakobsche, C.E., Parker, C.G., Tao, R.N. et al. (2013). Exploring binding and effector functions of natural human antibodies using synthetic immunomodulators. *ACS Chem. Biol.* 8: 2404–2411.

42 Oostindie, S.C., van der Horst, H.J., Lindorfer, M.A. et al. (2019). CD20 and CD37 antibodies synergize to activate complement by Fc-mediated clustering. *Haematologica* 104: 1841–1852.

43 Gray, M.A., Tao, R.N., DePorter, S.M. et al. (2016). A nanobody activation immunotherapeutic that selectively destroys HER2-positive breast cancer cells. *ChemBioChem* 17: 155–158.

44 Rathmann, S.M., Genady, A.R., Janzen, N. et al. (2021). A versatile platform for the development of radiolabeled antibody-recruiting small molecules. *Mol. Pharm.* 18: 2647–2656.

45 Castellano, G., Woltman, A.M., Schlagwein, N. et al. (2007). Immune modulation of human dendritic cells by complement. *Eur. J. Immunol.* 37: 2803–2811.

46 Junker, F., Gordon, J., and Qureshi, O. (2020). Fc gamma receptors and their role in antigen uptake, presentation, and t cell activation. *Front. Immunol.* 11: 1393.

47 Golay, J. and Taylor, R.P. (2020). The role of complement in the mechanism of action of therapeutic anticancer mAbs. *Antibodies* 9: 58.

48 Vogelpoel, L.T., Baeten, D.L., de Jong, E.C. et al. (2015). Control of cytokine production by human Fc gamma

receptors: implications for pathogen defense and autoimmunity. *Front. Immunol.* 6: 79.

49 Kazmierski, W.M., Kenakin, T.P., and Gudmundsson, K.S. (2006). Peptide, peptidomimetic and small-molecule drug discovery targeting HIV-1 host-cell attachment and entry through gp120, gp41, CCR5 and CXCR4. *Chem. Biol. Drug Des.* 67: 13–26.

50 Parker, C.G., Domaoal, R.A., Anderson, K.S. et al. (2009). An antibody-recruiting small molecule that targets HIV gp120. *J. Am. Chem. Soc.* 131: 16392–16394.

51 Parker, C.G., Dahlgren, M.K., Tao, R.N. et al. (2014). Illuminating HIV gp120-ligand recognition through computationally-driven optimization of antibody-recruiting molecules. *Chem. Sci.* 5: 2311–2317.

52 Liu, X., Zhang, B., Wang, Y. et al. (2020). A universal dual mechanism immunotherapy for the treatment of influenza virus infections. *Nat. Commun.* 11: 5597.

53 Hafeez, U., Parakh, S., Gan, H.K. et al. (2020). Antibody-drug conjugates for cancer therapy. *Molecules* 25: 4764.

54 Gavrilyuk, J., Uehara, H., Otsubo, N. et al. (2010). Potent inhibition of HIV-1 entry with a chemically programmed antibody aided by an efficient organocatalytic synthesis. *ChemBioChem* 11: 2113–2118.

55 Sato, S., Inokuma, T., Otsubo, N. et al. (2013). Chemically programmed antibodies AS HIV-1 attachment inhibitors. *ACS Med. Chem. Lett.* 4: 460–465.

56 Asano, S., Gavrilyuk, J., Burton, D.R. et al. (2014). Preparation and activities of macromolecule conjugates of the CCR5 antagonist Maraviroc. *ACS Med. Chem. Lett.* 5: 133–137.

57 Regeneron Pharmaceuticals (2021). Protein-antiviral compound conjugates. WO 2021/151031Al[P].

58 Cidara Therapeutics (2022). Protein-drug conjugates for antiviral therapy. WO 2022/192685Al[P].

59 DeGrace, M.M., Ghedin, E., Frieman, M.B. et al. (2022). Defining the risk of SARS-CoV-2 variants on immune protection. *Nature* 605: 640–652.

60 Maillard, P.V., van der Veen, A.G., Poirier, E.Z. et al. (2019). Slicing and dicing viruses: antiviral RNA interference in mammals. *EMBO J.* 38: e100941.

61 Alshaer, W., Zureigat, H., Al Karaki, A. et al. (2021). siRNA: mechanism of action, challenges, and therapeutic approaches. *Eur. J. Pharmacol.* 905: 174178.

62 Traube, F.R., Stern, M., Tolke, A.J. et al. (2022). Suppression of SARS-CoV-2 replication with stabilized and click-chemistry modified siRNAs. *Angew. Chem. Int. Ed. Engl.* 61: e202204556.

63 Tai, W., Zhao, P., and Gao, X. (2020). Cytosolic delivery of proteins by cholesterol tagging. *Sci. Adv.* 6: eabb0310.

64 von Itzstein, M. (2007). The war against influenza: discovery and development of sialidase inhibitors. *Nat. Rev. Drug Discov.* 6: 967–974.

65 Lv, X., Wang, P., Li, C. et al. (2021). Zanamivir-cholesterol conjugate: a long-acting neuraminidase inhibitor with potent efficacy against drug-resistant influenza viruses. *J. Med. Chem.* 64: 17403–17412.

66 Lazzarin, A. (2005). Enfuvirtide: the first HIV fusion inhibitor. *Expert. Opin. Pharmacother.* 6: 453–464.

67 Xue, J., Chong, H., Zhu, Y. et al. (2022). Efficient treatment and pre-exposure prophylaxis in rhesus macaques by an HIV fusion-inhibitory lipopeptide. *Cell* 185: 131–144.e18.

68 Matos, M.J. (2018). Learning from nature: the role of albumin in drug delivery. *Future Med. Chem.* 10: 983–985.

69 Liu, Z. and Chen, X. (2016). Simple bioconjugate chemistry serves great clinical advances: albumin as a versatile platform for diagnosis and precision therapy. *Chem. Soc. Rev.* 45: 1432–1456.

70 Chong, H., Yao, X., Zhang, C. et al. (2012). Biophysical property and broad anti-HIV activity of albuvirtide, a 3-maleimimidopropionic acid-modified peptide fusion inhibitor. *PLoS ONE* 7: e32599.

71 Su, B., Yao, C., Zhao, Q.X. et al. (2020). Efficacy and safety of the long-acting fusion inhibitor albuvirtide in antiretroviral-experienced adults with human immunodeficiency virus-1: interim analysis of the randomized, controlled, phase 3, non-inferiority TALENT study. *Chin. Med. J.* 133: 2919–2927.

72 Agrawal, N., Dasaradhi, P.V., Mohmmed, A. et al. (2003). RNA interference: biology, mechanism, and applications. *Microbiol. Mol. Biol. Rev.* 67: 657–685.

73 Yu, A.M. and Tu, M.J. (2022). Deliver the promise: RNAs as a new class of molecular entities for therapy and vaccination. *Pharmacol. Ther.* 230: 107967.

74 Setten, R.L., Rossi, J.J., and Han, S.P. (2019). The current state and future directions of RNAi-based therapeutics. *Nat. Rev. Drug Discov.* 18: 421–446.

75 Costales, M.G., Matsumoto, Y., Velagapudi, S.P. et al. (2018). Small molecule targeted recruitment of a nuclease to RNA. *J. Am. Chem. Soc.* 140: 6741–6744.

76 Costales, M.G., Suresh, B., Vishnu, K. et al. (2019). Targeted degradation of a hypoxia-associated non-coding RNA enhances the selectivity of a small molecule interacting with RNA. *Cell Chem. Biol.* 26: 1180–1186.e1185.

77 Costales, M.G., Aikawa, H., Li, Y. et al. (2020). Small-molecule targeted recruitment of a nuclease to cleave an oncogenic RNA in a mouse model of metastatic cancer. *Proc. Natl. Acad. Sci. USA* 117: 2406–2411.

78 Dey, S.K. and Jaffrey, S.R. (2019). RIBOTACs: small molecules target RNA for degradation. *Cell Chem. Biol.* 26: 1047–1049.

79 Haniff, H.S., Tong, Y., Liu, X. et al. (2020). Targeting the SARS-CoV-2 RNA genome with small molecule binders and ribonuclease targeting chimera (RIBOTAC) degraders. *ACS Cent. Sci.* 6: 1713–1721.

80 Mikutis, S., Rebelo, M., Yankova, E. et al. (2023). Proximity-induced nucleic acid degrader (PINAD) approach to targeted RNA degradation using small molecules. *ACS Cent. Sci.* 9: 892–904.

81 Rashad, A.A., Song, L.R., Holmes, A.P. et al. (2018). Bifunctional chimera that coordinately targets human immunodeficiency virus 1 envelope gp120 and the host-cell CCR5 coreceptor at the virus-cell interface. *J. Med. Chem.* 61: 5020–5033.

82 Gaffney, A., Nangarlia, A., Ang, C.G. et al. (2021). HIV-1 Env-dependent cell killing by bifunctional small-molecule/peptide conjugates. *ACS Chem. Biol.* 16: 193–204.

83 Wang, C., Wang, X., Wang, H. et al. (2021). A "two-birds-one-stone" approach toward the design of bifunctional human immunodeficiency virus type 1 entry inhibitors targeting the CCR5 coreceptor and gp41 N-terminal heptad repeat region. *J. Med. Chem.* 64: 11460–11471.

84 Xu, S., Ding, D., Zhang, X. et al. (2022). Newly emerging strategies in antiviral drug discovery. *Molecules* 27: 850.

85 O'Herin, C.B., Moriuchi, Y.W., Bemis, T.A. et al. (2023). Development of human carbonic anhydrase II heterobifunctional degraders. *J. Med. Chem.* 66: 2789–2803.

86 Chamakuri, S., Lu, S., Ucisik, M.N. et al. (2021). DNA-encoded chemistry technology yields expedient access to SARS-CoV-2 M(pro) inhibitors. *Proc. Natl. Acad. Sci. USA* 118: e2111172118.

87 Doudna, J.A. (2020). The promise and challenge of therapeutic genome editing. *Nature* 578: 229–236.

88 Chen, X.F., Zhao, X., and Yang, Z. (2021). Aptamer-based antibacterial and antiviral therapy against infectious diseases. *J. Med. Chem.* 64: 17601–17626.

89 Nguyen, P.D.M., Zheng, J., Gremminger, T.J. et al. (2020). Binding interface and impact on protease cleavage for an RNA aptamer to HIV-1 reverse transcriptase. *Nucleic Acids Res.* 48: 2709–2722.

90 Musumeci, D., Riccardi, C., and Montesarchio, D. (2015). G-Quadruplex forming oligonucleotides as anti-HIV agents. *Molecules* 20: 17511–17532.

91 Duclair, S., Gautam, A., Ellington, A. et al. (2015). High-affinity RNA aptamers against the HIV-1 protease inhibit both *in vitro* protease activity and late events of viral replication. *Mol. Ther. Nucl. Acids* 4: e228.

92 Tauber, C., Wamser, R., Arkona, C. et al. (2021). Chemical evolution of antivirals against enterovirus D68 through protein-templated knoevenagel reactions. *Angew. Chem. Int. Ed. Engl.* 60: 13294–13301.

93 Li, W., Zhang, J., Guo, L. et al. (2022). Importance of three-body problems and protein-protein interactions in proteolysis-targeting chimera modeling: insights from molecular dynamics simulations. *J. Chem. Inf. Model.* 62: 523–532.

94 Desantis, J., Mercorelli, B., Celegato, M. et al. (2021). Indomethacin-based PROTACs as pan-coronavirus antiviral agents. *Eur. J. Med. Chem.* 226: 113814.

95 Shaheer, M., Singh, R., and Sobhia, M.E. (2021). Protein degradation: a novel computational approach to design protein degrader probes for main protease of SARS-CoV-2. *J. Biomol. Struct. Dyn.* 40: 10905–10917.

96 Jin, H., Li, D., Lin, M.H. et al. (2020). Tat-based therapies as an adjuvant for an HIV-1 functional cure. *Viruses* 12: 415.

12

Polypharmacology for the Treatment of Major Depressive Disorder

Multitargeted Antidepressants

Tiffany Schwasinger-Schmidt

12.1 Introduction

Major depressive disorder is a complex medical condition that affects an estimated 265 million people worldwide [1, 2]. This disorder is associated with multiple comorbid conditions resulting in increased healthcare utilization, with accompanying individual- and societal-based social and economic burdens [3]. Depression results in substantial reductions in productivity and often leads to disability [2]. The challenge with drug development for this and other psychiatric conditions is that the symptoms are highly variable among affected individuals [1, 3]. Additionally, the underlying pathophysiology of this disorder is not well characterized and likely involves multiple genetic, environmental, and developmental influences [1, 3–5]. Signaling within the brain in individuals with depression is increasingly complex both within and between neurons across diverse areas of the brain including the hippocampus, amygdala, prefrontal cortex, cingulate cortex, thalamus, and basal ganglia [1, 3, 6]. This degree of intricacy highlights some of the challenges faced by pharmaceutical companies in producing medications for this therapeutic indication and also explains why many corporations have reduced or eliminated drug development in central nervous system (CNS) disorders [6].

The drug development process for psychiatric conditions has historically occurred through serendipity, with the medications used to treat major depressive disorder being discovered by chance [7]. This approach has been beneficial in developing numerous medications to treat psychiatric conditions but is limited in use as it employs a high-risk financial investment with limited financial reward, due to only a small number of compounds approved for clinical use [7]. More recent drug design approaches have focused on targeting specific neurotransmitters in the CNS to modulate their concentration within the synaptic cleft, thus altering signaling among the various neural pathways involved in emotional response and mood regulation [7].

Drug development for the treatment of major depressive disorder has largely centered on the monoamine hypothesis, which originated from the serendipitous discovery of tricyclic antidepressants (TCAs) and monoamine oxidase inhibitors (MAOIs) [8]. These were the first medications used to treat depression and were the only treatments available for major depressive disorder from the 1950s until the 1980s [8, 9]. TCAs were discovered through the investigation of H_1 antihistamine receptor modulators through the alteration of the chemical structure of chlorpromazine in an attempt to produce a medication to treat psychosis [6]. These drugs were not successful in treating psychosis but did show significant benefit in the treatment of depression [6]. MAOIs changed the field of psychiatry by identifying the role of serotonin and norepinephrine in the development of symptoms of depression [6]. This discovery led to the recognition of the beneficial role of neurogenesis and neuronal rescue in the brain that occurs during the treatment of major depressive disorder [6].

The monoamine hypothesis for drug development largely focuses on the modulation of catecholamines including serotonin, norepinephrine, and dopamine within the brain to treat the symptoms of depression [3, 8, 10]. This hypothesis suggests that the development of major depressive disorder is due to a reduction in the concentrations of catecholamines in the brain and was based on studies of reserpine conducted in the 1950s. This research showed that depletion of monoamines in the CNS resulted in symptoms of depression including increased sleep and reduced motor activities, while discontinuation of reserpine resulted in the resolution of symptoms [10, 11]. The monoamine hypothesis was further supported by the studies of the MAOI, iproniazid, that produced an antidepressant effect when it was given to patients with tuberculosis as a potential treatment for this respiratory disease [10]. Later studies with this molecule showed persistent antidepressant effects in patients without tuberculosis [10, 11].

Polypharmacology: Strategies for Multi-Target Drug Discovery, First Edition. Edited by Jens-Uwe Peters.

The monoamine hypothesis has been the basis for the development of numerous antidepressants designed to increase the concentration of monoamines in the synaptic cleft through the prevention of monoamine degradation or by blocking receptor reuptake of the neurotransmitters [11]. The first generation of antidepressants including TCAs and MAOIs had dual actions on norepinephrine and serotonin levels in the brain, but were limited in utility due to side effects and interactions with other medications and certain foods [3, 6, 11]. In an attempt to limit these significant side effects and to improve the treatment of major depressive disorder, selective serotonin reuptake inhibitors (SSRIs) were developed using a one-target, one-drug approach [3, 6, 8]. While the SSRIs have become the most frequently used medications for the treatment of depression worldwide due to effectiveness, tolerability, and reduced incidence of side effects, the clinical effect of these medications is delayed at 1–2 weeks with full efficacy reached at 4–6 weeks of therapy [6]. This delayed onset of efficacy raises concerns with both the monoamine hypothesis and the one-target, one-drug approach [10, 11]. If modulation of monoamine levels caused the development of major depressive disorder, then why is the onset of action of the targeted medications delayed? Additionally, if multiple monoamines were the cause of depression, then why is the patient not cured with the modulation of the monoamine levels in the synaptic cleft?

The one-target, one-drug approach of drug development limits the utility of established medications used for treating depression as there is no single genetic or neural pathway that defines this disorder [5]. There is increasing clinical evidence to suggest that single-target medications are insufficient to achieve remission in complex medical disorders, including depression [1, 2, 8, 12, 13]. Medications that act on multiple targets not only improve the potential therapeutic response but can also enhance safety, efficacy, tolerability, and medication compliance [2, 3, 8, 12, 13]. Ideally, multitargeted antidepressants would improve clinical outcomes in patients by regulating multiple neuronal pathways through different but complementary mechanisms of action allowing for decreased time to efficacy, significant symptom improvement, and reduced side effects [1, 3, 5, 12–14].

Medications that are rationally designed to modulate multiple targets within neural networks can alter disease states by inhibiting or activating multiple targets simultaneously to induce improved efficacy with reduced off-target side effects [3, 5, 15]. Synergism between compounds with complementary mechanisms of action that can be combined within a single medication has a better chance to control the symptoms of complex medical disorders [5]. Multitargeted medications should ideally result in significantly improved symptoms and increase the number of people achieving disease remission; improve patient functional status, thus improving treatment quality; reduce tachyphylaxis and the risk of relapse; accelerate efficacy with quicker onset and better sustained effects; broaden treatment indications to target commonly occurring comorbid conditions including anxiety; reduce side effects by preventing off-target effects; increase the therapeutic window for treatment; and enhance patient compliance through a single daily dose medication [1]. These and other benefits underscore the importance and need for the continued development of multitargeted medications for the treatment of complex medical conditions, including major depressive disorder [1].

The information presented in this chapter will not be a fully comprehensive list of all antidepressants with multitargeted therapeutic approaches, but instead will highlight some of the more notable mechanisms of action of common medications used in the treatment of major depressive disorder. This chapter will also discuss potential triple agonists and antagonists in development for the treatment of major depressive disorder and how these medications have the potential to improve therapeutic response and induce sustained remission.

12.2 Multitargeted Antidepressants

12.2.1 Tricyclic Antidepressants

TCAs were first discovered in the early 1950s and continue to be recognized as some of the most effective treatments for depression due to noted improvements in mood control, cognition, and motor function [1, 13]. TCAs increase the concentration of serotonin and norepinephrine in the synaptic cleft by inhibiting the reuptake of these molecules at post-synaptic neurons [1, 4, 13, 16, 17]. TCAs are typically highly protein bound with half-lives ranging from 13 to 51 hours [6]. The onset of action of these medications is 1–2 weeks with full benefit noted at 4–6 weeks of therapy (Figure 12.1) [6, 18–21].

The TCAs are one of the most effective treatments for major depressive disorder due to their interactions with the serotonin (5-hydroxytryptamine [5-HT]) receptor and the norepinephrine transporters [1, 4, 13, 16]. These medications additionally act at other receptors within the CNS including the 5-HT_{2A}, 5-HT_{2C}, 5-HT_3, 5-HT_7, α_1 adrenoceptors, histamine, and muscarinic acetylcholine receptors [1, 17]. The multiple targets across diverse regions of the brain result in numerous side effects including anticholinergic effects, drowsiness, tremor, reduced seizure threshold, hypotension, prolonged QT interval, cardiac

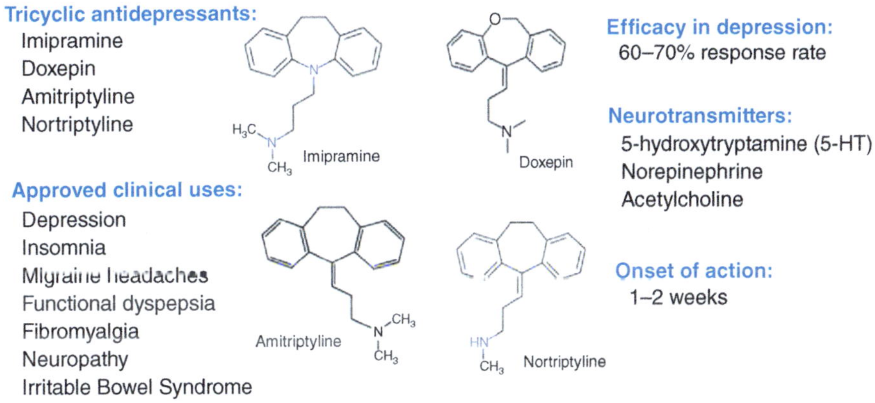

Figure 12.1 Description of tricyclic antidepressants including commonly used medications, clinical use, efficacy in depression, neurotransmitter targets, and onset of action [6, 18–21].

arrhythmias, constipation, urinary retention, sexual side effects, and increased weight gain [1, 4, 6, 18–21]. The numerous adverse effects observed with the TCAs including potential cardiac toxicity have limited their use in the treatment of major depressive disorder [1, 4, 17].

12.2.2 Monoamine Oxidase Inhibitors

MAOIs significantly advanced the field of psychiatry through the discovery of the role of serotonin and norepinephrine in the development of major depressive disorder [6]. These medications additionally aided in the measurement of amino acid levels in the brain during treatment, thus allowing for the recognition of the significant role of neurogenesis and neuronal rescue that occur with therapy [6, 22]. MAOIs increase serotonin and norepinephrine transmission in the synaptic cleft by preventing the degradation of these molecules through the blockade of mitochondrial enzymes located in the synaptic cleft [4, 16]. MAOIs have variable half-lives ranging from 90 minutes to 11.6 hours with an onset of action at 3–6 weeks for isocarboxazid and 1–2 weeks for the remaining medications within this class [6]. Full effects of the medications are noted at 4–6 weeks of treatment (Figure 12.2) [6, 23–26].

The utility of MAOIs for the treatment of major depressive disorder has been significantly limited due to food and drug interactions [6]. Tyramine-containing foods including cheese, red wine, beer, yeast extracts, aged meats, fava beans, Italian green beans, and liver must be avoided as they increase the risk of uncontrolled hypertension [6, 17]. Additional adverse effects noted with MAOI use include weight gain, orthostatic hypotension, sexual dysfunction, dry mouth, insomnia, and headache [6, 23–26]. Despite the significant drug and food interactions associated with MAOIs, these medications continue to be a third-line treatment for patients with treatment-resistant depression [6, 22].

12.2.3 Selective Serotonin Reuptake Inhibitors

SSRIs are currently the most prescribed medications for the treatment of major depressive disorder and remain first-line therapy due to efficacy, tolerability, and reduced side effects [6, 27]. SSRIs were the first class of psychiatric medications that were designed to modulate serotonin concentrations in the brain using receptor-binding affinity technology [6, 27, 28]. Fluoxetine was the first SSRI that was approved by the Food and Drug Administration (FDA) for use in the treatment of major depressive disorder in 1987, with additional SSRIs approved in the late 1980s and early 1990s [28]. The SSRIs are highly protein bound, excluding escitalopram, which is only 56% protein bound, with long half-lives ranging from 21 hours to 3

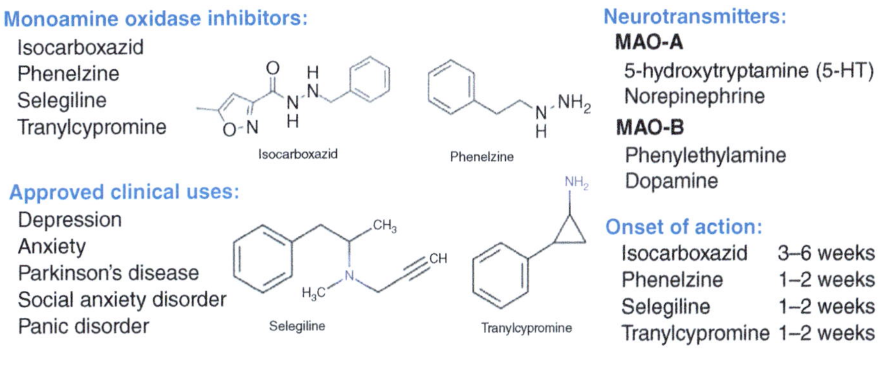

Figure 12.2 Description of monoamine oxidase inhibitors including commonly used medications, clinical use, neurotransmitter targets, and onset of action [6, 23–26].

days [6]. The onset of action for SSRIs is 1–2 weeks with full clinical efficacy achieved at 4–6 weeks of treatment [6]. Patients taking SSRIs have a 60%–80% response rate to the medications (Figure 12.3) [6, 29–33].

SSRIs were designed to selectively and effectively inhibit serotonin reuptake in the synaptic cleft [28]. These medications have additional receptor-binding affinities beyond just the serotonin (5-HT) receptor [8]. Paroxetine binds to norepinephrine transporters and the muscarinic cholinergic receptors, while sertraline binds to dopamine transporters [8]. Fluoxetine has increased affinity for the 5-HT_{2c} receptor, and citalopram has enhanced affinity for the $sigma_1$ and histamine H_1 receptors [8]. Escitalopram has additional affinity for the $sigma_1$ receptors [8]. These alterations in receptor binding can lead to differences in clinical efficacy and side effect profiles, which partially explains individual variability in the effectiveness and tolerability of the various SSRIs [8].

Several of the SSRIs also have inhibitory effects on the cytochrome P450 enzymes that mediate the majority of drug metabolism in the liver [28]. This needs to be considered when prescribing SSRIs due to potential drug–drug interactions and increased side effects of the medications [28]. SSRIs are noted to produce side effects including nausea, diarrhea, heartburn, sexual side effects, headaches, insomnia, dizziness, tremor, dry mouth, and anxiety [6, 29–33]. Withdrawal syndrome can also occur when stopping SSRIs and includes flu-like symptoms, lightheadedness, dizziness, restlessness, sleep and sensory disturbances, and headaches [28].

12.2.4 Selective Serotonin and Norepinephrine Reuptake Inhibitors

Selective serotonin and norepinephrine reuptake inhibitors (SNRIs) bind to both the serotonin (5-HT) receptor and the norepinephrine receptors with differing binding affinities and have been approved by the FDA to treat major depressive disorder since 1994 [6, 34]. The SNRIs show greater efficacy compared to the SSRIs and have a broader therapeutic range to treat multiple psychiatric conditions in addition to neuropathy and chronic pain [5, 6]. The SNRIs tend to be weekly protein bound between 13% and 27%, excluding duloxetine that is highly protein bound at greater than 90% [6, 34–39]. The half-life of these medications ranges from 5 to 12 hours with an onset of action at 1–2 weeks, with full effectiveness noted at 4–6 weeks of treatment [6, 34–39]. Approximately 40%–62% of patients with major depressive disorder will achieve remission while on these medications (Figure 12.4) [6].

Venlafaxine was the first SNRI approved for the treatment of major depressive disorder in 1994 [6]. This

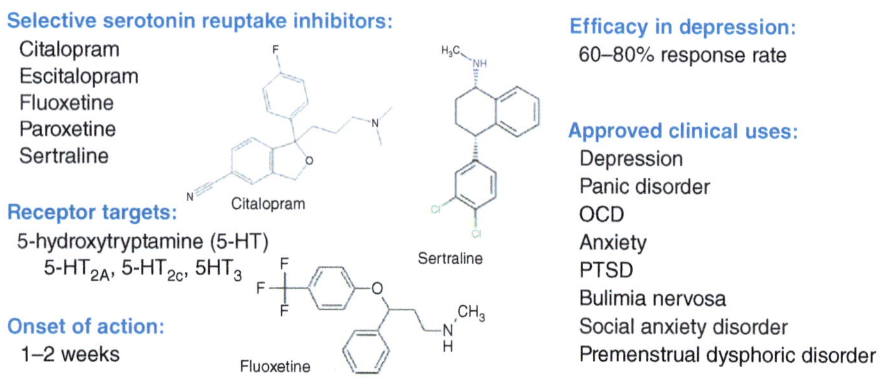

Figure 12.3 Description of selective serotonin reuptake inhibitors including commonly used medications, clinical use, neurotransmitter targets, efficacy in depression, and onset of action [6, 29–33].

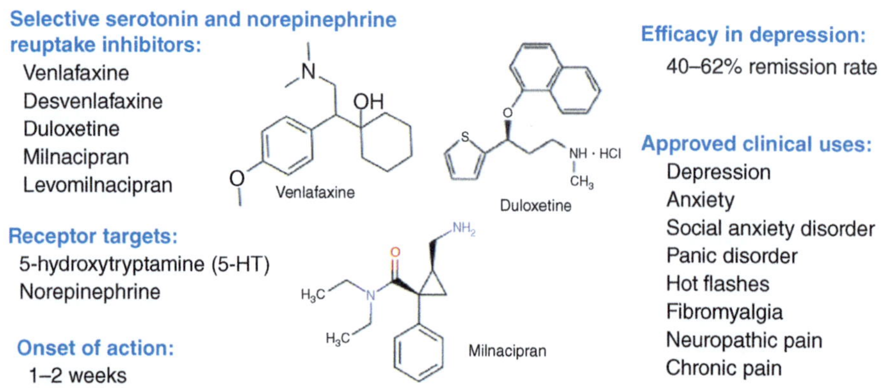

Figure 12.4 Description of selective serotonin and norepinephrine reuptake inhibitors including commonly used medications, clinical use, neurotransmitter targets, efficacy in depression, and onset of action [6, 34–39].

medication preferentially binds at low doses to the serotonin (5-HT) receptor and at higher doses binds with greater affinity to the norepinephrine receptors [8, 34]. Duloxetine was approved for the treatment of major depressive disorder in patients with chronic pain and treatment-resistant depression in 2004 [6]. This drug has a greater balance in receptor binding at both the serotonin and norepinephrine receptors with low affinity for off-target receptors including the histamine, muscarinic, adrenergic, and dopaminergic receptors, resulting in a more favorable side effect profile [8, 34]. Desvenlafaxine, a metabolite of venlafaxine, with increased binding at norepinephrine receptors was approved in 2008 with similar efficacy for the treatment of major depressive disorder compared to venlafaxine and duloxetine [6, 34]. Milnacipran was approved in 2009 for the treatment of fibromyalgia but is not currently approved by the FDA for the treatment of major depressive disorder in the United States, despite showing some benefit in depressive symptoms while patients are on therapy [6, 34]. This medication has moderate binding affinity for both the serotonin (5-HT) and norepinephrine receptors [34]. Levomilnacipran is an enantiomer of milnacipran with greater receptor binding at the serotonin (5-HT) and norepinephrine receptors and was FDA approved to treat depression in 2009 [34].

The side effects associated with SNRIs at low doses tend to be similar to SSRIs including nausea, diarrhea, sexual side effects, headaches, tremor, and dizziness [6]. The SNRIs do have an increased risk of clinically significant hypertension and tachycardia due to increased norepinephrine concentrations at augmented doses of the medications [6, 34]. Despite the known side effects of SNRIs, these medications have shown benefit in patients with major depressive disorder and chronic pain syndromes, thus indicating a possible area of focus for treatment in this specific population [34].

12.2.5 Bupropion

Bupropion is the only medication used to treat major depressive disorder in the United States that has no effect on serotonin concentrations in the CNS [6, 40]. This medication instead specifically targets norepinephrine and dopamine receptors [6, 40]. The initial formulation was FDA approved for treatment in 1989, and the extended-release formulation was approved in 2003 [17, 40]. The efficacy of bupropion is similar to SSRIs with a 60%–80% response rate and onset of action within 2 weeks, with full effectiveness achieved after 4 weeks of treatment [40]. The half-life of this medication is 14 hours with limited adverse effects including agitation, dry mouth, insomnia, headache, nausea, vomiting, constipation, and tremor [6, 40, 41]. Bupropion is contraindicated in patients with seizure disorders due to the reduced seizure threshold associated with this medication and must be dose reduced in patients with liver or kidney disease [40]. This medication has additionally been FDA approved for smoking cessation to reduce nicotine cravings and has shown benefit in patients with major depressive disorder who additionally use tobacco products [40]

12.2.6 Mirtazapine

Mirtazapine was approved by the FDA in 1996 for the treatment of major depressive disorder [40]. This medication is unique in that it increases norepinephrine and serotonin concentrations through direct interactions with the alpha-adrenergic and serotonin (5-HT) receptors [40]. Mirtazapine specifically blocks the α_2-adrenergic receptors, which increases norepinephrine through the activation of the α_1-adrenergic receptors on serotonergic neurons, thus increasing serotonin concentrations [40]. This medication also binds to the 5-HT_{2c}, 5-HT_3, and H_1 receptors, which prevents overexcitation of the serotonergic neurons and produces a noted sedative effect [4, 5, 40]. Mirtazapine has been used to potentiate the effects of SSRIs in clinical practice as a primarily adjunctive treatment option [1]. This medication is different from other drugs used in the treatment of major depressive disorder in that it decreases the release of corticotropin, thus reducing systemic cortisol levels and overall stress responses in patients with major depressive disorder [6, 40].

Mirtazapine has similar efficacy for the treatment of major depressive disorder compared to TCAs, SSRIs, and SNRIs with a 60%–80% response rate and reduced relapse rates compared to placebo at 40 weeks [6]. The half-life of the medication is longer at 20–40 hours with the onset of action occurring in 2 weeks of treatment [6, 40, 42]. Adverse effects associated with mirtazapine use include somnolence, weight gain, dizziness, agranulocytosis, and transient elevations in total cholesterol and liver function tests [6, 42]. The increased somnolence noted with the use of this medication can be attributed to the effects on the H_1 receptor, and it is unclear whether a tolerance to these side effects develops over time [40].

12.2.7 Trazadone

Trazodone was FDA approved for the treatment of major depressive disorder in 1981 and works as an antagonist at the 5-HT_2 receptors and as a serotonin reuptake inhibitor at the 5-HT receptor [4, 40]. The weak antagonistic effects of trazodone at the 5-HT_{2A} and 5-HT_{2c} receptors have been shown to enhance the release of dopamine and norepinephrine in the synaptic cleft of cortical neurons, resulting in subsequent reductions in anxiety and improved sleep [4]. Trazadone can be used as a monotherapy or in combination with other medications and is the second most prescribed medication for insomnia treatment, which is an off-label use, due to its antagonistic effects at the 5-HT_{2A}, H_1, and α_1-adrenergic receptors [6]. Trazadone has been shown to be as effective as TCAs for the treatment of major depressive disorder with a 60%–80% response rate with a shorter onset of action at 1 week [6]. This medication has a long half-life at 20–40 hours with side effects including drowsiness, dizziness, dry mouth, nervousness, bradycardia, hyponatremia, and acute angle glaucoma [6, 40, 43].

12.2.8 Vilazodone

Vilazodone is a SSRI and 5-HT_{1A} partial receptor agonist that was approved by the FDA in 2011 for the treatment of major depressive disorder [6, 8, 13, 40]. This medication is designed to increase serotonin levels in the brain without effects on norepinephrine and dopamine levels [6, 8, 17, 40]. Vilazodone has an onset of action within 1 week of treatment and has similar rates of remission compared to SSRIs [6, 40]. This medication is highly protein bound at 96%–99% and has a prolonged half-life of 25 hours [6, 40, 44]. Adverse effects from this medication are mild and include nausea, diarrhea, vomiting, insomnia, dizziness, dry mouth, fatigue, decreased libido, palpitations, arthralgias, and abnormal dreams [6, 8, 40, 44].

12.2.9 Vortioxetine

Vortioxetine was discovered through a drug development program that focused on the modulation of serotonin reuptake inhibition with additional modulation of the 5-HT_{1A} receptor to enhance the treatment of major depressive disorder [6, 40]. This medication was FDA approved for the treatment of major depressive disorder in 2013 in the United States and was found to function as a receptor antagonist, agonist, and partial agonist at multiple serotonin receptors [6, 8, 40]. Antagonism at the 5-HT_{1A} and 5-HT_3 receptors by vortioxetine has been shown to improve cognitive function including executive function, attention, and memory [8, 17]. Vortioxetine additionally acts as an antagonist at the 5-HT_7 and 5-HT_{1D} receptors, an agonist at the 5-HT_{1B} receptor, and a partial agonist at the 5-HT_{1A} receptor, resulting in increased concentrations of serotonin, norepinephrine, dopamine, acetylcholine, and histamine concentrations in the prefrontal cortex and ventral hippocampus [8, 13, 40]. Vortioxetine has been shown in studies to enhance glutamate transmission through the inhibition of the gamma-aminobutyric acid (GABA) neurons [8].

The diverse actions of vortioxetine on multiple receptor targets have led to reduced symptoms of depression and increased maintenance of remission [40]. Patients taking this medication for the treatment of major depressive disorder have noted substantial improvement in sleep, energy, and appetite within 1–2 weeks of starting the medication with full benefit at 4–6 weeks of therapy [6]. In animal models, vortioxetine has also been shown to increase neuroplasticity and cell proliferation in the hippocampus [8]. Vortioxetine has a long half-life at 57–66 hours and is well tolerated with adverse effects including nausea, sexual dysfunction, headache, dry mouth dizziness, constipation, and, in rare instances, pancreatitis and hypertensive crisis [6, 8, 45].

12.2.10 Triple Receptor Reuptake Agonists/Antagonists

Triple reuptake inhibitors that increase the concentrations of serotonin, norepinephrine, and dopamine in the synaptic cleft have the potential to better control symptoms of depression in patients who do not adequately respond to other available therapies [1, 16, 17]. These drugs are still under development but have the potential to improve the treatment of major depressive disorder by enhancing the known antidepressant effects of serotonin and norepinephrine in combination with enhanced dopamine effects on the reward systems of the brain [1, 8]. Given that anhedonia is a prominent feature of depression for most patients, the addition of dopamine modulation provides a promising avenue to alleviate this common symptom [8]. Additional modulation of multiple serotonin receptor subtypes in combination with serotonin reuptake inhibitors is another promising approach to the treatment of major depressive disorder that may mitigate some of the adverse effects associated with currently available therapies [16]. These medications have the potential to significantly improve treatment for patients with major depressive disorder.

12.3 Conclusions

Major depressive disorder is a complex medical condition that requires a multifaceted approach to treatment [2, 3, 8, 12, 13]. The historical approach to drug development for use in psychiatric conditions including the discovery of potential treatments by serendipity or the one-target, one-drug approach has not yielded significant sustained results for people living with depression

[2, 3, 5, 12]. Given the substantial personal and societal burden of this disease worldwide, the drug development process has transitioned to a multitargeted approach with a goal to modulate several neural pathways in diverse areas of the brain synergistically to improve symptoms and induce lasting disease remission [2, 5, 13, 15].

Multitargeted medications have the potential to modulate numerous targets within the CNS to enhance treatment efficacy and reduce off-target side effects [14–16]. The rational design of medications to treat depression that specifically target key receptors in the brain producing symptoms including anhedonia or sleep disturbances has the potential to enhance treatment for patients with this predominant symptomatology while limiting unwanted side effects [17]. Targeting drugs to specifically modulate neurotransmitters in key areas of the brain including the hippocampus, parietal cortex, or corticolimbic structures can substantially modify mood, motor behavior, cognition, and emotional responses that are common in major depressive disorder [4]. This polypharmacological approach to drug development can improve drug selection for providers and increase efficacy and safety for specific target populations [5, 13].

Major depressive disorder is a complex, heterogeneous disorder involving multiple genetic and environmental factors that combine to modulate the epigenetic mechanisms that control gene expression [1]. This complexity at the molecular level combined with neuroplasticity and diverse neuronal signaling within various structures of the brain highlights the need to control multiple mechanisms in a synergistic manner to better control the symptoms of depression and induce disease remission [1]. Multitargeted therapies that can manipulate these diverse neurochemical systems have the potential to produce substantial clinical efficacy with additional benefits in associated comorbid conditions and limited off-target side effects [3, 5]. Multitargeted drugs also have the benefit of preventing tachyphylaxis and the development of resistance to therapy that can often occur when patients are on a particular medication for an extended period of time [3].

Most medications that are developed to target multiple components of a clinical pathway are low-affinity, weak receptor binders [3]. This approach allows a single therapy to modulate multiple targets with fewer side effects due to the weak binding at off-target sites [3]. Since most signaling pathways have some degree of overlap, this weak binding mechanism allows the medication to exert a more profound effect at multiple receptors, thus synergistically controlling the pathways that lead to diverse symptoms in a complex medical condition like major depressive disorder [3]. Multitargeted therapies tend to be more potent and safer than single-targeted therapies due to the weak binding of these molecules to multiple receptor targets [3].

Treatment for major depressive disorder is increasingly complex, and it is clear that multitargeted medications will be the future of drug development within this therapeutic area. These medications offer the potential for greater efficacy, reduced time to remission, increased rates of remission, enhanced tolerability, greater patient safety, and improved patient compliance. The future of drug development in this and other complex medical conditions using multitargeted therapies holds great promise to improve clinical care and overall patient outcomes.

References

1 Millan, M.J. (2014). On 'polypharmacy' and multi-target agents, complementary strategies for improving the treatment of depression: a comparative appraisal. *Int. J. Neuropsychopharmacol.* 17 (7): 1009–1037.

2 Singh, K., Bhatia, R., Kumar, B. et al. (2022). Design strategies, chemistry and therapeutic insights of multi-target directed ligands as antidepressant agents. *Curr. Neuropharmacol.* 20 (7): 1329–1358.

3 Dessalew, N., Berhanu, W., and Hymete, A. (2008). Multitarget selective antidepressants design: latest developments, opportunities and challenges. *Curr. Med. Chem. - CNS Agents* 8 (2): 1–10.

4 Millan, M.J. (2004). The role of monoamines in the actions of established and "novel" antidepressant agents: a critical review. *Eur. J. Pharmacol.* 500 (1–3): 371–384.

5 Millan, M.J. (2009). Dual- and triple-acting agents for treating core and co-morbid symptoms of major depression: novel concepts, new drugs. *Neurotherapeutics* 6 (1): 53–77.

6 Schwasinger-Schmidt, T.E. and Macaluso, M. (2023). Biological treatments: psychopharmacology, brain stimulation, and innovations. In: *Atlas of Psychiatry* (ed. W.W. IsHak). Cham: Springer https://doi.org/10.1007/978-3-031-15401-0_10.

7 Schwasinger-Schmidt, T. and Preskorn, S.H. (2023). Discovery of new transmitter systems and hence new drug targets. In: *Drug Development in Psychiatry. Advances in Neurobiology*, vol. 30 (ed. M. Macaluso, S.H. Preskorn, and R.C. Shelton). Cham: Springer https://doi.org/10.1007/978-3-031-21054-9_7.

8 Dale, E., Bang-Andersen, B., and Sánchez, C. (2015). Emerging mechanisms and treatments for depression beyond SSRIs and SNRIs. *Biochem. Pharmacol.* 95 (2): 81–97.
9 Chockalingam, R., Gott, B.M., and Conway, C.R. (2019). Tricyclic antidepressants and monoamine oxidase inhibitors: are they too old for a new look? *Handb. Exp. Pharmacol.* 250: 37–48.
10 Delgado, P.L. (2000). Depression: the case for a monoamine deficiency. *J. Clin. Psychiatry* 61 (Suppl. 6): 7–11.
11 Hirschfeld, R.M. (2000). History and evolution of the monoamine hypothesis of depression. *J. Clin. Psychiatry* 61 (Suppl. 6): 4–6.
12 Ma, H., Huang, B., and Zhang, Y. (2020). Recent advances in multitarget-directed ligands targeting G-protein-coupled receptors. *Drug Discov. Today* 25 (9): 1682–1692.
13 Bawa, P., Pradeep, P., Kumar, P. et al. (2016). Multi-target therapeutics for neuropsychiatric and neurodegenerative disorders. *Drug Discov. Today* 21 (12): 1886–1914.
14 Si, T. and Wang, P. (2014). When is antidepressant polypharmacy appropriate in the treatment of depression? *Shanghai Arch. Psychiatry* 26 (6): 357–359.
15 Peters, J.-U. (2013). Polypharmacology–foe or friend? *J. Med. Chem.* 56 (22): 8955–8971.
16 Rodríguez-Lavado, J., Alarcón-Espósito, J., Mallea, M., and Lorente, A. (2022). A new paradigm shift in antidepressant therapy: from dual-action to multitarget-directed ligands. *Curr. Med. Chem.* 11 (29): 4896–4922.
17 O'Leary, O.F., Dinan, T.G., and Cryan, J.F. (2015). Faster, better, stronger: towards new antidepressant therapeutic strategies. *Eur. J. Pharmacol.* 753: 32–50.
18 Imipramine Pamoate Prescribing Information (2014). Mallinckrodt Pharmaceuticals. FDA approved. https://www.accessdata.fda.gov/drugsatfda_docs/label/2014/017090s078lbl.pdf (accessed 19 September 2024).
19 Doxepin HCL Prescribing Information (2007). Pfizer Inc. FDA approved. https://www.accessdata.fda.gov/drugsatfda_docs/label/2007/016798s054,017516s023lbl.pdf (accessed 19 September 2024).
20 Amitriptyline Hydrochloride Prescribing Information (2014). Sandoz Inc. FDA approved. https://www.accessdata.fda.gov/drugsatfda_docs/label/2014/085966s095,085969s084,085968s096,085971s075,085967s076,085970s072lbl.pdf (accessed 19 September 2024).
21 Nortriptyline Hydrochloride Prescribing Information (2012). Novartis Consumer Health Inc. FDA approved. https://www.accessdata.fda.gov/drugsatfda_docs/label/2007/018012s27lbl.pdf (accessed 19 September 2024).
22 Preskorn, S., Feighner, J., Stanga, C., and Ross, R. (2004). *Antidepressants: Past, Present and Future. Handbook of Experimental Pharmacology*. Switzerland: Springer.
23 Isocarboxazid Prescribing Information (2017). Neurocrine Biosciences Inc. FDA approved. https://www.accessdata.fda.gov/drugsatfda_docs/label/2017/209241lbl.pdf (accessed 19 September 2024).
24 Phenelzine Prescribing Information (2007). Pfizer Inc. FDA approved. https://www.accessdata.fda.gov/drugsatfda_docs/label/2007/011909s038lbl.pdf (accessed 19 September 2024).
25 Selegiline Prescribing Information (2007). Bristol-Myers Squibb Company. FDA approved. https://www.accessdata.fda.gov/drugsatfda_docs/label/2007/021336s002lbl.pdf (accessed 19 September 2024).
26 Tranylcypromine Prescribing Information (2015). Concordia Pharmaceuticals Inc. FDA approved. https://www.accessdata.fda.gov/drugsatfda_docs/label/2018/012342s064lbl.pdf (accessed 19 September 2024).
27 Macaluso, M. and Preskorn, S. (2019). *Antidepressants: From Biogenic Amines to New Mechanisms of Action.* Switzerland: Springer.
28 Lochmann, D. and Richardson, T. (2019). Selective serotonin reuptake inhibitors. *Handb. Exp. Pharmacol.* 250: 135–144.
29 Citalopram Prescribing Information (2009). Forest Pharmaceuticals Inc. FDA approved. https://www.accessdata.fda.gov/drugsatfda_docs/label/2009/020822s037,021046s015lbl.pdf (accessed 19 September 2024).
30 Escitalopram Prescribing Information (2009). Forest Pharmaceuticals Inc. FDA approved. https://www.accessdata.fda.gov/drugsatfda_docs/label/2009/021323s032,021365s023lbl.pdf (accessed 19 September 2024).
31 Fluoxetine Prescribing Information (1987). Eli Lilly and Company. FDA approved. https://www.accessdata.fda.gov/drugsatfda_docs/label/2011/018936s091lbl.pdf (accessed 19 September 2024).
32 Paroxetine Prescribing Information (2008). GlaxoSmithKline. FDA approved. https://www.accessdata.fda.gov/drugsatfda_docs/label/2008/020031s060,020936s037,020710s024lbl.pdf (accessed 19 September 2024).
33 Sertraline Prescribing Information (2009). Pfizer Inc. FDA approved. https://www.accessdata.fda.gov/drugsatfda_docs/label/2009/019839s070,020990s032lbl.pdf (accessed 19 September 2024).
34 Shelton, R.C. (2019). Serotonin and norepinephrine reuptake inhibitors. *Handb. Exp. Pharmacol.* 250: 145–180.
35 Venlafaxine Prescribing Information (1997). Wyeth Pharmaceuticals Inc. FDA approved. https://www

.accessdata.fda.gov/drugsatfda_docs/label/2017/020699s107lbl.pdf (accessed 19 September 2024).

36 Desvenlafaxine Prescribing Information (2008). Wyeth Pharmaceuticals Inc. FDA approved. https://www.accessdata.fda.gov/drugsatfda_docs/label/2012/021992s030lbl.pdf (accessed 19 September 2024).

37 Duloxetine Prescribing Information (2004). Eli Lilly and Company. FDA approved. https://www.accessdata.fda.gov/drugsatfda_docs/label/2010/022516lbl.pdf (accessed 19 September 2024).

38 Milnacipran Prescribing Information (2009). Forest Pharmaceuticals Inc. FDA approved. https://www.accessdata.fda.gov/drugsatfda_docs/label/2012/022256s013lbl.pdf (accessed 19 September 2024).

39 Levomilnacipran Prescribing Information (2019). Allergan USA, Inc. FDA approved. https://www.accessdata.fda.gov/drugsatfda_docs/label/2019/204168s006lbl.pdf (accessed 19 September 2024).

40 Schwasinger-Schmidt, T.E. and Macaluso, M. (2019). Other antidepressants. *Handb. Exp. Pharmacol.* 250: 325–355.

41 Bupropion Prescribing Information (2017). GlaxoSmithKline. FDA Approved. https://www.accessdata.fda.gov/drugsatfda_docs/label/2017/018644s052lbl.pdf (accessed 19 September 2024).

42 Mirtazapine Prescribing Information (2020). Merk and Co., Inc. FDA approved. https://www.accessdata.fda.gov/drugsatfda_docs/label/2020/020415s029,%20021208s019lbl.pdf (accessed 19 September 2024).

43 Trazadone Prescribing Information (2017). Pragma Pharmaceuticals. FDA approved. https://www.accessdata.fda.gov/drugsatfda_docs/label/2017/018207s032lbl.pdf (accessed 19 September 2024).

44 Vilazodone Prescribing Information (2021). Allergan sales, LLC. FDA approved. https://www.accessdata.fda.gov/drugsatfda_docs/label/2021/022567s022lbl.pdf (accessed 19 September 2024).

45 Vortioxetine Prescribing Information (2021). Takeda Pharmaceuticals America, Inc. FDA approved. https://www.accessdata.fda.gov/drugsatfda_docs/label/2021/204447s021s022lbl.pdf (accessed 19 September 2024).

13

Multi-target Drugs to Treat Metabolic Diseases

Felix F. Lillich, Samaneh Goorani, Ewgenij Proschak, and John D. Imig

13.1 Introduction

Metabolic diseases are increasing at an alarming rate since 2000 across all countries of the world [1–3]. This rising prevalence of metabolic diseases over the past two decades is a major global health burden [1, 4]. Metabolic diseases refer to the dysregulation of metabolic processes affecting obesity-linked insulin resistance, glucose homeostasis, lipid metabolism, and pro-inflammatory immune cells and cytokines [5, 6]. Major metabolic diseases include type 2 diabetes, hypertension, hyperlipidemia, obesity, and nonalcoholic fatty liver disease (NAFLD). These diseases are complex and commonly occur together [5, 6]. Several organs play a critical role in metabolic homeostasis by regulating synthesis, metabolism, redistribution, and utilization of carbohydrates, lipids, and amino acids as such metabolic diseases impact the whole body. Treatment approaches must act to improve the overall metabolic state rather than an individual disease, organ, or biological mechanism. Lifestyle modification that includes weight reduction and physical activity can improve the metabolic state of an individual [7, 8]. Likewise, pharmacological agents need to act to improve the overall metabolic state of an individual [9, 10]. Improvement of the overall metabolic state through drugs will likely mean that the drugs will need to act on multiple biological targets [11]. This review will focus on the development of multi-target drugs to improve the metabolic state to treat metabolic diseases.

13.2 Metabolic Diseases and Current Treatment Approaches

Metabolic diseases display common features including abdominal obesity, dyslipidemia, elevated blood pressure, and insulin resistance [6, 12]. In addition to the metabolic abnormalities, inflammation associated with metabolic diseases plays a key role in disease progression and associated comorbidities [5, 6]. Metabolic disease disrupts physiological regulatory systems due to excessive energy intake, which activates stressor stimuli to trigger inflammatory pathways [13–15]. Elevated levels of circulating free fatty acids also can result in inflammation during metabolic diseases [16, 17]. Likewise, adipose tissue expansion associated with metabolic diseases leads to adipokine and cytokine imbalances that result in a pro-inflammatory state [18, 19]. Inflammation during metabolic diseases also occurs in the pancreas, liver, and kidneys [19–21]. This low-grade inflammation is characterized by the activation of both the innate and adaptive immune systems [18, 19]. Plasma levels of tumor necrosis factor-α (TNF-α), interleukin 6 (IL 6), IL-1β, C-reactive protein, and other inflammatory mediators are elevated in humans with metabolic diseases [13–15]. Animal studies demonstrate that immune cell infiltration and increased cytokines in metabolic diseases could result in cardiovascular, neural, renal, and liver complications [13, 14, 22]. Although type 2 diabetes, hypertension, dyslipidemia, and liver disease are considered separate diseases, there is the commonality of inflammation for metabolic diseases (Figure 13.1) [13, 15].

Type 2 diabetes is a metabolic disease that is associated with obesity and decreased physical activity, which results in ineffective insulin utilization [23, 24]. The prevalence of this common disorder has increased at an alarming rate due to increased obesity and sedentary lifestyle [1, 3]. Currently, an estimated 462 million people worldwide, corresponding to 6% of the world's population, are affected by type 2 diabetes [1, 25]. Type 2 diabetes ranks as the ninth leading cause of mortality worldwide [25]. Patients with type 2 diabetes are at high risk for developing complications such as retinopathy, neuropathy, nephropathy, and cardiovascular diseases. As an example, diabetic nephropathy accounts for approximately one-third of all cases of advanced chronic kidney disease worldwide and is one of the leading causes

Polypharmacology: Strategies for Multi-Target Drug Discovery, First Edition. Edited by Jens-Uwe Peters.

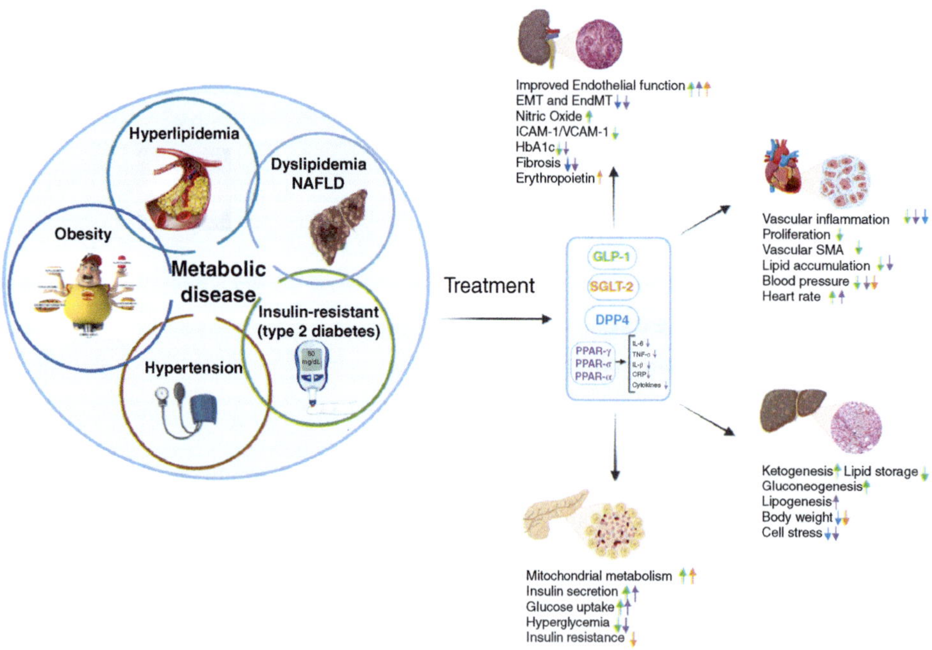

Figure 13.1 Metabolic diseases have characteristics that include obesity, hyperlipidemia, nonalcoholic fatty liver disease (NAFLD), hypertension, and insulin-resistant type 2 diabetes. Type 2 diabetes treatments can have effects on several characteristics of metabolic diseases. These treatments include glucagon-like peptide 1 (GLP-1) agonists, sodium–glucose cotransporter-2 (SGLT-2) inhibitors, dipeptidyl peptidase-4 (DPP4) inhibitors, and peroxisome proliferator-activated receptor (PPAR) agonists. These therapeutic targets lay the foundation for developing multi-target drugs to treat metabolic disease rather than an individual disease. Abbreviations: EMT = epithelial-to-mesenchymal transition, EndMT = endothelial-to-mesenchymal transition, ICAM-1 = intercellular adhesion molecule-1, VCAM-1 = vascular cell adhesion molecule-1, HbA1c = hemoglobin A1c, SMA = smooth muscle actin, IL-6 = interleukin-6, TNF-α = tumor necrosis factor alpha, IL-1β = interleukin-1 beta, CRP = C-reactive protein. With permission of BioRender.com.

of mortality in diabetic patients [26, 27]. Although current drugs lower blood glucose and hemoglobin A1c (HbA1c), complications due to inflammation remain common [28–30]. There is an urgent need for developing drugs that can treat diabetes as well as complications that occur in several organ systems.

Current antidiabetic drugs include peroxisome proliferator-activated receptor-γ (PPAR-γ) agonist insulin-sensitizing agents, glucagon-like peptide-1 (GLP-1) receptor agonists, dipeptidyl peptidase-4 (DPP4) inhibitors, and sodium–glucose cotransporter-2 (SGLT-2) inhibitors [10, 30]. As mentioned, these antidiabetic agents are effective in lowering blood glucose and HbA1c levels [10, 30]. Unfortunately, polypharmacy is a growing concern in patients with type 2 diabetes [31, 32]. Polypharmacy is defined by at least five drugs being prescribed and occurs in 60% of type 2 diabetic patients over 65 years old [32]. A large percentage of type 2 diabetics are on at least one antidiabetic drug, a lipid-lowering statin, and antihypertensive drugs [10, 31, 32]. There has been ongoing research into ways to limit the number of drugs needed to treat type 2 diabetes and associated complications and diseases.

The pleiotropic actions of several antidiabetic drugs could possibly limit the need for polypharmacy. As an example, PPAR-γ activation by thiazolidinediones (TZDs) has been demonstrated to lower blood pressure in humans and demonstrated to have vascular protective effects through unknown mechanisms [33–35]. However, the clinical use of TZDs is limited by excessive weight gain and fluid retention, which resulted in increased cardiovascular events and a warning label for patients with heart failure [35, 36]. Metformin that lowers hepatic glucose output through adenosine monophosphate (AMP) activation also prevents weight gain and lowers circulating plasminogen activator inhibitor-1 levels that could decrease cardiovascular events associated with type 2 diabetes [13, 30]. There has been recent excitement surrounding the use of GLP-1 receptor agonists and SGLT-2 inhibitors because of their potential to have organ protective effects beyond their antidiabetic actions [37, 38]. GLP-1 receptor agonists increase insulin secretion in response to glucose and suppress glucagon secretion from the pancreatic islet cells, which reduces postprandial glucose levels [38, 39]. In addition, a reduction in caloric intake and body weight occurs with GLP-1 receptor agonist treatment that appears to be related to effects on the gastrointestinal tract and central nervous system [38, 40]. Pleiotropic actions of GLP-1 receptor agonists on several organs have resulted in slowing diabetic nephropathy and reducing cardiovascular events [38, 39]. SGLT-2 inhibitors lower blood glucose and HbA1c levels by inhibiting glucose reabsorption in the kidneys [37, 41]. A reduction in body weight and blood pressure occurs with SGLT-2 inhibitor treatment [37, 42]. Like GLP-1 receptor agonists, SGLT-2 inhibitors have been demonstrated to slow the progression of diabetic nephropathy and significantly reduce cardiovascular events [37, 41]. Even with these promising actions of SGLT-2 inhibitors, their combination with renin–angiotensin system inhibitors is recommended as an approach to prevent the development and progression of diabetic kidney disease [43, 44]. Thus, the development of multi-target drugs for diabetes and associated complications and diseases is warranted.

Nonalcoholic liver disease (NAFLD), characterized by excessive fat buildup in the liver, is the most common liver disease and affects up to 30% of the adult population [1, 45]. Excessive caloric intake, a sedentary lifestyle, and insulin resistance lead to increased fatty acids and lipotoxic lipids to accumulate in the liver, resulting in NAFLD [45, 46]. People with obesity and type 2 diabetes will likely have NAFLD [1, 45]. A more advanced form of NAFLD is nonalcoholic steatohepatitis (NASH), which involves inflammation of the liver and fibrotic liver damage [45, 47]. NASH is emerging as the leading cause of cirrhosis, cirrhotic complications, hepatocellular carcinoma, and liver-related death [45, 48]. Fortunately, NASH makes up a small percentage of people with NAFLD [45]. There are currently limited options for treating NASH. Weight loss through caloric restriction, exercise, and healthy eating is the first-line treatment for NASH [49, 50]. Currently, there are no Food and Drug Administration (FDA)-approved medications to treat NAFLD or NASH.

Several therapeutic strategies and molecular targets have been proposed for the treatment of NAFLD and NASH. The bile acid (BA) transcription factor farnesoid X receptor (FXR) is one such molecular target [51, 52]. FXR activation prevents hepatocyte damage and reduces liver fat accumulation [51, 52]. FXR agonists have demonstrated therapeutic efficacy in humans with NASH; however, all have failed to get approval for various reasons [53, 54]. Other experimental strategies that decrease steatosis, fibrosis, and inflammation in NASH are in clinical trials [55]. A major challenge of NAFLD and NASH is that the disease often coexists with obesity and type 2 diabetes [1, 45]. Multi-target drugs could be the therapeutic approach needed for these multifactorial liver diseases.

Metabolic syndrome (MetS) also known as insulin resistance syndrome is highly prevalent and has a major impact on public health worldwide [1–3]. The risk for cardiovascular disease and other serious health problems is a MetS consequence [56, 57]. MetS is diagnosed when a patient has at least three of the following risk factors: abdominal obesity, elevated blood pressure, low high-density lipoprotein (HDL) cholesterol, hypertriglyceridemia, or glucose intolerance [10, 16]. As such, patients with type 2 diabetes or NAFLD and NASH will likely have MetS [1, 10]. Thus, a metabolic disease rarely occurs in isolation. Developing adequate preventive and therapeutic strategies for MetS and these multifactorial metabolic diseases has been challenging. Unfortunately, therapeutic approaches treat these risk factors and metabolic diseases as if they occur in isolation, which results in patients taking several drugs [31, 32]. Polypharmacy can easily be reached for MetS because for hypertension 70% will need at least two antihypertensive drugs, for type 2 diabetes 50% will need at least two antidiabetic drugs three years after initial diagnosis, for high cholesterol statins are prescribed, and for obesity and cardiovascular diseases additional drugs are needed [31, 32]. Intriguingly, inflammation is a common characteristic of these metabolic diseases and could serve as a starting point to develop multi-target drugs for metabolic diseases.

13.3 Strategies to Develop Multi-target Drugs for Metabolic Diseases

Therapeutic approaches to treat metabolic diseases have had varied success in improving patient outcomes [11, 52, 58]. Lifestyle modifications such as exercise, dieting, and cessation of smoking have been insufficient in mitigating disease risk factors [10, 30]. Multidrug regimens and polypharmacy are a major problem in metabolic diseases due to poor patient compliance, drug side effects, and drug–drug interactions [10, 11]. Presently, there are no approved drugs capable of consistently mitigating various conditions linked to metabolic diseases in the long term [10, 59]. The emergence of multi-target drugs in recent years has resulted in novel therapeutics that could treat multiple conditions in metabolic diseases [11, 60, 61]. Developing a multi-target drug that lowers plasma cholesterol, decreases inflammation, lowers blood pressure, and decreases HbA1c levels would be ideal for treating metabolic diseases.

Current drugs for metabolic diseases could serve as a foundation for developing multi-target drugs. Indeed, several of the drugs for metabolic diseases have multiple therapeutic actions or act on multiple molecular targets. Metformin, statins, PPAR agonists, FXR agonists, angiotensin receptor blockers (ARBs), and statins are examples [11, 62]. The multiple therapeutic actions for the biguanide metformin include lowering blood glucose, preventing weight gain, and decreasing cardiovascular events [13, 30]. Statins inhibit 3-hydroxy-3-methylglutaryl coenzyme A (HMG-CoA) reductase to lower lipids; however, statins also lower pro-inflammatory C-reactive protein levels and enhance liver insulin sensitivity through unknown mechanisms [3, 10]. ARBs such as telmisartan, irbesartan, and azilsartan possess pleiotropic effects beyond blood pressure reduction [11, 63]. These ARBs have effects on glucose metabolism that have been attributed to their ability to enhance PPAR-γ activity [63, 64]. Irbesartan and azilsartan lower blood pressure, decrease inflammation, improve endothelial function, and decrease kidney and heart injury in animal models and humans with MetS [64, 65]. PPAR agonists act as ligand-regulated transcription factors to impact gene expression. PPAR-γ agonists or TZDs are used as insulin-sensitizing agents to treat type 2 diabetes [30]. TZDs also have anti-inflammatory actions that decrease the progression of cardiovascular disease associated with type 2 diabetes [30]. Likewise, PPAR-α agonists, fibrates, have been approved for several decades to treat dyslipidemia [35]. Fibrates exhibit pleiotropic effects to lower blood pressure, decrease inflammation, improve endothelial function, and decrease heart attacks [35, 66]. There have been attempts to develop drugs that act on multiple PPARs to treat metabolic diseases [13, 30, 67]. Interestingly, the dual PPAR-α/PPAR-γ agonist, tesaglitazar, improves metabolic abnormalities and decreases renal in MetS animal models [68]. Unfortunately, these dual-acting PPAR agonists have failed in clinical trials. FXR activation is another target that could serve as a foundation for multi-target drugs to treat metabolic diseases [69, 70]. FXR agonists not only reduce hepatic fat accumulation but also have anti-inflammatory actions and decrease fibrosis in several organs [52, 71]. Several FXR agonists are in development and in clinical trials for biliary cholangitis, NASH, diabetic nephropathy, and type 2 diabetes [51, 71]. Unwanted side effects for FXR agonists include pruritus and increasing cholesterol levels [51]. These current therapeutic targets along with targeting inflammation provide the building blocks for developing bifunctional molecules that could combat multiple metabolic diseases.

There are major challenges for the development of multi-target drugs for the treatment of metabolic diseases. One challenge is the need to optimize the drug for multiple biological targets while keeping essential drug properties [72–74]. Multi-target drugs tend to have higher molecular weights and increased lipophilicity. Although the challenge of drug-like properties can be overcome, determining the correct biological targets is critical to the drug design. Many target combinations are amenable for developing multi-target drugs. Computational tools and structural information allow for pharmacophore modeling for the intended biological targets [72–74]. Chemical tractability is needed for the targets of interest to accommodate chemically similar ligands [72, 73]. On the other hand, there are examples of very different ligands that can be addressed in one drug. ARBs with PPAR-γ activity are a prominent example of dual modulators with very different ligands [63, 64], as exemplified by telmisartan (Figure 13.2, **1**). Defining desired activity with pharmacological properties and selectivity for biological targets is challenging. In addition, biological target combinations need to exhibit synergistic effects that contribute to the efficacy and safety of the multi-target drug. Synergistic target combinations have been successfully developed in the areas of antibiotics and cancer therapeutics [72]. There are several multi-target drugs with synergistic target combinations that are in early development stages for metabolic diseases [11, 72]. These multi-target drugs for metabolic diseases are demonstrating promising results in animal models of MetS, NASH, and type 2 diabetes [75–77].

Medicinal chemistry approaches to designing multi-target ligands fall into three categories: linked, fused, and merged pharmacophore drugs [72, 78]. Linked multi-target drugs have distinct pharmacophores for each target by a flexible linker. Although linked multi-target drugs typically exhibit sufficient *in vitro* potency at the targets, the larger molecular weight and increased number of free rotatable bonds lead to negative pharmacokinetic effects [79]. Eliminating the linker between two pharmacophores will result in them touching, which is known as a fused multi-target drug. This results in a lower molecular weight molecule but lacks the flexibility of a linked multi-target drug [78]. Identification of a merged pharmacophore for two targets is the most promising way to develop a multi-target ligand that will exhibit a promising pharmacodynamics and pharmacokinetic profile [72, 78]. Multi-target merged pharmacophore drugs are more challenging to design and optimize. Multi-target drugs in these three categories have been developed for metabolic diseases [11, 72].

Another major step in the development of multi-target drugs is to advance drugs with varying IC_{50} and EC_{50} *in vitro* into being effective in *in vivo* systems. An unanswered question with multi-target drugs *in vivo* is their ability to engage two biological targets at the same time. The engagement of a drug with two targets at the same time depends on the type of multi-target molecule, cell proximity of the targets, and binding properties to the targets. Linked multi-target drugs acting on two targets in close cellular proximity is the most likely scenario for the engagement

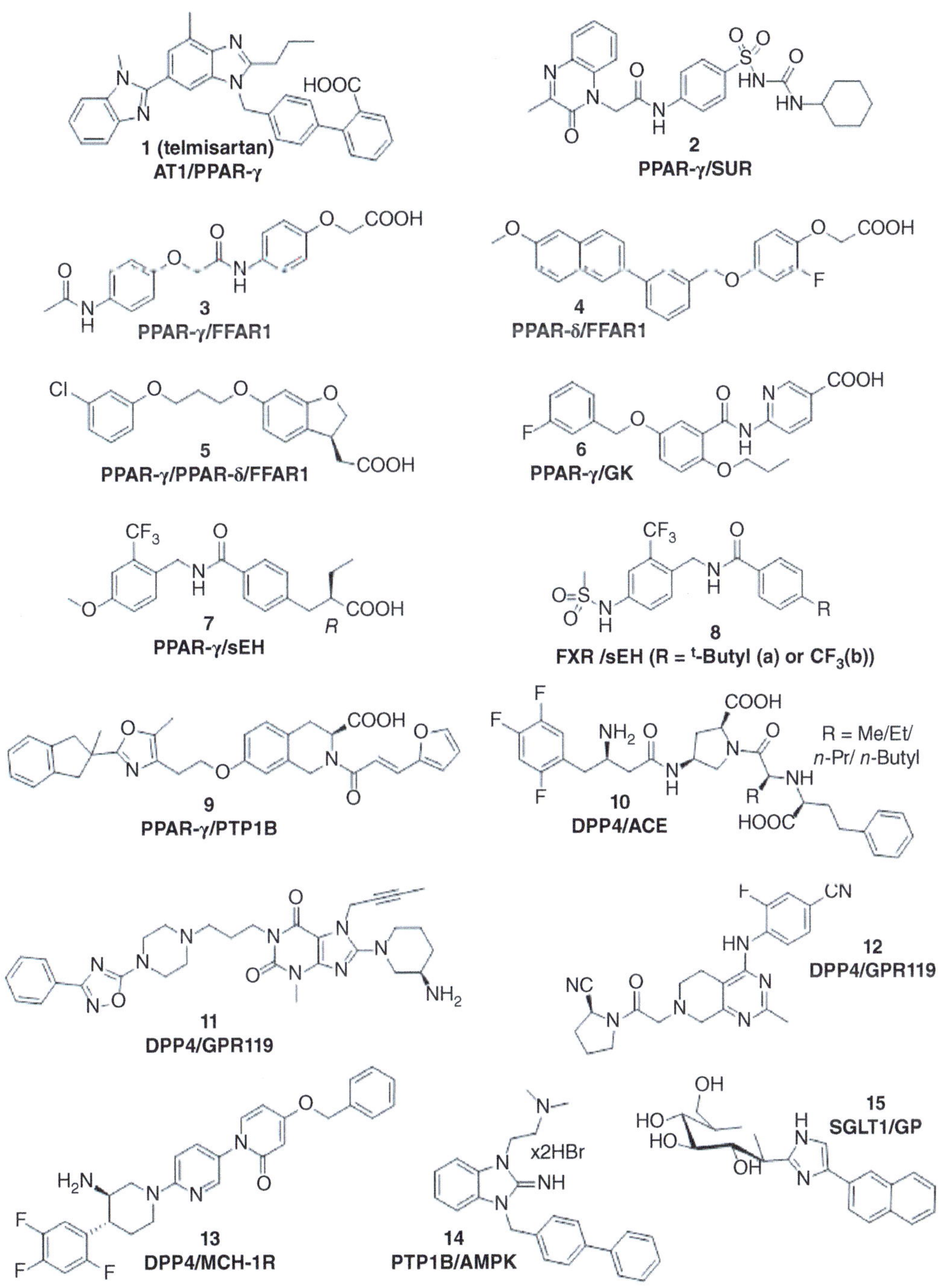

Figure 13.2 Dual modulators.

of two targets *in vivo* at the same time. A more plausible scenario would be that each target would be engaged separately, which would require a particular plasma level of the multi-target drug. Extensive *in vivo* evaluation is needed for assessing the proper target engagement by multi-target drugs. This book chapter discusses multi-target drugs that, in animal disease models, have been administered at doses verified to ensure effective *in vivo* target engagement.

Several examples of multi-target drugs for metabolic diseases will be summarized. A focus will be on explaining the rationale behind the choice of cellular and molecular target combinations. Moreover, challenges in the design and optimization of multi-target drugs and their potential treatment for metabolic diseases will be addressed.

13.4 Approaches Involving Modulation of PPARs and Other Metabolically Relevant Nuclear Receptors

PPARs and other nuclear receptors involved in metabolic processes (especially FXR) are activated by fatty acids or BAs. Given the chemical structure of their endogenous ligands, the synthetic modulators of these attractive drug targets are also known as fatty acid mimetics, and they often comprise an acidic head group and a large lipophilic tail [72]. This molecular blueprint offers a unique opportunity to combine the pharmacophores of PPAR or FXR ligands among each other or with other target classes that are modulated by fatty acids [80].

13.4.1 PPAR-γ/SUR

One of the first classes of oral antidiabetics was the sulfonylureas (SURs). SURs are binding to the adenosine triphosphate (ATP)-sensitive K^+ (K_{ATP}) channels on pancreatic beta cells, thereby stimulating insulin release. The pharmacophore of SURs consists of the name-giving acidic SUR group, flanked by two lipophilic moieties. Ibrahim et al. were able to combine the SUR and the PPAR-γ pharmacophore, resulting in the merged PPAR-γ/SUR ligand **2** (Figure 13.2) [81, 82]. **2** displays sub-micromolar PPAR-γ-binding affinity and insulin-secreting activity *in vitro* and reduced blood glucose levels in streptozotocin-induced hyperglycemic rats.

13.4.2 PPARs/FFAR1

Several G-protein-coupled receptors (GPCRs) are activated by free fatty acids. Especially, the free fatty acid receptor 1 (FFAR1) has been extensively studied in the context of diabetes, due to its stimulating effect on insulin release [83]. Due to the high similarity of the endogenous ligands, it was obvious to combine the pharmacophores to yield dual PPAR/FFAR1 modulators. Hidalgo-Figueroa et al. were able to design a dual PPAR-γ/FFAR1 ligand **3** (Figure 13.2), which exhibited a robust antihyperglycemic activity in several animal models of diabetes without adverse effects (hypoglycemia) [84]. Zhou et al. designed a dual PPAR-δ/FFAR1 ligand **4** (Figure 13.2), which was also exhibiting glucose-lowering effects *in vivo* without showing toxicity [85]. Finally, Li et al. developed a triple PPAR-δ/PPAR-γ/FFAR1 ligand **5**, which has also shown robust glucose-lowering effects in ob/ob mice [86].

13.4.3 PPAR-γ/GK

Glucokinase (GK) catalyzes the phosphorylation of glucose, which can be subsequently incorporated into glycogen. Thus, the activation of GK efficiently promotes the storage of glucose in hepatocytes. GK also functions as a glucose sensor in the islet cells of the pancreas and promotes insulin secretion [87]. Lei et al. prepared compound SHP289-03 (Figure 13.2, **6**), which modulates PPAR-γ and exhibits a 1.7-fold activation of GK at 50 μM. **6** acts as a partial PPAR-γ agonist, thereby activating the expression of PPAR-γ-dependent genes Adrp and aP2, without promoting adipocyte differentiation. In type 2 diabetic KK-A(y) mice, **6** exhibited dose-dependent improvement in hyperglycemia and insulin secretion. Surprisingly, treatment with **6** led to improvement of lipid parameters and lower body weight [88].

13.4.4 PPAR-γ/sEH

The metabolites, enzymes, and receptors of the arachidonic acid (AA) cascade are excellent therapeutic targets for the treatment of the MetS and inflammatory diseases, because it is one of the major pathways for the progression of pain and inflammation signals in the human body [89]. In this regard, important metabolites are the epoxyeicosatrienoic acids (EETs), whose accumulation is associated with anti-inflammatory effects [90–94]. However, in humans, EETs are predominantly metabolized via the soluble epoxide hydrolase and sEH gene (sEH and EPHX2) pathway [93]. sEH inhibition can stabilize EET levels [95]. PPAR-γ signaling and sEH metabolite EETs are tightly interconnected. While inhibition of sEH leads to increased EET levels, EETs were identified as endogenous ligands of PPAR-γ [90, 96].

In 2012, Imig and Hammock studied the effects of the selective PPAR-γ agonist rosiglitazone in combination with sEH inhibitor (t-AUCB) in an animal model of the MetS (SHROB rats). The authors concluded that the single application of an sEH inhibitor or a PPAR-γ modulator is beneficial, but in regard to the multifactorial characteristics of the MetS, the combination is even superior [97]. This and related studies emphasize the potential for the development of a dual sEH/PPAR-γ modulator for the treatment of MetS. Therefore, several studies have been conducted to design dual PPAR-γ/sEH modulators [98–100]. Blöcher et al. discovered *N*-benzyl benzamides as novel scaffolds of orally available dual PPAR-γ/sEH modulators. The lead compound RB394 was designed as a merged pharmacophore that modulates PPAR-γ and inhibits sEH at sub-micromolar concentrations [98]. In two animal models of the MetS (spontaniously hypertensive obese rat [SRHOB] and ZSF1 rats), RB394 showed its potential to improve all risk factors for MetS *in vivo* [75]. Subsequently, Hartmann et al. demonstrated that the eutomer of RB394 (Figure 13.2, **7**) is able to promote adipocyte browning in cell culture and exhibits cardioprotective activity in isolated buffer-perfused mouse hearts following ischemia [100]. These results underline that the combination of an sEH inhibitor with a selective PPAR-γ agonist in one molecule could be beneficial for the treatment of type 2 diabetes and hypertension, related complications, and, by this, the MetS in general.

13.4.5 FXR/sEH

FXR (NR1H4) is a nuclear BA receptor [101, 102]. FXR regulates the transcription of genes that, among others, are involved in inflammation and glucose and lipid homeostasis [103]. The physiological role of FXR is liver protective, as it is an enterohepatic regulator of BA homeostasis [104]. On the other hand, animal studies with induced fatty liver mice could show that reduced or lack of sEH activity has positive outcomes on the disease [105–107]. Taken together, the positive effects of FXR activation in NASH and the anti-inflammatory and anti-steatotic effects of sEH inhibition clearly show the potential of a dual FXR/sEH modulator.

The first dual modulator (Figure 13.2, **8a**) was published by Schmidt et al. in 2017, following a combinatorial approach. **8a** can stabilize EET levels and partially activate FXR target gene expression *in vitro* and *in vivo* [108]. In addition, the compound demonstrated excellent efficacy in the complex setting of NASH by reducing hepatic steatosis and fibrosis and exhibiting a pronounced anti-inflammatory effect in different mouse models of NASH [76, 77]. In 2021, Helmstädter et al. further optimized the compound with regard to its pharmacological properties. By replacing the tert-butyl against a trifluoromethyl motif (Figure 13.2, **8b**), they were able to significantly improve the metabolic stability and *in vivo* half-life [77]. Interestingly, dual FXR/sEH modulators had pronounced anti-inflammatory and anti-fibrotic effects not only in mice with toxin- and diet-induced obesity (DIO) [76, 77] but also in the model of kidney fibrosis [109].

13.4.6 PPAR-γ or PPAR-α/γ/PTP1B

Another extensively validated drug target for the treatment of obesity and type 2 diabetes is the enzyme protein tyrosine phosphatase 1B (PTP1B). This is due to its important role in inhibiting signaling through the insulin and leptin receptors. PTP1B controls the cellular response to insulin signaling by dephosphorylating phosphorylated tyrosine residues of insulin receptor substrate-1 (IRS-1) and the activated insulin receptor b (IRb) subunit. In the case of leptin signaling, an element located further downstream in the cascade, the tyrosine kinase Janus kinase 2 (JAK2), becomes the primary target of dephosphorylation [110–114]. Inhibition of PTP1B can lead to increased sensitivity of the insulin receptor and consequently to an improvement in diseases associated with insulin resistance [110]. Therefore, it is expected that the inhibition of PTP1B exerts synergistic effects with PPAR-γ activation on insulin sensitization [115].

Over the years, Hiroaki Shirahase's group has published various dual or triple modulators with PPAR-γ/α and PTP1B [116–120]. All modulators are based on the 1,2,3,4-tetrahydroisoquinoline-3-carboxylic acid scaffold, which was systematically investigated. Different acyl substituents were coupled to the nitrogen at position 2, while at position 7, the 5-methyloxazole-ethoxy substituent was varied. The first compounds of their structure–activity studies yielded triple modulators (PPAR-α/γ + PTP1B) [116–118]. They subsequently developed these further into dual PPAR-γ/PTP1B modulators [119]. The most potent compound was **9** (Figure 13.2), which proved to be a potent PPAR-γ partial agonist (EC_{50}: 0.85 μM [65% maximum activation relative to farglitazar]), while being inactive against the other PPAR subtypes. Human PTP1B is inhibited with an IC_{50} of 1.0 μM. In an initial animal study in KK-A(y) mice, the compound **9** showed safer antidiabetic effects (e.g., weaker hemodilution and hepatotoxicity) compared to rosiglitazone, which the authors attributed to its dual mode of action as a dual partial PPAR-γ agonist/PTP1B inhibitor [120].

13.5 Approaches Involving Inhibition of DPP4

A very prominent drug class for the treatment of type 2 diabetes mellitus (T2DM) is gliptins [121]. By inhibiting the DPP4 enzyme and preserving the action of GLP-1, these molecules have the potential to increase glucose uptake. The peptide hormone GLP-1, which belongs to the incretins, is able to liberate insulin in a dose-dependent manner after oral glucose uptake. Beyond that, GLP-1 suppresses the formation of glucagon [122]. GLP-1 is degraded rapidly after formation by DPP4 [123]. Semaglutide, a very effective GLP-1 agonist, already entered the market [124]. However, DPP4 as the GLP-1-degrading enzyme is as well a valuable target to face T2DM. DPP4, itself, is an exopeptidase, which cleaves proline or alanine dipeptides from the N-terminus of proteins and peptides, thereby changing their activity [125]. Several gliptins have already come onto the market, but this class is also associated with different side effects such as skin reactions or gastrointestinal problems [126–128]. Taken together, this underlines not only the potential of DPP4 inhibition as an effective way to prevent hyperglycemia but also the importance of developing new and safer inhibitors for this enzyme.

13.5.1 DPP4/ACE

Angiotensin-converting enzyme (ACE) inhibitors have been shown to be an efficient treatment of hypertension, used to medicate cardiovascular (e.g., congestive heart failure [CHF]) or renal diseases [129]. By inhibiting ACE, these molecules are able to decrease the activity of the renin–angiotensin–aldosterone system (RA[A]S). RAS controls the electrolyte balance, cardiovascular, adrenal, and renal functions and arterial blood pressure [129, 130]. To do so, in the kidney renin is released, for example, in response to low blood pressure, hyponatremia, or potassium depletion [129]. Renin then transforms its substrate angiotensinogen into angiotensin I (Ang I). Ang I is further cleaved by ACE to the physiologically active angiotensin II (Ang II) [131]. The inhibition of the conversion of Ang I to Ang II leads very generally to a decrease in blood pressure and vascular tone. In addition, it has been shown that ACE inhibitors reduce the progress of diabetic nephropathy and provide renovascular protection [132].

In 2012, Wang et al. reported the discovery of a peptide from an egg protein hydrolysate, which shows DDP4- and ACE-inhibitory activity [133]. The first rational approach toward a small-molecule dual ACE/DPP4 inhibitor for the treatment of MetS was described by Sattigeri et al. in 2017. The authors merged the most relevant functional groups of enalaprilat, the active metabolite of enalapril (ACE inhibitor), and sitagliptin (DDP4 inhibitor). *In vitro* characterization with enzymes from rat, mouse, and human plasma showed that the compounds (Figure 13.2, **10**) can inhibit both enzymes, especially the human isoforms with good potency; however, the compounds also partially showed a very large interspecies shift. In addition, they showed an excellent selectivity profile and metabolic stability [134].

13.5.2 DPP4/GPR119

As mentioned above, DPP4 inhibitors lead to the secretion of insulin in response to elevated blood glucose levels. A similar effect can also be achieved by the activation of GPCR GPR119 [135, 136]. Thus, GPR119 agonists could improve glucose homeostasis in patients with T2DM. Two main classes of endogenous ligands of GPR119 are described: (i) amide derivatives of fatty acids, such as oleoylethanolamide (OEA), and (ii) phospholipids, such as oleoyl-lysophosphatidylcholine [137]. Due to the complementary mechanisms of action of DPP4 inhibitors and GPR119 agonists, the synergistic effect of this target combination has been investigated [138]. Li et al. described the first dual DPP4/GPR119 modulator in 2016. In a pharmacophore merging approach, they combined the most important pharmacophore features of a DPP4 inhibitor (xanthine scaffold from linagliptin) and GPR119 agonists (4-piperidine moiety from AR231453, APD597, and PSN632408). Their investigations led to a compound, which is able to activate GPR119 (EC_{50}: 0.95 μM, 81.5% activity of AR231453 @1 μM) and inhibit DPP4 (IC_{50}: 0.22 μM) *in vitro*. In addition, it exhibits favorable selectivity against DPP8/9 as well as an acceptable *in vitro* pharmacological profile [139].

This compound was further optimized by Huan et al. HBK001 (Figure 13.2, **11**) can inhibit DPP4 *ex vivo* (IC_{50}: 0.066 μM) as well as *in vivo* (ICR mice, 30 mg/kg, 50% DPP4 inhibition for about 4 hours), and GPR119 is activated *in vitro* (EC_{50}: 0.03 μM, 73.6% activity of APD597 @10 μM). It is also selective over other DPPs and GPCRs. Further ex and *in vivo* experiments support the authors' statement that the dual DPP4/GPR119 modulator **11** could be a very promising candidate for the treatment of T2DM [140]. In 2020, the same group published the structure-activity relationships (SAR) of **11** and reported that by using the hydrochloride salt of **11** physicochemical properties (bioavailability, blood sugar-lowering effect in ICR, favorable pharmakokinetics [PK]/pharmcodynamics [PD]) could be improved [141].

In the same year, Fang et al. presented another dual modulator, designed by pharmacophore combination and scaffold hopping. A central tetrahydropyrimidine nucleus (GPR119 agonist scaffold) was combined with a substituted aniline and fragments from seven different DPP4 inhibitors from the literature. Their most promising compound **12** (Figure 13.2) was highly potent on DPP4 (74.5% @10 μM) and GPR119 (EC_{50}: 0.0087 μM, in a cell-based cAMP assay, 85.4% max; compared to maximum effect of GSK1292263). In an oral glucose tolerance test in C57BL/6N mice, **12** showed a stronger hypoglycemic effect than the DPP4 inhibitor vildagliptin [142].

13.5.3 DPP4/MCH-1R

One of the many approaches to treat obesity (a central risk factor of MetS) is appetite control [143]. In this regard, the melanin-concentrating hormone receptor 1 (MCH-1R/GPR24/ SLC-1) could play an important role. This GPCR is activated by the MCH, a cyclic neuropeptide. The important role of MCH for food intake and energy balance has been investigated and confirmed in numerous animal studies [144, 145]. Several MCH-1R antagonists have already been described in the literature [146]. It could be shown that they suppress MCH-induced food intake and thus lead to a reduction in weight gain [147, 148]. Up to now, only a few candidates were investigated in clinical trials, but, for various reasons, the concept of weight loss mediated by MCH-1R inhibition has not been confirmed in humans yet [149]. The development of selective MCH-1R antagonists is associated with several challenges resulting in an increased cardiovascular risk [150]. However, the combination with a DPP4 inhibitor could be an attractive combination resulting in an oral antidiabetic agent, which would also contribute to the reduction in body weight. Gattrell and colleagues presented the first approach to develop a dual DPP4/MCH-1R modulator. They developed a lead structure by determining molecules that share a common structural motif in ligand data from chemical literature and patents. Subsequently, they identified further MCH-1R ligands with a similar binding mode via molecular docking, fragmented them, and filtered them according to predetermined physicochemical properties. After further modifications, their best compound **13** (Figure 13.2) was almost equipotent on both targets (IC_{50} [MCH-1R]: 0.44 μM), IC_{50} [DPP4]: 0.35 μM) and also selective against DDP8/9. The human ether-a-go-go related gene (hERG) inhibition data of two analogs of 24 also suggest that the presented chemotype is a good starting point for further lead structure optimization [151].

13.6 Diverse Target Combinations for Polypharmacological Treatment of Metabolic Disorders

13.6.1 PTP1B/AMPK

AMP-activated protein kinase (AMPK) is a conserved and ubiquitously expressed sensor for the energy status of a cell. In response to energy stress (increasing AMP and/or adenosine diphosphate [ADP] together with decreasing ATP), the metabolism is adapted by AMPK-promoting ATP-producing catabolic pathways and inhibiting energy-consuming processes. The cell can thus be restored to a state of energy homeostasis. AMPK can therefore be regarded as a central regulator of metabolism at both the cellular and physiological levels [152–154].

Both proteins, PTP1B and AMPK, by themselves already represent attractive targets in the search for new antidiabetic agents. However, as various studies suggest, there is a connection between the signaling cascades of PTP1B and AMPK. Thus, the effect of inhibition of PTP1B is due to downstream phosphorylation of AMPK [155–157]. Babkov and colleagues concluded that the synergistic effect of PTP1B inhibition with direct and PTP1B-mediated indirect AMPK activation could be a novel starting point for the development of new dual antidiabetic agents. In previous work, they identified a potent AMPK activator consisting of a fused benzimidazole backbone further functionalized with a biphenyl residue [158]. They also found various PTP1B inhibitors in the literature that had a similar pharmacophore with a biphenyl unit bound to an aromatic core. Based on the strong similarity between the two structural motifs, they developed a new lead structure using the merged multiple targeting ligand approach. This molecule consists of a central 2-aminobenzimidazole core, which is crucial for AMPK activation. The biphenyl residue was retained at nitrogen atom N1 and a side chain was introduced at N3, which is intended to mimic structural motifs of known PTP1B inhibitors by adding H-bond donors and acceptors (e.g., tertiary nitrogen) [114]. Babkov et al. synthesized a total of 11 compounds and characterized them regarding their activity against both targets. They were able to show that the presence of the biphenyl residue correlates strongly with the activity toward both targets. Their SAR also showed that the activity toward both targets correlates inversely with the number of H-bond acceptors. They concluded that repulsive interactions could impair binding. The most active compound of this SAR is 1-(2-dimethylaminoethyl)-3-(biphenyl-4-yl)methyl-2-imino-benzimidazoline dihydrobromide (**15**). It is both a moderate PTP1B inhibitor (IC_{50}: 30.5 ± 70.7 μM, human) and AMPK activator (EC_{50}: 21.7 ± 8.4 μM, human). Due to the high structural similarity of the compound with the Ang II type 1 (AT1) receptor antagonists (sartans), it was also examined in this respect and **15** showed a statistically significant effect at a concentration of 10 μM (% AT1 receptor antagonist activity, 10 μM: 14.4 ± 0.3). In an animal model with obese, diabetic rats, the authors were able to demonstrate the *in vivo* activity of their dual modulator. The animals received **15** ip over 21 days at a dose of 30 mg/kg. As a reference drug, the approved biguanide antidiabetic metformin was tested at a dose of 450 mg/kg, which has a similar mechanism of action [159]. The animal study showed that an acute and chronic antihyperglycemic effect could be achieved with **15**. A further increase in body weight was also not observed. In addition, a positive influence on visceral fat distribution and mass was detected, although not quite as pronounced as in the metformin group. However, the improvement in glucose tolerance was similar in both groups. Compared to the vehicle group, an improvement in the lipid metabolism of the rats was observed after administration of both **15** and metformin. Although no significant reduction in triglycerides was observed for **15**, there was a clear reduction in low-density lipoproteins and an accumulation of HDLs. The effect of **15** was more pronounced than with metformin administration. This suggests that the mechanism of action of the dual PTP1P/AMPK modulators could have a positive effect on the treatment of type 2 diabetes and associated comorbidities [114].

13.6.2 SGLT/GP

Another promising approach for the treatment of type 2 diabetes is the suppression of glucose reabsorption [160]. This process is mediated by two sodium-dependent glucose cotransporters (SGLT-1/2), which are predominantly expressed in the kidneys. In the healthy state, less than 1% of the plasma glucose filtered by the kidneys is excreted via the urine; approximately 99% is reabsorbed. SGLT-2, which is located on the surface of the epithelial cells lining the S1 segment of the proximal tubule, is responsible for about 90% of this reabsorption under normoglycemic conditions. This is due to its high capacity and low affinity. SGLT-1 most likely accounts for the remaining 10%. S1 is localized in the S3 segment of the proximal tubule and has a low capacity and high affinity [161, 162]. Since the blood glucose level is lowered independently of insulin secretion, the inhibition of SGLT-2 represents a promising approach for the treatment of type 2 diabetes without the risk of hyperglycemia. SGLT-2 inhibitors, the so-called gliflozines, have a C-linked β-glycoside as a common structural feature and represent a further development of the natural substance phlorizin [161, 163].

Another approach to control blood glucose levels in type 2 diabetic patients is the inhibition of glycogen phosphorylase (GP). In the liver, this enzyme catalyzes the release of glucose from stored glycogen, a polymer of α-1,4- and α-1,6-linked glucose units [164]. GP is an allosteric enzyme that exists in two convertible forms: GPa (phosphorylated, active, high substrate affinity) and GPb (unphosphorylated, inactive, low substrate affinity). Seven different binding sites have been described at GP, which leads to a large molecular variety of potential structures that can inhibit GP. The current focus is particularly on various glucose derivatives that bind to the catalytic site of the enzyme [163, 165]. Sipos et al. analyzed that there is a large structural overlap between the known SGLT-2 inhibitors and analogous GP inhibitors, which also have a glucose moiety, and investigated whether dual-active compounds could be derived from this [166]. Initially, the authors characterized phlorizin and the approved SGLT-2 inhibitor empagliflozin. While phlorizin is still able to inhibit GP in the high nanomolar range, the authors were no longer able to demonstrate inhibition with empagliflozin. Subsequently, they investigated different (C-ß-D-glucopyranosylhetaryl)methyl arene-type compounds for their inhibition potential against both target structures. In the first step, they investigated 15 different arylmethyl-substituted glucosyl heterocycles and were able to characterize new SGLT-1/2 inhibitors, but most of them showed no activity against GP. In the second step, Sipos et al. selected 10 aryl-substituted glucopyranosyl azoles (inactive compounds as well as compounds that inhibit GP in the low nanomolar range), originally designed as GP inhibitors, to investigate their ability to inhibit both SGLTs. The best candidate from this series was 2-(ß-D-glucopyranosyl)-4(5)-(2-naphthyl)-imidazole (Figure 13.2, **15**). This compound inhibits GP with a K_i of 31 nM and SGLT-2 with an IC_{50} of 3.5 μM (IC_{50} [SGLT-1] = 11.89 μM). Furthermore, the authors were able to show that 24b showed no cytotoxicity in the CyQUANT assay [166]. Taken together, these compounds are a good starting point for the development of further new dual-active compounds.

13.7 Conclusion

In summary, the studies presented in this chapter show that designed multitarget ligands (DMLs) exhibit the potential to combat a multifactorial disease such as MetS by combining two selective inhibitors in one molecule with dual modulator activity. There are several reasons for introducing multi-target activity into agents for the treatment of MetS. First, polypharmaceutical treatment can be avoided by combining antihypertensive and diabetic-lowering activity, as has been shown for AT1/PPAR-γ, PPAR-γ/sEH, or DPP4/ACE ligands. *In vivo* evaluation of most of the advanced compounds shows great potential for efficacy; however, clinical studies are still lacking. Second, the enhancement of efficacy against certain MetS-associated diseases can be achieved by combining targets such as FXR/sEH or PPAR-γ/GK. This approach also holds great potential to reduce polypharmacologic treatment, as these diseases are currently treated with multiple drugs. Medicinal chemistry has shown that it is capable of potent DMLs with favorable pharmacokinetic properties. Moreover, innovative animal models of MetS are able to reflect the full spectrum of symptoms, so that the full potential of multi-target ligands can be evaluated.

References

1 Chew, N.W.S., Ng, C.H., Tan, D.J.H. et al. (2023). The global burden of metabolic disease: data from 2000 to 2019. *Cell Metab.* 35: 414–428.e3.

2 Potenza, M.V. and Mechanick, J.I. (2009). The metabolic syndrome: definition, global impact, and pathophysiology. *Nutr. Clin. Pract.* 24: 560–577.

3 Shaw, J.E., Sicree, R.A., and Zimmet, P.Z. (2010). Global estimates of the prevalence of diabetes for 2010 and 2030. *Diabetes Res. Clin. Pract.* 87: 4–14.

4 Zhang, P., Zhang, X., Brown, J. et al. (2010). Global healthcare expenditure on diabetes for 2010 and 2030. *Diabetes Res. Clin. Pract.* 87: 293–301.

5 Wu, H. and Ballantyne, C.M. (2020). Metabolic inflammation and insulin resistance in obesity. *Circ. Res.* 126: 1549–1564.

6 Saltiel, A.R. and Olefsky, J.M. Inflammatory mechanisms linking obesity and metabolic disease. *J. Clin. Invest.* 127: 1–4.

7 Wadden, T.A., Tronieri, J.S., and Butryn, M.L. (2020). Lifestyle modification approaches for the treatment of obesity in adults. *Am. Psychol.* 75: 235–251.

8 Grave, R.D., Calugi, S., Centis, E. et al. (2010). Lifestyle modification in the management of the metabolic syndrome: achievements and challenges. *Diabetes Metab. Syndr. Obes.* 3: 373–385.

9 Binesh Marvasti, T. and Adeli, K. (2010). Pharmacological management of metabolic syndrome and its lipid complications. *Daru* 18: 146–154.

10 Grundy, S.M. (2006). Drug therapy of the metabolic syndrome: minimizing the emerging crisis in polypharmacy. *Nat. Rev. Drug Discov.* 5: 295–309.

11 Lillich, F.F., Imig, J.D., and Proschak, E. (2021). Multi-target approaches in metabolic syndrome. *Front. Pharmacol.* 11.

12 Huang, P.L. (2009). A comprehensive definition for metabolic syndrome. *Dis. Model. Mech.* 2: 231–237.

13 Dandona, P., Aljada, A., Chaudhuri, A. et al. (2005). Metabolic syndrome: a comprehensive perspective based on interactions between obesity, diabetes, and inflammation. *Circulation* 111: 1448–1454.

14 Esser, N., Legrand-Poels, S., Piette, J. et al. (2014). Inflammation as a link between obesity, metabolic syndrome and type 2 diabetes. *Diabetes Res. Clin. Pract.* 105: 141–150.

15 Hotamisligil, G.S. (2006). Inflammation and metabolic disorders. *Nature* 444: 860–867.

16 Alberti, K.G.M.M., Zimmet, P., and Shaw, J. (2005). The metabolic syndrome—a new worldwide definition. *Lancet* 366: 1059–1062.

17 Forbes, J.M. and Cooper, M.E. (2013). Mechanisms of diabetic complications. *Physiol. Rev.* 93: 137–188.

18 Ouchi, N., Parker, J.L., Lugus, J.J., and Walsh, K. (2011). Adipokines in inflammation and metabolic disease. *Nat. Rev. Immunol.* 11: 85–97.

19 Kawai, T., Autieri, M.V., and Scalia, R. (2020). Adipose tissue inflammation and metabolic dysfunction in obesity. *Am. J. Physiol. Cell Physiol.* 320: C375–C391.

20 Catanzaro, R., Cuffari, B., Italia, A., and Marotta, F. (2016). Exploring the metabolic syndrome: nonalcoholic fatty pancreas disease. *World J. Gastroenterol.* 22: 7660–7675.

21 Lin, L., Tan, W., Pan, X. et al. (2022). Metabolic syndrome-related kidney injury: a review and update. *Front. Endocrinol.* 13, 904001.

22 Bernardi, S., Marcuzzi, A., Piscianz, E. et al. (2018). The complex interplay between lipids, immune system and interleukins in cardio-metabolic diseases. *Int. J. Mol. Sci.* 19.

23 Galicia-Garcia, U., Benito-Vicente, A., Jebari, S. et al. (2020). Pathophysiology of type 2 diabetes mellitus. *Int. J. Mol. Sci.* 21: 6275.

24 Jin, X., Qiu, T., Li, L. et al. (2023). Pathophysiology of obesity and its associated diseases. *Acta Pharm. Sin. B* 13: 2403–2424.

25 Khan, M.A.B., Hashim, M.J., King, J.K. et al. (2020). Epidemiology of type 2 diabetes–global burden of disease and forecasted trends. *J. Epidemiol. Glob. Health* 10: 107–111.

26 Gheith, O., Farouk, N., Nampoory, N. et al. (2015). Diabetic kidney disease: world wide difference of prevalence and risk factors. *J. Nephropharmacol.* 5: 49–56.

27 Thomas, M.C., Brownlee, M., Susztak, K. et al. (2015). Diabetic kidney disease. *Nat. Rev. Dis. Primers* 1: 15018.

28 Tsalamandris, S., Antonopoulos, A.S., Oikonomou, E. et al. (2019). The role of inflammation in diabetes: current concepts and future perspectives. *Eur. Cardiol.* 14: 50–59.

29 Donath, M.Y. and Shoelson, S.E. (2011). Type 2 diabetes as an inflammatory disease. *Nat. Rev. Immunol.* 11: 98–107.

30 Bianchi, C., Penno, G., Romero, F. et al. (2007). Treating the metabolic syndrome. *Expert. Rev. Cardiovasc. Ther.* 5: 491–506.

31 Grant, R.W., Devita, N.G., Singer, D.E., and Meigs, J.B. (2003). Polypharmacy and medication adherence in patients with type 2 diabetes. *Diabetes Care* 26: 1408–1412.

32 Alwhaibi, M., Balkhi, B., Alhawassi, T.M. et al. (2018). Polypharmacy among patients with diabetes: a cross-sectional retrospective study in a tertiary hospital in Saudi Arabia. *BMJ Open* 8: e020852.

33 Fang, S., Livergood, M.C., Nakagawa, P. et al. (2021). Role of the peroxisome proliferator activated receptors in hypertension. *Circ. Res.* 128: 1021–1039.

34 Kvandová, M., Majzúnová, M., and Dovinová, I. (2016). The role of PPARgamma in cardiovascular diseases. *Physiol. Res.* 65: S343–S363.

35 Lalloyer, F. and Staels, B. (2010). Fibrates, glitazones, and peroxisome proliferator-activated receptors. *Arterioscler. Thromb. Vasc. Biol.* 30: 894–899.

36 Lincoff, A.M., Wolski, K., Nicholls, S.J., and Nissen, S.E. (2007). Pioglitazone and risk of cardiovascular events in patients with type 2 diabetes mellitus: a meta-analysis of randomized trials. *JAMA* 298: 1180–1188.

37 DeFronzo, R.A., Reeves, W.B., and Awad, A.S. (2021). Pathophysiology of diabetic kidney disease: impact of SGLT2 inhibitors. *Nat. Rev. Nephrol.* 17: 319–334.

38 Nauck, M.A., Quast, D.R., Wefers, J., and Meier, J.J. (2021). GLP-1 receptor agonists in the treatment of type 2 diabetes – state-of-the-art. *Mol. Metab.* 46: 101102.

39 Górriz, J.L., Soler, M.J., Navarro-González, J.F. et al. (2020). GLP-1 receptor agonists and diabetic kidney disease: a call of attention to nephrologists. *J. Clin. Med.* 9: 947.

40 Shah, M. and Vella, A. (2014). Effects of GLP-1 on appetite and weight. *Rev. Endocr. Metab. Disord.* 15: 181–187.

41 Lupsa, B.C. and Inzucchi, S.E. (2018). Use of SGLT2 inhibitors in type 2 diabetes: weighing the risks and benefits. *Diabetologia* 61: 2118–2125.

42 Janež, A. and Fioretto, P. (2021). SGLT2 inhibitors and the clinical implications of associated weight loss in type 2 diabetes: a narrative review. *Diabetes Ther.* 12: 2249–2261.

43 Al Rashid, S., Elango, P., and Rahman, S.Z. (2023). SGLT2 inhibitors for cardioprotection. *Oman Med. J.* 38: e521.

44 Vart, P., Vaduganathan, M., Jongs, N. et al. (2022). Estimated lifetime benefit of combined RAAS and SGLT2 inhibitor therapy in patients with albuminuric CKD without diabetes. *Clin. J. Am. Soc. Nephrol.* 17: 1754–1762.

45 Wong, V.W.-S., Ekstedt, M., Wong, G.L.-H., and Hagström, H. (2023). Changing epidemiology, global trends and implications for outcomes of NAFLD. *J. Hepatol.* 79: 842–852.

46 Wasilewska, N. and Lebensztejn, D.M. (2021). Non-alcoholic fatty liver disease and lipotoxicity. *Clin. Exp. Hepatol.* 7: 1–6.

47 Heyens, L.J.M., Busschots, D., Koek, G.H. et al. (2021). Liver fibrosis in non-alcoholic fatty liver disease: from liver biopsy to non-invasive biomarkers in diagnosis and treatment. *Front. Med.* 8: 615978.

48 Younossi, Z.M. and Henry, L. (2021). Epidemiology of non-alcoholic fatty liver disease and hepatocellular carcinoma. *JHEP Rep.* 3: 100305.

49 Semmler, G., Datz, C., Reiberger, T., and Trauner, M. (2021). Diet and exercise in NAFLD/NASH: beyond the obvious. *Liver Int.* 41: 2249–2268.

50 Holmer, M., Lindqvist, C., Petersson, S. et al. (2021). Treatment of NAFLD with intermittent calorie restriction or low-carb high-fat diet – a randomised controlled trial. *JHEP Rep.* 3: 100256.

51 Ali, A.H., Carey, E.J., and Lindor, K.D. (2015). Recent advances in the development of farnesoid X receptor agonists. *Ann. Transl. Med.* 3: 5.

52 Oseini, A.M. and Sanyal, A.J. (2017). Therapies in non-alcoholic steatohepatitis (NASH). *Liver Int.* 37 (Suppl 1): 97–103.

53 Mudaliar, S., Henry, R.R., Sanyal, A.J. et al. (2013). Efficacy and safety of the farnesoid X receptor agonist obeticholic acid in patients with type 2 diabetes and nonalcoholic fatty liver disease. *Gastroenterology* 145: 574–82.e1.

54 Han, C.Y. (2018). Update on FXR biology: promising therapeutic target? *Int. J. Mol. Sci.* 19.

55 Harrison, S.A., Loomba, R., Dubourg, J. et al. (2023). Clinical trial landscape in NASH. *Clin. Gastroenterol. Hepatol.* 21: 2001–2014.

56 Guembe, M.J., Fernandez-Lazaro, C.I., Sayon-Orea, C. et al. (2020). Risk for cardiovascular disease associated with metabolic syndrome and its components: a 13-year prospective study in the RIVANA cohort. *Cardiovasc. Diabetol.* 19: 195.

57 Tune, J.D., Goodwill, A.G., Sassoon, D.J., and Mather, K.J. (2017). Cardiovascular consequences of metabolic syndrome. *Transl. Res.* 183: 57–70.

58 Clemente-Suárez, V.J., Martín-Rodríguez, A., Redondo-Flórez, L. et al. (2023). New insights and potential therapeutic interventions in metabolic diseases. *Int. J. Mol. Sci.* 24: 10672.

59 Handelsman, Y. and Jellinger, P.S. (2011). Overcoming obstacles in risk factor management in type 2 diabetes mellitus. *J. Clin. Hypertens.* 13: 613–620.

60 Imig, J.D., Merk, D., and Proschak, E. (2021). Multi-target drugs for kidney diseases. *Kidney* 360 (2): 1645–1653.

61 Makhoba, X.H., Viegas, C., Mosa, R.A. et al. (2020). Potential impact of the multi-target drug approach in the treatment of some complex diseases. *Drug Des. Devel. Ther.* 14: 3235–3249.

62 Imig, J.D. (2023). Peroxisome proliferator-activated receptors, farnesoid X receptor, and dual modulating drugs in hypertension. *Front. Physiol.* 14: 1186477.

63 Choung, W., Jung, H.J., Nam, E.H. et al. (2018). Discovery of the bifunctional modulator of angiotensin II type 1 receptor (AT1R) and PPARγ derived from the AT1R antagonist, Fimasartan. *Bioorg. Med. Chem. Lett.* 28: 3155–3160.

64 Schupp, M., Janke, J., Clasen, R. et al. (2004). Angiotensin type 1 receptor blockers induce peroxisome proliferator-activated receptor-gamma activity. *Circulation* 109: 2054–2057.

65 Hye Khan, M.A., Neckář, J., Cummens, B. et al. (2014). Azilsartan decreases renal and cardiovascular injury in the spontaneously hypertensive obese rat. *Cardiovasc. Drugs Ther.* 28: 313–322.

66 Jao, T.-M., Nangaku, M., Wu, C.-H. et al. (2019). ATF6α downregulation of PPARα promotes lipotoxicity-induced tubulointerstitial fibrosis. *Kidney Int.* 95: 577–589.

67 Jain, M.R., Giri, S.R., Bhoi, B. et al. (2018). Dual PPARα/γ agonist saroglitazar improves liver histopathology and biochemistry in experimental NASH models. *Liver Int.* 38: 1084–1094.

68 Liao, J., Soltani, Z., Ebenezer, P. et al. (2010). Tesaglitazar, a dual peroxisome proliferator-activated receptor agonist (PPAR alpha/gamma), improves metabolic abnormalities and reduces renal injury in obese Zucker rats. *Nephron Exp. Nephrol.* 114: e61–e68.

69 Kamata, S., Honda, A., and Ishii, I. (2023). Current clinical trial status and future prospects of PPAR-targeted drugs for treating nonalcoholic fatty liver disease. *Biomol. Ther.* 13.

70 Hong, F., Xu, P., and Zhai, Y. (2018). The opportunities and challenges of peroxisome proliferator-activated receptors ligands in clinical drug discovery and development. *Int. J. Mol. Sci.* 19 (8): 2189.

71 Li, C., Yang, J., Wang, Y. et al. (2020). Farnesoid X receptor agonists as therapeutic target for cardiometabolic diseases. *Front. Pharmacol.* 11: 1247.

72 Proschak, E., Stark, H., and Merk, D. (2019). Polypharmacology by design: a medicinal chemist's perspective on multitargeting compounds. *J. Med. Chem.* 62: 420–444.

73 Bansal, Y. and Silakari, O. (2014). Multifunctional compounds: smart molecules for multifactorial diseases. *Eur. J. Med. Chem.* 76: 31–42.

74 Talevi, A. (2015). Multi-target pharmacology: possibilities and limitations of the "skeleton key approach" from a medicinal chemist perspective. *Front. Pharmacol.* 6: 205.

75 Hye Khan, M.A., Kolb, L., Skibba, M. et al. (2018). A novel dual PPAR-γ agonist/sEH inhibitor treats diabetic complications in a rat model of type 2 diabetes. *Diabetologia* 61: 2235–2246.

76 Hye Khan, M.A., Schmidt, J., Stavniichuk, A. et al. (2019). A dual farnesoid X receptor/soluble epoxide hydrolase modulator treats non-alcoholic steatohepatitis in mice. *Biochem. Pharmacol.* 166: 212–221.

77 Helmstädter, M., Schmidt, J., Kaiser, A. et al. (2021). Differential therapeutic effects of FXR activation, sEH inhibition, and dual FXR/sEH modulation in NASH in diet-induced obese mice. *ACS Pharmacol. Transl. Sci.* 4: 966–979.

78 Morphy, R. and Rankovic, Z. (2009). Designing multiple ligands – medicinal chemistry strategies and challenges. *Curr. Pharm. Des.* 15: 587–600.

79 Morphy, R. and Rankovic, Z. (2006). The physicochemical challenges of designing multiple ligands. *J. Med. Chem.* 49: 4961–4970.

80 Ammazzalorso, A., Maccallini, C., Amoia, P., and Amoroso, R. (2019). Multitarget PPARγ agonists as innovative modulators of the metabolic syndrome. *Eur. J. Med. Chem.* 173: 261–273.

81 Ibrahim, M.K., Eissa, I.H., Alesawy, M.S. et al. (2017). Design, synthesis, molecular modeling and anti-hyperglycemic evaluation of quinazolin-4(3H)-one derivatives as potential PPARγ and SUR agonists. *Bioorg. Med. Chem.* 25: 4723–4744.

82 Ibrahim, M.K., Eissa, I.H., Abdallah, A.E. et al. (2017). Design, synthesis, molecular modeling and anti-hyperglycemic evaluation of novel quinoxaline derivatives as potential PPARγ and SUR agonists. *Bioorg. Med. Chem.* 25: 1496–1513.

83 Burant, C.F., Viswanathan, P., Marcinak, J. et al. (2012). TAK-875 versus placebo or glimepiride in type 2 diabetes mellitus: a phase 2, randomised, double-blind, placebo-controlled trial. *Lancet* 379: 1403–1411.

84 Hidalgo-Figueroa, S., Rodríguez-Luévano, A., Almanza-Pérez, J.C. et al. (2021). Synthesis, molecular docking, dynamic simulation and pharmacological characterization of potent multifunctional agent (dual

GPR40-PPARγ agonist) for the treatment of experimental type 2 diabetes. *Eur. J. Pharmacol.* 907: 174244.

85 Zhou, Z., Cai, Z., Zhang, C. et al. (2022). Design, synthesis, and biological evaluation of novel dual FFA1 and PPARδ agonists possessing phenoxyacetic acid scaffold. *Bioorg. Med. Chem.* 56: 116615.

86 Li, Z., Zhou, Z., Deng, F. et al. (2018). Design, synthesis, and biological evaluation of novel pan agonists of FFA1, PPARγ and PPARδ. *Eur. J. Med. Chem.* 159: 267–276.

87 Toulis, K.A., Nirantharakumar, K., Pourzitaki, C. et al. (2020). Glucokinase activators for type 2 diabetes: challenges and future developments. *Drugs* 80: 467–475.

88 Lei, L., Liu, Q., Liu, S. et al. (2015). Antidiabetic potential of a novel dual-target activator of glucokinase and peroxisome proliferator activated receptor-γ. *Metabolism* 64: 1250–1261.

89 Imig, J.D. and Hammock, B.D. (2009). Soluble epoxide hydrolase as a therapeutic target for cardiovascular diseases. *Nat. Rev. Drug Discov.* 8: 794–805.

90 Spector, A.A. (2009). Arachidonic acid cytochrome P450 epoxygenase pathway. *J. Lipid Res.* 50 (Suppl): S52–S56.

91 Spector, A.A. and Norris, A.W. (2007). Action of epoxyeicosatrienoic acids on cellular function. *Am. J. Physiol. Cell Physiol.* 292: C996–C1012.

92 Shen, H.C. and Hammock, B.D. (2012). Discovery of inhibitors of soluble epoxide hydrolase: a target with multiple potential therapeutic indications. *J. Med. Chem.* 55: 1789–1808.

93 Morisseau, C. and Hammock, B.D. (2013). Impact of soluble epoxide hydrolase and epoxyeicosanoids on human health. *Annu. Rev. Pharmacol. Toxicol.* 53: 37–58.

94 Campbell, W.B., Imig, J.D., Schmitz, J.M., and Falck, J.R. (2017). Orally active epoxyeicosatrienoic acid analogs. *J. Cardiovasc. Pharmacol.* 70: 211–224.

95 Sun, C.-P., Zhang, X.-Y., Morisseau, C. et al. (2021). Discovery of soluble epoxide hydrolase inhibitors from chemical synthesis and natural products. *J. Med. Chem.* 64: 184–215.

96 Liu, Y., Zhang, Y., Schmelzer, K. et al. (2005). The antiinflammatory effect of laminar flow: the role of PPARgamma, epoxyeicosatrienoic acids, and soluble epoxide hydrolase. *Proc. Natl. Acad. Sci. U.S.A* 102: 16747–16752.

97 Imig, J.D., Walsh, K.A., Hye Khan, M.A. et al. (2012). Soluble epoxide hydrolase inhibition and peroxisome proliferator activated receptor γ agonist improve vascular function and decrease renal injury in hypertensive obese rats. *Exp. Biol. Med.* 237: 1402–1412.

98 Blöcher, R., Lamers, C., Wittmann, S.K. et al. (2016). N-benzylbenzamides: a novel merged scaffold for orally available dual soluble epoxide hydrolase/peroxisome proliferator-activated receptor γ modulators. *J. Med. Chem.* 59: 61–81.

99 Lillich, F.F., Willems, S., Ni, X. et al. (2021). Structure-based design of dual partial peroxisome proliferator-activated receptor γ agonists/soluble epoxide hydrolase inhibitors. *J. Med. Chem.* 64: 17259–17276.

100 Hartmann, M., Bibli, S.-I., Tews, D. et al. (2021). Combined cardioprotective and adipocyte browning effects promoted by the eutomer of dual sEH/PPARγ modulator. *J. Med. Chem.* 64: 2815–2828.

101 Makishima, M., Okamoto, A.Y., Repa, J.J. et al. (1999). Identification of a nuclear receptor for bile acids. *Science* 284: 1362–1365.

102 Cariou, B. and Staels, B. (2007). FXR: a promising target for the metabolic syndrome? *Trends Pharmacol. Sci.* 28: 236–243.

103 Carr, R.M. and Reid, A.E. (2015). FXR agonists as therapeutic agents for non-alcoholic fatty liver disease. *Curr. Atheroscler. Rep.* 17: 500.

104 Massafra, V., Pellicciari, R., Gioiello, A., and van Mil, S.W.C. (2018). Progress and challenges of selective farnesoid X receptor modulation. *Pharmacol. Ther.* 191: 162–177.

105 Liu, Y., Dang, H., Li, D. et al. (2012). Inhibition of soluble epoxide hydrolase attenuates high-fat-diet-induced hepatic steatosis by reduced systemic inflammatory status in mice. *PLoS ONE* 7: e39165.

106 Mangels, N., Awwad, K., Wettenmann, A. et al. (2016). The soluble epoxide hydrolase determines cholesterol homeostasis by regulating AMPK and SREBP activity. *Prostaglandins Other Lipid Mediat.* 125: 30–39.

107 Schuck, R.N., Zha, W., Edin, M.L. et al. (2014). The cytochrome P450 epoxygenase pathway regulates the hepatic inflammatory response in fatty liver disease. *PLoS ONE* 9: e110162.

108 Schmidt, J., Rotter, M., Weiser, T. et al. (2017). A dual modulator of farnesoid X receptor and soluble epoxide hydrolase to counter nonalcoholic steatohepatitis. *J. Med. Chem.* 60: 7703–7724.

109 Khan, M.A.H., Nolan, B., Stavniichuk, A. et al. (2023). Dual soluble epoxide hydrolase inhibitor – farnesoid X receptor agonist interventional treatment attenuates renal inflammation and fibrosis. *Front. Immunol.* 14: 1269261.

110 Nandi, S. and Saxena, M. (2020). Potential inhibitors of protein tyrosine phosphatase (PTP1B) enzyme: promising target for Type-II diabetes mellitus. *Curr. Top. Med. Chem.* 20: 2692–2707.

111 Thomis, M.A., van Leemputte, M., Maes, H.H. et al. (1997, 1985). Multivariate genetic analysis of maximal isometric muscle force at different elbow angles. *J. Appl. Physiol.* 82: 959–967.

112 Yip, S.-C., Saha, S., and Chernoff, J. (2010). PTP1B: a double agent in metabolism and oncogenesis. *Trends Biochem. Sci.* 35: 442–449.

113 Chernoff, J. (1999). Protein tyrosine phosphatases as negative regulators of mitogenic signaling. *J. Cell. Physiol.* 180: 173–181.

114 Babkov, D.A., Zhukovskaya, O.N., Brigadirova, A.A. et al. (2023). Discovery and evaluation of biphenyl derivatives of 2-iminobenzimidazoles as prototype dual PTP1B inhibitors and AMPK activators with in vivo antidiabetic activity. *Chem. Biol. Drug Des.* 101: 896–914.

115 Morishita, K., Ito, Y., Otake, K. et al. (2021). Synthesis and evaluation of a novel series of 2,7-substituted-6-tetrazolyl-1,2,3,4-tetrahydroisoquinoline derivatives as selective peroxisome proliferator-activated receptor γ partial agonists. *Chem. Pharm. Bull.* 69: 333–351.

116 Otake, K., Azukizawa, S., Takahashi, K. et al. (2011). 2-Acyl-tetrahydroisoquinoline-3-carboxylic acids: lead compounds with triple actions, peroxisome proliferator-activated receptor α/γ agonist and protein-tyrosine phosphatase 1B inhibitory activities. *Chem. Pharm. Bull.* 59: 876–879.

117 Otake, K., Azukizawa, S., Fukui, M. et al. (2011). A novel series of (S)-2,7-substituted-1,2,3,4-tetrahydroisoquinoline-3-carboxylic acids: peroxisome proliferator-activated receptor α/γ dual agonists with protein-tyrosine phosphatase 1B inhibitory activity. *Chem. Pharm. Bull.* 59: 1233–1242.

118 Kubo, M., Fukui, M., Ito, Y. et al. (2014). Insulin sensitization by a novel partial peroxisome proliferator-activated receptor γ agonist with protein tyrosine phosphatase 1B inhibitory activity in experimental osteoporotic rats. *J. Pharmacol. Sci.* 124: 276–285.

119 Otake, K., Azukizawa, S., Fukui, M. et al. (2012). Novel (S)-1,2,3,4-tetrahydroisoquinoline-3-carboxylic acids: peroxisome proliferator-activated receptor γ selective agonists with protein-tyrosine phosphatase 1B inhibition. *Bioorg. Med. Chem.* 20: 1060–1075.

120 Otake, K., Azukizawa, S., Takeda, S. et al. (2015). Novel 2,7-substituted (S)-1,2,3,4-tetrahydroisoquinoline-3-carboxylic acids: peroxisome proliferator-activated receptor γ partial agonists with protein-tyrosine phosphatase 1B inhibition. *Chem. Pharm. Bull.* 63: 998–1014.

121 Scheen, A.J. (2015). A review of gliptins for 2014. *Expert. Opin. Pharmacother.* 16: 43–62.

122 Grigoropoulou, P., Eleftheriadou, I., Zoupas, C. et al. (2013). Incretin-based therapies for type 2 diabetes mellitus: effects on insulin resistance. *Curr. Diabetes Rev.* 9: 412–417.

123 Tomlinson, B., Hu, M., Zhang, Y. et al. (2016). Investigational glucagon-like peptide-1 agonists for the treatment of obesity. *Expert Opin. Investig. Drugs* 25: 1167–1179.

124 Dhillon, S. (2018). Semaglutide: first global approval. *Drugs* 78: 275–284.

125 Durinx, C., Lambeir, A.M., Bosmans, E. et al. (2000). Molecular characterization of dipeptidyl peptidase activity in serum: soluble CD26/dipeptidyl peptidase IV is responsible for the release of X-Pro dipeptides. *Eur. J. Biochem.* 267: 5608–5613.

126 Charbonnel, B., Karasik, A., Liu, J. et al. (2006). Efficacy and safety of the dipeptidyl peptidase-4 inhibitor sitagliptin added to ongoing metformin therapy in patients with type 2 diabetes inadequately controlled with metformin alone. *Diabetes Care* 29: 2638–2643.

127 Nelson, M., Bhandari, N., and Wener, J. (2014). Sitagliptin-induced pancreatitis – a longer road than expected. *Clin. Case Rep.* 2: 149–152.

128 Scheen, A.J. (2018). The safety of gliptins: updated data in 2018. *Expert Opin. Drug Saf.* 17: 387–405.

129 Regulski, M., Regulska, K., Stanisz, B.J. et al. (2015). Chemistry and pharmacology of Angiotensin-converting enzyme inhibitors. *Curr. Pharm. Des.* 21: 1764–1775.

130 Ferrario, C.M. and Strawn, W.B. (2006). Role of the renin-angiotensin-aldosterone system and proinflammatory mediators in cardiovascular disease. *Am. J. Cardiol.* 98: 121–128.

131 Pacurari, M., Kafoury, R., Tchounwou, P.B., and Ndebele, K. (2014). The Renin-Angiotensin-aldosterone system in vascular inflammation and remodeling. *Int. J. Inf. Secur.* 2014: 689360.

132 Feng, Y., Huang, R., Kavanagh, J. et al. (2019). Efficacy and safety of dual blockade of the renin-angiotensin-aldosterone system in diabetic kidney disease: a meta-analysis. *Am. J. Cardiovasc. Drugs* 19: 259–286.

133 Wang, Y., Landheer, S., van Gilst, W.H. et al. (2012). Attenuation of renovascular damage in Zucker diabetic fatty rat by NWT-03, an egg protein hydrolysate with ACE- and DPP4-inhibitory Activity. *PLoS ONE* 7: e46781.

134 Sattigeri, J.A., Sethi, S., Davis, J.A. et al. (2017). Approaches towards the development of chimeric DPP4/ACE inhibitors for treating metabolic syndrome. *Bioorg. Med. Chem. Lett.* 27: 2313–2318.

135 Yang, J.W., Kim, H.S., Choi, Y.-W. et al. (2018). Therapeutic application of GPR119 ligands in metabolic disorders. *Diabetes Obes. Metab.* 20: 257–269.

136 Odori, S., Hosoda, K., Tomita, T. et al. (2013). GPR119 expression in normal human tissues and islet cell tumors: evidence for its islet-gastrointestinal distribution, expression in pancreatic beta and alpha cells, and involvement in islet function. *Metabolism* 62: 70–78.

137 Hansen, H.S., Rosenkilde, M.M., Holst, J.J., and Schwartz, T.W. (2012). GPR119 as a fat sensor. *Trends Pharmacol. Sci.* 33: 374–381.

138 Ansarullah, Lu, Y., Holstein, M. et al. (2013). Stimulating β-cell regeneration by combining a GPR119 agonist with a DPP-IV inhibitor. *PLoS ONE* 8: e53345.

139 Li, G., Huan, Y., Yuan, B. et al. (2016). Discovery of novel xanthine compounds targeting DPP-IV and GPR119 as anti-diabetic agents. *Eur. J. Med. Chem.* 124: 103–116.

140 Huan, Y., Jiang, Q., Li, G. et al. (2017). The dual DPP4 inhibitor and GPR119 agonist HBK001 regulates glycemic control and beta cell function ex and in vivo. *Sci. Rep.* 7: 4351.

141 Li, G., Meng, B., Yuan, B. et al. (2020). The optimization of xanthine derivatives leading to HBK001 hydrochloride as a potent dual ligand targeting DPP-IV and GPR119. *Eur. J. Med. Chem.* 188: 112017.

142 Fang, Y., Zhang, S., Wu, W. et al. (2020). Design and synthesis of tetrahydropyridopyrimidine derivatives as dual GPR119 and DPP-4 modulators. *Bioorg. Chem.* 94: 103390.

143 Sargent, B.J. and Moore, N.A. (2009). New central targets for the treatment of obesity. *Br. J. Clin. Pharmacol.* 68: 852–860.

144 Luthin, D.R. (2007). Anti-obesity effects of small molecule melanin-concentrating hormone receptor 1 (MCHR1) antagonists. *Life Sci.* 81: 423–440.

145 Mihalic, J.T., Chen, X., Fan, P. et al. (2011). Discovery of a novel series of melanin-concentrating hormone receptor 1 antagonists for the treatment of obesity. *Bioorg. Med. Chem. Lett.* 21: 7001–7005.

146 Johansson, A. and Löfberg, C. (2015). Novel MCH1 receptor antagonists: a patent review. *Expert Opin. Ther. Pat.* 25: 193–207.

147 Borowsky, B., Durkin, M.M., Ogozalek, K. et al. (2002). Antidepressant, anxiolytic and anorectic effects of a melanin-concentrating hormone-1 receptor antagonist. *Nat. Med.* 8: 825–830.

148 Takekawa, S., Asami, A., Ishihara, Y. et al. (2002). T-226296: a novel, orally active and selective melanin-concentrating hormone receptor antagonist. *Eur. J. Pharmacol.* 438: 129–135.

149 Omran, Z. (2017). Obesity: current treatment and future horizons. *Mini Rev. Med. Chem.* 17: 51–61.

150 Högberg, T., Frimurer, T.M., and Sasmal, P.K. (2012). Melanin concentrating hormone receptor 1 (MCHR1) antagonists-still a viable approach for obesity treatment? *Bioorg. Med. Chem. Lett.* 22: 6039–6047.

151 Gattrell, W.T., Sambrook Smith, C.P., and Smith, A.J. (2012). An example of designed multiple ligands spanning protein classes: dual MCH-1R antagonists/DPPIV inhibitors. *Bioorg. Med. Chem. Lett.* 22: 2464–2469.

152 Steinberg, G.R. and Hardie, D.G. (2023). New insights into activation and function of the AMPK. *Nat. Rev. Mol. Cell Biol.* 24: 255–272.

153 Sharma, A., Anand, S.K., Singh, N. et al. (2023). AMP-activated protein kinase: an energy sensor and survival mechanism in the reinstatement of metabolic homeostasis. *Exp. Cell Res.* 428: 113614.

154 Lin, S.-C. and Hardie, D.G. (2018). AMPK: sensing glucose as well as cellular energy status. *Cell Metab.* 27: 299–313.

155 Yang, J.L., Ha, T.K.Q., Lee, B.W. et al. (2017). PTP1B inhibitors from the seeds of Iris sanguinea and their insulin mimetic activities via AMPK and ACC phosphorylation. *Bioorg. Med. Chem. Lett.* 27: 5076–5081.

156 Xu, Q., Wu, N., Li, X. et al. (2019). Inhibition of PTP1B blocks pancreatic cancer progression by targeting the PKM2/AMPK/mTOC1 pathway. *Cell Death Dis.* 10: 874.

157 Schultze, S.M., Hemmings, B.A., Niessen, M., and Tschopp, O. (2012). PI3K/AKT, MAPK and AMPK signalling: protein kinases in glucose homeostasis. *Expert Rev. Mol. Med.* 14: e1.

158 Babkov, D.A., Zhukowskaya, O.N., Borisov, A.V. et al. (2019). Towards multi-target antidiabetic agents: discovery of biphenyl-benzimidazole conjugates as AMPK activators. *Bioorg. Med. Chem. Lett.* 29: 2443–2447.

159 Zhou, G., Myers, R., Li, Y. et al. (2001). Role of AMP-activated protein kinase in mechanism of metformin action. *J. Clin. Invest.* 108: 1167–1174.

160 Abdul-Ghani, M.A., Norton, L., and DeFronzo, R.A. (2011). Role of sodium-glucose cotransporter 2 (SGLT 2) inhibitors in the treatment of type 2 diabetes. *Endocr. Rev.* 32: 515–531.

161 Choi, C.-I. (2016). Sodium-glucose cotransporter 2 (SGLT2) inhibitors from natural products: discovery of next-generation antihyperglycemic agents. *Molecules* 21: 1136.

162 Meng, W., Ellsworth, B.A., Nirschl, A.A. et al. (2008). Discovery of dapagliflozin: a potent, selective renal sodium-dependent glucose cotransporter 2 (SGLT2) inhibitor for the treatment of type 2 diabetes. *J. Med. Chem.* 51: 1145–1149.

163 Pałasz, A., Cież, D., Trzewik, B. et al. (2019). In the search of glycoside-based molecules as antidiabetic agents. *Top. Curr. Chem.* 377: 19.

164 Agius, L. (2015). Role of glycogen phosphorylase in liver glycogen metabolism. *Mol. Asp. Med.* 46: 34–45.

165 Leonidas, D.D., Zographos, S.E., Tsitsanou, K.E. et al. (2021). Glycogen phosphorylase revisited: extending the resolution of the R- and T-state structures of the free enzyme and in complex with allosteric activators. *Acta Crystallogr. F Struct. Biol. Commun.* 77: 303–311.

166 Sipos, Á., Szennyes, E., Hajnal, N.É. et al. (2021). Dual-target compounds against type 2 diabetes mellitus: proof of concept for sodium dependent glucose transporter (SGLT) and glycogen phosphorylase (GP) inhibitors. *Pharmaceuticals* 14: 364.

14

Overcoming the Challenges of Multi-Target-Directed Ligands for Alzheimer's Disease

Tips and Examples

Elisa Uliassi, Anna M. Pasieka, Eleonora Diamanti, and Maria Laura Bolognesi

14.1 Introduction

The emergence, now almost two decades ago, of the multi-target-directed ligand (MTDL) concept to combat neurodegenerative diseases has proven to be a landmark in the medicinal chemistry field. The Perspective article where this term was first coined was published in the Journal of Medicinal Chemistry in 2008 [1] and, as a highly cited paper, it is supposed to have influenced the research frontier. In a more objective way, a comprehensive bibliometric analysis performed by Munoz-Torrero and coworkers [2] within the 1990–2020 timeframe has revealed that the number of publications related to the multi-target Alzheimer's drug discovery field really boomed by 2011 and then continued ascending.

This undeniable growth trend has been driven primarily by a shared view that MTDLs can better address the multifactorial nature of Alzheimer's disease (AD) and that of other similarly complex neurodegenerative diseases. In the years, this view has even consolidated, and recent papers from leaders in neurology reinforce the urgent need for a paradigm shift in AD drug development from a single target to multiple targets [3]. In spite of that, as with many emerging concepts, a large dose of hype has accompanied this growth, and it appears that one of the most sought-after strategies isn't playing the major role we may have hoped for.

AD is still a largely unconquered therapeutic area by MTDLs, given that none of the thousands of hit compounds emerging from early drug discovery efforts have been translated to a drug for the millions of patients in need all over the world. More critically, when looking at the list of current multi-target agents in Phase 3 clinical trials [4], the impression is that MTDLs have still a long road ahead before they can successfully enter the market.

Some years ago, with the opportunity of a critical reflection on the field, we were motivated to provide the community with some basic prioritization concepts to potentially guide a complex process such as the MTDL pre-clinical development [5].

By considering the state of the art and lessons learned from our previous experience, we identified critical "points of attention," some of which are listed below:

- Target identification – a new MTDL should be directed to networked targets whose connectivity has been proven.
- The pharmacokinetic (PK) challenge – the physicochemical properties of MTDLs have been found to be less drug-like than single-target ones.
- Phenotypic screening – proof-of-concept in these assays should be provided as earlier as possible in the development workflow. Different from isolated protein assays, this system-based approach preserves molecular-pathway interactions and allows to disclosing actual polypharmacological profiles [5].

From the case studies discussed below, the field seems to have judiciously considered them and, by taking advantage of recent drug discovery advances and rationale medicinal chemistry strategies, is implementing effective countermeasures. These will hopefully contribute to bridging the valley of death across preclinical and clinical development in the near future.

14.2 Target Identification: In the Search for New Target Pairs

The selection of the right target pair is a fundamental aspect in MTDL drug discovery [5]. This is because, being the first step of the drug discovery pipeline, it likely affects the outcome of the entire campaign. Indeed, many MTDL drug discovery efforts failed because of the selection of the wrong targets [6]. In addition, over the past decades, we have witnessed that the canonical targets of AD drug

Polypharmacology: Strategies for Multi-Target Drug Discovery, First Edition. Edited by Jens-Uwe Peters.

discovery – acetylcholinesterase (AChE) and amyloid-beta (Aβ) – have been over-exploited, leading inevitably to a limited set of target pairs [5]. Most of the reported anti-AD MTDLs have been designed to hit AChE together with another target of interest [2]. Aβ aggregation, oxidative stress, and metal chelation have been the most pursued among AChE-based target combinations, whereas combinations with other key targets have been only slightly exploited [2]. This has consequently led to a high-level of chemical structure similarity (e.g., tacrine-based hybrids) [5]. To enhance the therapeutic potential, MTDL drug discovery should parallel the new trends of AD drug discovery, where a quite robust array of novel targets is being studied [7]. Interestingly, a recent report reveals that seventy percent of current trials are assessing non-Aβ and non-tau targets, with novel agents addressing inflammation and synaptic plasticity [7] as promising points of intervention. Such novel targets from ongoing research efforts should inspire novel target pairs for anti-AD MTDLs. To accelerate the development of MTDLs utilizing under-explored AD targets, several databases [8] and platforms [9, 10] have been proposed. Still, their systematic and routinely use is far to be applied. Nevertheless, in-silico network pharmacology approaches to identify interconnected pathways and targets relevant to AD pathogenesis are being increasingly useful tools for expanding the polypharmacology target space [11]. By analyzing molecular networks, key nodes, and signaling pathways that can be modulated by MTDLs to achieve the desired therapeutic effect, new targets may be found, hence accelerating the identification of new target pairs in a time- and cost-effective manner.

In the following, we will discuss two selected examples of MTDLs which, although maintaining constant AChE modulation, explore a second previously uncharted AD target. Importantly, in both examples, the critical step of validation of the target pair has been performed, and connectivity has been demonstrated through purposely addressed experiments.

An interesting example of anti-AD MTDLs designed to simultaneously modulate an "unconventional" target pair has been developed by Rochais and colleagues [12, 13]. The founding idea was based on combining AChE inhibition with serotonin subtype 4 receptor (5-HT$_4$R) activation [14]. AChE needs no introduction, since most of AD-marketed drugs act as AChE inhibitors improving cognitive functions [15]. On the other hand, 5-HT$_4$R involvement in memory processes has been recently documented, making it a suitable AD target [16]. Particularly, elevated cerebral 5-HT$_4$R density has been linked to the neurological amyloid status observed in AD patients [16]. To test the suitability of this new target pair, donepezil (Figure 14.1a), an AChE inhibitor, and RS67333 (Figure 14.1a), a selective 5-HT$_4$R agonist, were administered in combination to assess their effects on cognitive function in mice. Once the synergistic effects of AChE inhibition and 5-HT$_4$R activation were confirmed, new MTDLs acting as both AChE inhibitors and 5-HT$_4$R agonists were designed [12, 13]. Particularly, based on the structural similarity (highlighted in green in Figure 14.1), RS67333 and donepezil have been selected as starting points to design MTDLs displaying both activities. Intriguingly, thanks to the presence of the piperidine nitrogen establishing a Π-cation interaction with a Phe of the gorge, RS67333 displayed also a significant inhibitory effect toward AChE (IC_{50} = 403 nM) (Figure 14.1a). Several derivatives were synthesized, with donecopride being prioritized for its remarkable and balanced *in vitro* efficacy against both 5-HT$_4$R (EC_{50} = 11 nM) and AChE (IC_{50} = 16 nM). The introduction of the cyclohexyl group (highlighted in orange in Figure 14.1a) appeared to be a good compromise between the *N*-butyl substituent of RS67333 and the bulky benzyl group of donepezil. Preclinical studies demonstrated the compound's efficacy in animal models of AD, highlighting its ability to improve cognitive performance and mitigate disease-related neurodegeneration [17].

In summary, donecopride emerges as a promising MTDL candidate for AD treatment, offering a valuable therapeutic approach by simultaneously targeting key pathological mechanisms, one of which relatively underexplored in the MTDL drug discovery arena.

The group of Muñoz-Torrero proposed another interesting target combination [18]. The idea is to inhibit two key enzymes of the AD network: AChE and soluble epoxide hydrolase (sEH). sEH is an enzyme involved in the metabolism of arachidonic acid, especially epoxyicosatrienoic acids with remarkable anti-inflammatory potential [19]. sEH upregulation in AD patients leads to the progress of neuroinflammation. Moreover, sEH inhibition showed beneficial effects in different AD mouse models [20, 21]. More importantly, positive outcomes were observed following chronic co-administration of TPPU (a sEH inhibitor) and 6-chlorotacrine (an AChE inhibitor) in the senescence-accelerated mouse prone 8 (SAMP8) mouse model, as well as with TPPU and rivastigmine co-administration in the 5XFAD mouse model, when compared to corresponding monotherapy regimens [22]. Building on this strong target pair validation, the authors sought to develop MTDLs targeting both sEH and AChE [18]. The design of this novel set of MTDLs was based on linking two single-targeted scaffolds: 6-chlorotacrine/huprine Y – well-known AChE inhibitors – and TPPU [23] (Figure 14.1b). The linker attachment point on parent compounds and the proper length were selected based on docking studies performed

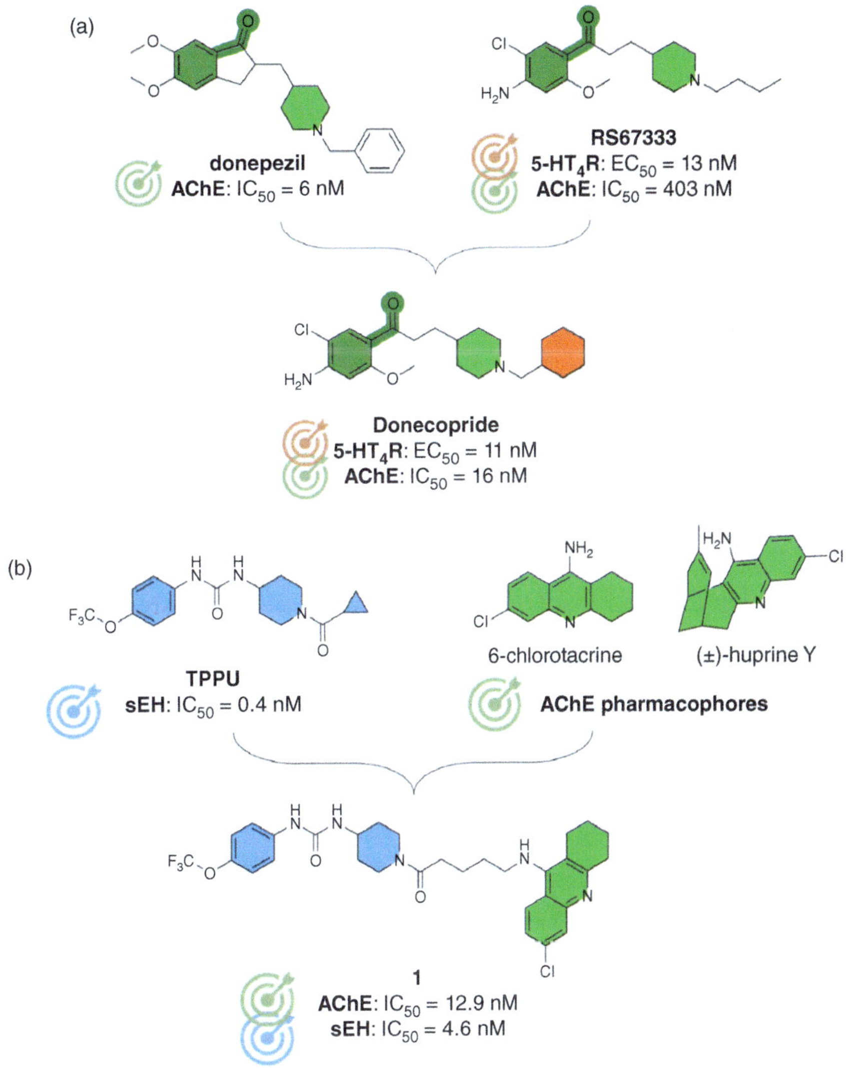

Figure 14.1 Design strategies to new anti-AD MTDLs donecopride (a) and **1** (b) directed to an innovative target pair.

at both targets, leading to the design and synthesis of 5 new MTDLs. The dual inhibitory activity against AChE and sEH of this small set was within the nanomolar range. The preliminary ADME studies (blood-brain barrier [BBB] permeability, aqueous solubility, cytotoxicity, and stability) allowed to select the most promising MTDL **1**, with a balanced inhibitory profile toward both AChE and sEH (IC_{50} of 12.9 nM and of 4.6 nM, respectively) for further *in vivo* evaluation (Figure 14.1b). A chronic *in vivo* efficacy study in SAMP8 mice confirmed the therapeutic potential of **1**. Notably, **1** after oral administration improved significantly short-term and long-term memory in SAMP8 mice. It is worth noting that together with a symptomatic effect on cognition, a concomitant reduction of neuroinflammation, sAPPβ, and tau phosphorylation levels was observed, thus accounting for a disease-modifying potential.

14.3 PK Challenges in MTDL Optimization

Framework combination is one of the most followed medicinal chemistry strategies in MTDL drug discovery [24, 25]. It is based on the combination of pharmacophoric moieties of single-target compounds, generally resulting in large molecules with evident PK drawbacks. Richard Morphy and Zoran Rankovic discussed thoroughly this aspect in a highly cited article introducing "The Physicochemical Challenges of Designing Multiple Ligands" [24]. Physicochemical properties that pertain to molecular size and complexity are indeed of particular significance in AD, considering their crucial association with the development of orally administered agents directed to the central nervous system (CNS). In addition, the linked-pharmacophore strategy has been favored over design methods relying on merging or fusing pharmacophores [2], making MTDL optimization even more challenging. This is especially true when the two pharmacophores are linked by polymethylene linkers.

In this respect, Cavalli and Rosini designed and synthesized compound **ARN14140** [26], an heterobifunctional molecule composed by galantamine, an AChE inhibitor, and memantine, a non-competitive NMDA antagonist, linked by a four-carbon chain (Figure 14.2). The rationale behind this is to have a MTDL with a neuroprotective profile arising from the simultaneous stimulation of the cholinergic system and suppression of glutamate neurotoxicity. However, despite **ARN14140** possesses good *in vitro* activity against AChE and NMDA receptors, it suffered of poor PK profile and low brain penetration that hampered further follow-up studies. Besides drug delivery strategies, the authors sought to pursue structural modification aimed to enhance BBB penetration. Considering that the CNS drugs are characterized by strong rigidity, with fused ring systems and limited rotatable bonds [27], they proposed to swap the polymethylene linker for more rigid alkene (**2**) and alkyne (**3**) chains (Figure 14.2) [28]. Interestingly, the newly synthesized compounds showed an improved *in vitro* IC_{50} value against hAChE, along with a better selectivity index between hBChE and hAChE compared to **ARN14140** and galantamine. The boost in activity was rationally explained by docking studies. Importantly, an extensive *in vitro* PK characterization was performed with the aim to effectively check how the difference in rigidity could impact the drug-like properties. Unfortunately, the introduction of an unsaturated chain gave rise to oxidation problems that normally occur in Phase I metabolism. In fact, alkenes and alkynes are much more likely to undergo oxidation respect to their alkane counterparts due to their π-electrons, but with a different extent [29]. In fact, compound **3** underwent a fast degradation, hampering its use for further biological studies, while compound **2** proved to be just slightly less stable than **ARN14140** ($t_{1/2}$ 45 versus >60 min), therefore feasible enough for *in vivo* administration. On the contrary, compound **ARN14140** and **2** were not affected during Phase II metabolism. Noteworthy, compound **2** showed improved brain permeation compared to **ARN14140**. In fact, at the t_{max} (5 min), the average amount of compound **2** in the brain is more than two times the amount of **ARN14140** (0.524 ng/mg versus 0.141 ng/mg). Therefore, despite the brain exposure is still low for a future clinical development of compound **2**, the authors successfully showed how proper modifications of the chemical structure can dramatically influence the PK properties.

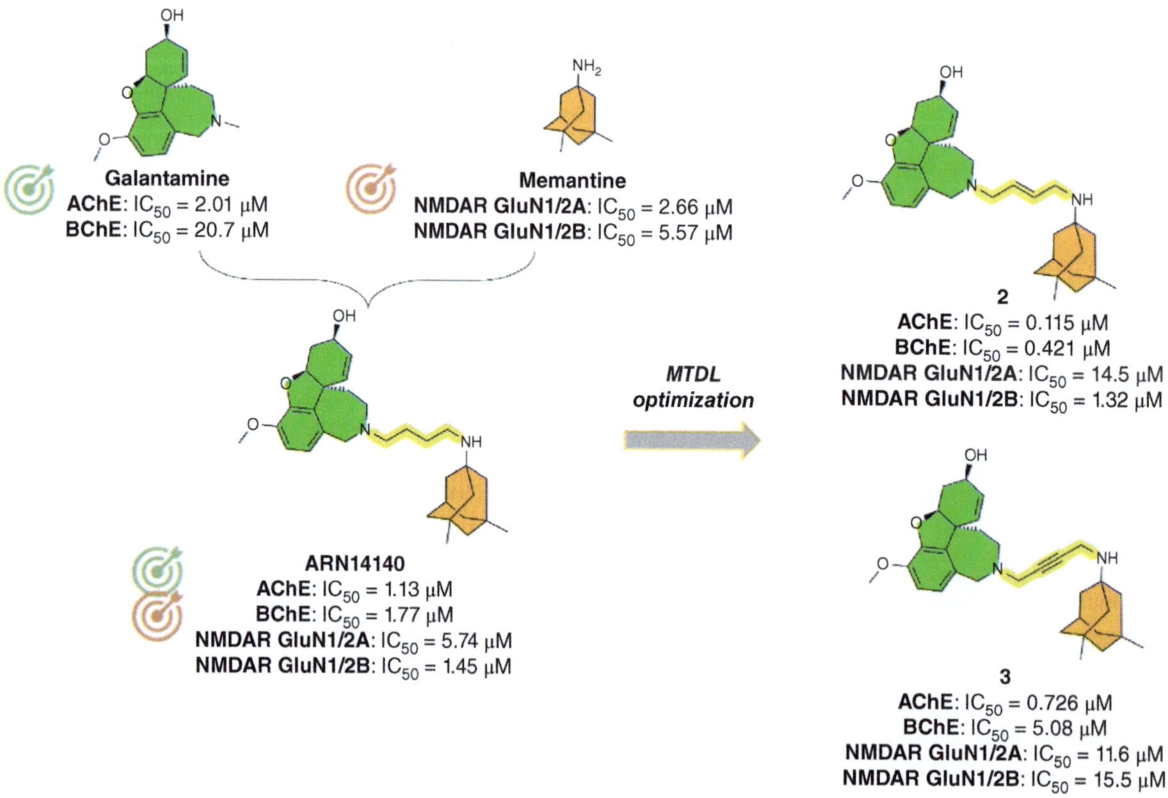

Figure 14.2 Optimization strategy to anti-AD MTDLs **2** and **3**.

14.4 Phenotypic Screening: In a Search for an Early Proof-of-Concept

As we discussed above, in AD multi-target drug discovery, the selection and combination of the right targets is a critical aspect. Separate, although clearly related to this, is to push forward the use of cellular and *in vivo* phenotypic assays in the screening pipeline. Different from isolated proteins, cell systems maintain a reasonably high throughput and reasonable costs, while preserving molecular-pathway interactions, which are crucial for disclosing a truly multi-target profile (with synergistic/additive effects). Even more relevant would be to obtain *in vivo* proof-of-concept of the MTDL profile as early as possible. This is especially pressing considering that a generally low correlation between *in vitro* and *in vivo* activities of MTDLs has been reported [30].

The use of *in vivo* models to accelerate MTDL development runs counter the fact that an animal model recapitulating all AD hallmarks and pathologic characteristics does not exist. The most common *in vivo* AD models are mice and rats, covering more than 50% of the available ones [31]. Some of them include streptozotocin- [32] or scopolamine-induced AD models [33, 34]. Other, like 3xTg, 5xFAD transgenic models, are more complex and not widely available for fast screening, especially in academic settings [35, 36]. Moreover, special attention is also paid to the number of animals used in research, emphasizing the need to apply the "3Rs rule" (replacement, reduction, and refinement) more strongly, routinely, and effectively [37]. Therefore, irrespective of the multi- or single-target approach, new "smart" alternative *in vivo* models to support early drug discovery are highly needed [31, 38]. The advent of sophisticated genomic tools applicable to a wide range of model organisms, combined with their typically short life cycles and the ease of generating large numbers of individuals, has stimulated the use of non-mammalian models such as *Caenorhabditis elegans, Drosophila melanogaster*, and *Danio rerio* (zebrafish) in AD drug discovery [38]. We believe that a wider use of these models for phenotype-based screening can facilitate MTDL drug discovery.

In the last 20 years, two Nobel Prizes were awarded to scientists working on *Drosophila* models [39]. First, *Drosophila* has a great gene homology with many species due to the conservation at genetic and molecular levels. Second, the organization of its brain into three parts and the presence of many common biological processes, such as differentiation, development, and other functions of the CNS, make *Drosophila* an alternative and validated neurological model [40, 41]. Another benefit is the ability to softly modulate gene expression. This has been broadly applied for creating AD models where different pathological proteins (Aβ_{42}, tau, APP, and BACE-1), mimicking AD pathological features, are overexpressed. In *Drosophila*, cognitive impairment is typically associated with a lifespan reduction. The second assessment is the evaluation of the locomotor function, i.e., climbing, which is connected to the neuronal loss. In some studies, immuno- and thioflavin staining to quantify the aggregation process has been also performed [40, 42, 43]. Various *Drosophila* models were successfully implemented in the development of anti-AD MTDLs. Albertini C. et al. presented a small series of cromolyn-ibuprofen conjugates as potential codrugs to modulate AD neuroinflammation/amyloid pathways [44]. The design was based on the idea of transforming the cromolyn-ibuprofen combination tested in clinical trials [45] into single-molecule "codrugs," where the two drugs are chemically linked through *in vivo* hydrolysable bonds. This aimed at (i) potentially reaching the CNS as single molecules and delivering the parent drugs into the brain, and (ii) enhancing the delivery and PK profile of cromolyn, which exhibits low membrane permeability [46]. Thus, these two parent drugs were combined via three different linkers (ethylene glycol, ethylenediamine, and ethanolamine), leading to neutral ester/amide codrugs **4–6** (Figure 14.3a, cromolyn in light blue, ibuprofen in red, and linker in gray). This modification should increase brain penetration, and consequently, once hydrolyzed, deliver to the CNS cromolyn and ibuprofen in a ratio 1:2. The good plasma stability and Aβ_{42} aggregation inhibitory profile (% inhibition @50 μM: 72%), along with immunomodulatory and neuroprotection properties have allowed to prioritize compound **6** for further phenotypic evaluation. The *in vivo* effect was evaluated in an Aβ_{42}-expressing *Drosophila* model [47]. Twenty-day experiment showed promising results: both lifespan and locomotor activity were improved compared to non-treated flies. This example shows the potential of *Drosophila* models in MTDL *in vivo* evaluation [44].

Another positive example refers to a library of 2,4-thiazolidinedione (TZD)-based bivalent derivatives directed to both Aβ and tau (Figure 14.3b, TZD in violet) [48]. Based on the idea of "palindromic compounds" possessing anti-aggregating properties, a new series of twenty-four symmetrical bifunctional structures was designed and synthetized [49]. As an amyloid protein recognition motif (PRM), the well-known antiaggregating TZD scaffold was selected and then decorated with different ionizable groups, such as a carboxylic acid or an amine. Two identical PRMs were connected via various planar, π-conjugated rings (phenyl, biphenyl, biphenylmethane, carbazole, fluorene, and bisthiophene), previously reported as Aβ and tau aggregation inhibitors

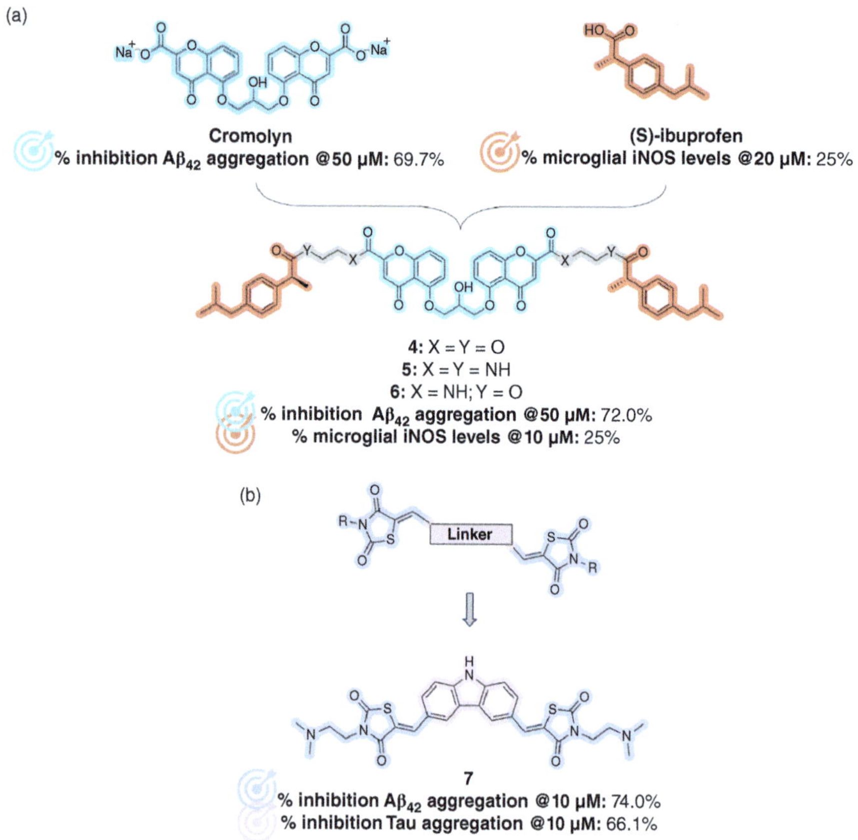

Figure 14.3 Design strategies to new anti-AD MTDLs **4–6** (a) and **7** (b) evaluated in alternative AD *in vivo* models.

[50–52]. The anti-aggregating profiles against $A\beta_{42}$ and tau proteins were determined using recombinant *Escherichia coli* cells [53]. The most promising compound **7**, featuring a central carbazole moiety (pink in Figure 14.3b), at 10 μM inhibited $A\beta_{42}$ and tau aggregation by 74% and 66%, respectively. **7** also displayed low neurotoxicity in primary neurons and was predicted to passively pass the BBB. This encouraging outcome prompted further *in vivo* studies. Similar to the previous example, *Drosophila* flies overexpressing $A\beta_{42}$ were used. Compound **7** improved both lifespan and climbing activities and reduced the presence of amyloid aggregates in the brain [48]. Notably, *Drosophila* models of AD were also successfully implemented for the identification of other MTLDs [54, 55].

Besides *Drosophila* species, there are other non-mammalian alternatives such as *D. rerio* or *C. elegans* for AD *in vivo* phenotypic screening [43, 56]. *D. rerio* (zebrafish) was at first exploited as an animal model to monitor toxic substances and water contamination. Over the years, thanks to its small body size, low cost of maintenance, and fast development cycle, it has become a widely used *in vivo* model, also for high-throughput screening. The ease of optical imaging and high homogeneity of the CNS compared to mammals are particularly important advantages in the context of neurodegenerative diseases [57, 58]. Similarly to *Drosophila*, the proper neural activity is scored mostly as a locomotor function, i.e., spontaneous swimming. There are two main types of AD models: neurotoxic (AD induced by okadaic acid, metal ions, or scopolamine) and genetic (APP/Aβ, MAPT/Tau, PSEN1, and PSEN2) [57]. Moreover, zebrafish models were reported as a promising phenotype-based drug screening for the identification of both "single-target" and "multi-target" anti-AD agents [59–62].

C. elegans is a small nematode which shows a great homology (at least 83%) with the human genome [63]. Alike to *Drosophila* it is also convenient for genetic manipulation, has a short lifespan and cost-effectiveness. Additionally, the possibility of the fluorescent dyeing, *in vivo* imaging, and neuronal signaling conservation makes it very attractive for neurodegenerative models [64, 65]. There are two main types of AD models: overexpressing amyloid-β (Aβ) or tau protein. Unfortunately, until now these *C. elegans* models are mostly applied in research focusing on disease mechanism elucidation rather than drug screening [65–67].

As discussed, there are now several options for alternative *in vivo* AD models (*Drosophila*, zebrafish, and *C. elegans*). Although none of them is perfect, they emerge as suitable tools for obtaining early proof-of-concept of the intended multi-target profile in a cost- and time-effective manner in model organisms able to recapitulate the complex molecular features underlying AD.

14.5 Conclusions

The MTDL concepts have been around since the early 2000s however, much to our regret, no new drug for AD has successfully emerged from the application of MTDL-based strategies. How might we progress this important field, still supported by the strong rationale of the complex nature of the disease? The translational disappointments experienced so far warrant a reappraisal of the types of preclinical questions asked of MTDLs. In this respect, this chapter aims to provide specific points of attention that should be dealt with when aiming at implementing MTDL strategies for AD. They focus on target identification, PK optimization, and phenotypic proof-of-concept. Although the list is clearly not exhaustive and the discussed case studies are limited to the early stages of drug discovery, we can conclude that medicinal chemists play an important role in steering MTDL early discovery. We also need experts from in silico network pharmacology who, by understanding the AD pathological complexity and analyzing the increasingly available omics data, can pinpoint unprecedented combinations of potential targets. Similarly critical is the support of pharmacologists that can deliver early *in vivo* proof-of-concept in "smart" models. Finally, we need out-of-the-box thinkers who push us to move beyond our assumptions and can see surprising connections!

References

1 Cavalli, A., Bolognesi, M.L., Minarini, A. et al. (2008). Multi-target-directed ligands to combat neurodegenerative diseases. *J. Med. Chem.* 51: 347–372.

2 Sampietro, A., Pérez-Areales, F.J., Martínez, P. et al. (2022). Unveiling the multitarget anti-Alzheimer drug discovery landscape: a bibliometric analysis. *Pharmaceuticals* 15: 545.

3 Gong, C.X., Dai, C.L., Liu, F., and Iqbal, K. (2022). Multi-targets: an unconventional drug development strategy for Alzheimer's disease. *Front. Aging Neurosci.* 14: 837649.

4 Alam, J., Kalash, A., Hassan, M.I., and Rahman, S.Z. (2024). Agents at the peak of US FDA approval for the treatment of Alzheimer's disease. *Neurol. Res.* 46 (4): 1–8.

5 Bolognesi, M.L. (2019). Harnessing polypharmacology with medicinal chemistry. *ACS Med. Chem. Lett.* 10: 273–275.

6 Zięba, A., Stępnicki, P., Matosiuk, D., and Kaczor, A.A. (2022). What are the challenges with multi-targeted drug design for complex diseases? *Expert Opin. Drug Discov.* 17: 673–683.

7 Cummings, J.L., Osse, A.M.L., and Kinney, J.W. (2023). Alzheimer's disease: novel targets and investigational drugs for disease modification. *Drugs* 83: 1387–1408.

8 Wang, Z. and Yang, B. (2022). Databases for rational design and discovery of multitarget drugs. In: *Polypharmacology: Principles and Methodologies* (ed. Z. Wang and B. Yang), 737–780. Switzerland AG: Springer Nature.

9 Zhou, Y., Fang, J., Bekris, L.M. et al. (2021). AlzGPS: a genome-wide positioning systems platform to catalyze multi-omics for Alzheimer's drug discovery. *Alzheimers Res. Ther.* 13: 1–13.

10 Liu, H., Wang, L., Lv, M. et al. (2014). AlzPlatform: an Alzheimer's disease domain-specific chemogenomics knowledgebase for polypharmacology and target identification research. *J. Chem. Inf. Model.* 54: 1050–1060.

11 Fang, J., Pieper, A.A., Nussinov, R. et al. (2020). Harnessing endophenotypes and network medicine for Alzheimer's drug repurposing. *Med. Res. Rev.* 40: 2386–2426.

12 Lecoutey, C., Hedou, D., Freret, T. et al. (2014). Design of donecopride, a dual serotonin subtype 4 receptor agonist/acetylcholinesterase inhibitor with potential interest for Alzheimer's disease treatment. *Proc. Natl. Acad. Sci. U.S.A* 111: 3825–3830.

13 Rochais, C., Lecoutey, C., Gaven, F. et al. (2015). Novel multitarget-directed ligands (MTDLs) with acetylcholinesterase (AChE) inhibitory and serotonergic subtype 4 receptor (5-HT4R) agonist activities as potential agents against Alzheimer's disease: the design of donecopride. *J. Med. Chem.* 58: 3172–3187.

14 Freret, T., Bouet, V., Quiedeville, A. et al. (2012). Synergistic effect of acetylcholinesterase inhibition (donepezil) and 5-HT(4) receptor activation (RS67333) on object recognition in mice. *Behav. Brain Res.* 230: 304–308.

15 Marucci, G., Buccioni, M., Ben, D.D. et al. (2021). Efficacy of acetylcholinesterase inhibitors in Alzheimer's disease. *Neuropharmacology* 190: 108352.

16 Lezoualc'h, F. (2007). 5-HT4 receptor and Alzheimer's disease: the amyloid connection. *Exp. Neurol.* 205: 325–329.

17 Rochais, C., Lecoutey, C., Hamidouche, K. et al. (2020). Donecopride, a Swiss army knife with potential against Alzheimer's disease. *Br. J. Pharmacol.* 177: 1988–2005.

18 Codony, S., Pont, C., Griñán-Ferré, C. et al. (2022). Discovery and in vivo proof of concept of a highly potent dual inhibitor of soluble epoxide hydrolase and acetylcholinesterase for the treatment of Alzheimer's disease. *J. Med. Chem.* 65: 4909–4925.

19 Griñán-Ferré, C., Codony, S., Pujol, E. et al. (2020). Pharmacological inhibition of soluble epoxide hydrolase as a new therapy for Alzheimer's disease. *Neurotherapeutics* 17: 1825–1835.

20 Ghosh, A., Comerota, M.M., Wan, D. et al. (2020). An epoxide hydrolase inhibitor reduces neuroinflammation in a mouse model of Alzheimer's disease. *Sci. Transl. Med.* 12: eabb1206.

21 Chen, W., Wang, M., Zhu, M. et al. (2020). 14, 15-Epoxyeicosatrienoic acid alleviates pathology in a mouse model of Alzheimer's disease. *J. Neurosci.* 40: 8188–8203.

22 Jarne-Ferrer, J., Griñán-Ferré, C., Bellver-Sanchis, A. et al. (2022). A combined chronic low-dose soluble epoxide hydrolase and acetylcholinesterase pharmacological inhibition promotes memory reinstatement in Alzheimer's disease mice models. *Pharmaceuticals* 15: 908.

23 Rose, T.E., Morisseau, C., Liu, J.-Y. et al. (2010). 1-Aryl-3-(1-acylpiperidin-4-yl)urea inhibitors of human and murine soluble epoxide hydrolase: structure–activity relationships, pharmacokinetics, and reduction of inflammatory pain. *J. Med. Chem.* 53: 7067–7075.

24 Morphy, R. and Rankovic, Z. (2006). The physicochemical challenges of designing multiple ligands. *J. Med. Chem.* 49: 4961–4970.

25 Uliassi, E., Prati, F., Bongarzone, S., and Bolognesi, M.L. (2017). 10 – Medicinal chemistry of hybrids for neurodegenerative diseases. In: *Design of Hybrid Molecules for Drug Development* (ed. M. Decker), 259–277. Amsterdam: Elsevier.

26 Simoni, E., Daniele, S., Bottegoni, G. et al. (2012). Combining galantamine and memantine in multitargeted, new chemical entities potentially useful in Alzheimer's disease. *J. Med. Chem.* 55: 9708–9721.

27 Xiong, B., Wang, Y., Chen, Y. et al. (2021). Strategies for structural modification of small molecules to improve blood-brain barrier penetration: a recent perspective. *J. Med. Chem.* 64: 13152–13173.

28 Basagni, F., Ortega, J.A., Bertozzi, S.M. et al. (2023). Galantamine-memantine hybrids for Alzheimer's disease: The influence of linker rigidity in biological activity and pharmacokinetic properties. *Eur. J. Med. Chem.* 261: 115803.

29 Obach, R.S. (2014). Functional group biotransformations. In: *Handbook of Metabolic Pathways of Xenobiotics*, 1–41. Chichester: John Wiley & Sons.

30 Muñoz-Torrero, D. (2017). 6 – Multitarget anti-Alzheimer hybrid compounds: do they work in vivo? In: *Design of Hybrid Molecules for Drug Development* (ed. M. Decker), 167–192. Amsterdam: Elsevier.

31 Cacabelos, R., Carrera, I., Martínez-Iglesias, O. et al. (2021). What is the gold standard model for Alzheimer's disease drug discovery and development? *Expert Opin. Drug Discov.* 16: 1415–1440.

32 Grieb, P. (2016). Intracerebroventricular streptozotocin injections as a model of Alzheimer's disease: in search of a relevant mechanism. *Mol. Neurobiol.* 53: 1741–1752.

33 Klinkenberg, I. and Blokland, A. (2011). A comparison of scopolamine and biperiden as a rodent model for cholinergic cognitive impairment. *Psychopharmacology* 215: 549–566.

34 Chen, W.N. and Yeong, K.Y. (2020). Scopolamine, a toxin-induced experimental model, used for research in Alzheimer's disease. *CNS Neurol. Disord. Drug Targets* 19: 85–93.

35 Dhapola, R., Kumari, S., Sharma, P., and HariKrishnaReddy, D. (2023). Insight into the emerging and common experimental in-vivo models of Alzheimer's disease. *Lab. Anim. Res.* 39: 1–18.

36 Louit, A., Galbraith, T., Berthod, F. et al. (2022). Animal models of neurodegenerative disease: recent advances in fly highlight innovative approaches to drug discovery. *Front. Mol. Neurosci.* 15: 1–19.

37 Gowans, J.L. (1974). Alternatives to animal experiments in medical research. *Br. Med. J.* 1: 557–559.

38 Dawson, T.M., Golde, T.E., and Lagier-Tourenne, C. (2018). Animal models of neurodegenerative diseases. *Nat. Neurosci.* 21: 1370–1379.

39 Fernandez-Funez, P., de Mena, L., and Rincon-Limas, D.E. (2015). Modeling the complex pathology of Alzheimer's disease in *Drosophila*. *Exp. Neurol.* 274: 58–71.

40 Bellen, H.J., Tong, C., and Tsuda, H. (2010). 100 years of *Drosophila* research and its impact on vertebrate neuroscience: a history lesson for the future. *Nat. Rev. Neurosci.* 11: 514–522.

41 Reichert, H. (2005). A tripartite organization of the urbilaterian brain: developmental genetic evidence from *Drosophila*. *Brain Res. Bull.* 66: 491–494.

42 Bolus, H., Crocker, K., Boekhoff-Falk, G., and Chtarbanova, S. (2020). Modeling neurodegenerative disorders in *Drosophila* melanogaster. *Int. J. Mol. Sci.* 21: 3055.

43 Tello, J.A., Williams, H.E., Eppler, R.M. et al. (2022). Animal models of neurodegenerative disease: recent advances in fly highlight innovative approaches to drug discovery. *Front. Mol. Neurosci.* 15: 1–21.

44 Albertini, C., Naldi, M., Petralla, S. et al. (2021). Polypharmacology – cromolyn-ibuprofen conjugates for Alzheimer's disease. *Molecules* 26: 1112.

45 Cummings, J., Lee, G., Ritter, A. et al. (2020). Alzheimer's disease drug development pipeline: 2020. *Alzheimers Dement.* 6: e12050.

46 Mori, T., Nishimura, K., Tamaki, S. et al. (1988). Pro-drugs for the oral delivery of disodium cromoglycate. *Chem. Pharm. Bull.* 36: 338–344.

47 Costa, R., Speretta, E., Crowther, D.C., and Cardoso, I. (2011). Testing the therapeutic potential of doxycycline in a *Drosophila* melanogaster model of Alzheimer disease. *J. Biol. Chem.* 286: 41647–41655.

48 Gandini, A., Gonçalves, A.E., Strocchi, S. et al. (2022). Discovery of Dual Aβ/Tau inhibitors and evaluation of their therapeutic effect on a *Drosophila* model of Alzheimer's disease. *ACS Chem. Neurosci.* 13: 3314–3329.

49 Kolstoe, S.E., Mangione, P.P., Bellotti, V. et al. (2010). Trapping of palindromic ligands within native transthyretin prevents amyloid formation. *Proc. Natl. Acad. Sci. U.S.A* 107: 20483–20488.

50 Petrlova, J., Kálai, T., Maezawa, I. et al. (2012). The influence of spin-labeled fluorene compounds on the assembly and toxicity of the Aβ peptide. *PLoS ONE* 7: e35443.

51 Reinke, A.A. and Gestwicki, J.E. (2007). Structure-activity relationships of amyloid beta-aggregation inhibitors based on curcumin: influence of linker length and flexibility. *Chem. Biol. Drug Des.* 70: 206–215.

52 Zhang, X., Wang, Y., Wang, S. et al. (2018). Discovery of a novel multifunctional carbazole–aminoquinoline dimer for Alzheimer's disease: copper selective chelation, anti-amyloid aggregation, and neuroprotection. *Med. Chem. Res.* 27: 777–784.

53 Pouplana, S., Espargaro, A., Galdeano, C. et al. (2014). Thioflavin-S staining of bacterial inclusion bodies for the fast, simple, and inexpensive screening of amyloid aggregation inhibitors. *Curr. Med. Chem.* 21: 1152–1159.

54 Kiran, P.V.R., Waiker, D.K., Verma, A. et al. (2023). Design and development of benzyl piperazine linked 5-phenyl-1,2,4-triazole-3-thione conjugates as potential agents to combat Alzheimer's disease. *Bioorg. Chem.* 139: 106749.

55 De Lorenzi, E., Seghetti, F., Tarozzi, A. et al. (2023). Targeting the multifaceted neurotoxicity of Alzheimer's disease by tailored functionalisation of the curcumin scaffold. *Eur. J. Med. Chem.* 252: 115297.

56 Trujillo-Estrada, L., Sanchez-Mejias, E., Sanchez-Varo, R. et al. (2022). Animal and cellular models of Alzheimer's disease: progress, promise, and future approaches. *Neuroscientist* 28: 572–593.

57 Wang, X., Zhang, J.B., He, K.J. et al. (2021). Advances of zebrafish in neurodegenerative disease: from models to drug discovery. *Front. Pharmacol.* 12: 1–19.

58 Chia, K., Klingseisen, A., Sieger, D., and Priller, J. (2022). Zebrafish as a model organism for neurodegenerative disease. *Front. Mol. Neurosci.* 15: 1–27.

59 Reinhardt, L., Kordes, S., Reinhardt, P. et al. (2019). Dual inhibition of GSK3β and CDK5 protects the cytoskeleton of neurons from neuroinflammatory-mediated degeneration in vitro and in vivo. *Stem Cell Rep.* 12: 502–517.

60 Wang, K., Shi, J., Zhou, Y. et al. (2021). Design, synthesis and evaluation of cinnamic acid hybrids as multi-target-directed agents for the treatment of Alzheimer's disease. *Bioorg. Chem.* 112: 104879.

61 Koehler, D., Shah, Z.A., and Williams, F.E. (2019). The GSK3β inhibitor, TDZD-8, rescues cognition in a zebrafish model of okadaic acid-induced Alzheimer's disease. *Neurochem. Int.* 122: 31–37.

62 Sang, Z., Wang, K., Han, X. et al. (2019). Design, synthesis, and evaluation of novel ferulic acid derivatives as multi-target-directed ligands for the treatment of Alzheimer's disease. *ACS Chem. Neurosci.* 10: 1008–1024.

63 Lai, C.H., Chou, C.Y., Ch'ang, L.Y. et al. (2000). Identification of novel human genes evolutionarily conserved in *Caenorhabditis elegans* by comparative proteomics. *Genome Res.* 10: 703–713.

64 Li, J. and Le, W. (2013). Modeling neurodegenerative diseases in *Caenorhabditis elegans*. *Exp. Neurol.* 250: 94–103.

65 Wu, Y., Chen, Y., Yu, X. et al. (2023). Towards understanding neurodegenerative diseases: insights from *Caenorhabditis elegans*. *Int. J. Mol. Sci.* 25: 443.

66 McColl, G., Roberts, B.R., Pukala, T.L. et al. (2012). Utility of an improved model of amyloid-beta ($A\beta_{1-42}$) toxicity in *Caenorhabditis elegans* for drug screening for Alzheimer's disease. *Mol. Neurodegener.* 7: 57.

67 Miyasaka, T., Xie, C., Yoshimura, S. et al. (2016). Curcumin improves tau-induced neuronal dysfunction of nematodes. *Neurobiol. Aging* 39: 69–81.

15

The Role of Polypharmacology in the History of Drug Discovery

Axel Helmstaedter

15.1 Introduction: Drug Discovery in the Twentieth Century

15.1.1 Medicinal Chemistry

Many therapy options we are using today have been developed throughout the twentieth century. In particular, the period from 1935 to 1965 was suggested as the "golden age of treatment discovery," while development slowed down thereafter overall, although important success was made in some areas, i.e., in the treatment of viral infections like human immunodeficiency virus (HIV) or hepatitis C [1]. In these thirty years, sufficient treatment options, for example, hypertension, Parkinson's disease, depression, and, most importantly, bacterial infections, became possible for the first time. Pharmaceutical chemistry played a predominant role in the development of new, mostly low molecular drugs evolving as blockbusters during the mid-twentieth century still remaining relevant up to the present day. In a similar way, Lombardino and Lowe in 2004 proposed two periods from a pharmaceutical chemist's point of view, the era from the 1950s to the 1980s ("then") and the following decades ("now") [2]. These periods provided different challenges for medicinal chemistry and for drug development in general. In the first period, chemists synthesized manually a variety of compounds to be tested pharmacologically in animal experiments. Successful compounds, i.e., those showing a particular effect on symptoms to be treated, were further developed, optimized, tested again, and so on. This is shown for the development of piroxicam, a nonsteroidal anti-inflammatory agent, which was launched by Pfizer in 1980 after almost 20 years of research and development. Research strategies were driven by chemist's experiences in the field and common knowledge suggesting important molecular structures of a successful drug candidate for a particular condition. Mechanistic considerations were not mandatory, as solely clinical effects were to be considered. This changed dramatically with the possibility to characterize particular drug targets serving as pharmacological models in the laboratory. Now, synthesizing developing substances precisely fitting the target preceded animal experiments. This is exemplified by the history of ziprasidone, a drug for the treatment of schizophrenia, launched in 2001. Initially intended as a supplementary approach to *in vivo* animal model, *in vitro* receptor-based pharmacology emerged in the 1980s and 1990s and became primarily dominant in the field of antipsychotic drug development. The characterization of a variety of different serotonin and dopamine receptor subtypes has enabled chemistry to tailor substances with highly specific receptor targeting profile [3].

15.1.2 The Evolution of the Receptor Concept

First ideas of what is now called "receptor" date back to the late nineteenth and early twentieth centuries and are closely connected to the names John Newport Langley (1852–1925) and Paul Ehrlich (1854–1915). Experimenting with pilocarpine and atropine, Langley anticipated specific structures in the body or on cells capable of being addressed by pharmacological agents in a dose-dependent way and even in competition with each other, so that actions of one can be erased by the other. Paul Ehrlich, a physician highly interested in chemistry, speculated in his PhD thesis already in 1878 about a specific chemical structure of cells where substances (dyes) can bind selectively. Roughly 20 years later and influenced by early immunological considerations around bacterial toxins and "antitoxins," he came up with the imagination of "side chains" of cells capable of specifically binding toxins. In 1900, he coined the term "receptor" instead of "side chain" [4, 5]. It was not until the early 1930s that the receptor concept received additional support based on quantitative analysis of drug action on cells by Alfred Joseph Clark (1885–1941). From calculations of molecular size and cell

Polypharmacology: Strategies for Multi-Target Drug Discovery, First Edition. Edited by Jens-Uwe Peters.

surface area, Clark realized that drugs, such as adrenaline and acetylcholine, at low concentrations needed for their biological effects, were unlikely to form a monomolecular layer over heart cells. Rather, such drugs were likely to, as Clark stated explicitly, "exert their action by uniting with certain specific receptors in or on the heart cells" [4]. Eventually, Raymond P. Ahlquist (1914–1983) made his famous distinction, in 1948, between α- and β-adrenoceptors, that receptor research began to provide a powerful basis for pharmaceutical innovation. Still, nobody had ever seen a receptor, and everything was just concluded from experimental data, though everybody strongly felt that such structures were highly likely to exist. This situation provoked the imagination of receptors as a "remote lady," as the Dutch scientist D.K. DeJong wrote in 1964: "To most of the modern pharmacologists the receptor is like a beautiful but remote lady. He has written her many a letter and quite often she has answered the letters. From these answers the pharmacologist has built himself an image of this fair lady. He cannot, however, truly claim ever to have seen her, although one day he may do so" [6]. In terms of pharmaceutical development, the following years bore the first innovative drugs based on receptor theory and receptor differentiation (Table 15.1).

15.1.3 The Race for Receptor Specificity

A historical landmark is the development of beta blockers connected with the Imperial Chemical Industries (ICI) researcher James Black (1924–2010) whose discoveries led to the launch of the first clinically useful beta-blocker propranolol in 1965 [7]. Black was awarded the Nobel Prize for Physiology or Medicine in 1988. This led to the run for compounds specifically and exclusively active on certain receptors; as specific receptors are preferably located at specific organs, this offered the opportunity to specifically address organ functions by pharmacological means. The following years, efforts were made to custom-tailor substances specifically binding to a receptor type or even subtype hoping to create remedies with rather specific activity and reduce side effects, which were proposed to be caused by unspecific receptor binding. So, a variety of (more or less) cardioselective beta blockers became available.

Phosphodiesterase (PDE) inhibitors may also serve as an example, because sildenafil's side effects like disturbed vision and unusual color sensations are caused by a still insufficient specificity of the compound to PDE5, also addressing the other known PDEs to some extent [8]. In terms of psychotherapeutic drugs, the characterization of a variety of serotonin receptor subtypes triggered the hope that specifically addressing these structures may lead to drugs suitable for the treatment of specific conditions [9, 10]. However, it became clear over time that not a complete receptor subtype specificity but a pattern of addressing a variety of receptors, a "receptor profile," made certain drugs useful for certain conditions [11]. Nevertheless, the second half of the nineteenth century was characterized by a "race for receptor specificity," leading to the attitude that everything that was not that "clean" enough to exclusively target one specific receptor has to be regarded as a "dirty

Table 15.1 Key events in the discovery and establishment of the concept of receptors.

1878	Ehrlich assumes specific chemical characteristics of cells as an explanation for the selective binding of dyes. Langley concludes: "There is some substance in the nerve endings or gland cells with which atropine and pilocarpine are capable of forming compounds according to some law of which their relative mass and chemical affinity for the substances are factors"
1897	Ehrlich proposes the "side chain theory" for the binding of toxins to cells
1905	Langley proposes that nicotine and curare act on the "receptive substance" of muscle cells
1907	Ehrlich proposes the existence of "chemoreceptors" for drugs
1908	Ehrlich is awarded the Nobel Prize in Medicine in recognition of his work on immunity
1933	Clark proposes the "receptor occupancy theory"
1948	Ahlquist proposes that adrenaline acts on two pharmacologically distinct receptor subtypes, which he terms α- and β-receptors
1954	Ariens modifies the receptor occupancy theory to account for the intrinsic activity of drugs
1956	Stephenson modifies the receptor occupancy theory to account for spare receptors and discrepancies between the binding and effect of drugs at receptors
1965	Propranolol was introduced as the first clinically useful β-receptor blocker
1980	The nicotinic acetylcholine receptor (nAChR) is isolated and purified, and its subunit stoichiometry is determined
1982	cDNA sequences for nAChR subunits are reported
1988	Black is awarded the Nobel Prize for his work on developing receptor-subtype-selective drugs
1994	Gilman and Rodbel are awarded the Nobel Prize for their work on the discovery of G-protein-coupled receptors and their role in signal transduction

Source: Adapted from Maehle et al. [4].

drug." This development might have also been triggered by the advent of monoclonal antibodies from 1975 onward, in fact being able to specifically target unique compounds present in the human body [12]. What has been, in this sense, not "clean," logically became "dirty." So, the term "dirty drugs" was coined; however, it seems unclear so far, by whom and in which context this terminology was introduced. An early document is dated from 1991 comprising the term "dirty drug" in the headline, however, without a proper definition [13]. The term is neither mentioned nor explained in the article itself, suggesting that it has already gained common acceptance around 1990. The paper deals with the pharmacological properties of licorice, a phytopharmaceutical mainly used against gastrointestinal disorders since antiquity. Components of licorice constituted "the first practical therapy for gastric and duodenal ulcers" [13] but show different and serious side effects, mainly caused by influences on glucocorticoid and mineralocorticoid pathways. Explaining the concept of "dirty drugs" by means of a phytopharmaceutical absolutely makes sense, as phytopharmaceuticals typically contain a great variety of active ingredients contributing to the overall effect of crude drugs and extracts.

15.2 Natural Products

Phytopharmaceuticals typically contain a wide variety of ingredients more or less contributing to the therapeutic effects observed. In many cases of plants and plant extracts traditionally and successfully used, not a single compound could be made responsible for the overall activity, so it becomes totally clear that therapeutic effects are achieved by the ingredient profile as a whole. Terms to describe this include "multidimensional," "multimodal," "multitargeting," or "pleiotropic" effects; this may lead to the establishment of "pleiotropic signatures" of phytopharmaceuticals instead of defining a single active component [14]. Interestingly, even the term "polypharmacology" was attributed to phytopharmaceuticals, and it has been discussed if natural products may serve as "the best contenders for the development of multitargeting drugs" [15]. In this sense, it was also proposed to research traditional Persian medicine in view of the long desired "polypill" against cardiovascular diseases [16].

Valeriana officinalis preparations, for example, have traditionally been used as sleeping aids since antiquity. Despite its widespread use and the characterization of several active components in the plant, no single compound could be made responsible for the observed effects so far. As known for decades now, Valerian contains different classes of compounds contributing to some extent to the therapeutic effects observed. Among the 15 major compounds isolated are the so-called valepotriates, valerenic acid, valerenol, valtrate, isovalerate, and many others interacting with different neurotransmitters in the human brain including gamma-aminobutyric acid (GABA), serotonin, and dopamine, overall resulting in sleep-inducing efficacy as well as some usefulness in the treatment of anxiety (Figure 15.1). As always in phytotherapy, extraction method and extraction solvent play a major role in the

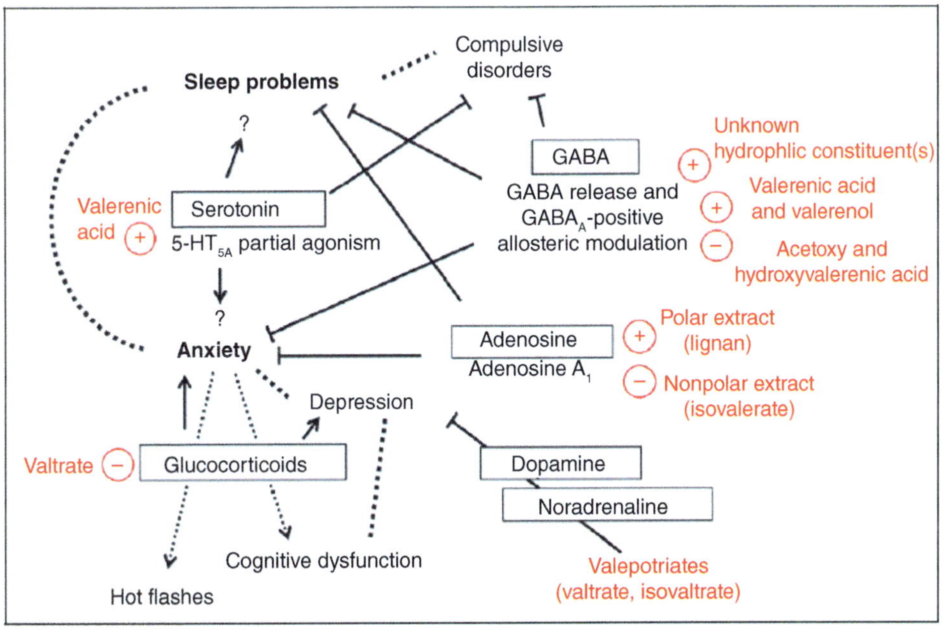

Figure 15.1 The multimodal action of valerian. Shinjyo et al. [17]/SAGE Publishing.

usefulness of the extract against certain conditions [17]. In other words, "Regarding the chemical basis for the observed sedative and tranquillizing activity it can be clearly seen that Valerian is a classical example of a herbal drug where the overall effect is due to several types of constituent and modes of activity" [18].

This also holds true for another member of the Valerianaceae family, *Nardostachys jatamansi*. Preparations have been used since decades against depression and again; no single component could be made responsible for the effects observed. By molecular docking studies, at least four components, i.e., acacin, cryptotanshinone, oleanolic acid, and ursolic acid, could be identified to significantly contribute to the antidepressant activity of *N. jatamansi* extracts [19]. In general, it has recently been suggested to include network pharmacology and systems biology methods to further study multicomponent herbal preparations and their therapeutic effects [20].

While a multimodal effect might, on one hand, be achieved by a variety of compounds addressing a single target each, the other possibility is that a specific compound itself is able to influence different structures and pathways. This is not only the case with natural products; history has shown a variety of examples where synthetic drugs initially introduced for the treatment of a single condition were proven to be also effective in totally different areas. In some cases, what was initially regarded as an adverse effect became a serious treatment option.

15.3 Historical Drugs with Multiple Actions

15.3.1 Aspirin

Certainly, one of the most prominent examples is acetylsalicylic acid (aspirin). Introduced as an analgesic agent in the late nineteenth century, it became one of the most successful drugs ever and is still widely used against headaches, migraines, rheumatic disorders, etc. Its mechanism of action has not been understood until the early 1970s, when John Vane (1927–2004) unveiled the cyclooxygenase (COX)-inhibiting effect of the drug [21]. This leads to the suppression of certain prostaglandins triggering pain. Vane could show that the acetyl group from aspirin is transferred to the hydroxy group of a serine residue located 70 amino acids from the C-terminal end of COX. This also explained the prolonged bleeding times in aspirin users already observed around 1950. The same time, it was already considered to use aspirin for cardiovascular diseases. In the first study, it was shown that out of 8000 patients taking aspirin, not a single one suffered from heart attacks. This remained widely unknown as results were not published in major journals; moreover, the investigator himself died, despite having taken aspirin, of a heart attack, which impaired the credibility of his work. In 1967, Harvey Weiss established the platelet-inhibiting effect of aspirin, before the underlying mechanisms were eventually clarified in 1971 [22]. As the drug addresses the same target but with different physiological consequences, this may not be an example of "polypharmacology" in its narrow sense. However, there are many others. Secondary drug uses, however, often arose from clinical observations, not from pharmacological considerations, and had their haydays in the 1960s. Spohrer gives 20 examples for "secondary indications" in this sense (Table 15.2) [23]. With sildenafil, an important example of recent years has certainly to be added.

15.3.2 Sulfonamides

Sulfonamides were introduced into therapy as the first bacteriostatic chemotherapeutic agents in 1935. German researcher Gerhard Domagk (1895–1964) had observed that certain dyestuffs were able to prevent bacterial growth. French researchers could show that only the sulfonamide part of Domagk's deeply red-colored pioneering substance (marketed as "Prontosil rubrum") was responsible for the bacteriostatic effect and came up with sulfanilamide ("Prontosil album") as the "mother substance" of all sulfonamides medicinally used. Already in the 1940s, Fildes and Woods established the "antimetabolite" concept following the observation that sulfonamides are competitive antagonists and structural analogs of *p*-aminobenzoic acid (PABA) in the synthesis of folic acid [24]. Sulfonamides inhibit and replace PABA in the enzyme dihydropteroate synthase (whose activity is important for the production of folate), eventually inhibit the formation of dihydrofolate and tetrahydrofolate, and also subsequently inhibit bacterial deoxyribonucleic acid (DNA) growth and cell division or replication [25]. While a great variety of sulfa compounds were synthesized and tested, it became obvious that certain derivatives had other than bacteriostatic effects.

So, in 1940, it was observed that sulfanilamide and other compounds were able to inhibit the enzyme carbonic anhydrase, leading to a diuretic effect. Initially regarded as a side effect of bacteriostatic therapy, the diuretic action of sulfanilamide was introduced as a treatment for congestive heart failure in the late 1940s. Medicinal chemistry then tried to optimize diuretic actions and soon came up with heterocyclic sulfonamides including acetazolamide, being manyfold more powerful as carbonic anhydrase inhibitor than sulfanilamide. The drug was launched in 1952. Clinical use, however, was severely limited due to side effects. Thus, the search for more useful sulfa

Table 15.2 "Secondary indications" derived from clinical observation for drugs introduced in the twentieth century.

Drug	Primary indication	Secondary indication
Acetylsalicylic acid	Analgesic	Antithrombotic
Allopurinol	Xanthine oxidase inhibitor	Uricostatic
Amantadine	Influenza prophylaxis	Parkinson's disease
Aminoglutethimide	Anticonvulsant	Antiestrogen
Carbutamide/tolbutamide	Bacteriostatic	Antidiabetic
Chlorpromazine	Antihistaminic	Neuroleptic
Chloroquine	Antimalarial	Antirheumatic
Clonidine	Rhinologic	Antihypertensive
Fenoterol	Bronchospasmolytic	Tocolytic
Imipramine	Hypnotic	Antidepressant
Iproniazid	Tuberculostatic	Antidepressant
Methotrexate	Cytotoxic	Antirheumatic
Minoxidil	Antihypertensive	Hair growth stimulator
Novasurol	Antisyphilitic	Diuretic
Phenobarbitone	Hypnotic	Antiepileptic
Piracetam	Antiemetic	Nootropic
Propranolol	ß-Sympathomimetic	ß-Blocker
Sulfasalazine	Antirheumatic	Against Colitis ulcerosa
Thalidomide	Hypnotic	Against leprosy
Verapamil	Coronary dilatator	Antiarrhythmic
Sildenafil	Erectile dysfunction	Pulmonary arterial hypertension

Source: Adapted from Spohrer [23].

derivatives continued and ended up with the well-known thiazide diuretics, chlorothiazide (introduced in 1957), hydrochlorothiazide (1959), and eventually furosemide [26], showing a totally different mechanism of action [26].

From clinical trials with sulfonamide antibiotics against typhoid in the early 1940s, it became obvious that some patients developed severe hypoglycemia. Initially ignored, this aspect became a matter of serious research in the mid-1950s, when the same effect was recognized by testing a sulfonylurea derivative, carbutamide, as a long-acting sulfonamide antibiotic against urinary tract infections. Important developments had been achieved in the German Democratic Republic before leading scientists moved to the west and restarted research there [27]. So, historians state that antidiabetic sulfonylureas "had to be discovered three times" [28]. Carbutamide was introduced into the German market in 1956, as was tolbutamide, another derivative being almost completely devoid of antibacterial activity. Antidiabetic sulfonylureas exert their activity by closing adenosine triphosphate (ATP)-sensitive K-channels in the beta-cell plasma membrane and so initiating a chain of events that result in insulin release [29].

As can be seen, sulfonamides address at least three totally different targets, and thus, as a drug class, they can certainly be regarded to act "polypharmacologically." Derivatives with considerable market success at the end, however, were more specific ones developed in the direction desired by means of medicinal chemistry.

15.3.3 Psychopharmaceuticals

The history of psychopharmaceuticals is closely connected to that of antihistamines. This holds especially true for chlorpromazine, known as the first clinically successful neuroleptic. History starts in the late 1940s in Tunisia with experiments done by the French Navy surgeon Henri Laborit (1914–1995) [30]. He searched for a potent antihistamine able to minimize the traumatic effects of surgery considering the possibility that antihistamines might prevent capillary hyperpermeability caused by histamine release in shock. Back to Paris, Laborit started systematic investigations with antihistamines in his "lytic cocktail," which should lower the risk of shock and reduce body temperature during surgical procedures. Experiments with mepyramine and promethazine turned out promising, and Laborit already noticed mood-stabilizing and anxiolytic effects. These were considerably enhanced with chlorpromazine, synthesized in 1951 at Rhone-Poulenc, which led to investigations into the usefulness of the compound in the treatment of psychoses, which turned out to be successful at the end. Licensed in France in 1952, chlorpromazine had not yet found its final place in therapy and was primarily marketed against diseases like seasickness and morning sickness in pregnancy. Trademark in France was "Largactil," suggesting the wide range of potential indications (large action) including vomiting, pruritus, or hiccups [31, 32]. Some years later, however, its outstanding value for the treatment of manic, agitated, and psychotic patients was recognized, and it became the first "neuroleptic," a term introduced by Delay and Deniker in 1955. Its synthesis was said to constitute "the begin of modern psychopharmacology" [32]. As is known today, chlorpromazine acts as an antagonist of a variety of dopamine and serotonin receptors and is, as initially intended, also active as H1 antihistamine.

Finding a new antihistamine also triggered research on iminodibenzyl derivatives at the Geigy company at the end of the 1940s. Some of them were active but showed strong sedation as an adverse effect. They were, thus, tested as hypnotics in the late 1950s but with dissatisfying results. However, significant improvement was seen with agitated schizophrenia patients, rather similar to what was already known from chlorpromazine. Geigy then followed this path and synthesized derivatives with structure analogies to that; as a result, imipramine was introduced as a potent antidepressant [32].

In parallel, also the first monoamine oxidase (MAO) inhibitor, iproniazid, was developed in the 1950s. Although the compound was marketed only for a few years, its appearance was said to be a historic achievement equally important to imipramine [33]. This compound was derived, like isoniazid, from a research program to find new antibiotics against tuberculosis. Both compounds turned out to be successful in this respect, but iproniazid showed a variety of central nervous side effects, absent under isoniazid treatment. The latter became a first-line therapy for tuberculosis, while iproniazid made a short career as a stimulating antidepressant and the first MAO inhibitor, before it was withdrawn because of hepatotoxicity [34, 35]. Its antidepressant effect became obvious, when tuberculosis patients treated with iproniazid began to sing and dance on the hospital floors [35]. It has to be mentioned here that many phenothiazines and thioxanthenes used as psychopharmaceuticals, most prominently chlorpromazine, also possess antibacterial activity [36]. Anthelminthic activities were also reported [37].

15.3.4 Amantadine

Amantadine was synthesized in 1960 at Munich University and was found to act virustatic in an industrial screening program. In 1966, the substance was introduced in the United States for the prophylaxis against influenza infections [38, 39]. It was still used against influenza in Asia in the 2000s [40]. Accidentally used in a Parkinson's patient in 1968, the substance was reported as considerably relieving symptoms of the disease, returning after discontinuation of the influenza prophylaxis. Clinical trials were initiated confirming therapeutic usefulness against Parkinson's disease, in combination with levodopa (l-DOPA) or without [41]. Although not widely used anymore, amantadine was regarded as one of the "milestones in Parkinson's disease therapeutics" [42]. In terms of influenza prophylaxis and therapy, amantadine inhibits the function of the M2 protein and thus stops the replication process. Amantadine is only effective against influenza A and not influenza B, because influenza B does not have a M2 protein [43]. Anti-Parkinson activity is explained by an indirect dopamine-releasing action and stimulation of dopamine receptors [44].

15.3.5 Minoxidil/Finasteride

In the late 1950s, Upjohn Company tried to develop an innovative antacid. A compound named U-7720 did only show a weak antacid effect, but impressed with strong antihypertensive action in animal models. The compound was not suitable for use in humans, but the antihypertensive effect was attributed to an animal metabolite comprising an N-oxide structure. From a program to develop N-oxide-containing structures as antihypertensives, eventually minoxidil emerged as the most promising compound. It was launched in the United States in 1979 and in Germany in 1983 as a second-line antihypertensive. Already in 1971, clinical researchers observed slight hypertrichosis in some patients and tested this effect successfully on themselves after topical application [45]. After additional testing, Upjohn started to market a 2% topical preparation of minoxidil in 1988 against alopecia. Minoxidil is a cellular potassium channel opener that contains nitric oxide (NO), i.e., a vasodilator, in its chemical structure and possibly acts as an NO agonist. This results in a direct relaxation of the arterioles and thus a drop in blood pressure. The mode of topical action has not yet been clarified, but it is assumed that minoxidil dilates the capillaries and increases local blood flow. Additionally, the hair growth telogen phase is shortened so that the growth (anagen) phase is reached more quickly. Along with the 5α-reductase inhibitor finasteride, minoxidil became the first-line treatment for hair loss. Finasteride's use is explained by the fact that 5α-reductase, which is responsible for the conversion of testosterone into dihydrotestosterone, is almost exclusively located in the body in the prostate and the scalp [46].

15.3.6 Sildenafil

Sildenafil arose out of a Pfizer screening program to find blood pressure-lowering compounds enhancing the activity of the atrial natriuretic peptide, which was known to increase the synthesis of cyclic guanosine monophosphate (cGMP). As an alternative way, reducing the breakdown of cGMP catalyzed by phosphodiesterases was suggested as a mechanism of action. In search of a suitable lead structure, research came up with zaprinast, a compound initially intended but never marketed as an antiallergic drug. It had been described as a weak PDE inhibitor in the literature. This compound was further developed into sildenafil being 100 times more potent than zaprinast as a PDE5 inhibitor. Clinical studies against coronary heart disease in view of the vasodilating effects of the compound brought rather disappointing results. However, unexpected effects of sildenafil turned up: "One of several side effects was only revealed when participants in a trial of sildenafil on 30 men in the Welsh town of Merthyr Tydfil in 1992 were questioned about their reluctance to turn unused tablets when the trial was stopped. On questioning by the physician in charge, it emerged that they had discovered that the drug was inducing an increase in erectile dysfunction" [47]. The company now concentrated on this "new" indication promising tremendous commercial success. Eventually, Viagra® appeared on the market in 1998. Thereafter, research moved back to cardiovascular indications, and in 2006, sildenafil was launched under a different trade name and dose against pulmonary hypertension [48]. As a nonmedical use, it was shown that small concentrations of the drug dissolved in a vase of water can double the shelf life of cut flowers, making them stand up straight for as long as a week beyond their natural life span [49]. Recently, it was discussed whether it could even be suitable for the treatment of Alzheimer's disease (AD) [50].

15.4 From Serendipity to Concept: Repurposing and Polypharmacology

Just these few historical examples show that the "one-compound–one-target–one-action approach" is much too simple. History shows that many, if not most, of the drugs are not acting as specifically as intended. This, in turn, makes it reasonable to systematically screen commercially available drugs for actions and indications outside the pharmacological field they were initially developed and licensed for. In case of success, this approach also saves a great deal of resources. The concept is known as "repurposing" or "repositioning"; it was firstly described and defined as "identifying and developing new uses for existing drugs" in 2004 [51]. This approach has been followed since then for a variety of unmet medical needs including Alzheimer's disease (AD), COVID-19, and cancer. Among the more than 100 clinically tested compounds against AD in 2017, 27 were already marketed for different purposes [52]. A variety of existing compounds proved to be active against the severe acute respiratory syndrome coronavirus 2 (SARS-CoV-2) *in vitro*, but clinical studies ended up disappointing in most cases, so for ivermectin and hydroxychloroquine [53]. Others like tocilizumab, baricitinib, or molnupiravir are proven to be somewhat more successful [54]. For the treatment of cancer, a variety of compounds originally introduced as antibiotics, antivirals, antidepressants, and others are clinically tested at the moment [55].

So, on the one hand, most pharmaceuticals seem to influence not only one but a variety of targets. On the other hand, diseases usually are not monocausal but develop out of a variety of physiological abnormalities. In the development of, for example, metabolic imbalances, central nervous system (CNS) disorders, and cancer, multiple signaling pathways and dysregulation of multiple physiological processes are involved. Thus, it seems totally reasonable to try to find "multi-target" compounds combating "multifactorial" diseases. From 2004 onward, this strategy became known as the "polypharmacology" approach (leading to drugs with "polypharmacological" activities) [56], a term somewhat unfortunately chosen from a linguistic point of view. "Pharmacology" is the scientific discipline investigating drug actions, and as such, only one "pharmacology" can exist. From this point of view, terms like "multi-target therapeutics" [57] seem preferable; more metaphoric descriptions have also been given [58]. Whatever called, the approach provides significant challenges for medicinal chemistry [56], because compounds have to be designed that are almost equally effective at different targets. Among many other examples, the concept has been proven by the development of dual action compounds against nonalcoholic steatohepatitis (NASH) simultaneously targeting the nuclear farnesoid X receptor (FXR) and the enzyme-soluble epoxide hydrolase (sEH) [59, 60]. These compounds are, although highly interesting from a chemical and theoretical point of view, still far from clinical use.

Nevertheless, the development constitutes some kind of paradigm shift. The development of the receptor concept and the characterization of receptor families with different but closely related members with their specific physiological relevance led to the desire to address targets most specifically. This approach proved effective in many cases; see the development of cardioselective beta blockers starting in the late 1960s. On the other hand, it became obvious that natural products and many synthetic drugs are addressing multiple targets and thus may exert a variety of actions. Serendipitous discoveries led to second and third indications for established drugs. In the early twenty-first century, this knowledge was transformed into the strategies of repurposing and polypharmacology. Known compounds were systematically tested for possible further uses, and medicinal chemistry tries to develop compounds able to address more than a single target. This seems particularly useful in diseases and syndromes known for their multifactorial causes.

References

1 Wortmann, R.J. and Bettiker, R.L. (1995). The slowing of treatment discovery, 1965–1995. *Nat. Med.* 1: 1122–1125.

2 Lombardino, J.G. and Lowe, J.A. III, (2004). The role of the medicinal chemist in drug discovery – then and now. *Nat. Rev. Drug Discov.* 3: 853–862.

3 David, D.J. and Gardier, A.M. (2016). Les bases de pharmacologie fondamentale du système sérotoninergique: application à la réponse antidepressive. *Encéphale* 42: 255–263.

4 Maehle, A.-H., Prüll, C.-R., and Halliwell, R.F. (2002). The emergence of the drug receptor theory. *Nat. Rev. Drug Discov.* 1: 638–641.

5 Maehle, A.-H., Prüll, C.-R., and Halliwell, R.F. (2003). Drugs and cells – pioneering the concept of receptors. *Pharm. Hist.* 18: 18–30.

6 Maehle, A.-H. (2009). A binding question: the evolution of the receptor concept. *Endeavour* 33: 135–140.

7 Quirke, V. (2006). Putting theory into practice: James Black, receptor theory, and the development of beta-blockers at ICI, 1958–1978. *Med. Hist.* 50: 69–92.

8 Helmstädter, M. and Schubert-Zsilavecz, M. (2021). Phosphodiesterases. In: *Pharmaceutical Biocatalysis. Important Enzymes, Novel Targets, and Enzymes* (Jerry Stanford Series on Biocatalysis, 6) (ed. P. Grunwald), 47–82. Singapore: Jenny Stanford Publishing.

9 Lucki, I. (1996). Serotonin receptor specificity in anxiety disorders. *J. Clin. Psychiatry* 57 (Suppl. 6): 5–10.

10 Montgomery, S.A. and Fineberg, N. (1989). Is there a relationship between serotonin receptor subtypes and selectivity of response in specific psychiatric illnesses? *Br. J. Psychiatry* 155 (S8): 63–69.

11 Miller, J.J. (2020). Drug receptor profiles matter. *Psychiatr. Times* 37 (April): 2–4.

12 Malaviya, A.N. and Mehra, N.K. (2018). A fascinating story and development of biologicals in clinical medicine. *Indian J. Med. Res.* 148: 236–278.

13 Molder, C. (1991). Corticosteroids, kidneys, sweet roots, and dirty drugs. *Mol. Cell. Endocrinol.* 78: C95–C98.

14 Schwabl, H., Vennos, C., and Saller, R. (2014). Pleiotrope Signaturen – eine Anregung zur Aufwertung traditioneller pflanzlicher Arzneimittel. *Forsch. Komplementmed.* 21 (Suppl. 1): 25–28.

15 Ho, T.T., Tran, Q.T., and Chai, C.L.L. (2018). The polypharmacology of natural products. *Future Med. Chem.* 10: 1361–1368.

16 Noroozi, S., Zargaran, A., and Karimi, M. (2021). Medicinal herbs: potential polypills in cardiovascular diseases. *J. Clin. Hypertens.* 23: 896–896.

17 Shinjyo, N., Waddell, G., and Green, J. (2020). Valerian root in treating sleep problems and associated disorders – a systematic review and meta analysis. *J. Evid. Based Integr. Med.* 25: 1–31.

18 Houghton, P.J. (1999). The scientific basis for the reputed activity of Valerian. *J. Pharm. Pharmacol.* 51: 505–512.

19 Jalali, S., Zarrinhaghighi, A., Sadraei, S. et al. (2018). A system pharmacology study for deciphering anti depression activity of Nardostachys jatamansi. *Curr. Drug Metab.* 19: 469–476.

20 Panossian, A. (2023). Challenges in phytotherapy research. *Front. Pharmacol.* 14: 1199516.

21 Botting, R.M. (2010). Vane's discovery of the mechanism of action of aspirin changed our understanding of its clinical pharmacology. *Pharmacol. Rep.* 62: 518–526.

22 Sneader, W. (2005). *Drug Discovery, a History*, 360. Chichester: Wiley.

23 Spohrer, B. (1996). Die Beobachtung am Krankenbett – ein Weg zu neuen Arzneimitteln. Diss. rer. nat. Jena.

24 Lesch, J.E. (2007). *The First Miracle Drugs: How the Sulfa Drugs Transformed Medicine.* Oxford: Oxford University Press.

25 Ovung, A. and Bhattacharyya, J. (2021). Sulfonamide drugs: structure, antibacterial property, toxicity, and biophysical interactions. *Biophys. Rev.* 13: 259–272.

26 Rau, S. (2006). Vom Coffein zum Furosemid. Die Geschichte der Diuretika. *Pharm. Unserer Zeit* 35: 286–292.

27 Kleinsorge, H. (1998). Carbutamide – the first oral antidiabetic. A retrospect. *Exp. Clin. Endocrinol. Diabetes* 106: 149–151.

28 Jörgens, V. (2022). *Die Geschichte der Diabetesforschung. Vom Opium zum Insulin*, 214–217. Mainz: Kirchheim.

29 Ashcroft, F.M. (1996). Mechanisms of the glycaemic effects of sulfonylureas. *Horm. Metab. Res.* 28: 456–463.

30 Sneader [22], pp. 407–408.

31 Balz, V. (2009). Megaphen wird zur Wirkung gebracht. Die klinische Konstitution eines erfolgreichen Behandlungsfalls an der Psychiatrischen Universitätsklinik Heidelberg. In: *Arzneimittel des 20. Jahrhunderts. Historische Skizzen von Lebertran bis Contergan* (ed. N. Eschenbruch, V. Balz, U. Klöppel, and M. Hulverscheidt), 167–198. Bielefeld: transcript.

32 Braslow, J.T. and Marder, S.R. (2019). History of psychopharmacology. *Annu. Rev. Clin. Psychol.* 15: 25–50.

33 Slattery, D.A., Hudson, A.L., and Nutt, D.J. (2004). Invited review: the evolution of antidepressant mechanisms. *Fundam. Clin. Pharmacol.* 18: 1–21.

34 Spohrer [23], pp. 45–53.

35 Lopez-Munoz, F. and Alamo, C. (2009). Monoaminergic neurotransmission: the history of the discovery of antidepressants from 1950s until today. *Curr. Pharm. Des.* 15: 1563–1586.

36 Nehme, H., Saulnier, P., Ramadan, A.A. et al. (2018). Antibacterial activity of antipsychotic agents, their association with lipid nanocapsules and its impact on the properties of the nanocarriers and on antibacterial activity. *PLoS ONE* 13: e0189950.

37 Weeks, J.C., Roberts, W.M., Leasure, C. et al. (2018). Sertraline, paroxetine, and chlorpromazine are rapidly acting anthelmintic drugs capable of clinical repurposing. *Sci. Rep.* 8: 975.

38 Hornick, R.B., Togo, Y., Dawkins, A.T., and Snyder, M.J. (1967). Clinical evaluation of amantadine-HCL in the prophylaxis of induced influenza. In: *Probleme der Verhütung von Viruserkrankungen* (ed. J. Ströder and W. Henle). Berlin, Heidelberg: Springer.

39 Delgado-Figueroa, N. (2016). 50 years ago in the journal of pediatrics: the suppression of A2 influenza in children by the chemoprophylactic use of amantadine. *J. Pediatr.* 177: 26.

40 Cowling, B.J., Chui, C.S.L., Lim, W.W. et al. (2018). Use of influenza antivirals in patients hospitalized in Hong Kong, 2000–2015. *PLoS ONE* 13: e0190306.

41 Hubsher, G., Haider, M., and Okun, M.S. (2012). The journey from fighting flu to treating Parkinson disease. *Neurology* 78: 1096–1099.

42 Rascol, O., Lozano, A., Stern, M., and Poewe, W. (2011). Milestones in Parkinson's disease therapeutics. *Mov. Disord.* 26: 1072–1082.

43 Stiver, G. (2003). The treatment of influenza with antiviral drugs. *Can. Med. Assoc. J.* 168: 49–57.

44 Bailey, E.V. and Stone, T.W. (1975). The mechanism of action of amantadine in parkinsonism: a review. *Arch. Int. Pharmacodyn. Ther.* 216: 246–262.

45 Zins, G.R. (1988). The history of the development of minoxidil. *Clin. Dermatol.* 6: 132–147.

46 Steiner, J.F. (1996). Clinical pharmacokinetics and pharmacodynamics of finasteride. *Clin. Pharmacokinet.* 30: 16–27.

47 Sneader [22], pp. 135–136.

48 Barnett, C.F. and Machado, R.F. (2006). Sildenafil in the treatment of pulmonary hypertension. *Vasc. Health Risk Manag.* 2: 411–422.

49 Siegel-Itzkovich, J. (1999). In brief: viagra makes flowers stand up straight. *Br. Med. J.* 319 (7205): 274B.

50 Huo, X. and Finkelstein, J. (2023). Using big data to uncover association between sildenafil use and reduced risk of Alzheimer's disease. *Stud. Health Technol. Inform.* 302: 866–870.

51 Ashburn, T.T. and Thor, K.B. (2004). Drug repositioning: identifying and developing new uses for existing drugs. *Nat. Rev. Drug Discov.* 3: 673–683.

52 Jourdan, J.-P., Bureau, R., Rochais, C., and Dallemagne, P. (2020). Drug repositioning: a brief overview. *J. Pharm. Pharmacol.* 72: 1145–1151.

53 Vaz, E.S., Vassiliades, S.V., Giarolla, J. et al. (2023). Drug repositioning in the COVID-19 pandemic: fundamentals, synthetic routes, and overview of clinical studies. *Eur. J. Clin. Pharmacol.* 79: 723–751.

54 Kumar, S. and Roy, V. (2023). Repurposing drugs: an empowering approach to drug discovery and development. *Drug Res.* 73: 481–490.

55 Pawar, V.A., Tyagi, A., Verma, C. et al. (2023). Unlocking therapeutic potential: integration of drug repurposing and immunotherapy for various disease targeting. *Am. J. Transl. Res.* 15: 4984–5006.

56 Proschak, E., Stark, H., and Merk, D. (2019). Polypharmacology by design. A medicinal chemist's perspective on multitargeting compounds. *J. Med. Chem.* 62: 420–444.

57 Zimmermann, G.R., Lehár, J., and Keith, C.T. (2007). Multi-target therapeutics: when the whole is greater than the sum of the parts. *Drug Discov. Today* 12: 34–42.

58 Stark, H. (2004). Turning from monogamy to strategic promiscuity. *Drug Discov. Today* 9: 736–737.

59 Schierle, S., Helmstädter, M., Schmidt, J. et al. (2020). Dual farnesoid X receptor/soluble epoxide hydrolase modulators derived from zafirlukast. *Chem. Med. Chem.* 15: 50–67.

60 Helmstädter, M., Kaiser, A., Brunst, S. et al. (2021). Second-generation dual FXR/sEH modulators with optimized pharmacokinetics. *J. Med. Chem.* 64: 9525–9536.

Part C

How to Discover Polypharmacological Drugs

16

Strategies for Multi-target Drug Discovery

Dayong Shi and Xiangqian Li

16.1 Introduction

Over the past few decades, the emphasis in drug development has primarily been on the creation of single drugs that exhibit high potency and selectivity toward specific biological targets [1]. Undoubtedly, this strategy of "one-molecule, one-target" has yielded noteworthy accomplishments, resulting in the development of numerous successful drugs [2, 3]. However, the use of highly selective ligands that target specific receptors may not always lead to the production of clinically effective medications. This is because the interaction with a single target often fails to sufficiently impact the diseased system, particularly in the case of complex multifactorial diseases like diabetes, neurodegenerative syndromes, cardiovascular diseases, and cancer [3].

In recent research, there has been a notable shift toward the modulation of multiple targets simultaneously [4, 5]. A proposed approach involves combining two distinct inhibitory activities into a single molecule, known as the multi-target drug strategy, to achieve combination therapy [6, 7]. It is important to differentiate this multi-target drug strategy from multiple-medication therapy, which entails the simultaneous use of two or more different drugs [8–10]. Multi-target drugs offer several potential advantages, including enhanced therapeutic effects, reduced risk of drug–drug interactions, and more predictable pharmacokinetics (PKs) and pharmacodynamics (PD) [11–13]. Moreover, the development of MTDLs represents a novel avenue to combat drug resistance. For instance, Choong-Kun Lee demonstrated that the multi-target angiokinase inhibitor TKI258 could overcome colorectal cancer resistance to both epidermal growth factor receptor (EGFR) and vascular endothelial growth factor (VEGF) inhibitors [14, 15].

In 2004, Morphy et al. introduced a groundbreaking concept in drug discovery known as designed multiple ligands (DMLs) [2, 4], which marked the beginning of a new era in the research and development of multi-target drugs. Anighoro et al. and Lu et al. conducted assessments to evaluate the advantages of multi-target drugs in comparison with single-target and combination therapies [1, 16]. Additionally, several successful examples of multitargeted agents were highlighted, encompassing various areas such as pharmacophore modes, anticancer agents, neurodegenerative diseases, Alzheimer's disease (AD), anti-inflammatory drugs, and multi-target therapeutics [3, 9, 11, 17–23]. Zhou et al. delved into discussions regarding the establishment of target combinations, as well as the identification, optimization, and synthesis of MTDLs [24]. Furthermore, Andrew Anighoro performed a comprehensive analysis of the advantages and disadvantages associated with MTDLs in comparison with combination therapies [1], while Ewgenij Proschak assessed the importance and feasibility of MTDL design [25].

The rational design of MTDLs continues to pose significant challenges, necessitating further implementation and methodological advancements [25, 26]. In contrast to previous reviews, the aim of this perspective is to illustrate the rational design of multi-target agents from the standpoint of medicinal chemists, leveraging existing information. This article provides a comprehensive review and analysis of the design strategy that involves incorporating a second activity into an existing ligand while preserving its affinity for the original target.

16.2 Rational Design of Multitargeted Ligands

Four commonly known strategies for obtaining MTDLs are merged-pharmacophore mode, fused-pharmacophore mode, non-cleavable linked-pharmacophore mode, and cleavable linked-pharmacophore mode [12, 25, 26]. Moreover, dozens of successful examples have been reported based on these strategies [9, 11, 19, 23]. Here, we discuss

Polypharmacology: Strategies for Multi-Target Drug Discovery, First Edition. Edited by Jens-Uwe Peters.

how to choose an appropriate strategy based on known information and imitate similar examples to design multitargeted ligands. The main design idea of multitargeted ligands is to obtain the second activity while maintaining the original activity. The key pharmacophores of MTDLs should be able to maintain the interaction with the original target, while being compatible with the second target [11]. First, the key pharmacophore elements of individual ligands should be identified and analyzed before the promiscuity of individual ligands. Then, an appropriate strategy should be employed to yield hybrid molecules according to the characteristics of key pharmacophore elements and scaffold structures. Moreover, the activities of the two targets must be adjusted to achieve an optimal ratio [4].

According to the similarities of pharmacophoric elements and scaffold structures, a dual-target inhibitor can be obtained from two lead molecules in several ways (Figure 16.1). If the binding sites of two targets can bind ligands with similar geometry and pharmacophore distribution, a dual-target inhibitor can be designed by merged-pharmacophore mode [27–30]. Two similar or partially similar compounds can be merged into one molecular skeleton with key pharmacophore elements, such as ionic bond, hydrogen bond, and hydrophobic interaction [27–31]. These dual-target inhibitors were often obtained with small structural changes but inspiring designs.

If the two lead compounds have no similar pharmacophoric elements or the binding regions of the two target proteins are not similar, the merged-pharmacophore strategy is difficult to implement. Under the premise of maintaining the original pharmacophore binding, the pharmacophores of the two lead compounds can be fused directly or connected with a linker. If the binding regions are large or adjacent to the solution, two key pharmacophores can be fused directly [32–34]. If the binding regions are small and deep inside the protein, the linked-pharmacophore strategy might be the only way to design dual-target inhibitors [6, 35].

16.2.1 Similar Pharmacophoric Elements

Ligands with clear binding modes can be preferentially selected as lead compounds. X-ray structure information is essential for identifying the key pharmacophore elements, such as crucial hydrogen bonds, charges, and hydrophobic groups [36, 37]. If the selected ligands have structural similarities, including similar functional groups or molecular backbones, the merged-pharmacophore method can be employed to make the pharmacophores overlap each other.

The merged-pharmacophore method is to superimpose the same structural parts of different ligands or to modify one of the ligands to obtain a new compound that integrates another pharmacophore. This strategy is the most elegant way to design MTDLs. Compared with the linked and fused molecules, the merged molecules have higher ligand efficiency, smaller molecular weight, and better physical and chemical properties. However, the design of merged molecules is a very difficult task. The merged sites and manners should be carefully evaluated and optimized to avoid mutual influence. Several successful examples of merged molecules with similar pharmacophoric elements are listed as follows.

16.2.1.1 HDAC and HMGR

Histone deacetylases (HDACs) are important targets for the treatment of cancer and neurodegenerative diseases [38]. Typically, the classical pharmacophore for

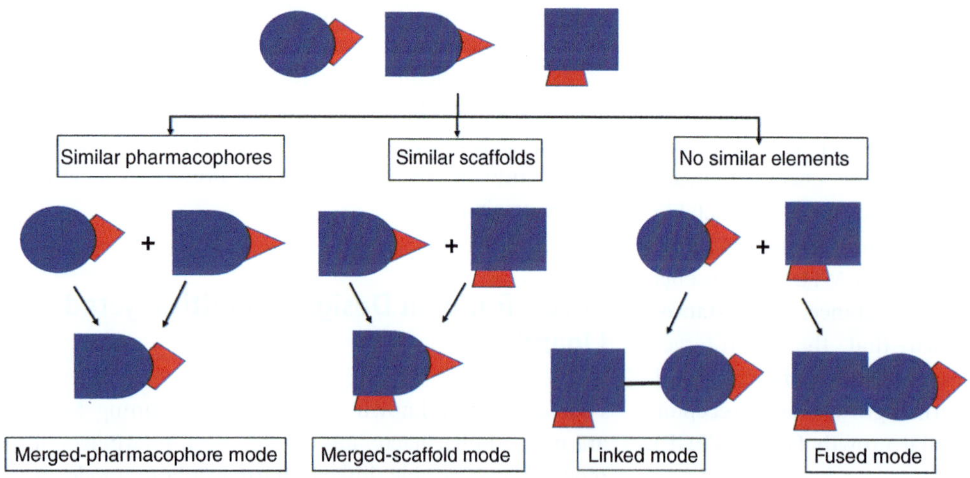

Figure 16.1 The design strategy of dual-target inhibitors. The blue graphics represent the molecular scaffold, and the red graphics represent the pharmacophoric elements.

HDACi, such as vorinostat (suberoylamilide hydroxamic acid [SAHA]; **1**), contains a cap group, a linker, and a hydroxamic acid group. Among them, the most critical one is the terminal hydroxamic acid group, which coordinates with zinc ion at the bottom of the HDAC pocket (Protein Data Bank [PDB] code 1C3S) [39]. Statins, such as lovastatin (**2**), can effectively reduce total cholesterol and low-density lipoprotein by inhibiting the activity of 3-hydroxy-3-methylglutaryl-CoA reductase (HMGR; PDB code 1HW9) [40]. The terminal carboxylate and hydroxyls (HMG moiety) of **2** form a hydrogen-bonding network with the enzyme active site of HMGR, which is characterized by a loop (residues 682–694). The rigid hydrophobic groups of the statins are situated in a shallow groove of HMGR [41]. *In vitro* experiments showed that the combination of HDACi and HMGRi can synergistically induce HeLa cell apoptosis [42].

SAHA (**1**) and lovastatin (**2**) share similar pharmacophore elements, including a hydrophilic head, a flexible long chain, and a hydrophobic tail. The hydroxamic acid of **1** is a bioisostere of carboxylic acid of **2**. By merging these similar pharmacophores of two compounds, the combined molecule **3** can act as a dual-targeting inhibitor of HDAC–HMGR (Figure 16.2a) [27]. The hydroxamate group of **3** consistently inserts into the catalytic pocket of HDAC, while the hydrophobic moiety can be used as a cap group. *In vitro* results showed that compound **3** can selectively kill A549 lung cancer cells (half-maximal inhibitory concentration [IC_{50}] = 18.2 μM).

16.2.2 Similar Scaffold Structures

More often, two ligands might only have "compatibility districts" that have little effect on activity instead of similar crucial functional groups. These compatibility districts usually have similar skeleton structures, and their changes have little effect on the binding of ligands to proteins. After these similar compatibility districts are merged, the functional groups of the two ligands can be added at appropriate positions, respectively.

16.2.2.1 FXR and sEH

In view of the clinical efficacy of farnesol X receptor (FXR) activation in reducing lipid accumulation and the important role of soluble epoxide hydrolase (sEH) inhibition in anti-hepatitis and anti-steatosis, dual regulation of FXR and sEH may be a new strategy for the treatment of nonalcoholic fatty liver and nonalcoholic steatohepatitis. Schmidt et al. combined sEH inhibitor **4** and FXR agonist **5** to produce a dual modulator (**6**), which can be used to treat nonalcoholic steatohepatitis and related metabolic diseases (Figure 16.2) [30]. The active site in sEH is formed by a channel between two large, solvent-accessible pockets (PDB code 3I28) [43]. The amide group of sEH inhibitor **4** forms two hydrogen bonds with the residues Tyr383 and Asp335 [44]. The 4-cyano-phenyl group forms an π-stacking interaction with Trp336. The 4-fluorophenyl group occupies a deep hydrophobic pocket, whereas the 4-methanesulphonyl substituted phenyl group is directed toward the solvent.

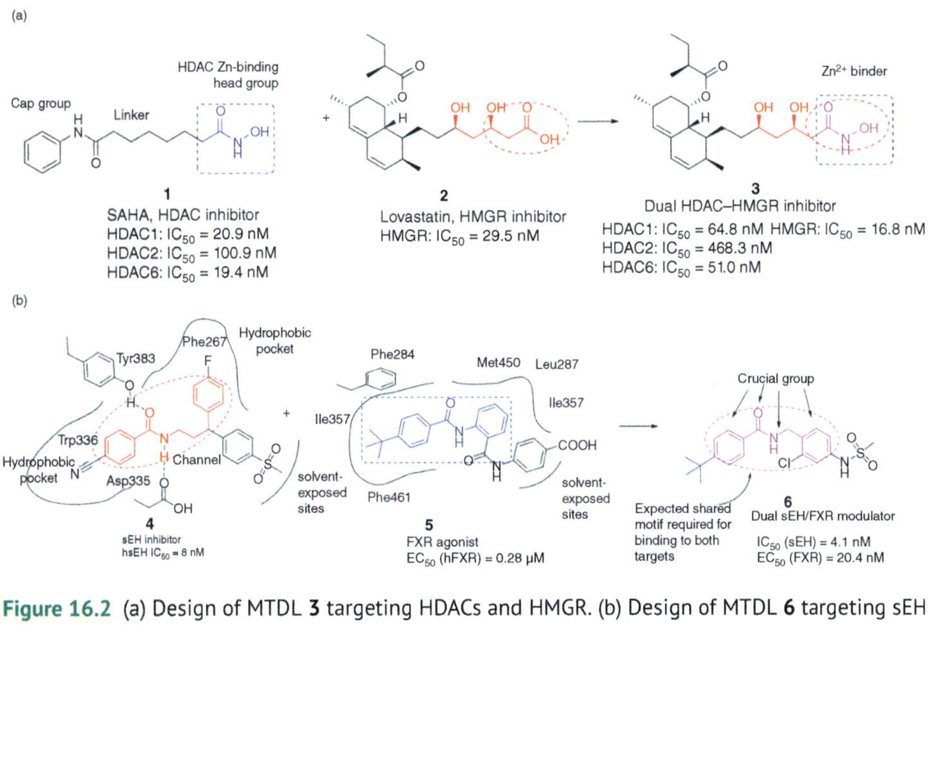

Figure 16.2 (a) Design of MTDL **3** targeting HDACs and HMGR. (b) Design of MTDL **6** targeting sEH and FXR.

The ligand binding pocket in the full agonistic conformation of FXR is narrow and long (PDB code 4QE8). The critical lipophilic substituents of **5**, the 4-tert-butylbenzoyl group and the central aromatic ring, were buried in two hydrophobic pockets, respectively. The benzoic acid is exposed to the solvent without any significant interaction with the protein [45]. The sEH inhibitor **4** containing the *N*-benzylamide group has some structural similarities with the FXR agonist **5** [46]. These similar scaffold structures were extracted and merged as the key structure of the dual modulator. Other critical pharmacophore elements were added to obtain a potent dual modulator (**6**), which could activate FXR and inhibit sEH *in vitro* and *in vivo* [30].

16.2.3 Targets Lacking Similar Ligands or Similar Endogenous Substrates

In these merged MTDLs, multiple functional groups are integrated into a compatible molecular scaffold. In order to allow such pharmacophore hybridization, similar pharmacophores or scaffolds should be present in two single-target ligands. However, most selected ligands that bind different targets do not have structural similarities. Without similar pharmacophores or scaffold structures, two selected lead compounds can be connected directly (fused) or by using a linker to obtain a new molecule that retains most of the original key pharmacophore elements.

For linked-pharmacophore mode, the biggest advantage is that such linked molecules are relatively easy to obtain. The key pharmacophores of the original ligands and their activities on the multiple targets can almost be retained with only a linker. However, this method also has obvious shortcomings. If the linker is a stable, non-cleavable fragment, the dual-target compound can recognize and bind to two or multiple targets separately as an integral molecule *in vivo*. In this case, the properties of these linkers, such as length, hydrophobicity, and rigidity, may affect the overall efficacy of MTDLs, including physical and chemical properties, and the degree of adaptation to the two targets. These linked molecules often have an increase in volume and molecular weight and a decrease in ligand efficiency, which might make their properties deviate from Lipinski's "rule of five" [47].

The active binding regions of many targets penetrate deep into the proteins and connect to the peripheral binding regions through long channels or narrow gorges. For example, the active region of acetylcholinesterase (AChE) spans 20 Å from the outer peripheral binding site to the catalytic triad, deep within the molecule [48, 49]. The catalytic site of HDAC is an 11-Å tubular channel with a zinc ion at the bottom [50]. Targeting such protein with narrow throat area, the principles of dual-target molecular design are that (a) the designed hybrid can reproduce the interaction between the key pharmacophore and the original catalytically active site, (b) the flexible linker spans across the gorge, and (c) the added fragment acts as a binder for the peripheral binding site.

16.2.3.1 AChE and GSK-3β

Tacrine (**7**) is the first AChE inhibitor used in clinical trials for the treatment of AD patients [19]. In the complex with AChE, the acridine of **7** is sandwiched between the aromatic rings of Trp-84 and Phe-330 in the "anionic" subsite of the active site (PDB code 1ACJ). Its primary amine group is located at the solvent interface and points toward the solvent through a narrow gorge [48]. The pyridine carboxamide of glycogen synthase kinase-3β inhibitor (GSK-3βI, **8**) acts as a hinge to form two hydrogen bonds with V135. Thiazolyl primary amide is another key pharmacophore that forms a hydrogen bond with Lys85 (PDB code 4PTC) [51]. The primary amide exposed to the solvent is the best position to connect the primary amine of **7** via an appropriate linker.

As the main pathogenesis of AD, GSK-3β and AChE are ideal target combinations for multi-target methods. Jiang et al. designed a new multi-target ligand **9** by connecting **7** to the solvent-exposed area of **8** with an alkylene diamine tether as a linker (Figure 16.3) [35]. Docking studies showed that the position and orientation of the tacrine fragment of **9** are consistent with those of **7** in the AChE active-site gorge. The flexible chain threads through the narrow gorge, and the pyridoxathiazole fragment is located in the peripheral anion site (PAS) of AChE binding groove. In the complex with GSK-3β, pyridine carboxamide pharmacophore maintains GSK-3β kinase domain critical binding components, occupying the adenosine triphosphate (ATP)-binding site. Compound **9** significantly improved the cognitive disorder of Institute for Cancer Research (ICR) mice treated with scopolamine and showed lower liver toxicity than **7**.

16.2.3.2 IN and RT

Rational design of dual-target inhibitors of integrase (IN) and reverse transcriptase (RT) is an effective strategy for the treatment of human immunodeficiency virus (HIV). IN binding requires minimally a diabetic ketoacidosis (DKA) type of metal-chelating functionality and a hydrophobic aromatic ring directly connected to the chelator [52, 53]. However, the binding pocket of RT is largely hydrophobic and is not a suitable environment for the binding of the DKA group. A hydrogen bond network is formed among the pyrimidine ring of tenecteplase (TNK)-651 (**10**), a water molecule, and Lys101, Glu138 in the RT inhibitor binding pocket of RT (PDB code 1RT2) [54]. A large

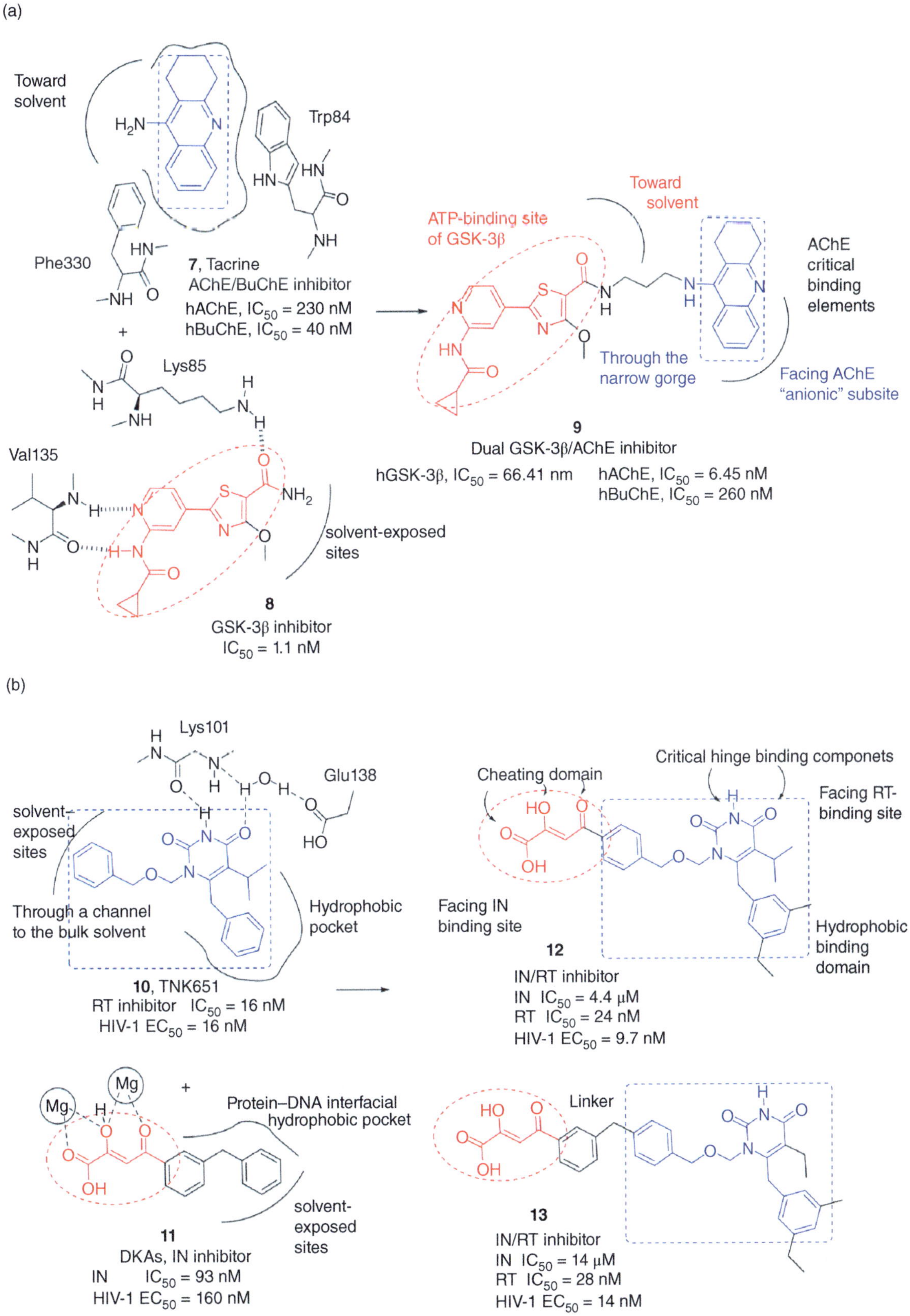

Figure 16.3 (a) Design of MTDL **9** targeting AChE and GSK-3β. (b) Design of MTDLs **12** and **13** targeting RT and IN.

(benzyloxy)methyl tail extends to the protein/solvent interface through a narrow channel. This phenyl group provides a suitable site for linking the DKA moiety to reproduce the IN inhibitory activity.

Based on RT inhibitor **10** and IN inhibitor DKAs (**11**), bifunctional inhibitors were directly fused (**12**) or combined by a linker (**13**; Figure 16.3b) [34, 55]. Both **12** and **13** showed higher potency against HIV than **10** and **11**. Docking studies showed that the DKA functional groups are located at the protein/solvent interface and may be solvated without hydrogen bonding with RT. Similarly, the methyl sulfonamide group of the RT inhibitor delavirdine is located in an open area that can tolerate structural modifications (PDB code 1KLM) [56]. Using the DKA functional group instead of the sulfonamide group can not only affect the RT inhibitory activity, but also obtain the inhibitory activity of IN [57].

16.3 Discussion and Conclusion

Incorporating the second activity into an existing single-active ligand, while maintaining affinity for the original target, is a challenging task. In order to overcome this challenge, we must identify the common characteristics, key pharmacophores, and substitutable sites of two single-target ligands to perform molecular hybridization without interfering with the original binding activity. First, each pharmacophore of the multitargeted agents should retain the ability to interact with its specific site(s) on original targets. The crucial hydrogen bonds need to be maintained, and the structure alterations of hydrophobic groups should be accommodated by the binding domain. Second, the added pharmacophores of second targets should not affect the binding of the original ones. These added groups can (i) be accommodated by the catalytic active sites; (ii) bind to the peripheral binding sites; and (iii) bind to the surface of the protein or be exposed to solution.

The design of multi-target molecules should accurately grasp the key pharmacophore of the target and look for similar pharmacophores as shared fragments. Under the premise of maintaining the orientation and spatial distance of original pharmacophores, two lead ligands can be merged, fused, or linked together. Duplicate or unnecessary fragments are removed to reduce molecular size and relative molecular mass. The flexibility and lipophilicity should not be too large to prevent the influence on PK problems such as drug absorption and distribution.

References

1 Andrew, A., Jürgen, B., and Giulio, R. (2014). Polypharmacology: challenges and opportunities in drug discovery. *J. Med. Chem.* 57: 7874–7887.

2 Morphy, R., Kay, C., and Rankovic, Z. (2004). From magic bullets to designed multiple ligands. *Drug Discov. Today* 9: 641–651.

3 Cavalli, A., Bolognesi, M.L., Minarini, A. et al. (2008). Multi-target-directed ligands to combat neurodegenerative diseases. *J. Med. Chem.* 51: 347–372.

4 Morphy, R. and Rankovic, Z. (2005). Designed multiple ligands. An emerging drug discovery paradigm. *J. Med. Chem.* 48: 6523–6543.

5 Jérémy, B., Gian Filippo, R., Vincent, S. et al. (2012). Automated design of ligands to polypharmacological profiles. *Nature* 492: 215–220.

6 Yao, L., Mustafa, N., Tan, E.C. et al. (2017). Design and synthesis of ligand efficient dual inhibitors of Janus kinase (JAK) and histone deacetylase (HDAC) based on ruxolitinib and vorinostat. *J. Med. Chem.* 60: 8336–8357.

7 Petrelli, A. and Giordano, S. (2008). From single- to multi-target drugs in cancer therapy: when aspecificity becomes an advantage. *Curr. Med. Chem.* 15: 422–432.

8 Abdel-Rahman, O. and Fouad, M. (2015). Temozolomide-based combination for advanced neuroendocrine neoplasms: a systematic review of the literature. *Future Oncol.* 11: 1275–1290.

9 Sun, G., Fan, T., Zhao, L. et al. (2017). The potential of combi-molecules with DNA-damaging function as anticancer agents. *Future Med. Chem.* 9: 403–435.

10 García, M., Virgili, M., Alonso, M. et al. (2020). Discovery of EST73502, a dual μ-opioid receptor agonist and σ 1 receptor antagonist clinical candidate for the treatment of pain. *J. Med. Chem.* 63: 15508–15526.

11 Fu, R.G., Sun, Y., Sheng, W.B., and Liao, D.F. (2017). Designing multi-targeted agents: an emerging anticancer drug discovery paradigm. *Eur. J. Med. Chem.* 136: 195–211.

12 Mondello, P., Derenzini, E., Asgari, Z. et al. (2017). Dual inhibition of histone deacetylases and phosphoinositide 3-kinase enhances therapeutic activity against B cell lymphoma. *Oncotarget* 8: 14017–14028.

13 Hartmann, M., Bibli, S.I., Tews, D. et al. (2021). Combined cardioprotective and adipocyte browning effects promoted by the eutomer of dual sEH/PPARγ modulator. *J. Med. Chem.* 64: 2815–2828.

14 Ko, K.S., Steffey, M.E., Brandvold, K.R., and Soellner, M.B. (2013). Development of a chimeric c-Src kinase and HDAC inhibitor. *ACS Med. Chem. Lett.* 4: 779–783.

15 Choong-Kun, L., Myung Eun, L., Won Suk, L. et al. (2015). Dovitinib (TKI258), a multi-target angiokinase inhibitor, is effective regardless of KRAS or BRAF mutation status in colorectal cancer. *Am. J. Cancer Res.* 5: 72–86.

16 Lu, J.J., Pan, W., Hu, Y.J., and Wang, Y.T. (2012). Multi-target drugs: the trend of drug research and development. *PLoS ONE* 7: e40262.

17 Jin, G., Yan, M., Liu, K. et al. (2020). Discovery of a novel dual-target inhibitor against RSK1 and MSK2 to suppress growth of human colon cancer. *Oncogene* 39: 6733–6746.

18 Liang, Q., Wang, J., Zhao, L. et al. (2021). Recent advances of dual FGFR inhibitors as a novel therapy for cancer. *Eur. J. Med. Chem.* 214: 113205.

19 Tumiatti, V., Minarini, A., Bolognesi, M.L. et al. (2010). Tacrine derivatives and Alzheimer's disease. *Curr. Med. Chem.* 17: 1825–1838.

20 Ferreira, J.P.S., Albuquerque, H.M.T., Cardoso, S.M. et al. (2021). Dual-target compounds for Alzheimer's disease: natural and synthetic AChE and BACE-1 dual-inhibitors and their structure-activity relationship (SAR). *Eur. J. Med. Chem.* 221: 113492.

21 Malafaia, D., Albuquerque, H.M.T., and Silva, A.M.S. (2021). Amyloid-β and tau aggregation dual-inhibitors: a synthetic and structure-activity relationship focused review. *Eur. J. Med. Chem.* 214: 113209.

22 Qandil, A.M. (2012). Prodrugs of nonsteroidal anti-inflammatory drugs (NSAIDs), more than meets the eye: a critical review. *Int. J. Mol. Sci.* 13: 17244–17274.

23 Zimmermann, G.R., Lehár, J., and Keith, C.T. (2007). Multi-target therapeutics: when the whole is greater than the sum of the parts. *Drug Discov. Today* 12: 34–42.

24 Zhou, J., Jiang, X., He, S. et al. (2019). Rational design of multi-target-directed ligands: strategies and emerging paradigms. *J. Med. Chem.* 62: 8881–8914.

25 Proschak, E., Stark, H., and Merk, D. (2019). Polypharmacology by design: a medicinal chemist's perspective on multitargeting compounds. *J. Med. Chem.* 62: 420–444.

26 Hopkins, A.L., Mason, J.S., and Overington, J.P. (2006). Can we rationally design promiscuous drugs? *Curr. Opin. Struct. Biol.* 16: 127–136.

27 Chen, J.-B., Chern, T.-R., Wie, T.-T. et al. (2013). Design and synthesis of dual-action inhibitors targeting histone deacetylases and 3-hydroxy-3-methylglutaryl coenzyme A reductase for cancer treatment. *J. Med. Chem.* 56: 3645–3655.

28 Tavera-Mendoza, L.E., Quach, T.D., Dabbas, B. et al. (2008). Incorporation of histone deacetylase inhibition into the structure of a nuclear receptor agonist. *Proc. Natl. Acad. Sci. U.S.A* 105: 8250–8255.

29 Jing, T., Maddali, K., Dreis, C.D. et al. (2011). N-3 hydroxylation of pyrimidine-2,4-diones yields dual inhibitors of HIV reverse transcriptase and integrase. *ACS Med. Chem. Lett.* 2: 63–67.

30 Schmidt, J., Rotter, M., Weiser, T. et al. (2017). A dual modulator of farnesoid x receptor and soluble epoxide hydrolase to counter nonalcoholic steatohepatitis. *J. Med. Chem.* 60: 7703–7724.

31 Huang, W., Lu, D., Yu, H. et al. (2010). Dual-target-directed 1,3-diphenylurea derivatives: BACE 1 inhibitor and metal chelator against Alzheimer's disease. *Bioorg. Med. Chem.* 18: 5610–5615.

32 Kogen, H., Toda, N., Tago, K. et al. (2002). Design and synthesis of dual inhibitors of acetylcholinesterase and serotonin transporter targeting potential agents for Alzheimer's disease. *Org. Lett.* 4: 3359–3362.

33 Siavosh, M., Stefan, D., Andreas, S. et al. (2009). Design of chimeric histone deacetylase- and tyrosine kinase-inhibitors: a series of imatinib hybrides as potent inhibitors of wild-type and mutant BCR-ABL, PDGF-Rbeta, and histone deacetylases. *J. Med. Chem.* 52: 2265–2279.

34 Wang, Z., Bennett, E.M., Wilson, D.J. et al. (2007). Rationally designed dual inhibitors of HIV reverse transcriptase and integrase. *J. Med. Chem.* 50: 3416–3419.

35 Jiang, X.Y., Chen, T.K., Zhou, J.T. et al. (2018). Dual GSK-3β/AChE inhibitors as a new strategy for multi-targeting anti-Alzheimer's disease drug discovery. *ACS Med. Chem. Lett.* 9: 171–176.

36 Sun, B., Dong, Y., An, Y. et al. (2020). Design, synthesis and bioactivity evaluation of novel arylalkene-amide derivatives as dual-target antifungal inhibitors. *Eur. J. Med. Chem.* 205: 112645.

37 An, Y., Dong, Y., Liu, M. et al. (2020). Novel naphthylamide derivatives as dual-target antifungal inhibitors: Design, synthesis and biological evaluation. *Eur. J. Med. Chem.* 210: 112991.

38 Falkenberg, K.J. and Johnstone, R.W. (2014). Histone deacetylases and their inhibitors in cancer, neurological diseases and immune disorders. *Nat. Rev. Drug Discov.* 13: 673.

39 Marks, P.A. and Breslow, R. (2007). Dimethyl sulfoxide to vorinostat: development of this histone deacetylase inhibitor as an anticancer drug. *Nat. Biotechnol.* 25: 84–90.

40 Hawk, E. and Viner, J.L. (2005). Statins and cancer--beyond the "one drug, one disease" model. *N. Engl. J. Med.* 352: 2238–2239.

41 Istvan, E.S. and Deisenhofer, J. (2001). Structural mechanism for statin inhibition of HMG-CoA reductase. *Science* 292: 1160–1164.

42 Lin, Y., Lin, J., Chou, C. et al. (2008). Statins increase p21 through inhibition of histone deacetylase activity and release of promoter-associated HDAC1/2. *Cancer Res.* 68: 2375–2383.

43 Gomez, G.A., Morisseau, C., Hammock, B.D., and Christianson, D.W. (2004). Structure of human epoxide hydrolase reveals mechanistic inferences on bifunctional catalysis in epoxide and phosphate ester hydrolysis. *Biochemistry* 43: 4716–4723.

44 Eldrup, A.B., Soleymanzadeh, F., Taylor, S.J. et al. (2009). Structure-based optimization of arylamides as inhibitors of soluble epoxide hydrolase. *J. Med. Chem.* 52: 5880–5895.

45 Merk, D., Lamers, C., Ahmad, K. et al. (2014). Extending the structure-activity relationship of anthranilic acid derivatives as farnesoid X receptor modulators: development of a highly potent partial farnesoid X receptor agonist. *J. Med. Chem.* 57: 8035–8055.

46 Kompa, A.R., Wang, B.H., Xu, G. et al. (2013). Soluble epoxide hydrolase inhibition exerts beneficial anti-remodeling actions post-myocardial infarction. *Int. J. Cardiol.* 167: 210–219.

47 Lipinski, C.A., Lombardo, F., Dominy, B.W., and Feeney, P.J. (1997). Experimental and computational approaches to estimate solubility and permeability in drug discovery and development settings. *Adv. Drug Deliv. Rev.* 23: 3–25.

48 Harel, M., Kryger, G., Rosenberry, T.L. et al. (2000). Three-dimensional structures of Drosophila melanogaster acetylcholinesterase and of its complexes with two potent inhibitors. *Protein Sci.* 9: 1063–1072.

49 Harel, M., Schalk, I., Ehret-Sabatier, L. et al. (1993). Quaternary ligand binding to aromatic residues in the active-site gorge of acetylcholinesterase. *Proc. Natl. Acad. Sci. U.S.A* 90: 9031–9035.

50 Rabal, O., Sanchez-Arias, J.A., Cuadrado-Tejedor, M.D.M. et al. (2016). Design, synthesis, and biological evaluation of first-in-class dual acting HDAC and PDE5 inhibitors for the treatment of Alzheimer's disease. *J. Med. Chem.* 59: 8967–9004.

51 Sivaprakasam, P., Han, X., Civiello, R.L. et al. (2015). Discovery of new acylaminopyridines as GSK-3 inhibitors by a structure guided in-depth exploration of chemical space around a pyrrolopyridinone core. *Bioorg. Med. Chem. Lett.* 25: 1856–1863.

52 Dayam, R., Sanchez, T., Clement, O. et al. (2005). β-diketo acid pharmacophore hypothesis. 1. Discovery of a novel class of HIV-1 integrase inhibitors. *J. Med. Chem.* 48: 111–120.

53 Hazuda, D.J., Felock, P., Witmer, M. et al. (2000). Inhibitors of strand transfer that prevent integration and inhibit HIV-1 replication in cells. *Science* 287: 646–650.

54 Hopkins, A.L., Ren, J., Esnouf, R.M. et al. (1996). Complexes of HIV-1 reverse transcriptase with inhibitors of the HEPT series reveal conformational changes relevant to the design of potent non-nucleoside inhibitors. *J. Med. Chem.* 39: 1589–1600.

55 Wang, Z., Tang, J., Salomon, C.E. et al. (2010). Pharmacophore and structure–activity relationships of integrase inhibition within a dual inhibitor scaffold of HIV reverse transcriptase and integrase. *Bioorg. Med. Chem.* 18: 4202–4211.

56 Esnouf, R.M., Ren, J., Hopkins, A.L. et al. (1997). Unique features in the structure of the complex between HIV-1 reverse transcriptase and the bis (heteroaryl) piperazine (BHAP) U-90152 explain resistance mutations for this nonnucleoside inhibitor. *Proc. Natl. Acad. Sci.* 94: 3984–3989.

57 Wang, Z. and Vince, R. (2008). Design and synthesis of dual inhibitors of HIV reverse transcriptase and integrase: introducing a diketoacid functionality into delavirdine. *Bioorg. Med. Chem.* 16: 3587–3595.

17

Predicting Polypharmacology with Web-Based Tools

Maedeh Darsaraee, Sacha Javor, and Jean-Louis Reymond

17.1 Introduction

While a very large number of approaches have been reported to perform target prediction for small molecules [1, 2], only a few have resulted in tools that are freely available online for use. In this chapter, we review a series of such web-based tools that are currently active. All of these tools take a query molecule, usually drawn in a browser using an online drawing tool such as JSME [3], or entered as SMILES [4], and assign possible targets to this query based on its molecular similarity to known, target-annotated molecules, which are taken from public datasets such as ChEMBL [5] and BindingDB [6]. These datasets contain better accessible molecule-target associations than the very general archive PubChem [7].

Molecular similarities are quantified using various molecular fingerprints (Table 17.1) [18–20]. As for ligand-based virtual screening methods [21], target prediction generally works best when using the classical Morgan fingerprints, also called extended connectivity fingerprints (ECFP) or circular fingerprints [8, 22, 23], reflecting the fact that biological activity often depends on the presence of very specific substructural features in molecules. Other substructure fingerprints used in web-based target prediction include multilevel neighborhood of atoms (MNA) [9] and FP2, an OpenBabel version of the Daylight substructure fingerprint [10] Substructure similarity, however, sometimes focuses too much on details and misses overall similarities in molecular shape and pharmacophores as important determinants of biological activity [24, 25]. Accordingly, many target prediction tools also measure molecular similarities using shape and pharmacophore fingerprints, such as those based on atom pairs [11, 12, 26–28]. Even more global descriptions of molecules such as Molecular ACCess System (MACCSkeys) [13], Electrotopological State (E-state) [14], molecular quantum numbers (MQN) [15] or SMILES fingerprint (SMIfp) [16] may provide alternative, relevant comparisons in terms of target prediction. Fusion fingerprints including, multiple fingerprints simultaneously, can also be employed [17].

Identifying a series of molecules with high molecular fingerprint similarity to the query molecule from a reference database (ChEMBL or BindingDB in the present tools) is computationally very fast. Each tool then uses specific data integration methods and statistical models to deduce the most probable targets for the query molecule based on the targets associated with the molecules identified by the similarity search. These methods are detailed below for each of the reviewed web-based tools presented in this chapter (Table 17.2), which cover, in chronological order of their first release, prediction of activity spectra for substances (PASS), [29, 30] Similarity Ensemble Approach (SEA), [31, 32] Super-PRED [33, 34], TargetHunter [35], SwissTargetPrediction [36, 37], TargetNet [38], Polypharmacology Browser (PPB) [17], and Polypharmacology Browser 2 (PPB2) [39].

17.2 PASS

17.2.1 Data and Methodology

PASS was one of the very first target prediction tools to be reported [29], and is currently available as a free online tool [30]. It performs a target prediction using data from ChEMBL database version 19. For the filtration of the retrieved data, only those compounds with measured bioactivity types of AC_{50}, EC_{50}, IC_{50}, K_i, K_d, and target's activity percentage were used. The next step of data filtration involves the evaluation of activity, relying on median values of different activity types. For this determination, a cut-off value of 10 μM in compound activity is considered. In cases where this specific concentration data is unavailable, the authors utilized the median values for activity types expressed as a percentage of the target's

Polypharmacology: Strategies for Multi-Target Drug Discovery, First Edition. Edited by Jens-Uwe Peters.

Table 17.1 Brief description of the molecular fingerprints mentioned in this chapter.

Fingerprint type	Fingerprint	Description
Substructure	ECFP2, ECFP4, ECFP6 [8]	Extended-connectivity circular fingerprints with bond diameters 2, 4, and 6 with a length of 1024 or 2048 bits.
	MNA [9]	A set of 2D molecular descriptors generated from multilevel neighborhoods of atom.
	FP2 [10]	A 1024-bit binary path-based daylight-type substructure fingerprint that captures structural features with a path length of 7 in a molecule.
Molecular shape and pharmacophore	ElectroShape [11]	An 18-dimensional molecular descriptor that encapsulates information about shape, electrostatics, and lipophilicity in a molecule.
	APfp [12]	A 21-dimensional atom-pair fingerprint which recognizes molecular shape characteristics.
	Xfp [12]	A 55-dimensional atom category extended atom-pair fingerprint that discerns pharmacophores by considering atom pairs in its representation.
Indices	MACCSkeys [13]	Molecular ACCess System fingerprint which is a binary molecular descriptor that encodes the presence or absence of predefined structural keys.
	E-state [14]	A topological fingerprint based on Electrotopological State indices which encodes the intrinsic electronic state of the atom as perturbed by the electronic influence of all other atoms in the molecule.
	MQN [15]	A 42-dimensional molecular quantum numbers scalar fingerprint which quantifies atoms, bonds, polarity, and ring features, discerning constitution, topology, and molecular shape.
	SMIfp [16]	A 34-dimensional scalar fingerprint, SMIfp counts the occurrence of characters in SMILES representation while perceives features such as rings, aromaticity, and polarity.
Combined	Ffp1 [17]	Fusion fingerprint: combination of Xfp, SMIfp, and Sfp.
	Ffp2 [17]	Fusion fingerprint: combination of Xfp, SMIfp, and MQN.
	Ffp3 [17]	Fusion fingerprint: combination of Xfp, SMIfp, Sfp, and ECFP4.
	Ffp4 [17]	Fusion fingerprint: combination of Xfp, SMIfp, Sfp, and MQN.

activity, using a cut-off value set at 50%. Furthermore, only targets with at least 10 active compounds are considered. Eventually, the final dataset consists of 2507 unique protein targets, 589,107 unique compounds, and 1,563,481 target-compound relations.

To qualify for PASS analysis, a query molecule must have at least three carbon atoms, a molecular weight less or equal to 1250 Da, be uncharged, and consist of only one component. The compound chemical structure is represented as a set of MNA representing substructures present in the molecule [9]. PASS uses Naïve Bayesian (NB) classifier as its mathematical approach. For a given query molecule, NB uses the Bayesian theorem of conditional probability and calculates and ranks the possibility of the presence of targets in the predictions. The formula for the Bayesian theorem is as follows:

$$P(A|B) = \frac{P(B|A) \cdot P(A)}{P(B)} \tag{17.1}$$

where $P(A|B)$ is the probability of event A (target) occurring given that event B (individual bits in a molecular fingerprint) has occurred (posterior probability: a compound to be active on target A), $P(B|A)$ shows the probability of event B occurring given that event A has occurred (prior probabilities calculated based on fingerprints of known bioactive compounds active against target A), $P(A)$ represents the prior probability of event A (the relative frequency of target A considering all the other targets in the database), and $P(B)$ is the probability of event B occurring.

Considering multiple evidence (fingerprint bits: $B_{i=1} \dots B_n$), the probability for each piece of evidence given that event A is true ($P(B_i|A)$) can be calculated using the following formula in Eq. (17.2), where $\alpha = 1$ is the Laplace smoothing:

$$P(B_i|A) = \frac{((count\ of\ evidence\ B_i\ given\ A) + \alpha)}{((total\ count\ of\ evidence\ A) + \alpha\ (number\ of\ evidence))} \tag{17.2}$$

For evaluation of PASS target prediction performance, two main validation methods are performed. The evaluation of prediction accuracy involves the use of Area under

Table 17.2 Summary of the datasets, methods, and final outputs of eight web-based target prediction tools that are discussed in this chapter. These tools are ranked chronologically, based on the releasing year of their first version.

Web servers (URL)	Data	Method	Output
PASS [29, 30] (http://www.way2drug.com/PASSonline)	ChEMBL 19: 2507 protein targets, 589,107 unique compounds, and 1,563,481 datapoints.	MNA atomic neighborhood descriptors using Bayesian analysis	A list of proteins predicted as possible direct targets consisting of direct interactions and probable indirect interactions.
SEA [31, 32] (https://sea.bkslab.org/)	ChEMBL 21: 4161 targets and 65,241 unique ligands.	Cross-target similarity maps based on an extreme value distribution using Daylight fingerprints	A list of predicted targets annotated with *p*-values and Tanimoto coefficients of the most similar ligand.
Super-PRED [33, 34] (https://prediction.charite.de/index.php)	ChEMBL 29: 365,719 compounds, 2353 targets, and 500,979 datapoints.	Logistic regression machine learning models based on Morgan fingerprints.	A table containing a list of strong binders displayed with their corresponding assay binding data and two accuracy scores and visualization links.
TargetHunter [35] (https://www.cbligand.org/TargetHunter/)	ChEMBL 11: 117,535 unique compounds against 794 targets.	A similarity search using ECFP4, ECFP6, and FP2 fingerprints.	A list of predicted targets based on the compound's chemical structure and its similarity to known compounds in the chemogenomic database.
SwissTargetPrediction [36, 37] (http://www.swisstargetprediction.ch)	ChEMBL 23: 376,42 small molecules against 3,068 targets and 580,496 interactions.	Multiple logistics regression models based on the combination of 2D and 3D similarity.	A list of the maximum 100 most probable predicted targets.
TargetNet [38] (http://targetnet.scbdd.com/)	Binding DB: 109,061 compounds against 623 protein targets and 115,257 datapoints.	Seven NB models built with FP2, MACCS, E-State, FP4, Daylight, ECFP2, ECFP4, and ECFP6 fingerprints.	A list of all 623 protein targets in the database sorted by the probability of the ligand-target interactions.
Polypharmacology Browser (PPB) [17] (https://webtools.gdb.tools/PPB/)	ChEMBL 21: 871,673 bioactive compounds against 4613 targets.	Similarity searching using APfp, SMIfp, MQN, Xfp, Sfp, and ECFP4 fingerprints.	A merged list of the top 20 targets for each fingerprint.
Polypharmacology Browser 2.0 (PPB2) [39] (https://ppb2.gdb.tools/)	ChEMBL 22: 344,163 compounds, 1720 protein targets.	Machine learning techniques such as DNN, NB, NN, and a combination of NN + NB using APfp, SMIfp, MQN, Xfp, Sfp, and ECFP4 fingerprints.	A list of maximum 20 most probable predicted targets.

the curve (AUC) measurements through leave-one-out cross-validation and 20-fold cross-validation.

17.2.2 Evaluation

To evaluate PASS, chemical structures of 758 compounds for 206 targets retrieved from ChEMBL [5] were used as the external test set, in which the number of active compounds per target ranged from 1 to 59. The results showed the average accuracy and specificity around 92%, whereas the average value of ROC AUC and sensitivity is 89.7% and 75.9%, respectively. Diving more into the details, for 25 targets with at least 10 known drugs, the average values of ROC AUC were around 93%. Moreover, in the case of 55 targets possessing a minimum of five drugs, both the mean and median ROC AUC values were 91% and 93%, respectively.

In a second evaluation of this tool, benchmarking datasets containing the information of seven targets and their corresponding active and inactive compounds were used. Analyzing the number of active and inactive compounds, the number of inactive compounds was noticeably higher than the number of active ones. In this case, active compounds with their corresponding data were excluded from the training set. Then, the quality of prediction for the seven targets was estimated, in which the average value of ROC AUC, sensitivity, specificity, and accuracy for these targets was 90.84%, 73.13%, 88.63%, and 88.62%, respectively.

One of the main applications of PASS Targets is the prediction of the interaction spectrum for kinase inhibitors, which often have low selectivity across the kinome [40], as well as the identification of potential off-targets that share structural characteristics with kinases, as discussed in the updated version of the website [30].

17.3 SEA

17.3.1 Data and Methodology

SEA [31] uses compounds and targets extracted from ChEMBL version 21. In the data filtration, duplicated compounds, salts, and fragments, as well as non-molecular targets and targets with less than five active compounds, were excluded, resulting in a subset of 4161 target proteins and 65,241 unique ligands.

Employing a 2D ligand-based similarity ensemble approach, SEA characterizes each target by examining its active ligand set. To compare the query molecules with the ligand sets, $T_{c(A,B)}$ (Tanimoto similarity coefficient of ECFP4 fingerprints of the query molecule A and compound B in the database) is calculated using the formula in Eq. (17.3):

$$T_{c(A,B)} = \frac{N_C}{(N_A + N_B - N_C)} \tag{17.3}$$

where N_C is the number of ON bits common between A and B, and N_A and N_B are the number of ON bits in the query and a compound in the database.

As input, SEA takes a SMILES representation of the query molecule, and it gives the user a list of predicted targets ranked by the *p*-values and T_c (Tanimoto coefficient) of the most similar compounds related to the predicted targets to a query molecule. Also, by clicking on the corresponding maximum T_c value, the user can access information about the most similar ligand associated with each target.

The potential protein targets are predicted by summing up the Tanimoto similarities above 0.57 and obtaining the raw scores, calculating *Z-score,* and converting *Z-scores* to *p*-values. For each target, the raw score is converted to a *Z-score* using the formula in Eq. (4):

$$Z - score = \frac{(Raw\ score\ (A,B) - Mean_{(s)})}{SD_{(s)}} \tag{17.4}$$

where *Raw score* (A,B) is the raw score for ligand set A and ligand B pair and s is the product of size of ligand set A and ligand set B. $Mean_{(s)}$ and $SD_{(s)}$ can be retrieved by plotting the means and standard deviations of the raw score. The *p*-value for a target indicates the probability of by chance similarity between ligands of a target and a query molecule, and it is calculated following the formula in Eq. (17.5):

$$p - value(z) = 1 - e^{x(z)}\ considering\ z \le 28\ and\ x(z) = -e^{\frac{-z\pi}{\sqrt{6}-\Gamma'(1)}} \tag{17.5}$$

where $\Gamma'(1)$ is the Euler–Mascheroni constant (≈0.577215665). When $z > 28$, to avoid exceeding numerical precision Taylor expansion approximation in Eq. (17.6) can be used:

$$p - value(z) = -x(z) - \frac{x(z)^2}{2} - \frac{x(z)^3}{6}\ considering\ z > 28\ and\ x(z) = -e^{\frac{-z\pi}{\sqrt{6}-\Gamma'(1)}} \tag{17.6}$$

17.3.2 Evaluation

Not only was SEA initially employed to find relationships among different target proteins by comparing their ligand sets [31], but it was also later used to identify new targets for known drugs [32]. Among 30 newly proposed drug-target interactions identified by SEA, 77% were confirmed. For instance, SEA confirmed interactions like the adrenergic β1-receptor (β1-AR) with the drug Prozac (a transporter inhibitor), the 5-hydroxytryptamine (5-HT) transporter with Vadilex (an ion channel drug), and the histamine H4 receptor with Rescriptor (an enzyme inhibitor). Additionally, SEA was used in a comprehensive study to assess the similarity of 695 drugs binding to 73 targets associated with adverse drug reactions, and approximately 50% of the predictions were confirmed [41].

17.4 Super-PRED

17.4.1 Data and Methodology

Super-PRED [33] offers a knowledge-based method for target prediction of various compounds which is based on machine learning models. The dataset that is used in Super-PRED is retrieved from ChEMBL database version 29, focusing on binding values for *Homo sapiens* binding or functional assay types. Activity values were limited to measured IC_{50}, EC_{50}, K_i, K_d, and potency of 1000 nM or less. Data filtration resulted in 500,979 unique compound-target relations between 365,719 strong binders and 2353 targets.

To predict the targets, it is necessary to input a molecular structure into the ChemDoodle [42] web interface. Structures can be obtained by entering a PubChem [7] name, a SMILES string, loading a structure file, or using the provided drawing tools. Compared to its previous versions released in 2008 [43] and 2014 [33] the latest version of

Super-PRED, released in 2019 [34], uses instead of overall similarity searching various machine learning models, including logistic regression, linear discriminant analysis, *k*-nearest neighbors, decision trees, support vector machines, Gaussian Naïve Bayes, and random forests. One machine learning model was trained for each predictable target.

17.4.2 Evaluation

Model accuracy was evaluated through 10-fold cross-validation of the ChEMBL 29 dataset, with 82% of the target models achieving at least 85% accuracy and only 5% scoring below 70% accuracy. Due to varying performance across targets, two reported scores include the probability of interaction between the input structure and the target determined by the corresponding machine learning model, as well as the overall accuracy of the respective model. Predictions made by logistic regression based on Morgan fingerprints of length 2048 obtained the highest accuracy amongst all the machine learning models.

With Super-PRED, the identification of therapeutic targets has evolved beyond relying solely on active binders; it now incorporates experimentally confirmed non-binders extracted from the ChEMBL database. This, coupled with the machine learning approach, allows for a more accurate evaluation of structural groups influencing protein binding. Emphasizing substructure features over overall structural similarity provides an additional advantage [34].

17.5 TargetHunter

17.5.1 Data and Methodology

TargetHunter [35] performs a target prediction utilizing the data from ChEMBL version 11, consisting of 117,535 unique compounds from 794 targets. Targets were associated with more than 30 compounds and the compounds, exhibited activity values of IC_{50}, EC_{50}, K_i, and ED_{50} below 10 μM at a high confidence level for direct interactions. Consequently, the number of compounds related to each target varied between 30 and 2180. These targets encompassed diverse proteins, including enzymes, kinases, and receptors. Also, single compounds associated with multiple targets were included within this subset.

TargetHunter searches through 2D similarity based on ECFP4, ECFP6, and FP2 OpenBabel fingerprints. It assigns a target to a query molecule using multiple category Bayesian models (MCM) [44], which is based on the NBalgorithm using a Laplacian-corrected estimator, which is similar to the approach used in PASS discussed above. This model uses individual bits of molecular fingerprints as the molecular descriptors, and the contribution score of molecular descriptors in compounds (C_{d_i}) is calculated using the formula in Eq. (17.7):

$$C_{d_i} = \frac{(A_{d_i} + 1)}{\left((T_{d_i}) \times \left(\frac{A}{T}\right) + 1\right)} \tag{17.7}$$

where A_{d_i} is the total number of active compounds possessing d_i descriptor, T_{d_i} is the total number of active and inactive compounds possessing d_i descriptor, and 1 is the Laplacian-corrected estimator. With the summation of n descriptor contribution scores, the activity score of a given compound related to a certain target (C_{active}) is calculated as follows in Eq. (17.8):

$$C_{active} = \sum_{i=1}^{n} log(C_{d_i}) \tag{17.8}$$

17.5.2 Evaluation

For model evaluation, seven-fold cross-validation was performed by assigning the compounds to seven subsets and using the subsets as test sets in seven rounds. The prediction accuracy was calculated using the following formulas in Eqs. (17.9 and 17.10):

$$P_i = \sum_{j=1}^{m=3} \frac{N_{ij}}{T_i} \tag{17.9}$$

$$P_{total} = \sum_{i=1}^{794} \sum_{j=1}^{m=3} \frac{N_{ij}}{T} \tag{17.10}$$

where P_i is the prediction accuracy of the ith target, P_{total} is the prediction accuracy for 794 targets, m is the number of total three predictions made by the TargetHunter for each case, N_{ij} is the number of true predictions of ligand-target pairs for the ith target by the jth prediction, T_i is the number of testing ligand-target, pairs from the ith target and T is the total number of testing ligand-target pairs. The average accuracy of seven-fold cross-validation of MCM is 74.8% in the top 3 predictions.

17.6 SwissTargetPrediction

17.6.1 Data and Methodology

First released in 2014 [36], SwissTargetPrediction is a ligand-based target prediction web-based tool that predicts the targets based on the similarity principle. In the latest version released in 2019 [37], SwissTargetPrediction uses the dataset from ChEMBL 23 for human, mouse and rat, considering only molecules smaller than 80 heavy atoms reported to act on single protein targets with bioactivity values of K_i, K_d, IC_{50}, or EC_{50} better than 10 μM.

Compared to the first version of SwissTargetPrediction, the number of unique compounds in the entire dataset of the latest version increased to 376,342, which is 34% more than the number of compounds extracted from ChEMBL 16 in the initial dataset. Moreover, the number of protein targets and datapoints (ligand-target interactions) reached 3068 and 580,496 respectively, which shows 19% and 32% of an increase in the number of unique targets and datapoints compared to the dataset used in the initial release of SwissTargetPrediction in 2014.

SwissTargetPrediction uses ligand-based methodology based on similarity searching to find the most similar bioactive molecules to the query molecule using both 2D and 3D similarity comparisons. The 2D similarity searching compares the query molecule (*A*) with each dataset compound (*B*) using Tanimoto coefficients for the Open Babel FP2 fingerprint [10] which is defined as similarity value S_1'. For the 3D-comparison between *A* and *B*, 20 conformers are generated for each molecule, and an 18-dimensional Electroshape fingerprint in computed in each case [11, 45]. The Electroshape fingerprint vectors (*x* and *y*) are compared using the Manhattan distance (*d*), which is calculated using the following formula:

$$d = \sum_{i=1}^{18} |x_i - y_i| \tag{17.11}$$

The smallest Manhattan distance ($d_{\min(A,B)}$) across the 20 × 20 conformer comparison is retained and converted to a 3D similarity value between *A* and *B*. For each dataset compound, the similarity value S_2' is computed using the following equation:

$$S_2' = \frac{1}{\left(\left(\frac{1}{18} d_{\min(A,B)}\right) + 1\right)} \tag{17.12}$$

For each target, the most similar ligand according to each similarity measure is retrieved. The two similarity values S_1' and S_2', with S_1' corresponding to the 3D similarity and S_2' to the 2D similarity, are then normalized using the formulas in Eqs. (17.13 and 17.14):

$$S_1 = \frac{(S_1' - 0.65)}{(1 - 0.65)} \tag{17.13}$$

$$S_2 = \frac{(S_2' - 0.3)}{(1 - 0.3)} \tag{17.14}$$

The normalized similarity values of S_1 and S_2 are then combined using a logistic regression in Eq. (17.15) (a_0, a_1, and a_2 are molecule size-dependent coefficients) to obtain a final similarity score $f(S_1,S_2)$ for the corresponding target combining 2D and 3D similarity, ranging between 0 (no similarity found between query molecules and the compounds in the dataset for this target) and 1 (query molecule is found in the dataset as a known compound for this target).

$$f(S_1, S_2) = \frac{1}{\left(1 + e^{(-a_0 - a_1 S_1 - a_2 S_2)}\right)} \tag{17.15}$$

The input for the query molecule can be in the form of SMILES, and the structural representation of the molecule can be drawn using the built-in tool. Possible targets are assigned and ranked according to the combined score $f(S_1, S_2)$. The Swisstargetprediction website offers a variety of options to visualize and download. The prediction results target list is provided for the top-N assigned targets as pie chart of the predicted classes alongside with a list of predictions containing a maximum of the top 100 predicted targets, as illustrated here for the target search for triazine **1** discussed in Section 10 below (Figure 17.1).

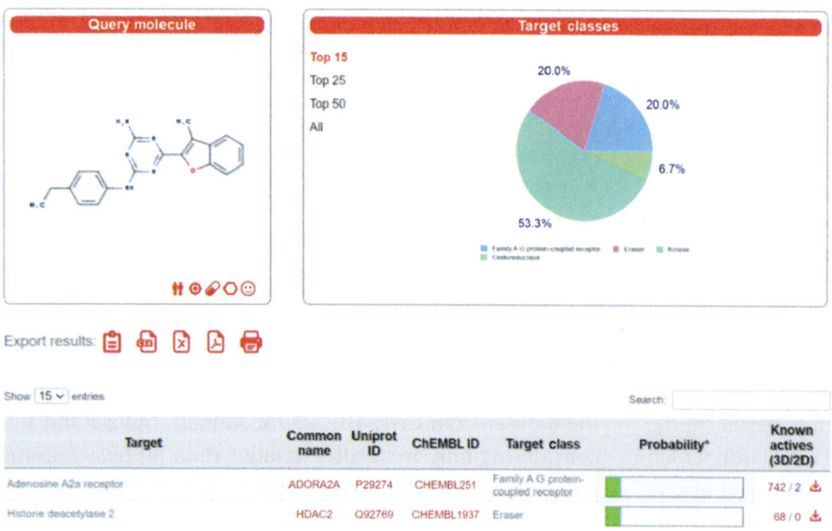

Figure 17.1 Swisstargetprediction results window as shown for triazine **1** as query molecule. The top target listed is the adenosine receptor A2aR due to a 2D-similarity with CHEMBL604535 (see also Figure 17.3 and Section 10).

17.6.2 Evaluation

The model performance of SwissTargetPrediction was evaluated by the average precision and validated by using an external test set of 500 experimentally strong binders associated with at least two human protein targets, measured K_i, K_d, IC_{50} or EC_{50} less than 1 nM, and less than 80 heavy atoms. These binders were extracted from ChEMBL 24 by emphasizing on not being recorded in ChEMBL 23. By running the model on the external test set, the new SwissTargetPrediction model could predict at least one of the expected targets in the first 15 predictions for 72% of the compounds.

Utilizing computational predictions grounded in the similarity to known ligands is effective in reducing the number of potential targets and providing insights into the side effects of established molecules. SwissTargetPrediction online tool proves valuable for comprehending the molecular mechanisms associated with a specific phenotype or bioactivity, elucidating potential side effects, anticipating off-target effects, and evaluating the feasibility of repurposing therapeutically significant compounds.

17.7 TargetNet

17.7.1 Data and Methodology

TargetNet [38] uses Binding Database [6] as its source of information. The final dataset of TargetNet contains compounds with only three types of activity: IC_{50}, EC_{50}, or K_i. Additionally, only targets with more than 200 biological activity data were considered. As a result, the final dataset consists of 109,061 compounds associated with 623 unique target proteins, representing enzymes (276), receptors (255), kinases (85), transporters (14), ion channels (9) and other protein classes (69). The threshold for activity value (or mean activity value for the compounds with more than one measured activity value) of 10 μM was considered.

After data filtration, TargetNet uses seven fingerprints (FP2, MACCS, Estate, FP4, ECFP2, ECFP4, and ECFP6) to build in each case a NBmodel as described above for PASS.

17.7.2 Evaluation

The evaluation of the performance of each model was determined by calculating the accuracy, Matthew's correlation coefficient (MCC), F1-score, and AUC. With the prediction accuracies of 70–99.8%, the MCC scores of 0.42–1.00 and F1 scores of 68.7–100%, the NBmodel using ECFP4 showed the best prediction results. Additionally, the AUC scores were in the range 75.9–100% for all the models. More specifically, the AUC scores for the models based on ECFP4 fingerprints were higher than 90%. TargetNet is also used for identifying new targets for known drugs. As an example, by analyzing the top 30 TargetNet predictions for drug bromocriptine, it is anticipated that bromocriptine might have an interaction with the dopamine receptor D_1A.

17.8 PPB

17.8.1 Data and Methodology

Polypharmacology Browser) [17] uses data on targets and their corresponding ligands from ChEMBL version 21. For each target, the compounds having IC_{50}, EC_{50}, GI_{50}, K_i, K_d, or potency values equal or less than 10 μM, or an inhibitory percentage greater than 50, were retained. All molecules were processed as non-stereo SMILES and ionized at pH 7.4. Subsequently, duplicate molecules were removed within the context of each target. Finally, only targets with at least 10 compounds were used in the selection, comprising 4613 targets represented by 871,673 unique bioactive compounds and 2.7 million ligand-target interactions. Among these targets, 60% are of a single protein type, 55% are human targets, and 45% have fewer than 50 bioactive compounds.

In ligand-based methodologies, each target protein is characterized by a set of known active ligands, and the ranking of targets is based on the molecular similarity of the query compound to the representative ligands of the target proteins. Manhattan distance was used as the similarity measure, and each fingerprint was scaled by analyzing the distance distribution of 50 million random pairs of compounds in each fingerprint space.

The similarity search is executed employing six distinct fingerprints: Xfp, MQN, SMIfp, APfp, Sfp, and ECFP4, and involves four fingerprint combinations: Ffp1 = Xfp + SMIfp + Sfp, Ffp2 = Xfp + MQN + SMIfp, Ffp3 = Xfp + SMIfp + Sfp + ECFP4, and Ffp4 = Xfp + MQN + SMIfp + Sfp + ECFP4. Conducting similarity searches within the chemical space associated with the target would identify the closest analogs and nearest neighbors to the query molecule, establishing a connection to the respective target proteins.

The query molecule can be drawn, provided as SMILES, or as an SDF file. Additionally, users can extract the query molecule from Protein Data Bank [46] (PDB) using the protein-ligand corresponding PDB ID. PPB calculates all the 10 mentioned fingerprints for the entered query compound and sorts the 871,673 unique compounds retrieved from ChEMBL 21 by the similarity to the query compound based on each fingerprint individually and creates a list of targets in each case. The *p*-value [47] for each target is calculated from the nearest neighbor found for that target, using the same strategy for calculating the *p*-value in SEA as discussed in the corresponding section. The *p*-value for each target was determined by calculating the best match compound of the target protein to the query molecule, ranging from a *p*-value of 0.01 to 0. Subsequently, the top 20 targets for each fingerprint space are identified by PPB, and lists of targets are represented on the web page. The results are presented as a list of targets annotated with a bar that shows the probability of the predicted target based on each fingerprint.

17.8.2 Evaluation

The performance of PPB was evaluated by testing its ability to identify targets for 670 approved drugs and drugs in clinical trials. These drugs were associated with a maximum of 20 targets each, which led to the total number of 4794 drug-target interactions from ChEMBL 21. In the evaluation, 670 individual searches were conducted with PPB, and analyzing predicted targets for each fingerprint was based on different criteria, including all targets, targets with *p*-value ≤ 0.01, and targets with *p*-value > 0.01. Additionally, metrics such as the fraction of known targets, total predicted targets, and hit rate were calculated for each drug.

While methods based on MQN, SMIfp, and APfp had a performance below 50%, the pharmacophore fingerprint Xfp and the binary fingerprints Sfp and ECFP4 have an accuracy exceeding 55%. The fusion fingerprints Ffp1, Ffp3, and Ffp4 demonstrated the highest average accuracy with correctly predicting around 65% of known targets. The combined method, which only considers targets voted by at least two fingerprints, performed exceptionally well in identifying known targets within the top 5 predictions, with a notable contribution from a low *p*-value for accurate target predictions.

Despite statistically similar overall performance among many fingerprints, pairwise overlap analysis revealed that, on average, any two fingerprints shared less than 45% of common targets. Interestingly, shape pharmacophore fingerprints, APfp, and Xfp, identified a considerable number of exclusive targets with the highest percentages of 52% and 31%, respectively, whereas fused fingerprints displayed only 3–10% of unique targets due to notable overlap among themselves. In the final analysis, a significant overlap, averaging 47%, was observed when comparing high-confidence targets characterized by low *p*-values across various fingerprints. Conversely, for low-confidence targets with high *p*-values, there was a smaller overlap, averaging at 24%, which can be attributed to the reduced apparent structural similarity between a query and its nearest neighbor compound associated with the target at higher *p*-values.

17.9 PPB2

17.9.1 Data and Methodology

Polypharmacology browser version 2 [39] was constructed by compiling a dataset of bioactive compounds with IC_{50}, EC_{50}, K_D, or K_i values less than 10 μM on a single protein target from ChEMBL database version 22. Considering only single protein targets from either rat or human as a source organism and with at least 10 bioactive compounds resulted in a dataset of 344,163 unique compounds linked to 1720 distinct protein targets across eight target families, corresponding in 555,346 target-compound associations.

PPB2 offers eight different methods for target assignment from similarity analysis. The first three are Nearest Neighbor (NN) searches using the fingerprints MQN, Xfp, or ECFP4, assigning targets to the query molecule following the order of the most similar database compounds. The next three methods extract the top 2000 NN for either MQN, Xfp, or ECFP4, and then perform a Laplacian-modified multinomial NB classification similar to the PASS method, focusing only on these 2000 NN and using ECFP4, to assign targets. The idea of this approach is that substructure comparisons are only performed on molecules with a certain degree of similarity, which avoids comparing unrelated compounds. Method number seven is a multinomial NB model like the one used for PASS, and method eight is a Deep Neural Network (DNN), both built using ECFP4 fingerprints calculated from the entire dataset of 344,163 molecules.

The PPB2 webpage opens with a JSME window to input the query molecule and the list of the eight possible search methods listed above. Upon choosing one of the methods, a results panel opens displaying the query structure and the top-20 predicted targets, both as pie chart by protein class and by decreasing order of priority as illustrated here for the target prediction of triazine **1** (Figure 17.2).

17.9.2 Evaluation

The performance of eight target prediction models was evaluated by calculating overall and average recall and precision using a 10-fold cross-validation procedure. The *overall recall* was calculated by the division of the number of times a target was correctly predicted (*TP*: true positive) to the sum of the number of true positives and the number of interactions of the given target which were not predicted (*FN*: false negatives). The *overall recall* of top n predictions

Figure 17.2 PPB2 results window as shown for triazine **1** as query molecules, using NN-search with the molecular fingerprint Xfp as similarity measure. This search method lists the experimentally correct target, LPAAT-β, at rank 1 (see also Section 10 and Figure 17.3). The molecules corresponding to each target can be inspected by clicking each 'shown NN' button.

for the targets across the whole dataset (*i*) and all the 10 cross-validation runs (*j*) is calculated as follows:

$$\textit{overall recall} = \frac{\sum_{i=1}^{n}\sum_{j=1}^{10} TP_{i,j}}{\sum_{i=1}^{n}\sum_{j=1}^{10} TP_{i,j} + \sum_{i=1}^{n}\sum_{j=1}^{10} FN_{i,j}} \times 100 \tag{17.16}$$

Subsequently, the *overall precision* was calculated by the division of true positives to the summation of true positives and the number of times when the desired target was found among the top n predicted targets but not annotated for the compound (*FP*: false positive). The *overall precision* of top n predictions for the targets across the whole dataset (*i*) and all the 10 cross-validation runs (*j*) is calculated as follows:

$$\textit{overall precision} = \frac{\sum_{i=1}^{n}\sum_{j=1}^{10} TP_{i,j}}{\sum_{i=1}^{n}\sum_{j=1}^{10} TP_{i,j} + \sum_{i=1}^{n}\sum_{j=1}^{10} FP_{i,j}} \times 100 \tag{17.17}$$

The average recall and precision for each target were calculated for 10 cross-validation runs. In terms of recall, the NN method using ECFP4 fingerprints showed the best performance considering the top 10 predictions with the *overall recall* of 86%, whereas the NN + NB model using ECfp4 showed the highest *overall precision* of 41%.

17.10 Comparison of Different Web-Based Tools

A very thorough comparison of different target prediction tools has recently been reported, discussing quantitative performances across 17 different target-centric models similar to those reviewed in this chapter [48]. Here, we focus our attention on two case studies involving inhibitors discovered in our own laboratory. The first example concerns the estimation of off-target effects of an ion channel inhibitor [49] and the second example is a target identification problem for an active compound discovered in a phenotypic screen [50]. These represent typical use cases for target prediction tools in medicinal chemistry.

In the first example, we consider the prediction of off-targets for compound *cis*-**22a** [49], a molecule which was developed as an inhibitor of TRPV6 [51], a calcium channel known for its overexpression in hyperparathyroidism, transient neonatal and hyperparathyroidism, neonatal severe, breast, and prostate cancer (Figure 17.3a) [52]. The compound was originally tested for 24 possible off-targets selected from a safety panel and was reported as active (>50% inhibition at 10 μM) for twelve of these off-targets. Further engineering of *cis*-**22a** to a photoswichtable and a more natural product-like analog allowed to mostly overcome the off-targets effects of *cis*-**22a** and obtain an equally potent but more selective analog [53, 54]. Electrophysiological, X-ray, and cryo-EM studies recently established that *cis*-**22a** and its optimized analogs all bind and inhibit TRPV6 in a similar manner [55].

From the eight target prediction tools discussed in this chapter, the best performance for predicting the off-targets of *cis*-**22a** is achieved by PPB2, which identifies 8 out of 12 active off-targets (using the ECFP4 DNN option). TargetNet successfully predicts seven experimentally active off-targets, including the important hERG channel also pointed out by PPB2 and Super-PRED. SwissTargetPrediction and PPB fail to predict hERG but nevertheless predict five, respectively, six of the proven off-targets for *cis*-**22a**. Several tools fall rather short, namely TargetHunter only predicts two off-targets, Super-PRED and SEA one each, and PASS misses all of them. Note that none of the eight prediction tools predicts any of the 12 experimentally verified inactive off-targets of *cis*-**22a**.

Considering that all tools tested rely on similarities to known inhibitors to assign off-targets, one can take note of the high structural similarity between *cis*-**22a** and CHEMBL2112778, known as active on ADRA1A, HTR1A, and DRD2. These two molecules only differ by a Me→OMe substituent change at the *meta*-position of the terminal benzene ring and the *meta*→*ortho* position change for the

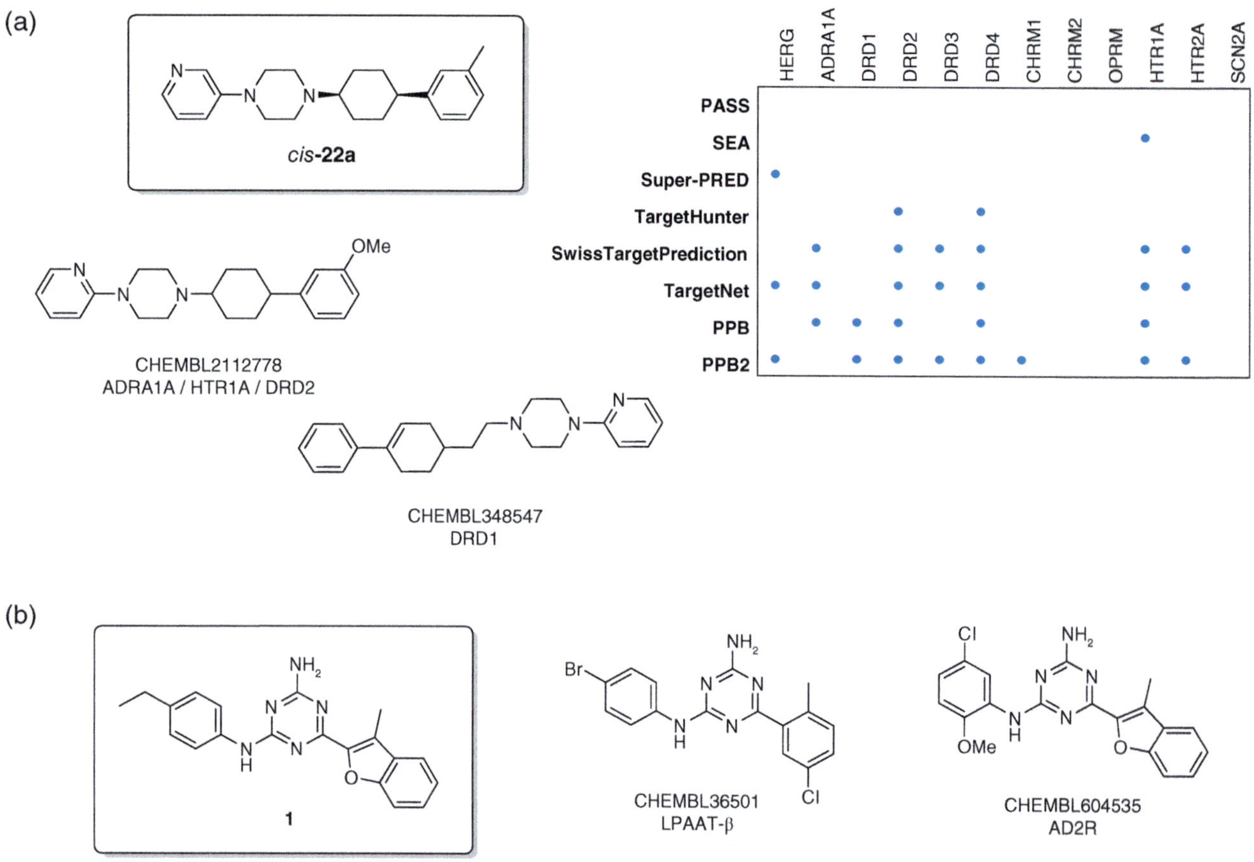

Figure 17.3 (a) Structure of the TRPV6 inhibitor *cis*-**22a** with the table outcome of target predictions of eight target prediction tools using their latest online version (matches indicated by blue circles) for twelve experimentally verified off-targets of *cis*-**22a**: voltage gated potassium channel subfamily H member 2 (hERG), adrenergic α1A (ADRA1A), dopamine receptor subtypes D1–4 (DRD1–4), cholinergic muscarinic receptor 1 (CHRM1) and 2 (CHRM2), μ opioid receptor (OPRM), 5-hydroxytryptamine (serotonin) receptor 1A (HTR1A), 2A (HTR2A), and voltage gated Na^+ channel (SCN2A). The structure of two CHEMBL molecules with similarity to *cis*-**22a** and the associated targets are also shown. None of the prediction tools predicts any of the verified inactive off-targets of *cis*-**22a**, which are the adrenergic α2A receptor, 5-hydroxytryptamine receptors 1B and 2B, the voltage dependent L- and N-type Ca^{2+} channel, the 5-hydroxytryptamine receptor 3, the *N*-methyl-D-aspartate receptor, the γ-aminobutyric acid receptor, the adrenergic β1 and β2 receptors, and the cannabinoid 1 and 2 receptors. (b) Structure of the angiogenesis phenotypic screening hit **1** later identified to inhibit LPAAT-β (lysophosphatidic acid acyl transferase β) and the related ChEMBL molecules with their known targets.

pyridine nitrogen atom. On the other hand, the similarity between *cis*-**22a** and the DRD1 inhibitor CHEMBL348547 is less precise and seemingly only recognized as a relevant comparison by PPB and PPB2, since these are the only tools correctly assigning DRD1 to *cis*-**22a**.

The second example is that of aminotriazine **1**, which was identified in a phenotypic screen for angiogenesis focused on kinase inhibitor discovery but turned out to be entirely inactive in a kinome-wide activity profile (Figure 17.3b) [50]. Investigation of possible targets of **1** using PPB2 pointed to three possible main targets, namely VEGFR-2 (vasoendothelical growth factor receptor 2), A2aR (adenosine receptor 2a), and LPAAT-β (Lysophosphatidic Acid Acyltransferase β). Experimental testing showed that aminotriazine **1** was inactive on VEGFR-2 and A2aR but had a low nanomolar activity on LPAAT-β, matching its observed toxicity on cancer cells. The key analogs pointing to these targets were triazines CHEMBL36501, known as LPAAT-β inhibitor and investigated as an anti-cancer drug [56–58], and CHEMBL604535, known as an A2aR inhibitor.

To test the ability of the target prediction tools discussed in this chapter, we have retested them for their ability to predict LPAAT-β as the target of aminotriazine **1**. Only PPB2 identifies LPAAT-β as a possible target for aminotriazine **1**, while the other seven web-based tools do not list LPAAT-β in the predicted targets. On the other hand, all the target prediction tools except Super-PRED list LPAAT-β as one of the predicted targets for the anti-cancer drug CHEMBL36501, indicating that they indeed have this target in their dataset. Therefore, it seems that only PPB2 identified the similarity between aminotriazine **1** and

CHEMBL36501 as a relevant comparison. This similarity originates from the pharmacophore fingerprint Xfp, which encodes molecular shape and pharmacophore based on counts of atom-pairs [12]. Note that PASS, SwissTargetPrediction, TargetNet, and PPB2 predict A2aR (Adenosine A2a Receptor) as a possible target for aminotriazine **1**, most likely by similarity to the related triazine CHEMBL604535.

17.11 Conclusion

Here we reviewed eight web-based target prediction tools that are currently active. All of them capitalize on the availability of public datasets of compound-target associations, which is one of the most valuable assets to support drug discovery. Indeed, the question when developing a drug is not whether it has off-targets, but which these off-targets might be considering the most recent state-of-the-art.

Almost all of the tools described in this chapter rely on the time-honored Morgan fingerprint (ECFP4) to encode molecular features [22] and the exploitation of Naïve Bayes statistics interpreting each fingerprint bit as an (almost) independent information, an approach which was first reported by the online tool PASS [29]. Strikingly, this simple approach remains competitive in terms of compute time and prediction recall and accuracy compared to more advanced models such as deep neural networks, as discussed in the performance evaluation of PPB2 [39]. Nevertheless, other molecular structure encoding methods such as molecular shape analysis or simple structural keys can provide decisive alternative insights, and many of these are implemented alongside ECFP4 in the various online tools. It must be pointed out that a key difference distinguishing the different online tools available is the selection of the source data, which are subsets obtained by applying different types of activity thresholds and target types. One must hope that all these tools will constantly be updated by the latest releases of public datasets such as ChEMBL.

Acknowledgement

This work was supported financially by the Swiss National Science Foundation, Grant no. 200020_207976.

References

1 Chen, X., Yan, C.C., Zhang, X. et al. (2016). Drug–target interaction prediction: databases, web servers and computational models. *Brief. Bioinform.* 17 (4): 696–712. https://doi.org/10.1093/bib/bbv066.

2 Yang, S.-Q., Ye, Q., Ding, J.-J. et al. (2021). Current advances in ligand-based target prediction. *WIREs Comput. Mol. Sci.* 11 (3): e1504. https://doi.org/10.1002/wcms.1504.

3 Bienfait, B. and Ertl, P. (2013). JSME: a free molecule editor in JavaScript. *Aust. J. Chem.* 5 (1): 24. https://doi.org/10.1186/1758-2946-5-24.

4 Weininger, D. (1988). SMILES a chemical language and information system. 1. Introduction to methodology and encoding rules. *J. Chem. Inf. Comput. Sci.* 28 (1): 31–36. https://doi.org/10.1021/ci00057a005.

5 Zdrazil, B., Felix, E., Hunter, F. et al. (2024). The ChEMBL database in 2023: a drug discovery platform spanning multiple bioactivity data types and time periods. *Nucleic Acids Res.* 52 (D1): D1180–D1192. https://doi.org/10.1093/nar/gkad1004.

6 Liu, T., Lin, Y., Wen, X. et al. (2007). BindingDB: a web-accessible database of experimentally determined protein–ligand binding affinities. *Nucleic Acids Res.* 35 (Database issue): D198–D201. https://doi.org/10.1093/nar/gkl999.

7 Kim, S., Chen, J., Cheng, T. et al. (2023). PubChem 2023 update. *Nucleic Acids Res.* 51 (D1): D1373–D1380. https://doi.org/10.1093/nar/gkac956.

8 Rogers, D. and Hahn, M. (2010). Extended-connectivity fingerprints. *J. Chem. Inf. Model.* 50 (5): 742–754. https://doi.org/10.1021/ci100050t.

9 Filimonov, D., Poroikov, V., Borodina, Y., and Gloriozova, T. (1999). Chemical similarity assessment through multilevel neighborhoods of atoms: definition and comparison with the other descriptors. *J. Chem. Inf. Comput. Sci.* 39 (4): 666–670. https://doi.org/10.1021/ci980335o.

10 O'Boyle, N.M., Banck, M., James, C.A. et al. (2011). Open babel: an open chemical toolbox. *J. Chem.* 3 (1): 33. https://doi.org/10.1186/1758-2946-3-33.

11 Armstrong, M.S., Morris, G.M., Finn, P.W. et al. (2010). ElectroShape: fast molecular similarity calculations incorporating shape, chirality and electrostatics. *J. Comput. Aided Mol. Des.* 24 (9): 789–801. https://doi.org/10.1007/s10822-010-9374-0.

12 Awale, M. and Reymond, J.L. (2014). Atom pair 2D-fingerprints perceive 3D-molecular shape and pharmacophores for very fast virtual screening of ZINC and GDB-17. *J. Chem. Inf. Model.* 54: 1892–1897. https://doi.org/10.1021/ci500232g.

13 Durant, J.L., Leland, B.A., Henry, D.R., and Nourse, J.G. (2002). Reoptimization of MDL keys for use in drug discovery. *J. Chem. Inf. Comput. Sci.* 42 (6): 1273–1280. https://doi.org/10.1021/ci010132r.

14 Hall, L.H. and Kier, L.B. (1995). Electrotopological state indices for atom types: a novel combination of electronic, topological, and valence state information. *J. Chem. Inf. Comput. Sci.* 35 (6): 1039–1045. https://doi.org/10.1021/ci00028a014.

15 Nguyen, K.T., Blum, L.C., van Deursen, R., and Reymond, J.-L. (2009). Classification of organic molecules by molecular quantum numbers. *ChemMedChem* 4 (11): 1803–1805. https://doi.org/10.1002/cmdc.200900317.

16 Schwartz, J., Awale, M., and Reymond, J.-L. (2013). SMIfp (SMILES fingerprint) chemical space for virtual screening and visualization of large databases of organic molecules. *J. Chem. Inf. Model.* 53 (8): 1979–1989. https://doi.org/10.1021/ci400206h.

17 Awale, M. and Reymond, J.-L. (2017). The polypharmacology browser: a web-based multi-fingerprint target prediction tool using ChEMBL bioactivity data. *J. Chem.* 9 (1): 11. https://doi.org/10.1186/s13321-017-0199-x.

18 Willett, P., Barnard, J.M., and Downs, G.M. (1998). Chemical similarity searching. *J. Chem. Inf. Comput. Sci.* 38 (6): 983–996. https://doi.org/10.1021/ci9800211.

19 Willett, P. (2006). Similarity-based virtual screening using 2D fingerprints. *Drug Discov. Today* 11 (23–24): 1046–1053.

20 Stumpfe, D. and Bajorath, J. (2011). Similarity Searching. *WIREs Comput. Mol. Sci.* 1 (2): 260–282. https://doi.org/10.1002/wcms.23.

21 Ripphausen, P., Nisius, B., and Bajorath, J. (2011). State-of-the-art in ligand-based virtual screening. *Drug Discov. Today* 16 (9): 372–376. https://doi.org/10.1016/j.drudis.2011.02.011.

22 Morgan, H.L. (1965). The generation of a unique machine description for chemical structures-a technique developed at chemical abstracts service. *J. Chem. Doc.* 5 (2): 107–113. https://doi.org/10.1021/c160017a018.

23 Glem, R.C., Bender, A., Arnby, C.H. et al. (2006). Circular fingerprints: flexible molecular descriptors with applications from physical chemistry to ADME. *IDrugs* 9 (3): 199–204.

24 Nicholls, A., McGaughey, G.B., Sheridan, R.P. et al. (2010). Molecular shape and medicinal chemistry: a perspective. *J. Med. Chem.* 53 (10): 3862–3886. https://doi.org/10.1021/jm900818s.

25 Scior, T., Bender, A., Tresadern, G. et al. (2012). Recognizing pitfalls in virtual screening: a critical review. *J. Chem. Inf. Model.* 52 (4): 867–881. https://doi.org/10.1021/ci200528d.

26 Carhart, R.E., Smith, D.H., and Venkataraghavan, R. (1985). Atom pairs as molecular features in structure-activity studies: definition and applications. *J. Chem. Inf. Comput. Sci.* 25 (2): 64–73. https://doi.org/10.1021/ci00046a002.

27 Schneider, G., Neidhart, W., Giller, T., and Schmid, G. (1999). "Scaffold-hopping" by topological pharmacophore search: a contribution to virtual screening. *Angew. Chem. Int. Ed. Engl.* 38 (19): 2894–2896.

28 Capecchi, A., Probst, D., and Reymond, J.-L. (2020). One molecular fingerprint to rule them all: drugs, biomolecules, and the metabolome. *J. Chem.* 12 (1): 43. https://doi.org/10.1186/s13321-020-00445-4.

29 Lagunin, A., Stepanchikova, A., Filimonov, D., and Poroikov, V. (2000). PASS: prediction of activity spectra for biologically active substances. *Bioinformatics* 16 (8): 747–748. https://doi.org/10.1093/bioinformatics/16.8.747.

30 Pogodin, P.V., Lagunin, A.A., Filimonov, D.A., and Poroikov, V.V. (2015). PASS targets: ligand-based multi-target computational system based on a public data and naive Bayes approach. *SAR QSAR Environ. Res.* 26 (10): 783–793. https://doi.org/10.1080/1062936x.2015.1078407.

31 Keiser, M.J., Roth, B.L., Armbruster, B.N. et al. (2007). Relating protein pharmacology by ligand chemistry. *Nat. Biotechnol.* 25 (2): 197–206. https://doi.org/10.1038/nbt1284.

32 Keiser, M.J., Setola, V., Irwin, J.J. et al. (2009). Predicting new molecular targets for known drugs. *Nature* 462 (7270): 175–181. https://doi.org/10.1038/nature08506.

33 Nickel, J., Gohlke, B.-O., Erehman, J. et al. (2014). SuperPred: update on drug classification and target prediction. *Nucleic Acids Res.* 42 (Web Server issue): W26–W31. https://doi.org/10.1093/nar/gku477.

34 Gallo, K., Goede, A., Preissner, R., and Gohlke, B.-O. (2022). SuperPred 3.0: drug classification and target prediction – a machine learning approach. *Nucleic Acids Res.* 50 (W1): W726–W731. https://doi.org/10.1093/nar/gkac297.

35 Wang, L., Ma, C., Wipf, P. et al. (2013). TargetHunter: an in silico target identification tool for predicting therapeutic potential of small organic molecules based on chemogenomic database. *AAPS J.* 15 (2): 395–406. https://doi.org/10.1208/s12248-012-9449-z.

36 Gfeller, D., Grosdidier, A., Wirth, M. et al. (2014). SwissTargetPrediction: a web server for target prediction of bioactive small molecules. *Nucleic Acids Res.* 42 (Web Server issue): W32–W38. https://doi.org/10.1093/nar/gku293.

37 Daina, A., Michielin, O., and Zoete, V. (2019). SwissTargetPrediction: updated data and new features for efficient prediction of protein targets of small molecules.

Nucleic Acids Res. 47 (W1): W357–W364. https://doi.org/10.1093/nar/gkz382.

38 Yao, Z.-J., Dong, J., Che, Y.-J. et al. (2016). TargetNet: a web service for predicting potential drug-target interaction profiling via multi-target SAR models. *J. Comput. Aided Mol. Des.* 30 (5): 413–424. https://doi.org/10.1007/s10822-016-9915-2.

39 Awale, M. and Reymond, J.-L. (2019). Polypharmacology browser PPB2: target prediction combining nearest neighbors with machine learning. *J. Chem. Inf. Model.* 59 (1): 10–17. https://doi.org/10.1021/acs.jcim.8b00524.

40 Attwood, M.M., Fabbro, D., Sokolov, A.V. et al. (2021). Trends in kinase drug discovery: targets, indications and inhibitor design. *Nat. Rev. Drug Discov.* 20 (11): 839–861. https://doi.org/10.1038/s41573-021-00252-y.

41 Lounkine, E., Keiser, M.J., Whitebread, S. et al. (2012). Large-scale prediction and testing of drug activity on side-effect targets. *Nature* 486 (7403): 361–367. https://doi.org/10.1038/nature11159.

42 Burger, M.C. (2015). ChemDoodle web components: HTML5 toolkit for chemical graphics, interfaces, and informatics. *J. Chem.* 7 (1): 35. https://doi.org/10.1186/s13321-015-0085-3.

43 Dunkel, M., Günther, S., Ahmed, J. et al. (2008). SuperPred: drug classification and target prediction. *Nucleic Acids Res.* 36 (Web Server issue): W55–W59. https://doi.org/10.1093/nar/gkn307.

44 Nidhi, Glick, M., Davies, J.W., and Jenkins, J.L. (2006). Prediction of biological targets for compounds using multiple-category Bayesian models trained on chemogenomics databases. *J. Chem. Inf. Model.* 46 (3): 1124–1133. https://doi.org/10.1021/ci060003g.

45 Armstrong, M.S., Finn, P.W., Morris, G.M., and Richards, W.G. (2011). Improving the accuracy of ultrafast ligand-based screening: incorporating lipophilicity into electroshape as an extra dimension. *J. Comput. Aided Mol. Des.* 25 (8): 785–790. https://doi.org/10.1007/s10822-011-9463-8.

46 Kirchmair, J., Markt, P., Distinto, S. et al. (2008). The protein data bank (PDB), its related services and software tools as key components for in silico guided drug discovery. *J. Med. Chem.* 51 (22): 7021–7040. https://doi.org/10.1021/jm8005977.

47 Baldi, P. and Nasr, R. (2010). When is chemical similarity significant? The statistical distribution of chemical similarity scores and its extreme values. *J. Chem. Inf. Model.* 50 (7): 1205–1222. https://doi.org/10.1021/ci100010v.

48 Jimenes-Vargas, K., Pazos, A., Munteanu, C.R. et al. (2024). Prediction of compound-target interaction using several artificial intelligence algorithms and comparison with a consensus-based strategy. *Aust. J. Chem.* 16 (1): 27. https://doi.org/10.1186/s13321-024-00816-1.

49 Simonin, C., Awale, M., Brand, M. et al. (2015). Optimization of TRPV6 calcium channel inhibitors using a 3D ligand-based virtual screening method. *Angew. Chem. Int. Ed. Engl.* 54 (49): 14748–14752. https://doi.org/10.1002/anie.201507320.

50 Poirier, M., Awale, M., Roelli, M.A. et al. (2019). Identifying lysophosphatidic acid acyltransferase β (LPAAT-β) as the target of a nanomolar angiogenesis inhibitor from a phenotypic screen using the polypharmacology browser PPB2. *ChemMedChem* 14 (2): 224–236. https://doi.org/10.1002/cmdc.201800554.

51 Yue, L., Peng, J.B., Hediger, M.A., and Clapham, D.E. (2001). CaT1 manifests the pore properties of the calcium-release-activated calcium channel. *Nature* 410 (6829): 705–709. https://doi.org/10.1038/35070596.

52 Khattar, V., Wang, L., and Peng, J.-B. (2022). Calcium selective channel TRPV6: structure, function, and implications in health and disease. *Gene* 817: 146192. https://doi.org/10.1016/j.gene.2022.146192.

53 Cunha, M.R., Bhardwaj, R., Lindinger, S. et al. (2019). Photoswitchable inhibitor of the calcium channel TRPV6. *ACS Med. Chem. Lett.* 10 (9): 1341–1345. https://doi.org/10.1021/acsmedchemlett.9b00298.

54 Cunha, M.R., Bhardwaj, R., Carrel, A.L. et al. (2020). Natural product inspired optimization of a selective TRPV6 calcium channel inhibitor. *RSC Med. Chem.* 11 (9): 1032–1040. https://doi.org/10.1039/D0MD00145G.

55 Bhardwaj, R., Lindinger, S., Neuberger, A. et al. Inactivation-mimicking block of the epithelial calcium channel TRPV6. *Sci. Adv.* 6 (48): eabe1508. https://doi.org/10.1126/sciadv.abe1508.

56 Hideshima, T., Chauhan, D., Hayashi, T. et al. (2003). Antitumor activity of lysophosphatidic acid acyltransferase-β inhibitors, a novel class of agents, in multiple myeloma. *Cancer Res.* 63 (23): 8428–8436.

57 Hong, F., Hollenback, D., Singer, J.W., and Klein, P. (2005). Diamino-C,N-diarylpyridine positional isomers as inhibitors of lysophosphatidic acid acyltransferase-beta. *Bioorg. Med. Chem. Lett.* 15 (21): 4703–4707. https://doi.org/10.1016/j.bmcl.2005.07.055.

58 Pagel, J.M., Laugen, C., Bonham, L. et al. (2005). Induction of apoptosis using inhibitors of lysophosphatidic acid acyltransferase-beta and anti-CD20 monoclonal antibodies for treatment of human non-Hodgkin's lymphomas. *Clin. Cancer Res.* 11 (13): 4857–4866. https://doi.org/10.1158/1078-0432.ccr-04-2352.

18

Using Phenotypic Screening to Uncover the Full Potential of Polypharmacology

Arsenio Nueda

18.1 Introduction: Phenotypic Screening and Phenotypic Drug Discovery

Phenotypic screening is a strategy for the identification of chemical and biological drug discovery starting points that uses a cell, tissue, organ, or *in vivo* model system recapitulating a disease-relevant functional endpoint (the phenotype) to screen a library of perturbagens such as small molecules, extracts containing natural products, peptides, siRNAs, or antibodies. A phenotypic screen is the first stage of what is otherwise known as phenotypic drug discovery (PDD) [1–3]. Complementary to PDD, target-based drug discovery (TDD) relies on a target-centric "target-first" endpoint referred to the function of a previously validated protein target for library screening. A target-based screen aims to identify hits engaging the selected target with the highest possible activity and selectivity.

The concept of PDD is not new, as it was the basis for the discovery of many early drugs well before the introduction of molecular biology tools to drug discovery at the end of the last century, which supported the adoption of TDD approaches still dominant today [4]. The emergence of systems biology in the early 2000s facilitated the renaissance of PDD as an approach that provided an experimental strategy to interrogate disease models from a systems biology perspective [5].

In phenotypic screening, the library and the model system screen each other to identify the combination of actives and mechanisms of action (MoA) with the highest impact in the selected disease-linked phenotype, placing no restrictions on the mechanism/s and number of target engagements required to achieve activity. In the context of polypharmacology, phenotypic screening allows the integration of the combined modulation of any number of target proteins (on-targets) into a single measurable endpoint. Subsequent mapping of these individual contributions can reveal a polypharmacology signature without a previous hypothesis on the MoA.

PDD has proved successful in delivering novel drugs. This was initially shown in a seminal publication by Swinney and Anthony analyzing drugs approved between 1999 and 2008 [1] and in a subsequent assessment for the period 1999–2013 that considered only high-throughput phenotypic screening as PDD [6]. A more recent estimation has suggested that PDD approaches may have contributed directly or indirectly to the discovery of up to 61% of drugs approved between 1995 and 2022 [7].

In PDD, the concept of target is not denied. Yet, it is not considered the starting point of the drug discovery process, allowing the possibility to capture both single and multi-target MoAs. Identification of the MoA of a PDD candidate molecule is not a regulatory requirement but is a significant advantage [4, 8]. It can simplify hit and lead optimization, offer new target opportunities (that can be re-routed to a TDD strategy), and allow early identification of potential safety liabilities and biomarkers which can be key to supporting more informed decision-making in the progression to clinical development.

The success of a phenotypic screen largely relies on an *in vitro*, *ex vivo*, or *in vivo* model system recapitulating one or more aspects of the pathophysiology of the disease of interest based on what was described as the "chain of translatability" by Moffat et al. [2]. In phenotypic screening, the disease model offers access to a biology space spanning dimensions that include target classes (proteins, nucleic acids, or other biomolecules), their combinations in functional macromolecular complexes, and their dynamics, without any assumption on what is needed for activity across any of those dimensions.

Achieving sufficient throughput has traditionally been a challenge for phenotypic screening, as the most relevant model systems, particularly *ex vivo* and *in vivo* models, have limitations when screening large libraries. Remarkably, *in vivo* models have proven suitable for phenotypic screening [9, 10] and were at the origin of the discovery of many classical drugs, which typically relied on focused chemical

Polypharmacology: Strategies for Multi-Target Drug Discovery, First Edition. Edited by Jens-Uwe Peters.

library strategies with a limited number of molecules being screened [1].

The scope of phenotypic screens can be narrowed down to restricted library subsets or selected biological pathways. A small molecule library could be as diverse as technically possible or custom-built to include a limited number of molecules related by their chemical structure or by the target families they engage. Similarly, a phenotypic endpoint could be selected to capture inputs converging on a single gene, protein, or signaling pathway-based readout (e.g., increased protein expression) or related to a complex cellular phenotype (e.g., cellular differentiation), with the latter being more likely to capture a multi-target polypharmacology MoA. Therefore, a phenotypic screen can range from a focused exercise with limited risk but a lower chance of finding a novel MoA to an exploratory fishing expedition with higher risk and higher potential reward. Selecting between one of these approaches depends on the final aim of the project. For example, the use of kinase-focused libraries in phenotypic screens led to the discovery of drugs such as deucravacitinib for the treatment of psoriasis [11] or the identification of investigational polypharmacology compounds inducing neurite growth [12].

Phenotypic screening is the most defining phase of PDD, as this is the stage where active molecules that will be the starting point for further optimization are identified. Optimization of phenotypic hits must be initially based on their phenotypic activity unless the molecule identified as a hit has a previously annotated (known) target or targets and these targets are proven to be related to the modulation of the phenotypic endpoint. This can be verified using tool molecules or probes selective for those targets. If the targets engaged by the hit are considered of interest, the PDD process can divert to a TDD project where the primary phenotypic model can be used as a secondary assay during optimization. The opposite is also possible: a disease-relevant model used in TDD as a secondary assay could be used for phenotypic screening in a target-agnostic way. This illustrates the interconnectivity between PDD and TDD (Figure 18.1) and speaks to the fact that a phenotypic screen can be used for target identification when screening annotated or molecular probe libraries.

In the absence of any knowledge of the on-target signature of phenotypic hits, assessment of PDD selectivity must initially rely on a phenotype as close as possible to the primary phenotypic endpoint but outside the chain of translatability. For more advanced PDD compounds, information about the target or targets mediating activity can become available, allowing a more conventional approach to selectivity optimization based on reducing interactions with targets not related to the modulation

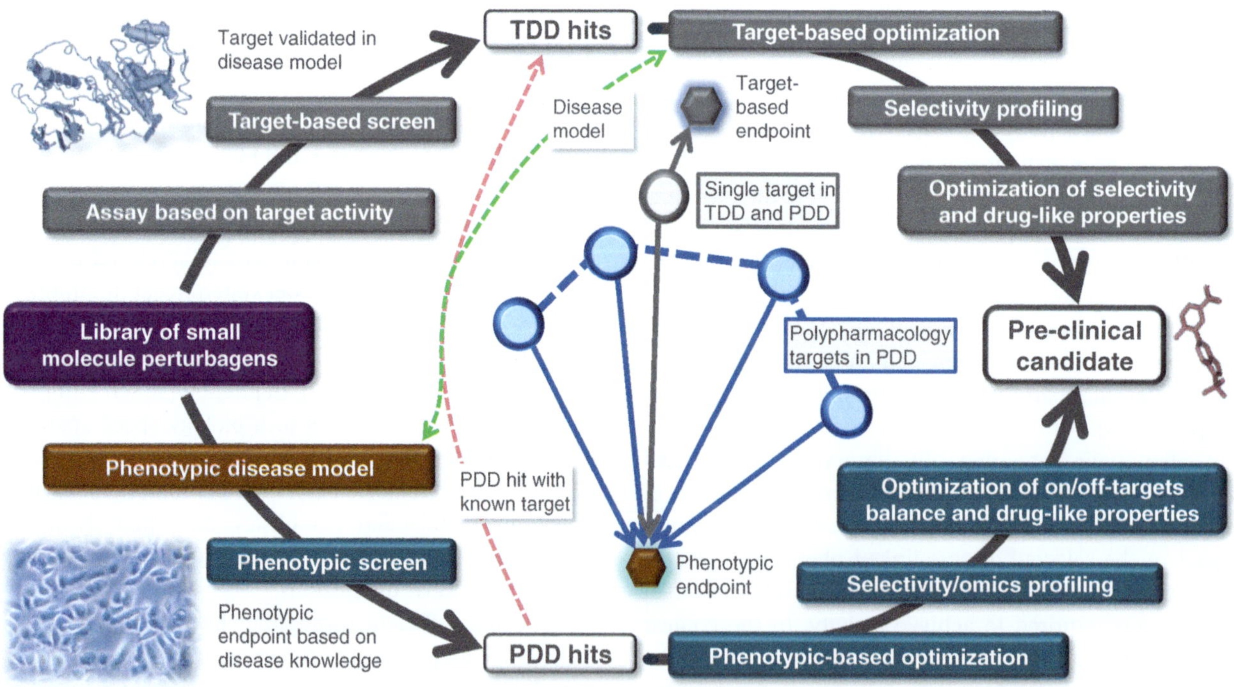

Figure 18.1 Polypharmacology in PDD and interconnectivity between PDD and TDD. Schematic overview of generic PDD and TDD processes highlighting the potential of PDD to capture polypharmacology (blue lines and circles) and the interconnectivity between both approaches (red and green dashed arrows). The process is exemplified for a cell-based phenotypic screening using a library of small molecules. Parts of this Figure have been adapted from [51] under the terms of a Creative Commons Attribution License.

of the phenotypic endpoint (off-targets). In cases where target identification is not obvious even after some level of characterization, partial knowledge of the MoA (e.g., signaling pathways modulated by the hit) can be enough to support progressing the candidate molecule to the pre-clinical stage while additional target/s identification efforts are undertaken [8].

The following sections will highlight aspects and strategies of phenotypic screening and PDD that can be used to uncover novel polypharmacology.

18.2 Polypharmacology Discovered Using Phenotypic Screening

Polypharmacology was a common trait in many of the drugs discovered in the pre-molecular biology era when, in the absence of target molecular knowledge, evaluating the activity in a cell-based, tissue, or *in vivo* phenotypic model was the only possible approach to searching for novel drug candidates [4]. In many cases, in-depth MoA determination and target identification have only been possible decades later, oftentimes revealing extensive on-target polypharmacology [7].

A list of investigational compounds, clinical candidates, and drugs discovered using PDD and showing a polypharmacology MoA was compiled by Vincent et al. [3]. Two of those examples, imipramine, and SEP-363856, representing classical phenotypic screening in human subjects and more modern systematic screening in *in vivo* models, are highlighted below (Figure 18.2).

Imipramine is an anti-depressant identified serendipitously in the 1950s while exploring the antipsychotic activity of a selection of anti-histaminic compounds in patients suffering from schizophrenia [13, 14]. This drug shows a complex multi-target signature as an antagonist of histamine receptors, an inhibitor of serotonin and norepinephrine transporters, and an activator of TrkB neurotrophin receptor signaling [15, 16], among other activities. These include off-target mechanisms not related to its anti-depressant activity, such as inhibition of estrogen receptor signaling and modulation of the activity of the yes associated protein YAP oncogene [17, 18].

SEP-363856, for schizophrenia and psychosis, is proof of how modern *in vivo* phenotypic model platforms can be used to discover novel drug candidates. This molecule, identified using a battery of *in vivo* mouse behavioral models [9, 19], and currently being evaluated in phase 3 clinical trials [20], shows dual target polypharmacology, with trace-amine-associated receptor-1 (TAAR1) and serotonin-1A receptor (5-HT1A) agonist activities as on-targets. Several off-targets that are not involved in its therapeutic effect have been identified, with 5-HT1B, 5-HT1D, and 5-HT7 being the most prominent [19–21].

A more recent investigational compound, BMS-794833, is an example of how a focused phenotypic screening approach can uncover unexpected polypharmacology (Figure 18.2) [22]. The compound showed inhibition of protumoral polarization of tumor-associated macrophages in a phenotypic screen using a library of 85 previously known kinase inhibitors targeting a representation of the human kinome. BMS-794833, initially annotated as a dual inhibitor of c-MET and VEGFR2, and reported as a MERTK kinase inhibitor [23], revealed a polypharmacology signature that included FAK, SRC, STAT3, and p38 MAPK, where c-MET and VEGFR2 were off-targets not involved in the phenotypic effect.

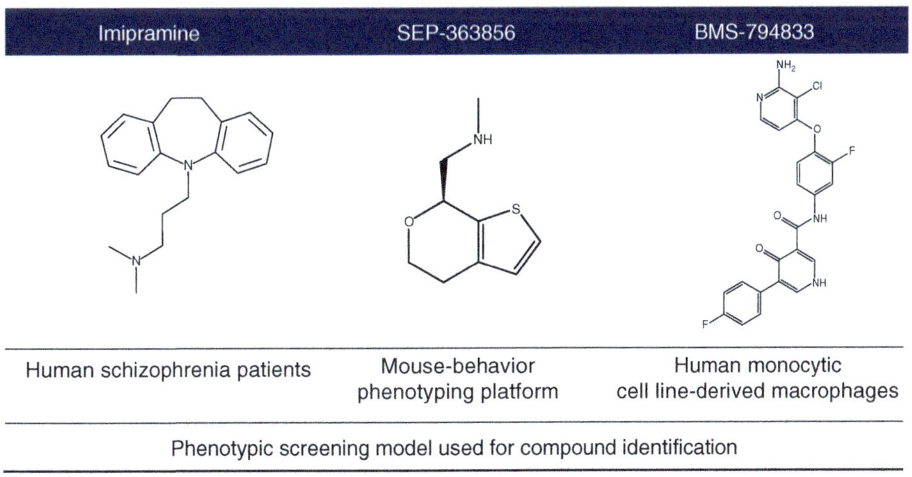

Figure 18.2 Representative examples of PDD-derived polypharmacology. Structure of imipramine, clinical candidate SEP-363856, and investigational compound BMS-794833. The phenotypic model used for their identification is indicated for each molecule.

18.3 PDD Strategies to Discover Novel Polypharmacology

Phenotypic screening provides a vantage point to explore multi-target MoAs not approachable with TDD, where the endpoint selected for the primary screen is locked into the function of a single target. Both approaches can identify non-selective molecules, but only PDD is geared to capture active polypharmacology signatures.

Selecting the best possible phenotypic disease model, a fit-for-purpose library compatible with engaging multiple targets, and identifying a sufficiently integrative translational endpoint are key aspects of a phenotypic screening campaign aiming to take advantage of the benefits of polypharmacology.

18.3.1 Disease Model

A description of the most relevant drivers for successful phenotypic assays from a disease model and assay perspectives was provided by Vincent et al. in their 2015 publication [24] describing the "rule of three": the right disease-relevant model system, the right stimulus mimicking a disease context (when required), and the right phenotypic endpoint.

In the context of a disease model, the phenotypic endpoint is one of the keys to increasing the chances of capturing a polypharmacology MoA. Phenotypic endpoints can range from gene- or protein-based readouts (e.g., increased gene transcription or protein expression) to more holistic endpoints in the form of omics signatures and cellular phenotypes such as viability, growth, differentiation, or migration. While all of them are open to multi-target MoAs in a phenotypic screen, the latter holistic endpoints are more likely to capture polypharmacology.

As large *in vitro* primary screens can only be conducted using cells or tissue samples from a limited number of donors, typically one or two, selecting representative donors is a critical aspect. This requires looking at a minimum number of key properties which, in addition to the phenotype to be evaluated, should include cell viability, expression of tissue-specific markers, and, if possible, some level of general expression profiling (which can be benchmarked with information from the literature or public domain datasets).

From a general standpoint, three key aspects are contributing to enabling access to increasingly translational disease models suitable for phenotypic screening:

1) *Knowledge-based sourcing of patient samples*: Clinician researchers are a valuable source of in-depth patient-based disease knowledge that, in combination with the access to human samples and associated clinical information that they can facilitate, allows establishing correlations between molecular features such as expression of relevant markers or omics signatures, and clinical phenotypes like disease severity or response to existing treatments.
2) *Ethico-legal environment*: Over the past decade, a solid ethical and data protection regulatory environment fostered by national and international legislation like the European GDPR [25, 26] has been developed. This has contributed to creating a culture of trust between research study volunteers, ethics committees, and the biomedical research community from industry, academia, and medical stakeholders.
3) *Industrialization*: The increased access to human samples from trusted commercial vendors and the development of dependable and well-regulated biobanking repositories supported by national and international consortiums such as the European Research Infrastructure for Biobanking (BBMRI-ERIC) [27] are examples of industrialization with direct impact in PDD.

Despite these advances, the use of animal or human-derived model systems for phenotypic screening still faces limitations when screening large chemical libraries, which can range from several hundred to millions of molecules.

18.3.2 Library

From a polypharmacology perspective, the capability of phenotypic screening to offer access to multi-target signatures is defined by the combination of the target space accessible in the disease model and the potential pharmacological activities in each of the molecules contained within the library.

Ideally, the library must be sufficiently extensive to take full advantage of the universe of potential MoA offered by the disease model, but also a realistically selected subset of the theoretical chemical space so that patient-derived samples are not wasted screening molecules unlikely to have any relevant bioactivity (e.g., highly cytotoxic). Basic criteria for molecule selection in small molecule libraries are assumed: lack of structural alerts and chemical diversity based on selecting molecules representative of the chemical families contained in the original library.

The type of library used will determine the wealth in hit classes resulting from a phenotypic screen. We may be interested in testing approved kinase inhibitors which are a typical source of polypharmacology [28], to repurpose them for another indication, or in evaluating a restricted number of potential target combinations, for example, using bi-specific antibodies. Three relevant examples of bi-specific antibody-based phenotypic screening have been provided by Bhatta et al. [29]: identification of inhibitors of B-cell receptor function screening 300 bi-specific combinations of 23 targets in peripheral blood mononuclear cells, identification of inhibitors of extracellular matrix accumulation in a co-culture of lung epithelial cells and fibroblasts, screening 239 bi-specific combinations of 23 targets, and identification of modulators of T cell activation using antibodies representing 969 bi-specific combinations of 49 targets.

The use of well-annotated compound or molecular probe subsets [30–32] and libraries of bioactive molecules based on high-dimensional transcriptomics and image-based profiling [33] can be a rich source of both tools and bioactive hits for phenotypic screening.

DNA-encoded libraries (DEL) are a formidable source of potential diversity based on the sheer number of molecules that can be screened [34]. The technology was initially limited to using purified protein targets, which prevented its application for phenotypic screening, but several examples show that it can also be adapted to cell-based assays which can include antibacterial screens [35–37], increasing the opportunities for identifying multi-target ligands in phenotypic screening applications. However, as mentioned above, screening such large libraries would only be possible in some *in vitro* phenotypic models, for example, those based on transformed cell lines.

Exploring diversity beyond conventional small-molecule libraries is also an attractive possibility. Libraries based on extracts containing natural products are a good example of how bioactivity without initial knowledge of chemical structure could expand the possibilities to identify hits with multi-target MoA [38]. However, it should be kept in mind that the presence of multiple unknown components in the extracts evaluated could interfere with phenotypic assays both from a biology (e.g., cytotoxicity) or technology (e.g., assay interference) perspectives, at the concentrations needed to uncover sufficient activity.

Very importantly, the value of libraries and library annotations stored within the walls of public and private biomedical research organizations as a result of decades of legacy drug discovery and research projects is an asset that should not be underestimated when considering a phenotypic screening campaign. Efforts to tap into this potential are underway, such as the European Lead Factory [39] or EU-Openscreen [40] consortiums, but common sense dictates that a large fraction of the existing chemical space will continue to remain within institutions and organizations' silos. Typically, these libraries have undergone multiple high throughput screening (HTS) campaigns and hit or lead optimization bioassays resulting not only in positive (active) but also in negative (non-active) annotations which are a highly valuable source of selectivity information. Therefore, the first analysis that must be conducted when assessing a list of hit compounds resulting from a phenotypic screen is a comprehensive in-house data mining exercise that can be complemented with data retrieved from external resources like PubChem [41] or ChEMBL [42], for cases when the hit of interest has been available in the public domain. This can provide a first indication of the MoA of phenotypic hits, including the potential for a multi-target signature, without the need for any experimental activity.

18.3.3 Running a Phenotypic Screening and Selecting Hits

Running a phenotypic screening campaign is a challenging endeavor. Compared with biochemical endpoint-based target screens, the use of a live cellular or *in vivo* model introduces added sources of variability resulting from the subject's individual background and the experimental conditions (e.g., type of growth medium). This demands strict supervision of screening conditions and the use of proper controls, including reference compounds that need to be consistent across the screen.

Multiple technologies can be applied to translate a phenotypic endpoint into a robust readout in the form of fluorescence, luminescence, or more complex imaging or omics signature-based endpoints. A phenotypic endpoint can be a negative loss-of-function trait (for example, inhibition of viability or downregulation of the expression of a gene) or a positive gain-of-function endpoint (for example, upregulation of the expression of a protein or induction of cellular differentiation). Screens based on positive gain-of-function endpoints have intrinsic advantages, as they are more likely to avoid false actives resulting from a non-specific negative impact on cell fitness that can complicate the interpretation of results from loss-of-function screens [43].

Using a loss-of-function endpoint requires not only setting up strict viability/cytotoxicity controls but also relying heavily on a robust phenotypic selectivity counter-screen to assess how specific is the inhibition of the phenotypic endpoint. This becomes particularly important if polypharmacology is to be explored, as impaired cell viability is highly likely when considering a yet-to-be optimized combination of targets. The idea of phenotypic selectivity can be counterintuitive when considering target-agnostic PDD approaches. However, in much the same way that structurally similar proteins to the target of interest are the first-line selectivity choice in TDD, we can think of endpoints close enough to the selected phenotypic endpoint but still outside the chain of translatability [2].

For gain-of-function screens, cytotoxic hits tend to rule themselves out from the active hits list. In some cases, hits that have an impact on viability can be sufficiently active to overcome their detrimental effect. In these cases, further progression of the hits is advisable, as they may show other promising properties worth assessing that could be potentially optimized to reduce cytotoxicity at later stages. Setting up a phenotypic selectivity assay is also important when using gain-of-function screens but, in general, is less critical for initial hit selection than in loss-of-function screens.

When possible, running the primary or secondary screens in samples from other representative donors can provide valuable information on the subject-specific activity of the hits. Being active in samples derived from only one donor is not necessarily a negative trait, and it may be representative of a restricted patient population that could be of interest (e.g., good responders to a drug).

According to standard practice, hits resulting from a phenotypic screen must be (i) confirmed in the same assay, (ii) validated using an alternative detection technology, and (iii) evaluated in dose-response as added proof of activity. Verification of potential hit interference with the chemistry of the detection assay in the absence of a biological matrix is also essential.

Finally, once all the proper controls have been conducted, evaluation of the hits to see whether they can be grouped into structural families is an important aspect of final hit selection. Having a structure-activity relationship (SAR), i.e., more than one active hit within a structural class, is more proof of activity than having a singleton or isolated active when other structurally similar molecules do not show activity.

18.4 Optimizing Polypharmacology in Phenotypic Screening Hits

18.4.1 Hit Expansion and Hit Characterization

In the absence of target annotation information, phenotypic screening-derived hits are selected based on their activity in the phenotypic assay using a "no-questions-asked" criteria regarding targets or selectivity. This can of course change when using annotated libraries to find single target starting points in phenotypic screening campaigns, reverting to TDD (Figure 18.1) [44].

For small molecules, the initial list of hits obtained in a phenotypic screening can be expanded based on structural similarity to compounds not initially included in the library subset selected for the screen but easily accessible in the original library or from commercial sources [45]. This provides a powerful capability to zoom in on a chemical family, saving valuable biological material and resources at the initial primary screen. This technique, known as hit expansion, offers the opportunity to have the first SAR information without additional chemical synthesis. Hit expansion can also be based on similarity in the bioactivity signature, further increasing the possibility of finding other chemical scaffolds sharing the activities of interest [46].

For molecules with little or no bioactivity annotation, information on target signatures can be initially captured using conventional selectivity panels which, despite their limitation, as they include only a finite number of pre-selected activities, can be useful to get a first snapshot of the targets engaged by representative hits [47, 48]. In contrast to TDD, where the focus of this preliminary profiling is to explore hit selectivity, in PDD this represents an opportunity to identify potential targets. In this context, as important as uncovering the MoA is getting early warnings of potential safety liabilities based either on direct hints from safety-oriented selectivity panels or on structural and/or bioactivity similarity to reference compounds having known safety/toxicity issues [49, 50].

In some cases, characterization of representative hit compounds that do not yet meet the minimal target candidate profile requirements can be a strategy to gather information on potential weaknesses of a hit series which can be useful to select key assays that will be used during hit and lead optimization. In addition, characterizing, and progressing compounds that may otherwise be discarded based on some of their properties can be highly informative, as they may harbor other valuable potential features required by a preclinical candidate: suitable physicochemical and drug metabolism and pharmacokinetics (DMPK) properties, potency, phenotypic selectivity, and absence of known liabilities.

18.4.2 Optimizing Polypharmacology

From a polypharmacology perspective, the key advantage of phenotypic activity-based optimization is that the phenotypic endpoint can integrate the effects of the combined modulation of all the targets engaged by the compound without introducing any external bias and without the need for a priori knowledge of the MoA at the screening and hit selection phases. As for any functional endpoint in cell-based assays, information integrated into a phenotypic endpoint would also include properties that may be affecting activity, such as poor solubility or cell membrane permeability, cytotoxicity, and indirect modulation through off-targets. This should also be factored in when interpreting structure-activity results and may require implementing added controls.

A representative example of this concept was the characterization of the main hit (compound **4**) described in Orellana et al. [51]. This compound was an inhibitor of thymic stromal lymphopoietin (TSLP) production identified in a phenotypic screen using human primary bronchial epithelial cells (NHBE) from 2 donors and a library of 44,974 small molecules. Extensive characterization of this hit evaluating >400 activities in 2 selectivity panels and a BioMAP® phenotypic platform [52], showed inhibition of multiple targets in the 0.1–10 μM IC_{50} range (Figure 18.3a). Among those activities, mTOR was identified as a potential on-target based on the similarity of the bioactivity signature of the compound to reference mTOR inhibitors in the BioMAP® platform, where it showed exquisite phenotypic selectivity and lack of cytotoxicity in NHBE cells. Inhibition of TSLP by reference selective mTOR inhibitors confirmed this was an on-target in these cells. However, the inhibition of mTOR activity by this hit (IC_{50} ~1 μM) was >10-fold less potent than its inhibition of TSLP production in the phenotypic assay (IC_{50} = 75 nM). In the absence of cytotoxicity and given its good phenotypic selectivity, this suggested the contribution of other targets in addition to mTOR to the activity of the compound in NHBE. Other potential contributions to TSLP inhibition were explored among the targets identified with IC_{50} < 5 μM using IBM Watson for Drug Discovery, a cognitive literature mining tool [53]. This led to the hypothesis that PIK3CD, PRKDC, and PIK3CG kinases could also be involved in TSLP downregulation, while the other targets identified were more likely to be off-targets as they were either not related to mTOR/TSLP or were functionally downstream from them. This is an example of the information that can be gathered for representative hits and that can be used to guide optimization and MoA characterization efforts in the presence of a polypharmacology signature.

Including sets of similar molecules with previously annotated targets in a phenotypic screen can uncover enriched targets in the more active hits. This can be helpful as a first assessment to discriminate potential on-targets from off-targets. A simplified example reproduced in Figure 18.3b shows the correlation of p38 and JAK but not Syk kinase inhibition with the phenotypic activity of screened compounds derived from legacy drug discovery programs for each of those targets and that were included in the library used for the phenotypic screen described in Orellana et al. [51].

The use of known drugs or reference molecules or probes engaging the targets identified can provide information on the role of these targets in mediating phenotypic activity at the single target or polypharmacology level. Having more than one probe for each target with different selectivity

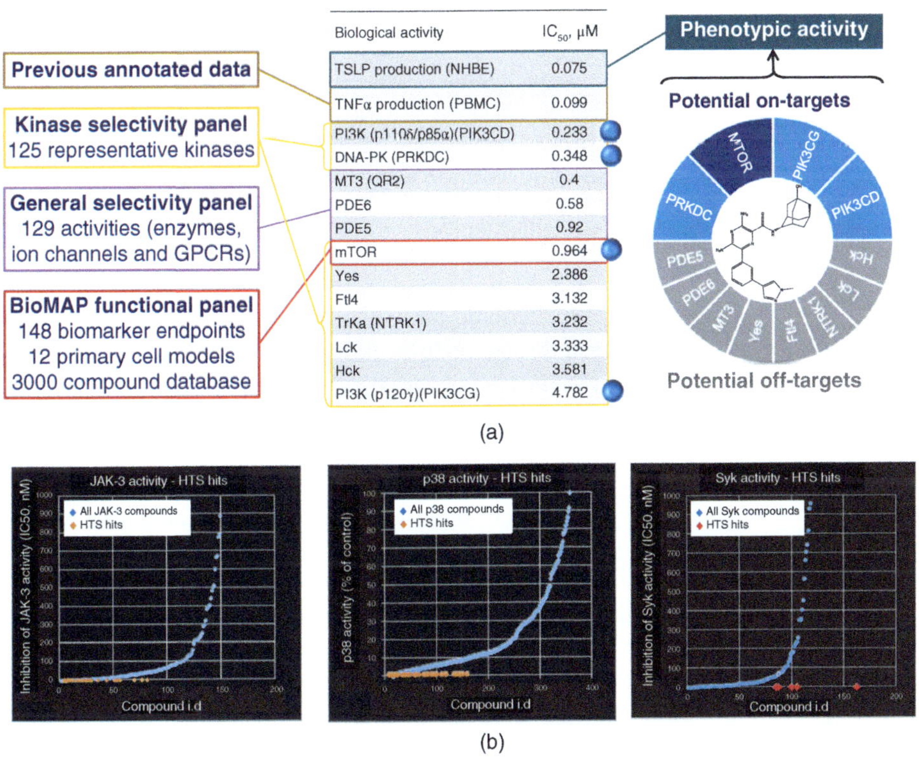

Figure 18.3 Characterization of phenotypic hits and potential targets in phenotypic screening. (a) Top 14 activities modulated by hit compound **4** among >400 activities evaluated in 2 selectivity panels and 12 primary cell-based systems (BioMAP® platform) [51, 52]. Potential on-targets are shown in blue and marked with blue circles (mTOR highlighted in darker blue)*. (b) Global view of the relative activity of compounds derived from p38, JAK, and Syk kinase inhibitors discovery programs present in the library and represented as blue dots, with potency for the indicated targets shown as IC_{50} or % of control activity (lower values in the vertical axis indicate more potent activity). Actives in the phenotypic screen are shown as orange-red dots in the compound i.d. axis. *Adapted from [51] under the terms of a Creative Commons Attribution License.

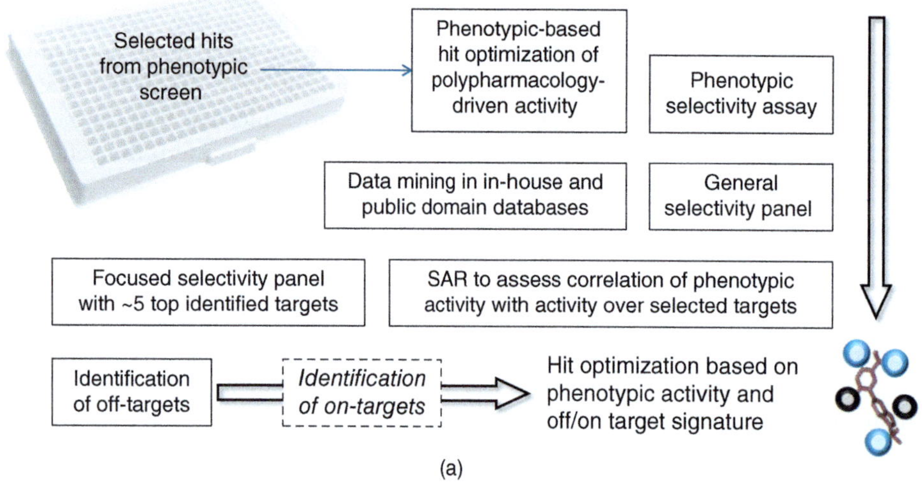

(a)

Hit and structural variants	Structural features				Relative modulation of phenotypic activity	Relative modulation of target activity				
	S1	S2	S3	S4		Target A	Target B	Target C	Target D	Target E
Starting hit #1	1	1	1	1	100	100	100	100	100	100
Compound #1.1	1	1	0	0	5	100	90	15	20	95
Compound #1.2	0	0	1	1	20	25	15	100	85	40
Compound #1.3	1	0	1	1	95	90	4	100	100	7
Compound #1.4	0	1	0	1	10	10	85	15	95	93
Compound #1.5	1	0	1	0	85	90	8	96	15	2
	Target A	Targets B & E	Target C	Target D		**on target**	off target	**on target**	off target	off target

(b)

Figure 18.4 Prototypical drug discovery optimization process in PDD. (a) Reference top to bottom and left to right screening cascade for hit optimization aimed at identifying on- and off-targets represented by blue and grey circles, respectively. (b) Prototypical SAR strategy to map the on-target and off-target signatures of a hit molecule, highlighting relative modulation of target activities (green) and phenotypic activity (blue).

signatures, and structurally unrelated, reduces the chances of potential confounding effects of poorly characterized reference molecules often considered selective at concentrations relevant for assessing the target of interest when they are not [54]. In Orellana et al., the weak relative activity of selective p38 reference inhibitors in the phenotypic TSLP production assay suggested that other targets in addition to p38 were likely contributing to the phenotypic activity of hits initially annotated as highly potent and selective p38 inhibitors [51].

As mentioned above, one key aspect during optimization in phenotypic screening is to assess phenotypic selectivity without full knowledge of the target or MoA. An example of the practical application of this concept is the evaluation of the phenotypic selectivity of translational read-through inducers in premature termination codon (PTC) mutations (translational stop codons appearing in a coding sequence) which, without full knowledge of the molecular MoA underlying the activity (e.g., interactions with ribosome proteins), relies on verifying non-specific read-through activity in natural stop codons [55]. In this case, the phenotypic endpoint is a gain-of-function readout based on measuring the expression of the full-length form of a protein whose coding RNA has a PTC mutation [56].

Without knowledge of the MoA, optimizing activity and selectivity of multi-target hits requires combining phenotypic endpoint-based optimization of polypharmacology-driven activity with a reverse engineering approach to identify, and attempt to reduce off-target interactions not necessary for the phenotypic activity. This is important to mitigate the risk of an unexpected safety liability appearing during subsequent molecule progression. Implementing this strategy demands a systematic target mapping and SAR approach that is outlined below and is summarized in Figure 18.4a:

1) Phenotypic activity-based optimization of hits and compounds identified by hit expansion or generated

in a first limited medicinal chemistry campaign. A reasonable effort at this stage to find the best possible starting points for further optimization is essential before embarking on more resource-intensive profiling and lead optimization activities.

2) Mapping of the targets engaged by the partially optimized hit derivatives using selectivity panels, previous annotations (in-house data), or information available in the public domain if the hits are previously described compounds.
3) Setting up a panel of assays to assess the correlation between the main targets identified in the previous step and the phenotypic activity of the hits. Typically, this can be a 2-concentrations-in-duplicate panel evaluating ~3–5 targets.
4) Generation of analogs with structural modifications and evaluation of their activity in the panel described above to conduct a systematic SAR analysis to map on-/off-targets based on their correlation with the phenotypic activity. As an example, this type of approach was used to improve the selectivity over A2A adenosine receptor, p38 kinase, and benzodiazepine receptor BZD of a translational read-through inducer identified in a phenotypic screen using expression of collagen VII as gain-of-function endpoint [57].
5) Selection of molecules that, while maintaining phenotypic activity, have the least number of off-target activities, even if the on-targets and MoA cannot yet be fully identified. A generic example of this process for a hypothetical hit showing dual-target polypharmacology is shown in Figure 18.4b.

In summary, in the presence of a multi-target MoA, phenotypic and target-based SAR can be combined to optimize a polypharmacology signature, reduce off-targets, and attempt identification of on-targets.

18.5 Understanding the MoA from a PDD and Polypharmacology Perspectives

Standard selectivity studies in representative hits derived from phenotypic screening can be extremely helpful for first optimization efforts but may not identify the complete on-target signature underlying the MoA of candidate molecules. For more advanced hits, gaining a sufficient understanding of the MoA is important for progression to the preclinical and clinical phases, allowing more informed decisions when significant resources, time, and patient well-being are at stake. However, this must be put into perspective so that efforts to unveil the MoA do not become a hurdle to access new drug opportunities [8] while, at the same time, sufficient information is gathered to reduce the risk of failure of a clinical candidate.

Beyond the limited selectivity assessment needed for the initial hit optimization effort, more in-depth profiling can be conducted for advanced compounds using focused selectivity panels which can be customized to include specific target classes, such as transcription factors or ion channels, based on the knowledge gained in the preliminary selectivity assessment of the hits. This type of analysis can identify potential liabilities when no clear MoA is yet obvious.

To complement the information obtained with selectivity panels, bulk transcriptomics combined with a gene ontology analysis [58, 59] in relevant cells or tissue exposed to the candidate molecules can provide a non-biased quick snapshot of the signaling pathways they may mobilize which can offer the first clues of the target/s being engaged.

Additional approaches can be used to gain further insight into the MoA. For example, on-/off-targets can be explored in the context of polypharmacology using methodologies such as the transcription factor reporter systems-based profiling methodology described by Medvedev et al. [60], but perhaps the most representative example is the Library of Integrated Network-Based Cellular Signatures (LINCS) platform [61, 62]. In LINCS, the data from molecules of interest can be compared with omics signatures obtained from cell lines treated with different classes of chemical and biological perturbagens. Similar omics-based profiling approaches that can be applied to advanced hits are machine learning-powered cell imaging [63 65] and, as exemplified above, platforms like the BioMAP® or BioPrint® panels [47, 52, 66], where the bioactivity signature of the molecule of interest can be compared with the signatures stored in a database of drugs and reference molecules previously tested. More recently developed methodologies based on systematic profiling of proteomic signatures also hold high potential to support MoA characterization studies [67].

Profiling based on comparing with previously compiled bioactivity signatures is a powerful approach that can provide a robust starting point to dissect the MoA of a candidate molecule based on the additional information available for the reference molecules. For example, the use of the BioMAP® platform contributed to the identification of the MoA of phenotypic hits inducing the differentiation of acute myeloid leukemia cell lines as anti-mitotic tubulin disruptors [68]. A potential pitfall of this strategy is the possibility that the reference molecules and drugs used to build these databases may not be selective or fully annotated at the concentrations where similarities with test compounds may be found. In this context, CRISPR-generated gene-selective knock-outs can offer another complementary source of evidence [69].

Selectivity panels, omics-based platforms, and phenotypic profiling tools are powerful but may not capture all the possible target engagements of a candidate drug. Despite this limitation, the combination of pharmacology data, transcriptomics, gene ontology, and similarity-based MoA identification can provide enough information on off-targets and potential on-targets/MoA to optimize and progress a candidate molecule without the pre-requisite of comprehensive target/s identification at earlier stages.

18.6 The Path to Virtual PDD-Derived Polypharmacology

Will it ever be possible to recreate disease biology with enough depth to build in silico models that could be used to run virtual phenotypic screens? This possibility may be close but is yet to become a reality [70, 71]. If it ever does, we will need to master the combination of all available accumulated knowledge that can be considered relevant for this purpose: scientific literature, bioactivity, drug, protein, genetic, disease, and patient data repositories (e.g., real-world data).

This concept was explored in work published by Sidders et al. [72], using scientific literature-based text mining to generate a chronic pain-specific target network based on information from reported protein-protein interactions. This network was then used to run a virtual screen with a library of annotated compounds to find a set of potentially active molecules engaging multiple targets. Additional polypharmacology space was explored using pairwise combinations of these compounds. Interestingly, the actives identified showed correlation with those obtained using an experimental *in vitro* phenotypic model of pain based on rat dorsal root ganglion neurons, one example being chlorpromazine. As also exemplified above [53], the application of existing literature text mining platforms for the identification of non-obvious functional links between biological entities (proteins, nucleic acids, macromolecular complexes), perturbagens, and diseases, is a highly attractive approach given the large quantity of information stored in PubMed, containing more than 36 million citations, many of them in open access format [73], provided that some level of appropriate filtering is applied (e.g., evidence from more than one research group). Sourcing other less structured information repositories like blogs or scientific discussion groups could be envisioned in the future, also assuming that strict filtering criteria are applied (e.g., authored by scientists with a publication track record) and leveraging the support of language-based artificial intelligence (AI) tools, which are proving particularly useful for text-based applications.

The above concept could be expanded and used in a systematic (i.e., industrialized) way to generate a digital library of disease models recapitulating relevant aspects of disease conditions and their variability and severity, integrating information from genome-wide association studies [74–76], transcriptomics [77, 78] and proteomics [67, 79] among other data sources. These models could be interrogated using bioactivity data from existing compounds with experimental and computationally predicted target annotations [80], to identify candidate molecules likely to have an impact on their molecular disease features.

The bioactivity signature of existing compounds could be further enriched using protein structural information to include predicted pharmacological activities even for proteins whose structure has not been solved, taking advantage of the latest advances in machine learning-guided 3D structure determination such as the Alphafold technology [81].

From a more general perspective, one additional hurdle must be considered. Key drug discovery stakeholders from industry and academia will need to accept the output from virtual and digital approaches so that later investments in experimental studies to confirm these results are supported. In this context, developing convincing quality standards for in silico disease models will be important to build the confidence needed to invest in the costly experimental resources required for proof-of-concept validation of in silico work. Otherwise, the vicious cycle of lack of experimental evidence blocking the progression of in silico-generated chemistry and biology that never gets validated will be perpetuated.

Given that many of the current *in vitro* and *in vivo* experimental models of disease are considered to only recapitulate partial aspects of disease complexity, and yet they are considered acceptable for their intended purpose, similar criteria should be applied to in silico models. Experimental models will still be required to support the progression of any candidate molecule resulting from virtual phenotypic screening. However, being able to run the primary screen in silico combining computational resources with publicly available databases and available in-house data would be a major step forward in the industrialization of PDD and the exploration of new polypharmacology landscapes.

18.7 Conclusions and Future Directions

Understanding the complexity of disease is a phenomenal challenge that can be overwhelming or just impossible to fully solve in many cases. Even for monogenic rare diseases, where the first trigger can be traced to single base mutations, symptoms typically vary between affected individuals due to different genetic and environmental backgrounds. For multifactorial polygenic pathologies such as immune-mediated CNS, dermatological, or respiratory diseases, the picture can be much more complex, with final symptoms resulting from the contribution of many genes along various stages of development and aging and subject to the influence of environmental factors.

Following the adoption of single target-friendly molecular biology approaches that revolutionized drug discovery at the end of the twentieth century, the re-introduction of phenotypic screening by the academic and industrial drug discovery communities over the last 20 years [1–5] has facilitated the acceptance of polypharmacology in addition to single target MoAs as a valid drug discovery strategy. Identification of hits boasting single target-based MoAs is just one of the potential outcomes of phenotypic screening. Often, complex, and not fully elucidated MoAs are being described for phenotypic drugs in reminiscence of the early drugs identified by classical phenotypic screening, where MoA determination was limited by available profiling technology. However, in contrast with the limitations encountered for the discovery of these early phenotypic drugs, we now have a toolbox of powerful profiling and target identification techniques that can deliver enough understanding of the MoA of a phenotypic hit to allow its progression to preclinical testing within a reasonable risk framework, even if its molecular basis cannot be fully elucidated [3].

For many existing long-proven drugs discovered in the last few decades either through phenotypic or target-based approaches, it has become evident that efficacy is based on engaging multiple targets [3, 7]. These polypharmacology MoAs have been found years after the drugs were approved, when the technology for in-depth profiling and target identification became standard practice.

Further standardization of the PDD process is still needed. This refers to the possibility of increasing model translatability, reproducibility, and throughput in a cost-effective manner and can take advantage of the increased availability of well-characterized human patient samples in addition to the improved automation and detection technologies that continue to evolve. The development of guidelines to ensure that minimum quality standards are met is also a critical aspect for the reproducibility, predictability, and credibility of the new drug candidates and their biology [24, 82].

In the future, the discovery of novel polypharmacology derived from phenotypic screening will rely on increasingly relevant disease models and advanced multi-omics profiling tools coupled with computational and machine learning/AI technologies to map relevant on-targets and dispensable off-target activities into a starting target signature. Optimization of this target signature must follow a systematic SAR approach to generate hits with the best possible on-/off-targets balance to reduce potential safety liabilities as they are progressed to preclinical characterization.

However, some key challenges remain before the drug discovery community can take full advantage of PDD-derived polypharmacology.

First, further integration of existing bioactivity datasets and a more quantitative correlation between annotated targets pharmacology of reference drugs and their activity, efficacy, and potential safety/toxicity are needed to better the accuracy of predictions based on bioactivity and omics signatures similarity.

Second, will machine learning/AI approaches live up to current expectations? Is it going to be possible to model disease biology with sufficient complexity to make in silico phenotypic screening a realistic alternative? Will multidimensional profiling technologies be capable of predicting the activity and potential side effects of new phenotypic drug candidates? These are all still open questions, but answers may be coming sooner than we expect with the exponential growth in biological data and AI methodologies that we are currently witnessing. This could further improve the predictive capabilities of profiling systems and the integration of existing knowledge already available across multiple public and private databases.

One other challenge remains to be fully addressed. Will all key stakeholders accept phenotypic screening-derived polypharmacology MoAs after decades of single target-based drugs? The answer must be yes. We should be able to overcome any negative perception by providing sufficient molecular knowledge of the omics landscape carved by multi-target candidates using well-characterized marketed drugs originally identified using phenotypic screening and with post hoc uncovered polypharmacology MoA, as success examples and proof of feasibility.

Phenotypic screening offers a unique vantage point to unleash the full potential of polypharmacology using multi-target "grade" libraries, ever more disease-relevant models, and state-of-the-art profiling coupled with AI-based tools, leveraging a translational approach that recognizes the high complexity of biological systems and disease.

References

1 Swinney, D.C. and Anthony, J. (2011). How were new medicines discovered? *Nat. Rev. Drug Discov.* 10: 507–519.

2 Moffat, J.G., Vincent, F., Lee, J.A. et al. (2017). Opportunities and challenges in phenotypic drug discovery: an industry perspective. *Nat. Rev. Drug Discov.* 16: 531–543.

3 Vincent, F., Nueda, A., Lee, J. et al. (2022). Phenotypic drug discovery: recent successes, lessons learned and new directions. *Nat. Rev. Drug Discov.* 21: 899–914.

4 Swinney, D.C. and Lee, J.A. (2020). Recent advances in phenotypic drug discovery. *F1000Research* 9: F1000 Faculty Rev-944.

5 Butcher, E.C. (2005). Can cell systems biology rescue drug discovery? *Nat. Rev. Drug Discov.* 4: 461–467.

6 Eder, J., Sedrani, R., and Wiesmann, C. (2014). The discovery of first-in-class drugs: origins and evolution. *Nat. Rev. Drug Discov.* 13: 577–587.

7 Sadri, A. (2023). is target-based drug discovery efficient? Discovery and "off-target" mechanisms of all drugs. *Chem* 66: 12651–12677.

8 Davis, R.L. (2020). Mechanism of action and target identification: a matter of timing in drug Discovery. *iScience* 23: 101487.

9 Alexandrov, V., Brunner, D., Hanania, T., and Leahy, E. (2015). High-throughput analysis of behavior for drug discovery. *Eur. J. Pharmacol.* 750: 82–89.

10 Lipinski, C.A. and Reaume, A.G. (2020). High throughput in vivo phenotypic screening for drug repurposing: discovery of MLR-1023 a novel insulin sensitizer and novel Lyn kinase activator with clinical proof of concept. *Bioorg. Med. Chem.* 28: 115425.

11 Tokarski, J.S., Zupa-Fernandez, A., Tredup, J.A. et al. (2015). Tyrosine kinase 2-mediated signal transduction in T lymphocytes is blocked by pharmacological stabilization of its pseudokinase domain. *J. Biol. Chem.* 24: 11061–11074.

12 Al-Ali, H., Bixby, J.L., and Lemmon, V.P. (2016). Exploiting kinase polypharmacology for nerve regeneration. *Neural Regen. Res.* 11: 71–72.

13 Brown, W.A. and Rosdolsky, M. (2015). The clinical discovery of imipramine. *Am. J. Psychiatry* 172: 426–429.

14 Helmchen, H. (2023). Alltägliche Grenzüberschreitungen: zur Skandalisierung der klinischen Arzneimittelprüfungen des Psychiaters Roland Kuhn [Everyday transgressions of borderlines: the scandalization of clinical drug trials of the psychiatrist Roland Kuhn]. *Nervenarzt* 94: 243–249.

15 Gillman, P.K. (2007). Tricyclic antidepressant pharmacology and therapeutic drug interactions updated. *Br. J. Pharmacol.* 151: 737–748.

16 Casarotto, P.C., Girych, M., Fred, S.M. et al. (2021). Antidepressant drugs act by directly binding to TRKB neurotrophin receptors. *Cell* 184: 1299–1313.

17 Timilsina, S., Rajamanickam, S., Rao, A. et al. (2022). The antidepressant imipramine inhibits breast cancer growth by targeting estrogen receptor signaling and DNA repair events. *Cancer Lett.* 540: 215717.

18 Wang, Y., Wang, X., Wang, X. et al. (2021). Imipramine impedes glioma progression by inhibiting YAP as a Hippo pathway independent manner and synergizes with temozolomide. *J. Cell. Mol. Med.* 25: 9350–9363.

19 Dedic, N., Jones, P.G., Hopkins, S.C. et al. (2019). SEP-363856, a novel psychotropic agent with a unique, non-D2 receptor mechanism of action. *J. Pharmacol. Exp. Ther.* 371: 1–14.

20 Achtyes, E.D., Hopkins, S.C., Dedic, N. et al. (2023). Ulotaront: review of preliminary evidence for the efficacy and safety of a TAAR1 agonist in schizophrenia. *Eur. Arch. Psychiatry Clin. Neurosci.* 273: 1543–1556.

21 Nair, P.C., Miners, J.O., McKinnon, R.A. et al. (2022). Binding of SEP-363856 within TAAR1 and the 5HT1A receptor: implications for the design of novel antipsychotic drugs. *Mol. Psychiatry* 27: 88–94.

22 Nishida-Aoki, N. and Gujral, T.S. (2022). Polypharmacologic reprogramming of tumor-associated macrophages toward an inflammatory phenotype. *Cancer Res.* 82: 433–446.

23 Bae, S.H., Kim, J.H., Park, T.H. et al. (2022). BMS794833 inhibits macrophage efferocytosis by directly binding to MERTK and inhibiting its activity. *Exp. Mol. Med.* 54: 1450–1460.

24 Vincent, F., Loria, P., Pregel, M. et al. (2015). Developing predictive assays: the phenotypic screening "rule of 3". *Sci. Transl. Med.* 7: 293ps15.

25 Mondschein, C.F. and Monda, C. (2019). The EU's general data protection regulation (GDPR) in a research context. In: *Fundamentals of Clinical Data Science* (ed. P. Kubben, M. Dumontier, and A. Dekker), 55–71. Cham: Springer; Chapter 5.

26 Staunton, C., Slokenberga, S., Parziale, A., and Mascalzoni, D. (2022). Appropriate safeguards and article 89 of the GDPR: considerations for biobank, databank and genetic research. *Front. Genet.* 18: 719317.

27 Mayrhofer, M.T., Holub, P., Wutte, A., and Litton, J.E. (2016). BBMRI-ERIC: the novel gateway to biobanks. From humans to humans. *Bundesgesundheitsblatt Gesundheitsforschung Gesundheitsschutz* 59: 379–384.

28 Reinecke, M., Brear, P., Vornholz, L. et al. (2024). Chemical proteomics reveals the target landscape of

1000 kinase inhibitors. *Nat. Chem. Biol.* 20: 577–585. https://doi.org/10.1038/s41589-023-1459-3.

29 Bhatta, P., Whale, K.D., Sawtell, A.K. et al. (2021). Bispecific antibody target pair discovery by high-throughput phenotypic screening using in vitro combinatorial Fab libraries. *MAbs* 13: 1859049.

30 Wassermann, A.M., Camargo, L.M., and Auld, D.S. (2014). Composition and applications of focus libraries to phenotypic assays. *Front. Pharmacol.* 5: 164.

31 Jörg, M. and Madden, K.S. (2021). The right tools for the job: the central role for next generation chemical probes and chemistry-based target deconvolution methods in phenotypic drug discovery. *RSC Med. Chem.* 12: 646–665.

32 Antolin, A.A., Sanfelice, D., Crisp, A. et al. (2023). The chemical probes portal: an expert review-based public resource to empower chemical probe assessment, selection and use. *Nucleic Acids Res.* 51: D1492–D1502.

33 Wawer, M.J., Li, K., Gustafsdottir, S.M. et al. (2014). Toward performance-diverse small-molecule libraries for cell-based phenotypic screening using multiplexed high-dimensional profiling. *Proc. Natl. Acad. Sci. USA* 111: 10911–10916.

34 Peterson, A.A. and Liu, D.R. (2023). Small-molecule discovery through DNA-encoded libraries. *Nat. Rev. Drug Discov.* 22: 699–722.

35 Cai, B., Kim, D., Akhand, S. et al. (2019). Selection of DNA-encoded libraries to protein targets within and on living cells. *J. Am. Chem. Soc.* 141: 17057–17061.

36 Petersen, L.K., Christensen, A.B., Andersen, J. et al. (2021). Screening of DNA-encoded small molecule libraries inside a living cell. *J. Am. Chem. Soc.* 143: 2751–2756.

37 Cochrane, W.G., Fitzgerald, P.R., and Paegel, B.M. (2021). Antibacterial discovery via phenotypic DNA-encoded library screening. *ACS Chem. Biol.* 16: 2752–2756.

38 Miethke, M., Pieroni, M., Weber, T. et al. (2021). Towards the sustainable discovery and development of new antibiotics. *Nat. Rev. Chem.* 5: 726–749.

39 van Vlijmen, H., Pannifer, A.D., Cochrane, P. et al. (2024). The European Lead Factory: Results from a decade of collaborative, public-private, drug discovery programs. *Drug Discov. Today* 29: 103886.

40 Brennecke, P., Rasina, D., Aubi, O. et al. (2019). EU-OPENSCREEN: a novel collaborative approach to facilitate chemical biology. *SLAS Discov* 24: 398–413.

41 Kim, S., Chen, J., Cheng, T. et al. (2023). PubChem 2023 update. *Nucleic Acids Res.* 6: D1373–D1380.

42 Zdrazil, B., Felix, E., Hunter, F. et al. (2024). The ChEMBL Database in 2023: a drug discovery platform spanning multiple bioactivity data types and time periods. *Nucleic Acids Res.* 5: D1180–D1192.

43 Dahlin, J.L., Auld, D.S., Rothenaigner, I. et al. (2021). Nuisance compounds in cellular assays. *Cell Chem. Biol.* 28: 356–370.

44 Jones, L.H. and Bunnage, M.E. (2017). Applications of chemogenomic library screening in drug discovery. *Nat. Rev. Drug Discov.* 16: 285–296.

45 Klutzny, S., Kornhuber, M., Morger, A. et al. (2022). Quantitative high-throughput phenotypic screening for environmental estrogens using the E-Morph Screening Assay in combination with in silico predictions. *Environ. Int.* 158: 106947.

46 Riniker, S., Wang, Y., Jenkins, J.L., and Landrum, G.A. (2014). Using information from historical high-throughput screens to predict active compounds. *J. Chem. Inf. Model.* 54: 1880–1891.

47 Bendels, S., Bissantz, C., Fasching, B. et al. (2019). Safety screening in early drug discovery: an optimized assay panel. *J. Pharmacol. Toxicol. Methods* 99: 106609.

48 Bowes, J., Brown, A.J., Hamon, J. et al. (2012). Reducing safety-related drug attrition: the use of in vitro pharmacological profiling. *Nat. Rev. Drug Discov.* 11: 909–922.

49 Wassermann, A.M., Lounkine, E., and Glick, M. (2013). Bioturbo similarity searching: combining chemical and biological similarity to discover structurally diverse bioactive molecules. *J. Chem. Inf. Model.* 53: 692–703.

50 Berg, E.L. (2019). Human cell-based in vitro phenotypic profiling for drug safety-related attrition. *Front. Big Data* 2: 47.

51 Orellana, A., García-González, V., López, R. et al. (2018). Application of a phenotypic drug discovery strategy to identify biological and chemical starting points for inhibition of TSLP production in lung epithelial cells. *PLoS ONE* 10: e0189247.

52 Berg, E.L. (2017). Phenotypic chemical biology for predicting safety and efficacy. *Drug Discov. Today Technol.* 23: 53–60.

53 Chen, Y., Elenee Argentinis, J.D., and Weber, G. (2016). IBM Watson: how cognitive computing can be applied to big data challenges in life sciences research. *Clin. Ther.* 38: 688–701.

54 Sterling, J., Baker, J.R., McCluskey, A., and Munoz, L. (2022). Systematic literature review reveals suboptimal use of chemical probes in cell-based biomedical research. *Nat. Commun.* 14: 3228.

55 Crawford, D.K., Alroy, I., Sharpe, N. et al. (2020). ELX-02 generates protein via premature stop codon read-through without inducing native stop codon

read-through proteins. *J. Pharmacol. Exp. Ther.* 374: 264–227.

56 Nagel-Wolfrum, K., Möller, F., Penner, I. et al. (2016). Targeting nonsense mutations in diseases with translational read-through-inducing drugs (TRIDs). *BioDrugs* 30: 49–74.

57 Jover, I., Ramos, M.C., Escámez, M.J. et al. (2024). Identification of novel small molecule-based strategies of COL7A1 upregulation and readthrough activity for the treatment of recessive dystrophic epidermolysis bullosa. *Sci. Rep.* 14: 18969.

58 Gene Ontology Consortium (2021). The gene ontology resource: enriching a gold mine. *Nucleic Acids Res.* 49: D325–D334.

59 Subramanian, A., Tamayo, P., Mootha, V.K. et al. (2005). Gene set enrichment analysis: a knowledge-based approach for interpreting genome-wide expression profiles. *Proc. Natl. Acad. Sci. U.S.A* 102: 15545–15550.

60 Medvedev, A., Moeser, M., Medvedeva, L. et al. (2019). Evaluating biological activity of compounds by transcription factor activity profiling. *Sci. Adv.* 4: eaar4666.

61 Iwata, M., Sawada, R., Iwata, H. et al. (2017). Elucidating the modes of action for bioactive compounds in a cell-specific manner by large-scale chemically-induced transcriptomics. *Sci. Rep.* 7: 40164.

62 Xie, Z., Kropiwnicki, E., Wojciechowicz, M.L. et al. (2022). Getting started with LINCS datasets and tools. *Curr Protoc.* 2: e487.

63 Zoffmann, S., Vercruysse, M., Benmansour, F. et al. (2019). Machine learning-powered antibiotics phenotypic drug discovery. *Sci. Rep.* 21: 5013.

64 Chandrasekaran, S.N., Ceulemans, H., Boyd, J.D., and Carpenter, A.E. (2020). Image-based profiling for drug discovery: due for a machine-learning upgrade? *Nat. Rev. Drug Discov.* 20: 145–159.

65 Rohban, M.H., Fuller, A.M., Tan, C. et al. (2022). Virtual screening for small-molecule pathway regulators by image-profile matching. *Cell Syst.* 13: 724–736.

66 Plavec, I., Sirenko, O., Privat, S. et al. (2004). Method for analyzing signaling networks in complex cellular systems. *Proc. Natl. Acad. Sci. USA* 101: 1223–1228.

67 Mitchell, D.C., Kuljanin, M., Li, J. et al. (2023). A proteome-wide atlas of drug mechanism of action. *Nat. Biotechnol.* 41: 845–857.

68 Jackson, T.R., Vuorinen, A., Josa-Culleré, L. et al. (2022). A tubulin binding molecule drives differentiation of acute myeloid leukemia cells. *iScience* 25: 104787.

69 Matsumoto, K. and Yoshida, M. (2022). Mammalian chemical genomics towards identifying targets and elucidating modes-of-action of bioactive compounds. *ChemBioChem* 5: e202100561.

70 Barh, D., Yiannakopoulou, E.C., Salawu, E.O. et al. (2014). In silico disease models: from simple networks to complex diseases. In: *Anim. Biotechnol.* (ed. A.S. Verma and A. Singh), 385–404.

71 Moingeon, P., Chenel, M., Rousseau, C. et al. (2023). Virtual patients, digital twins and causal disease models: paving the ground for in silico clinical trials. *Drug Discov. Today* 28: 103605.

72 Sidders, B., Karlsson, A., Kitching, L. et al. (2018). Network-based drug discovery: coupling network pharmacology with phenotypic screening for neuronal excitability. *J. Mol. Biol.* 430: 3005–3015.

73 He, F., Liu, K., Yang, Z. et al. (2023). Applications of cutting-edge artificial intelligence technologies in biomedical literature and document mining. *Med. Rev.* 3: 200–204.

74 Mailman, M.D., Feolo, M., Jin, Y. et al. (2007). The NCBI dbGaP database of genotypes and phenotypes. *Nat. Genet.* 39: 1181–1186.

75 Landrum, M.J., Lee, J.M., Benson, M. et al. (2016). ClinVar: public archive of interpretations of clinically relevant variants. *Nucleic Acids Res.* 44: D862–D868.

76 Sollis, E., Mosaku, A., Abid, A. et al. (2023). The NHGRI-EBI GWAS Catalog: knowledgebase and deposition resource. *Nucleic Acids Res.* 51: D977–D985.

77 Barrett, T., Wilhite, S.E., Ledoux, P. et al. (2013). NCBI GEO: archive for functional genomics data sets—update. *Nucleic Acids Res.* 41: D991–D995.

78 Parkinson, H., Kapushesky, M., Kolesnikov, N. et al. (2009). ArrayExpress update—from an archive of functional genomics experiments to the atlas of gene expression. *Nucleic Acids Res.* 37: D868–D872.

79 Uhlén, M., Fagerberg, L., Hallström, B.M. et al. (2015). Proteomics. Tissue-based map of the human proteome. *Science* 347: 1260419.

80 Falaguera, M.J. and Mestres, J. (2023). Illuminating the chemical space of untargeted proteins. *J. Chem. Inf. Model.* 8: 2689–2698.

81 Jumper, J., Evans, R., Pritzel, A. et al. (2021). Highly accurate protein structure prediction with AlphaFold. *Nature* 596: 583.

82 Quancard, J., Bach, A., Cox, B. et al. (2021). The european federation for medicinal chemistry and chemical biology (EFMC) best practice initiative: phenotypic drug discovery. *ChemMedChem* 16: 1736–1739.

19

Phenotypic Polypharmacology Drug Discovery for CNS Applications

Alberto Ambesi-Impiombato, Lee McDermott, Alan Lars Pehrson, and Daniela Brunner

19.1 Introduction

We would like to believe that as our understanding of biological systems and the molecular basis of diseases improves over time, we are better equipped to rationally design new, more effective, even patient-customized drug therapies that rectify the malfunctioning cellular processes causing disease. This is certainly the case in some areas of medicine, such as oncology, for example, where our understanding of the molecular processes underlying the disease has advanced greatly. Technological progress in molecular diagnostics and improved understanding of the mechanisms of oncogenesis, cancer progression, and resistance have empowered researchers to craft therapies targeted toward the specific genetic composition of cancer cells, and often to tailor them at the individual patient level. In this context, the design and discovery of new drugs is guided by knowledge of what molecular target needs to be suppressed or modulated. This drug discovery approach, where compounds are developed based on their activity on their molecular target, is collectively referred to as target-based drug discovery (TDD). Unfortunately, in the realm of central nervous system (CNS) diseases, our understanding of the disease processes is much more limited, in part because of the immense complexity of the brain. Moreover, it has become clear that more than one target may need to be modulated for a drug to be effective in psychiatric patients, and TDD is not well suited to address polypharmacology, i.e., the ability of a drug to achieve its therapeutic activity by concurrently modulating multiple targets [1]. Unsurprisingly, successful attempts to rationally design CNS drugs with novel mechanisms through TDD have been rare [2].

In this chapter, we will explore approaches to drug discovery that may be better suited for tackling the specific challenges of CNS drug discovery. After briefly reviewing the history of psychopharmacology, we will argue that methods that do not require assumptions on the molecular underpinnings of brain disease are better positioned to facilitate the discovery of new CNS drugs with completely novel mechanism of actions (MOAs). We believe that approaching CNS drug discovery without preconceived notions on receptor/protein targets not only will help address the unmet need of more effective neuropsychiatric drugs, but also help us gain new insights on the pathophysiology of brain diseases. Approaching CNS drug discovery in a mechanism-agnostic fashion can be achieved through phenotypic drug discovery (PDD), where compounds are screened by measuring a *phenotypic* outcome, rather than their effects on one or more brain receptors. The readout of PDD can be a biomarker, or other *in vitro* or *ex vivo* data, or *in vivo* data from the whole animal. We will focus and provide data from an *in vivo* behavioral PDD (BPDD). We propose that one of the most promising BPDD strategies is based on measuring the compounds' global effects on behavior in laboratory animals rather than a single outcome of a behavioral assay. We will illustrate, with examples, that if the behavioral readout is comprehensive, automated, and therefore standardized, it is possible to expedite the drug discovery process in CNS. This acceleration can be achieved in a cost-effective manner, while inherently fostering polypharmacology and even facilitating the identification of entirely novel MOAs.

19.2 BPDD Lessons from the History of Psychopharmacology

The first step in selecting the best approach to tackle a complex challenge is to look back and identify historical patterns of strategies that have worked and those that have not. A critical review of the history of CNS pharmacology,

Polypharmacology: Strategies for Multi-Target Drug Discovery, First Edition. Edited by Jens-Uwe Peters.

and psychopharmacology in particular, reveals that phenotypic observations, often accidental, have typically advanced our knowledge. It has been often the case that it was only after studying the pharmacology of novel drugs in laboratory animals or in patients that insights into the molecular mechanisms underlying brain pathophysiology were drawn (and not vice versa, with biological insights driving drug discovery). Indeed, the "psychopharmacology revolution" began in the 1950s as physicians and scientists observed that drugs used for treating non-psychiatric conditions had noticeable effects on psychiatric symptoms [3]. Calling it a *revolution* is by no means an exaggeration, as the introduction of chlorpromazine, the first antipsychotic brought to market, triggered a profound societal shift in our approach to mental illness. This shift dramatically improved the approaches to psychiatric treatment and their effectiveness to such an extent that many schizophrenic patients could be reintegrated into society, thus contributing to rendering long-term psychiatric institutionalization obsolete [4].

19.2.1 Antipsychotics and Antidepressants

Chlorpromazine was discovered by noticing its neuroleptic effects when administered as an adjunct anesthetic [3]. Similarly, positive effects on mood by iproniazid, a monoaminoxidase inhibitor (MAOI), were first discovered in patients otherwise receiving it for tuberculosis [5]. Imipramine, a tricyclic antidepressant (TCA), was also discovered serendipitously, as it improved symptoms in patients with severe depression, at a time when it was considered a putative antipsychotic [6]. It was the investigation into the mechanism of action (MOA) of these early psychotropic drugs and specifically their modulation of serotonin (5HT), norepinephrine (NE), and dopamine (DA) that led to the formulation of the various theories of "neurochemical imbalance" in psychiatric disorders [7]. As such, the finding that antidepressants enhanced the concentration of monoamine neurotransmitters led to the "monoaminergic hypothesis" of mood disorders. On the other hand, the observation that antipsychotic drugs shared antagonism of D2 receptors – while displaying diverse monoamine receptor binding profiles – led to the "dopamine hypothesis" of schizophrenia [8]. Following the introduction of chlorpromazine, many other D2 antagonists were developed, becoming what are now referred to as *typical antipsychotics* [9]. Haloperidol, for instance, was discovered in 1958 by the Belgian pharmacologist Paul Janssen while investigating the structure-activity relationship (SAR) of analogs of meperidine as putative analgesic compounds [10]. Janssen noticed that one of the compounds in the series, haloperidol, induced an unexpected trance-like behavior in mice, which resembled the effects of chlorpromazine, effectively pioneering a BPDD approach. Following the success of the first-generation antipsychotics, the discovery of clozapine in 1972 [9], with a MOA extending well beyond dopamine (including important action at 5HT receptors), spearheaded a second wave of development of antipsychotics (the so-called *atypical* antipsychotics). These newer drugs promised fewer neurological side effects (i.e., extrapyramidal side effects, or EPSEs) and increased effectiveness, especially in terms of the difficult-to-treat negative symptoms (e.g., social withdrawal, anhedonia, avolition, and blunted affect). Atypical antipsychotics, including clozapine, olanzapine, risperidone, and aripiprazole, among others, were developed and brought to market, on the polypharmacological hypothesis that a clozapine-like 5HT/DA binding profile would result in improved tolerability and efficaciousness on negative symptoms. These second-generation antipsychotics are widely used for the treatment of schizophrenia and other psychotic disorders. The initial enthusiasm for atypical antipsychotics, however, has waned over time, as evidence supporting their improved tolerability [9] and efficacy on negative symptoms [11] may not be as robust as initially thought, except for clozapine, which continues to offer distinct beneficial effects despite the complexities associated with its use [12].

19.2.2 Other Antidepressants

Following the introduction of the TCAs, the development of the selective serotonin reuptake inhibitors (SSRIs) is an exception to the BPDD trend. As the role of serotonin began to be understood to be the most relevant among other monoamine neurotransmitters' role in depression, a TDD strategy led to the development of fluoxetine, the first SSRI, developed to selectively block the serotonin transporter (SERT) [13]. Similarly, citalopram, and particularly its S-enantiomer escitalopram, were subsequently developed because of their even greater selectivity for SERT [14]. The introduction of fluoxetine was so successful that it spearheaded a cultural phenomenon, sometimes termed "Prozac Nation" (from fluoxetine's trade name), promoting societal awareness of mental health and acceptance of mental health issues. The resulting enthusiasm within pharma possibly had the effect of validating the TDD approach that led to fluoxetine's discovery. It can be argued, however, that such enthusiasm for TDD strategies in psychopharmacology was misguided for a few reasons. First, by targeting the reuptake transporter of a neurotransmitter, as is the case for SSRIs, the increased availability of the neurotransmitter in the synaptic cleft will affect the various pre- and post-synaptic subtypes of the neurotransmitter, rendering the resulting pharmacology effectively akin to polypharmacology. Second, while SSRIs may have a better tolerability profile than other antidepressants with broader polypharmacology, such as TCAs or MAOIs, they are not superior in terms of their therapeutic effects [15]. Indeed, patients that do not respond to SSRIs are treated with antidepressants with broader polypharmacology, or even with augmentation strategies adding a second drug (i.e., *polypharmacy*). Furthermore, treatment-resistant patients can benefit from electroconvulsive therapy (ETC), arguably the least specific treatment for depression [15, 16]. Thus, for schizophrenia and depression, it is plausible that greater efficacy might depend on polypharmacology, rather than single-target pharmacology, as "dirty" or "promiscuous" drugs are often more effective than "clean" highly specific drugs [17].

19.2.3 Mood Stabilizers

Another example of a "dirty" but effective drug is lithium, still used today for the treatment of bipolar disorders. This drug has the simplest chemical structure possible, consisting of a single ion, Li^+, which contrasts with the complexity of its MOA [18]. The mood-stabilizing properties of lithium were recognized accidentally by Australian scientist John Cade in the late 1940s, while he was studying the toxicity of urine from bipolar patients [19]. Hypothesizing a toxic metabolite as the cause of mania, he began testing the effects of intraperitoneal injections of patients' urine in guinea pigs, which indeed appeared to be more toxic than control urine samples. However, when he injected uric acid solutions to investigate its role in urine's toxicity, he noticed that lithium salts unexpectedly reduced its toxicity. He then tested the effects of lithium salts alone in guinea pigs and noticed it's calming properties [19], in another example of target-agnostic BPDD. After self-experimenting to assess toxicity in humans, a widespread practice at the time, he decided to test it on psychiatric patients and found it to be beneficial in manic patients [19].

19.2.4 Anxiolytics and Sedatives

As with the SSRIs, target discovery for anxiolytics and sedatives started with serendipitous BPDD findings. As it happens, many substances found in nature have been sought and consumed throughout humanity's history, due to their beneficial effects counteracting stress and anxiety. These substances include alcohol, opium alkaloids, and other narcotic plants [20]. At the beginning of modern pharmacology, the first sedative drug was barbital (diethyl-barbiturate), discovered in 1903 by the German pharmacologist Von Mering, who was led by the intuition that ethyl groups had hypnotic properties [20]. Meprobamate, which became the first "blockbuster" psychotropic drug, at the time called a "tranquilizer" (and later "minor tranquilizer," to distinguish it from neuroleptics or "major tranquilizers," now called antipsychotics) [21] was developed to overcome the shortcomings of mephenesin, whose calming effects were first noticed in rodents while searching for a preservative for penicillin [2, 22]. Finally, chlordiazepoxide (Librium), the first benzodiazepine, was "officially" discovered in 1955 by Dr. Sternbach during a laboratory cleanup, although the truth is that he had synthesized the drug several months earlier just because he enjoyed working on the benzodiazepine privileged structure [23]. The drug proved to be an anxiolytic after observing its effects on muscle tone and alertness in mice and cats. The discovery of this pharmacophore [2] constituted one of the major breakthroughs for both CNS pharmacology and the understanding of mental disorders, even if initially the MOA was completely unknown [24]. Nowadays, benzodiazepines are commonly prescribed for the short-term treatment of anxiety disorders, with antidepressants, calcium modulators, and 5HT1A agonists preferred for long-term treatments [25].

19.3 Current Trends in Psychopharmacology

19.3.1 Lessons from History

As discussed above, great advances in our knowledge of mental health over the past decades were largely due to target-agnostic *in vivo* PDD, often fortuitous, followed by careful elucidation of CNS drug's mechanisms of action. It is possible, therefore, that the next breakthrough that will expand our pharmacological arsenal for CNS disorders will also stem from *in vivo* phenotypic observations. Supporting this view is the observation that mental health disorders are typically polygenic, often modulated by the upbring environment and living conditions. As we have learned from the pharmacological characterization of many drugs routinely and widely prescribed, effective drugs often require simultaneous modulation of multiple receptor targets [26–28]. As such, small molecule *de novo* design efforts (either receptor- or ligand-based) that rely on a single target have not been proven to be optimal for CNS drug discovery [29–31].

Examining the drug discovery efforts of "big pharma" in the more recent past, it seems that TDD approaches have largely failed to bring to market new and better first-in-class drugs. Indeed, the low approval rate for CNS drugs over the past few decades has prompted numerous big pharma companies to reduce or abandon psychiatric drug discovery efforts altogether [32]. The huge unmet medical need and sheer scale of the psychotropic drug market, however, is pushing pharmaceutical companies to reconsider the pursuit of these indications by embracing novel solutions. At the very least, a multi-target pharmacology approach would seem to be superior to the classical single-target pharmacology strategy. While multi-target drug design is possible with receptor- and ligand-based approaches [22, 33], it is limited by considering only a subset of all possible receptor/protein targets that may impact the disease phenotype if one knew which target set to consider and in which specific proportion each target needs to be modulated. Target-agnostic PDD approaches, on the other hand, measure the phenotypic outcome of a compound's activity rather than its effect on one or more molecular targets. This is particularly effective when animal models exist that recapitulate the pathophysiology underlying particular diseases, that is, animal models that have construct and etiological validity [34]. PDD seems, therefore, to be the best way to move forward in discovering novel CNS therapies where polypharmacology is key to efficacy [35]. In fact, as the history of modern psychopharmacology shows, more than 80% of new, first-in-class CNS treatments were discovered phenotypically [1, 36, 37], rather than by pursuing a TDD approach. We believe that, rather than wait to be helped by serendipity, industry needs to strategically and methodically embrace BPDD approaches for the discovery of novel CNS treatments. Such BPDD systems should be high-throughput, unbiased, and target-agnostic, and should not overtly rely on our limited knowledge of brain disorders. Indeed, target-agnostic approaches can enable the discovery of novel mechanisms of action, including complex polypharmacology action.

19.3.2 New Trends in Industry and Biotech

Recent efforts focused on the development of novel psychedelic/entheogen drugs for unmet needs (such as treatment-resistant depression (TRD), post-traumatic stress disorder (PTSD), and substance use disorders (SUD)) constitute a revival of sorts of the investigation of the psychedelic's therapeutic potential during the 1950s and 1960s. Regardless of early studies corroborating a remarkable potential for psychedelics, a legally mandated, decades-long global "war on drugs" all but blocked any further development or scientific research for this class of compounds [38]. Pushing back against the misguided "war on drugs," the recent interest in psychedelic research, sometimes referred to as the *psychedelic renaissance* [39], resulted in a series of landmark achievements, including positive outcomes of clinical trials for psilocybin-assisted therapy for TRD (Phase II) [40], and methylenedioxymethamphetamine (MDMA) for PTSD (Phase III) [41], as well as the FDA approval of Esketamine (the *S* enantiomer of ketamine) for the treatment of depression in adults (albeit with some controversy) [42]. Many psychedelics have complex pharmacological profiles, but the exact mechanism driving their putative therapeutic properties is not yet understood. Indeed, psychedelics pose special challenges, for instance, in how to even assess their efficacy, as it is difficult to design studies that are truly blind, as the psychedelic effects are easily recognizable by patients and experimenters. In addition, there is no consensus as to whether the psychedelic experience itself is an integral part of the therapeutic mechanism.

Scientists believing that the psychedelic experience is not necessarily required for the therapeutic efficacy are actively pursuing programs with derivative compounds that do not induce hallucinations or similar effects, with the goal of identifying equally effective compounds with improved tolerability. At the molecular level, these different approaches track different downstream effects, the efficacy of which can only be determined phenotypically. The recently approved esketamine has a superior therapeutic profile compared to classical antidepressants not only because it is effective in patient populations that did not respond to other treatments (TRD), but also because its antidepressant effects are rapid and sustained. In contrast, traditional antidepressants typically require at least two to three weeks of daily administration to begin showing any clinical effect. Ketamine is a N-methyl-D-aspartate receptor (NMDAR) blocker, among other pharmacological actions. The precise mechanisms of its antidepressant action are not fully understood, and several theories have been proposed [43]. Interestingly, only a few decades ago, before its antidepressant properties were known, ketamine was widely used in rodents as a psychotomimetic, i.e., a "tool" compound to model psychosis, because of the many similarities between its effects in humans and the symptoms of schizophrenia. Once again, this highlights how the study of the pharmacology of psychotropic drugs has shaped our understanding of biological psychiatry, but also how limited this understanding often is.

19.3.3 Novel Antipsychotics

Despite the various therapeutic options available today, there still is a great need for more effective and tolerable treatments for schizophrenia. After a decades-long hiatus of innovation, the recent development of two promising experimental drugs is finally sparking hope for a new generation of drugs: ulotaront and KarXT. What is remarkable and unique about these two experimental antipsychotics is their complete lack of affinity for DA receptors. In contrast, all available antipsychotics work by blocking DA receptors, which are responsible for severe neurological symptoms (extrapyramidal side effects, EPSEs) and various metabolic disturbances, including hyperprolactinemia, weight gain, and insulin resistance. Ulotaront is a $TAAR_1/5\text{-}HT_{1A}$ agonist, a novel MOA discovered through BPDD that will be discussed below [44]. KarXT, which the FDA has recently approved for the treatment of schizophrenia in adults [45], is a combination therapy of xanomeline, an M_1/M_4 muscarinic acetylcholine receptor agonist, and trospium, a peripheral muscarinic antagonist [46]. The role of trospium, which does not pass the blood-brain barrier, is to counter the cholinergic side effects of xanomeline in the periphery. While the use of this combination is an example of a very clever therapeutic design, aimed at improving the tolerability of a promising novel MOA for the treatment of psychosis, the antipsychotic effects of xanomeline were discovered observationally in a Phase II clinical trial for Alzheimer's patients [11]. In that trial, in addition to a modest improvement in cognitive symptoms, xanomeline unexpectedly demonstrated significant and dose-dependent reductions in various psychotic symptoms. While the role of the neurotransmitter acetylcholine (ACh) in cognition is well established (cholinesterase inhibitors improve cognition by increasing ACh in the brain), the unexpected amelioration of psychotic symptoms by agonism on the specific subset of ACh receptors targeted by xanomeline suggests previously unknown aspects of the pathophysiology of psychosis. The simultaneous pro-cognitive effects of M_1/M_4 agonism by KarXT [46] may alleviate the cognitive impairment commonly seen in psychotic patients, in contrast to current antipsychotics that, in some cases, may aggravate it. This is a remarkable clinical benefit over current antipsychotics which often require the addition of anticholinergic drugs to counter EPSEs, resulting in a further worsening of cognitive function [47].

19.4 A Machine Learning-Based System for Global Behavior Profiling for CNS Drug Discovery

19.4.1 Standardization of Screening Tools

We have provided evidence suggesting that a target-agnostic PDD approach is the best strategy for pursuing polypharmacology for CNS disorders. Agnostic phenotypic screening can be accomplished by monitoring a biological parameter, or biomarker, *in vitro* or *in vivo* (Table 19.1) for a desired change thought to be correlated to a therapeutic outcome, rather than assessing specific binding profiles defined *a priori*. Moreover, when a suitable animal model of disease or a predictive *in vivo* platform exists, the behavioral version of PDD, BPDD, is preferable. The choice of the experimental readout in PDD approaches still relies on assumptions or expectations based on previous scientific knowledge. This choice may depend on prior knowledge of its association with disease and the reliability of such association. The simplest *in vitro* phenotypic screening systems employ cell lines to monitor biomarkers such as expression of specific proteins or genes. More complex neuronal physiology can also be studied *in vitro* with brain slices or cell cultures and organoids generated from induced pluripotent stem cells (iPSCs) with various techniques ranging from biochemical,

Table 19.1 Different approaches to drug discovery and their reliance on prior knowledge.

Drug discovery approach	Experimental setting	Readout type	Comments
Target-based Drug Discovery (TDD)	*in vitro*	Target affinity	Relies on knowledge of target(s), limited polypharmacology[a)]
Phenotypic Drug Discovery (PDD)	*in vitro*	Biomarker	Choice of biomarker bias, and likely translatable
	in vivo	Biomarker	Choice of biomarker bias, and increased translatability
		EEG	Pathophysiology agnostic, unbiased, and translatable
		Single behavioral assay	Pathophysiology agnostic, requires disease association, and translatable
		Behavioral Phenotyping (BPDD)	Pathophysiology and target agnostic, comprehensive, translatable, and high-throughput

a) Some targets pursued by TDD may effectively impact multiple brain receptors indirectly: examples include monoamine transporters (as discussed for SSRIs).

imaging, and electrophysiological and genomics assays drug screening applications, especially in neurological diseases [48]. Whole-organism *in vivo* systems are irreplaceable in their ability to assess even more complex brain functions and test the polypharmacological effects of compounds as they affect the whole brain activity. Assessing polypharmacology *in vivo* has the foremost advantage of evaluating the effects of a compound in controlled experimental settings, resulting in high translational value to humans.

Phenotypic profiles in rodents can be acquired by directly measuring brain activity through electroencephalogram (EEG) recordings. This type of readout can be used for drug discovery as it provides valuable insights into the effects of potential treatments on brain activity, and it has been shown to be predictive in therapeutic classes of psychotropic drugs. Systematic and standardized EEG recordings in free-moving mice can be acquired with PsychoGenics' eCube® platform in a way that is amenable to drug screening. Alternatively, a phenotypic readout obtained by measuring animal behavior (as in BPDD), sometimes referred to as *chemobehavioral phenomics* [49], is particularly useful in the context of drug discovery for neuropsychiatric conditions. This approach has been carried out in simple organisms such as worms [48] and zebrafish [49, 50], or in rodents, which are phylogenetically closer to humans [51]. The advantage of studying behavior in whole organisms is that it can capture functional relationships that would be missed by more reductionist approaches. The use of rodents, for instance, can model complex brain functions from mood, anxiety, and cognition to social behavior. Commonly used single behavior assays in rodents include the forced swim test (FST) and tail suspension test (TST), modeling depressive behavior, and the open field test (OFT) and elevated plus maze (EPM), for anxiety. The ability of drugs such as antidepressants or anxiolytics to modify animal behavior in these tests has translational value, even if we do not fully understand the complex neuronal mechanisms involved. Those behavioral assays are useful tools for guiding identification and optimization of drug leads, having been the fundamental discovery tools for early psychiatric drugs (such as the benzodiazepines, as described above). Pushing the notion of using behavioral assessment in CNS drug discovery even further, behavioral profiles that capture enough features in a systematic and standardized fashion can provide low variability, high quality data in a high throughput fashion.

We posit that quantifying the overall drug-induced behavior in rodents rather than using biomarker or single-parameter behavioral assays may be most useful for early drug discovery programs. Compared to biomarker-based and even single-outcome behavioral assay methods, BPDD methods based on global behavior profiling give us the opportunity to immediately gain insights on a drug's potency, tolerability, and even therapeutic potential by observing how it impacts global animal behavior. This type of extensive behavioral assessment can be done systematically to generate behavioral profiles amenable to quantitative analyses.

19.4.2 The SmartCube System®

One such approach, developed in the early 2000s [52], based on unsupervised identification of behavioral modules of pose dynamics captured through 3D computer vision, has proven to be predictive of psychoactive drug treatments [53]. In general, BPDD can be accomplished with commercially available platforms for measuring and testing behaviors of laboratory rodents, such as Noldus' PhenoTyper. Alternatively, proprietary platforms can offer behavioral phenotyping and analysis as a service, such as PsychoGenics' SmartCube® platform [51]. This technology employs computer vision and deep learning to generate high-content behavioral readouts in a high-throughput and quantitative fashion. The resulting behavioral signatures can be used for various aspects of drug discovery programs [51, 54–58] (Figure 19.1). In this system, animal behavior is assessed after a fixed pre-treatment time over a 45-minute-long standard protocol of automated stress and motor challenges. The key to the system is the use of this standardized protocol for all reference drugs and compounds being screened. Computer vision is used to identify the animal characteristics and posture and to derive granular frame-by-frame data that can be used to also identify target behaviors (e.g., rearing, locomotion, etc.), temporal dynamics, and behavioral motifs [52]. Machine learning (ML)-based classifiers can be trained on drug-induced behavioral profiles of marketed drugs and tool compounds with proven therapeutic indications in the literature and the clinic. Thus, a novel compound's minimum and maximum effective dose, behavioral toxicity, and CNS activity, can be evaluated by SmartCube® in an unbiased manner and without any knowledge of the putative biological mechanisms (novel or otherwise) involved. The core ability of the system is to provide a prediction for the compounds' therapeutic potential, focused on the main CNS drug classes: antidepressants, antipsychotics, anxiolytics, analgesics, mood stabilizers, psychedelics/hallucinogens, psychostimulants, and cognitive enhancers (see Figures 19.4 and 19.5 for examples). The system is maximally automated, allowing the screening of 100 compound doses/week or more.

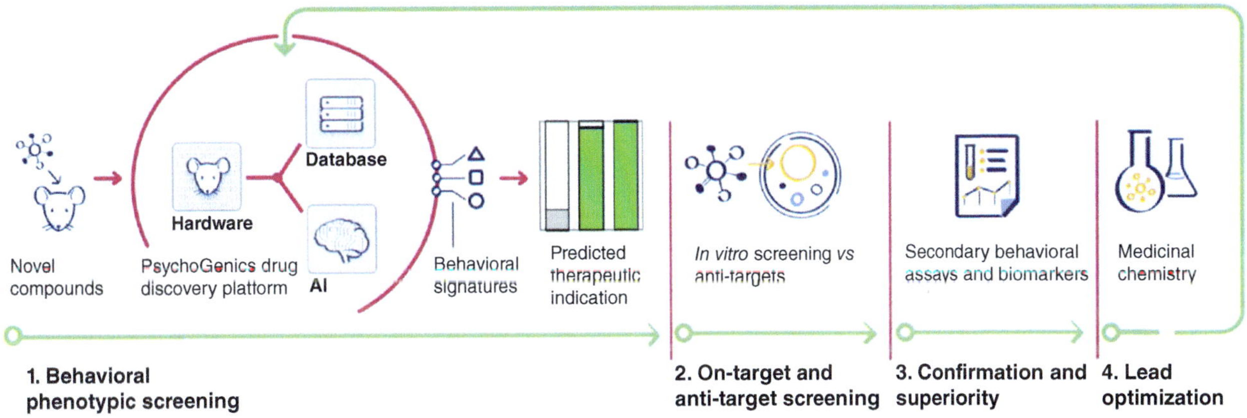

Figure 19.1 Drug discovery phases implementing BPDD: (a) during the initial screening, a novel compound or a whole compound library is assessed using the SmartCube® system, and a measure of overall behavioral activity and therapeutic potential is found; (b) to ensure that a hit is not a me-too drug, and it does not have an undesired risky profile, *in vitro* panels can be used; (c) a second-tier set of assays is used to confirm efficacy for the particular indication chosen; and (d) novel analogs can be synthesized and screened in the system, if either the hit needs optimization or if a backup compound is desired.

19.5 Modeling Chemical and Phenotypic Relationships of Compounds Screened in SmartCube®

The systematic use of a standardized BPDD profiling system over time can allow for the collection of behavioral profiles of diverse chemical classes. As an example, PsychoGenics and its partners collected over the course of almost 2 decades an extensive dataset of phenotypic drug profiles, at several doses, of compounds from commercial libraries (fragment, GPCR, and kinase compound libraries), as well as thousands of compounds synthesized around hits identified during drug discovery projects. In addition to this collection, totaling over 8500 compounds, hundreds of reference CNS compounds, such as marketed drugs, "tool" compounds, and compounds in clinical trials, have also been characterized using the same standardized platform, SmartCube®, at several doses. This rich and diverse dataset allows for additional downstream analyses to find potential drugs for repurposing [59], or to identify structural descriptors that predict *in vivo* activity, and many other uses [55]. To exemplify the type of data and analyses that can be pursued, we present below an analysis of the behavioral profiles of reference compounds, focusing on a comparison of the structural, binding, and *in vivo* phenotypic profiles, with the hope that this analysis will highlight the value of the BPDD assessment.

19.5.1 Antidepressants

To illustrate how a BPDD approach can help identify therapeutic potential of different psychotropic drugs, several drugs marketed as antidepressants were screened in SmartCube®, including classic, rapid-acting antidepressants, and potential novel antidepressants (e.g., psychedelics). Each drug screen resulted in the standard feature set describing its phenotypic behavioral readout. Such data was then used for clustering and visualization to explore similarities and differences between drug profiles. A t-distributed Stochastic Neighbor Embedding (t-SNE) analysis of the dataset was used to visualize both the distinct clusters driven by compounds' structure (Figure 19.2) and by their phenotypic BPDD profile (Figure 19.3). Of note, these analyses are unsupervised, meaning that the algorithm does not know *a priori* the therapeutic classes each test drug belongs to. As expected, drug treatments at lower doses clustered closer to the vehicle control groups, and higher doses clustered together consistent to their therapeutic category (Figure 19.3). Among antidepressants, clearly SSRIs display a robust cluster, quite distinct from tricyclics and MAOIs.

19.5.2 Psychedelics

Another robust cluster is formed by classic serotonergic psychedelics, including psilocybin, psilocin, and dimethyltryptamine (DMT). Compounds that block the NMDA receptor, such as phencyclidine (PCP), dizocilpine (MK801), and high doses of ketamine and dextromethorphan, also grouped tightly together, while the entactogen compound 3,4-methylenedioxymethamphetamine (MDMA) and the psychedelic DOI each formed a distinct, separate cluster of their own. This demonstrates the exciting potential of BPDD to separate drugs by their therapeutic mode of action/profile using the drug-induced behavioral readout alone, in a data-driven fashion.

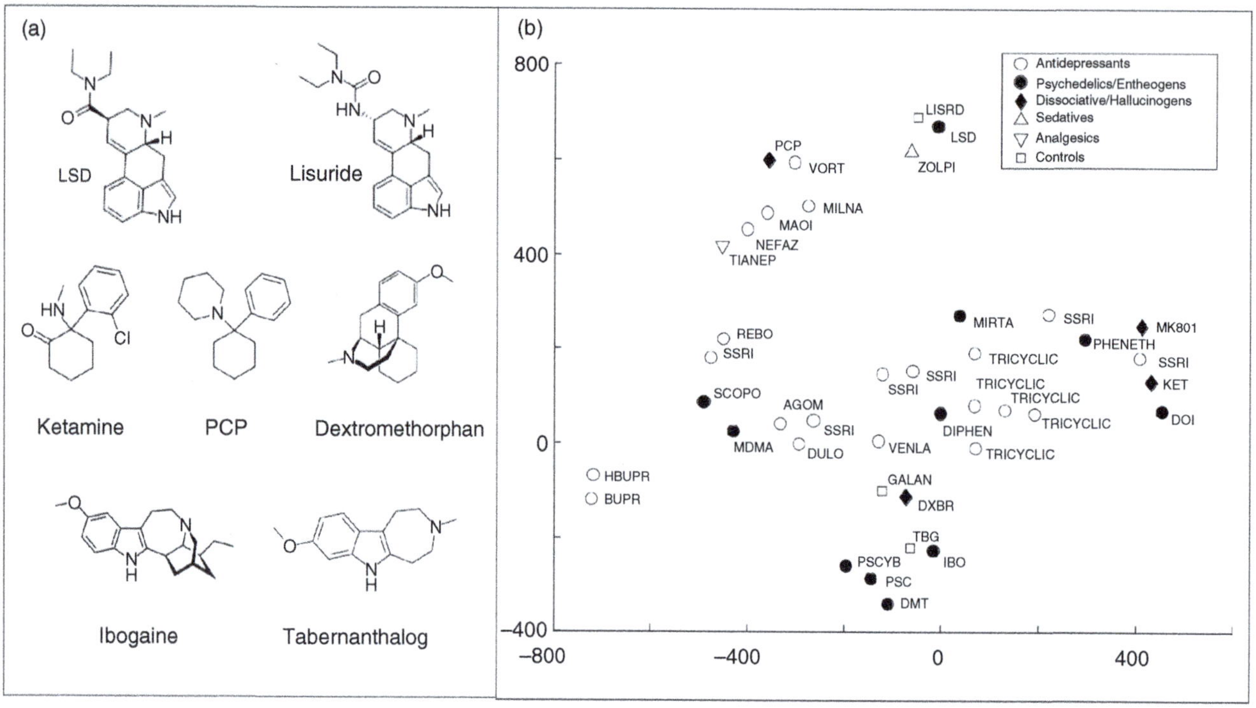

Figure 19.2 (a) Structures of compounds with proven or putative antidepressant action. (b) Unsupervised two-dimensional t-SNE analysis (MATLAB 2020b; Statistics and Machine Learning Toolbox Version 12.0.) based on Tanimoto similarity scores used as pairwise distances of the different structures' MACC fingerprints. BUPR: bupropion; DIPHEN: diphenhydramine; DMT: dimethyltryptamine; DOI: 2,5-Dimethoxy-4-iodoamphetamine; DULO: duloxetine; DXBR: dextromethorphan; HBUPR: hydroxybupropion; IBO: ibogaine; GALAN: galantamine; KET: ketamine; LSD: lysergic acid diethylamide; MAOI: moclobemide; MDMA: 3,4-methylenedioxymethamphetamine; MK801: dizocilpine; PHENETH: phenethylamine; PCP: phencyclidine; PSC: psilocin; PSCYB: psilocybin; REBO: reboxetine; SSRIs: fluoxetine, sertraline, and others; TIANEP: tianeptine; TBG: tabernanthalog; TRICYCLIC: imipramine, desipramine, and others; VORT: vortioxetine; ZOLPI: zolpidem.

The topological organization of drugs identified by this method, including the profiles of different doses of some drugs (e.g., ketamine), would not be possible with data solely based on chemical structure. To illustrate this point, let us consider a well-known example of structurally related compounds with different pharmacological effects. Lysergic acid diethylamide (LSD) and lisuride are two ergoline derivatives with a remarkably similar chemical structure (Figure 19.2a,b). Because they share a high affinity to the serotoninergic 5-HT_{2A} receptor [Ki = 0.73, and 1.4 nM, respectively], they would hypothetically be picked up by a drug screen targeted at 5-HT_{2A}. Their pharmacological effects in humans are however quite different, with LSD notoriously inducing hallucinations while lisuride does not, at least not at therapeutic doses. This pharmacological difference is well captured in their behavioral profile (Figure 19.3), as they cluster in different areas in the t-SNE space. Any approach driven by the molecular structure, on the other hand, would most likely clump them into the same group (as in Figure 19.2b).

In a separate example, the antagonistic activity at the NMDA receptor of certain drugs is robustly captured by their phenotypic profiles, even when brought by chemically different drugs, such as dextromethorphan, a morphinan, as compared to PCP, an arylcyclohexylamine, or ketamine. Indeed, note the structural differences shown in Figures 19.2a,b and their phenotypic similarity in Figure 19.3. Notably, dextromethorphan itself lacks the euphoric and analgesic properties and behavioral profile of the structurally related opioid drugs.

A final example illustrating the power of phenotypic profiling to discriminate between distinct pharmacology of compounds that have common structural elements comes from ibogaine and tabernanthalog (TBG) (Figures 19.2a,b). Ibogaine is a natural psychedelic with dissociative properties that is thought to act through kappa opioid receptors [Ki at kappa opioid receptors = 206 nM versus >10 μM at 5-HT_{2A} receptors] [60]. Anecdotal evidence suggests it may be highly effective in curbing cravings for opioids, alcohol, and other drugs of abuse in humans, echoing similar

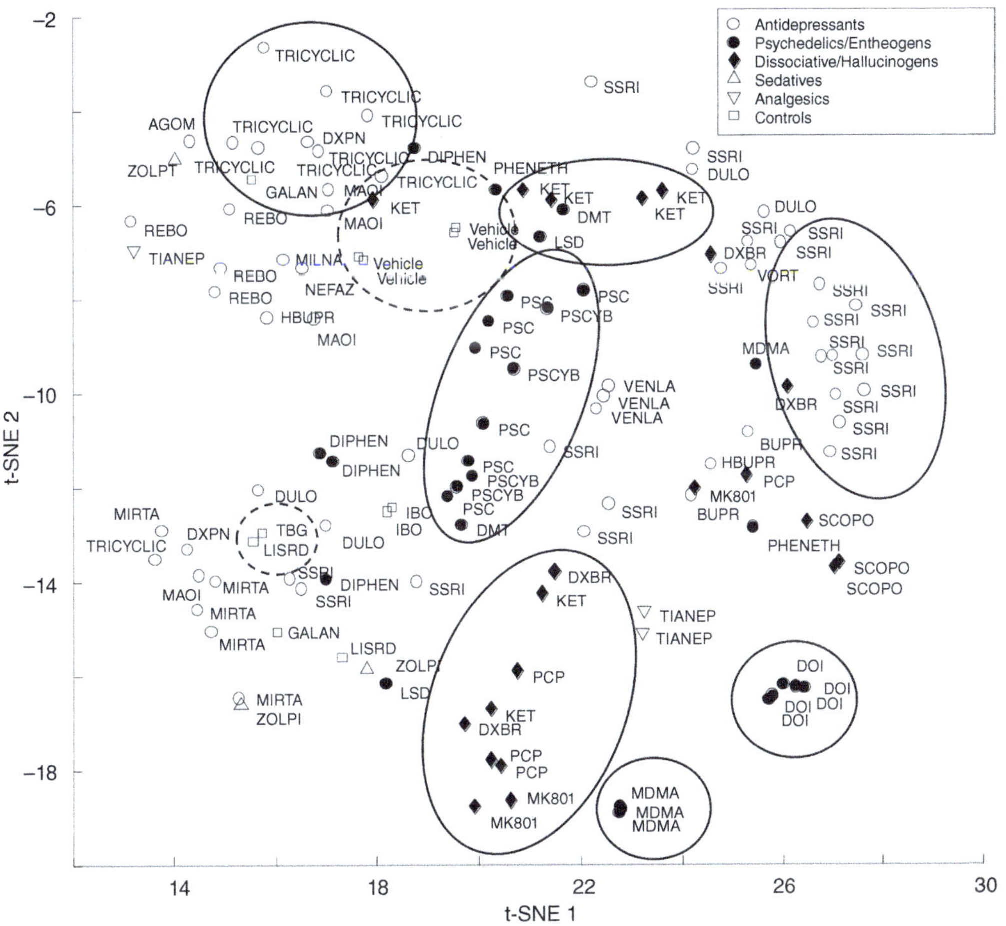

Figure 19.3 Unsupervised two-dimensional t-SNE analysis (MATLAB 2020b; Statistics and Machine Learning Toolbox Version 12.0.) based on similarity scores quantified as pairwise distances of the phenotypic profiles obtained with the PDD platform SmartCube®. Each point in this graph represents an experimental group of mice treated with a particular dose of a drug. To quantify overall behavioral similarity, each experimental group's behavioral profile was compared, at the dose level, with all other groups using machine learning. Clusters of compounds are highlighted by ellipses (the two dashed lines ellipses contain vehicle controls/low doses and lisuride/tabernanthalog controls). BUPR: bupropion; DIPHEN: diphenhydramine; DMT: dimethyltryptamine; DOI: 2,5-Dimethoxy-4-iodoamphetamine; DULO: duloxetine; DXBR: dextromethorphan; HBUPR: hydroxybupropion; IBO: ibogaine; GALAN: galantamine; KET: ketamine; LSD: lysergic acid diethylamide; MAOI: moclobemide; MDMA: 3,4-methylenedioxymethamphetamine; MK801: dizocilpine; PHENETH: phenethylamine; PCP: phencyclidine; PSC: psilocin; PSCYB: psilocybin; REBO: reboxetine; SSRIs: fluoxetine, sertraline, and others; TIANEP: tianeptine; TBG: tabernanthalog; TRICYCLIC: imipramine, desipramine, and others; VORT: vortioxetine; ZOLPI: zolpidem.

reports for serotoninergic psychedelics, including LSD and psilocybin [61]. However, ibogaine has been associated with serious adverse effects, including cardiotoxicity. The interest in its potential therapeutic effect led to the synthesis of the less toxic azepinoindole derivative TBG. Animal studies suggest that TBG lacks the toxicity and hallucinogenic effects of ibogaine, while maintaining its desired properties, such as reducing drug-seeking behavior, promoting antidepressant-like behavioral responses, and neural plasticity [62].

Interestingly, TBG clusters tightly to lisuride, a non-hallucinogenic derivative of LSD, rather than ibogaine (Figure 19.3), again pointing to the ability of behavioral profiling to distinguish structurally related compounds with different pharmacologic properties.

These examples highlight the power of unbiased BPDD to capture behavioral signatures of neurotropic drugs that are much more predictive of their therapeutic class than their chemical structure, even over and above their mechanisms of action.

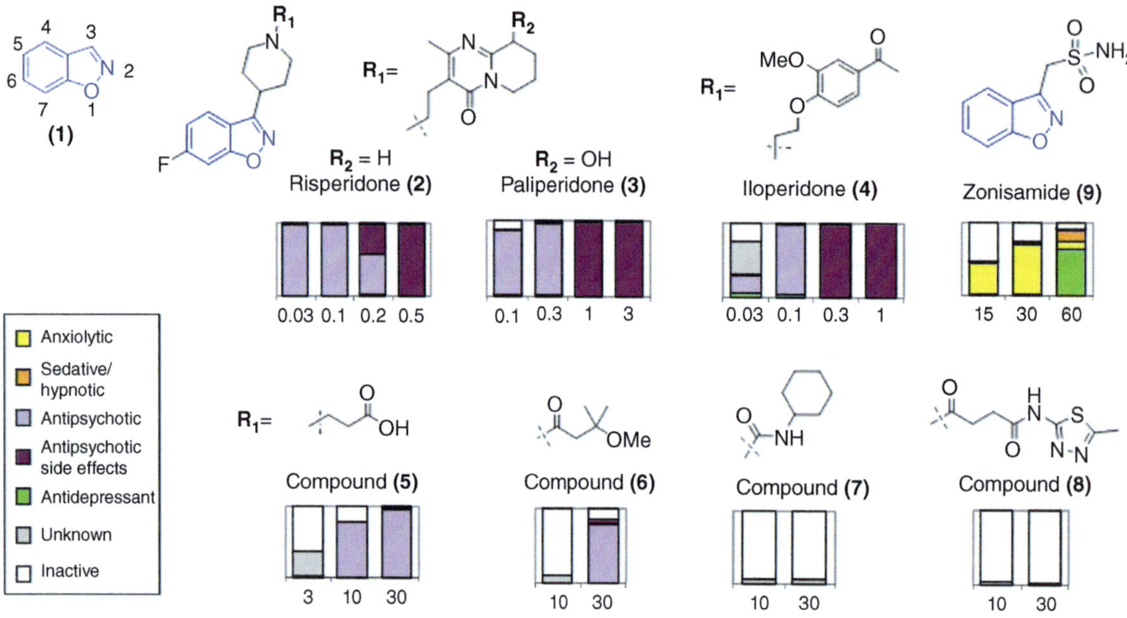

Figure 19.4 Structures and SmartCube signatures of selected 1,2-benzisoxazole compounds.

19.6 Privileged Scaffolds and BPDD with SmartCube®

As mentioned above, benzodiazepines triggered a revolution in pharmacology, in part for the richness of their pharmacophore, recognized as a privileged structure or scaffold, i.e., a structural motif that can be a common substructure in compounds that bind a variety of receptors or proteins. In other words, a privileged scaffold is a structural motif that provides substitution vectors that can be exploited to produce compounds that have the ability to bind a diverse set of biological targets [63]. 1,2-benzisoxazole **(1)** is an example of a privileged heterocyclic motif that can be seen in several compounds with CNS activity, such as the atypical antipsychotics risperidone (RISP, **2**), paliperidone (PALI, **3**) and iloperidone (ILOPE, **4**) and the anticonvulsant zonisamide (ZONI, **9**), for example [64]. The pharmacological profile of risperidone, paliperidone, and iloperidone features broadly similar activities at serotonergic, dopaminergic, and adrenergic receptor targets (see Table 19.3), but quite different from that of zonisamide, which interacts with voltage gated sodium and calcium channels to have an anticonvulsant effect [65]. SmartCube® can capture the phenotypic differences and similarities of these compounds and by extension, their underlying pharmacology (Figure 19.4). In addition to being able to classify compounds based on biological/phenotypic activity, SmartCube® is also able to provide the dose range in which compounds are effective, thus allowing not only for the assessment of a therapeutic index but also for the development of SAR within a series of similar compounds. The behavioral activities of risperidone, paliperidone, iloperidone, and compounds **5–8** in Figure 19.4 illustrate this point. Appending different side chains on the common 6-fluoro-3-(piperidin-4-yl)benzisoxazole core seen in these compounds imparts potency and therapeutic index changes that are easily captured by the platform, including those modifications that render the compounds rather inactive (Figure 19.4, compounds **7** and **8**).

The neurotransmitter tryptamine (TRYP, **10**; Figure 19.5) is another example of a privileged moiety that can be seen in several CNS active molecules with varying pharmacology (see Table 19.4) and indications. For instance, this moiety is found in melatonin (MELATON, **12**), dimethyl tryptamine (DMT, **11**), tabernanthalog (TBG, **13**), and mitragynine (MITRA, **14**), to name a few (Figure 19.5). Simple changes in the core structure of tryptamine as seen between DMT and melatonin or larger changes, like tying the tryptamine ethyl amine chain in a ring, as seen in tabernanthalog and mitragynine, provide for profound differences in pharmacology, differences that can be seen in the phenotypic behavior captured and output provided by SmartCube®, with either mixed signatures for tryptamine (**10**) and TBG (**13**), antidepressant for DMT (**11**), anxiolytic at low doses moving to sedative at high doses for melatonin (**12**), and analgesic for mitragynine (**14**) (Figure 19.5).

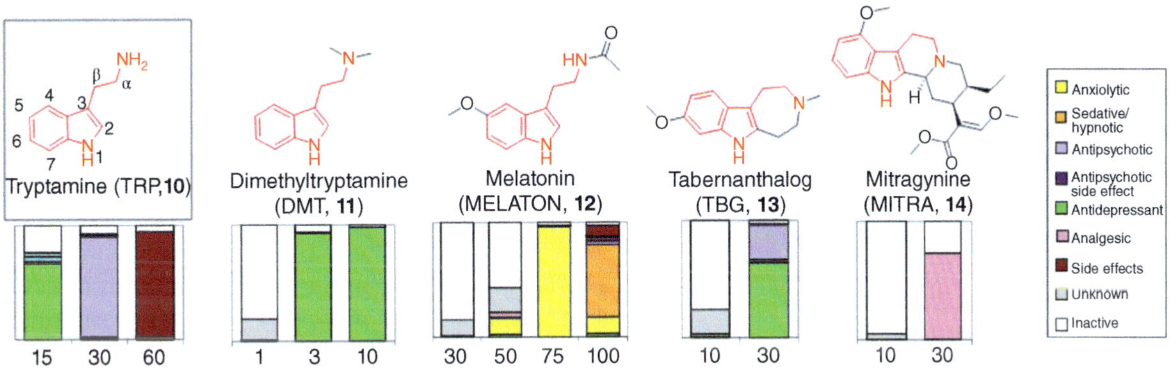

Figure 19.5 Structures and SmartCube® signatures of selected 1,2-benzisoxazole compounds.

19.7 Ulotaront (SEP-363856) a BPDD Case Study

Ulotaront (SEP-363856) is a putative antipsychotic with agonistic action at 5HT1A receptor and the trace amine-associated receptor 1 (TAAR1), with no affinity for DA receptors (Figure 19.6). Ulotaront was the result of a phenotypic lead identification and lead optimization project based on SmartCube® screening aimed at designing safer antipsychotic compounds with novel mechanisms of action. A key aspect of the project was an "anti-target" *in vitro* panel that ensured that chosen compounds with potent antipsychotic signatures had no 5HT2A or D2 receptor antagonism, mechanisms traditionally seen in antipsychotics such as haloperidol, risperidone, iloperidone, clozapine, and others. The medicinal chemistry effort also included testing of compounds to prioritize candidates with good drug metabolism and pharmacokinetics (DMPK) profiles and *in vivo* efficacy confirmation [26]. Following this paradigm, ulotaront emerged as the lead candidate with an antipsychotic profile (with some anxiolytic signal at the lowest dose of 0.3 mg/kg; Figure 19.6). Ulotaront was first disclosed in patent application WO2011069063A1 that covered approximately 300 derivatives and analogs. The cryo-electron microscopy (cryoEM) structure of the compound in complex with human TAAR1 was recently reported [66]. Full pharmacological characterization of ulotaront has been reported by Dedic *et al.* [67], while SAR against human TAAR1 for the compound and a selected number of its derivatives was recently disclosed by Heffernan *et al.* [68]. Whereas ulotaront has a very weak affinity towards 5HT2A (Ki = 17 μM) and D2 (Ki = 21 μM) receptors, it potently binds and activates the human TAAR1 receptor (EC_{50} = 38 nM, E_{max} = 109%; 140 nM, E_{max} 101%) [67, 68] and to a lesser extent, 5HT1A receptor (EC_{50} = 2.3 μM, E_{max} = 75%) [68]. Interestingly, ulotaront is active in SmartCube® but its enantiomer, SEP-363855 is not. Dosed at 10 mg/kg SEP-363855 fails to show any appreciable activity while ulotaront is fully active at 0.3 mg/kg (Figure 19.6).

Consistently, phenotypic profiles of ulotaront and its enantiomer correlate with the human TAAR1 and 5HT1A pharmacologies thought to be the drivers of its *in vivo* activity (Table 19.2). Specifically, SEP-363855 is 12- to 17-fold less active with respect to TAAR1 activation (EC_{50} = 645 nM, E_{max} = 108%; 1700 nM, E_{max} = 70%)

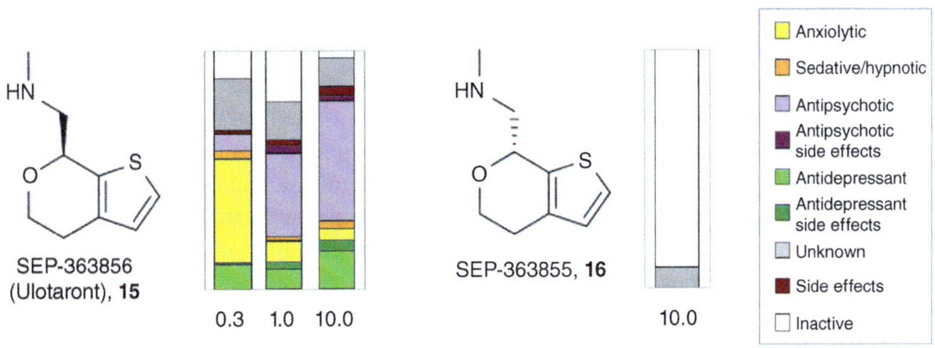

Figure 19.6 Structures of ulotaront and its enantiomer SEP-363855. Screening in SmartCube® showed a strong antipsychotic signature for ulotaront but lack of activity for its enantiomer up to 10 mg/kg.

Table 19.2 Activities of ulotaront and its enantiomer SEP-363855 at the TAAR1 and 5HT1A receptors.

Compound	Receptor	EC_{50} (nM)/(E_{max})	Ki (nM)
Ulotaront	TAAR1	38/109%; 140/101%	N/A
	5HT1A	2300/75%	284
SEP-363855	TAAR1	645/108%; 1700/70%	N/A
	5HT1A	N/A	10,500

and over 36,000-fold weaker in binding assays toward 5HT1A (ulotaront: 5HT1A Ki = 284 nM; SEP-363855: 5HT1A Ki = 10,500 nM) [67, 68]. It is also of note that although TAAR1 had been proposed in the past as target for schizophrenia, the SmartCube® classifiers used during the ulotaront medicinal chemistry campaign were not trained using TAAR1 agonists. The TAAR1/5HT1A target combination responsible for ulotaront's efficacy originated empirically through phenotypic SmartCube® screening results.

Ulotaront is currently in Phase III clinical development in several clinical trials, after promising results of two Phase II trials (4-week acute study and 26-week open-label extension) which led to Breakthrough Therapy Designation from the US Food and Drug Administration (FDA) for the treatment of schizophrenia. Overall, the Phase II studies demonstrated statistically significant improvements in positive and negative symptoms of schizophrenia with a placebo-like tolerability profile [69]. However, two highly anticipated Phase III studies were recently announced to be unsuccessful in meeting the primary efficacy endpoint of a reduction from baseline of PANSS (Positive and Negative Syndrome Scale) total score at six weeks, in part due to an unexpectedly high placebo response. The occurrence of the COVID-19 coronavirus pandemic in 2020 may have impacted the studies, as demonstrated by a pooled analysis of the two studies considering only subjects enrolled before March 13, 2020. In this analysis, ulotaront essentially replicated the findings previously reported in the Phase III studies [70]. Ulotaront is still actively being developed in other Phase III studies.

19.8 Conclusions

The complex interplay among multiple neurotransmitter systems, particularly in th>e context of the pathophysiology of CNS diseases where the underlying biological mechanisms are often poorly understood, presents a formidable challenge for the development of new drugs for psychiatric disorders. The most effective drugs used in the clinic typically exhibit polypharmacology, modulating multiple targets simultaneously. As described earlier in this chapter, most first-in-class CNS drugs were discovered phenotypically, often by serendipitous observations of treated subjects rather than by target-based design that followed some hypothetical disease processes. There is a serious challenge pursuing the identification of novel drugs for mental disorders using TDD. Many companies withdrew from this space, but there seems to be a renaissance triggered by innovative strategies that go beyond the classic target-based approaches. The most promising and cost-effective strategies are target-agnostic, such as those based on *in vivo* phenotypic profiling. We showed that BPDD assays, like SmartCube®, can capture often unpredicted and unexpected therapeutic profile changes that can result from relatively minor changes in a molecule's structure. We also described the development of ulotaront, a putative antipsychotic with a novel polypharmacology MOA, to highlight the power and promise of BPDD platforms.

In this chapter, we highlighted the advantages of BPDD approaches that are high-throughput, of high-informational content, standardized, and combined with innovative ML methodologies and artificial intelligence (AI). BPDD methods are mechanism-agnostic and inherently sensitive to polypharmacologic action, therefore, they can facilitate the discovery of new, first-in-class breakthrough drugs with novel MOAs. They provide (directly or indirectly) multiple readouts at once, including overall drug potency, therapeutic window, brain exposure, and toxicity. Considering all its advantages, BPDD is extremely efficient and cost-effective once its required infrastructure is in place. This is not to say that BPDD does not have some drawbacks. The infrastructure required for the standardization and collection of high information phenotypic profiles requires considerable initial investments and time to set up. There are, however, contract research organizations that already have such infrastructure in place and can conduct BPDD studies as a service. BPDD does not directly provide insights into the MOAs of compounds, apart from the quantification of the similarity of the pharmacological action. However, MOAs can be deciphered once compounds that provide a desired phenotype have been identified, and along with behavioral data, they can guide decisions in the discovery and lead optimization process. A limitation of in *vivo* PDD that is shared with all *in vivo* methodologies is that experimental observations may not fully translate to humans, due to differences between humans and model animals. Genetic, physiological, and anatomical differences between species may lead to pharmacokinetic or pharmacodynamic differences in the response to treatment. Another limitation of BPDD stems from the necessity of standardization of the experimental design for drug screening. Standardization inherently imposes constraints on several parameters, such as route of administration, time and duration of the measurement period, and type of vehicle used. On balance, however, despite these limitations, as we delineated in this chapter, BPDD approaches are best suited for discovering first-in-class/best-in-class medicines.

In conclusion, we have shown how PDD methods (and BPDD in particular) can assist in several aspects of the drug discovery pipeline, from compound screening, lead optimization, and SAR development. We provided evidence for the immense potential these methods have to facilitate the discovery of drugs that are effective through unique multi-target profiles (polypharmacology). We also highlighted the potential of BPDD to lead to the discovery of breakthrough therapies that can revolutionize the treatment of neuropsychiatric disorders, while providing new insights into the brain's inner workings.

References

1 Vincent, F., Nueda, A., Lee, J. et al. (2022). Phenotypic drug discovery: recent successes, lessons learned and new directions. *Nat. Rev. Drug Discov.* 21: 899–914.

2 Ban, T.A. (2006). The role of serendipity in drug discovery. *Dialogues Clin. Neurosci.* 8 (3): 335–344.

3 Lopez-Munoz, F., Alamo, C., Cuenca, E. et al. (2005). History of the discovery and clinical introduction of chlorpromazine. *Ann. Clin. Psychiatry* 17 (3): 113–135.

4 Chow, W.S. and Priebe, S. (2013). Understanding psychiatric institutionalization: a conceptual review. *BMC Psychiatry* 13: 169.

5 Suchting, R., Tirumalaraju, V., Gareeb, R. et al. (2021). Revisiting monoamine oxidase inhibitors for the treatment of depressive disorders: a systematic review and network meta-analysis. *J. Affect. Disord.* 282: 1153–1160.

6 Hillhouse, T.M. and Porter, J.H. (2015). A brief history of the development of antidepressant drugs: from monoamines to glutamate. *Exp. Clin. Psychopharmacol.* 23 (1): 1–21.

7 Ryznar, E. and Meltzer, H.Y. (2020). The neurochemical basis of psychiatric disorders. In: *Landmark Papers in Psychiatry* (ed. E. Ryznar, A.B. Pederson, M.A. Reinecke, et al.), 43–58. Oxford: Oxford University Press.

8 Creese, I., Burt, D.R., and Snyder, S.H. (1976). Dopamine receptors and average clinical doses. *Science* 194 (4264): 546.

9 Fischer-Barnicol, D., Lanquillon, S., Haen, E. et al. (2008). Typical and atypical antipsychotics—the misleading dichotomy. results from the working group 'drugs in psychiatry' (AGATE). *Neuropsychobiology* 57 (1–2): 80–87.

10 Granger, B. and Albu, S. (2005). The haloperidol story. *Ann. Clin. Psychiatry* 17 (3): 137–140.

11 Bodick, N.C., Offen, W.W., Levey, A.I. et al. (1997). Effects of xanomeline, a selective muscarinic receptor agonist, on cognitive function and behavioral symptoms in Alzheimer disease. *Arch. Neurol.* 54 (4): 465–473.

12 Qubad, M. and Bittner, R.A. (2023). Second to none: rationale, timing, and clinical management of clozapine use in schizophrenia. *Ther. Adv. Psychopharmacol.* 13: 20451253231158152.

13 Wong, D.T., Horng, J.S., Bymaster, F.P. et al. (1974). A selective inhibitor of serotonin uptake: Lilly 110140, 3-(p-trifluoromethylphenoxy)-N-methyl-3-phenylpropylamine. *Life Sci.* 15 (3): 471–479.

14 Owens, M.J., Knight, D.L., and Nemeroff, C.B. (2001). Second-generation SSRIs: human monoamine transporter binding profile of escitalopram and R-fluoxetine. *Biol. Psychiatry* 50 (5): 345–350.

15 Thase, M.E. (2003). Effectiveness of antidepressants: comparative remission rates. *J. Clin. Psychiatry* 64 (Suppl 2): 3–7.

16 Pagnin, D., de Queiroz, V., Pini, S., and Cassano, G.B. (2004). Efficacy of ECT in depression: a meta-analytic review. *J. ECT* 20 (1): 13–20.

17 Mencher, S.K. and Wang, L.G. (2005). Promiscuous drugs compared to selective drugs (promiscuity can be a virtue). *BMC Clin. Pharmacol.* 5: 3.

18 Jakobsson, E., Argüello-Miranda, O., Chiu, S.W. et al. (2017). Towards a unified understanding of lithium action in basic biology and its significance for applied biology. *J. Membr. Biol.* 250 (6): 587–604.

19 Schioldann, J. (2013). From guinea pigs to manic patients: cade's 'story of lithium'. *Aust. N. Z. J. Psychiatry* 47 (5): 484–486.

20 Lopez-Munoz, F., Ucha-Udabe, R., and Alamo, C. (2005). The history of barbiturates a century after their

clinical introduction. *Neuropsychiatr. Dis. Treat.* 1 (4): 329–343.

21 Balon, R., Starcevic, V., Silberman, E. et al. (2020). The rise and fall and rise of benzodiazepines: a return of the stigmatized and repressed. *Braz. J. Psychiatry* 42 (3): 243–244.

22 Besnard, J., Ruda, G.F., Setola, V. et al. (2012). Automated design of ligands to polypharmacological profiles. *Nature* 492 (7428): 215–220.

23 Colen, B.D. (1980). Adventurous chemist and his pill. *The Washington Post* (19 January). https://www.washingtonpost.com/archive/politics/1980/01/20/adventurous-chemist-and-his-pill/16cd98f6-00a6-4f5c-8e86-08451ed3451b/ (accessed 22 September 2024).

24 Lopez-Munoz, F., Alamo, C., and Garcia-Garcia, P. (2011). The discovery of chlordiazepoxide and the clinical introduction of benzodiazepines: half a century of anxiolytic drugs. *J. Anxiety Disord.* 25 (4): 554–562.

25 Bandelow, B., Michaelis, S., and Wedekind, D. (2017). Treatment of anxiety disorders. *Dialogues Clin. Neurosci.* 19 (2): 93–107.

26 Leahy, E., Varney, M., and Brunner, D. (2021). Use of phenotypic screening in mice in the development of a novel non-D2-receptor-targeting drug for the treatment of schizophrenia. In: *Phenotypic Drug Discovery* (ed. B. Isherwood and A. Augustin), 175–197. London: The Royal Society of Chemistry.

27 Brunner, D., Nestler, E., and Leahy, E. (2002). In need of high-throughput behavioral systems. *Drug Discov. Today* 7 (18): S107–S112.

28 Hopkins, A.L. (2012). Introduction: the case for polypharmacology. In: *Polypharmacology in Drug Discovery* (ed. J.-W. Peters), 1–6. Hoboken, NJ: Wiley.

29 Bung, N., Krishnan, S.R., and Roy, A. (2022). An in silico explainable multiparameter optimization approach for de novo drug design against proteins from the central nervous system. *J. Chem. Inf. Model.* 62 (11): 2685–2695.

30 Gusman, D.H. and Shoemake, C. (2017). Evaluation and optimization of in silico designed sphingosine-1-phosphate (S1P) receptor subtype 1 modulators for the management of multiple sclerosis. *Yale J. Biol. Med.* 90 (1): 15–23.

31 Malik, R., Mehta, P., Srivastava, S. et al. (2017). Structure-based screening, ADMET profiling, and molecular dynamic studies on mGlu2 receptor for identification of newer antiepileptic agents. *J. Biomol. Struct. Dyn.* 35 (16): 3433–3448.

32 Miller, G. (2010). Is pharma running out of brainy ideas? *Science* 329 (5991): 502–504.

33 Paolini, G.V., Shapland, R.H., van Hoorn, W.P. et al. (2006). Global mapping of pharmacological space. *Nat. Biotechnol.* 24 (7): 805–815.

34 Brunner, D., Balci, F., and Ludvig, E.A. (2012). Comparative psychology and the grand challenge of drug discovery in psychiatry and neurodegeneration. *Behav. Process.* 89 (2): 187–195.

35 Roth, B.L., Sheffler, D.J., and Kroeze, W.K. (2004). Magic shotguns versus magic bullets: selectively non-selective drugs for mood disorders and schizophrenia. *Nat. Rev. Drug Discov.* 3 (4): 353–359.

36 Swinney, D.C. and Anthony, J. (2011). How were new medicines discovered? *Nat. Rev. Drug Discov.* 10 (7): 507–519.

37 Sadri, A. (2023). Is target-based drug discovery efficient? Discovery and "off-target" mechanisms of all drugs. *J. Med. Chem.* 66 (18): 12651–12677.

38 Nutt, D. (2015). Illegal drugs laws: clearing a 50-year-old obstacle to research. *PLoS Biol.* 13 (1): e1002047.

39 Sessa, B. (2012). Shaping the renaissance of psychedelic research. *Lancet* 380 (9838): 200–201.

40 Goodwin, G.M., Aaronson, S.T., Alvarez, O. et al. (2022). Single-dose psilocybin for a treatment-resistant episode of major depression. *N. Engl. J. Med.* 387 (18): 1637–1648.

41 Mitchell, J.M., Ot'alora, G.M., van der Kolk, B. et al. (2023). MDMA-assisted therapy for moderate to severe PTSD: a randomized, placebo-controlled phase 3 trial. *Nat. Med.* 29 (10): 2473–2480.

42 Cristea, I.A. and Naudet, F. (2019). U.S. Food and Drug administration approval of esketamine and brexanolone. *Lancet Psychiatry* 6 (12): 975–977.

43 Zanos, P. and Gould, T.D. (2018). Mechanisms of ketamine action as an antidepressant. *Mol. Psychiatry* 23 (4): 801–811.

44 Hopkins, S.C., Ogirala, A., Worden, M. et al. (2021). Depicting safety profile of TAAR1 agonist ulotaront relative to reactions anticipated for a dopamine D2-based pharmacological class in FAERS. *Clin. Drug Investig.* 41 (12): 1067–1073.

45 PureTech-Invented KarXT Receives U.S. Food and Drug Administration Approval for the Treatment of Schizophrenia in Adults. *BusinessWire* (27 September 2024).

46 Sauder, C., Allen, L.A., Baker, E. et al. (2022). Effectiveness of KarXT (xanomeline-trospium) for cognitive

impairment in schizophrenia: post hoc analyses from a randomised, double-blind, placebo-controlled phase 2 study. *Transl. Psychiatry* 12 (1): 491.
47 Georgiou, R., Lamnisos, D., and Giannakou, K. (2021). Anticholinergic burden and cognitive performance in patients with schizophrenia: a systematic literature review. *Front. Psych.* 12: 779607.
48 Ji, P., Li, H., Jin, Y. et al. (2022). *C. elegans* as an *in vivo* model system for the phenotypic drug discovery for treating paraquat poisoning. *PeerJ* 10: e12866.
49 Kokel, D. and Peterson, R.T. (2008). Chemobehavioural phenomics and behaviour-based psychiatric drug discovery in the zebrafish. *Brief. Funct. Genomic. Proteomic.* 7 (6): 483–490.
50 McCarroll, M.N., Gendelev, L., Keiser, M.J., and Kokel, D. (2016). Leveraging large-scale behavioral profiling in zebrafish to explore neuroactive polypharmacology. *ACS Chem. Biol.* 11 (4): 842–849.
51 Brunner, D., Gondhalekar, V., Leahy, E. et al. (2002). *Systems and Methods for Monitoring Behavior Informatics* (ed. USPTO). Pittsburgh, PA: Carnegie Mellon University.
52 Brunner, D., Gondhalekar, V., Leahy, E. et al. (2009). *Method for Predicting Treatment Classes Using Animal Behavior Informatics* (ed. USPTO). Pittsburgh, PA: Carnegie Mellon University, Psychogenics Inc.
53 Wiltschko, A.B., Tsukahara, T., Zeine, A. et al. (2020). Revealing the structure of pharmacobehavioral space through motion sequencing. *Nat. Neurosci.* 23 (11): 1433–1443.
54 Alexandrov, V., Brunner, D., Hanania, T., and Leahy, E. (2015). High-throughput analysis of behavior for drug discovery. *Eur. J. Pharmacol.* 750: 82–89.
55 Brunner, D., Alexandrov, V., Caldarone, B. et al. (2012). Behavior-based screening as an approach to polypharmacological ligands. In: *Polypharmacology in Drug Discovery* (ed. J.-U. Peters), 303–313. Hoboken, New Jersey: John Wiley & Sons, Inc.
56 Prinssen E, Moreau, J.-L., Mueggler, T. et al. (2016). Screening for antipsychotic drugs which do not interact with monoamine receptors. *5th Biennial Schizophrenia International Research Society Conference*, Florence, Italy (2–6 April 2016).
57 Alexandrov, V., Brunner, D., Menalled, L.B. et al. (2016). Large-scale phenome analysis defines a behavioral signature for Huntington's disease genotype in mice. *Nat. Biotechnol.* 34 (8): 838–844.
58 KevináFang, Q. (2016). *In vivo* phenotypic drug discovery: applying a behavioral assay to the discovery and optimization of novel antipsychotic agents. *MedChemComm* 7 (6): 1093–1101.
59 Ambesi-Impiombato, A., Cox, K., Ramboz, S. et al. (2023). Enrichment analysis of phenotypic data for drug repurposing in rare diseases. *Front. Pharmacol.* 14: 1128562.
60 Glick, S.D., Maisonneuve, I.M., and Szumlinski, K.K. (2001). Mechanisms of action of ibogaine: relevance to putative therapeutic effects and development of a safer *iboga* alkaloid congener. In: *The Alkaloids* (ed. G.A. Cordell), 39–53. San Diego: Academic Press.
61 Olson, D.E. (2018). Psychoplastogens: a promising class of plasticity-promoting neurotherapeutics. *J. Exp. Neurosci.* 12: 1179069518800508.
62 Cameron, L.P., Tombari, R.J., Lu, J. et al. (2021). A non-hallucinogenic psychedelic analogue with therapeutic potential. *Nature* 589 (7842): 474–479.
63 Welsch, M.E., Snyder, S.A., and Stockwell, B.R. (2010). Privileged scaffolds for library design and drug discovery. *Curr. Opin. Chem. Biol.* 14 (3): 347–361.
64 Uto, Y. (2016). 1,2-benzisoxazole: a privileged structure with a potential for polypharmacology. *Curr. Pharm. Des.* 22 (21): 3201–3211.
65 Leppik, I.E. (2004). Zonisamide: chemistry, mechanism of action, and pharmacokinetics. *Seizure* 13 (Suppl 1): S5–S9. discussion S10.
66 Liu, H., Zheng, Y., Wang, Y. et al. (2023). Recognition of methamphetamine and other amines by trace amine receptor TAAR1. *Nature* 624: 663–671.
67 Dedic, N., Jones, P.G., Hopkins, S.C. et al. (2019). SEP-363856, a novel psychotropic agent with a unique, non-D(2) receptor mechanism of action. *J. Pharmacol. Exp. Ther.* 371 (1): 1–14.
68 Heffernan, M.L.R., Herman, L.W., Brown, S. et al. (2022). Ulotaront: a TAAR1 agonist for the treatment of schizophrenia. *ACS Med. Chem. Lett.* 13 (1): 92–98.
69 Achtyes, E.D., Hopkins, S.C., Dedic, N. et al. (2023). Ulotaront: review of preliminary evidence for the efficacy and safety of a TAAR1 agonist in schizophrenia. *Eur. Arch. Psychiatry Clin. Neurosci.* 273 (7): 1543–1556.
70 (2023). ACNP 62nd annual meeting: poster abstracts P251–P500. *Neuropsychopharmacology* 48 (Suppl 1): 211–354.

Appendix

Table 19.3 Affinity profile of 1,2 benzisoxazole compounds.

	Receptor Ki in nM (Range)		
Receptor target	**Risperidone**	**Paliperidone**	**Illoperidone**
5-HT_{1A}	497 (190–1170)	569 (480–638)	73 (33–93)
5-HT_{1B}	249 (9.8–718)	304 (12–876)	94 (89–99)
5-HT_{1D}	50 (3.9–140)	69 (19–170)	12.5 (10–15)
5-HT_{1E}	1320	ND (1020 – >10000)	—
5-HT_{1F}	1240	—	—
5-HT_{2A}	0.31 (0.2–0.42)	1.1 (0.4–1.9)	1.7 (0.2–5.6)
5-HT_{2B}	30.4 (16–42)	40 (18–62)	27
5-HT_{2C}	25.1 (32–64)	43 (38–48)	14
5-HT_5	200 (195–206)	353 (278–428)	472 (394–550)
5-HT_6	1666 (962–2400)	1547 (681–2414)	112 (63–189)
5-HT_7	4.2 (0.4–6.9)	2.75 (2.7–2.8)	7.8
α_{1A} adrenergic	5	2.5	1.4
α_{1B} adrenergic	9	0.7	224
α_{1D} adrenergic	—	—	1.2
α_{2A} adrenergic	151 (150.8–151)	86 (4.7–222)	111 (8.3–162)
α_{2B} adrenergic	108 (107.6–108)	33 (9.4–57)	112 (11–162)
α_{2C} adrenergic	1.3	7.4	16.2
D_1 Dopamine	334 (60–580)	41	129 (43–216)
D_2 Dopamine	3.7 (0.3–6.4)	3.8 (0.4–9.4)	131 (3.3–369)
D_3 Dopamine	8.7 (1.6–16)	2.6 (0.1–6.9)	7.1
D_4 Dopamine	13 (0.25–45)	54	25
D_5 Dopamine	289.5 (16–563)	29	319
H_1 Histamine	28.3 (3.5–54)	14 (3.4–32)	933 (12–1854)
H_2 Histamine	120	121	—
M_1 Muscarinic	2800	>10000	—
M_2 Muscarinic	2900	>10000	—
hERG	167	—	—
SERT	1000	3717	443
DAT	—	>10000	2950
NET	—	>10000	1007 (535–1479)

Ki data was derived from the PDSP database [22]. Ki values represent affinity at human receptors, and averages were derived from studies using agonist radioligands, where available.
Abbreviations: ND, not determined.

Table 19.4 Affinity profile of tryptamine-like compounds.

	Receptor Ki (nM)						
Receptor target	Tryptamine	N-acetyl-5-HT	Melatonin	5MeODMT	Psilocin	Mitragynine	Ibogaine
5-HT$_{1A}$	105	—	—	1.9	49	161	>10000
5-HT$_{1B}$	305 (86–525)	—	—	74		>10000	>10000
5-HT$_{1D}$	173 (42–521)	—	—	6.3	—	>10000	>10000
5-HT$_{1E}$	2559	—	—	360	—	>10000	—
5-HT$_{2A}$	ND (89->10000)	—	>10000	1109 (207–2011)	−18.5 (12–25)[a]	ND (4393- >10000)	>10000
5-HT$_{2B}$	67 (25–102)	—	ND (5754->10000	3884	450[a]	2046 (304–3788)	—
5-HT$_{2C}$	1320 (295–2344)	—	>10000	538	10[a]	>10000	>10000
5-HT$_5$	—	—	—	277	—	>10000	—
5-HT$_6$	438	—	—	36	—	>10000	—
5-HT$_7$	148.5 (148-149)	—	—	4	—	865	—
α_{2A} adrenergic	—	—	—	1890	—	ND (3672->10000)	—
α_{2B} adrenergic	—	—	—	2640	—	ND (1030->10000)	—
α_{2C} adrenergic	—	—	—	508	—	ND (1454->10000)	—
D_1 Dopamine	—	—	—	80	—	ND (2736->10000)	>10000
D_2 Dopamine	—	—	—	3562	—	ND (1683->10000)	>10000
D_3 Dopamine	—	—	—	498	—	>10000	—
D_4 dopamine	—	—	—	3120	—	>10000	—
Kappa opioid	—	—	—	>10000	—	823 (386-1260)	206
Melatonin$_{1A}$	—	735 (484-986)	0.66 (0.08-1.5)	—	—	—	—
Melatonin$_{1B}$	—	595	0.41 (0.18-1.1)	—	—	—	—
Mu opioid	—	—	—	—	—	624 (149–1100)	4362
Sigma$_2$	—	—	—	>10000	—	—	2717
TAAR$_1$	1084	—	—	—	—	—	—
SERT	—	—	—	2032	—	>10000	—
DAT	—	—	—	—	—	ND (5362->10000)	—
NET	—	—	—	2859	—	>10000	—

Ki data was derived from the PDSP database [22]. Ki values represent affinity at human receptors, unless noted otherwise. Ki averages were derived from studies using agonist radioligands, where available.

a) Represent cases in which Ki values were derived from non-human receptors. These cases were included where no human Ki values were available for important targets.

Abbreviations: ND, not determined.

20

Multi-target Peptides for the Treatment of Metabolic Diseases

Martin Bossart and Gerhard Hessler

20.1 Introduction

The modern sedentary lifestyle has led to a worrying increase in obesity and type 2 diabetes (T2D) in the last decades. Lifestyle modifications such as a healthy diet and regular physical activity can help to improve overall metabolic health. Unfortunately, such interventions are often not successful and pharmacological treatment is necessary.

A number of drug classes are available for the therapy of T2D, and it is common in a patient journey that single drug treatment is followed by combination treatments of oral options like metformin with sulfonylureas, non-sulfonylurea secretagogues (metiglinide derivatives), alpha-glucosidase inhibitors, and for some time thiazolidinediones (TZDs) and later by combinations of short- and long-acting insulins [1]. Thus, the concept of polypharmacology is deeply rooted in the treatment paradigm.

For many years, there were no adequate options available for the treatment of obesity. The first three drugs approved by the Food and Drug Administration (FDA), orlistat (Xenical, Alli), phentermine–topiramate (Qsymia), and naltrexone–bupropion (Contrave), have limited efficacy and often quite severe side effects [2]. Remarkably, among these three products, two are co-formulations of two active entities, underlining that addressing multiple mechanisms simultaneously is often necessary to achieve significant weight loss.

Peptide drugs have several well-established advantages, including high affinity and specificity, as well as good safety properties. However, peptides often encounter two major drawbacks: membrane impermeability and poor *in vivo* stability [3]. Despite these limitations, peptide drugs are well-established options for the treatment of diabetes and obesity. Especially, the introduction of GCG-like peptide-1 (GLP-1) receptor (GLP-1R) agonists first for the treatment of T2D and later also obesity has dramatically improved treatment success. In addition, it has inspired a wide search for suitable pharmacological combinations with GLP-1 to further improve treatment outcomes, which will be the topic of this chapter. Metabolically active peptide drugs are good examples to illustrate the benefit of pharmacological combinations.

Fusing different peptides presents the opportunity to integrate multiple therapeutic principles into a single molecule, thus enhancing efficacy. This can be achieved through various approaches, such as designing a single peptide sequence that targets two distinct therapeutic targets or linking two peptide sequences or biologically active principles together (Figure 1a). The former approach is limited to the combination of closely related peptides that share a similar mode of action at two targets, whereas the latter strategy allows to merge very different pharmaceutically active principles. In the realm of metabolically active peptides, the latter approach is commonly employed to extend the half-life of peptide drugs by combining them with albumin-binding motifs or antibody-derived Fc fragments. The field of incretins provides rich examples for the use of these different strategies.

20.2 Glucagon-like Peptide-1 (GLP-1) Receptor Agonists

GLP-1 is a peptide hormone secreted by the L cells in the gut upon food intake [4]. Its main effects are glucose-dependent stimulation of insulin secretion, a delay of gastric emptying, and an increase in satiety leading to decreased food intake. GLP-1 also has cardioprotective and neuroprotective effects, decreases inflammation and apoptosis, and has implications for learning and memory, reward behavior, and palatability, so additional applications of GLP-1R agonists for the treatment of metabolic dysfunction-associated steatohepatitis (MASH, formerly called nonalcoholic steatohepatitis [NASH]) and

Polypharmacology: Strategies for Multi-Target Drug Discovery, First Edition. Edited by Jens-Uwe Peters.

Alzheimer's or Parkinson's disease are currently under investigation. Marketed GLP-1R agonists are either based on the sequence of GLP-1 itself or based on exendin-4, a GLP-1R agonist identified in the saliva of the Gila monster [5]. Various modifications have been introduced to stabilize the peptide and particularly to extend *in vivo* half-life. An overview of different marketed GLP-1R agonists can be found in Figure 20.1b [6].

Exenatide, containing exendin-4 in its unmodified form as an active ingredient, was the first GLP-1R agonist to be approved by the FDA in 2005, administered as a twice-daily injection. A prolonged-release suspension via microsphere encapsulation of exenatide was later approved for once-weekly administration and marketed as Bydureon. Lixisenatide, also based on the exendin-4 sequence with a C-terminal lysine stretch to achieve a pharmacokinetic (PK) profile allowing once-daily administration, was FDA approved in 2016. Fatty acid conjugation as protraction strategy via binding to human serum albumin (HSA) was used for the first marketed GLP-1R agonist liraglutide based on the GLP-1 sequence. Peptides bound to HSA are sterically shielded from proteolytic degradation and protected from rapid renal filtration due to the relatively large size of HSA of approximately 66 kDa as well as neonatal Fc receptor (FcRn)–mediated recycling of albumin [7]. A 1.8 mg version was approved for the treatment of T2D as Victoza in 2010 and a higher-dose 3 mg version as Saxenda in 2020 for the treatment of obesity. Two drugs with once-weekly profile, which are produced by recombinant synthesis, were approved in 2014. Dulaglutide achieves its PK profile via conjugation of a GLP-1 analog peptide sequence to an immunoglobulin G4 (IgG4) Fc fusion fragment and is marketed as Trulicity. The Fc fusion fragment enables FcRn receptor–mediated recycling of the peptide and thus increases its half-life significantly [8, 9]. The second benefit of the Fc fusion fragment is the increased size of the molecule, which reduces renal filtration [10]. Albiglutide contains two directly connected GLP-1-based peptide sequences linked to an albumin as protraction unit ensuring reduced renal clearance. It was marketed as Tanzeum but discontinued in 2017 due to limited prescriptions. All these treatments showed decent glucose control combined with limited weight loss of a maximum of 5–10%. GLP-1R agonists are also used in combination with long-acting insulins to achieve improved outcomes and are even available as co-formulations to reduce the injection burden for patients [11].

A breakthrough in the field of selective GLP-1R agonists was achieved with the identification of semaglutide [12]. Based on the GLP-1 sequence, it carries a 2-aminoisobutyric acid (Aib) in position 2 to suppress dipeptidyl peptidase-4 (DPP-IV) cleavage and a fatty acid side chain conjugation at position 20. The side chain consisting of octadecanedioic acid and a linker of a γ-glutamate and two 2-[2-(2-aminoethoxy)ethoxy]acetate (AEEAc) units improved albumin binding to achieve a once-weekly PK profile via fatty acid conjugation [7]. The resulting PK profile, in combination with carefully developed uptitration schemes to limit the transient adverse effects of nausea and vomiting, allows to reach higher active plasma levels of the GLP-1R agonist. Semaglutide in a dose of 1 mg once weekly is approved as Ozempic for the treatment of T2D and at the dose of 2.4 mg once weekly as Wegovy for the treatment of obesity. For T2D, even a once-daily oral formulation with 14 mg of semaglutide and sodium salcaprozate (SNAC) as gastrointestinal permeation enhancer is available as Rybelsus.

Selective GLP-1R agonists already offer a promising approach to treat obesity and related metabolic disorders, but further improved effects on glucose control and weight loss and enhanced effects on β-cell function and liver lipids potentially combined with a more favorable side-effect profile are still urgently needed for the benefit of patients.

20.3 Unimolecular Multiagonists Based on Glucagon-like Peptide-1 (GLP-1) Following the One-pharmacophore Approach

The sequence of GLP-1 (7–37) shows a high similarity to other members of the secretin-like family, a group of peptide hormones that play a critical role in energy metabolism, especially glucagon (GCG), its C-terminal elongated relative oxyntomodulin (OXM), and glucose-dependent insulinotropic polypeptide (GIP). Receptors of these peptides belong to the class B GPCRs, which have an extracellular binding domain. The C-terminal part of the peptidic ligand interacts with the extracellular domain (ECD) triggering receptor activation by binding of the N-terminus to the transmembrane region [13–16]. The determination of the receptor–ligand complex structures for a triagonistic peptide ligand with the three receptors of GLP-1, GCG, and GIP highlights the similar binding mode, demonstrating that these peptides form a helical structure in the complexes and interact with a similar binding site (Figure 20.2a). Binding is driven by conserved key interactions with the corresponding receptors in the lower half of the transmembrane domain (TMD) pocket [13, 17, 18]. The similar binding modes of the peptide ligands at three different receptors underscore the potential to merge three distinct biological activities into a single chimeric molecule. The synergy between the complementary pharmacological effects of the individual peptide hormones

(a) Different options to design multi-targeting peptides

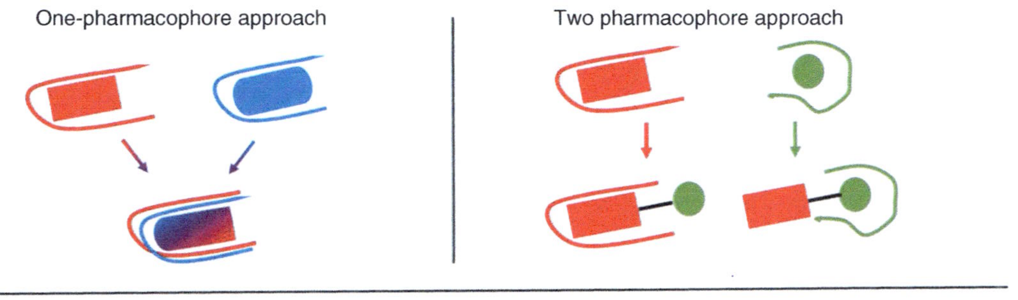

(b) Marketed selective GLP-1R agonists

name	1	2	3	4	5	6	7	8	9	10	11	12	13	14	15	16	17	18	19	20
GLP-1(7-36)NH_2	H	A	E	G	T	F	T	S	D	V	S	S	Y	L	E	G	Q	A	A	K
Exenatide	H	G	E	G	T	F	T	S	D	L	S	K	Q	M	E	E	E	A	V	R
Lixisenatide	H	G	E	G	T	F	T	S	D	L	S	K	Q	M	E	E	E	A	V	R
Liraglutide	H	A	E	G	T	F	T	S	D	V	S	S	Y	L	E	G	Q	A	A	K[gGlu-Palm]
Dulaglutide	H	G	E	G	T	F	T	S	D	V	S	S	Y	L	E	E	Q	A	A	K
Albiglutide	H	G	E	G	T	F	T	S	D	V	S	S	Y	L	E	G	Q	A	A	K
Semaglutide	H	Aib	E	G	T	F	T	S	D	V	S	S	Y	L	E	G	Q	A	A	K[AEEAc-AEEAc-gGlu-C18diacid]

name	21	22	23	24	25	26	27	28	29	30	31	32	33	34	35	36	37	38	39	40	41	42	43	44	C-term.
GLP-1(7-36)NH_2	E	F	I	A	W	L	V	K	G	R															NH_2
Exenatide	L	F	I	E	W	L	K	N	G	G	P	S	S	G	A	P	P	P	S						NH_2
Lixisenatide	L	F	I	E	W	L	K	N	G	G	P	S	S	G	A	P	P	S	K	K	K	K	K	K	NH_2
Liraglutide	E	F	I	A	W	L	V	R	G	R	G														OH
Dulaglutide	E	F	I	A	W	L	V	K	G	G*															
Albiglutide	E	F	I	A	W	L	V	K	G	R§															
Semaglutide	E	F	I	A	W	L	V	R	G	R	G														OH

Violet: Amino acids identical in GLP-1 and exendin-4
White: Amino acids from GLP-1
Grey: Amino acids from exendin-4

Orange: New amino acid mutations
Red font: Protraction unit

G*: GS-Linker to Fc part of IgG4
R§: GLP-1 sequence is repeated as linker to human albumin as protraction unit

Figure 20.1 (a) Different strategies employed for metabolically active peptides to target multiple targets: one-pharmacophore approach – two related receptors addressed by the same peptide sequence. Two-pharmacophore approach: Two molecules are linked together; each part of the linked molecule addressing their respective pharmacophore. (b): Overview of marketed selective GLP-1R agonists.

has inspired numerous research initiatives that investigate unimolecular combinations of two or three components, seeking to optimize therapeutic efficacy for obesity and/or T2D [19].

20.4 GLP-1 Receptor/Glucagon Receptor Dual Agonists

GCG is a 29-amino acid peptide from the secretin-like family of peptide hormones [20]. It is secreted by the α-cells of the pancreatic islets of Langerhans. Hypoglycemia is physiologically the most potent secretory trigger, and the best-known action of GCG is the stimulation of glucose production via activation of the GCG receptor (GCGR) in the liver, leading to glycogenolysis and gluconeogenesis and thereby maintenance of adequate plasma glucose concentrations. Already in the 1950s, GCG was described to diminish feelings of hunger [21] and to decrease food intake [22]. Nair et al. reported in 1987 that GCG also increases energy expenditure [23]. These early observations were, however, never pursued in the context of diabetes or obesity due to GCG's role as the counter-regulatory hormone to insulin that increases blood glucose. Only recently, the effects of GCG in combination with GLP-1 were explored [24, 25]. According to theory, GLP-1R activation triggers glucose lowering and body weight reduction. Additional activation of the GCGR then should lead to more significant weight loss, with the risk that excessive GCGR activation could lead to overall blood glucose elevation in case the GLP-1R-mediated glucose-lowering effect can't counterbalance the GCGR-stimulated glucose anymore.

OXM is a 37-amino acid peptide hormone having the GCG sequence at the N-terminus and an 8-amino acid C-terminal extension (see Figure 20.2b). It was first described in 1981 [26]. Like GLP-1, OXM is mainly produced in gut endocrine L cells by processing the preproglucagon precursor and secreted together with GLP-1 following nutrient intake [27]. OXM is an endogenous unimolecular dual GLP-1R/GCGR agonist, showing *in vitro* activity at both the GLP-1R and the GCGR although its activity is a factor 50–100 diminished compared to the two more selective agonists GLP-1 and GCG on their respective receptors. Bloom and coworkers studied OXM in overweight-to-obese subjects [28]. OXM shows in humans a very short half-life of only 12 minutes. Therefore, 400 nmol OXM was administered subcutaneously three times daily prior to meals. After four weeks of treatment, an average, placebo-corrected body weight loss of 1.8 kg was reported. Later work showed that both reduced food intake and increased energy expenditure contribute to the observed weight loss [29]. Following these results, Bloom and coworkers identified a stabilized OXM analog TKS1125/OAP-189/PF5212389 (Thiakis/Wyeth/Pfizer), which was tested in a phase 1 study starting in 2009. Unfortunately, no further information has been disclosed on this compound.

Pocai et al. published their preclinical work on a close OXM analog in 2009 [30]. Introducing a D-serine in position 2 to increase proteolytic stability and attaching an albumin-binding cholesterol moiety via a C-terminal cysteine to extend half-life, they obtained their GLP-1R/GCGR dual agonist DualAG. Changing a single amino acid (Q->E) in position 3, they prepared a GLP-1R selective comparator molecule GLPAG. In diet-induced obese (DIO) mice, DualAG exhibited superior weight loss and higher lipid-lowering activity than GLPAG while showing comparable antihyperglycemic efficacy, thereby proving the therapeutic concept. In addition, improvements in plasma metabolic parameters including insulin, leptin, and adiponectin were more pronounced upon chronic treatment with DualAG than with GLPAG. Dual receptor agonism also increased fatty acid oxidation and reduced hepatic steatosis in DIO mice.

Two further OXM analogs reached human studies. Pegapamodutide (TT-401/LY2944876/OPK88003; Figure 20.2b), originally developed by Lilly, contains seven amino acid changes compared to native OXM. At the C-terminus, it carries two cysteines linked to a 20 kDa polyethylene glycol (PEG) unit each to increase half-life to allow for once-weekly administration. In a 24-week phase 2 trial in patients with T2D, pegapamodutide (50 mg, qw) administration resulted in a hemoglobin A1c (HbA1C) reduction of 1.3%, which was similar to once-weekly GLP-1R agonist exenatide long-acting release (LAR) (1.4%) and statistically superior to placebo (0.4%) [31]. The observed weight loss of 3.3 kg at week 24 was limited, although superior to both placebo (1.6 kg) and exenatide LAR (2.1 kg).

With a very similar design, JNJ54728518 carries an Aib in position 2 to improve metabolic stability and a cysteine mutation at position 30 with a PEG conjugation for half-life extension in combination with a truncation of residues 32–37, producing a dual agonist with good activity on both receptors [32]. It was tested in phase 1 but stopped afterward.

At the same time, the groups of DiMarchi and Tschöp worked on GLP-1R/GCGR dual agonists based on the structure of GCG. In a first paper in 2009, they described a PEGylated dual agonist "Aib2 chimera lactam*" [33] (see Figure 20.2b). The mutations in comparison with GCG comprise an Aib in position 2 to block DPP-IV-mediated cleavage, integration of the amino acids from GLP-1 in positions 17–21 and 23, and a single-chain lactam bridge between positions 16 and 20 (i, i+4) to stabilize the backbone α-helix. "Aib2 chimera lactam*" showed a balanced

(a) Structural similarity of GLP-1R, GCGR and GIPR, allowing interaction by all 3 receptors with a single peptide

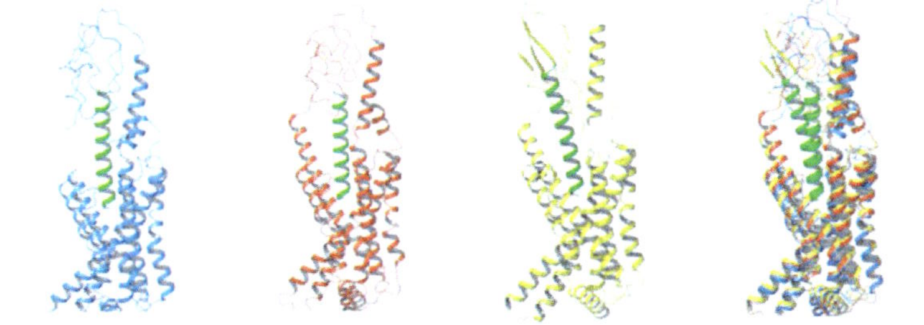

Peptide 20 – GLP-1R Peptide 20 – GCGR Peptide 20 – GIPR Superposition of all three receptors

(b) GLP-1R / GCGR dual agonists

name	1	2	3	4	5	6	7	8	9	10	11	12	13	14	15	16	17	18	19
GLP1(7-36)NH_2	H	A	E	G	T	F	T	S	D	V	S	S	Y	L	E	G	Q	A	A
Exenatide	H	G	E	G	T	F	T	S	D	L	S	K	Q	M	E	E	E	A	V
Glucagon	H	S	Q	G	T	F	T	S	D	Y	S	K	Y	L	D	S	R	R	A
GIP(1-42)	Y	A	E	G	T	F	I	S	D	Y	S	I	A	M	D	K	I	H	Q
Oxyntomodulin (hu)	H	S	Q	G	T	F	T	S	D	Y	S	K	Y	L	D	S	R	R	A
DualAG	H	D Ser	Q	G	T	F	T	S	D	Y	S	K	Y	L	D	S	R	R	A
GLPAG	H	D Ser	E	G	T	F	T	S	D	Y	S	K	Y	L	D	S	R	R	A
Pegapamodutide	H	Aib	Q	G	T	F	T	S	D	Y	S	K	Y	L	D	S	K	K	A
JNJ54728518	H	Aib	Q	G	T	F	T	S	D	Y	S	K	Y	L	D	S	R	R	A
Aib2 chimera lactam	H	Aib	Q	G	T	F	T	S	D	Y	S	K	Y	L	D	E(1)	Q	A	A
Aib2 chimera	H	Aib	Q	G	T	F	T	S	D	Y	S	K	Y	L	D	E	Q	A	A
SAR425899	H	D Ser	Q	G	T	F	T	S	D	L	S	K	Q	K[gGlu-Palm]	E	S	K	A	A
DiMarchi seq.17	H	Aib	Q	G	T	F	T	S	D	K[gGlu-gGlu-Palm]	S	K	Y	L	D	Aib	R	A	A
MK-1462	H	D Ser	Q	G	T	F	T	S	D	K[gGlu-gGlu-Palm]	S	K	Y	L	D	A	R	A	A
cotadutide	H	S	Q	G	T	F	T	S	D	K[gGlu-Palm]	S	E	Y	L	D	S	E	R	A
NN1177	H	Aib	Q	G	T	F	T	S	D	L	S	K	Y	L	E	S	K	R	A
Mazdutide	H	Aib	Q	G	T	F	T	S	D	Y	S	K	Y	L	D	E	K	K	A
survodutide	H	Cyb	Q	G	T	F	T	S	D	Y	S	K	Y	L	D	E	R	A	A
Efinopegdutide	H	Aib	Q	G	T	F	T	S	D	Y	S	K	Y	L	D	E(1)	K	R	A
Pemvidutide	H	Aib	Q	G	T	F	T	S	D	Y	S	K	Y	L	D	E	K[Sugar]	A	A

name	20	21	22	23	24	25	26	27	28
GLP1(7-36)NH_2	K	E	F	I	A	W	L	V	K
Exenatide	R	L	F	I	E	W	L	K	N
Glucagon	Q	D	F	V	Q	W	L	M	N
GIP(1-42)	Q	D	F	V	N	W	L	L	A
Oxyntomodulin (hu)	Q	D	F	V	Q	W	L	M	N
DualAG	Q	D	F	V	Q	W	L	M	N
GLPAG	Q	D	F	V	Q	W	L	M	N
Pegapamodutide	Q	E	F	V	Q	W	L	L	N
JNJ54728518	Q	D	F	V	Q	W	L	M	N
Aib2 chimera lactam	K(1)	E	F	I	C^{Δ}	W	L	M	N
Aib2 chimera	K	E	F	I	C^{Δ}	W	L	M	N
SAR425899	Q	D	F	I	E	W	L	K	A
DiMarchi seq.17	Q	D	F	V	Q	W	L	M	N
MK-1462	Q	D	F	V	Q	W	L	L	D
cotadutide	R	D	F	V	A	W	L	E	A
NN1177	R	E	F	V	Q	W	L	L	K[gGlu-gGlu-AEEAc-AEEAc-gGlu-C18diacid]
Mazdutide	K(AEEAc-AEEAc-gGlu-C20diacid)	E	F	V	E	W	L	L	E
survodutide	K	D	F	I	K[GGSGSGgGlu-C18diacid]	W	L	E	S
Efinopegdutide	K(1)	E	F	V	Q	W	L	M	N
Pemvidutide	K	E	F	I	Q	W	L	L	Q

name	29	30	31	32	33	34	35	36	37	38	39	40	41	42	C-term
GLP1(7-36)NH_2	G	R													NH_2
Exenatide	G	G	P	S	S	G	A	P	P	P	S				NH_2
Glucagon	T														OH
GIP(1-42)	Q	K	G	K	K	N	D	W	K	H	N	I	T	Q	OH
Oxyntomodulin (hu)	T	K	R	N	R	N	N	I	A						OH
DualAG	T	K	R	N	R	N	N	I	A	C[Chol]					NH_2
GLPAG	T	K	R	N	R	N	N	I	A	C[Chol]					NH_2
Pegapamodutide	Aib	G	R	N	R	N	N	I	A	$C^{\S}$	$C^{\S}$				NH_2
JNJ54728518	T	$C^{\$}$	R												NH_2
Aib2 chimera lactam	T														NH_2
Aib2 chimera	T														NH_2
SAR425899	G	G	P	S	S	G	A	P	P	P	S				NH_2
DiMarchi seq.17	T	K	gE												NH_2
MK-1462	T														NH_2
cotadutide	G	G													OH
NN1177	T														NH_2
Mazdutide	G	G	P	S	S	G									NH_2
survodutide	A														NH_2
Efinopegdutide	T	C^{*}													
Pemvidutide	T														NH_2

Green: Amino acids identical in GCG and exendin-4
Grey: Amino acids from exendin-4
Yellow: Amino acids from GCG/OXM
White: Amino acids from GLP-1

Violet: Amino acids from GIP
Orange: New amino acid mutations
Red font: Protraction unit

$C^{\S}$: Cysteine with 20kPEG as protraction unit
$C^{\$}$: Cysteine with 30kPEG as protraction unit
C^{Δ}: Cysteine with 40kPEG as protraction unit
C^{*}: l0kDa PEGILinker to Fc fragment of IgG4

C[Chol]:

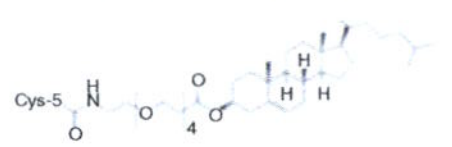

K[Sugar]:

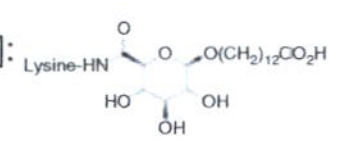

Figure 20.2 (a) Receptor overlay for GLP-1R, GCGR, and GIPR and interaction with triagonist 20. (b) GLP-1R/GCGR dual agonists.

in vitro activity on both receptors in contrast to the related analog without the lactam bridge, which showed seven times diminished activity at the GCGR. After treatment of DIO mice for four weeks, the balanced compound showed higher body weight loss driven in part by increased energy expenditure in addition to improved glucose control; decreased plasma triglycerides, low-density lipoprotein (LDL) cholesterol, and total cholesterol; and normalized liver lipid content. These results strongly supported the potential of GLP-1R/GCGR dual agonists not only for T2D and obesity but also for MASH.

Based on these results, several clinical candidates were developed by a number of companies, using a variety of half-life extending techniques. These GLP-1R/GCGR dual agonists show a series of different *in vitro* activity profiles, some having a higher relative activity on the GLP-1R and some showing a balanced profile on both receptors with others showing a preferred activation of the GCGR to investigate the optimal receptor activation ratio for maximum weight loss without deteriorated glucose control. In an experiment published in 2012 with carefully designed GLP-1R/GCGR dual agonists with different activity profiles, Day et al. showed that the balanced dual agonist demonstrated intermediate weight loss but with glucose control comparable to that of the most GLP-1R-selective peptide in mice [34]. Nevertheless, it was concluded that the optimal GLP-1R/GCGR activity ratio could be different for the various species and will need to be carefully evaluated [35].

Sanofi's SAR425899 is a GLP-1R/GCGR dual agonist structurally based on the selective GLP-1R agonist exendin-4 (see Figure 20.2b). Exendin-4 was chosen as a starting point due to its improved physicochemical profile compared to GCG, mainly driven by the C-terminal exendin tail, which helps to stabilize the α-helical fold of the peptide. After careful analysis of the structures of the selective peptides with the extracellular domains of their corresponding receptors, the positions were identified to introduce amino acids from GCG into the exendin-4 sequence to achieve dual agonism. In addition to GCG amino acids in positions 3, 16, and 19–21, at position 2, a D-Ser was introduced instead of GCG's L-Ser to stabilize against DPP-IV cleavage. Mutations were added in positions 17 and 28 to improve the overall profile as well as a lysine carrying a C16 fatty acid side chain linked via γ-glutamate to the lysine side chain nitrogen in position 14 to extend the half-life. SAR425899 shows a half-maximal effective concentration (EC50) of 1.1 pM at the human GLP-1R and of 5.5 pM at the human GCGR in overexpressing human embryonic kidney 293 (HEK293) cells. This bias for the GLP-1R was chosen to ensure proper glucose control in T2D patients. In a phase 1 study in healthy volunteers, a human half-life of 11–17 hours was observed, suitable for q.d. dosing [36, 37]. In a 4-week study in T2D patients, a 180 μg daily subcutaneous injection of SAR425899 led to a mean body weight loss of 5.46 kg versus 2.37 kg in the placebo group and an HbA1c change of −0.59% compared to 0.06% for placebo. The bias for the GLP-1R could be confirmed in humans in a positron emission tomography (PET) tracer target occupancy study where SAR425899 at the clinical dose of 200 μg once daily demonstrated an average GLP-1R occupancy of 49.9 ± 13.3% and a GCGR occupancy of 11.2 ± 14.4% in $N = 5$ patients [38]. In 2019, further development of SAR425899 was halted.

Merck tested the once-daily subcutaneously dosed GLP-1R/GCGR dual agonist MK-8521 in clinical trials. The sequence has not been disclosed so far. DiMarchi and colleagues described in one of the related patents a dual agonist with four amino acid mutations in comparison with GCG with an Aib in position 2 to suppress DPP-IV cleavage, a C16 fatty acid side chain attached to the lysine in position 10 via two γ-glutamate linkers to extend half-life, a second Aib amino acid in position 16 to stabilize the α-helix as well as an alanine from GLP-1 at 18. [39]. Additionally, a lysine and a γ-glutamate amino acid were added to the C-terminus. It provided higher body weight loss and improved glucose control in comparison with a structurally related selective GLP-1R agonist in a 21-day experiment in rhesus monkeys [40]. For MK-8521, results of a 12-week phase 2 study in T2D patients were released [41]. Once-daily subcutaneous administration of 300 μg MK-8521 led to body weight loss of 3.0 kg, superior to placebo (1.3 kg), but not different from the open comparator arm liraglutide, 1.8 mg q.d., which showed a body weight loss of 2.9 kg. MK-8521 showed an HbA1c reduction of 1.05%, again superior to placebo (0.44%), but inferior to the reduction in the liraglutide arm with 1.42%.

Merck recently described a second clinical GLP-1R/GCGR dual agonist MK-1462 with a balanced receptor activity profile on the two receptors. Starting from the GCG sequence, four amino acids at positions 16, 18, 27, and 28 were mutated in addition to the introduction of a D-Ser in position 2 and the C16 fatty acid side chain, linked via two γGlu units to the side chain nitrogen of the lysine, in position 10 to achieve the balanced profile. In a 22-day study in T2D rhesus monkeys, MK-1462 at a dose of 3 μg/kg q.d. showed superior body weight loss compared to liraglutide (30 μg/kg q.d), whereas liraglutide showed better glucose control than the dual agonist. Based on the favorable PK and pharmacodynamics (PD) results, MK-1462 was advanced to the clinic, but no results are available so far [42].

The sequence of AstraZeneca/MedImmune's once-daily GLP-1R/GCGR dual agonist cotadutide (MEDI0382;

Figure 20.2b) is also based on GCG and contains nine amino acid mutations to increase GLP-1R agonism [43]. It is completely designed with natural L-amino acids with only one lysine at position 10 in the peptide chain to enable semisynthetic access. On top of the typical C16 fatty acid side chain linked via γ-glutamate to the lysine nitrogen, amino acids from exendin-4 at positions 17, 20, 29, and 30; an amino acid from GLP-1 at position 24; and further mutations at positions 12, 27, and 28 were introduced. This led to *in vitro* EC50 values of 6.9 nM at the GLP-1R and 10.2 pM at the GCGR in overexpressing Chinese hamster ovary (CHO) cells. In a 41-day phase 2a study in T2D patients, s.c. administration of 200 µg q.d. cotadutide showed significantly greater weight loss of 3.84 kg versus 1.70 kg for placebo. This weight loss was accompanied by a reduction in subcutaneous and visceral adipose tissue. An HbA1c decrease of 0.9% was observed, which was significantly higher than for placebo (0.6%). In addition, treatment with cotadutide led to a significantly higher liver fat reduction of 39% compared to the placebo group (20%), underlining the potential of GLP-1R/GCGR dual agonism for the treatment of MASH [44]. Based on these results and follow-up phase 2 studies, cotadutide was advanced to a phase 2/3 study in non-cirrhotic MASH patients with fibrosis, but shortly after initiation, AstraZeneca decided in April 2023 to stop enrollment after recruiting 45 patients, well short of the targeted 1860 patients, in favor of a once-weekly analog AZD9550 whose structure is still unknown.

Novo Nordisk's GLP-1R/GCGR dual agonist NN1177 (NNC9204-1177) is based on the human GCG sequence [45]. Changes were done by introducing amino acids from exendin-4 at positions 10, 15, and 20; GLP-1 at position 21; and additional substitutions at positions 2, 17, and 27–28 to obtain dual receptor activity and to ensure long-term formulation stability. A longer plasma half-life was secured by the attachment of a C18 fatty diacid, in line with the once-weekly GLP-1R agonist semaglutide [22]. However, in contrast to the semaglutide side chain, the linker between the peptide backbone and the fatty acid contains several negatively charged residues. Incorporation of such residues was generally shown to improve the physical stability of GCG-based co-agonists when evaluated in a thioflavin T-based fibrillation assay. A human mean half-life between 77 and 111 hours was observed [46]. In a 12-week multiple dosing study, clinically relevant weight loss of 12.6% for the highest weekly dose of 4.2 mg was obtained. Although NN1177 appeared tolerable across trials, several unexpected treatment-related safety signals like increased heart rate, decreased reticulocyte count, increased markers of inflammation, increased aspartate and alanine aminotransferase, and impaired glucose tolerance were observed. Due to these observed safety signals, the clinical development of NN1177 was terminated.

20.5 Clinical Advanced GLP-1/GCGR Dual Agonists

The currently clinically most advanced GLP-1R/GCGR dual agonists have recently reached phase 3. Most advanced is Innovent's mazdutide (IBI362 or LY3305677). Based on the structure of GCG, the team of the original inventors at Lilly used an Aib in position 2 to block DPP-IV cleavage and a C20 diacid fatty acid side chain at position 20 linked via a semaglutide-type linker for half-life extension. In addition, in positions 29–34, part of the exendin tail was introduced, potentially to stabilize the α-helical structure. Further mutations were three glutamates from exendin-4 in positions 16 and 24 and from GLP-1 in position 21 to introduce GLP-1R agonism. Additionally, two lysines were introduced in positions 17 and 18 together with a leucine in position 27 and an aspartate in position 28. In phase 1 studies, both in healthy subjects and in T2D patients, dose escalation schemes up to 10 mg were established and the once-weekly administration profile was confirmed with a terminal half-life of 8.5–11.6 days. In overweight-to-obese Chinese adults, 6 mg weekly doses of mazdutide for 26 weeks led to a body weight reduction of 11.3%, compared to a 1% weight gain with placebo [47]. Similarly, a body weight loss of 7.1% was observed in type 2 diabetic Chinese patients after 20 weeks of dosing of 6 mg in comparison with a loss of 1.4% with placebo and 2.7% with selective GLP-1R agonist comparator dulaglutide [48]. High HbA1c reductions of −1.41%, −1.67%, and −1.55% were achieved for all three dose groups of mazdutide (3, 4.5, and 6 mg). This corresponds to a numerically greater HbA1c reduction with 4.5 and 6 mg mazdutide versus dulaglutide (−1.35%), although superiority was not established. These results underline the potential of the GLP-1R/GCGR dual agonist mazdutide to confer additional weight loss and potentially cumulative benefits in cardiometabolic risk factors, without significantly compromising glycemic effect in comparison with selective GLP-1R agonists. Currently, four phase 3 studies of mazdutide in Chinese patients with overweight or obese (GLORY-1 and GLORY-2) and type 2 diabetic (DREAMS-1 and DREAMS-2) subjects are underway. High-level results of the first of these studies, GLORY-1, have recently been shown where mazdutide met the primary and all key secondary endpoints versus placebo [49].

Boehringer Ingelheim's survodutide is currently undergoing a phase 3 study program as well. After the return of the once-daily dual agonist ZP2929 from Boehringer back to their collaboration partner Zealand Pharma, survodutide (BI456906) was selected as follow-up compound in the collaboration. Based on the structure of GCG, it carries a 1-amino-1-cyclobutanecarboxylic acid in position 2 to increase metabolic stability and a C18 diacid fatty acid side chain in position 24, linked via a glycyl-glycyl-seryl-glycyl-seryl-glycyl-gamma-glutamyl-(GGSGSG-γ-Glu) linker to the nitrogen of the corresponding lysine to extend half-life [50]. Introduction of aspartate in position 16, alanine in position 18, and isoleucine in position 23 from exendin-4 and lysine from GLP-1 in position 20 in combination with the new C-terminus glutamyl-seryl-aspartyl (ESA) in positions 27–29 provided a balanced dual agonist with a profile suitable for clinical development. A first phase 2 study assessed survodutide in overweight-to-obese participants with T2D on a background of metformin therapy with doses up to 2.7 mg once weekly and 1.8 mg twice weekly in comparison with placebo and an open-arm 1.0 mg semaglutide group over 16 weeks [51]. The two highest once-weekly doses of 1.8 and 2.7 mg showed superior body weight loss compared to semaglutide (−6.6% versus −5.3%), with similar reductions of HbA1c. In a 46-week follow-up study, weekly doses of 4.8 mg survodutide led to a body weight loss of 14.9% in comparison with 2.8% with placebo. Two parallel phase 3 studies assess now weekly doses of 3.6 mg and 6 mg over 76 weeks in obese patients either with T2D or without T2D. In addition, a long-term cardiovascular (CV) safety study and a phase 2 program in MASH were initiated.

Efinopegdutide (HM12525A, JNJ-64565111, and MK-6024) has undergone a journey with a lot of ups and downs. Based on Hanmi's lapscovery platform technology, a GCG-based peptide sequence is conjugated via the C-terminal cysteine and a maleimide to a 10 kDa PEG linking it to an IgG4 Fc region to extend half-life. The dual agonist peptide contains an Aib in position 2 to improve metabolic stability, an glutamate–lysine lactam in positions 16 and 20 to stabilize the α-helix, and a lysine in position 17 and a glutamate from GLP-1 in position 21. Being licensed by J&J, it underwent several phase 2 studies. In a 26-week study in obese subjects of body mass index (BMI) 35–50 kg/m^2 without T2D, once-weekly dosing provided placebo-subtracted body weight changes of −6.8, −8.1, and −10.0% for the JNJ-64565111 dose groups (5.0, 7.4, and 10.0 mg, respectively) versus −5.8% for the 3 mg liraglutide group [52]. The incidence of treatment-emergent adverse events, especially nausea and vomiting, was higher in each JNJ-64565111 treatment group compared to placebo and liraglutide. Based on these and similar results from the other studies, J&J ended the collaboration with Hanmi in 2019. But already in 2020, Merck and Co. licensed the drug for MASH and started a phase 2 clinical trial. In an 8-week study, a significantly higher reduction in liver fat content was obtained with efinopegdutide (10 mg, −72.2%) than with semaglutide (1 mg, −42.3%) [53]. Based on these results, efinopegdutide was granted fast-track designation by the FDA.

Pemvidutide (ALT801, SP1373, and VPD1070) is an additional GLP-1R/GCGR dual agonist with first results from phase 2. Based on the GCG sequence, amino acids from GLP-1 or exendin-4 were introduced at positions 16, 18, 20, 21, and 23. It contains the well-known Aib mutation in position 2 as well as two further changes at the C-terminus in positions 27 and 28. As a protraction moiety, a C18 fatty diacid linked via glucopyranose unit to the lysine in position 10 led to a once-weekly profile. In a first 48-week phase 2 study in overweight and obese subjects without T2D, a body weight loss of 15.6% was observed for the once-weekly 2.4 mg dose in comparison with 2.2 % for placebo [54]. While a number of serum lipids were reduced as well, glucose homeostasis was maintained. A phase 2 study for the treatment of MASH is ongoing.

Another GLP-1R/GCGR dual agonist is still in clinical trials without exact structural information being disclosed. G3215 is reported to be an OXM analog without protraction moiety and is currently being developed by the team at Imperial College, which ran the original clinical studies with OXM [55]. To achieve a faster uptitration regime, they are exploring a dual agonist for continuous subcutaneous infusion via pump in an adaptive design based on the occurrence of nausea and vomiting. In a 14-day phase 1 trial, the treatment group showed a 2.39 kg body weight loss in comparison with 0.84 kg in the placebo group without negative impact on glucose control. Further studies are needed to better explore the potential of this approach.

GLP-1R agonists have been extensively studied structurally to elucidate their mode of action. Peptidic agonists bind to the ECD and TMD of the GLP-1R, triggering conformational changes. The ECD rotates upward and clockwise relative to the TMD, while the transmembrane region of the TMD rearranges to stabilize a kink in TM6 [56]. The peptidic agonists bind in a helical conformation, while the overall binding modes are very similar for the different ligands, with differences in the dynamics of the protein–ligand complex between the different ligands. Molecular dynamic simulations of OXM and GLP-1 complexes revealed extensive interactions between N-terminal residues and amino acids located deep in the receptor pocket for GLP-1, while OXM forms more transient interactions at these positions. The different dynamical behavior for the peptides GLP-1, exendin-4, OXM, and exendin-P5 has been associated with different kinetics of G-protein binding, which might affect the efficacy of cyclic adenosine monophosphate (cAMP) coupling [56]. Structural analysis of GLP-1R structures also provides the first evidence that biased signaling might be associated with the TM7 and TM5 helices [57].

Semaglutide contains a fatty acid derivative to extend its half-life via albumin binding. The structure of semaglutide does not reveal the molecular details of this interaction but suggests two different conformations for the fatty acid substituent, one pointing toward TM1 and the other toward the ECD [58]. Molecular dynamic simulations of different acylated liraglutide derivatives indicate interactions of the fatty acid side chain with the ECD [59]. The interactions between the fatty acid and the receptor contribute significantly to the binding affinity, while significant differences in the interactions depend on the position and type of the fatty acid substituents [60].

These structural studies reveal that while the overall binding mode of GLP-1R ligands is similar, the receptor exhibits different interactions and dynamics depending on the specific agonist. The structural understanding of how these differences influence the pharmacology of the different ligands starts to emerge.

The similarity between GLP-1 and GCG sequences as well as the similar mode of action at their respective receptors allows for achieving dual target active peptide sequences with a strong pharmacological effect on metabolic parameters. The addition of an albumin-binding moiety was successfully used to extend half-life for some molecules adding a third binding feature to the ligands. While no GLP-1R/GCGR dual agonists have been approved so far, several compounds are now in phase 3 for the treatment of T2D, obesity, and MASH with mazdutide having even completed a first study. It will be very exciting to see the upcoming results to be able to compare how the different activity profiles will compare with the selective GLP-1R agonists and other dual agonists. It is likely that different profiles will be suitable for different indications.

20.6 GLP-1 Receptor/Glucose-dependent Insulinotropic Polypeptide (GIP) Receptor Dual Agonists

The incretin effect refers to enhanced insulin secretion following an oral glucose load relative to an intravenous glucose administration and is caused by the two incretin hormones, GLP-1 and GIP. Both hormones account equally for this insulin secretion in humans [61], and therefore, an improved antidiabetic effect of such dual agonistic peptides compared to sole GLP-1R agonists could be expected. The role of GIP with respect to energy homeostasis is unclear, as there is conflicting data reported in favor of and abrogating GIP action in animal models of obesity [62–64].

The groups of DiMarchi and Tschöp presented the first highly active GLP-1R/GIP receptor (GIPR) dual agonists [65]. Based on the sequence of their GLP-1R/GCGR co-agonist "Aib2 chimera lactam*" (Figure 20.2b), seven amino acids from endogenous GIP were introduced, most importantly a tyrosine in position 1 and an isoleucine in position 12. Additionally, an Aib was introduced in position 20 to stabilize the α-helix instead of the lactam bridge between positions 16 and 20, and the exendin tail was added to the C-terminus to further stabilize the peptide backbone. As protraction unit, a 40 kDa PEG unit was attached via a cysteine at position 24. These mutations provided the GLP-1R/GIPR dual agonist MAR701 (Marcadia Biotech – later licensed to Roche as RG-7685/RO-6807952; Figure 20.3) showing balanced activity at both receptors. Extensive profiling in rodent models showed high body weight loss in DIO mice without an increase in energy expenditure and excellent glucose control in db/db mice. In a 6-week phase 1 study in T2D patients, a mean elimination half-life of 120–141 hours was observed. An HbA1c reduction of 1.11% was obtained for the highest tested dose of once-weekly 30 mg compared to 0.16% for placebo. Effects on body weight were not statistically significant. After the discontinuation of MAR701, the once-daily follow-up compound MAR709 (RG7697/RO6811135/NNC0090-2746; Figure 20.3), also originally developed by Marcadia, was pursued by Roche and later Novo Nordisk [65]. Being a close analog of MAR701, it carries a C16 fatty acid side chain attached to the lysine in position 40 to extend half-life via albumin binding. In the single ascending dose study in humans, a rapid s.c. absorption and a flat PK profile with a terminal half-life of 15–25 hours for doses ≥0.2 mg were observed [66]. MAR709 was tested at 1.8 mg q.d. without any uptitration regimen in a 12-week phase 2 trial in T2D patients and compared to 1.8 mg liraglutide as an open-label comparator, showing a placebo-corrected HbA1c reduction of 0.96%, which was similar to liraglutide (1.12%) [67]. Body weight decreased by 1.67% compared to placebo during the 12 weeks, again similar to liraglutide (1.23%).

The best-known GLP-1R/GIPR dual agonist is tirzepatide (LY3298176; Figure 20.3) from Eli Lilly, which was approved by the FDA for the treatment of T2D in 2022 as Mounjaro and for obesity in 2023 as Zepbound. This hybrid of both endogenous ligands contains two Aib amino acids in position 2 for DPP-IV protection and position 13 as well as the C-terminal exendin tail to stabilize the

GLP-1R/GIPR dual agonists

name	1	2	3	4	5	6	7	8	9	10	11	12	13	14	15	16	17	18	19	20	21	22	23	24	25	26	27	28	29	30	31	32	33	34	35	36	37	38	39	40	41	42	C-term
GLP1(7-36)NH_2	H	A	E	G	T	F	T	S	D	V	S	S	Y	L	E	G	Q	A	A	K	E	F	I	A	W	L	V	K	G	R													NH_2
Exenatide	H	G	E	G	T	F	T	S	D	L	S	K	Q	M	E	E	E	A	V	R	E	F	I	E	W	L	K	N	G	G	P	S	S	G	A	P	P	P	S				NH_2
Glucagon	H	S	Q	G	T	F	T	S	D	Y	S	K	Y	L	D	S	R	R	A	Q	D	F	V	Q	W	L	M	N	T														OH
GIP(1-42)	Y	A	E	G	T	F	I	S	D	Y	S	I	A	M	D	K	I	H	Q	Q	D	F	V	N	W	L	L	A	Q	K	G	K	K	N	D	W	K	H	N	I	T	Q	OH
MAR701	Y	Aib	E	G	T	F	T	S	D	Y	S	I	Y	L	D	K	Q	A	A	Aib	E	F	V	C[PEG40k]	W	L	L	A	G	G	P	S	S	G	A	P	P	P	S	K			NH_2
MAR709	Y	Aib	E	G	T	F	T	S	D	Y	S	I	Y	L	D	K	Q	A	A	Aib	E	F	V	N	W	L	L	A	G	G	P	S	S	G	A	P	P	P	S	K(Palm)			NH_2
tirzepatide	Y	Aib	E	G	T	F	T	S	D	Y	S	I	Aib	L	D	K	I	A	Q	K[AEEAc-AEEAc-gGlu-C20diacid]	A	F	V	Q	W	L	I	A	G	G	P	S	S	G	A	P	P	P	S				NH_2
Viking Cpd.4	Y	Aib	E	G	T	F	T	S	D	Y	S	I	Aib	L	D	K	I	A	Q	K[AEEAc-AEEAc-gGlu-C19phosphonic acid]	A	F	V	Q	W	L	I	A	G	G	P	S	S	G	A	P	P	P	S				NH_2

GLP-1R/GCGR/GIPR triple agonists

name	1	2	3	4	5	6	7	8	9	10	11	12	13	14	15	16	17	18	19	20	21	22	23	24	25	26	27	28	29	30	31	32	33	34	35	36	37	38	39	40	41	42	C-term
GLP1(7-36)NH_2	H	A	E	G	T	F	T	S	D	V	S	S	Y	L	E	G	Q	A	A	K	E	F	I	A	W	L	V	K	G	R													NH_2
Exenatide	H	G	E	G	T	F	T	S	D	L	S	K	Q	M	E	E	E	A	V	R	L	F	I	E	W	L	K	N	G	G	P	S	S	G	A	P	P	P	S				NH_2
Glucagon	H	S	Q	G	T	F	T	S	D	Y	S	K	Y	L	D	S	R	R	A	Q	D	F	V	Q	W	L	M	N	T														OH
GIP(1-42)	Y	A	E	G	T	F	T	S	D	Y	S	I	A	M	D	K	I	H	Q	Q	D	F	V	N	W	L	L	A	Q	K	G	K	K	N	D	W	K	H	N	I	T	Q	OH
Triagonist 20	H	Aib	Q	G	T	F	T	S	D	K[gGlu-Palm]	S	K	Y	L	D	E	R	A	A	Q	D	F	V	Q	W	L	L	D	G	G	P	S	S	G	A	P	P	P	S				NH_2
SAR441255	H	Aib	H	G	T	F	T	S	D	L	S	K	L	K[gGlu-gGlu-Palm]	E	E	Q	R	Q	Aib	E	F	I	E	W	L	K	A	DAla	G	P	P	S	Aib	K	P	P	P	K				NH_2
Retatrutide	Y	Aib	Q	G	T	F	T	S	D	Y	S	I	MeL	L	D	K	K[AEEAc-gGlu-C20-diacid]	A	Q	Aib	A	F	I	E	Y	L	L	E	G	G	P	S	S	G	A	P	P	P	S				NH_2
efocipegtrutide	Y	Aib	Q	G	T	F	T	S	D	Y	S	K	Y	L	D	E	K	R	A	K	E	F	V	Q	W	L	L	D	H	H	P	S	S	G	Q	P	P	P	S	C*			

Green: Amino acids identical in GCG and exendin-4
Grey: Amino acids from exendin-4
Yellow: Amino acids from GCG/OXM
White: Amino acids from GLP-1

Violet: Amino acids from GIP
Orange: New amino acid mutations
Red font: Protraction unit

C*: 10kDa PEGILinker to Fc fragment of IgG4

Figure 20.3 GLP-1R/GIPR dual agonists and GLP-1R/GCGR/GIPR triple agonists

α-helical fold [68]. In addition, it carries a lipophilic side chain (AEEAc)2-γGlu-C20-diacid, on the lysine in position 20 to achieve a long half-life. The fatty acid side chain has an analogous design as the side chain in semaglutide with a C2-elongated diacid part. Tirzepatide has high *in vitro* activity at both human receptors with GIPR activity comparable to endogenous GIP and a fivefold weaker activity than native GLP-1 at the GLP-1R. A human half-life of 106–124 hours was observed suitable for once-weekly dosing. In a 40-week head-to-head phase 3 comparison between dual agonist tirzepatide and the selective GLP-1R agonist semaglutide in T2D patients, all doses of tirzepatide (5, 10, and 15 mg) surpassed semaglutide (1 mg) in both glucose lowering (−2.01 to −2.3% HbA1c lowering versus −1.86%) and body weight loss (7.6 kg to 11.2 kg versus 5.7 kg) [69]. In obese patients without T2D, higher body weight losses up to −26% were observed [70]. The CV outcome study "A Study of Tirzepatide (LY3298176) Compared With Dulaglutide on Major Cardiovascular Events in Participants With Type 2 Diabetes" (SURPASS-CVOT) is ongoing and expected to read out in 2024.

Protein–ligand complexes could be obtained by cryo-electron microscopy (cryo-EM) illustrating the binding of tirzepatide to the GLP-1R as well as the GIPR [18, 71]. Tirzepatide binds to both receptors in a similar manner with the N-terminus binding into the TMD and the C-terminus interacting with the ECD. Interestingly, the C-terminal part of tirzepatide tilts to the extracellular loop 1 (ECL1). Such conformations have been observed in other peptide–GLP-1R complexes before, like for EX–P5 [72] and for taspoglutide [58], and can be attributed to the flexibility of GLP-1R ECL1. The conformation of the half-life protraction fatty acid group was not resolved in the structure, but molecular dynamic simulations indicate conformational diversity.

Viking Therapeutics recently announced top-line results from clinical trials of their GLP-1R/GIPR dual agonist VK2735. In the VENTURE phase 2 study, obese patients lost up to 14.7% body weight upon once-weekly administration of 15 mg s.c. after 13 weeks compared to 1.7% weight loss for placebo-treated patients [73]. An oral formulation of VK2735 showed dose-dependent decreases in body weight in a 28-day phase 1 study with the highest daily dose of 40 mg achieving 3.3% placebo-corrected weight loss [74]. While the exact structure of VK2735 has not been disclosed yet, Viking Therapeutics recently published patent applications highlighting compound 4 (see Figure 20.3) [75, 76]. It contains the identical peptide sequence as tirzepatide but differs in the terminal acidic moiety in the side chain where a phosphonic acid group replaces the carboxylic acid. The inventors claim that they surprisingly observe a longer half-life in cynomolgus monkeys compared to tirzepatide although a comparison of *in vitro* activities with and without HSA for the two molecules would suggest lower albumin-binding affinity for compound 4.

Additional GLP-1R/GIPR agonists are or were in clinical phases, but their structures are not known. Most advanced is Roche's RG6641 (CT-868) in phase 2. Originating from Carmot's chemotype evolution platform, the once-daily compound was designed with biased signaling to preferentially activate cAMP while having minimal to no ß-arrestin recruitment and therefore receptor internalization, which is claimed to improve the therapeutic window. The once-weekly follow-up compound RG6640 (CT-388) recently completed a phase 1 study. Additional dual agonists currently reported in phase 1 are HZ-010, SCO094, HRS9531, and HISHS-2001. SAR438335 was in phase 1 but was since discontinued.

20.7 GLP-1 Receptor/Glucagon Receptor/GIP Receptor Triple Agonists

The groups of DiMarchi and Tschöp were even able to generate unimolecular triple agonists, which interact with all three receptors GLP-1R, GCGR, and GIPR [77]. This is possible because the three endogenous hormones GLP-1, GCG, and GIP share a high sequence overlap, especially with identical amino acids in positions 4, 5, 6, 8, 9, 11, 22, 25, and 26 and related amino acids in positions 10, 14, 15, 21, 22, and 27 (see Figure 20.3). Based on this observation, triagonist 20 was designed by carefully selecting amino acids from the sequences of the three endogenous agonists. In the stretch from positions 11 to 26, only two amino acid mutations were done to the GCG sequence, achieving high activity on the GCGR. An Aib in position 2 was introduced, not only to block DPP-IV-mediated cleavage but in combination with the introduction of the exendin-4 tail at the C-terminus also to obtain a high activity at the GIPR without observing any change in activity for the GLP-1R or GCGR. A fatty acid side chain–conjugated lysine in position 10 was added to extend half-life. The resulting triagonist was broadly profiled in a number of rodent models. Superior results were observed both on body weight loss and on glucose control in comparison with the GLP-1R/GIPR dual agonist MAR709.

Structural studies illustrated the binding mode of triagonist 20 (MAR423) to all three receptors [18]. Triagonist 20 adopts a conformation that is similar to the binding of its cognate ligands for all three receptors. While the ligands bind to the ECD in a similar fashion for all three receptors, some conformational divergence can be observed in the region of ECL1. The fatty acid side chain forms significant interactions in a cleft between TM1 and TM2 in simulations for GIPR and GCCR, which is in line with a drop of activity for a delipidated form for the GIPR and the GCCR.

Based on these results, a number of further triple agonists have reached clinical trials.

Sanofi's triple agonist SAR441255 is based on the exendin-4 structure. Amino acids of GLP-1 at positions 17 and 21, of GCG at position 18, and of GIP at positions 19 and 28 were introduced to enhance receptor activation at the GCGR and GIPR. In position 2, an Aib was introduced to further strengthen the activation of all three receptors as well as to protect against DPP-IV cleavage. Four amino acids were mutated in the exendin tail (positions 32, 34, 35, and 39) to improve the physicochemical profile in the presence of antimicrobial preservatives, particularly to avoid aggregation issues at acidic pH in the presence of phenolic preservatives. Further changes at positions 3, 13, 20, and 29 were introduced to achieve high and balanced activation at each receptor and to ameliorate chemical stability in buffer formulations. At position 14, a fatty acid side chain was introduced at the ε-amino group of lysine using two γ-glutamic acid spacers to prolong *in vivo* half-life via albumin binding [78]. SAR441255 not only showed impressive body weight loss and improved glucose control in obese diabetic cynomolgus monkeys but also showed simultaneous target engagement of GLP-1R, GCGR, and GIPRs in humans in a phase 1 study. Nevertheless, the compound was stopped for strategic reasons.

The most advanced GLP-1R/GCGR/GIPR triple agonist is Lilly's retatrutide (LY3437943), which is currently undergoing phase 3 clinical trials. It is based on the GIP peptide backbone combined with amino acids from GCG in positions 3, 7, and 14 as well as from GLP-1 in positions 18, 23, and 24. To further improve GIPR activity, the exendin tail and an Aib amino acid in position 20 were added, which also stabilized the helical fold. To achieve the balanced activity on all three receptors, an α-methyl leucine in position 13 was introduced in addition to the Aib in position 2 and the fatty acid side chain on lysine in position 17 to prolong half-life. Further mutations were done with an alanine at position 21, a tyrosine in position 25 replacing the tryptophan in the core of the tryptophan cage as well as a glutamate in position 28 to achieve the desired overall profile [79]. In healthy subjects, a mean half-life of approximately six days was obtained. Phase 2 data were recently reported. After 36 weeks of treatment, HbA1c lowering of over 2% and a body weight loss of more than 17% in patients with T2D and a body weight loss of 24% in patients without T2D were obtained, clearly underlying the potential of triple agonists to outcompete the dual agonists [80, 81].

Hanmi's GLP-1R/GCGR/GIPR efocipegtrutide (HM-15211) is in analogy to their GLP-1R/GCGR dual agonist formed via conjugation of a peptide agonist to an IgG Fc fragment to extend half-life. The peptide part is based on the GIP sequence with amino acids from GCG at positions 3, 7, 12–14, 18–19, and 24; from GLP-1 at positions 20–21; and exendin-4 at positions 16 as well as 31–34 and 36–39 in the exendin tail. To achieve the desired overall profile, further mutations with an Aib in position 2, a lysine at position 17, and a aspartyl-histidyl-histidyl (DHH)-tripeptide at positions 28–30 were introduced. It is currently profiled in phase 2 for the treatment of MASH [82].

A further GLP-1/GIP/GCGR triple agonist NNC9204-1706 or NN9423 by Novo Nordisk with an unknown structure was evaluated in a phase 1 trial but has been terminated [83].

20.8 Further Unimolecular Multiagonists Based on Glucagon-like Peptide-1 (GLP-1) Following the One-pharmacophore Approach

Following the one-pharmacophore approach, GLP-1R agonists have been combined with additional endogenous ligands of members of the secretin-like GPCR family.

Zealand Pharma designed GLP-1R/GLP-2R dual agonist dapiglutide (ZP7570), which is currently tested in clinical phase 2 for the treatment of short bowel syndrome (see Figure 20.4). Based on the sequence of GLP-2, amino acids of GLP-1 were introduced in positions 3, 7–8, 17, and 24 and from exendin–4 in position 10. In combination with the well-known mutations of Aib in position 2 and a fatty acid side chain carrying lysine 16, an alanine in position 11 and a histidine in position 29 were chosen to provide the desired profile. Dapiglutide significantly improved oral glucose tolerance, reduced intestinal transit time, and promoted intestinal growth in rodent models [84]. Similar results were published by Gubra for another GLP-1R/GLP-2R co-agonist GUB09-145 [85].

Based on the observation that in addition to GLP-1 secretin is also upregulated in a rat model of bariatric surgery, the same group recently described a secretin/GLP-1 co-agonist GUB06-046 (see Figure 20.4). The N-terminus of GLP-1 with positions 1–9 with a mutated Aib in position 2 is combined with the C-terminus of secretin at positions 10–27. To get a balanced profile, amino acids from GLP-1 at positions 14, 18, and 22–23 were introduced [86]. GUB06-046 showed reduced food intake in lean mice and improved glucose control and β-cell mass in db/db mice after eight weeks of treatment.

Multi agonists combining GLP-IR agonism with other pharmacology

name	1	2	3	4	5	6	7	8	9	10	11	12	13	14	15	16	17	18	19	20	21	22	23
GLP-1(7-36)NH$_2$	H	A	E	G	T	F	T	S	D	V	S	S	Y	L	E	G	Q	A	A	K	E	F	I
Exenatide	H	G	E	G	T	F	T	S	D	L	S	K	Q	M	E	E	E	A	V	R	L	F	I
Glucagon	H	S	Q	G	T	F	T	S	D	Y	S	K	Y	L	D	S	R	R	A	Q	D	F	V
GIP(1-42)	Y	A	E	G	T	F	I	S	D	Y	S	I	A	M	D	K	I	H	Q	Q	D	F	V
GLP-2	H	A	D	G	S	F	S	D	E	M	N	T	I	L	D	N	L	A	A	R	D	F	I
Dapiglutide	H	Aib	E	G	S	F	T	S	E	L	A	T	I	L	D	K[gGlu-C18diacid]	Q	A	A	R	D	F	I
GUB09-145	H	G	D	G	S	F	T	S	E	L	S	T	Y	K[gGlu-Palm]	D	A	L	A	A	R	D	F	I
Secretin	H	S	D	G	T	F	T	S	E	L	S	R	L	R	E	G	A	R	L	Q	R	L	L
GUB06-046	H	Aib	E	G	T	F	T	S	D	L	S	R	L	L	E	G	A	A	L	Q	R	E	I
ZP3022	H	G	E	G	T	F	T	S	D	L	S	K	Q	M	E	E	E	A	V	R	L	F	I
C2816	H	G	E	G	T	F	T	S	D	L	S	K	Q	M	E	E	E	A	V	R	L	F	I
xGLP/GCG/gastrin	H	DSer	Q	G	T	Y	T	N	D	V	S	K	Y	K[gGluPalm]	D	S	R	R	A	Q	D	F	I
Exendin-4/xenin-8-Gln	H	G	E	G	T	F	T	S	D	L	S	K	Q	M	E	E	E	A	V	R	L	F	I
Exendin-4/gastrin/xenin-8-Gln	H	G	E	G	T	F	T	S	D	L	S	K	Q	M	E	E	E	A	V	R	L	F	I
EP45	H	G	E	G	T	F	T	S	D	L	S	K	Q	M	E	E	E	A	V	R	L	F	I
GEP44	H	DSer	Q	G	T	F	T	S	D	L	S	K	Y	L	E	E	E	A	V	R	E	F	I
GEP12	F	DSer	Q	G	T	F	T	S	D	L	S	K	Y	L	E	E	E	A	V	R	E	F	I
Peptide19	H	Aib	E	G	T	F	T	S	D	V	S	S	Y	L	E	G	Q	A	L	R	H	Y	I
MEDI4166	H	G	E	G	T	F	T	S	D	L	S	K	Q	M	E	E	E	C(1)	A	R	L	F	I
Insulin/GLP-1/GCG triagonist	H	DSer	Q	G	T	F	T	S	D	K[gGlu-gGlu-Palm]	S	K	Y	L	D	E	R	A	A	Q	D	F	V
Maridebart cafraglutide	H	Aib	E	G	T	F	T	S	D	Y	S	S	Y	L	E	E	Q	A	A	K	E	F	I

name	24	25	26	27	28	29	30	31	32	33	34	35	36	37	38	39	40	41	42	43	44	45	46	47	C-term
GLP-1(7-36)NH$_2$	A	W	L	V	K	G	R																		NH$_2$
Exenatide	E	W	L	K	N	G	G	P	S	S	G	A	P	P	P	S									NH$_2$
Glucagon	Q	W	L	M	N	T																			OH
GIP(1-42)	N	W	L	L	A	Q	K	G	K	K	N	D	W	K	H	N	I	T	Q						OH
GLP-2	N	W	L	I	Q	T	K	I	T	D															OH
Dapiglutide	A	W	L	I	Q	H	K	I	T	D															OH
GUB09-145	A	W	L	I	Q	T	K	I	T	D															NH$_2$
Secretin	Q	G	L	V																					NH$_2$
GUB06-046	Q	W	L	V																					NH$_2$
ZP3022	E	W	L	K	N	AEEAc	AEEAc	Y	G	W	L	D	F												NH$_2$
C2816	E	W	L	K	N	PEG4	Nle	G	W	K[Tac]	D	NMeF													NH$_2$
xGLP/GCG/gastrin	E	W	L	K	N	G	G	P	S	S	G	A	P	P	P	S	AEEAc	AEEAc	Y	G	W	L	D	F	NH$_2$
Exendin-4/xenin-8-Gln	E	W	L	K	N	AEEAc	AEEAc	H	P	Q	Q	P	W	I	L										OH
Exendin-4/gastrin/xenin-8-Gln	E	W	L	K*	N	AEEAc	AEEAc	Y	G	W	L	D	F	AEEAc	AEEAc	H	P	Q	Q	P	W	I	L		OH
EP45	E	W	L	K	N	G	G	P	S	S	R	H	Y	L	N	L	V	T	R	Q	R	Y			NH$_2$
GEP44	A	W	L	K	N	G	G	P	S	R	H	Y	L	N	L	V	T	R	Q	R	Y				NH$_2$
GEP12	A	W	L	K	N	G	G	P	S	R	H	Y	L	N	L	V	T	R	Q	R	Y				NH$_2$
Peptide19	N	W	L	T	R	Q	R	Y																	NH$_2$
MEDI4166	E	W	L	K	N	Q	G	P	S	S	G	A	P	P	P	G	C(1)	Linker-PCSK9 Ab							
Insulin/GLP-1/GCG triagonist	K[click-PEG4-B1 F-insulin]	W	L	L	D	T																			NH$_2$
Maridebart cafraglutide	A	W	L	V	K	G	G	G	(GGGGS)3	K[BrAc)															NH$_2$

name	1	2	3	4	5	6	7	8	9	10	11	12	13	14	15	16	17	18	19	20	21	22	23	24	25	26	27	28	29	30	31
Example 0011	H	Aib	E	G	T	F	T	S	D	V	S	S	Y	L	E	E	Q	A	A	R	E	F	I	A	W	L	V	R	G	R	K[AEEAc-AEEAc-gGlu-C18diacid]
salmon calcitonin																															

name	32	33	34	35	36	37	38	39	40	41	42	43	44	45	46	47	48	49	50	51	52	53	54	55	56	57	58	59	60	61	62	63	64	65	66	67	68	C-term
Example 0011	G	G	G	G	E	A	S	E	L	S	T	A	A	L	G	R	L	S	A	E	L	H	E	L	A	T	L	P	R	T	E	T	G	S	G	S	P	NH$_2$
salmon calcitonin						C(1)	S	N	L	S	T	C(1)	V	L	G	K	L	S	Q	E	L	H	K	L	Q	T	Y	P	R	T	N	T	G	S	G	T	P	NH$_2$

Violet: Amino acids identical in GLP-1 and exendin-4
White: Amino acids from GLP-1
Grey: Amino acids from exendin-4
Yellow: Amino acids from glucagon

Dark violet: Amino acids from GIP
Orange: New amino acid mutations
Red font: Protraction unit

C*: 10kDa PEGILinker to Fc fragment of IgG4

Figure 20.4 Multiagonists combining GLP-1R with other pharmacology

20.9 Unimolecular Multiagonists Based on Glucagon-like Peptide-1 (GLP-1) Following the Two-pharmacophore Approach

In the two-pharmacophore approach, two or more bioactive peptides are conjugated, usually via a short linker, to provide a single peptidic entity where the individual parts are able to bind to their respective receptors. In recent years, more and more different designs have appeared in the literature, but the group is still considerably less mature as only three members thereof have yet reached the clinical phase.

A unimolecular GLP-1R/gastrin receptor dual agonist ZP3022 was described by Zealand Pharma [87]. It contains the amino acids 1–28 of exendin-4 at the N-terminus combined with gastrin-6, a gastrin receptor (also called CCK2 receptor) agonist, at the C-terminus. ZP3022 improved glycemic control in db/db mice and showed increased pancreatic islet and β-cell mass compared to vehicle.

In a similar approach, a GLP-1R/GCGR dual agonist, based on a potent GLP-1 analog derived from the African frog *Xenopus laevis*, was conjugated with gastrin-6. xGLP/GCG/gastrin is a potent and long-acting triple agonist with a fatty acid side chain at position 14 [88]. It showed higher body weight loss in DIO mice than ZP3022 and liraglutide and sustained improvement in glucose control compared to ZP3022, liraglutide, and the GLP-1R/GCGR dual agonist cotadutide.

GLP-1R agonist combinations with an agonist of the related CCK1 receptor were also described to have beneficial metabolic effects. C2816 was constructed such that a stabilized GLP-1R agonist (AC3174) was linked to the N-terminus of a CCKR1-selective agonist (AC170222), thus preserving the C-terminal amide of the CCK moiety, which is important for activity [89]. C2816 retained full agonism at both receptors, but at lower potency compared to parent molecules. In sub-chronic studies in DIO mice, C2816 exerted superior reduction in body weight compared to co-administration of AC3174 and AC170222 albeit at a higher molar dose. These data suggest that the synergistic pharmacological effects of GLP-1 and CCK pathways can be harnessed in a single therapeutic peptide.

A unimolecular combination of an exendin-4 (amino acids 1–28) with an insulin secretion-enhancing xenin analog linked by two AEEAc units was presented by Irwin and colleagues to reduce non-fasting glucose and increase circulating insulin levels in high-fat-fed mice [90]. The same group then presented a next-generation triple agonist by adding a gastrin agonist to the mix. Combining the exendin-4 sequence with gastrin-6 and xenin-8-Gln, both linkers again consisting of two AEEAc units and conjugating a C16 fatty acid side chain to extend half-life provided a triagonist, which showed improved glucose-lowering actions of insulin compared to an agonist missing the xenin activity [91].

Fibroblast growth factor 21 (FGF21) is another hormone that was studied for the treatment of obesity and MASH. A unimolecular combination of GLP-1 and FGF21 was presented by Yuhan Corporation. YH-25723 is a novel IgG Fc-fused protein comprising a GLP-1 variant and a FGF21 variant; however, the exact structure is not known. It combines GLP-1's well-known pharmacology with FGF21's effects as a metabolic regulator, which can improve insulin sensitivity and lipid metabolism in addition to having an antifibrotic effect. YH-25723 showed promising results in a DIO mouse model of MASH [92].

Peptide YY (PYY) is a 36-amino acid peptide released from L cells in the ileum and colon in response to feeding. PYY_{1-36} is rapidly processed by the enzyme DPP-IV to the 34-amino acid peptide PYY_{3-36}. PYY_{1-36} binds to all known Y receptors with differing affinities, and PYY_{3-36} preferentially binds to the Y_2 receptor and to a lesser extent to the Y_5 receptor. PYY inhibits gastric motility, increases water and electrolyte absorption in the colon, and has been shown to reduce appetite. Co-infusion of GLP-1 and PYY_{3-36} exerts a synergistic effect on energy intake in healthy overweight men [93].

A unimolecular approach was presented by the groups of Doyle and Holz. By combining the N-terminal 33 amino acids of exendin-4 with the 12 C-terminal amino acids of PYY_{3-36}, the dual GLP-1/PYY agonist EP45 (see Figure 20.4) was obtained. *In vitro*, EP45 shows high activity on both the GLP-1 and the Y_2 receptor; however, no *in vivo* data were reported [94]. Two related unimolecular GLP-1/PYY hybrids were designed by introducing GLP-1 amino acids to EP45. GEP44 was obtained by mutations in positions 13–14, 21, and 24 as well as a D-seryl-asparaginyl (DSer-Gln) dipeptide in position 2/3. GEP12 carries an additional phenylalanine mutation in position 1 [95]. Two biased agonists of the GLP-1, neuropeptide Y1, and neuropeptide Y2 receptors were obtained, which showed more profound reductions in food intake and body weight than liraglutide when administered to DIO rats.

Colleagues from Novo Nordisk showed that a GLP-1/PYY combo cannot only be done by conjugating two active parts of the endogenous ligands, but that by careful analysis of the key binding interactions, a shorter 31 amino acid dual agonist peptide 19 can be achieved, which has overlapping stretches of the peptide binding to the two receptors. The N-terminal 19 amino acids from GLP-1 are combined with the 13 C-terminal amino acids from PYY introducing three mutations to ensure dual activity. Leucines were used in positions 23 and 25 to replace an isoleucine and a tryptophan as well as a valine in position 26 to substitute a leucine. Peptide 19 showed strongly reduced food intake in an acute model in comparison with both an analogous selective GLP-1R and selective Y2 receptor (Y2R) agonist [96].

Following the co-formulations of selective GLP-1R agonists with insulins, colleagues from Merck designed a unimolecular insulin/GLP-1/GCG triple agonist [97]. Based on their GLP-1R/GCGR dual agonist MK-1462 (see Figure 20.2b), they identified position 24 as the ideal point for the introduction of the azidolysine to enable the click reaction. The alkyne as reaction partner was introduced at position B29 of the insulin. A PEG4 linker between insulin and alkyne was shown to provide balanced activity on the GLP-1R and GCGR, while longer linkers showed preferential activation at the GLP-1R. The insulin/GLP-1R/GCGR triagonist was tested in mini-pigs and showed an i.v. half-life of four hours. Further studies in T2D-relevant animal models are ongoing.

A combination of GLP-1 with a proprotein convertase subtilisin/kexin type 9 (PCSK9)-neutralizing antibody was explored by MedImmune [98]. To match activity with the PCSK9 antibody part, a GLP-1R agonist with reduced activity was designed with the guidance of an in silico PK/PD model to limit the expected dose-limiting side effects of nausea and vomiting. Exendin-4 was selected as base sequence as fusions starting from GLP-1 were metabolically labile and proved to be unstable at Trp25 during the manufacturing process. This instability could be avoided as the exendin tail stabilizes this region by forming a protein mini-fold, the so-called tryptophan cage. For further stabilization, a disulfide bridge between cysteines 18 and 40 was introduced (see Figure 20.4). To fine-tune activity to the desired level, Val19 was replaced by Ala. In cynomolgus monkeys, MEDI4166 (100 mg/kg, s.c.) achieved a half-life of more than 100 hours for both the GLP-1 and the PCSK9 part. A single dose of 100 mg/kg in healthy cynomolgus monkeys suppressed PCSK9 levels completely for at least 30 days, leading to an LDL-C reduction of greater than 70%, while food intake was not affected. In healthy human subjects, a half-life of 4.5–5 days was observed, confirming the suitability for once-weekly administration [99]. In a 5-week study in T2D patients, s.c. doses of 50, 200, and 400 mg q.w. MEDI4166 showed LDL-C lowering of 1.25, 1.97, and 1.96 mmol/L, respectively. Further clinical testing was discontinued because no improvement in postprandial glucose control was observed during a mixed meal tolerance test (MMTT), showing that the selected GLP-1R activity was not sufficient to trigger glucose-lowering effects.

The second unimolecular combination product reaching human clinical trials is the bispecific molecule maridebart cafraglutide (AMG133) engineered by conjugation of two GLP-1 analogs to a fully human monoclonal anti-human GIPR antagonist antibody using amino acid linkers [100]. While this combination is surprising considering the success of the GLP-1R/GIPR dual agonists like tirzepatide, it is based on reports in the literature that GIPR KO mice are protected from DIO obesity and that anti-GIPR antibodies prevented body weight gain in DIO mice and obese cynos. The GLP-1R agonistic part is based on the GLP-1 sequence with only three mutations, an Aib in position 2 to stabilize against DPP-IV cleavage, a tyrosine from GCG in position 10, and a glutamate from exendin-4 in position 16. It carries a C-terminal amino acid linker ending with a lysine carrying a bromoacetamide group at the side chain nitrogen during synthesis, which is reacted with a specific cysteine mutation at the E384 position of the anti-GIPR antibody. Intact AMG133 has a human half-life of 14–16 days. In a multiple-dose study in adult participants with obesity, three doses of 420 mg every four weeks provided a body weight loss of −14.5% after 78 days and body weight remained reduced 150 days after the last dose. AMG133 treatment was also associated with changes in glucose homeostasis markers as well as a dose-dependent reduction in levels of the inflammation marker high-sensitivity C-reactive protein (hs-CRP). It is currently undergoing phase 2 clinical studies.

The latest addition to the list of clinical compounds is a unimolecular combination of a GLP-1R agonist with an amylin receptor agonist called amycretin (NN9487) from Novo Nordisk. It is precedented by co-formulation studies of the selective GLP-1R agonist semaglutide and the amylin receptor agonist cagrilintide. This combination called CagriSema, which is currently studied in phase 3, showed not only impressive HbA1c reduction in T2D patients upon once-weekly subcutaneous dosing but also a mean change in body weight from baseline to week 32 of −15.6% compared to −5.1% semaglutide and −8.1% for cagrilintide [101]. The unimolecular version of amycretin was recently reported to give a 12% placebo-corrected weight loss upon daily oral dosing over 12 weeks [102]. A clinical trial studying an injectable formulation is ongoing, and longer studies are planned. The structure of amycretin has not been disclosed so far, but Novo Nordisk published a patent application disclosing GLP-1R/Amylin-R dual agonists. A highlighted compound example 0011 comprises a typical GLP-1-based sequence in positions 1–30 linked via a protraction unit carrying amino acid linker to an amylin receptor agonist, which is based on the sequence of salmon calcitonin (see Figure 20.4). Eleven mutations were introduced to salmon calcitonin, among them most importantly two alanines in positions 37 and 43 to replace the two cysteines, which form the N-terminal cycle in calcitonin [103].

20.10 Conclusion

Peptide drug discovery has seen a huge surge in the last two decades. Therapeutic peptides address a unique chemical space with molecular weights between small molecules and biologics like nano- or antibodies. They show often remarkable potency, selectivity, and low toxicity. At the same time, however, their limitations, including low oral bioavailability, low plasma stability, and short circulation time, were recognized, but many strategies to address these shortcomings were developed like the use of albumin-binding side chains or formulations with absorption enhancer.

T2D and obesity are on the rise worldwide, but the introduction of GLP-1R agonists has provided powerful options to improve the treatment of these diseases. In addition, GLP-1 and exendin-4 act as ideal partners for unimolecular combinations, which enable the incorporation of additional mechanisms of action to further strengthen pharmacological effects. Two general strategies are pursued. In the one-pharmacophore approach, the similarity of endogenous agonists and receptors in the secretin-like GPCR family allowed the identification of key interactions to design dual agonists with overlapping pharmacophore regions. The first marketed member of this class, the GLP-1R/GIPR dual agonist tirzepatide, is showing impressive efficacy both on body weight loss and on glucose control. Further combinations like GLP-1R/GCGR dual agonists and GLP-1R/GCGR/GIPR triple agonists are undergoing phase 3 testing and should soon add further options for the treatment of type 2 diabetic and obese patients and quite likely also patients with MASH. The two-pharmacophore approach is applied for combinations where these similarities are not given. GLP-1R agonistic peptide sequences are then connected mostly via a short amino acid linker to a second pharmacologically active sequence. These approaches are clinically less advanced, but it will be exciting to track their progress. Most recent additions like Novo Nordisk's amycretin combination also highlight the new maturity level of the field, which allows peptides with over sixty amino acids not only to reach clinical trials with subcutaneous administration but also to be applied orally. This should allow even more exciting new combinations in the future.

References

1 Derosa, G. and Sibilla, S. (2007). Optimizing combination treatment in the management of type 2 diabetes. *Vasc. Health Risk Manag.* 3: 665–671.

2 Ruban, A., Stoenchev, K., Ashrafian, H., and Teare, J. (2019). Current treatments for obesity. *Clin. Med.* 19: 205–212. https://doi.org/10.7861/clinmedicine.19-3-205.

3 Wang, L., Wang, N., Zhang, W. et al. (2022). Therapeutic peptides: current applications and future directions. *Sig. Transduct. Target Ther.* 7: 48.

4 Müller, T.D., Finan, B., Bloom, S.R. et al. (2019). Glucagon-like peptide 1 (GLP-1). *Mol Metab.* 30: 72–130.

5 Eng, J., Kleinman, W.A., Singh, L. et al. (1992). Isolation and characterization of exendin-4, an exendin-3 analogue, from Heloderma suspectum venom. Further evidence for an exendin receptor on dispersed acini from guinea pig pancreas. *J. Biol. Chem.* 267: 7402–7405.

6 Nauck, M.A., Quast, D.R., Wefers, J., and Meier, J.J. (2021). GLP-1 receptor agonists in the treatment of type 2 diabetes – state-of-the-art. *Molecular Metab.* 46: 101102.

7 Bech, E.M., Pederson, S.L., and Jensen, K.J. (2018). Chemical strategies for half-life extension of biopharmaceuticals: lipidation and its alternatives. *ACS Med. Chem. Lett.* 9: 577–580.

8 Glaesner, W., Vick, A.M., Millican, R. et al. (2010). Engineering and characterization of the long-acting glucagon-like peptide-1 analogue LY2189265, an Fc fusion protein. *Diabetes Metab. Res. Rev.* 26: 287–296.

9 Ko, S., Jo, M., and Jung, S.T. (2021). Recent achievements and challenges in prolonging the serum half-lives of therapeutic IgG antibodies through Fc engineering. *BioDrugs* 35: 147–157.

10 Yu, M., Benjamin, M.M., Srinivasan, S. et al. (2018). Battle of GLP-1 delivery technologies. *Adv. Drug Deliv. Rev.* 130: 113–130.

11 Nuffer, W., Guesnier, A., and Trujillo, J.M. (2018). A review of the new GLP-1 receptor agonist/basal insulin fixed-ratio combination products. *Ther. Adv. Endocrinol. Metab.* 9: 69–79.

12 Bergmann, N.C., Davies, M.J., Lingvay, I., and Knop, F.K. (2023). Semaglutide for the treatment of overweight and obesity: a review. *Diabetes Obes. Metab.* 25: 18–35.

13 El Eid, L., Reynolds, C.A., Tomas, A., and Jones, B. (2022). Biased agonism and polymorphic variation at the GLP-1 receptor: implications for the development of personalised therapeutics. *Pharmacol. Res.* 184: 106411.

14 Zhang, Y., Sun, B., Feng, D. et al. (2017). Cryo-EM structure of the activated GLP-1 receptor in complex with G protein. *Nature* 546: 248–253.

15 Mann, R., Nasr, N., Hadden, D. et al. (2007). Peptide binding at the GLP-1 receptor. *Biochem. Soc. Trans.* 35: 713–716.

16 Brubaker, P.L. and Drucker, D.J. (2002). Structure-function of the glucagon receptor family of G protein-coupled receptors: the glucagon, GIP, GLP-1, and GLP-2 receptors. *Receptors Channels* 8: 179–188.

17 Krumm, B. and Roth, B.L. (2020). A structural understanding of class B GPCR selectivity and activation revealed. *Structure* 28: 277–279.

18 Zhao, F., Zhou, Q., Cong, Z. et al. (2022). Structural insights into multiplexed pharmacological actions of tirzepatide and peptide 20 at the GIP, GLP-1 or glucagon receptors. *Nat. Commun.* 13: 1057.

19 Hasib, A. (2020). Multiagonist unimolecular peptides for obesity and type 2 diabetes: current advances and future directions. *Clin. Med. Insights: Endocrinol. Diabetes* 13: 1179551420905844.

20 Müller, T.D., Clemmensen, C., Finan, B. et al. (2018). Anti-obesity therapy: from rainbow pills to polyagonists. *Pharmacol. Rev.* 70: 712–746.

21 Stunkard, A.J., Van Itallie, T.B., and Reis, B.B. (1955). The mechanism of satiety: effect of glucagon on gastric hunger contractions in man. *Proc. Soc. Exp. Biol. Med. Soc. Exp. Biol. Med. N. Y. N.* 89: 258–261.

22 Schulman, J.L., Carleton, J.L., Whitney, G., and Whitehorn, J.C. (1957). Effect of glucagon on food intake and body weight in man. *J. Appl. Physiol.* 11: 419–421.

23 Nair, K.S. (1987). Hyperglucagonemia increases resting metabolic rate in man during insulin deficiency. *J. Clin. Endocrinol. Metab.* 64: 896–901.

24 Habegger, K.M., Heppner, K.M., Geary, N. et al. (2010). The metabolic actions of glucagon revisited. *Nat. Rev. Endocrinol.* 6: 689–697.

25 Heppner, K.M., Habegger, K.M., Day, J. et al. (2010). Glucagon regulation of energy metabolism. *Physiol. Behav.* 100: 545–548.

26 Unger, R.H., Eisentraut, A.M., McCall, M.S. et al. (1959). Glucagon antibodies and their use for immunoassay for glucagon. *Proc. Soc. Exp. Biol. Med.* 102: 621–623.

27 Zhihong, Y., Chen, W., Qianqian, Z. et al. (2023). Emerging roles of oxyntomodulin-based glucagon-like peptide-1/glucagon co-agonist analogs in diabetes and obesity. *Peptides* 162: 170955.

28 Wynne, K., Park, A.J., Small, C.J. et al. (2005). Subcutaneous oxyntomodulin reduces body weight in overweight and obese subjects: a double-blind, randomized, controlled trial. *Diabetes* 54: 2390–2395.

29 Wynne, K., Park, A.J., Small, C.J. et al. (2006). Oxyntomodulin increases energy expenditure in addition to decreasing energy intake in overweight and obese humans: a randomized controlled trial. *Int. J. Obes.* 30: 1729–1736.

30 Pocai, A., Carrington, P.E., Adams, J.R. et al. (2009). Glucagon-like peptide 1/glucagon receptor dual agonism reverses obesity in mice. *Diabetes* 58: 2258–2266.

31 Sheetz, M., Bray, R., Tham, L.S.M. et al. (2017). Results of a phase 2 study of the oxyntomodulin (OXM) analogue LY2944876 in patients with type 2 diabetes (T2DM). *77th Scientific Session American Diabetes Association*, San Diego, CA (10 June 2017). Wiley.

32 You, S., McDonald, M., Case, M. et al. (2016). Long-acting GLP-1 and glucagon receptor dual agonists for the treatment of type 2 diabetes. *Diabetes* 65 (Supplement1): A221–A360.

33 Day, J.W., Ottaway, N., Patterson, J.T. et al. (2009). A new glucagon and GLP-1 co-agonist eliminates obesity in rodents. *Nat. Chem. Biol.* 5: 749–757.

34 Day, J.W., Gelfanov, V., Smiley, D. et al. (2012). Optimization of co-agonism at GLP-1 and glucagon receptors to safely maximize weight reduction in DIO-rodents. *Biopolymers* 98: 443–450.

35 Hope, D.C.D., Vincent, M.L., and Tan, T.M.M. (2021). Striking the balance: GLP-1/glucagon co-agonism as a treatment strategy for obesity. *Front. Endocrinol.* 12: 735019.

36 Tillner, J., Posch, M.G., Wagner, F. et al. (2019). A novel dual glucagon-like peptide and glucagon receptor agonist SAR425899: results of randomized, placebo-controlled first-in-human and first-in-patient trials. *Diabetes Obes. Metab.* 21: 120–128.

37 Tillner, J., Posch, M.G., Hueser, A. et al. (2016). Safety, pharmacokinetics, and pharmacodynamics of the novel dual GLP-1/glucagon agonist SAR425899 in healthy subjects and diabetes patients. *Diabetes* 65 (Supplement 1): A6.

38 Eriksson, O., Haack, T., Hijazi, Y. et al. (2020). Receptor occupancy of dual glucagon-like peptide 1/glucagon receptor agonist SAR425899 in individuals with type 2 diabetes. *Sci. Rep.* 10: 16758.

39 DiMarchi, R.D. and Smiley, D.L. (2012). Glucagon/glp-1 receptor co-agonists. US Patent 10,730,923, filed 12 June 2012 and issued 4 April 2020.

40 Tschöp, M.H., Finan, B., Clemmensen, C. et al. (2016). Unimolecular polypharmacy for treatment of diabetes and obesity. *Cell Metab.* 24: 51–62.

41 ClinicalTrials.gov. (2018) A preliminary study of the efficacy and safety of MK-8521 for type 2 diabetes (MK-8521-004). Study Results.

42 Palani, A., Nawrocki, A.R., Orvieto, F. et al. (2022). Discovery of MK-1462: GLP-1 and glucagon receptor dual agonist for the treatment of obesity and diabetes. *ACS Med. Chem. Lett.* 13: 1248–1254.

43 Henderson, S.J., Konkar, A., Hornigold, D.C. et al. (2016). Robust anti-obesity and metabolic effects of a dual GLP-1/glucagon receptor peptide agonist in rodents and non-human primates. *Diabetes Obes. Metab.* 18: 1176–1190.

44 Ambery, P., Parker, V.E., Stumvoll, M. et al. (2018). MEDI0382, a GLP-1 and glucagon receptor dual agonist, in obese or overweight patients with type 2 diabetes: a randomised, controlled, double-blind, ascending dose and phase 2a study. *Lancet* 391: 2607–2618.

45 Simonsen, L., Lau, J., Kruse, T. et al. (2022). Preclinical evaluation of a protracted GLP-1/glucagon receptor co-agonist: translational difficulties and pitfalls. *PLoS ONE* 17: e0264974.

46 Friedrichsen, M.H., Endahl, L., Kreiner, F.F. et al. (2023). Results from three phase 1 trials of NNC9204-1177, a glucagon/GLP-1 receptor co-agonist: effects on weight loss and safety in adults with overweight or obesity. *Mol Metab.* 78: 101801.

47 Ji, L., Jiang, H., Cheng, Z. et al. (2023). A phase 2 randomised controlled trial of mazdutide in Chinese overweight adults or adults with obesity. *Nat. Commun.* 14: 8289.

48 Zhang, B., Cheng, Z., Chen, J. et al. (2024). Efficacy and safety of mazdutide in Chinese patients with type 2 diabetes: a randomized, double-blind, placebo-controlled phase 2 trial. *Diabetes Care* 47: 160–168.

49 InnoventBio (2024). Innovent announces the first phase 3 clinical trial of mazdutide in Chinese adults with overweight or obesity met the primary and all key secondary endpoints, and plans to submit NDA of mazdutide to the NMPA. *Press Release* (9 January).

50 Zimmermann, T., Thomas, L., Baader-Pagler, T. et al. (2022). BI 456906: discovery and preclinical pharmacology of a novel GCGR/GLP-1R dual agonist with robust anti-obesity efficacy. *Mol Metab.* 66: 101633.

51 Blüher, M., Rosenstock, J., Hoefler, J. et al. (2024). Dose–response effects on HbA1c and bodyweight reduction of survodutide, a dual glucagon/GLP-1 receptor agonist, compared with placebo and open-label semaglutide in people with type 2 diabetes: a randomised clinical trial. *Diabetologia* 67: 470–482.

52 Alba, M., Yee, J., Frustaci, M.E. et al. (2021). Efficacy and safety of glucagon-like peptide-1/glucagon receptor co-agonist JNJ-64565111 in individuals with obesity without type 2 diabetes mellitus: a randomized dose-ranging study. *Clin. Obes.* 11: e12432.

53 Romero-Gómez, M., Lawitz, E., Shankar, R.R. et al. (2023). A phase IIa active-comparator-controlled study to evaluate the efficacy and safety of efinopegdutide in patients with non-alcoholic fatty liver disease. *J. Hepatol.* 79: P888–P897.

54 Altimmune Presentation. MOMENTUM — Pemvidutide Phase 2 Obesity Trial, Topline Week 48 Results. 30 November 2023 https://ir.altimmune.com/static-files/0ff26b19-1b74-491a-8a37-8d6f1c8f12d4 (assessed 17 March 2024).

55 Hope, D.C.D., Ansari, S., Choudhury, S. et al. (2024). Adaptive infusion of a glucagon-like peptide-1/glucagon receptor co-agonist G3215, in adults with overweight or obesity: Results from a phase 1 randomized clinical trial. *Diabetes Obes. Metab.* 26: 1479–1491.

56 Deganutti, G., Liang, Y.L., Zhang, X. et al. (2022). Dynamics of GLP-1R peptide agonist engagement are correlated with kinetics of G protein activation. *Nat. Commun.* 13: 92.

57 Kawai, T., Sun, B., Yoshino, H. et al. (2020). Structural basis for GLP-1 receptor activation by LY3502970, an orally active nonpeptide agonist. *Proc. Natl. Acad. Sci. USA* 117: 29959–29967.

58 Zhang, X., Belousoff, M.J., Liang, Y.L. et al. (2021). Structure and dynamics of semaglutide- and taspoglutide-bound GLP-1R-Gs complexes. *Cell Rep.* 36: 109374.

59 Frimann, T.M., Ko, S.K., Harris, P. et al. (2023). *In-silico* study of the interactions between acylated glucagon like-peptide-1 analogues and the native receptor. *J. Biomol. Struct. Dyn.* 41: 5007–5021.

60 Knudsen, L.B., Nielsen, P.F., Huusfeldt, P.O. et al. (2000). Potent derivatives of glucagon-like peptide-1 with pharmacokinetic properties suitable for once daily administration. *J. Med. Chem.* 43: 1664–1669.

61 Nauck, M.A., Homberger, E., Siegel, E.G. et al. (1986). Incretin effects of increasing glucose loads in man calculated from venous insulin and C-peptide responses. *J. Clin. Endocrinol. Metab.* 63: 492–498.

62 Holst, J.J. and Rosenkilde, M.M. (2021). GIP as a therapeutic target in diabetes and obesity: insight from incretin co-agonists. *J. Clin. Endocrinol. Metab.* 105: e2710–e2716.

63 Campell, J. (2021). Targeting the GIPR for obesity: to agonize or antagonize? Potential mechanisms. *Mol. Metabol.* 46: 101139.

64 Liskiewicz, A., Khalil, A., Liskiewicz, D. et al. (2023). Glucose-dependent insulinotropic polypeptide regulates body weight and food intake via GABAergic neurons in mice. *Nat. Metab.* 5: 2075–2085.

65 Finan, B., Ma, T., Ottaway, N. et al. (2013). Unimolecular dual incretins maximize metabolic benefits in rodents, monkeys, and humans. *Sci. Transl. Med.* 5: 209ra151.

66 Portron, A., Jadıdi, S., Sarkar, N. et al. (2017). Pharmacodynamics, pharmacokinetics, safety and tolerability of the novel dual glucose-dependent insulinotropic polypeptide/glucagon-like peptide-1 agonist RG7697 after single subcutaneous administration in healthy subjects. *Diabetes Obes. Metab.* 19: 1446–1453.

67 Frias, J.P., Bastyr, E.J., Vignati, L. et al. (2017). The sustained effects of a dual GIP/GLP-1 receptor agonist, NNC0090-2746, in patients with type 2 diabetes. *Cell Metab.* 26: 343–352.e2.

68 Coskun, T., Sloop, K.W., Loghin, C. et al. (2018). LY3298176, a novel dual GIP and GLP-1 receptor agonist for the treatment of type 2 diabetes mellitus: From discovery to clinical proof of concept. *Mol. Metab.* 18: 3–14.

69 Frías, J.P., Davies, M.J., Rosenstock, J. et al. (2021). Tirzepatide versus semaglutide once weekly in patients with type 2 diabetes. *N. Engl. J. Med.* 385: 503–515.

70 Aronne, L.J., Sattar, N., Horn, D.B. et al. (2024). Continued treatment with tirzepatide for maintenance of weight reduction in adults with obesity: the SURMOUNT-4 randomized clinical trial. *JAMA* 331: 38–48.

71 Sun, B., Willard, F.S., Feng, D. et al. (2022). Structural determinants of dual incretin receptor agonism by tirzepatide. *Proc. Natl. Acad. Sci. USA* 119: e2116506119.

72 Liang, Y.L., Khoshouei, M., Glukhova, A. et al. (2018). Phase-plate cryo-EM structure of a biased agonist-bound human GLP-1 receptor-Gs complex. *Nature* 555: 121–125.

73 Viking Therapeutics (2024). Viking therapeutics announces positive top-line results from phase 2 VENTURE trial of dual GLP-1/GIP receptor agonist VK2735 in patients with obesity. *Press Release* (27 February).

74 Viking Therapeutics (2024). Viking therapeutics announces results from phase 1 clinical trial of oral tablet formulation of dual GLP-1/GIP receptor agonist VK2735. *Press Release* (26 March).

75 Lian, B., Barker, G.E., Barnes, M., and Yagiz, K. (2022). Pharmaceutical formulations and methods for the treatment of metabolic and liver disorders. World Intellectual Property Organization WO2023141044, filed 18 January 2022 and published 27 July 2023.

76 Lian, B., Barker, G.E., Barnes, M. et al. (2022). Pharmaceutical formulations and methods for the treatment of metabolic and liver disorders. World Intellectual Property Organization WO2024020388, filed 20 July 2022 and published 25 January 2024.

77 Finan, B., Yang, B., Ottaway, N. et al. (2015). A rationally designed monomeric peptide triagonist corrects obesity and diabetes in rodents. *Nat. Med.* 21: 27–36.

78 Bossart, M., Wagner, M., Elvert, R. et al. (2022). Effects on weight loss and glycemic control with SAR441255, a potent unimolecular peptide GLP-1/GIP/GCG receptor triagonist. *Cell Metab.* 34: 59–74.e10.

79 Coskun, T., Urva, S., Roell, W.C. et al. (2022). LY3437943, a novel triple glucagon, GIP, and GLP-1 receptor agonist for glycemic control and weight loss: From discovery to clinical proof of concept. *Cell Metab.* 34: 1234–1247.e9.

80 Jastreboff, A.M., Kaplan, L.M., Frías, J.P. et al. (2023). Retatrutide phase 2 obesity trial investigators. triple-hormone-receptor agonist retatrutide for obesity - a phase 2 trial. *N. Engl. J. Med.* 389: 514–526.

81 Rosenstock, J., Frias, J., Jastreboff, A.M. et al. (2023). Retatrutide, a GIP, GLP-1 and glucagon receptor agonist, for people with type 2 diabetes: a randomised, double-blind, placebo and active-controlled, parallel-group, phase 2 trial conducted in the USA. *Lancet* 402: 529–544.

82 World Health Organization (2021). WHO drug information, Vol. 35, No. 4, Proposed INN List 126, p. 960.

83 Novo Nordisk (2020). Financial report for the period 1 January 2020 to 30 June 2020. https://www.novonordisk.com/content/dam/nncorp/global/en/investors/pdfs/financial-results/2020/Company%20Announcement_Q2%202020.pdf (assessed 17 March 2024).

84 Reiner, J., Berlin, P., Held, J. et al. (2022). Dapiglutide, a novel dual GLP-1 and GLP-2 receptor agonist, attenuates intestinal insufficiency in a murine model of short bowel. *J. Parenter. Enter. Nutr.* 46: 1107–1118.

85 Wismann, P., Pedersen, S.L., Hansen, G. et al. (2018). Novel GLP-1/GLP-2 co-agonists display marked effects on gut volume and improves glycemic control in mice. *Physiol. Behav.* 192: 72–81.

86 van Witteloostuijn, S.B., Dalbøge, L.S., Hansen, G. et al. (2017). GUB06-046, a novel secretin/glucagon-like peptide 1 co-agonist, decreases food intake, improves glycemic control, and preserves beta cell mass in diabetic mice. *J. Pept. Sci.* 23: 845–854.

87 Fosgerau, K., Jessen, L., Lind Tolborg, J. et al. (2013). The novel GLP-1-gastrin dual agonist, ZP3022,

increases β-cell mass and prevents diabetes in db/db mice. *Diabetes Obes. Metab.* 15: 62–71.

88 Zhao, S., Yan, Z., Du, Y. et al. (2022). A GLP-1/glucagon (GCG)/CCK2 receptors tri-agonist provides new therapy for obesity and diabetes. *Br. J. Pharmacol.* 179: 4360–4377.

89 Hornigold, D.C., Roth, E., Howard, V. et al. (2018). A GLP-1:CCK fusion peptide harnesses the synergistic effects on metabolism of CCK-1 and GLP-1 receptor agonism in mice. *Appetite* 127: 334–340.

90 Hasib, A., Ng, M.T., Khan, D. et al. (2018, 2018). A novel GLP-1/xenin hybrid peptide improves glucose homeostasis, circulating lipids and restores GIP sensitivity in high fat fed mice. *Peptides* 100: 202–211.

91 Hasib, A., Ng, M.T., Tanday, N. et al. (2019). Exendin-4(Lys27 PAL)/gastrin/xenin-8-Gln: a novel acylated GLP-1/gastrin/xenin hybrid peptide that improves metabolic status in obese-diabetic (ob/ob) mice. *Diabetes Metab. Res. Rev.* 35: e3106.

92 Hong, H.N., Kim, J.H., Choi, H.H. et al. (2017). PS-024 - YH25724, a novel long-acting GLP-1/FGF21 dual agonist lowers both non-alcoholic fatty liver disease activity score and fibrosis stage in a diet-induced obese mouse model of biopsy-confirmed non-alcoholic steatohepatitis. *J. Hepatol.* 66 (1, Supplement): S16–S17.

93 Schmidt, J.B., Gregersen, N.T., Pedersen, S.D. et al. (2014). Effects of PYY3-36 and GLP-1 on energy intake, energy expenditure, and appetite in overweight men. *Am. J. Physiol. Endocrinol. Metab.* 306: E1248–E1256.

94 Chepurny, O.G., Bonaccorso, R.L., Leech, C.A. et al. (2018). Chimeric peptide EP45 as a dual agonist at GLP-1 and NPY2R receptors. *Sci. Rep.* 8: 3749.

95 Chichura, K.S., Elfers, C.T., Salameh, T.S. et al. (2023). A peptide triple agonist of GLP-1, neuropeptide Y1, and neuropeptide Y2 receptors promotes glycemic control and weight loss. *Sci. Rep.* 13: 9554.

96 Østergaard, S., Paulsson, J.F., Kjaergaard Gerstenberg, M., and Wulff, B.S. (2021). The design of a GLP-1/PYY dual acting agonist. *Angew. Chem. Int. Ed. Engl.* 60: 8268–8275.

97 Huang, C., Palani, A., Yang, Z. et al. (2022). Discovery of insulin/GLP-1/glucagon triagonists for the treatment of diabetes and obesity. *ACS Med. Chem. Lett.* 13: 1255–1261.

98 Chodorge, M., Celeste, A.J., Grimsby, J. et al. (2018). Engineering of a GLP-1 analogue peptide/anti-PCSK9 antibody fusion for type 2 diabetes treatment. *Sci. Rep.* 8: 17545.

99 Jain, M., Carlson, G., Cook, W. et al. (2018). Randomised, phase 1, dose-finding study of MEDI4166, a PCSK9 antibody and GLP-1 analogue fusion molecule, in overweight or obese patients with type 2 diabetes mellitus. *Diabetologia* 62: 373–386.

100 Véniant, M.M., Lu, S.C., Atangan, L. et al. (2024). A GIPR antagonist conjugated to GLP-1 analogues promotes weight loss with improved metabolic parameters in preclinical and phase 1 settings. *Nat. Metab.* 6: 290–303.

101 Frias, J.P., Deenadayalan, S., Erichsen, L. et al. (2023). Efficacy and safety of co-administered once-weekly cagrilintide 2.4 mg with once-weekly semaglutide 2.4 mg in type 2 diabetes: a multicentre, randomised, double-blind, active-controlled, phase 2 trial. *Lancet* 402: 720–730.

102 Marcus Schindler (2024) Novo Nordisk capital market day 2024. https://www.novonordisk.com/content/dam/nncorp/global/en/investors/irmaterial/cmd/2024/P2-r-and-d-updated.pdf (assessed 17 March 2024).

103 Kruse, T., Kodal, A.L.B., Madsen, J., et al. (2021). C-agonists of the GLP-1 and amylin receptors. World Intellectual Property Organization WO2022129526, filed 17 December 2021 and published 23 June 2022.

21

The SOSA Approach to Drug Discovery

Norbert Handler, Michal Poznik, and Helmut Buschmann

21.1 Introduction

Drug research and development (R&D) is a highly complicated process requiring investment of capital, human resources, and various technological expertise, whereas attrition rates have been notoriously high. More importantly, drug development requires strict adherence to regulations on testing and manufacturing standards; thus, market introduction of a novel drug is time-consuming and costly. According to a 2016 analysis of 106 randomly selected new drugs, it costs well over two billion dollars to bring a single drug from a bench to a patient's bedside [1, 2].

In the search for new bioactive molecules as starting points for a drug development program, high-throughput screening (HTS) of large compound libraries has been the mainstay at most pharmaceutical companies for several decades, but considering the enormous investment in labor and machinery, its outcome has been sometimes meager, especially with difficult drug targets.

> "*The standard modus operandi of drug discovery is the identification of biologically active molecules via the high throughput screening (HTS) of large, diverse compound collections against specific biological targets, in low complexity assays systems.*
> *The specific biological targets of interest are selected from the results of basic scientific research on the aetiology of the disease of interest. Despite the a priori scientific appeal of the current standard approach, the rise in its popularity has also been concurrent with a reported decline in the productivity of the pharmaceutical industry, although of course this relationship is not necessarily causal and can be challenged.*
> *Additionally, the HTS approach has often over-shadowed a wider variety of alternative and complementary drug discovery strategies that have been very successfully employed in the discovery of new medicines*" [3].

During the last twenty years, pharmaceutical companies have massively invested in hit-and-lead generation and optimization technologies, and in the early 2000s, the industry began to use various computational approaches. Despite this effort, the productivity in terms of new marketed therapeutic entities was low and the attrition rates were high. Most of the compounds were removed from further development for unsatisfactory absorption, distribution, metabolism, and excretion (ADME) properties and, more recently, for toxicity reasons [4].

The abovementioned challenges have prompted several paradigm changes in modern lead generation, and one current trend is to use smaller and smarter libraries, especially focused libraries with increased drug likeness or libraries of small "fragment" molecules (fragment-based lead generation). The selective optimization of side activities (SOSA) approach represents another very compelling alternative to large-scale screening approaches, generating per se drug-like hits. SOSA makes use of already existing drugs or drug candidates and aims at optimizing their side activities, thus generating novel leads for targets and indications *different* to the ones targeted by the original drug or drug candidate. This re-evaluation of old drugs has emerged as one of the most rapid and efficient ways to counterbalance the escalating high cost and to increase the ebbing marketing rate in the R&D of new drugs [5, 6].

21.1.1 Drug Selectivity and Unwanted or Desired Side Effects

The rationale behind the SOSA approach lies in the fact that, in addition to their main activity, almost all drugs used in human therapy show one or several side effects. Virtually all drugs have side effects, either target-related ones or non-target-related ones. Side effects are carefully tracked during clinical trials and include any undesired effects of the drug. Side effects can often arise for many reasons, including binding to more than one target (lack

Polypharmacology: Strategies for Multi-Target Drug Discovery, First Edition. Edited by Jens-Uwe Peters.

of specificity) and metabolites that bind additional targets. As an example, the pharmaceutical industry withdrew a large number of drugs during the years 1996–2001. Some of these drugs interacted with nontherapeutic targets and thereby caused severe side effects. For instance, the widely used anorexigen, fenfluramine, led to pulmonary hypertension and heart valve damage, because its metabolite activated the serotonin 5-HT_{2B} receptor [7–9]. When fenfluramine was withdrawn in 1997, many patients were already permanently injured, and the case sparked a public debate on drug safety and vigilance.

Many established drugs are not selective [10], and a significant percentage of compounds in large databases bind to more than one target [11]: 33% of the compounds in the BioPrint database [12], 35% in a large proprietary database [13], 38% in the ChEMBL database [14], 50% in the PubChem database [15], and 52% of oral drugs in the ChEMBL database [16] (target binding is defined differently in the cited publications).

The Psychoactive Drug Screening Program (PDSP) database contains an even larger number of promiscuous compounds [17]. These databases cover only a fraction of the estimated 3000 druggable targets of the human genome [18], so the actual prevalence of promiscuity is likely higher than the indicated 33–52% [10, 19–21].

Most of the drugs bind to multiple proteins in an organism and can therefore produce a variety of unwanted effects (concerns between 35% and 62% of the chemicals – sometimes described as drug promiscuity). This characteristic is well-known and referred to as off-target effect or polypharmacology. Identifying the off-target proteins interacting with known drugs can provide repurposing opportunities [21].

In other words, if drugs are able to exert a strong interaction with the main target, they also exert less strong interactions with some other biological targets. Most of these targets are unrelated to the primary therapeutic activity of the compound. A drug with poor target specificity can be modified to enhance the side effect and decrease the original effect.

The objective is then to proceed to a reversal of the affinities; the identified side effect is becoming the main effect and vice versa. By contrast, there is only a limited chemical universe of small molecules that can be safely administered to humans.

Many recent analyses of marketed oral drugs provide clues to the medicinal chemist for the development of new pharmaceuticals [21–27]. A lot of this research is focused on defining and using drug-like property space, distinguishing drugs from nondrugs, and the evolution of drug properties over time [23, 24, 26]. The world of "drug-like" compounds (definitions see [28]) astonishingly only comprises some 10,000 molecules.

The SOSA approach applied to existing drug molecules is an efficient strategy for the generation of drug candidates with new biological activities [29, 30]. Only a limited number of highly diverse drug molecules need to be screened, for which bioavailability and toxicity studies have already been performed and efficacy in humans has been confirmed. Once the screening has generated a hit, it can be used either in a drug-repurposing approach or as a starting point for a new drug discovery program. As quoted by Raju, "The most fruitful basis for the discovery of a new drug is to start with an old drug, Sir James Black, Nobel Laureate 1988 in Physiology and Medicine" [31].

In this context [32], it is stated that some 90% of all the new drugs are provided by the modification of already known (lead) compounds. During the last several years, this relationship appears not to have changed considerably [33].

The similarity principle that similar compounds should exert similar biological activities has always been a most successful approach in drug research, despite many exceptions to this general concept.

21.2 Definition, Rational, and Concept of the SOSA Approach

Most drugs used in human therapy possess more than one active principle and display major and/or minor actions, thus delivering one or several pharmacological side effects in addition to their main activity.

The SOSA approach [34–38] uses old drugs for new pharmacological targets. The aim is to screen only a limited number of drug molecules that are structurally and therapeutically diverse and have known safety and bioavailability in humans. As a result, it is expected that such an approach might reduce the time and the cost compared with a standard hit identification process. The SOSA approach proceeds in two steps:

i) Screening of a limited set of carefully chosen structurally diverse drug molecules (a focused library of around 1000 compounds). For all these drugs, bioavailability and toxicity studies will have been performed already, and because they all have demonstrated efficacy and safety in human therapy, all hits developed by this process will be, by definition, drug-like.
ii) The hits are optimized (by means of traditional, parallel, combinatorial chemistry and various *in silico* methods) to give increased affinity for the new target and decreased affinity for the original target(s). The

objective is to prepare analogs of the hit molecule to transform the observed side activity into the main effect and strongly reduce or abolish the initial main pharmacological activity.

The SOSA approach appears to be an efficient strategy for drug discovery, particularly when it is based on the screening of drug molecules, thus automatically yielding drug-like hits. Before starting a costly HTS campaign, SOSA can represent a seductive alternative. Once the initial screening has provided a hit, it will be used as the starting point for a drug discovery program. Using traditional medicinal chemistry as well as parallel synthesis, the initial "side activity" is transformed into the main activity, and conversely, the initial main activity is strongly reduced or abolished. This strategy leads with a high probability to safe, bioavailable, original, and patentable analogs.

Choosing a known drug as the lead compound for a side effect has the advantage that the compound is already drug-like, and it should be more feasible to develop a clinically useful drug with the required pharmacodynamic and pharmacokinetic (PK) properties.

In contrast to the SOSA approach, many of the "hits" obtained from HTS do not have a drug-like structure and it may require far more effort to optimize them. Indeed, it has been argued that modifications of known drug structures should provide lead compounds in several areas of medicinal chemistry [37]. Many research groups are now screening compounds that are either in clinical use or have reached late-stage clinical trials to see whether they have side activities that would make them suitable lead compounds.

21.2.1 Multiple Ligands and Polypharmacology

In certain diseases and afflictions, there can be an advantage in "hitting" a number of different targets selectively, as this can be more beneficial than hitting just one (multiple ligands).

Polypharmacology describes the activity of compounds at multiple targets. Current research focuses on two aspects of polypharmacology:

i) Unintended polypharmacology can lead to adverse effects.
ii) Polypharmacology across several disease-relevant targets can improve therapeutic efficacy, prevent drug resistance, or reduce therapeutic target-related adverse effects.

Combination therapy is normally used to achieve this by administering two or more drugs showing selectivity against different targets. This is particularly the case in the treatment of cancer and human immunodeficiency virus (HIV) infection. However, combination therapies are also used in a variety of other situations. The disadvantage of combination therapies is the number of different medications and the associated dose regimens. Therefore, there are benefits in designing a single drug that can act selectively at different targets [21].

Many research projects now set out to discover new drugs with a defined profile of activity against a range of specific targets. Also, for multiple ligands, the SOSA approach may offer several advantages for future lead generation and optimization projects. For a multiple ligand, the optimization of a desired receptor profile has to be optimized using the same technologies as for a selective ligand, however, in more than one dimension.

21.2.2 Safety and Bioavailability

During years of practicing SOSA approaches, several examples are documented in the literature showing that starting with a drug molecule as the lead substance in performing analog synthesis increased notably the probability of obtaining safe new chemical entities [37].

In addition, most of the SOSA-generated leads satisfy Lipinski's [39], Veber's [40], Bergström's [41], and Wenlock's [25] recommendations in terms of solubility, oral bioavailability, and drug likeness.

21.3 Drugs in Other Drugs: Drug as Fragments

Siegel and Vieth [42] published a review entitled "*Drugs in other drugs: a new look at drugs as fragments.*" The authors of this review have examined the complete set of marketed drugs, with regard to looking for structural similarities between drugs. By comparing the structures of all drugs, it has been established how many times one marketed drug occurred as a substructure within another marketed drug. A total of 209 from 1386 marketed drugs sized between 100 and 1500 Da (i.e., 15% of the 1386 total) are contained within other drugs, differing by one or more continuous chemical fragments, and as many as 418 drugs from the total of 1386 (i.e., 30%) contain other drugs as substructure fragments. Many smaller drugs occur in multiple larger drugs. Most of the small changes tend to retain primary indicated pharmacology, whereas larger changes more often lead to different primary pharmacology. The authors identify a subset of drugs that can be used in fragment-based drug discovery strategies. In addition, the analysis enhances the understanding of marketed drug space from the chemical building-block perspective. The recent review, "*Twenty years on: the impact of fragments on drug discovery,"* summarizes the current state of fragment-based drug discovery and illustrates its growing application in academia and industry by presenting more than 30 drug candidates who entered the clinic [43].

21.3.1 Drug Repositioning and Drug Repurposing

In some cases, such "side" effects are therapeutic for other diseases. To extend such old drugs, a new term "drug repositioning" or "drug repurposing" has emerged and new uses for old drugs have raised great interest among drug scientists and clinicians. Actually, some old drugs or clinical candidates have been developed and successfully marketed for new indications, such as aspirin for stroke, sildenafil (Viagra) for erectile dysfunction, and methotrexate for rheumatoid arthritis.

21.4 Old Drugs

While the term "old drugs" is often used to explain the fundamental principles of SOSA, some of the drugs used in this lead generation approach especially for drug repurposing are in fact not that "old." An "old" drug could be defined as one of the following categories [2]:

i) *Off-Patent Drugs.* These kinds of drugs have been used for a certain period and become generic and are not protected by the patent agreement, for example, ampicillin.
ii) Those drugs are withdrawn from the market due to their latent toxicity, e.g., thalidomide.
iii) *Ancient Drugs.* These kinds of drugs are traditional in most Eastern or African countries and have been used for hundreds of years or longer but have not been scientifically developed in the modern era.
iv) To a significant extent, old drugs also include those candidates aborted during the clinical evaluation because of insufficient efficacy or safety liabilities.
v) Currently, most researchers also regard patented drugs as old drug resources because of their established safety evaluation and potential application for new indications.

21.5 The SOSA Approach and Analog Design

The most popular strategy in drug design is the synthesis of analogs of existing active molecules in order to further improve desired properties and alleviate liabilities. The objective is to start with known active principles and, by various chemical transformations, prepare new molecules (in certain cases referred to as "*me-too compounds*") for which an increase in potency, a better specific activity profile, improved safety, or a formulation that is easier to handle by physicians and nurses or more acceptable to the patient are claimed [44, 45].

In the pharmaceutical industry, motivations for analog design are often driven by competitive and economic factors. Indeed, if the sales of a given medicine are high and the company is found in a monopolistic situation, protected by patents and trademarks, other companies will want to produce similar medicines, if possible with some therapeutic improvements. They will therefore use the already commercialized drug as a lead compound and search for ways to modify its structure and some of its physical and chemical properties while retaining or improving its therapeutic properties.

The term analogy, derived from the Latin and Greek *analogia,* has been used in natural sciences since 1791 to describe structural *and* functional similarity. Extended to drugs, this definition implies that the analog of an existing drug molecule shares chemical and therapeutical similarities with the original compound. Formally, this definition allows anticipating three categories of drug analogs based on the level of similarity of the chemical structure and the level of common pharmacophoric features, and in some cases, there is a certain overlap between the different categories:

i) Analogs presenting chemical and pharmacological similarity ("full" or "true" analogs).
ii) Analogs presenting only chemical similarity ("structural analogs").
iii) Analogs displaying similar pharmacological properties but presenting totally different chemical structures ("functional analogs").

Representative examples of full analogs are shown in Figure 21.1 [46].

Structural analogs are shown by an example of classical isosteric and bioisosteric replacement of antipsychotic drugs (Figure 21.2) [47].

In Figure 21.3, structural analogs are shown by an example of nonclassical bioisosteric replacement of anxiolytic drugs. Structural analogs, which possess different activity types, are shown below as well.

Functional analogs exemplified by gamma-aminobutyric acid (GABA)-A agonists are represented in Figure 21.4 [47].

Chemical similarity does not necessarily mean biological similarity, as found for steroid hormones: Testosterone and progesterone, although being chemically very close, have totally different biological functions. Similarly, minaprine is a dopaminergic drug, whereas its cyano analog SR 95191

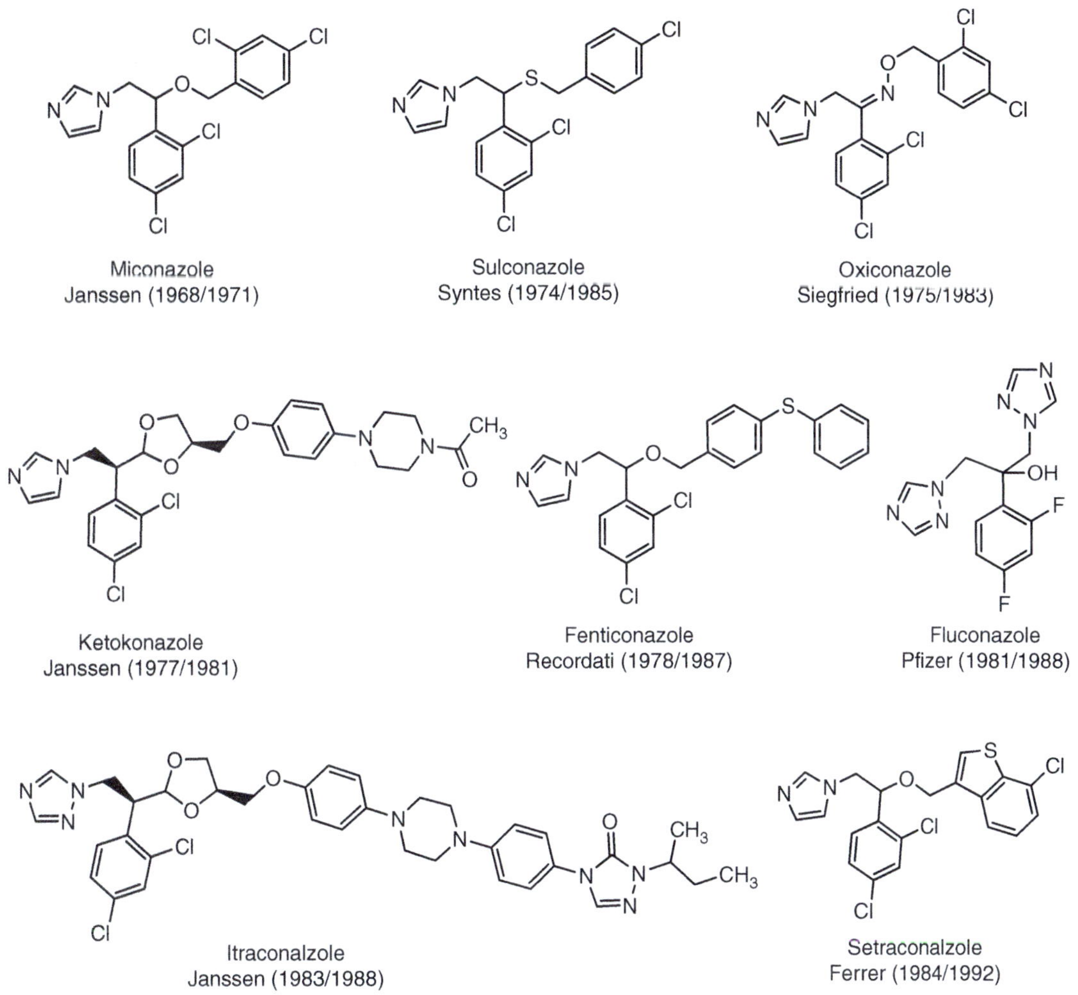

Figure 21.1 Examples of full analogs.

is a potent monoamine oxidase A (MAO-A) inhibitor (Figure 21.5) [48].

21.6 Patentability and Interference Risk of the SOSA Approach

According to the European Patent Convention (EPC), chemical compounds (or mixtures/compositions) can be patented if they are new (novelty), if the inventive step is involved, and if they are industrially applicable [49]. For example, a new chemical compound is always inventive if unexpected effects are implicated when industrially applied. With an increasingly broad and more general definition of the symbols and substituents used in the Markush formula [50, 51], the use of parameters and functional features in patent applications, even the straightforward assessment of novelty [52] of compound claims, is rendered difficult.

However, more difficult than novelty is a unified, clear, and objective assessment of the inventive step. The original idea was that not every new but logically deducible obvious modification of the state of the art that would occur to the skilled person as a simple further development of his or her daily routine work should be protected automatically pursuant to its mere novelty. The third and last requirement, industrial applicability, does as a rule not raise any problem when assessing the patentability of biologically active compounds.

Today, if you look through a copy of a journal referring to biologically/pharmacologically active compounds in the broadest sense, you will have difficulty in finding an article not referring to structure–activity–relationships. The correlations between molecular structure and (biological) activity are one of the most fundamental principles of modern medicinal and pharmaceutical chemistry and as such represent the basis for their industrial applicability.

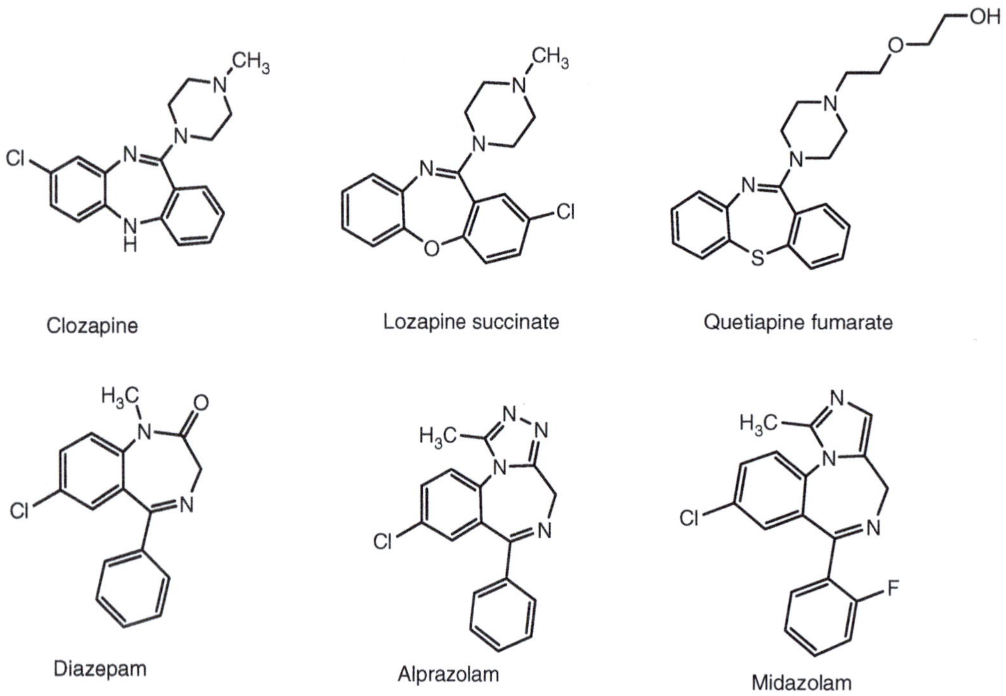

Figure 21.2 Structural analogs of classical isosteric (clozapine, lozapine succinate, and quetiapine fumarate) and bioisosteric (diazepam, alprazolam, and midazolam) replacement of antipsychotic drugs.

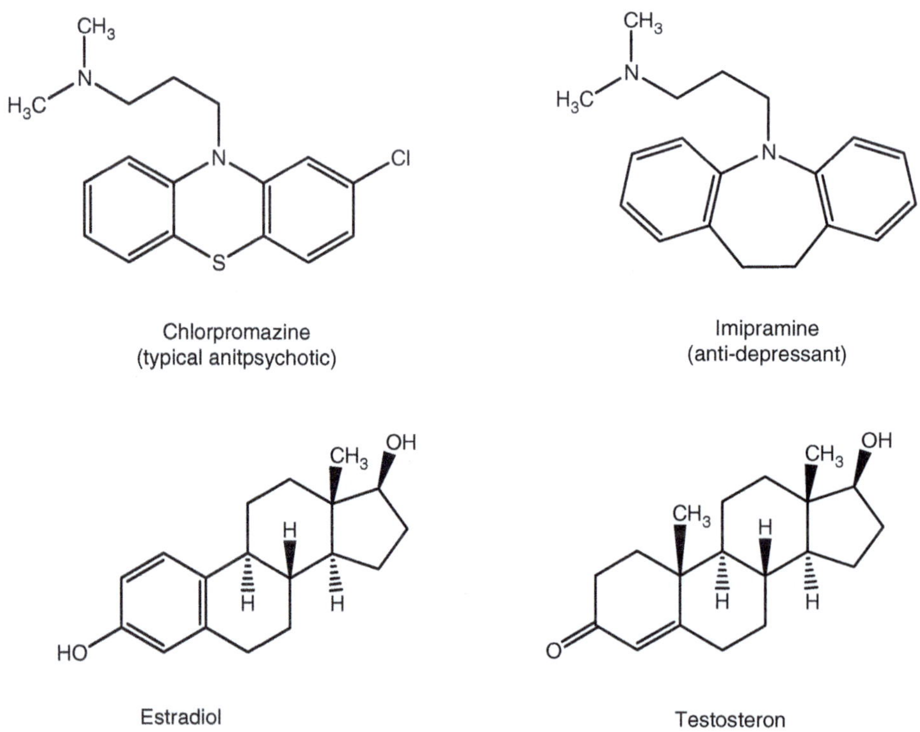

Figure 21.3 Structural analogs of nonclassical bioisosteric replacement.

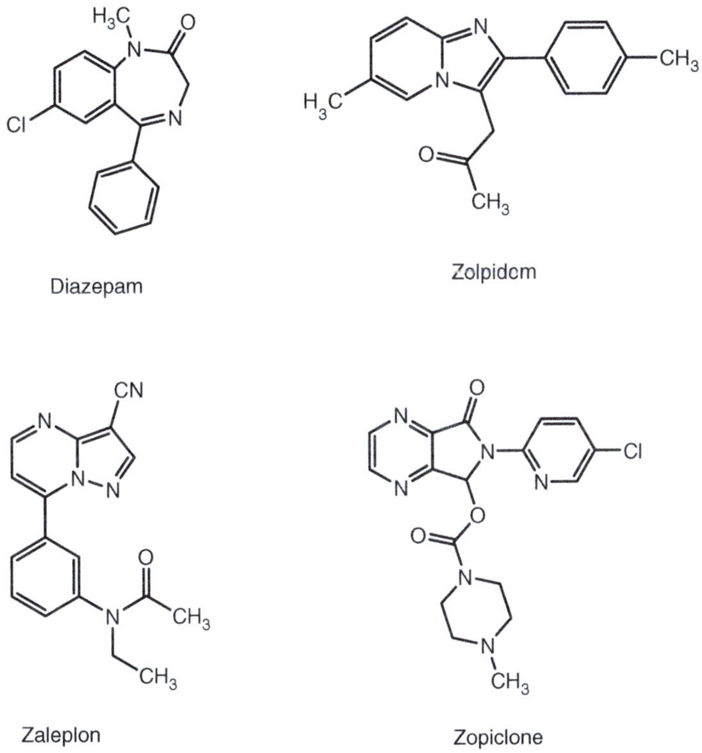

Figure 21.4 Functional GABA-A agonist analogs with similar pharmacological properties but different chemical structures.

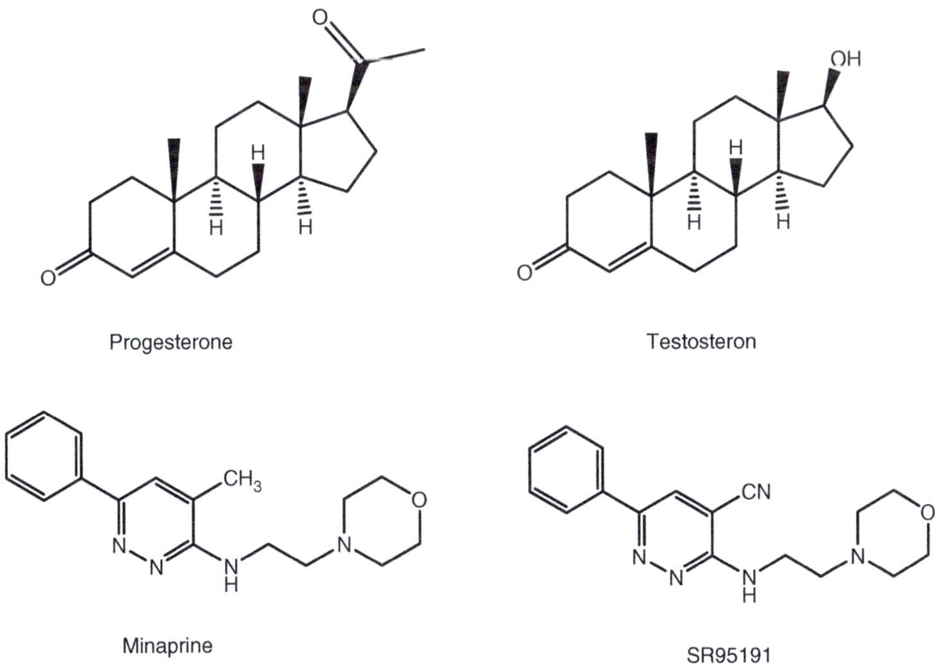

Figure 21.5 Structural analogs with different biological activity.

21.6.1 Analogization, Optimization, and Isosterism

Analogs falling within the scope of a generic formula are "equivalents" in the sense of patent law. This is the consequence of the existence of structure–activity–relationships. Equivalents do not have to be taken into account for novelty but for the assessment of inventive step [53].

When a well-known drug interacts with a new target, there is a risk that several hundreds or thousands of analogs of the original drug molecule have already been synthesized by the original inventors and/or their competitors. These molecules are usually protected by patents or belong to the public domain.

The aim is to modify the structure sufficiently such that it avoids patent restrictions, retains activity, and ideally has improved therapeutic properties.

At first glance, a high risk of interference thus appears probable.

However, when optimizing a therapeutic profile other than that of the original inventors during the SOSA approach, the medicinal chemist will rapidly prepare analogs with distinct chemical structures, thus avoiding overlap from the original hit in the sense of patent terms.

For example, a medicinal chemist interested in phosphodiesterases (PDEs) and using diazepam as a lead structure will rapidly prepare compounds (CI-1044), which are out of scope of the original patents, precisely because they exhibit dominantly PDE-inhibiting properties and almost no more affinity for the benzodiazepine receptor (Figure 21.6) [54].

21.7 Case Studies and Examples

Many drugs of the past resulted from the experimental or clinical observation of side effects.

As a consequence, several drugs have been developed by enhancing the side effect of another drug. As an example, the mood-improving effect of iproniazid was discovered when it was tested as an antituberculous drug; antidepressant inhibitors of neurotransmitter reuptake, like imipramine and desipramine, stem from the antipsychotic dopamine antagonist chlorpromazine, which itself was derived from H_1 antihistamines [36, 37, 55].

21.7.1 Sulfonamides

Diuretic, antihypertonic, antiglaucoma, and antidiabetic drugs were derived from the bacteriostatic sulfonamides, and most of them have been used as antibacterial agents. However, some sulfonamides with antibacterial activity could not be used clinically because they had convulsive side effects brought on by hypoglycemia (lowered glucose levels in the blood). Clearly, this is an undesirable side effect for an antibacterial agent, but the ability to lower blood glucose levels would be useful in the treatment of diabetes. Therefore, structural alterations were made to the sulfonamides concerned in order to eliminate the antibacterial activity and to enhance the hypoglycemic activity. This led to the antidiabetic agent tolbutamide (Figure 21.7) [55].

21.7.2 Morphine Analogs

Another historical example of the validity of this approach is the antitussive and constipating morphine analogs [55].

21.7.3 Warfarin

Another example was the discovery that the anticoagulant warfarin is also a weak inhibitor of a viral enzyme that is important in the life cycle of HIV. Warfarin was used as the lead compound in the development of an anti-HIV drug called tipranavir (Figure 21.7) [56].

21.7.4 Sildenafil (Viagra)

In some cases, the side effect may be strong enough that the drug can be used without modification. For example, the drug to treat erectile dysfunction, sildenafil (Viagra), was originally designed as a vasodilator to treat angina and hypertension (Figure 21.7) [2, 55].

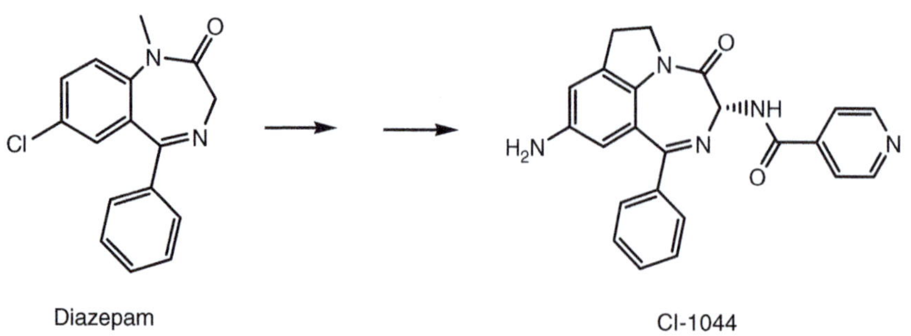

Figure 21.6 Diazepam as a lead structure leading to compounds with phosphodiesterase activity.

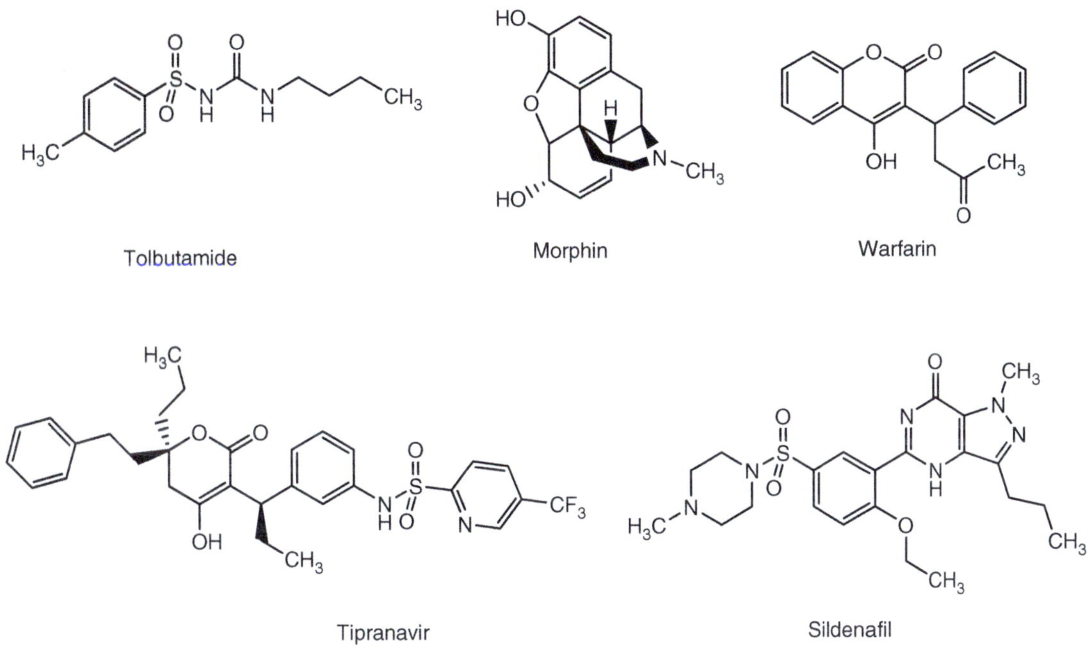

Figure 21.7 Structures of tolbutamide, morphine, warfarin, tipranavir, and sildenafil.

During clinical trials, it was found that it acted as a vasodilator more effectively in the penis than in the heart, resulting in increased erectile function. The drug is now used to treat erectile dysfunction and sexual impotence.

In the late 1980s, a Pfizer research team in England developed a promising drug for treating angina, a condition characterized by tightness, in phase I trials in the United States. Although UK-92,480 appeared to be safe, it had little impact on blood pressure, heart rate, and cardiac output. In a 1992 study, some volunteers receiving multiple doses reported an increased incidence of penile erections. Preliminary data were not overwhelming, but Pfizer spent another two years investigating the potential market for an erectile dysfunction drug. A study involving 300 patients in 1994 and 1995 showed excellent results for UK-92,480, which had been the chest because of limited blood flow to the heart. The compound, known at the time as UK-92,480, entered clinical trials in healthy volunteers, the equivalent of renamed sildenafil. The trials were so successful that, when the trials were complete, patients were reluctant to turn in their unused medication. Some patients even falsely claimed to have flushed leftover pills down the toilet in order to keep the unused medication. Sildenafil was approved by the Food and Drug Administration (FDA) in early 1997 under the trade name Viagra. Despite competition in the marketplace from other erectile dysfunction drugs, Viagra has been a blockbuster for Pfizer, with a peak in sales of nearly US$2 billion worldwide in 2008 [2].

21.7.5 Thalidomide Analogs

Thalidomide was originally developed in the 1950s for the treatment of pregnancy-associated morning sickness and was first marketed in 1957 as a "wonder drug" for insomnia, coughs, colds, and headaches because of its sedative–hypnotic effects. However, thalidomide was soon found to be associated with teratogenicity [2]. In the late 1950s and early 1960s, more than 10,000 children in 46 countries were born with deformities such as phocomelia as a consequence of thalidomide use associated with its inhibition on angiogenesis [57, 58]. Therefore, thalidomide was immediately withdrawn from the market in 1961. Research on thalidomide slowed down in the 1960s, but never stopped. Several lines of mechanisms of teratogenesis were proposed including angiogenesis inhibition [57, 59], free radical-mediated oxidative deoxyribonucleic acid (DNA) damage [60], and inhibition on insulin-like growth factor 1 (IGF-1) and fibroblast growth factor 2 (FGF2) expression [59, 61]. All these molecular events are associated with cancer development. Thalidomide was first tested as a single agent for the treatment of multiple myeloma in 1999 for its anti-angiogenesis activity [58], followed by a series of successful multicenter clinical trials that ensured that thalidomide was approved by the FDA. Thalidomide analogs such as lenalidomide were also developed and are more powerful, but with fewer side effects [62]. Thalidomide and lenalidomide are now used as a first-line therapy for multiple myeloma. The discovery of thalidomide was believed to be a breakthrough in the therapy of multiple myeloma in the last 30 years. Thalidomide has also been evaluated in a number of other hematological disorders, including acute myeloid leukemia (AML) [63, 64], myelofibrosis with myeloid metaplasia and myelodysplastic syndrome (MDS), and some solid tumors [65].

Thalidomide is teratogenic by inhibiting vascular endothelial growth factor (VEGF) signaling, thus disrupting angiogenesis during the formation of limbs [59]. Angiogenesis is proposed as an important factor in myeloma development because increased bone marrow vascularity and angiogenesis-associated signals of interleukin (IL)-6, VEGF, and fibroblast growth factor receptor (FGFR) are frequently seen in myeloma [66]. Thus, thalidomide was challenged for myeloma treatment in a clinical trial in 1999 using its antiangiogenic activity [58]. It turned out great remission after myeloma administration, which finally led to the wide application in myeloma.

Based on thalidomide, several derivatives were successfully synthesized such as lenalidomide and pomalidomide (Figure 21.8). More potent lenalidomide and pomalidomide are derived from thalidomide by adding an amino group. This secondary generation of thalidomide has been fully developed and displays more powerful therapeutic efficacy but less toxicity than thalidomide [67]. Of them, lenalidomide was approved for clinical use earlier than thalidomide by the FDA.

21.7.6 Bupropion

Another example is the antidepressant drug bupropion. Patients taking this drug reported that it helped them give up smoking, and so the drug is now marketed as an antismoking aid (Zyban). Astemizole is a medication used in the treatment of allergy but has been found to be a potent antimalarial agent (Figure 21.9) [68].

21.7.7 Chlorpromazine

Chlorpromazine is used as a neuroleptic agent in psychiatry but was developed from the antihistamine agent promethazine. This might appear an unexpected thing to do, but it is known that promethazine has sedative side effects, and so medicinal chemists modified the structure to enhance the sedative effects at the expense of antihistamine activity (Figure 21.10) [69].

21.7.8 Chlorothiazide

Similarly, the development of sulfonamide diuretics such as chlorothiazide arose from the observation that sulfanilamide has a diuretic effect in large doses due to its action on an enzyme called carbonic anhydrase (Figure 21.11) [70].

21.7.9 Propranolol

Propranolol, the first successful nonselective beta blocker, was derived from the early β-adrenergic antagonist's

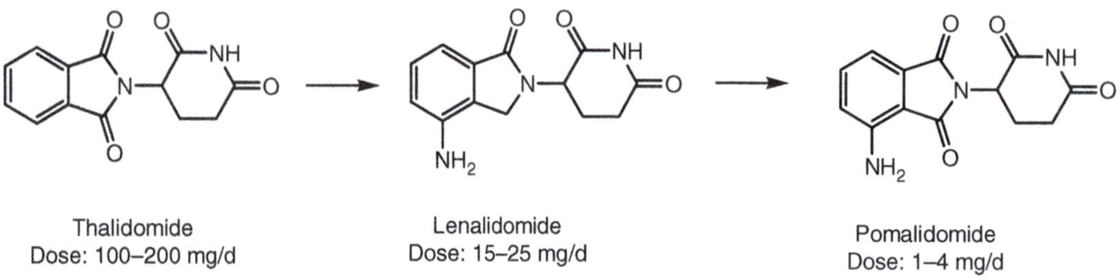

Figure 21.8 More potent lenalidomide and pomalidomide are derived from thalidomide by adding amino group.

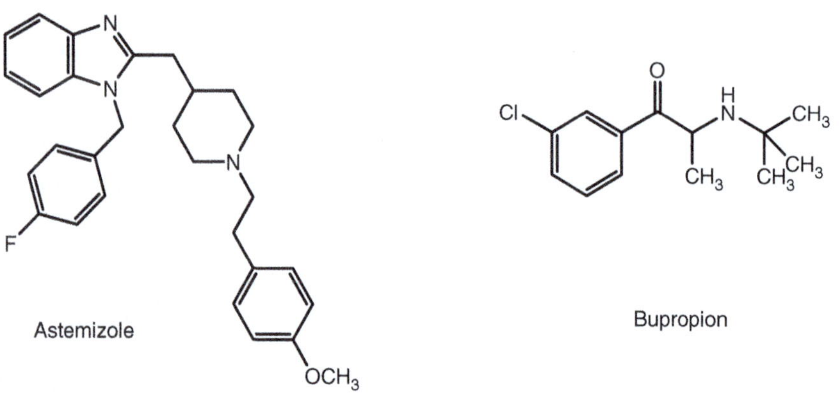

Figure 21.9 From the antidepressant drug bupropion used as an antismoking agent and astemizole to treat allergy, which also has some antimalarial activities.

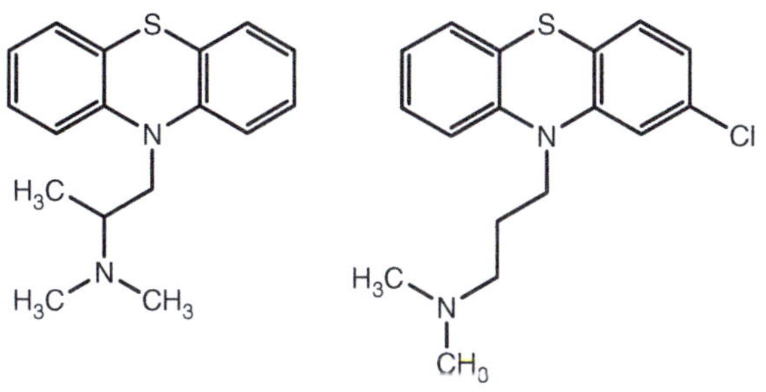

Figure 21.10 Chlorpromazine is used as a neuroleptic agent in psychiatry but was developed from the antihistamine agent promethazine.

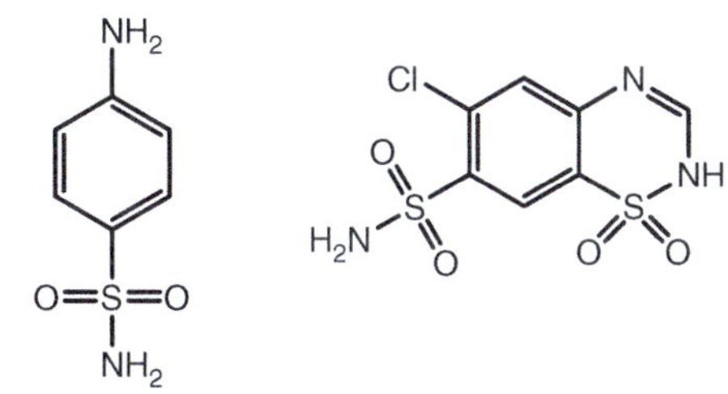

Figure 21.11 Development of sulfonamide diuretics such as chlorothiazide starting from sulfanilamide.

dichloroisoprenaline (DCI) and pronethalol [71] (Figure 21.12). DCI is a nonselective blocker for the β1-adrenergic and β2-adrenergic receptors, but it has low potency. Pronethalol was a promising clinical candidate with a similar structure to DCI but showed carcinogenicity in mice. However, the insertion of a -CH_2O- linkage into the arylethanolamine structure of pronethalol has increased the potency of the compound and has eliminated the carcinogenicity being found with pronethalol in animal models [72]. Similar efforts have brought newer and more selective beta-blockers (such as nebivolol) in the treatment of hypertension [73].

21.7.10 Minaprine Analogs

Sometimes slight changes to a structure can result in significant changes in pharmacological activity. For example, minaprine is an antidepressant agent that acts as a serotonin agonist. Adding a phenolic substituent resulted in 4-hydroxyminaprine, which is a potent dopamine agonist, whereas adding a cyano substituent gave bazinaprine, which is a potent inhibitor of the enzyme monoamine oxidase-A (Figure 21.13). Minaprine also binds weakly to muscarine receptors (Ki = 17 pM), and modifications were successfully carried out to give structure I with nanomolar activity (Ki = 3 nM) for the muscarinic receptor and negligible activity for dopamine and serotonin receptors. Minaprine also has weak affinity for the cholinesterase enzyme (half-maximal inhibitory concentration [IC_{50}] = 600 pM). Modifications led to structure II, which has strong affinity (IC_{50} = 10 nM) for this target [74].

21.7.11 Viloxazine Analogs

In the 1970s, the standard treatment for hypertension (high blood pressure) included β-blockers and diuretics, such as propranolol and hydrochlorothiazide, respectively. β-Blockers possess many other different activities, including sedative and anticonvulsant effects. To enhance some of the secondary effects, the central core, or pharmacophore, of β-blockers was modified to form a new ring. A thorough study of derivatives yielded the antidepressant viloxazine (Figure 21.14). The research on viloxazine implies that different conformations of β-blockers are responsible for their poor target selectivity. Restriction of the flexibility by a standard method, tying the molecule with a ring, boosts selectivity for one of the targets [37, 38].

21.7.12 Methylation in the SOSA Strategy of Drug Design [75]

A typical case in the use of methyls in the SOSA approach was the discovery of endothelin receptor antagonists (ETAs) by Bristol–Myers–Squibb (BMS) scientists [37, 43]. From sulfathiazole trials, a drug from BMS itself, they discovered that this antibiotic was a weak ligand for ETA receptors [76]. By testing isosteric modifications on the thiazole ring and mainly through the introduction of two methyls (at C-3 and C-4) to the isoxazolic compound sulfisoxazole [77], researchers discovered compound BMS-193884 [78]. This molecule exhibits optimized affinity for the ETA receptor, demonstrating that the methyl at C-4 is essential for potency, whereas the methyl at C-3 is responsible for selectivity (Figure 21.15) [75]. Additional modifications on the para-aminobenzenesulfonamide ring to increase its potency and bioavailability led to the discovery of BMS-207940 [79], a powerful and selective ETA receptor antagonist with *in vivo* hypotensive activity when taken orally.

In addition to the introduction of a methyl group to a bioactive compound, its regioisomers can also be important for optimizing its pharmacological activity [37]. For example, Wermuth and associates moved the methyl from C-4 to C-5 in minaprine, a drug that acts on dopaminergic

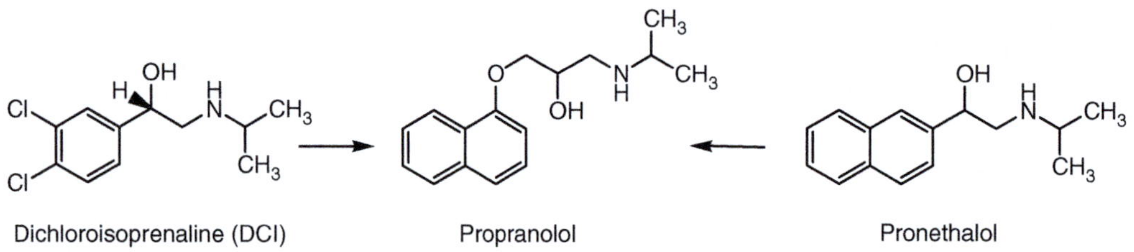

Figure 21.12 Propranolol was derived from the early β-adrenergic antagonist's dichloroisoprenaline (DCI) and pronethalol.

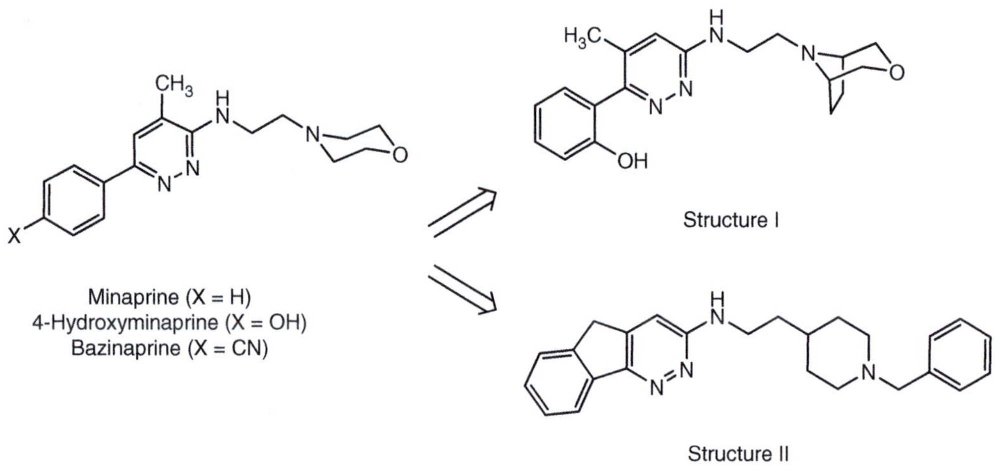

Figure 21.13 Minaprine analogs.

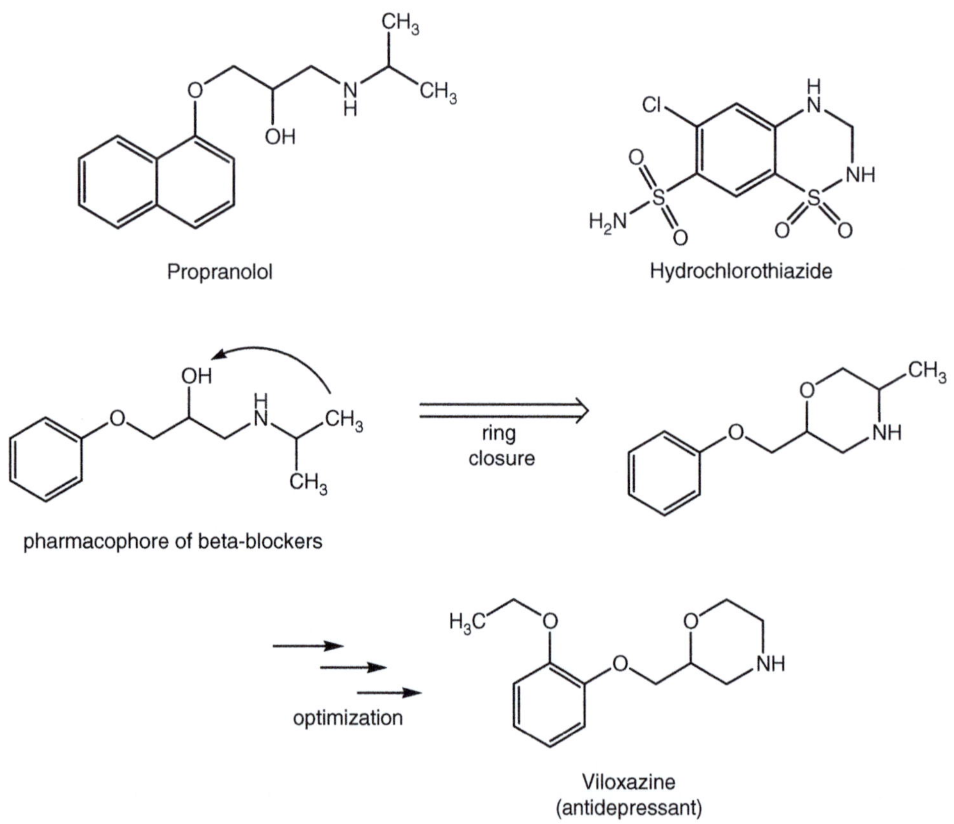

Figure 21.14 Viloxazine analogs.

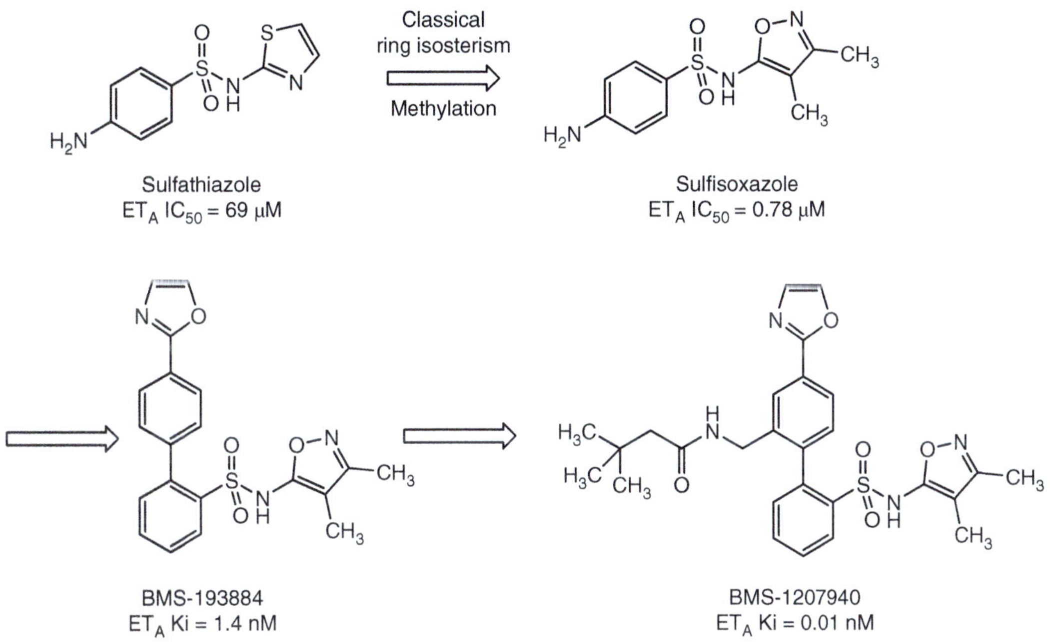

Figure 21.15 Endothelin receptor antagonists as a representative example of methylation in the SOSA strategy of drug design.

and serotoninergic transmission and also has a weak affinity for muscarinic M1 (Ki = 17 μM) receptors. This generated a molecule, which shows a 35-fold increase in affinity for M1 receptors (Figure 21.16). Additional modifications to the morpholine and the addition of an ortho-hydroxyl group to the phenyl group led to a potent partial agonist of M1 devoid of activity on dopaminergic or serotoninergic transmission (Figure 21.16) [74, 80–82].

21.7.13 Discovery of New Antiplasmodial Compounds [83]

Malaria remains one of the most challenging diseases. Over three billion people live under the threat of malaria across the world, and one million are killed each year, mostly children. *Plasmodium falciparum* accounts for the most severe and fatal form of the disease.

The dominant strategy for the discovery of new antimalarial drugs relies on cell-free assays on specific biochemical pathways of *P. falciparum*. However, it appears that screening directly on the parasite is a more rewarding approach. The applied "drug-to-genome-to-drug" approach consists of testing a small set of structural analogs of a certain drug acting on human proteins that have plasmodial orthologs. Both man and plasmodium possess cyclic nucleotide PDEs that are key players of cell homeostasis. The authors synthesized and tested 40 analogs of tadalafil, a human phosphodiesterase type 5 (PDE5) inhibitor, on *P. falciparum* in culture and obtained potent inhibitors of parasite growth. In addition, the authors also prove that antiplasmodial derivatives inhibit the hydrolysis of cyclic nucleotides of the parasite, validating the cyclic adenosine monophosphate (cAMP)/cyclic guanosine monophosphate (cGMP) pathways as therapeutic targets against *P. falciparum* (Figure 21.17).

21.7.14 Drugs Acting on Central Nervous System (CNS) Targets as Leads for Non-CNS Targets [84]

Kharkar [84] recently described the use of central nervous system (CNS) drugs as leads for non-CNS targets. The approach is related to the use of existing drugs for new indications. Suitable chemical modifications of CNS drugs abolish their CNS penetration. These novel analogs may then be screened for activity against non-CNS targets. Careful selection of the appropriate structural modifications remains the key to success.

The design and development of drugs that cross the blood-brain barrier (BBB) and act at some target site(s) in the CNS is a formidable task. In contrast, for the drugs to act on peripheral targets, it is important to restrict their passage through the BBB in order to avoid unwanted CNS side effects. Several physicochemical and molecular properties of CNS drugs differ from peripherally acting drugs; the former have lower molecular weights, are more lipophilic, have a smaller polar surface area (PSA), a fewer H-bond acceptors and donors, and fewer rotatable bond [85]. There is a fine balance between the physicochemical properties of CNS and non-CNS drugs.

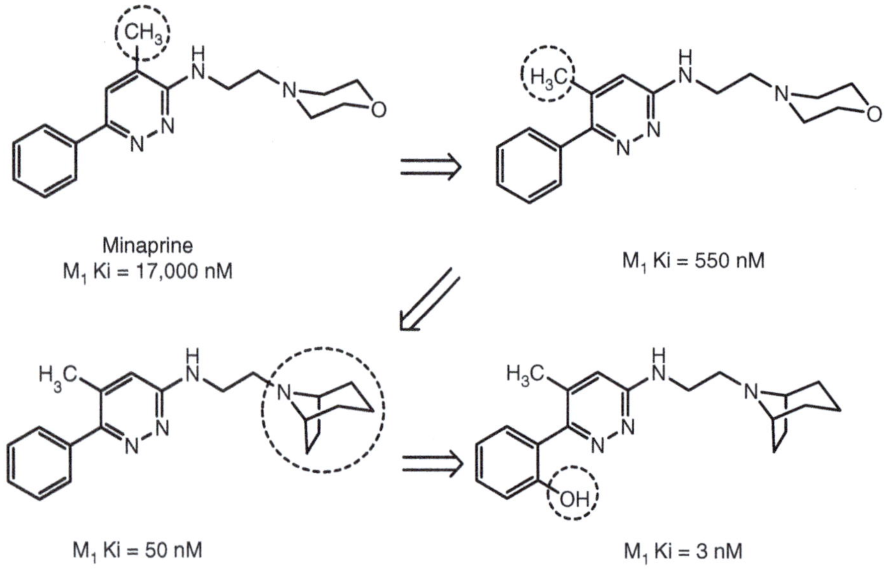

Figure 21.16 Modification of minaprine leading to compounds with M1 receptor affinity without a significant dopaminergic or serotoninergic activity.

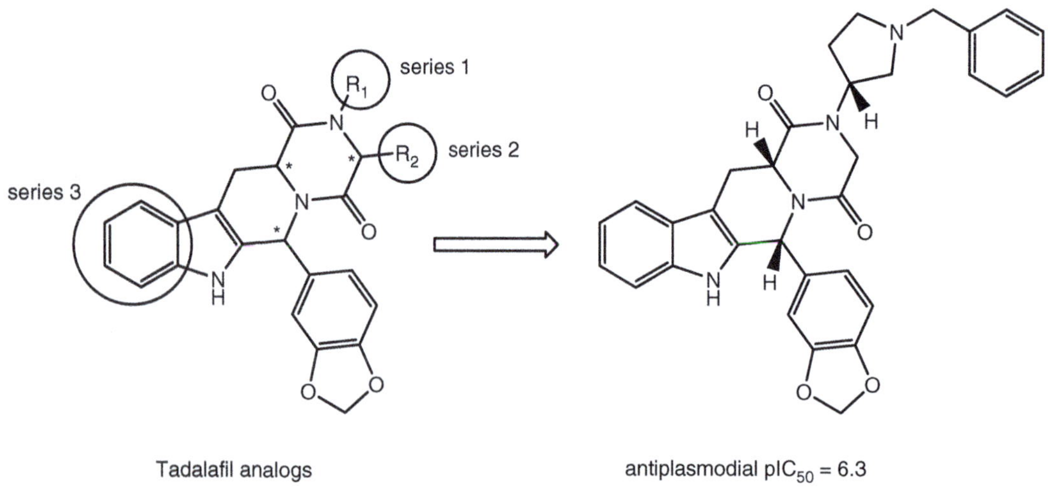

Figure 21.17 Tadalafil analogs leading to novel antiplasmodial compounds.

During typical lead optimization cycles in the discovery phase, the lead molecules undergo several chemical modifications in order to improve their potency and PK properties, which usually lead to an increased (i) molecular weight, (ii) lipophilicity, (iii) molecular complexity, (iv) number of rotatable bonds, and (v) number of H-bond donors and acceptors. In general, the drugs are more complex than the leads and exhibit higher values for the majority of the associated molecular properties listed above. Based on these findings, it can be hypothesized that "CNS drugs which are smaller and lower ranges of the aforementioned molecular properties make excellent starting points (as leads) for the development of non-CNS drugs" (Figure 21.18).

The use of CNS drugs as a starting point for developing non-CNS leads seems interesting with reference to potentially altered molecular, PK, pharmacodynamic, and/or toxicity properties. The combinatorial libraries based on CNS drug scaffolds may perturb novel chemical space not accessed previously by the non-CNS small-molecule drugs. Curious researchers interested in the above strategy may help in demonstrating its potential utility.

21.7.15 Mexiletine Derivatives as Orally Bioavailable Inhibitors of Urokinase-type Plasminogen Activator [86]

Urokinase-type plasminogen activator [87] (uPA*a*) is a trypsin-like serine protease [88] that, when bound to

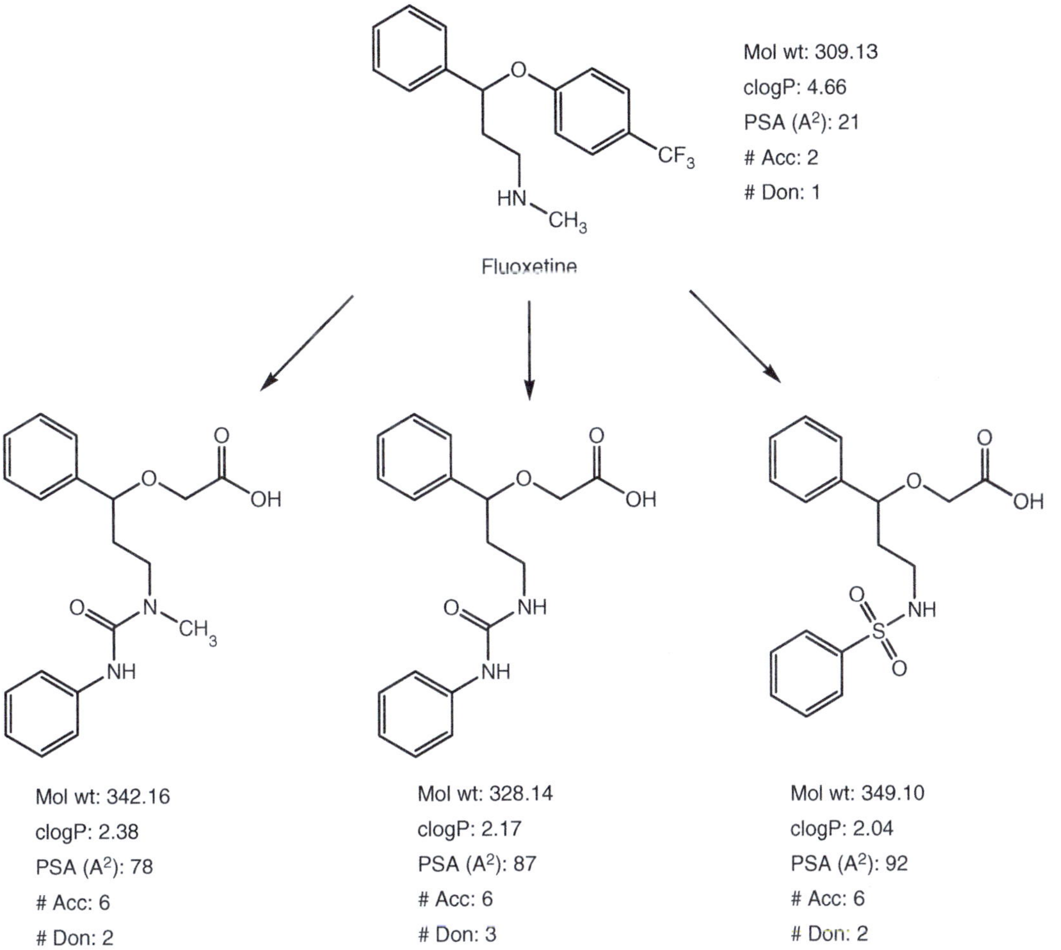

Figure 21.18 An illustration of the concept of using CNS drugs for non-CNS targets. Fluoxetine, an antidepressant agent, is used as a template. The synthesis of the analogs can be achieved by modifying the synthetic route of fluoxetine itself. The major structural modification includes abolishing the basicity of the secondary N in fluoxetine by converting it to urea or sulfonamide. Similarly, the 4-CF_3Ph ring is replaced with an acetic acid side chain. The resultant structural modifications lead to increased PSA, decreased calculated logarithm of partition coefficient (cLogP), and an increased number of H-bond donors and acceptors.

the extracellular membrane-associated uPA receptor [89] (uPAR), catalyzes the conversion of plasminogen to plasmin through the cleavage of the amide bond of an arginine/valine motif. Plasmin is responsible for a number of proteolytic processes that degrade various components of the extracellular matrix [90], thereby triggering the induction of cell migration. As a result, uPA has been implicated in the progression of a variety of disease states associated with abnormal tissue destruction and cell infiltration including aortic aneurism [91], multiple sclerosis [92], and metastasis processes in cancer [93]. Advanced inhibitors of uPA (compounds **1–4**) are shown in Figure 21.19.

Frederickson *et al.* [86] reported on a novel series of weakly basic orally bioavailable inhibitors of uPA derived from mexiletine **5**, as shown in Figure 21.20.

Mexiletine **5** is a class Ib antiarrhythmic licensed for use in patients with life-threatening irregular heartbeat and may also be of use in the control of refractory pain.

The 2-ethoxy derivative **6** (measured p*K*a of 8.75) is a potent inhibitor of uPA (IC_{50} 72 nM; ligand efficiency [LE] 0.31) and is comparable to **1–4** (figure before; reported K_i values of 410, 37, 12, and 20 nM [94], respectively). As expected, (*R*)-**7** (IC_{50} 460 nM; LE 0.27) had markedly less *in vitro* activity than **6** [86].

The selectivity of **6** for uPA was determined by reference to IC_{50} values (data not shown) measured against a panel of closely related proteases (plasmin, factor VIIa, factor Xa [FXa], tissue-type plasminogen activator, and trypsin). Amine **6** was found to be selective for uPA (>50-fold) over all but one of the enzymes but was only moderately so (three- to fourfold) against trypsin, the enzyme to which (in terms of the similarity of the S1 subside) it is most

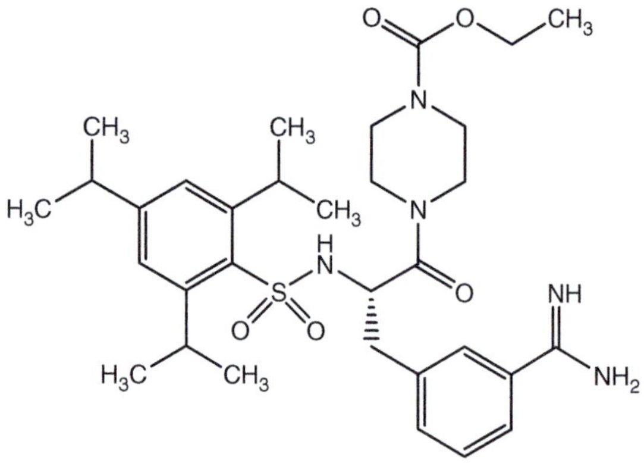

1

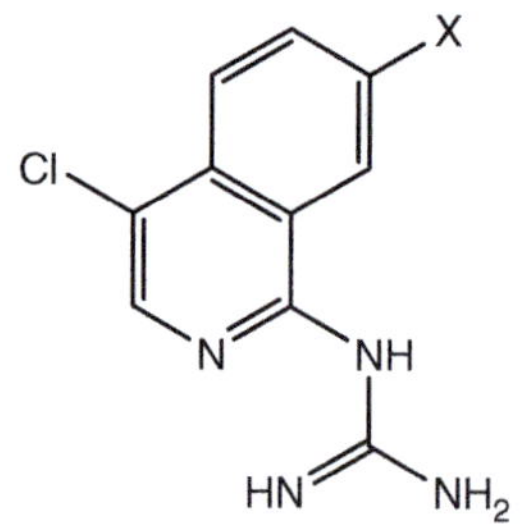

2: X = 3-COO_2H-Ph

3: X = $SO_2NHCMe_2CO_2H$

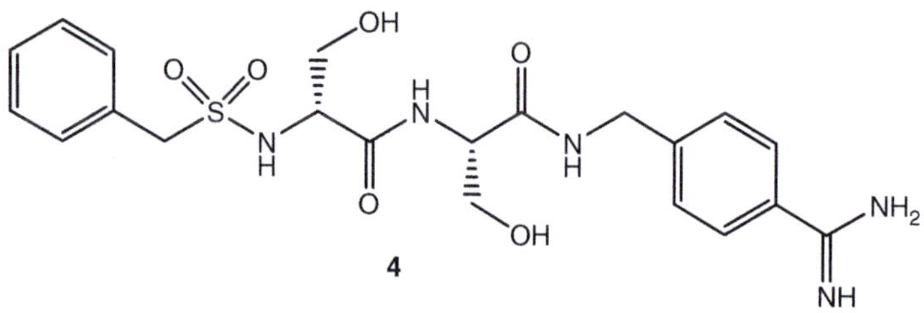

4

Figure 21.19 Advanced inhibitors of urokinase-type plasminogen activator.

closely related and over which selectivity for uPA is most difficult to obtain (Figure 21.20) [95]. That such limited selectivity over trypsin was not an inherent property of all compounds akin to **6** was shown by reference to compounds **7** and **8**, which showed somewhat greater selectivity for uPA (around 10-fold). On the basis of the observed high potency and selectivity, the PK profile of compound **6** was studied in the rat. When dosed orally at 4.2 mg/kg, **6** showed well-defined and reproducible PK data (Table 21.1); importantly, amine **6** had high oral bioavailability (*F*oral [rat] 60%). The moderate clearance level, high volume of distribution, and long half-life (7½ hours) observed for **6** are consistent with those observed for mexiletine **5** in man [96]. Compound **6** thus displays many of the highly desirable molecular properties expected of a good "lead-like" molecule, including high potency and selectivity over a series of closely related proteases and, importantly, high oral bioavailability, a molecular property that has so far remained relatively elusive [97] in advanced inhibitors of this heavily studied therapeutic target.

21.7.16 Amiloride Analogs as Inhibitors of the Urokinase-type Plasminogen Activator (uPA) [98, 99]

A known side activity of the oral potassium-sparing diuretic drug amiloride is the inhibition of the enzyme uPA ($K_i = 7\,\mu M$), a promising anticancer target. Several studies have demonstrated significant antitumor/metastasis properties for amiloride in animal cancer models, and it would appear that these arise, at least in part, through inhibition of uPA. Selective optimization of amiloride's structure for more potent inhibition of uPA and loss of diuretic effects would thus appear as an attractive strategy toward novel anticancer agents. A key finding was that the well-studied 5-substituted analogs ethyl isopropyl amiloride **9** and hexamethylene amiloride **10** are approximately twofold more potent than amiloride as uPA inhibitors (Figure 21.21).

Orally active noncytotoxic agents capable of inhibiting the progression of primary tumors to metastasis have long been a goal in anticancer drug discovery [100]. Upregulation of the plasminogen activation system (PAS) is known to play a critical role in tumor invasion and metastasis [87, 101], and several components of the PAS offer attractive drug targets in this context [102, 103]. In its simplest form, the PAS comprises the serine protease uPA, its cognate glycosylphosphatidylinositol (GPI)-anchored cell surface–bound receptor (uPAR), and two endogenous serpins; and plasminogen activator inhibitor 1 (PAI-1) and PAI 2 (PAI-2).

Receptor-bound uPA efficiently cleaves and thus activates co-localized plasminogen at the cell surface, which reveals the broad-spectrum serine protease plasmin.

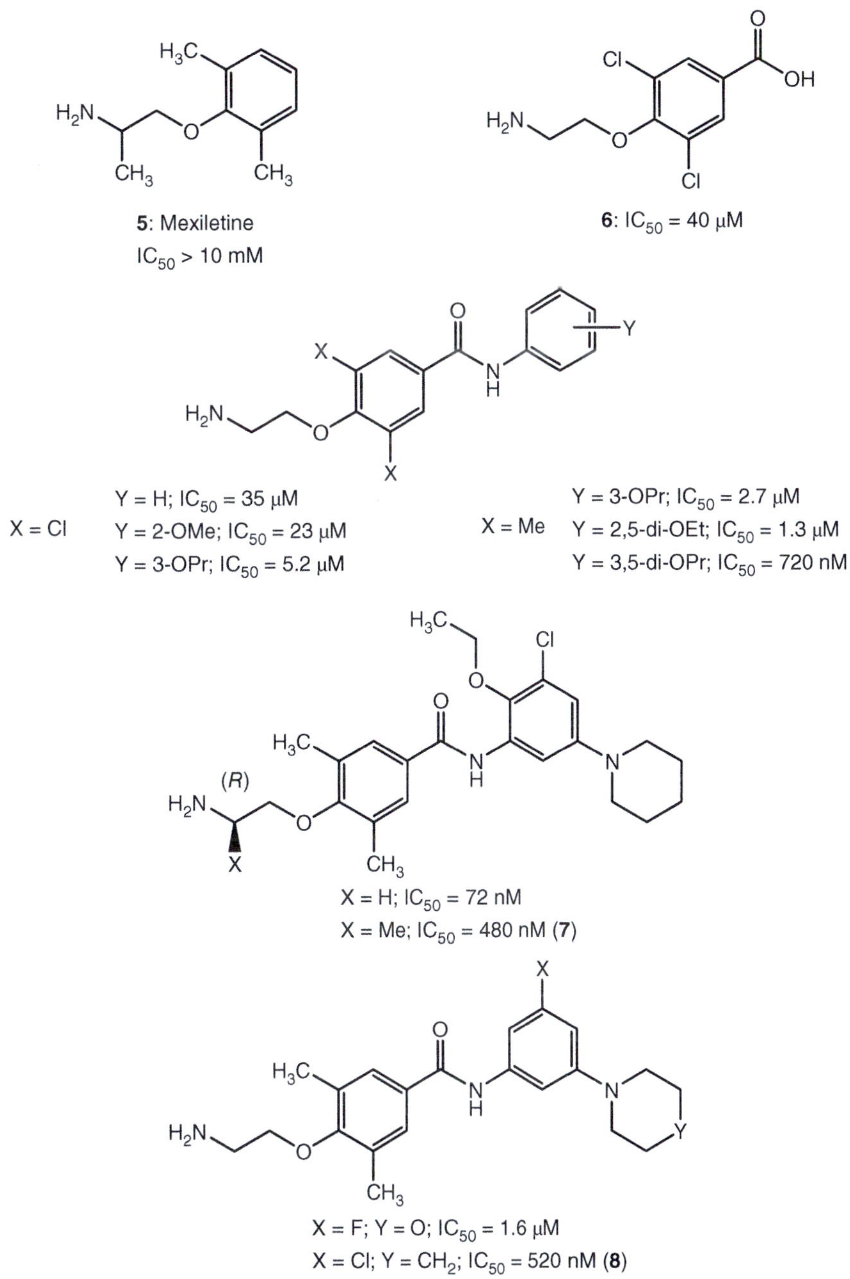

Figure 21.20 The 2-ethoxy derivative **6** (measured pKa of 8.75) as an analog of mexiletine **5** as a potent inhibitor of uPA.

Table 21.1 Pharmacokinetic profile for **6** (in rat).

Cl (mL/min/kg)[a)]	V_d[a)] (l/kg)	t1/2[a)](h)	F(%)[b)]
43	27	7.5	60

a) A 0.95 mg/kg i.v. dose (solution in 100% H_2O).
b) A 4.2 mg/kg oral dose (solution in 100% H_2O).

Plasmin activates downstream extracellular proteases (e.g., matrix metalloproteases) and latent growth factors, and together, these lead to pericellular proteolysis and remodeling of the local extracellular environment – key events required for metastasis.

Amiloride is an orally administered potassium-sparing diuretic that has been used clinically for several decades, most commonly in combination with thiazide (e.g., hydrochlorothiazide) or loop diuretics (e.g., frusemide) as an antikaliuretic. Studies have shown that amiloride is also a moderately potent competitive inhibitor of uPA (K_i = 7 μM) and that uPA inhibition correlates with significant antitumor/metastasis effects of the compound *in vivo* [99].

Analogs of **9** and **10** that retain the respective ethyl isopropyl and hexamethylene amines at C-5 while replacing the 6-chloro group with substituents that are able to

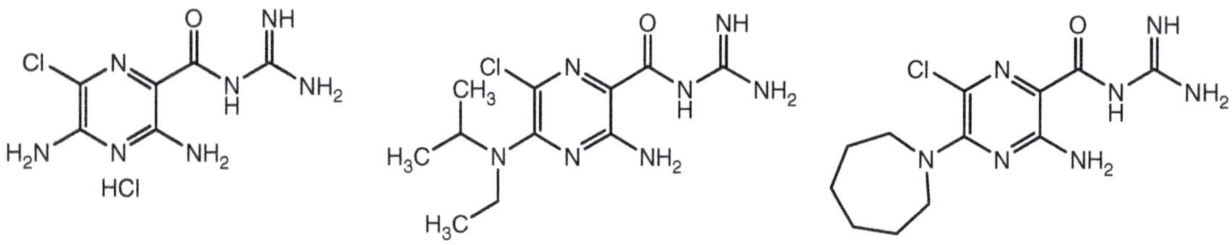

Amiloride HCl
uPA inhib. IC_{50} 11 μM

5-(*N*-ethyl-*N*-isopropyl) amiloride **9**
uPA inhib. IC_{50} 6 μM

Hexamethylene amiloride **10**
uPA inhib. IC_{50} 6 μM

Figure 21.21 Amiloride analogs as inhibitors of the urokinase-type plasminogen activator (uPA).

make more favorable contacts in the S1β site may lead to more potent uPA inhibitors that retain potency against sodium–hydrogen exchanger isoform 1 (NHE1). It is worth noting that 6-iodo-**9** and 6-bromo-**10** are 313- and 566-fold more potent NHE1 inhibitors than amiloride, respectively, and 1.40- and 1.08-fold higher in potency than their 6-chloro counterparts **9** and **10** [104].

The current study has shown that converting amiloride to the 6-bromo or 6-iodo derivatives enhances uPA inhibition. The compounds 6-iodo-**9**, 6-bromo-**10** and derivatives of **9** and **10** that carry other C-6 substituents are currently being examined in these laboratories in pursuit of dual uPA/NHE1 inhibitors.

21.7.17 Flavonoids with an Oligopolysulfated Moiety: A New Class of Anticoagulant Agents [105]

In the work of Correia-da-Silva *et al.* [105], polysulfated (oligo)flavonoids were synthesized and assayed for their *in vitro* and *in vivo* anticoagulant activities. The approach was based on the molecular hybridization of two classes of anticoagulants, sulfated polysaccharides and sulfated flavonoids. In their work, the synthesis, the *in vitro* and *in vivo* anticoagulant profiles, and the preliminary toxicological investigation of several polysulfated flavonoids/flavonosides are described (Figure 21.22).

The oligo-O-sulfated moiety is present in potent anticoagulant polysaccharides, suggesting the incorporation of an oligo-O-sulfated moiety into a flavonoid scaffold in order to improve the anticoagulant potency by increasing both molecular size and number of sulfate groups (Figure 21.23a,b).

Modification of molecules that have already been safely used in humans leads to compounds with more predictable and less complex PKs, lower incidence of side effects, and less-demanding clinical studies [37]. On the basis of these considerations, the authors selected for sulfation several flavonoid glycosides already used in human therapy as venotropic agents [106] (Figure 21.23c). This strategy follows the principles of the SOSA approach [37, 38].

The incorporation of an oligo-O-sulfated moiety into a flavonoid scaffold was a successful strategy to improve the anticoagulant potency of sulfated flavonoids. Polysulfated flavonosides proved to be active *in vitro* both in plasma and in whole human blood as well as *in* vivo, surpassing existing anticoagulant flavonoids. This study provided interesting structure-activity relationship (SAR) by revealing sulfated 7-O-rutinosides as indirect FXa inhibitors and sulfated 3-O-rutinosides as direct FXa inhibitors. In contrast to unfractionated and low-molecular-weight (LMW) heparins, the polysulfated oligoflavonoids correspond to compounds with a defined composition and with less charge density that allowed a systemic anticoagulant action after intraperitoneal (ip) administration and are not expected to induce hemorrhagic complications and hepatic toxicity, as it was preliminarily proved with the *in vivo* studies in mice.

The obtained polysulfated flavonosides were highly effective in increasing clotting times and able to completely block the clotting process, in contrast to their corresponding aglycones. The thromboelastography proved that polysulfated flavonosides possess good whole-blood anticoagulation activity.

The derivatives resulting from chemical molecular modifications of known therapeutic agents have high chances to produce safer anticoagulant agents, and this study provided interesting data to be explored by rational design in the future.

21.7.18 Clioquinol

Through this screen, Gan and coworkers identified the off-patented drugs clioquinol [107] and cyproheptadine [108] as potential therapeutics for both myeloma and leukemia.

Clioquinol is an antifungal and antiprotozoal drug developed in the 1930s and was very effective in anti-infective, intestinal antiamebic, and vaginal trichomonacide and soon was marketed worldwide. However, clioquinol was suspected to be highly associated with subacute

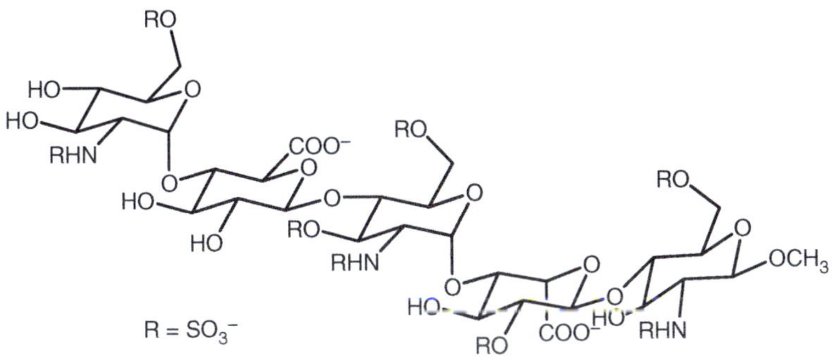

fondaparinux

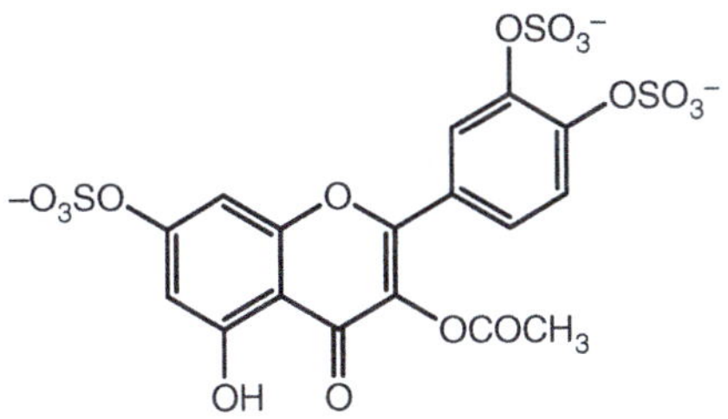

quercetin 3-acetyl-3',4',7-*O*-trisulfate

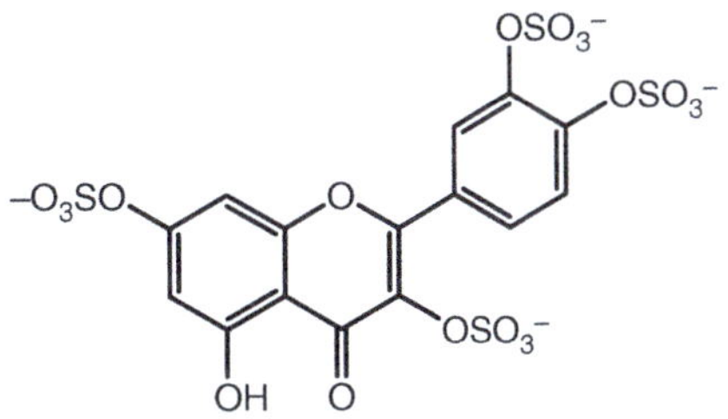

quercetin 3,3',4',7-*O*-tetrasulfate

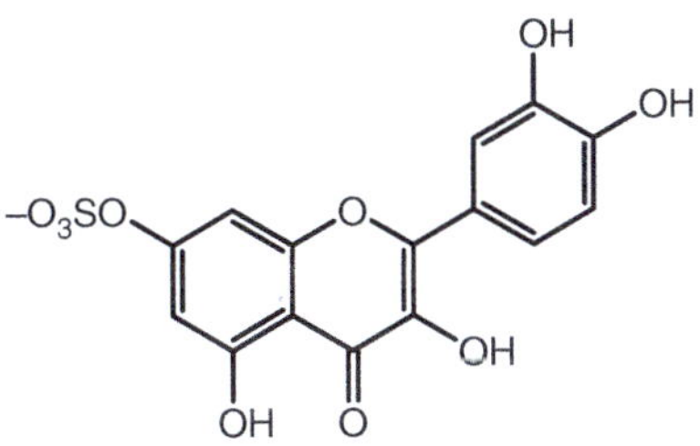

quercetin 7-*O*-sulfate

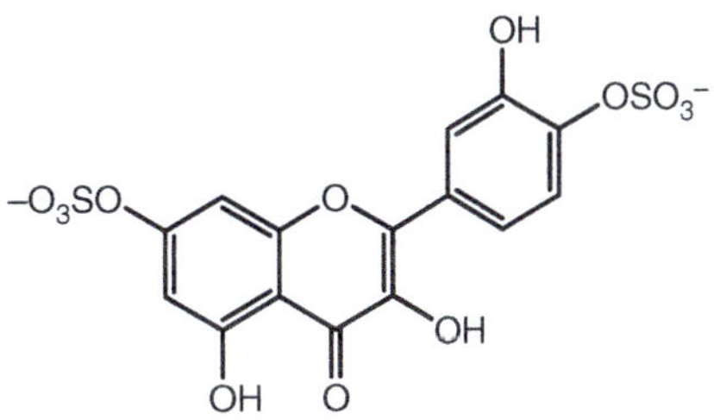

quercetin 4',7-*O*-disulfate

Figure 21.22 Sulfated polysaccharide fondaparinux and flavonoids with an oligopolysulfated moiety as new class of anticoagulant agents.

myelo-optic neuropathy (SMON) between 1957 and 1970 [109]. It has thus been restricted or discontinued in some countries. Recently, clioquinol was revitalized as a novel agent in the treatment of Alzheimer's disease [110] and Huntington's disease [111]. In 2005, it was reported to be a potential drug to induce breast and prostate cancer apoptosis in the presence of copper [112, 113]. Further investigations suggested that clioquinol alone can induce cell apoptosis of leukemia or myeloma cells and can delay tumor growth in leukemia cell xenografted mice, which is different from solid tumors such as breast and prostate cancer xenografts [107]. This kind of apoptosis induction is critical for the treatment because if clioquinol-induced cancer cell death is dependent on copper addition, it will add additional toxicity to patients due to the accumulative toxicity of copper. Clioquinol has been approved by Health Canada for clinical trial in the treatment of refractory and/or relapsed acute myelogenous leukemia (Figure 21.24).

21.7.19 From Antimalarial Drugs to Antifungals

Mefloquine is an FDA-approved antimalarial drug, however, when 400 of its analogs were tested, some exhibited antimicrobial activity [114]. Mefloquine itself showed only limited activity, but the potency of some derivatives like compound WR166391 against *C. albicans* and *C. neoformans* reached the levels of amphotericin, a drug used to treat serious fungal infections. The piperidine methanol side chain in mefloquine was identified as a crucial moiety for antimicrobial efficacy.

Work continued in 2020 by the Krysan group when they reported a mefloquine derivative with activity against pathogenic yeasts and molds. Out of 13 tested

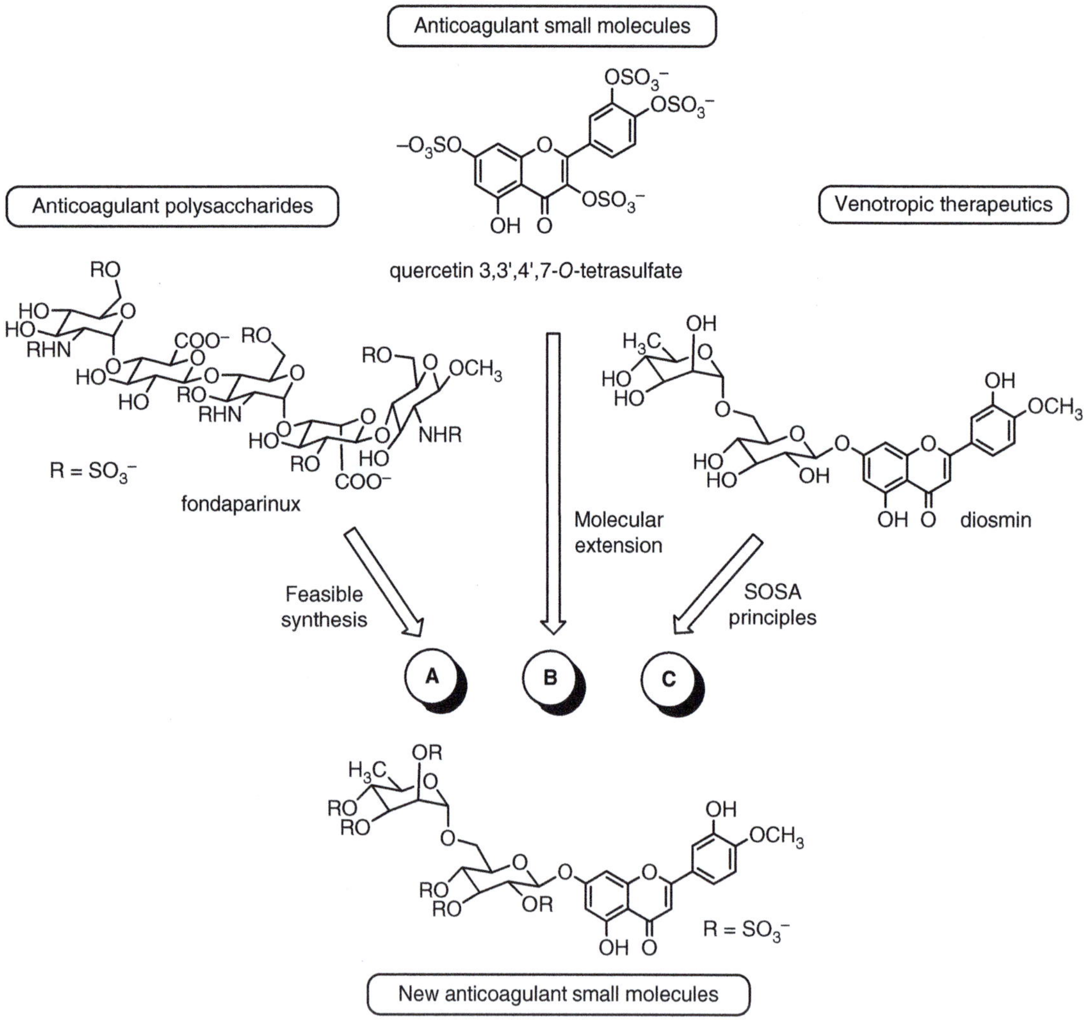

Figure 21.23 Schematic representation of the strategies used herein for the development of new anticoagulant small molecules based on three examples, namely fondaparinux, quercetin 3,3',4',7-O-tetrasulfate, and diosmin.

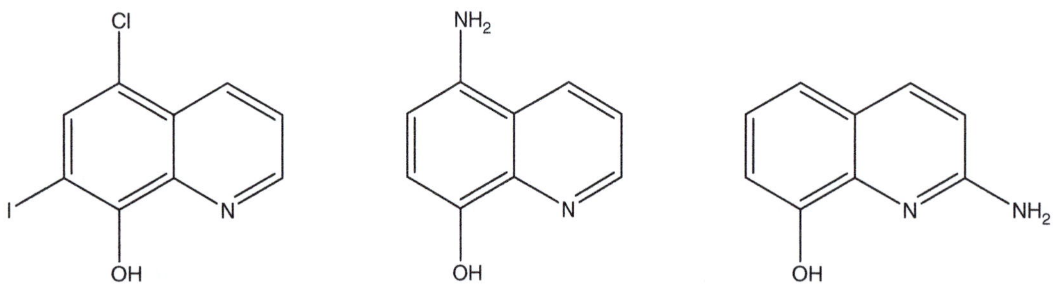

Clioquinol
IC_{50} = 31.5 μM

5-Amino-8-hydroxyquinoline
IC_{50} = 12.5 μM

2-Amino-8-hydroxyquinoline
IC_{50} = 125 μM

Figure 21.24 5-Amino-8-hydroxyquinoline was developed from clioquinol by replacing the chloride with amino at position 5 and removing the iodine.

analogs, they identified compound 4377 as the most potent (Figure 21.25), displaying broad-spectrum antifungal activity. Crucially, the *in vitro* toxicity profiles observed with human cell lines suggested that toxicity profiles of the mefloquine derivatives are very similar to those of the parent mefloquine, even though the derivatives exhibit up to a 64-fold increase in activity against fungal cells [115, 116].

21.7.20 From Immunosuppressives to Antifungals

Potential antifungal properties of the immunosuppressive drug FK506 have been known already for over 20 years [117]. Some approaches were examined to address this promising activity, but its complex natural structure rendered the development of derivatives laborious. However, exploiting biosynthetic engineering, Boem *et al.* were able to biosynthesize seven new FK506 analogs modified at both the FK506-binding protein 12- and calcineurin-binding regions. One of these analogs was identified to reduce more than 900-fold its immunosuppressive activity while retaining the antifungal activity [118].

In 2019, crystal structures of calcineurin subunits complexed with FK506 and the FK506-binding protein (FKBP12) were reported [119]. This revealed that a single phenylalanine residue as critical for binding and inhibition of fungal calcineurin led to the development of new derivative APX879 (Figure 21.26), with an acetohydrazide substitution of the C22 carbonyl of FK506. This derivative (APX879) exhibits 70-fold reduction in immunosuppressive activity while retaining broad-spectrum antifungal activity and efficacy in a murine model of invasive fungal infection.

21.7.21 From Antipsychotics to Antifungals

It is known that antipsychotic drugs can have antifungal side activity [116, 120]. Especially, their ability to cross BBB was shown to be desirable for treating certain fungal infections like cryptococcal meningitis.

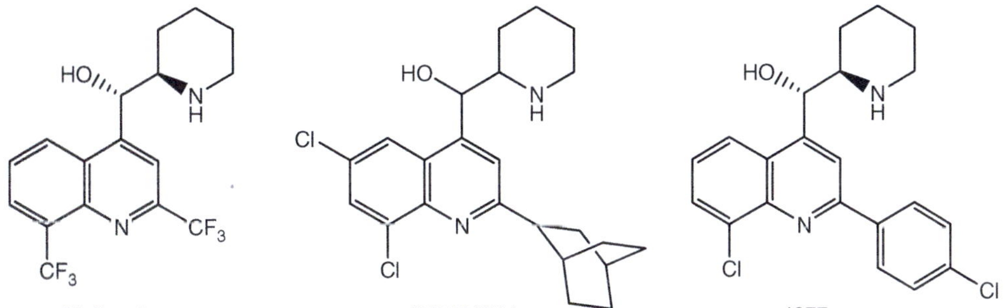

Mefloquine	WR166391	4377
C. albicans MIC >128 μg/mL	*C. albicans* MIC 1.5 μg/mL	*C. albicans* MIC 4 μg/mL
C. neoformans MIC 32 μg/mL	*C. neoformans* MIC 0.4 μg/mL	*C. neoformans* MIC 1 μg/mL

Figure 21.25 Antimalarial drug mefloquine and approach to its modification to yield derivative with improved antifungal properties.

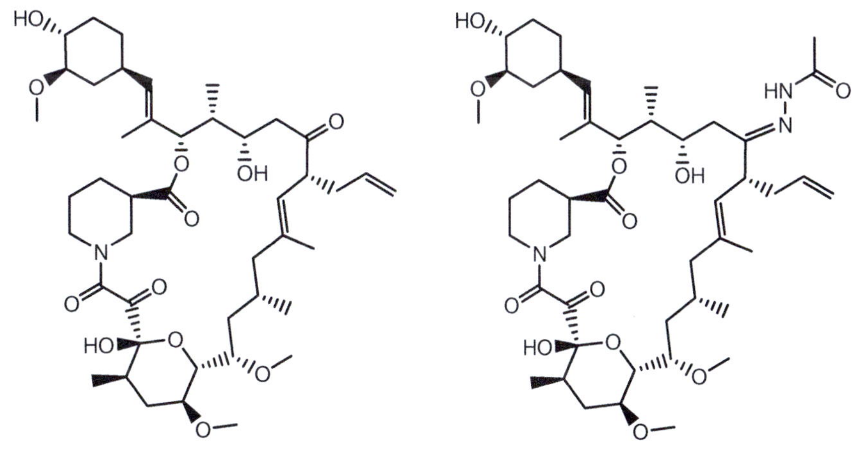

FK506
IL-2+ T cells IC_{50} 0.19 nM

APX879
IL-2 + T cells IC_{50} 13.48 nM

Figure 21.26 Modification of FK506 to semisynthetic derivative APX879 with reduced immunosuppressive activity against naive primary murine $CD4^+$ T cells.

Krysan *et al.* identified thioridazine and trifluoperazine as potential anticryptococcal agents, but modest potency would require the use of concentrations not considered safe in humans. Further SAR analysis of existing phenothiazines identified fluphenazine (FPZ) as a potential derivative, and limited screening of 11 analogs identified compound CWHM-974 to have improved potency while exhibiting reduced (compared to FPZ) binding affinity to dopamine D1-5 and six serotonin receptors (Figure 21.27) [121].

Another example is haloperidol used as an initial structure leading to the development of the derivate B10 (Figure 21.27) [122, 123].

Sertraline is another example of antipsychotic drug that was even tested and failed in clinical trials as antifungal agent. However, its structure optimization led to the development of the derivative D16 improving original potency significantly (Figure 21.27) [124].

21.7.22 Diclofenac as Inhibitor of the Transthyretin Amyloid Formation

Transthyretin (TTR) is a tetrameric protein. Under acidic conditions, TTR undergoes dissociation into an alternatively folded, monomeric intermediate that self-assembles into amyloid fibrils. These insoluble protein fibrils are implicated to be the culprit behind several human amyloid diseases like senile systemic amyloidosis (SSA), familial amyloid cardiomyopathy (FAC), and familial amyloid polyneuropathy (FAP) [116, 125, 126].

Through limited screening, the nonsteroidal anti-inflammatory drug diclofenac was identified as a potent inhibitor of TTR amyloid formation. Seeking compounds with both high inhibition capacities and selective binding to TTR, the optimization of diclofenac led to the development of 3,5-disubstituted positional isomers **11** and **12** (Figure 21.28) and substituted anthranilic acid.

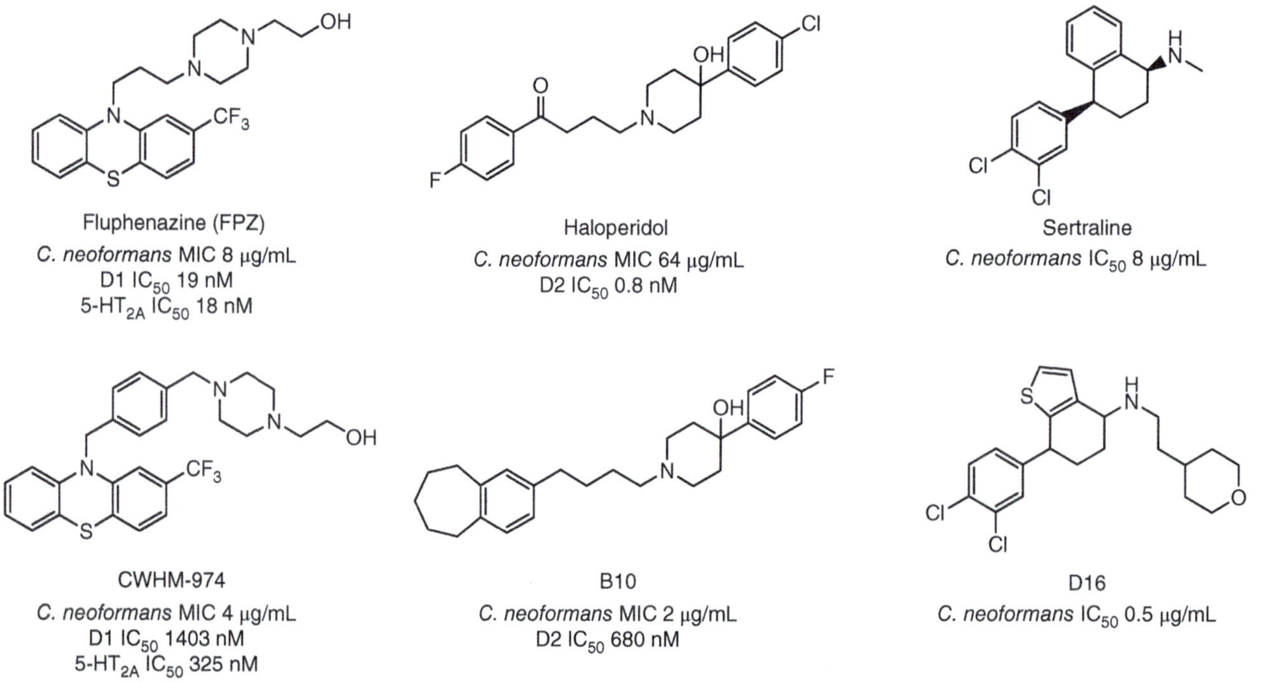

Figure 21.27 Recently developed derivatives of antipsychotic drugs with improved antifungal activity (CWHM-974, B10, and D16) and reduced affinity for dopamine D1 (CWHM-974), D2 (B10), and serotonin (CWHM-974) receptors.

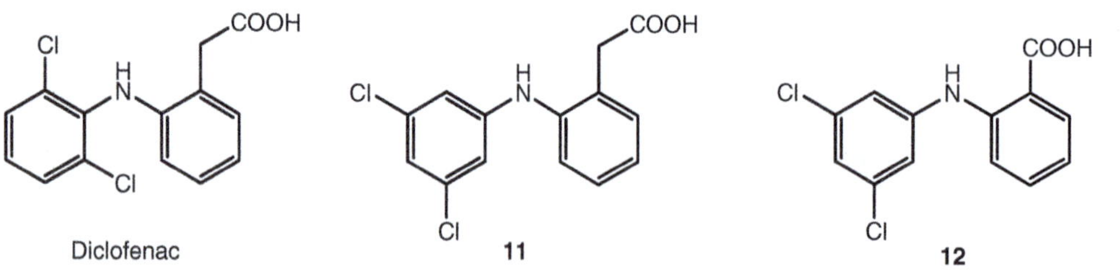

Figure 21.28 Structure of diclofenac and its analogs **11** and **12** that exhibited activity, an inhibitor of TTR amyloid formation.

Structure–activity relationships revealed that a carboxylic acid is required for activity, but changes in its position as well as the positions of other substituents are tolerated [127, 128].

21.7.23 Pranlukast

Pranlukast is a cysteinyl leukotriene receptor 1 (CysLT1R) antagonist. Its moderate side activity has been identified for the farnesoid X receptor (FXR) [129]. Through systematic structural modification of the drug, significant optimization of its partial FXR agonism to sub-nanomolar potency has been achieved [130]. The resulting FXR modulator **13** (Figure 21.29) exhibits no activity on CysLT1R, demonstrating high selectivity, metabolic stability, and low toxicity. These FXR modulators, derived from SOSA, present a novel chemotype for FXR ligands, showcasing a favorable *in vitro* profile and warranting further preclinical evaluation.

21.7.24 Loxapine [131]

Various disorders are accompanied by histamine-independent itching, which is often resistant to the currently available therapies. The pharmacological activation of Slack (Kcnt1 and KNa 1.1), a potassium channel highly expressed in itch-sensitive sensory neurons, has therapeutic potential for the treatment of itching. Based on the Slack-activating antipsychotic drug, loxapine, Balzulat *et al.* synthesized a series of new derivatives with improved pharmacodynamic and PK profiles that validate Slack as a pharmacological target *in vivo*. One of these new Slack activators, compound **14**, exhibits negligible dopamine D2 and D3 receptor binding, unlike loxapine (Figure 21.30). Notably, compound **14** displays potent on-target antipruritic activity in multiple mouse models of acute histamine-independent and chronic itch without motor side effects.

21.7.25 Computer-assisted SOSA

In recent years, computational approaches have become a norm in various facets of chemistry and SOSA is not an exception. Automated computational design of analogs can speed up hit-to-lead optimization, especially when working with a limited diversity of potential analogs.

21.7.26 Cinalukast

This approach was examined during the optimization of side activities of CysLT1R antagonist, as it was identified to have partial agonistic activity on peroxisome proliferator-activated receptor α (PPARa) [132, 133].

A virtual library of around 8000 compounds was automatically generated [134]. The classification of these compounds for PPARa agonism and CysLT1R antagonism was accomplished using automated affinity scoring and machine learning. The computationally favored analog for SOSA derivative **15** (Figure 21.31) was synthesized, with subsequent *in vitro* characterization disclosing a noteworthy activity shift, specifically, an enhanced PPARa activation and a concurrent reduction in CysLT1R antagonism. This study underscores the promising potential of automating the evaluation of SOSA.

Figure 21.29 Pranlukast and its derivative **13** developed as FXR modulator.

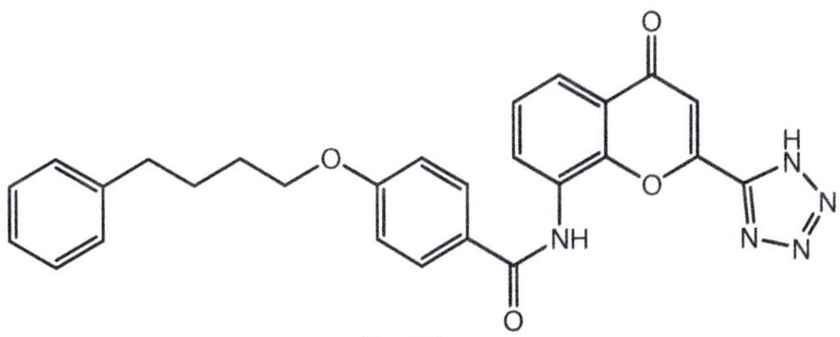

Pranlukast

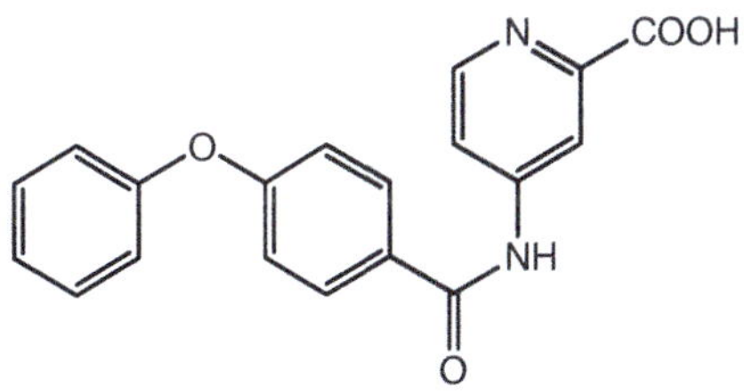

13

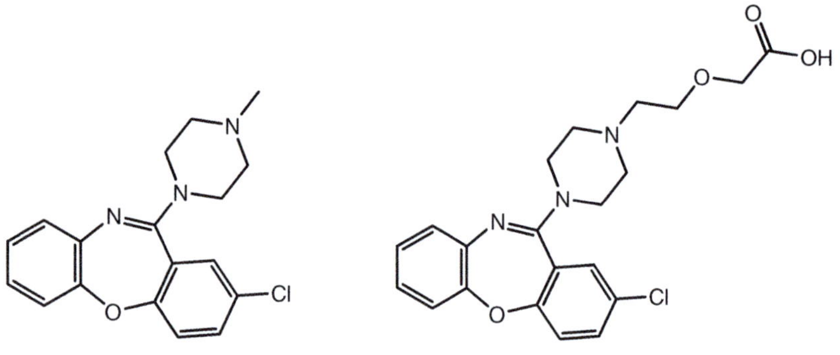

Loxapine
EC_{50} 20.7 μM
E_{MAX} 100%
Ki D_2 32 nM
Ki D_3 112 nM

14
EC_{50} 30.3 μM
E_{MAX} 96.2%
Ki D_2 1281 nM
Ki D_3 3520 nM

Figure 21.30 Structure of loxapine and derivative **14** and their Slack-activating properties. Potency (half-maximal effective concentration [EC_{50}]) and efficacy (E_{Max}; % relative to loxapine) values are from a FluxOR assay in HEK-Slack cells and dopamine D_2/D_3 receptor affinity.

21.7.27 Talinolol

Another example of computer-aided SOSA is demonstrated on talinol, which is an unselective antagonist of the beta-adrenergic receptors (βARs) but was also shown to reduce low-density lipoprotein cholesterol (LDL) cholesterol [135]. Therefore, the SOSA approach was used to deorphanize its pleiotropic cholesterol-lowering effects, which result from the inhibition of the enzyme soluble epoxide hydrolase (sEH) [136]. Talinol was co-crystallized with the C-terminal domain of sEH and analyzed via X-ray. The binding mode was used for the optimization of inhibitory potency. The target structure (morpholino-talinol; Figure 21.32) was synthesized and exhibited an increased *in vivo* activity in the rat model.

21.8 Conclusion

The SOSA approach has demonstrated over several decades to be an efficient lead generation strategy in drug discovery, particularly because – as it is based on screening drugs and clinical candidates for side activities – the method will deliver a priori drug-like hits. Before starting a costly HTS campaign, SOSA can represent an attractive alternative. Once the initial screening has provided a hit against a biological target different from the one responsible for the main activity, that molecule will be used as the starting point for a drug discovery program.

Using traditional medicinal chemistry as well as parallel synthesis and a growing number of *in silico* methods, the initial side activity is transformed into the main activity, and conversely, the initial main activity is strongly reduced or abolished.

This strategy has demonstrated a higher probability of yielding safe, bioavailable, original, and patentable analogs than conventional de novo screening techniques like HTS.

The SOSA approach can be compared with other approaches, such as hit or lead generation from known drug metabolites or the design of drug analogs, because it makes use of old drugs to generate new hits or leads. As a general rule, the activities of metabolites are often similar or close to the activity of the corresponding active molecule.

Contrary to this, the SOSA approach is based on the optimization of side activities that are totally different from the original activity of the molecule.

SOSA together with parallel pharmacophore-based virtual screening or other informatic-based molecular building tools is a straightforward and fairly inexpensive method for the generation of new lead compounds. In addition,

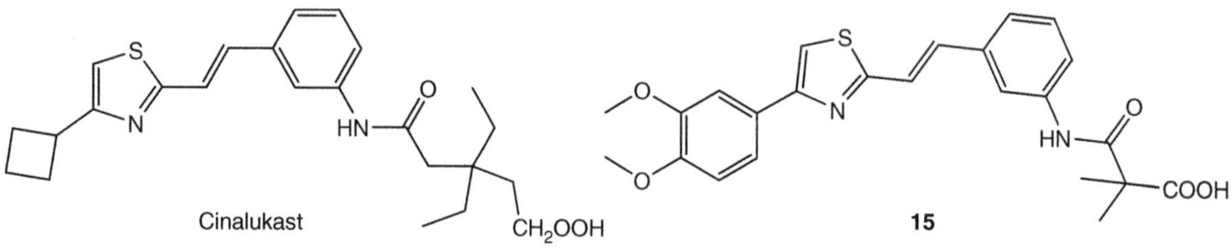

Figure 21.31 Cinalukast and its derivative **15** optimized for enhanced PPARa activation.

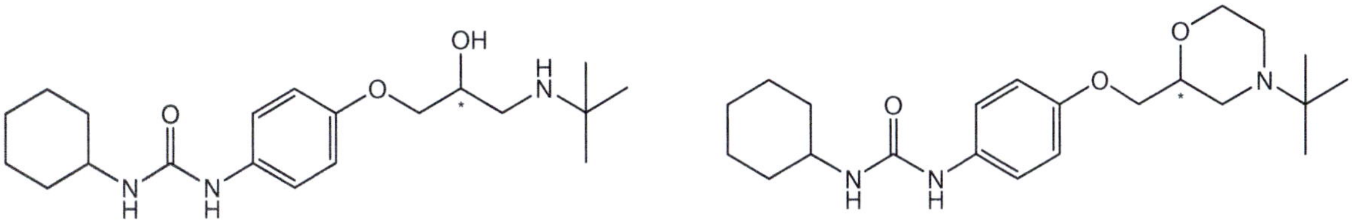

Figure 21.32 Talinolol and its structural analog with increased cholesterol reducing activity.

using methods of computer-aided molecular design, such as pharmacophore-based modeling, virtual screening, or affinity profiling, can provide valuable decision support for medicinal chemists when following a SOSA approach.

However, in recent years, the pure SOSA approach as originally defined is less common as it is being slowly merged with other drug discovery approaches as demonstrated in the latest examples provided in this chapter. The method as such is also slowly disappearing from the vocabulary of medicinal chemists in published papers, while its core advantages are still being utilized.

Credit

This work is based on the original chapter published in: Handler, N., Wolkerstorfer, A. and Buschmann, H./, J. Holenz (Ed.): Selective optimization of side activities: an alternative and promising strategy for lead generation. In Lead Generation. 221–257. 2016

References

1 DiMasi, J.A., Grabowski, H., and Hansen, R. (2016). Innovation in the pharmaceutical industry: new estimates of R&D costs. *J. Health Econ.* 47: 20–33.

2 Gan, F., Cao, B., Wu, D. et al. (2011). Exploring old drugs for the treatment of hematological malignancies. *Curr. Med. Chem.* 18: 1509–1514.

3 Hopkins, A., Lanfear, J., Lipinski, C., and Beetey, L. (2005). chemical tools for indications discovery. *Annu. Rep. Med. Chem.* 40: 339–348.

4 Ekins, S., Lane, T.R., Urbina, F., and Puhl, A.C. (2023). In silico ADME/tox comes of age: twenty years later. *Xenobiotica* 1–7.

5 Aronson, J.K. (2007). Old drugs-new uses. *Br. J. Clin. Pharmacol.* 64: 563–565.

6 Chong, C.R. and Sullivan, D.J. Jr., (2007). New uses for old drugs. *Nature* 448: 645–646.

7 Rothman, R.B., Baumann, M.H., Savage, J.E. et al. (2000). Evidence for possible involvement of 5-HT2B receptors in the cardiac valvulopathy associated with fenfluramine and other serotonergic medications. *Circulation* 102 (23): 2836–2841.

8 Edwards, B.S., Edwards, W.D., and Schaff, H.V. (1997). Valvular heart disease associated with fenfluramine-phentermine. *N. Engl. J. Med.* 337 (9): 581–588.

9 Hutcheson, J.D., Setola, V., Roth, B.L., and Merryman, W.D. (2011). Serotonin receptors and heart valve disease-It was meant 2B. *Pharmacol. Ther.* 132 (2): 146–157.

10 Mestres, J., Gregori-Puigjane, E., Valverde, S., and Sole, R.V. (2008). Data completeness-the Achilles heel of drug-target networks. *Nat. Biotechnol.* 26 (9): 983–984.

11 Hu, Y. and Bajorath, J. (2013). Compound promiscuity: what can we learn from current data? *Drug Discov. Today* 18 (13/14): 644–650.

12 Peters, J.-U. (2012). Opportunities and challenges in polypharmacological, drug discovery. *AAPS Annual Meeting and Exposition*, Chicago (14–18 October 2012).

13 Paolini, G.V., Shapland, R.H.B., van Hoorn, W.P. et al. (2006). Global mapping of pharmacological space. *Nat. Biotechnol.* 24 (7): 805–815.

14 Hu, Y. and Bajorath, J. (2013). How promiscuous are pharmaceutically relevant compounds? A data-driven assessment. *AAPS J.* 15 (1): 104–111.

15 Hu, Y. and Bajorath, J. (2013). What is the likelihood of an active compound to be promiscuous? Systematic assessment of compound promiscuity on the basis of PubChem confirmatory bioassay data. *AAPS J.* 15 (3): 808–815.

16 Gleeson, M.P., Hersey, A., Montanari, D., and Overington, J. (2011). Probing the links between

in vitro potency, ADMET and physicochemical parameters. *Nat. Rev. Drug Discov.* 10 (3): 197–208.

17 Kroeze, W.K. and Roth, B.L. (2012). Polypharmacological drugs: magic shotguns for psychiatric diseases. In: *Polypharmacology in Drug Discovery* (ed. J.-U. Peters), 135–148. Hoboken, NJ: Wiley.

18 Overington, J.P., Al-Lazikani, B., and Hopkins, A.L. (2006). How many drug targets are there? *Nat. Rev. Drug Discov.* 5 (12): 993–996.

19 Jalencas, X. and Mestres, J. (2013). On the origins of drug polypharmacology. *MedChemComm* 4 (1): 80–87.

20 Garcia-Serna, R. and Mestres, J. (2011). Chemical probes for biological systems. *Drug Discov. Today* 16 (3/4): 99–106.

21 Peters, J.-U. (2013). Polypharmacology – foe or friend? *J. Med. Chem.* 56: 8955–8971.

22 Lipinski, C. and Hopkins, A. (2004). Navigating chemical space for biology and medicine. *Nature* 432: 855–861.

23 Leeson, P.D. and Davis, A.M. (2004). Time-related differences in the physical property profiles of oral drugs. *J. Med. Chem.* 47: 6338–6348.

24 Vieth, M., Siegel, M.G., Higgs, R.E. et al. (2004). Characteristic physical properties and structural fragments of marketed oral drugs. *J. Med. Chem.* 47: 224–232.

25 Wenlock, M.C., Austin, R.P., Barton, P. et al. (2003). A comparison of physiochemical property profiles of development and marketed oral drugs. *J. Med. Chem.* 46: 1250–1256.

26 Vieth, M. and Sutherland, J.J. (2006). Dependence of molecular properties on proteomic family for marketed oral drugs. *J. Med. Chem.* 49: 3451–3453.

27 Proudfoot, J.R. (2005). The evolution of synthetic oral drug properties. *Bioorg. Med. Chem. Lett.* 15: 1087–1090.

28 Lipinski, C.A. (2000). Drug-like properties and the causes of poor solubility and poor permeability. *J. Pharmacol. Toxicol. Methods* 44: 235–249.

29 Kubinyi, H. (1998). Similarity and dissimilarity: a medicinal chemist's view. *Perspec.Drug Discov. Design* 9 (11): 225–252.

30 Martin, Y.C., Kofron, J.L., and Traphagen, L.M. (2002). *J. Med. Chem.* 45: 4350–4358.

31 Raju, T.N.K. (2000). The nobel chronicles. *Lancet* 255: 1022–1024.

32 Mager, P.P. (1987). Zur Entwicklung von bioaktiven Leitstrukturen. *Pharm. Unserer Zeit* 4: 97–113.

33 Bronson, J., Black, A., Murali Dhar, T.G. et al. (2013). To market, to market – 2012. *Annu. Rep. Med. Chem.* 48: 471–546.

34 Wermuth, C.G. (1998). Search for new lead compounds: the example of the chemical and pharmacological dissection of aminopyridazines. *J. Heterocyclic Chem.* 35: 1091–1100.

35 Wermuth, C.G. and Clarence-Smith, K. (2000). "Drug-like" leads: bigger is not always better. *Pharm. News* 7: 53–57.

36 Wermuth, C.G. (2001). The "SOSA" approach: an alternative to high-throughput screening. *Med. Chem. Res.* 10: 431–439.

37 Wermuth, C.G. (2004). Selective optimization of side activities: another way for drug discovery. *J. Med. Chem.* 47: 1303–1314.

38 Wermuth, C.G. (2006). Selective optimization of side activities: the SOSA approach. *Drug Discov. Today* 11: 160–164.

39 Lipinski, C.A., Lombardo, F., Dominy, B.W., and Feeney, P.J. (2001). Experimental and computational approaches to estimate solubility and permeability in drug discovery and development settings. *Adv. Drug Deliv. Rev.* 46: 3–26.

40 Veber, D.F., Johnson, S.R., Cheng, H.Y. et al. (2002). Molecular properties that influence the oral bioavailability of drug candidates. *J. Med. Chem.* 45: 2615–2623.

41 Bergström, C.A., Strafford, M., Lazorova, L. et al. (2003). Absorption classification of oral drugs based on molecular surface properties. *J. Med. Chem.* 46: 558–570.

42 Siegel, M.G. and Vieth, M. (2007). Drugs in other drugs: a new look at drugs as fragments. *Drug Discov. Today* 71–79.

43 Erlanson, D., Fesik, S., Hubbard, R. et al. (2016). Twenty years on: the impact of fragments on drug discovery. *Nat. Rev. Drug Discov.* 15: 605–619.

44 Fischer, J. and Ganellin, C.R. (ed.) (2006). *Analogue-based Drug Discovery*. Weinheim: Wiley-VCH Verlag ISBN 978-3-527-31257-3.

45 Wermuth, C.G. (2006). Analogues as a means of discovering new drugs. In: *Analogue-based Drug Discovery* (ed. J. Fischer and C.R. Ganellin), 3–23. Weinheim: Wiley-VCH Verlag ISBN 978-3-527-31257-3.

46 Wermuth, C.G., Villoutreix, B., Grisoni, S. et al. (2015). Chapter 4 – Strategies in the search for new lead compounds or original working hypotheses. In: *The Practice of Medicinal Chemistry (Fourth Edition)* (ed. C.G. Wermuth, D. Aldous, P. Raboisson, and D. Rognan), 73–99. San Diego, CA: Academic Press.

47 Ciapetti, P. and Giethlen, B. (2008). Chapter 8 – Molecular variations based on isosteric replacements. In: *The Practice of Medicinal Chemistry (Fourth Edition)* (ed. C.G. Wermuth, D. Aldous, P. Raboisson, and D. Rognan), 181–241. San Diego, CA: Academic Press.

48 Worms, P., Kan, J.P., Wermuth, C.G. et al. (1987). SR 95191, a selective inhibitor of type A monoamine oxidase with dopaminergic properties, I. Psychopharmacological profile in rodents. *J. Pharmacol. Exp. Ther.* 240 (1): 241–250.

49 Stellmach, J.A. (2009). The influences of the structure and activity of biologically active compounds on the assessment of inventive step. *World Patent Inf.* 31: 226–234.

50 Kaback, A.S.M. (1985). Access all the information. *ChemTech* 15 (3): 146–151.

51 Berks, H. (2001). Current state of the art of Markush topological search systems. *World Pat. Inform.* 23: 5–13.

52 European Patent Office (2006). *Case Law of the Boards of Appeal of the European Patent Office*, European Patent Office Directorate-General 3 (Appeals) Directorate 3.0 Legal Research and Administration, 5the ed., Par. 4.1, 88–95. Munich, Germany: European Patent Office https://link.epo.org/web/case_law_of_the_boards_of_appeal_2006_en.pdf.

53 European Patent Office (2006). *Case Law of the Boards of Appeal of the European Patent Office*, European Patent Office Directorate-General 3 (Appeals) Directorate 3.0 Legal Research and Administration, 5the. Munich, Germany: European Patent Office Par. 2.5, p. 73/74, Par. 8.9, p. 154.

54 Ouagued, M., Martin-Chouly, C.A.E., Brinchault, G. et al. (2005). The novel phosphodiesterase 4 inhibitor, CI-1044, inhibits LPS-induced TNF-α production in whole blood from COPD patients. *Pulm. Pharmacol. Ther.* 18 (1): 49–54.

55 Kubinyi, H. (2004). Chemogenomics in drug discovery – a medicinal chemistry perspective. In: *Methods and Principles in Medicinal Chemistry 22* (ed. H. Kubinyi and G. Müller), 43–67. Weinheim: Wiley-VCH.

56 Streeck, H. and Rockstroh, J.K. (2007). Review of tipranavir in the treatment of drug-resistant HIV. *Ther. Clin. Risk Manag.* 3: 641–651.

57 D'Amato, R.J., Loughnan, M.S., Flynn, E., and Folkman, J. (1994). Thalidomide is an inhibitor of angiogenesis. *Proc. Natl. Acad. Sci. USA* 91: 4082–4085.

58 Singhal, S., Mehta, J., Desikan, R. et al. (1999). Antitumor activity of thalidomide in refractory multiple myeloma. *N. Engl. J. Med.* 341: 1565–1571.

59 Therapontos, C., Erskine, L., Gardner, E.R. et al. (2009). Thalidomide induces limb defects by preventing angiogenic outgrowth during early limb formation. *Proc. Natl. Acad. Sci. USA* 106 (21): 8573–8578.

60 Parman, T., Wiley, M.J., and Wells, P.G. (1999). Free radical-mediated oxidative DNA damage in the mechanism of thalidomide teratogenicity. *Nat. Med.* 5: 582–585.

61 Stephens, T.D., Bunde, C.J., and Fillmore, B.J. (2000). Mechanism of action in thalidomide teratogenesis. *Biochem. Pharmacol.* 59: 1489–1499.

62 Chanan-Khan, A.A. and Cheson, B.D. (2008). Lenalidomide for the treatment of B-cell malignancies. *J. Clin. Oncol.* 26: 1544–1552.

63 Gore, S.D. (2008). New agents for the treatment of AML recent study findings. *Clin. Adv. Hematol. Oncol.* 6: 6–8.

64 Fehniger, T.A., Byrd, J.C., Marcucci, G. et al. (2009). Single-agent lenalidomide induces complete remission of acute myeloid leukemia in patients with isolated trisomy 13. *Blood* 113: 1002–1005.

65 Teo, S.K., Stirling, D.I., and Zeldis, J.B. (2005). Thalidomide as a novel therapeutic agent: new uses for an old product. *Drug. Discov. Today* 10: 107–114.

66 Li, W.W., Hutnik, M., and Gehr, G. (2008). Antiangiogenesis in haematological malignancies. *Br. J. Haematol.* 143: 622–631.

67 Stewart, A.K. (2009). Novel therapies for relapsed myeloma. *Hematology Am. Soc. Hematol. Educ. Program* 2009: 578–586.

68 Holenz, J., Mannhold, R., Kubinyi, H., and Folkers, G. (2016). *Lead Generation: Methods and Strategies.* Weinheim, Germany: Wiley-VCH Verlag GmbH & Co. KGaA.

69 Cookson, J. (2019). Histamine in psychiatry: promethazine as a sedative anticholinergic. *BJPsych Adv.* 25: 265–268.

70 Campos Rosa, J. (2024). *Pharmaceutical Chemistry.* Boston, MA: De Gruyter.

71 Barrett, A.M. (1985). Cardiac beta-adrenoceptor blockade: the quest for selectivity. *J. Pharmacol.* 16 (Suppl 2): 95–108.

72 Black, J.W., Duncan, W.A.M., and Shanks, R.G. (1997). Comparison of some properties of pronethalol and propranolol. *Br. J. Pharmacol.* 120: 285–299.

73 de Boer, R.A., Voors, A.A., and Veldhuisen, D.J. (2007). Nebivolol: third-generation beta-blockade. *Expert. Opin. Pharmacother.* 8: 1539–1550.

74 Wermuth, C.G. (1993). Aminopyridazines-an alternative route to potent muscarinic agonists with no cholinergic syndrome. *Il Farmaco* 48: 253–274.

75 Barreiro, E.J., Kümmerle, A.E., and Fraga, C.A.M. (2011). The methylation effect in medicinal chemistry. *Chem. Rev.* 111: 5215–5246.

76 Stein, P.D., Hunt, J.T., Floyd, D.M. et al. (1994). The discovery of sulfonamide endothelin antagonists and the development of the orally active ETA antagonist 5-(Dimethylamino)-N-(3,4-dimethyl-5-isoxazolyl)-l-naphthalenesulfonamide. *J. Med. Chem.* 37: 329–331.

77 Rudy, B. and Senkowski, Z.B. (1973). Sulfisoxazole. *Anal. Profiles Drug Subst.* 2: 487–506.

78 Miyauchi, T., Sato, R., Sakai, S. et al. (2000). Endothelin-1 and right-sided heart failure in rats: effects of an endothelin receptor antagonist on the failing right ventricle. *J. Cardiovasc. Pharmacol.* 36: S327–S330.

79 Hulpke-Wette, M. and Buchhorn, R. (2002). BMS-193884 and BMS-207940 Bristol-Myers Squibb. *Curr. Opin. Investig. Drugs* 3 (7): 1057–1061.

80 Wermuth, C.G., Schlewer, G., Bourguignon, J.-J. et al. (1989). 3-Aminopyridazine derivatives with atypical antidepressant, serotonergic and dopaminergic activities. *J. Med. Chem.* 32: 528–537.

81 Wermuth, C.G., Bourguignon, J.J., Hoffmann, R. et al. (1992). SR 46559 A and related aminopyridazines are potent muscarinic agonists with no cholinergic syndrome. *Bioorg. Med. Chem. Lett.* 2: 833–838.

82 Kan, J.-P., Steinberg, R., Oury-Donat, F. et al. (1993). SR 46559A: a novel and potent muscarinic compound with no cholinergic syndrome. *Psychopharmacology* 112: 219–227.

83 Beghyn, T.B., Charton, J., Leroux, F. et al. (2011). Drug to genome to drug: discovery of new antiplasmodial compounds. *J. Med. Chem.* 54: 3222–3240.

84 Kharkar, P.S. (2014). Drugs acting on central nervous system (CNS) targets as leads for non-CNS targets [version 2; peer review: 2 approved, 1 approved with reservations]. *F1000Research* 3: 40.

85 Pajouhesh, H. and Lenz, G.R. (2005). Medicinal chemical properties of successful central nervous system drugs. *NeuroRx* 2 (4): 541–553.

86 Frederickson, M., Callaghan, O., Chessari, G. et al. (2008). Fragment-based discovery of mexiletine derivatives as orally bioavailable inhibitors of urokinase-type plasminogen activator. *J. Med. Chem.* 51: 183–186.

87 Duffy, M.J. (2004). The urokinase plasminogen activator system: role in malignancy. *Curr. Pharm. Des.* 10: 39–49.

88 Ilies, M.A., Scozzafava, A., and Supuran, C.T. (2002). Therapeutic applications of serine protease inhibitors. *Expert Opin. Ther. Pat.* 12: 1181–1214.

89 Ge, Y. and Elghetany, T. (2003). Urokinase plasminogen activator receptor (CD87): something old, something new. *Lab. Hematol.* 9: 67–71.

90 Wong, A.P., Cortez, S.L., and Baricos, W.H. (1992). Role of plasmin and gelatinase in extracellular matrix degradation by cultured rat mesangial cells. *Am. J. Physiol. Renal Physiol.* 263: 1112–1118.

91 Jean-Claude, J., Newman, K.M., Li, H. et al. (1994). Possible key role for plasmin in the pathogenesis of abdominal aortic aneurysms. *Surgery* 116: 472–478.

92 Gveric, D., Hanemaaijer, R., Newcombe, J. et al. (2001). Plasminogen activators in multiple sclerosis legions. *Brain* 124: 1978–1988.

93 Almholt, K., Lund, L.R., Rygaard, J. et al. (2005). Reduced metastasis of transgenic mammary cancer in urokinase-deficient mice. *Int. J. Cancer* 113: 522–532.

94 Schweinitz, A., Steinmetzer, T., Banke, I.J. et al. (2004). Design of novel and selective inhibitors of urokinase-type plasminogen activator with improved pharmacokinetic properties for use as antimetastatic agents. *J. Biol. Chem.* 279: 33613–33622.

95 Mackman, R.L., Katz, B.A., Breitenbucher, J.G. et al. (2001). Exploiting subsite S1 of trypsin-like serine proteases for selectivity: potent and selective inhibitors of urokinase-type plasminogen activator. *J. Med. Chem.* 44: 3856–3871.

96 Monk, J.P. and Brogden, R.N. (1990). Mexiletine. A review of its pharmacodynamic and pharmacokinetic properties, and therapeutic use in the treatment of arrhythmias. *Drugs* 40: 374–411.

97 Bruncko, M., McClellan, W.J., Wendt, M.D. et al. (2005). Naphthamidine urokinase plasminogen activator inhibitors with improved pharmacokinetic properties. *Bioorg. Med. Chem. Lett.* 15: 93–98.

98 Matthews, H., Ranson, M., Tyndall, J.D.A., and Kelso, M.J. (2011). Synthesis and preliminary evaluation of amiloride analogs as inhibitors of the urokinase-type plasminogen activator (uPA). *Bioorg. Med. Chem. Lett.* 21 (22): 6760–6766.

99 Matthews, H., Ranson, M., and Kelso, M.J. (2011). Anti-tumour/metastasis effects of the potassium-sparing diuretic amiloride: an orally active anti-cancer drug waiting for its call-of-duty? *Int. J. Cancer* 129: 2051–2061.

100 Duffy, M. (1992). Inhibiting tissue invasion and metastasis as targets for cancer therapy. *J. Biotherapy* 4: 45–52.

101 Dass, K., Ahmad, A., Azmi, A.S. et al. (2008). Evolving role of uPA/uPAR system in human cancers. *Cancer Treat. Rev.* 34: 122–136.

102 Romer, J., Nielsen, B.S., and Ploug, M. (2004). The urokinase receptor as a potential target in cancer therapy. *Curr. Pharm. Des.* 10: 2359–2376.

103 Nozaki, S., Endo, Y., Nakahara, H. et al. (2006). Targeting urokinase-type plasminogen activator and its receptor for cancer therapy. *Anti-Cancer Drugs* 17: 1109–1117.

104 Kleyman, T.R. and Cragoe, E.J.J. Jr., (1988). Amiloride and its analogs as tools in the study of ion transport. *Membr. Biol.* 105: 1–21.

105 Correia-da-Silva, M., Sousa, E., Duarte, B. et al. (2011). Flavonoids with an oligopolysulfated moiety: a new class of anticoagulant agents. *J. Med. Chem.* 54: 95–106.

106 Lyseng-Williamson, K.A. and Perry, C.M. (2003). Micronised purified flavonoid fraction: a review of its use in chronic venous insufficiency, venous ulcers and haemorrhoids. *Drugs* 63: 71–100.

107 Mao, X., Li, X., Sprangers, R. et al. (2009). Clioquinol inhibits the proteasome and displays preclinical activity in leukemia and myeloma. *Leukemia* 23: 585–590.

108 Mao, X., Liang, S.B., Hurren, R. et al. (2008). Cyproheptadine displays preclinical activity in myeloma and leukemia. *Blood* 112: 760–769.

109 Mao, X. and Schimmer, A.D. (2008). The toxicology of clioquinol. *Toxicol. Lett.* 182: 1–6.

110 Adlard, P.A., Cherny, R.A., Finkelstein, D.I. et al. (2008). Rapid restoration of cognition in Alzheimer's transgenic mice with 8-hydroxy quinoline analogs is associated with decreased interstitial Abeta. *Neuron* 59: 43–55.

111 Nguyen, T., Hamby, A., and Massa, S.M. (2005). Clioquinol down-regulates mutant huntingtin expression in vitro and mitigates pathology in a Huntington's disease mouse model. *Proc. Natl. Acad. Sci. USA* 102: 11840–11845.

112 Daniel, K.G., Chen, D., Orlu, S. et al. (2005). Clioquinol and pyrrolidine dithiocarbamate complex with copper to form proteasome inhibitors and apoptosis inducers in human breast cancer cells. *Breast Cancer Res.* 7: R897–R908.

113 Chen, D., Cui, Q.C., Yang, H. et al. (2007). Clioquinol, a therapeutic agent for Alzheimer's disease, has proteasome-inhibitory, androgen receptor-suppressing, apoptosis- inducing, and antitumor activities in human prostate cancer cells and xenografts. *Cancer Res.* 67: 1636–1644.

114 Kunin, C.M. and Ellis, W.Y. (2000). Antimicrobial activities of mefloquine and a series of related compounds. *Antimicrob. Agents Chemother.* 44 (4): 848–852.

115 Montoya, M.C., Beattie, S., Alden, K.M., and Krysan, D.J. (2020). Derivatives of the antimalarial drug mefloquine are broad-spectrum antifungal molecules with activity against drug-resistant clinical isolates. *Antimicrob. Agents Chemother.* 64 (3): e02331–e02419.

116 Donlin, M.J. and Meyers, M.J. (2022). Repurposing and optimization of drugs for discovery of novel antifungals. *Drug Discov. Today* 27 (7): 2008–2014.

117 Odom, A., Del Poeta, M., Perfect, J., and Heitman, J. (1997). The immunosuppressant FK506 and its nonimmunosuppressive analog L-685,818 are toxic to Cryptococcus neoformans by inhibition of a common target protein. *Antimicrob. Agents Chemother.* 41 (1): 156–161.

118 Beom, J.Y., Jung, J.A., Lee, K.T. et al. (2009). Biosynthesis of Nonimmunosuppressive FK506 Analogues with Antifungal Activity. *J. Nat. Prod.* 82 (8): 2078–2086.

119 Juvvadi, P.R., Fox, D. 3rd., Bobay, B.G. et al. (2019). Harnessing calcineurin-FK506-FKBP12 crystal structures from invasive fungal pathogens to develop antifungal agents. *Nat. Commun.* 10 (1): 4275.

120 Spitzer, M., Griffiths, E., Blakely, K.M. et al. (2011). Cross-species discovery of syncretic drug combinations that potentiate the antifungal fluconazole. *Mol. Syst. Biol.* 7 (1): 499.

121 Montoya, M.C., DiDone, L., Heier, R.F. et al. (2018). Antifungal phenothiazines: optimization, characterization of mechanism, and modulation of neuroreceptor activity. *ACS Infect Dis.* 4 (4): 499–507.

122 Ji, C., Liu, N., Tu, J. et al. (2020). Drug repurposing of haloperidol: discovery of new benzocyclane derivatives as potent antifungal agents against cryptococcosis and candidiasis. *ACS Infect Dis.* 6 (5): 768–786.

123 Yang, W., Tu, J., Ji, C. et al. (2021). Discovery of piperidol derivatives for combinational treatment of azole-resistant candidiasis. *ACS Infect Dis.* 7 (3): 650–660.

124 Li, W., Yun, Z., Ji, C. et al. (2022). Discovery of novel sertraline derivatives as potent anti-cryptococcus agents. *J. Med. Chem.* 65 (9): 6541–6554.

125 Buxbaum, J.N. and Tagoe, C.E. (2000). The genetics of the amyloidosis. *Annu. Rev. Med.* 51: 543–569.

126 Plante-Bordeneuve, V. and Said, G. (2000). Transthyretin related familial amyloid polyneuropathy. *Curr. Opin. Neurol.* 13: 569–573.

127 Baures, P.W., Peterson, S.A., and Kelly, J.W. (1998). Discovering transthyretin amyloid fibril inhibitors by limited screening. *Bioorg. Med. Chem.* 6 (8): 1389–1401.

128 Oza, V.B., Smith, C., Raman, P. et al. (2002). Synthesis, structure, and activity of diclofenac analogues as transthyretin amyloid fibril formation inhibitors. *J. Med. Chem.* 45 (2): 321–332.

129 Proschak, E., Heitel, P., Kalinowsky, L., and Merk, D. (2017). Opportunities and challenges for fatty acid mimetics in drug discovery. *J. Med. Chem.* 60 (13): 5235–5266.

130 Schierle, S., Schmidt, J., Kaiser, A., and Merk, D. (2018). Selective optimization of pranlukast to farnesoid x receptor modulators. *ChemMedChem* 13 (23): 2530–2545.

131 Balzulat, A., Zhu, W.F., Flauaus, C. et al. (2024). Discovery of a small molecule activator of slack (Kcnt1)

potassium channels that significantly reduces scratching in mouse models of histamine-independent and chronic itch. *Adv. Sci.* 11: 2307237. https://doi.org/10.1002/advs.202307237.

132 Lamers, C., Schubert-Zsilavecz, M., and Merk, D. (2012). Therapeutic modulators of peroxisome proliferator-activated receptors (PPAR): a patent review (2008–present). *Expert Opin. Ther. Pat.* 22 (7): 803–841.

133 Michalik, L., Auwerx, J., Berger, J.P. et al. (2006). International union of pharmacology. LXI. Peroxisome proliferator-activated receptors. *Pharmacol. Rev.* 58 (4): 726–741.

134 Pollinger, J., Schierle, S., Neumann, S. et al. (2019). Computer-assisted selective optimization of side-activities-from cinalukast to a PPARα modulator. *ChemMedChem* 14 (14): 1343–1348.

135 Sourgens, H., Schmidt, J., and Derendorf, H. (2003). Comparison of talinolol and atenolol effects on blood pressure in relation to lipid and glucose metabolic parameters. Results from the TALIP study. *Int. J. Clin. Pharmacol. Ther.* 41 (1): 2–29.

136 Hiesinger, K., Kramer, J.S., Achenbach, J. et al. (2019). Computer-aided selective optimization of side activities of talinolol. *ACS Med. Chem. Lett.* 10 (6): 899–903.

Part D

Polypharmacology, Classic Case Studies and Recent Research

22

Dual Inhibitors of CDK4/6 for Treating Cancer

Peter L. Toogood

22.1 Introduction

The cell division cycle (cell cycle; Figure 22.1) requires a highly coordinated series of steps that enables eukaryotic cells to divide with high fidelity such that each daughter cell receives one complete and accurate copy of the genomic deoxyribonucleic acid (DNA) [1–3]. Upon receipt of a growth signal, cells exit a quiescent state (G_0) and enter the G1 phase of the cell cycle in which they prepare to replicate their DNA. Once all of the requisite criteria have been met, cells pass a restriction point and enter the S phase where the DNA is duplicated. A second gap phase (G2) follows, during which cells continue to grow and prepare to divide. Then, cell division occurs by mitosis (M phase). The regulation of this intricate ballet is conducted in part through the action of a family of holoenzymes that comprise a catalytic cyclin-dependent kinase (CDK) protein partnered with an activating cyclin [4]. While the levels of the CDKs are largely constant, the levels of the cyclins rise and fall throughout the cell cycle, resulting in the synchronized activation of successive CDKs as the cycle proceeds. CDKs 1–4 and CDK6, together with their partner cyclins, have been identified as cell cycle regulators, along with CDKs 14–18 [4]. CDK4 and CDK6 are 71% homologous (>90% homologous in the adenosine triphosphate [ATP]-binding site), and in partnership with cyclins D1, D2, and D3, they perform almost identical functions [4–6]. As a consequence, they are typically referred to simply as CDK4/6. This closely related pair of CDKs controls the G1 checkpoint by phosphorylating the retinoblastoma protein (pRb), a tumor suppressor that in its unphosphorylated state sequesters E2F transcription factors, preventing the expression of S-phase genes. The activity of CDK4/6 in turn is controlled by the endogenous **IN**hibitors of CD**K4** (INK4) proteins (p16, p18, and p19) and by CDK interacting protein/kinase inhibitory proteins (CIP/KIP) including p21 and p27, which can act both as scaffolding proteins and as endogenous CDK inhibitors (CDKi) [7].

Early cell cycle studies identified the complex of CDK4/6 and D-type cyclins as a potential target for drug discovery [8–10]. The pRb pathway is dysregulated in more than 80% of human cancers [9] with the most common aberrations including loss of the CDK4/6 substrate protein Rb, loss of p16, or overexpression of cyclin D1. Interestingly, these alterations tend to occur individually rather than in combination, suggesting that any one of these changes on its own is sufficient to cause uncontrolled cell proliferation. Inhibition of CDK4/6 activity was predicted to restore the G1 checkpoint.

The first small-molecule CDKi to be tested clinically included compounds like flavopiridol (alvocidib) that did not discriminate between different members of the CDK family [11, 12]. Broad-based CDK inhibition was associated with unacceptable levels of toxicity in patients, and these compounds failed to advance. A second generation of CDKi exhibited higher selectivity for CDKs versus other serine/threonine kinases. However, these compounds also lacked sufficient discrimination between key cell cycle CDKs such as CDK4, CDK2, and CDK1. Success was ultimately achieved in the clinic with the discovery of compounds that display high selectivity for inhibition of CDK4 and CDK6 versus other CDKs [13].

22.2 Selectivity Profile of Approved CDK4/6 Inhibitors

The U.S. approval of palbociclib [14, 15], ribociclib [16], abemaciclib [17], and now trilaciclib [18] (Figure 22.2) is generally attributed to their superior selectivity profile compared to earlier generation CDK inhibitors (CDKi). These "third-generation" CDKi were developed as selective CDK4/6 inhibitors (CDK4/6i), and despite ongoing studies to examine the extent of this selectivity, this initial assessment has largely stood up well to further scrutiny [19]. Interestingly, the basis for this selectivity

Polypharmacology: Strategies for Multi-Target Drug Discovery, First Edition. Edited by Jens-Uwe Peters.

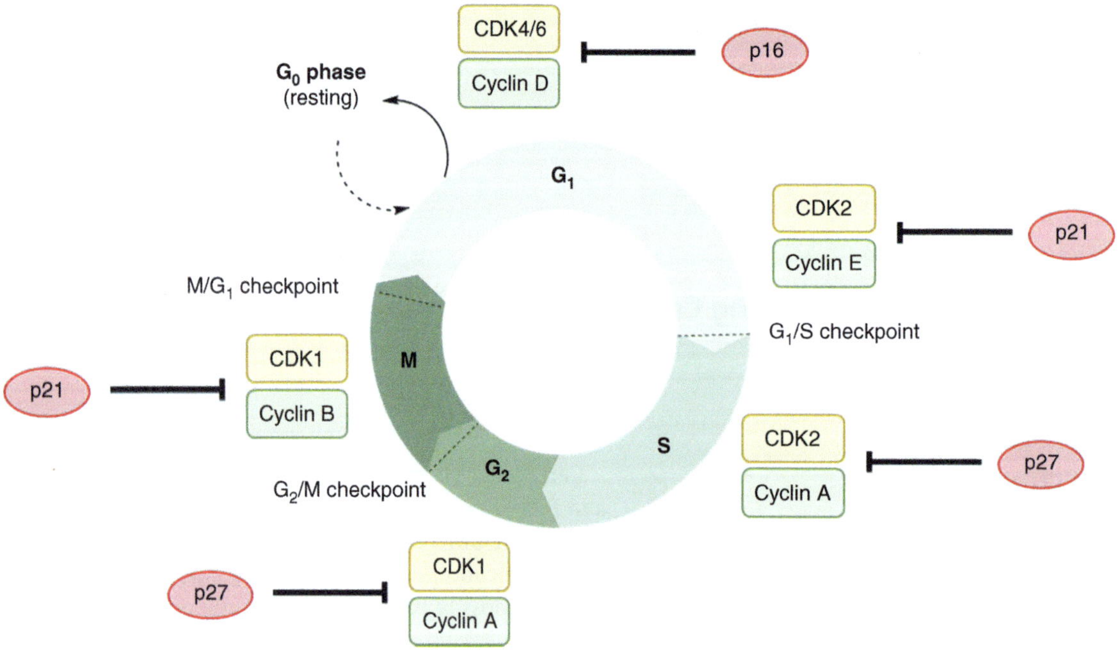

Figure 22.1 Cyclin-dependent kinases and the cell cycle.

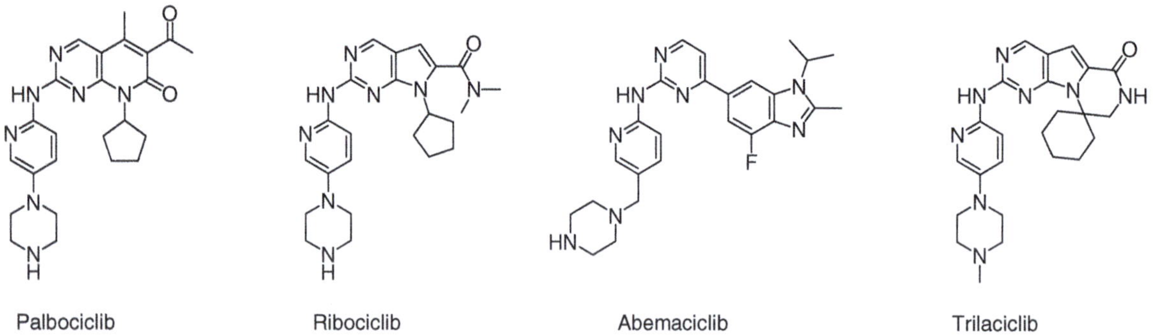

Figure 22.2 The four CDK4/6 inhibitors currently approved by the US Food and Drug Administration (FDA).

was initially unclear. All of the approved CDK4/6i incorporate an aminopyridine side chain at the C2 position of a pyrimidine ring. The pyridine nitrogen is an important contributor to the selectivity profile of these compounds [15]. However, the first available palbociclib/CDK6 crystal structure failed to explain how this critical nitrogen contributed to the compound's selectivity for CDK4/6 versus other CDKs, especially since it did not appear positioned to form a hydrogen bond with the protein [20]. A decade later, a crystal structure of abemaciclib bound to CDK4 identified a water molecule bridging the pyridine nitrogen and a histidine residue that is unique to CDK4/6 versus other CDKs [21]. This water-mediated hydrogen bond incorporates the pyridine ring into a hydrogen bonding network that extends and complements the hinge-binding hydrogen bond donor–accepter motif created by the aminopyrimidine.

While the initial selectivity claims, especially for palbociclib, were based on results obtained by counter-screening against a limited number of other kinases in biochemical assays, more recent studies have expanded upon these reports as larger panels of kinases have become available for screening. In addition, several groups have employed multiple complementary techniques in profiling CDK4/6i, making them one of the best-understood classes of kinase inhibitors on the market. Hendrychova et al. [19] provide an excellent comparison of the selectivity profiles for palbociclib, ribociclib, and abemaciclib from four different studies that employed different approaches to measuring inhibition or affinity of the test compounds with 274 kinases and up to 587 total proteins. There is remarkable agreement between these studies, which all support the overall impression that ribociclib binds to the fewest proteins overall, with palbociclib binding to marginally more and abemaciclib showing notably higher promiscuity.

22.2.1 Assessing CDK4/6 Inhibitor Selectivity

One popular method for screening kinase inhibitors against a large panel of kinases is called KINOME*scan*®, which is offered as a service by the large contract research organization (CRO) Eurofins. In KINOME*scan*, a test ligand is evaluated for its ability to prevent the binding of DNA-tagged kinase proteins to immobilized broad-spectrum kinase inhibitors. The difference in the amount of protein captured by the immobilized ligand in the presence or absence of the test compound can be precisely measured using a quantitative polymerase chain reaction (qPCR). Kim et al. [16] used KINOME*scan* to evaluate the binding of ribociclib to 465 kinases and confirmed its high selectivity for CDK4 (CDK6 was not in the panel). At 0.1 μM ribociclib, only CDK4/D1 and CDK4/D3 were inhibited, and at 1 μM ribociclib, only eight additional hits were identified. Similar KINOME*scan* experiments with palbociclib showed it to be similarly selective with only two and nine additional hits besides CDK4 identified at 0.1 and 1.0 μM palbociclib, respectively. In contrast, abemaciclib was less selective with 53 and 115 off-targets identified at the two test concentrations. In this study, palbociclib and ribociclib displayed comparable affinity for CDK4/D1 (Kd = 1.9 and 1.5 nM, respectively), and abemaciclib was 24-fold more potent (Kd = 0.078 nM).

A somewhat related but complementary method to KINOME*scan* is a technique referred to as Kinobeads. This approach features broad-spectrum kinase inhibitors immobilized on (polystyrene) beads that are then incubated with the test medium, e.g., a cell lysate, thoroughly rinsed, and analyzed by mass spectrometry. Proteins that bind with high affinity to the beads can be identified from their mass and peptide fragmentation pattern. When performed as a competition assay, Kinobeads can be used to examine the selectivity profile of kinase inhibitors. Interestingly, in using Kinobeads to examine the selectivity of 243 clinically tested kinase inhibitors, Klaeger et al. [22] identified some off-target proteins that were not kinases, indicating the risks of focusing too narrowly on a target class when considering the full extent of polypharmacology that a compound may exhibit. In contrast to other studies, the Kinobeads analysis did not characterize palbociclib as a particularly selective inhibitor for CDK4/6 with binding also seen for at least 13 additional proteins. While CDK6 was identified as the highest affinity target for palbociclib (Kd^{app} = 76 nM), it bound CDK4 more weakly, with Kd^{app} = 572 nM, and six other proteins, including five kinases and eukaryotic translation initiation factor 3J (EIF3J), were bound with Kd <1 μM. In contrast, ribociclib appeared to be less promiscuous in this assay, although CDK6 was not the highest scoring target (Kd^{app} = 497 nM) and CDK4 did not feature at all among the top nine hits. The highest affinity targets for ribociclib were calcium–calmodulin-dependent kinases 2G and 2D (CAMK2G Kd = 247 and CAMK2D Kd = 293 nM). Consistent with other studies, abemaciclib was found to be more potent and less selective than ribociclib or palbociclib with 22 proteins identified as binding with Kd^{app} < 200 nM. Based on the aggregate of available data, this particular study suggests that the kinobeads approach is a less reliable method for probing kinase selectivity, possibly due to the way that the small molecules are immobilized.

Uitdehaag and coworkers [23] used electrophoretic mobility shift assay (EMSA), enzyme-linked immunosorbent assay (ELISA), immobilized ion affinity particle (IMAP) assays, and surface plasmon resonance (SPR) to study the interaction of a panel of approved kinase inhibitor drugs with 280 kinases at ATP concentrations close to the Michaelis constant (K_M) for each kinase. In this work, abemaciclib was the most potent inhibitor of CDK4 and palbociclib was the most potent inhibitor of CDK6 (Table 22.1). When all three compounds were tested at 1 μM, ribociclib was the most selective CDK4/6i, inhibiting no other kinases; palbociclib inhibited CDK4/D3, CDK6/D3, and the dual-specificity kinase, cell-division cycle protein-like kinase 1 (CLK1); and abemaciclib inhibited 20 kinases by >95% at this concentration.

Table 22.1 Comparison of biochemical IC_{50} values published for three approved CDK4/6i.

	Palbociclib		Ribociclib		Abemaciclib	
IC_{50} (nM)	*Chen*	*Uitdehaag*	*Chen*	*Uitdehaag*	*Chen*	*Uitdehaag*
CDK4/D3	0.26	1.1	0.53	2.5	0.07	0.7
CDK6/D1	0.26		2.3		0.52	
CDK6/D3		3.9		29		8.4

Source: Chen et al. [21]; Uitdehaag et al. [24].

Similar results were reported by Chen and coworkers [21] based on data obtained using a mobility shift assay (Table 22.1).

A potential concern with biochemical assays is their physiological relevance. These assays are often conducted in the absence of substrate or using a short peptide rather than full-length protein, and the concentration of ATP may not reflect physiological ATP concentrations or even be matched to the K_M for the enzyme. Moreover, biochemical assays generally only test one posttranslational state of the kinase, ignore subcellular localization and higher-order protein complexes, and generally do not account for time-dependent inhibition or slow-binding kinetics. For all of these reasons, cell-based systems likely provide more relevant selectivity profiling data, and among cell-based systems, functional assays are likely superior. However, the technical challenge of evaluating the activity of >518 kinases in a cell-based environment as a function of inhibitor concentration makes this level of characterization prohibitive. Nonetheless, several informative cell-based studies have been described.

Sumi et al. [25] used whole-cell proteomics to identify direct and indirect protein partners for palbociclib and ribociclib in H157 cells and tumor tissue from patients diagnosed with squamous cell lung cancer. In this study, the CDK4/6i were first immobilized on an affinity resin and protein lysates were panned for binding partners, which were identified by mass spectrometry. Competition with non-immobilized CDK4/6i was used as a control to validate the observed binding partners. A primary conclusion from this study was that both palbociclib and ribociclib bound to and inhibited CDK9. In addition, palbociclib (but not ribociclib) bound to casein kinase 2α, adenosine monophosphate-dependent protein kinase (AMPK), and the lipid kinase phosphoinositide-3-kinase regulatory subunit 4 (PIK3R4). Although the inhibition of these proteins required more than 70-fold higher concentration of palbociclib than inhibition of CDK4/6, it was proposed that these activities may contribute to the pharmacological profile of palbociclib in some tumors. In contrast, Nomanbhoy and coworkers [26] employed an ATP derivative capable of functionally labeling ATP-binding proteins with desthiobiotin (KiNativ) in the absence of an ATP-competitive inhibitor. Using this approach, they showed that palbociclib most potently inhibited labeling of CDK4, CDK6, and myristoylated and palmitoylated serine/threonine protein kinase (MPSK1 – also known as STK16) in Colo-205 cell lysates, with significant inhibition also of the phosphatidylinositol-5-phosphate 4-kinase type 2γ (PIP4K2C) and phosphatidylinositol 3-kinase catalytic subunit type 3 (PIK3C3). These results were replicated in intact Colo-205 cells at shorter time points, while at longer timepoints, additional, presumably indirect, targets also were inhibited.

Hafner et al. [27] examined the similarities and differences between CDK4/6i-induced changes in gene expression profiles. They defined a gene signature associated with G1 arrest that was induced by palbociclib, ribociclib, and abemaciclib in Rb-positive breast cancer cell lines. Using this signature, they were able to determine an order of selectivity where ribociclib/palbociclib > abemaciclib > alvocidib (flavopiridol). However, for abemaciclib and alvocidib, the specific gene set affected was a function of drug concentration. For abemaciclib, at concentrations below 0.3 μM, the genes modulated were consistent with selective CDK4/6 inhibition, and the effect was cytostatic, similar to palbociclib and ribociclib. In contrast, at concentrations above 0.3 μM, abemaciclib but not palbociclib or ribociclib affected additional genes, and the result was cytotoxic. In Rb-deficient cells, only four genes were differentially affected by ribociclib consistent with ribociclib lacking activity in Rb-negative cell lines. The difference between abemaciclib and palbociclib could be discriminated by effects on as few as four genes: CCNE1 (cyclin E1), CDKN1a (cyclin-dependent kinase inhibitor 1a, p21), CDK9, and CDK-like 5 (CDKL5). Another interesting observation to arise from this study was that any differences between CDK4 inhibition and CDK6 inhibition for the three drugs were small and unlikely to be meaningful in a clinical setting. This conclusion contradicts earlier claims, in particular that the clinical profile of abemaciclib is attributable to higher activity versus CDK4 than CDK6 [28].

Overall, as newer data have become available, they have generally supported the initial characterization of palbociclib, ribociclib, and abemaciclib as potent and selective CDK4/6i [27]. All three compounds, when applied at clinically achievable concentrations, impose a G1 cell cycle block on cells that corresponds with inhibition of pRb phosphorylation consistent with the intended mechanism. Early speculation that the CDK4 versus CDK6 selectivity profile of abemaciclib may provide less risk for hematological side effects (e.g., neutropenia) appears not to be supported by more recent data [28]. At the exposures achieved clinically, palbociclib, ribociclib, and abemaciclib are likely to fully inhibit both CDK4 and CDK6. The single-agent efficacy displayed by abemaciclib in the clinic most likely reflects its broader off-target profile as demonstrated in multiple studies [21, 22, 24–27]. Preclinically, abemaciclib induces transcriptional changes akin to those seen with the pan-CDK inhibitor alvocidib (flavopiridol), and it produces more extensive phosphoproteome changes than either palbociclib or ribociclib [19]. Abemaciclib binds (and inhibits) multiple kinases besides CDK4/6, including

Table 22.2 Comparison of PFS and OS from randomized, controlled clinical trials [31–43].

Trial details	PALOMA-2	PALOMA-3	MONALEESA-2	MONALEESA-3	MONALEESA-7	MONARCH-3
CDK4/6i	Palbociclib	Palbociclib	Ribociclib	Ribociclib	Ribociclib	Abemaciclib
Endocrine therapy	Letrozole	Fulvestrant	Letrozole	Fulvestrant	Goserelin + tamox. or NSAI	NSAI
No. of patients	666	521	668	365	672	493
Med. follow-up	90	72	80	70.8	53.5	70.2
mPFS	24.8	9.2	25.3	33.6	23.8	28.2
mPFS control	14.5	3.8	16	19.2	13	14.8
mOS (months)	53.9	19.1%	63.9	67.6	58.7	67.1
mOS control	51.2	12.9%	51.4	51.8	47.7	54.5

NSAI = nonsteroidal aromatase inhibitor.

CDK2 and CDK9, and it is cytotoxic, whereas ribociclib and palbociclib primarily cause cytostasis in cell culture. Also, abemaciclib is cytotoxic toward Rb-deficient cells that are not dependent on CDK4/6 for cell cycle progression, which is a strong indication of non-CDK4/6-mediated pharmacology.

The most recently approved CDK4/6i trilaciclib (G1T28), from G1 Therapeutics, displays a similar profile to palbociclib and ribociclib in cell-based assays, imposing a reversible G1 block at concentrations up to >3000-fold the half-maximal inhibitory concentration (IC_{50}) for CDK4 [18]. Trilaciclib exhibits sub-nanomolar inhibition of CDK4 and CDK6 and is highly selective across a panel of 468 kinases. Interestingly, its most sensitive off-targets appear to be quite different from those seen with either ribociclib or palbociclib with a never-in-mitosis (NIMA)-related kinase 10 (NEK10) and sucrose non-fermentable 1/adenosine monophosphate (SNF1/AMP)-related kinase (SNARK) both inhibited with a sub-nanomolar IC_{50}. The implications of these observations are unclear but do not appear to manifest as cell death or increased DNA damage. Similar to ribociclib and palbociclib, trilaciclib has no effect on the growth of Rb-deficient cells.

22.3 Clinical Experience with CDK4/6 Inhibitors

22.3.1 Indications and Use for Approved CDK4/6 Inhibitors

The primary indication for CDK4/6i is hormone receptor-positive (HR^+) HER2 receptor-negative ($HER2^-$) breast cancer [29, 30]. Approximately 75% of stage IV breast cancer patients meet this definition, which is also referred to as luminal breast cancer. The use of palbociclib, ribociclib, and abemaciclib in clinical practice has been established based on the results of multiple pivotal, prospective, randomized, controlled clinical trials (Table 22.2) [31–37]. All three drugs are used in combination with endocrine therapy (ET). In first-line therapy for post-menopausal women with HR^+ ER^- advanced or metastatic breast cancer, palbociclib is approved for use in combination with the aromatase inhibitor, letrozole, and ribociclib is approved for use with a nonsteroidal aromatase inhibitor (NSAI) or the selective estrogen receptor degrader (SERD) fulvestrant. Ribociclib is also approved for use in pre- or perimenopausal women in combination with the luteinizing hormone blocker goserelin plus the selective estrogen receptor modulator (SERM) tamoxifen as a first-line therapy for advanced or metastatic breast cancer. Abemaciclib is approved in the first-line setting for use in combination with a NSAI in the treatment of advanced or metastatic breast cancer post-menopause. All three of these CDK4/6i provide similar benefits in median progression-free survival (mPFS – Table 22.2) when used in the treatment of advanced breast cancer in combination with ET. However, to date, only ribociclib and abemaciclib have provided statistically significant improvements in overall survival (OS) among this patient population [13, 33, 34, 36–39]. Data supporting a survival benefit for abemaciclib alone are still maturing. In the second-line setting, combination of abemaciclib with fulvestrant provided statistically significant improvements in OS upon long-term follow-up [40]. In the first-line setting, an interim analysis of the MONARCH-3 trial provided some support for a benefit in OS for abemaciclib as a single agent [35, 41]. However, these results did not reach statistical significance. In contrast, palbociclib in combination with either letrozole or fulvestrant did not produce statistical improvement in OS benefit in the intention-to-treat population, although benefit was seen in the subset of patients with endocrine-sensitive disease [42–44]. Interestingly, an analysis of outcomes for US-based patients receiving first-line endocrine-based therapy paints a slightly different picture for palbociclib with median OS increasing from 40.4 months to 53.4 months upon the addition of palbociclib to ET [45, 46].

Palbociclib, ribociclib, and abemaciclib are all approved for use as second-line therapy in combination with fulvestrant [47]. In the Palbociclib: Ongoing Trials in the Management of Breast Cancer-3 (PALOMA-3) trial, patients receiving palbociclib plus fulvestrant achieved a mPFS of 9.2 months versus patients receiving fulvestrant plus placebo who experienced a mPFS of 3.8 months. In the Mammary Oncology assessment of LEE11's (ribociclib) Efficacy and Safety-3 (MONALEESA-3) trials, patients receiving ribociclib plus fulvestrant achieved a mPFS of 20.5 months versus patients receiving fulvestrant plus placebo who experienced a mPFS of 12.8 months. In the MONARCH-2 trial, patients receiving abemaciclib plus fulvestrant achieved a median PFS of 16.4 months versus patients receiving fulvestrant plus placebo who experienced a mPFS of 9.3 months. In a small trial of 132 post-menopausal patients with HR^+/HER^- advanced or metastatic breast cancer, abemaciclib as a single agent produced a 19.7% response rate in patients that had experienced disease progression following prior treatment with ET and with chemotherapy [36, 39, 40].

CDK4/6i have also been investigated for their potential in the treatment of early-stage breast cancer. In this patient population, improvements in invasive disease-free survival (iDSF) or distant relapse-free survival (DRFS) produced by either palbociclib [48, 49], ribociclib [50, 51], or abemaciclib [52] in combination with ET have all been small with perhaps the best outcomes to date being seen in the MONARCH-E trial in which abemaciclib plus ET decreased the risk of cancer recurrence following two years of treatment and an additional two years of follow-up (iDSF 85.5% versus 78.6%) [52].

The standard dose of palbociclib is 125 mg, once per day for 21 days in a 28-day cycle. A seven-day dosing holiday is included to mitigate the neutropenia, which arises as a result of CDK6 inhibition in hematopoietic cells. The main toxicities observed with palbociclib include neutropenia, anemia, thrombocytopenia, diarrhea, fatigue, increased alanine transferase, and infections [13, 47, 53]. Ribociclib is used on a similar schedule as palbociclib, three weeks on and one week off; however, the clinical dose of ribociclib is higher at 600 mg, once per day. In addition to the adverse effects observed with palbociclib, including dose-limiting neutropenia, ribociclib carries a risk for QTc prolongation and requires monitoring for the first two cycles of dosing; it cannot be co-administered with other drugs that present a similar risk. Abemaciclib displays a lower risk for neutropenia, enabling continuous dosing, but a higher incidence of gastrointestinal (GI) toxicity. The highest recommended dose of abemaciclib is 150 mg twice per day in combination with ET. Abemaciclib is also approved for use as a single agent in a third-line setting, and the recommended dose in this case is 200 mg twice per day. Both ribociclib and abemaciclib, but not palbociclib, require that patients be monitored for liver function.

A large number of additional clinical trials have been initiated to examine various aspects of CDK4/6i clinical use, including various combination therapies and their effectiveness in different breast cancer patient subsets as well as indications outside of breast cancer [29, 54]. As of July 31, 2023, ClinicalTrials.Gov listed 360 trials for palbociclib, 151 for ribociclib, and 190 for abemaciclib. Two trials are particularly noteworthy. In the Study of Non-Inferiority Approach (SONIA) trial (NCT03425838), which is a phase 3 investigator-initiated trial being conducted in the Netherlands, treatment with a NSAI plus one of palbociclib, ribociclib, or abemaciclib in the first line followed by fulvestrant alone upon disease progression will be compared with treatment with a NSAI in the first line followed by fulvestrant plus palbociclib, ribociclib, or abemaciclib upon disease progression. This trial has enrolled 1050 patients and will be completed in December 2025. It is expected to provide PFS and OS data comparing palbociclib, ribociclib, and abemaciclib in the same trial. The HARMONIA trial (NCT0507709) is a head-to-head comparison of palbociclib and ribociclib in patients with advanced $HR^+/HER2^-$ and HER2-enriched breast cancer. The patient cohort comprises pre-menopausal women and men, and the study is unblinded. All patients will receive letrozole or fulvestrant and a luteinizing hormone-releasing hormone agonist plus either ribociclib or palbociclib. The primary stated objective of this study is to demonstrate superior PFS for ribociclib over palbociclib.

In contrast to the first three approved CDK4/6i, trilaciclib is approved for use as a chemoprotectant [18]. By blocking the cycling of CDK4/6-dependent hematopoietic stem and progenitor cells (HSPCs), while leaving Rb-negative tumors unaffected, trilaciclib is able to reduce the myelosuppression commonly associated with standard cytotoxic therapies such as 5-fluorouracil (5-FU) and topotecan. The use of trilaciclib for this clinical purpose is aided by its relatively short half-life and suitability for IV administration, such that bone marrow suppression can be tightly controlled and released once cytotoxic therapy is complete.

22.3.2 Mechanisms of Drug Resistance to CDK4/6 Inhibitors

As for all cancer drugs, the use of CDK4/6i ultimately results in the development of drug resistance [47, 55–59].

There is some evidence that once patients develop resistance to one CDK4/6i, they may see further benefit by switching to a different CDK4/6i and ET [60]; however, this does not appear to be universally the case [13, 53]. One known resistance mechanism toward selective CDK4/6 inhibition is an elevated expression of cyclin E, which can partner with CDK2 and phosphorylate pRb independently of CDK4/6. Thus, the inhibition of CDK2 and CDK4/6 at the same time by abemaciclib may represent a potential advantage. The premise that patients resistant to ribociclib or palbociclib treatment may still benefit from treatment with abemaciclib is being tested in the ongoing postMONARCH trial [61]. The selective inhibition of CDK4/6 and CDK2 activity by a single molecule is one of the profiles being targeted in fourth-generation CDKi (see below).

Multiple studies have confirmed that CDK4/6 selective inhibitors are only effective in preventing the growth of pRb-positive cells and tumors as anticipated based on their mechanism of action. However, in a surprising result, Nomanbhoy et al. showed that palbociclib and ribociclib fail even to bind to CDK4 in resistant cells [26]. Thus, the lack of potency appears to stem at least in part from a failure to inhibit the target as much as from a reduced requirement for CDK4/6 activity in Rb-negative cells. This inability of palbociclib to bind CDK4/6 in resistant cells is not completely understood but likely reflects an inactive conformational state of the protein(s), e.g., due to lack of posttranslation phosphorylation at threonine-172, or incorporation into an inactive complex with p27 that is unphosphorylated on tyrosine-88/89 [62].

Given the widespread dependence of different tumors on the CDK4/6-Rb-E2F pathway [9], it is surprising that CDK4/6i have not exhibited broader efficacy in the clinic. Despite large numbers of trials in different indications, the only cancer for which the use of CDK4/6i has been approved is $HR^+/HER2^-$ breast cancer. Recent work from Fassl et al. provided one potential explanation for the resistance of triple-negative breast cancer (TNBC) to palbociclib and ribociclib [63]. Despite CDK4/6 being essential for survival in some TNBC cells, these cells also display enhanced lysosomal biogenesis and the positively charged nature of palbociclib and ribociclib leads these compounds to be sequestered inside lysosomes reducing their efficacy. Co-administration of lysosomotropic or lysosome-destabilizing compounds limited palbociclib/ribociclib lysosomal sequestration and increased TNBC cell sensitivity toward these agents. Replacement of the piperazine group in ribociclib with a non-ionizable group that also lowered the pKa of the pyridine nitrogen mitigated lysosomal accumulation and resulted in improved cytotoxicity toward TNBC cells in culture. It is not currently known whether lysosomal sequestration also accounts for the lack of CDK4/6i sensitivity in other tumors.

22.3.3 CDK4/6 Biomarkers

One major limitation in identifying additional patients who may benefit from CDK4/6i treatment is a lack of appropriate patient selection biomarkers. At the current time, the primary way that patients are selected for treatment using a CDK4/6i is based on whether or not a tumor biopsy is positive for the expression of receptors for the hormones progesterone or estrogen. This is an imperfect criterion that leads to some patients being treated who have little chance of benefit and some patients who may benefit being neglected. More specific biomarkers predictive of response are highly desirable to identify the most sensitive patient population. Unfortunately, no single genetic aberration is indicative of CDK4/6i sensitivity. However, several potential biomarkers for predicting response to CDK4/6i have been suggested based on mechanism of action considerations [64–66], and recently, the Food and Drug Administration (FDA) approved the use of abemaciclib for the treatment of high-risk early-stage breast cancer with a Ki-67 score ≥20% based on results from the MONARCH-E study [67]. Preclinically, the presence of functional pRb can be used to discriminate between sensitive and resistant cell lines. However, surprisingly, pRb has not proven to be as robust a biomarker in clinical studies [64–66], possibly because of a correlation between HR status and Rb expression. Similarly, cyclin D levels, while predictive of response in cells, do not display useful predictive value in patients. Lower cyclin E1 but not cyclin E2 messenger ribonucleic acid (mRNA) expression was associated with higher efficacy in the PALOMA-3 trial (palbociclib plus fulvestrant) [68], but not in PALOMA-2 (palbociclib plus letrozole). Posttranslational phosphorylation of CDK4 on Thr172 is required for full activity, and this step is thought to be rate limiting for CDK4 activation. Unphosphorylated CDK4 was found by Raspé to be less sensitive to palbociclib inhibition, suggesting that CDK phosphorylation status may be useful as a clinical biomarker [69]. However, protein phosphorylation state can be difficult to assess in clinical samples, so Raspé identified a gene expression profile that correlated with CDK4 pThr172. This profile predicted palbociclib sensitivity in 49 of 52 breast cancer cell lines, but has yet to be tested clinically. In ER^- (but not ER^+) cell lines, high CDKN2 (p16) inhibits CDK4/6 activity and also prevents CDK4/6i target engagement. Based on these observations, CDKN2A is proposed to be an independent predictor of palbociclib sensitivity in cancer types where ER status is not relevant [70]. This hypothesis also has yet to be confirmed clinically. Finally, the phosphorylation status of p27 has potential utility as a patient selection biomarker. p27 phosphorylation on residue Y88 or Y89 is required to form a fully active CDK4/6/cyclin D/p27 complex. In cell culture experiments, phosphorylation of p27 on Y88 directly correlated with palbociclib responsiveness. Gottesman obtained further support for this hypothesis by testing explants from patients who had undergone surgery to remove their tumor for the proliferation marker Ki-67 [71]. These data then were correlated with results from archival biopsies stained for p27 and pY88-p27, which showed a decrease in Ki-67 in explants from the pY88-positive group and unchanged Ki-67 in the explants from patients whose biopsy samples had stained negatively for pY88. These results indicate that the p27 Y88 phosphorylation status correlated with inhibition of proliferation and hence sensitivity to palbociclib. Some of the palbociclib-sensitive patients were found in the $HER2^+$ subgroup and even in the triple-negative subgroup.

A second class of biomarkers that would be useful in supporting clinical studies are those that predict CDK4/6i resistance. These may be considered in two categories, those that are cell cycle-related and those that are non-cell-cycle-related. In the cell-cycle-related category, both Rb deletion and cyclin E1 overexpression are established resistance pathways and predictive of poorer outcomes in the clinic. From work with cell lines, Guarducci showed that a high CCNE1/Rb1 ratio was an even stronger predictor of palbociclib resistance than either measure on its own [57]. Low-molecular-weight versions of cyclin E (LMWE) have been identified in aggressive tumor subtypes [72]. Compared with full-length protein, LMWE/CDK2 is more resistant to inhibition by p21 and p27. LMWE promotes CDK activity and pRb hyperphosphorylation, bypassing CDK4/6. Consequently, high levels of LMWE correlate with insensitivity to palbociclib and with a worse prognosis in general. Multiple resistance mechanisms appear to converge on amplification of CDK6 and CDK2 [55, 56]. In contrast to CDK4, increased CDK6 can sequester p21 and p27 away from CDK2 [73]. In addition, loss of the tumor suppressor protein FAT atypical cadherin 1 (FAT1) significantly increases CDK6 expression via downregulation of the Hippo pathway [64, 74].

Non-cell cycle-related resistance pathways include thymidine kinase-1, the phosphoinositide-3-kinase/Ak strain transforming/mammalian target of rapamycin (PI3K/Akt/mTOR) pathway, and growth factors including fibroblast growth factor receptor (FGFr) and epidermal growth factor receptor (EGFr), which are upstream of cyclin D expression [57]. Thymidine kinase 1 (TK1) is expressed in S/G2 phase and catalyzes the synthesis of DNA precursors. TK1 expression was found to increase significantly after three months of treatment with letrozole plus palbociclib and elevated TK1 expression correlated with shorter PFS [75]. The PI3K/Akt/mTOR pathway is activated in ~30–40% of breast cancer, and this same pathway is activated in ribociclib-resistant breast cancer cells [76]. The PI3K inhibitor alpelisib can overcome resistance to ribociclib in a patient-derived xenograft (PDX) model [65]. Many of these resistance pathways are being targeted with potential combination therapies.

22.3.4 CDK4/6 Inhibitors in Combination Therapy

A common approach to achieving clinically meaningful efficacy in cancer is through combination therapy [77, 78]. CDK4/6i are approved for use in combination with antihormonal drugs such as the aromatase inhibitor letrozole or the SERD fulvestrant (see above). However, no other combination therapies containing CDK4/6i have yet been successfully advanced to approval, despite many completed and ongoing clinical investigations [79, 80]. The empirical approach of combining CDK4/6i with standard of care is complicated by questions of drug sequencing [81]. Indeed, concurrent treatment with CDK4/6i and chemotherapy can be ineffective, or even antagonistic, if cells are held up in G1 by CDK4/6i, since most DNA-targeted chemotherapy and antimitotic agents act upon cells in later stages of the cell cycle. However, sequencing treatment such that cells are coordinately released from G1 to continue through cell division can lead to increased chemotherapy effectiveness. Such combinations are based on an appreciation of CDK4/6i mechanism of action and resistance pathways. A primary mechanism of CDK4/6i resistance is reactivation of CDK2, e.g., through downregulation of p27 by protein degradation [79]. Since phospho-p27 has a shorter half-life than the unphosphorylated protein, the inhibition of kinases that regulate p27 offers a potential mechanism for combating CDK4/6i inhibitor resistance. One protein implicated in p27 phosphorylation is the Src family kinase Brk [82]. Administration of the Src family inhibitor saracatinib restored sensitivity to palbociclib in k-Ras mutant colorectal cancer cells *in vitro* and *in vivo* [83].

The rat sarcoma viral oncogene/dual specificity mitogen-activated protein kinase kinase 1/extracellular signal-related kinase (Ras-MEK-ERK) signaling pathway acts downstream of several important growth factor receptors and as such plays a role in the regulation of D-type cyclins that activate CDK4/6 activity. It has been estimated that about 10% of patients who become resistant to CDK4/6i harbor an activating Ras mutation. In contrast, Ras mutations are absent from CDK4/6i-sensitive tumors [84]. The combination of MEK and CDK4/6i was found to be synergistic in PDX models of colorectal cancer [85, 86]. There is an ongoing clinical trial to evaluate the combination of the MEK inhibitor binimetinib and palbociclib in patients with advanced Kirsten rat sarcoma viral oncogene (KRAS) mutant non–small cell lung cancer (NCT03170206). In other work, it has been found that CDK6/D3 stabilizes the transcription factor forkhead box protein O3 (FOXO3) and promotes transcription of the DNA damage response kinase ataxia telangiectasia and Rad3-related protein (ATR). Activation of ATR is a mechanism of resistance to DNA-damaging cytotoxic agents such as cisplatin. Accord ingly, CDK6 inhibition with ribociclib in combination with cisplatin was found to induce synthetic lethality [87].

The PI3K-Akt-mTOR pathway is frequently upregulated in cancer and is also implicated in CDK4/6i resistance [76]. The combination of a CDK4/6i and PI3K pathway inhibition has been shown to produce enhanced tumor cell toxicity as well as immunogenic effects. For example, ribociclib and the PI3K inhibitor alpelisib were found to act synergistically in murine models of TNBC, inducing cell cycle arrest and apoptosis as well as an increase in tumor-infiltrating T cells [88]. In pancreatic cancer cells, resistance to palbociclib was mitigated by treatment with an inhibitor of mTOR, and the mTOR inhibitor vistusertib (AZD2104) in combination with palbociclib produced durable growth arrest in HR^+ breast cancer cells [89].

Other potential combination therapies are based on targeting receptor tyrosine kinase (RTK) pathways including EGFr, FGFr, anaplastic lymphoma kinase (ALK), and erythropoietin-producing human hepatocellular receptors 1 and 2 (Eph1/2) [79]. Studies of RTK receptor inhibitors in combination with CDK4/6i have identified FGFrs in particular for the role they play in mediating resistance [65]. Amplification of FGFr1 sustains resistance to hormonal therapy, and FGFr signaling also induces cross-resistance to palbociclib. This effect could be overcome by inhibitors of FGFr or MEK [90], and further studies along these lines are warranted.

Given the time and expense of clinical trials, a quick way of identifying which of the many possible CDK4/6i plus chemotherapy/targeted therapy combinations are most likely to be successful would be valuable [91]. Indeed, the discovery that palbociclib was particularly effective against ER^+ cells in culture was a key observation that enabled the development of this first-in-class drug for the most appropriate patient population [92]. In an heroic effort testing 64 drugs against 1000 human PDX models of different origins and genotypes, Gao et al. [93] identified the combination of ribociclib and the BRAF inhibitor encorafenib as being particularly effective in inhibiting the growth of melanoma cells in culture. Melanoma is an indication for which no CDK4/6i are yet approved, despite inactivating mutations in p16 being common in this cancer [94]. Unfortunately, clinical assessments of ribociclib/encorafenib combinations could not be completed due to the frequency and severity of adverse events [95]. Nonetheless, this study illustrates a potentially effective way of prioritizing clinical studies and suggests that the further exploration of CDK4/6i and serine/threonine-protein kinase B-raf (BRAF) inhibitors may be warranted. In a separate study, Franco and coworkers examined the combination of palbociclib with a panel of 304 drugs and identified combinations containing MEK or PI3K/mTOR inhibitors as particularly promising for the treatment of pancreatic cancer [96].

22.3.5 CDK4/6 Inhibitors and Immunotherapy

The introduction of anti-cytotoxic T-cell antigen 4 (anti-CTLA-4) and anti-programmed cell death 1/programmed cell death ligand (anti-PD1/PD-L1) antibodies has transformed cancer treatment and care [97]. The remarkable results observed in some patients treated with these immunotherapies have spawned over a thousand clinical trials, many of them seeking to explore combinations of immunotherapy with other cancer drugs. However, there have been relatively few investigations of CDK4/6i with immunotherapy. Interestingly, CDK4/6 inhibition appears to exert its own effect on the immune response. A 2017 report by Goel et al. described how abemaciclib or palbociclib elevated antitumor immunity through a combination of T-regulatory cell suppression and enhanced tumor antigen presentation [98]. Both of these effects are believed to occur as a result of reduced E2F activity. Reduced activity of the E2F target DNA methyltransferase 1 activates expression of retroviral genes, which stimulates production of type III interferons. CDK4/6 inhibition also induces interferon/signal transducer and activator of transcription 1 (STAT1) signaling, which helps drive T-cell-mediated tumor clearance [98]. CDK4/6i also promote the secretion of inflammatory cytokines including chemokine (C-C motif) ligand 5 (CCL5) and C-X-C motif chemokine ligand 10 (CXCL10), which improve T cell homing to tumors [99]. Interestingly, this effect appears to be dependent on both Rb and reactive oxygen species (ROS). However, upon removal of the CDK4/6i, even though Rb phosphorylation was restored, chemokine levels remained high, indicating that this effect was not reversible.

In 2021, Heckler et al showed that the treatment of mice with palbociclib promoted $CD8^+$ T-cell persistence and immunologic memory [100]. Analysis of patients treated with either palbociclib or abemaciclib showed that these compounds similarly elevated $CD8^+$ memory precursors in patients. This effect was shown not to be solely dependent on cell cycle arrest but rather due to an increase in the levels of the transcription factor Max-interacting transcriptional repressor MAD4 (Mxd4). The Mxd4 protein associates with the myc partner myc-associated factor X (MAX), resulting in a decrease in the transcription of myc target genes. The mechanism through which CDK4/6 regulates Mxd4 is not fully established, but Heckler et al speculate that it occurs through phosphorylation of FOXO1 since the promoter region for mxd4 contains FOXO1 recognition sequences.

In a 2018 study published in Nature, Zhang et al described how the inhibition of CDK4/6 can increase PD-L1 levels by inhibiting PD-L1 ubiquitination [101]. This effect is mediated by the inhibition of CDK4/6/D1-mediated phosphorylation of speckle-type POZ protein (SPOP). Unphosphorylated SPOP is degraded, which interferes with its ability to promote cullin-mediated PD-L1 ubiquitination and proteolysis. These same authors demonstrated that co-treatment of tumor-bearing mice with palbociclib and an anti-PD1 antibody produced substantial tumor regression, and 8 out of 12 mice treated in this study achieved a complete response, with no measurable tumor remaining after treatment for three weeks.

Taken together, these studies support exploring the combination of CDK4/6 inhibition and checkpoint inhibition in clinical studies. Current studies are looking at sequencing the drugs such that CDK4/6 inhibition is deployed first to prime the immune system by increasing PD-L1 expression and antigen presentation, followed by a checkpoint inhibitor to combat T-cell exhaustion (e.g., NCT04075604). It will also be important to identify biomarkers, possibly including the cytokines CCL5 and CXCL10, which can be used to identify the patients most likely to benefit from co-treatment with a CDK4/6i and immunotherapy.

22.4 New Approaches and Agents for CDK4/6 Inhibition

22.4.1 Reversible Inhibitors

Among the newer CDK4/6i to enter development, two of them are notable for being structurally similar to approved compounds (Figure 22.3). Lerociclib (G1T38) [102] differs from trilaciclib (G1T28) only in the substituent on the terminal piperazine nitrogen, which is methyl in trilaciclib and isopropyl in lerociclib. This structural difference confers a shorter human plasma half-life on lerociclib, which otherwise appears to retain an almost unchanged potency and selectivity profile. In contrast to trilaciclib, lerociclib is being developed as an oral therapy. By contrast to palbociclib, lerociclib has a shorter half-life, which leads to less accumulation in plasma with multiple dosing, while at the same time lerociclib appears to accumulate in tumors, leading to a prolonged pharmacodynamic effect with sustained inhibition of pRb phosphorylation even after the plasma drug concentration has fallen to zero. This profile has the potential advantage of minimizing myelosuppression while allowing continuous daily dosing of the CDK4/6i. This expectation is somewhat borne out in dogs; however, the realization of a clinical benefit in patients is still to be established.

Dalpiciclib (SHR6390; Figure 22.3) [103] differs from palbociclib only in the replacement of the tertiary nitrogen in the side chain piperazine by a carbon atom. The preclinical profile of dalpiciclib matches very closely the profile of palbociclib with only minor differences in degree [104]. Dalpiciclib is approved in China for use in combination with fulvestrant for the treatment of $ER^+/HER2^-$ breast cancer and is being investigated for use in combination with antihormonal therapy such as letrozole or anastrozole [105], although other combinations also are being considered, including combinations with tyrosine kinase inhibitors or checkpoint inhibitors. BPI-16350 is related to abemaciclib, but is described as showing less inhibition of CDK9 [106]. A phase 1 clinical trial of BPI-16350 in combination with fulvestrant is ongoing in patients with advanced solid tumors [106]. In contrast, PF-07220060 is a structurally distinct CDK4/6i that is more potent versus CDK4/D1 (Ki = 0.6 nM) than CDK6/D3 (Ki = 13.9 nM), a profile predicted to present a lower risk for neutropenia. The amino-tetrahydropyranol side chain attached to the pyrimidine C2 carbon in PF-07220060 is likely acting as an isostere for the bridging water believed to mediate binding to the pyridine in earlier CDK4/6i (see above). Phase 1/2a clinical trial results indicate that PF-07220060 can be efficacious in patients with HR^+/HER^- metastatic breast cancer that have previously progressed on prior CDK4/6i plus ET treatment, and phase 3 trials commenced at the end of 2023.

In order to combat resistance to CDK4/6i, Pfizer also is developing a new CDK2/4/6 inhibitor, PF-6873600 [107]. This compound was recently advanced into human studies. PF-6873600 is based on the same pyridopyrimidine scaffold as palbociclib, but all of the substituents have been changed to recover potency versus CDK2. PF-6873600 inhibits CDK2/E1 and CDK2/A1 with Ki = 0.13 and 0.08 nM, CDK4/D with Ki = 1.25 nM, CDK4/D3 with Ki = 1.4 Nm, and CDK6/D3 with Ki = 2.5 nM. However, CDK1/A2 and CDK5/p35 are similarly inhibited by PF-6873600 (Ki = 4.5 nM and 0.1 nM, respectively), although the Ki for CDK9/T1 is higher at 19.6 nM [107]. Thus, PF-6873600 is perhaps better described as a CDK1/2/4/5/6 inhibitor, illustrating the challenge of trying to build compounds

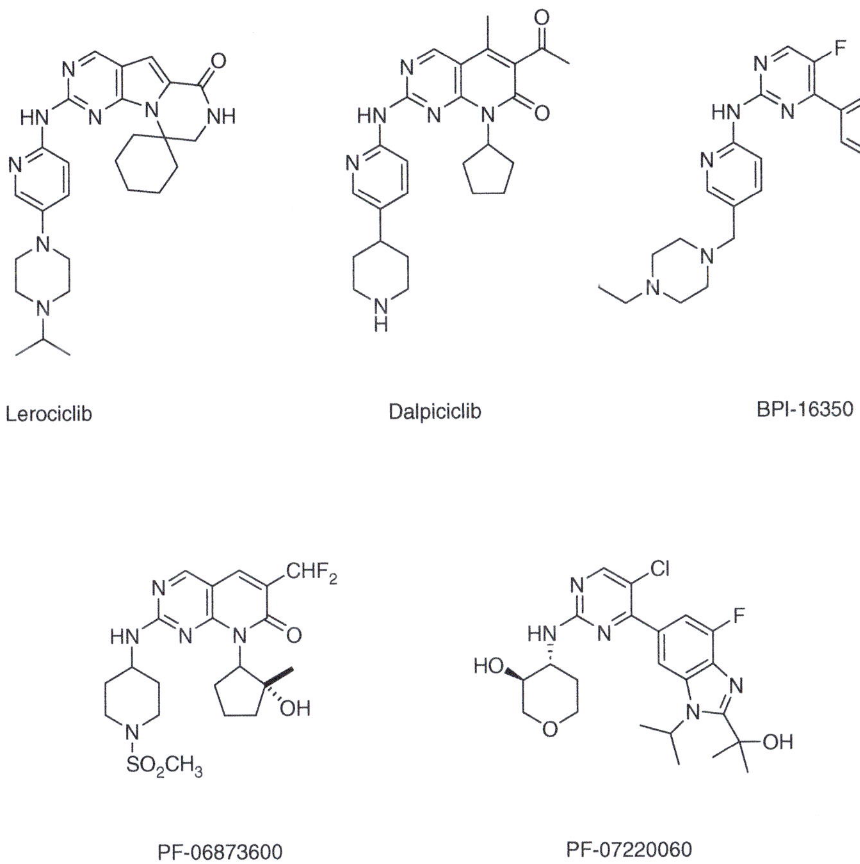

Figure 22.3 Investigational CDK4/6 inhibitors currently in clinical trials.

that only inhibit a defined subset of CDKs. A complementary strategy would be to combine an approved CDK4/6i with a highly selective CDK2i, several of which are now advancing through development [108–112].

22.4.2 PROTACs

Proteolysis-targeting chimeras (PROTACs) are heterobifunctional molecules in which a ligand for the target protein is attached via a linker to a ligand for an E3 ligase [113, 114]. The ternary complex formed between the target protein, the PROTAC, and the ligase facilitates ubiquitination of the target protein, resulting in its subsequent proteolysis by the proteasome. The PROTAC itself remains intact and is able to recycle and form additional ternary complexes, resulting in sub-stoichiometric activity. The chemical structures and binding modes of palbociclib, ribociclib, and abemaciclib are well suited for elaboration into PROTACs. Specifically, the side chain piperazine that is common to all three compounds extends out of the binding pocket and provides an ideal attachment point for a linker to the requisite E3 ligase ligand. Multiple groups have described versions of this chemistry with variations in the choice of CDK4/6 ligand, E3 ligase ligand, and tether length and composition [12, 115]. Some of these compounds induce degradation of CDK4 and CDK6, others appear to selectively degrade CDK4, and yet others induce selective degradation of CDK6. A summary of recently disclosed CDK4/6 degraders was provided by Rana et al. [115]. These initial CDK4/6 PROTACs are all first-generation efforts with highly flexible linkers that are not optimized for oral delivery. As a consequence, the pharmacological and toxicological consequences of targeting CDK4/6 proteolysis *in vivo* are not yet known.

22.5 Conclusion

CDK4/6i have been successful in the treatment of women with advanced or metastatic $HR^+/HER2^-$ breast cancer. While this represents a significant patient population, it is only a small fraction of all patients with tumors that might have been expected to benefit from targeting such a fundamental mechanism at the heart of cell proliferation. While it has been suggested that other cell cycle CDKs can functionally replace CDK4/6 in most cancers other than luminal breast cancer, more recent discoveries may help to identify additional indications where CDK4/6i can provide benefit. The identification and clinical validation of biomarkers will substantially aid in this effort, and here also, newer developments provide a basis for optimism. New combination treatments either together with ET or employing CDK4/6i in entirely new combinations without the need for ET also appear poised for advancement. In particular, combinations of CDK4/6i and inhibitors of the PI3-kinase family warrant further clinical study, as do combinations with immunotherapy. Additional combinations are likely to be explored with appropriate sequencing.

An interesting and important question to address for physicians is which CDK4/6i to recommend for patients who now have multiple options. Safety and tolerability considerations tend to favor palbociclib, which lacks the cardiovascular risk and GI toxicity of ribociclib and abemaciclib, respectively. However, physicians must weigh these concerns against potential differences in OS benefit. A third consideration is whether one or other of the CDK4/6i currently available is more likely to incur resistance. Mechanistic considerations might suggest that resistance will be least likely to emerge with abemaciclib due to its less selective mechanism of action. Certainly, cross-resistance between abemaciclib and palbociclib or ribociclib appears to be less compared to between ribociclib and palbociclib in preclinical studies, suggesting that patients whose disease progresses on either palbociclib or ribociclib may still receive some benefit from switching to abemaciclib. Switching between ribociclib and palbociclib may also provide benefit, but the evidence in support of this outcome is weaker.

In conclusion, CDK4/6i have been transformative for the treatment of breast cancer and are likely to continue to provide benefit for patients for many years. New approaches to the selective inhibition of CDK4 may mitigate the major adverse effect of CDK4/6i, which is neutropenia. Meanwhile, advances in our understanding of patient selection biomarkers will enable better targeting of therapy, and response biomarkers will inform ongoing efforts to develop combination therapies to address intrinsic and acquired resistance. Strategies for patient care once CDK4/6i resistance develops will become more sophisticated and likely include options for switching the specific CDK4/6i used for therapy. Finally, opportunities to expand the application of CDK4/6i beyond ER^+/$HER2^-$ breast cancer represent an exciting future for the field.

Acknowledgment

Thanks to all the researchers and patients who have contributed to our knowledge of this interesting and useful class of drugs. This chapter represents my personal perspective on the topic and is not intended to be a systematic review. I apologize to the many authors whose excellent work was not included. I would like to thank Michigan Drug Discovery and the University of Michigan College of Pharmacy for support, Dr. Judy Leopold for her insightful review of the manuscript, Dr. Stephen Eck for encouragement and feedback, Rajani Arora for help with the figures, and Nancy Kraus for her strength and inspiration.

References

1 Lodish, H., Berk, A., Kaiser, C.A. et al. (2021). *Molecular Cell Biology (Ninth Edition)*. New York: MacMillan.
2 Sherr, C.J. (2000). The pezcoller lecture: cancer cell cycles revisited. *Cancer Res.* 60: 3689–3695.
3 Nurse, P. (2000). A long twentieth century of the cell cycle and beyond. *Cell* 100: 71–78.
4 Malumbres, M. (2014). Cyclin-dependent kinases. *Genome Biol.* 15: 122.
5 Malumbres, M. and Barbacid, M. (2005). Mammalian cyclin-dependent kinases. *Trends Biochem. Sci.* 30: 630–641.
6 Pavletich, N.P. (1999). Mechanisms of cyclin-dependent kinase regulation: structures of Cdks, their cyclin activators, and Cip and INK4 inhibitors. *J. Mol. Biol.* 287: 821–828.
7 Sherr, C.J. and Roberts, J.M. (1999). CDK inhibitors: positive and negative regulators of G1-phase progression. *Genes Dev.* 13: 1501–1512.
8 Malumbres, M. and Barbacid, M. (2001). To cycle or not to cycle: a critical decision in cancer. *Nat. Rev. Cancer* 1: 222–231.
9 Fry, D. and Garrett, M. (2000). Inhibitors of cyclin-dependent kinases as therapeutic agents for the treatment of cancer. *Current Opin. Oncol. Endoc. & Metab. Investig.* 2: 2–40.
10 Diaz-Moralli, S., Tarrado-Castellarnau, M., Miranda, A., and Cascante, M. (2013). Targeting cell cycle regulation in cancer therapy. *Pharmacol. Ther.* 138: 255–271.
11 Mughal, M.J., Bhadresha, K., and Kwok, H.F. (2023). CDK inhibitors from past to present: a new wave of cancer therapy. *Semin. Cancer Biol.* 88: 106–122.

12 Xie, Z., Hou, S., Yang, X. et al. (2022). Lessons learned from past cyclin-dependent kinase drug discovery efforts. *J. Med. Chem.* 65: 6356–6389.

13 Johnston, S., Emde, A., Barrios, C. et al. (2023). Cyclin-dependent kinase 4 and 6 (CDK4/6) inhibitors: existing and emerging differences. *JNCI Cancer Spectr.* 7: 1–11.

14 Fry, D.W., Harvey, P.J., Keller, P.R. et al. (2004). Specific inhibition of cyclin-dependent kinase 4/6 by PD 0332991 and associated antitumor activity in human tumor xenografts. *Mol. Cancer Ther.* 3: 1427–1438.

15 Toogood, P.L., Harvey, P.J., Repine, J.T. et al. (2005). Discovery of a potent and selective inhibitor of cyclin-dependent kinase 4/6. *J. Med. Chem.* 48: 2388–2406.

16 Kim, S., Tiedt, R., Loo, A. et al. (2018). The potent and selective cyclin-dependent kinases 4 and 6 inhibitor ribociclib (LEE011) is a versatile combination partner in preclinical cancer models. *Oncotarget* 9: 35226–35240.

17 Gharbi, S.I., Pelletier, L.A., Espada, A. et al. (2022). Crystal structure of active CDK4-cyclin D and mechanistic basis for abemaciclib efficacy. *NPJ Breast Cancer* 8: 126.

18 Bisi, J.E., Sorrentino, J.A., Roberts, P.J. et al. (2016). Preclinical characterization of G1T28: a novel CDK4/6 inhibitor for reduction of chemotherapy-induced myelosuppression. *Mol. Cancer Ther.* 15: 783–793.

19 Hendrychova, D., Jorda, R., and Krystof, V. (2021). How selective are clinical CDK4/6 inhibitors? *Med. Res. Rev.* 41: 1578–1598.

20 Lu, H. and Schulze-Gahmen, U. (2006). Toward understanding the structural basis of cyclin-dependent kinase 6 specific inhibition. *J. Med. Chem.* 49: 3826–3831.

21 Chen, P., Lee, N.V., Hu, W. et al. (2016). Spectrum and degree of CDK drug interactions predicts clinical performance. *Mol. Cancer Ther.* 15: 2273–2281.

22 Klaeger, S., Heinzlmeir, S., Wilhelm, M. et al. (2017). The target landscape of clinical kinase drugs. *Science* 358: eaan4368.

23 Uitdehaag, J.C., de Roos, J.A., van Doornmalen, A.M. et al. (2014). Comparison of the cancer gene targeting and biochemical selectivities of all targeted kinase inhibitors approved for clinical use. *PLoS ONE* 9: e92146.

24 Uitdehaag, J.C.M., Kooijman, J.J., de Roos, J. et al. (2019). Combined cellular and biochemical profiling to identify predictive drug response biomarkers for kinase inhibitors approved for clinical use between 2013 and 2017. *Mol. Cancer Ther.* 18: 470–481.

25 Sumi, N.J., Kuenzi, B.M., Knezevic, C.E. et al. (2015). Chemoproteomics reveals novel protein and lipid kinase targets of clinical CDK4/6 inhibitors in lung cancer. *ACS Chem. Biol.* 10: 2680–2686.

26 Nomanbhoy, T.K., Sharma, G., Brown, H. et al. (2016). Chemoproteomic evaluation of target engagement by the cyclin-dependent kinase 4 and 6 inhibitor palbociclib correlates with cancer cell response. *Biochemistry* 55: 5434–5441.

27 Hafner, M., Mills, C.E., Subramanian, K. et al. (2019). Multiomics profiling establishes the polypharmacology of FDA-approved CDK4/6 inhibitors and the potential for differential clinical activity. *Cell Chem. Biol.* 26 (1067–1080): e1068.

28 Patnaik, A., Rosen, L.S., Tolaney, S.M. et al. (2016). Efficacy and safety of abemaciclib, an inhibitor of CDK4 and CDK6, for patients with breast cancer, non-small cell lung cancer, and other solid tumors. *Cancer Discov.* 6: 740–753.

29 Schettini, F., De Santo, I., Rea, C.G. et al. (2018). CDK 4/6 inhibitors as single agent in advanced solid tumors. *Front. Oncol.* 8: 608.

30 Morrison, L., Loibl, S., and Turner, N.C. (2023). The CDK4/6 inhibitor revolution – a game-changing era for breast cancer treatment. *Nat. Rev. Clin. Oncol.* 21: 89–105.

31 Finn, R.S., Martin, M., Rugo, H.S. et al. (2016). Palbociclib and letrozole in advanced breast cancer. *N. Engl. J. Med.* 375: 1925–1936.

32 Cristofanilli, M., Turner, N.C., Bondarenko, I. et al. (2016). Fulvestrant plus palbociclib versus fulvestrant plus placebo for treatment of hormone-receptor-positive, HER2-negative metastatic breast cancer that progressed on previous endocrine therapy (PALOMA-3): final analysis of the multicentre, double-blind, phase 3 randomised controlled trial. *Lancet Oncol.* 17: 425–439.

33 Hortobagyi, G.N., Stemmer, S.M., Burris, H.A. et al. (2016). Ribociclib as first-line therapy for HR-positive, advanced breast cancer. *N. Engl. J. Med.* 375: 1738–1748.

34 Slamon, D.J., Neven, P., Chia, S. et al. (2018). Phase III randomized study of ribociclib and fulvestrant in hormone receptor–positive, human epidermal growth factor receptor 2–negative advanced breast cancer: MONALEESA-3. *J. Clin. Oncol.* 36: 2465–2472.

35 Goetz, M.P., Toi, M., Campone, M. et al. (2017). MONARCH 3: abemaciclib as initial therapy for advanced breast cancer. *J. Clin. Oncol.* 35: 3638–3646.

36 Sledge, G.W., Toi, M., Neven, P. et al. (2017). MONARCH 2: abemaciclib in combination with fulvestrant in women with HR+/HER2− advanced

breast cancer who had progressed while receiving endocrine therapy. *J. Clin. Oncol.* 35: 2875–2884.

37 Tripathy, D., Im, S.-A., Colleoni, M. et al. (2018). Ribociclib plus endocrine therapy for premenopausal women with hormone-receptor-positive, advanced breast cancer (MONALEESA-7): a randomised phase 3 trial. *Lancet Oncol.* 19: 904–915.

38 Lu, Y.S., Im, S.A., Colleoni, M. et al. (2022). Updated overall survival of ribociclib plus endocrine therapy versus endocrine therapy alone in pre- and peri-menopausal patients with HR+/HER2- advanced breast cancer in MONALEESA-7: a phase III randomized clinical trial. *Clin. Cancer Res.* 28: 851–859.

39 Sledge, G.W. Jr., Toi, M., Neven, P. et al. (2020). The effect of abemaciclib plus fulvestrant on overall survival in hormone receptor-positive, ERBB2-negative breast cancer that progressed on endocrine therapy-monarch 2: a randomized clinical trial. *JAMA Oncol.* 6: 116–124.

40 Llombart-Cussac, A., Sledge, G., Toi, M. et al. (2023). Abstract PD13-11: PD13-11 final overall survival analysis of monarch 2: a phase 3 trial of abemaciclib plus fulvestrant in patients with hormone receptor-positive, HER2-negative advanced breast cancer. *Cancer Res.* 83: PD13-11.

41 Goetz, M.P., Toi, M., Huober, J. et al. (2022). LBA15 MONARCH 3: interim overall survival (OS) results of abemaciclib plus a nonsteroidal aromatase inhibitor (NSAI) in patients (pts) with HR+, HER2- advanced breast cancer (ABC). *Ann. Oncol.* 33: S1384.

42 Cristofanilli, M., Rugo, H.S., Im, S.A. et al. (2022). Overall survival with palbociclib and fulvestrant in women with HR+/HER2- ABC: updated exploratory analyses of PALOMA-3, a double-blind, phase iii randomized study. *Clin. Cancer Res.* 28: 3433–3442.

43 Finn, R.S., Rugo, H.S., Dieras, V.C. et al. (2022). Overall survival (OS) with first-line palbociclib plus letrozole (PAL+LET) versus placebo plus letrozole (PBO+LET) in women with estrogen receptor–positive/human epidermal growth factor receptor 2–negative advanced breast cancer (ER+/HER2– ABC): analyses from PALOMA-2. *J. Clin. Oncol.* 40: LBA1003.

44 Finn, R.S., Boer, K., Bondarenko, I. et al. (2020). Overall survival results from the randomized phase 2 study of palbociclib in combination with letrozole versus letrozole alone for first-line treatment of ER+/HER2- advanced breast cancer (PALOMA-1, TRIO-18). *Breast Cancer Res. Treat.* 183: 419–428.

45 DeMichele, A., Cristofanilli, M., Brufsky, A. et al. (2021). Comparative effectiveness of first-line palbociclib plus letrozole versus letrozole alone for HR+/HER2– metastatic breast cancer in US real-world clinical practice. *Breast Cancer Res.* 23: 37.

46 Rugo, H.S., Brufsky, A., Liu, X. et al. (2022). Real-world study of overall survival with palbociclib plus aromatase inhibitor in HR+/HER2– metastatic breast cancer. *npj Breast Cancer* 8: 114.

47 Abdelmalak, M., Singh, R., Anwer, M. et al. (2022). The renaissance of CDK inhibitors in breast cancer therapy: an update on clinical trials and therapy resistance. *Cancers (Basel)* 14: 5388.

48 Gnant, M., Dueck, A.C., Frantal, S. et al. (2022). Adjuvant palbociclib for early breast cancer: the PALLAS trial results (ABCSG-42/AFT-05/BIG-14-03). *J. Clin. Oncol.* 40: 282–293.

49 Loibl, S., Marmé, F., Martin, M. et al. (2021). Palbociclib for residual high-risk invasive HR-positive and HER2-negative early breast cancer-the penelope-B Trial. *J. Clin. Oncol.* 39: 1518–1530.

50 Slamon, D.J., Fasching, P.A., Hurvitz, S. et al. (2023). Rationale and trial design of NATALEE: a Phase III trial of adjuvant ribociclib + endocrine therapy versus endocrine therapy alone in patients with HR+/HER2- early breast cancer. *Ther. Adv. Med. Oncol.* 15: 17588359231178125.

51 Slamon, D.J., Stroyakovskiy, D., Yardley, D.A. et al. (2023). Ribociclib and endocrine therapy as adjuvant treatment in patients with HR+/HER2- early breast cancer: Primary results from the phase III NATALEE trial. *J. Clin. Oncol.* 41: LBA500.

52 Johnston, S.R.D., Toi, M., O'Shaughnessy, J. et al. (2023). Abemaciclib plus endocrine therapy for hormone receptor-positive, HER2-negative, node-positive, high-risk early breast cancer (monarchE): results from a preplanned interim analysis of a randomised, open-label, phase 3 trial. *Lancet Oncol.* 24: 77–90.

53 Grinshpun, A., Tolaney, S.M., Burstein, H.J. et al. (2023). The dilemma of selecting a first line CDK4/6 inhibitor for hormone receptor-positive/HER2-negative metastatic breast cancer. *npj Breast Cancer* 9: 15.

54 Panagiotou, E., Gomatou, G., Trontzas, I.P. et al. (2022). Cyclin-dependent kinase (CDK) inhibitors in solid tumors: a review of clinical trials. *Clin. Transl. Oncol.* 24: 161–192.

55 Brett, J.O., Herman, P.E., Mayer, E.L. et al. (2022). Mechanisms of resistance to CDK4/6 inhibitors in hormone receptor-positive (HR +) breast cancer: spotlight on convergent CDK6 upregulation and immune signaling. *Curr. Breast Cancer Rep.* 14: 222–232.

56 Gomatou, G., Trontzas, I., Ioannou, S. et al. (2021). Mechanisms of resistance to cyclin-dependent kinase 4/6 inhibitors. *Mol. Biol. Rep.* 48: 915–925.

57 Guarducci, C., Bonechi, M., Boccalini, G. et al. (2017). Mechanisms of resistance to CDK4/6 inhibitors in breast cancer and potential biomarkers of response. *Breast Care* 12: 304–308.

58 Ono, M., Oba, T., Shibata, T., and Ito, K.I. (2021). The mechanisms involved in the resistance of estrogen receptor-positive breast cancer cells to palbociclib are multiple and change over time. *J. Cancer Res. Clin. Oncol.* 147: 3211–3224.

59 Xu, X.Q., Pan, X.H., Wang, T.T. et al. (2021). Intrinsic and acquired resistance to CDK4/6 inhibitors and potential overcoming strategies. *Acta Pharmacol. Sin.* 42: 171–178.

60 Kalinsky, K., Accordino, M.K., Chiuzan, C. et al. (2023). Randomized phase II trial of endocrine therapy with or without ribociclib after progression on cyclin-dependent kinase 4/6 inhibition in hormone receptor–positive, human epidermal growth factor receptor 2–negative metastatic breast cancer: maintain trial. *J. Clin. Oncol.* 41: 4004–4013.

61 Kalinsky, K., Layman, R.M., Kaufman, P.A. et al. (2022). postMONARCH: a phase 3 study of abemaciclib plus fulvestrant versus placebo plus fulvestrant in patients with HR+, HER2-, metastatic breast cancer following progression on a CDK4 & 6 inhibitor and endocrine therapy. *J. Clin. Oncol.* 40: TPS1117.

62 Guiley, K.Z., Stevenson, J.W., Lou, K. et al. (2019). p27 allosterically activates cyclin-dependent kinase 4 and antagonizes palbociclib inhibition. *Science* 366: eaaw2106.

63 Fassl, A., Brain, C., Abu-Remaileh, M. et al. (2020). Increased lysosomal biomass is responsible for the resistance of triple-negative breast cancers to CDK4/6 inhibition. *Sci. Adv.* 6: eabb2210.

64 Schoninger, S.F. and Blain, S.W. (2020). The ongoing search for biomarkers of CDK4/6 inhibitor responsiveness in breast cancer. *Mol. Cancer Ther.* 19: 3–12.

65 Huang, W. and Wang, H. (2021). Potential biomarkers of resistance to CDK4/6 inhibitors: a narrative review of preclinical and clinical studies. *Transl. Breast Cancer Res.* 2: 12–12.

66 Bui, T.B.V., Burgering, B.M.T., Goga, A. et al. (2022). Biomarkers for cyclin-dependent kinase 4/6 inhibitors in the treatment of hormone receptor-positive/human epidermal growth factor receptor 2-negative advanced/metastatic breast cancer: translation to clinical practice. *JCO Precis. Oncol.* 6: e2100473.

67 Harbeck, N., Rastogi, P., Martin, M. et al. (2021). Adjuvant abemaciclib combined with endocrine therapy for high-risk early breast cancer: updated efficacy and Ki-67 analysis from the monarchE study. *Ann. Oncol.* 32: 1571–1581.

68 Turner, N.C., Liu, Y., Zhu, Z. et al. (2019). Cyclin E1 expression and palbociclib efficacy in previously treated hormone receptor-positive metastatic breast cancer. *J. Clin. Oncol.* 37: 1169–1178.

69 Raspé, E., Coulonval, K., Pita, J.M. et al. (2017). CDK4 phosphorylation status and a linked gene expression profile predict sensitivity to palbociclib. *EMBO Mol. Med.* 9: 1052–1066.

70 Green, J.L., Okerberg, E.S., Sejd, J. et al. (2019). Direct CDKN2 modulation of CDK4 alters target engagement of CDK4 inhibitor drugs. *Mol. Cancer Ther.* 18: 771–779.

71 Gottesman, S.R.S., Somma, J., Tsiperson, V. et al. (2019). Tyrosine phosphorylation of p27Kip1 correlates with palbociclib responsiveness in breast cancer tumor cells grown in explant culture. *Mol. Cancer Res.* 17: 669–675.

72 Caruso, J.A., Duong, M.T., Carey, J.P.W. et al. (2018). Low-Molecular-Weight cyclin E in human cancer: cellular consequences and opportunities for targeted therapies. *Cancer Res.* 78: 5481–5491.

73 Pack, L.R., Daigh, L.H., Chung, M., and Meyer, T. (2021). Clinical CDK4/6 inhibitors induce selective and immediate dissociation of p21 from cyclin D-CDK4 to inhibit CDK2. *Nat. Commun.* 12: 3356.

74 Li, Z., Razavi, P., Li, Q. et al. (2018). Loss of the FAT1 tumor suppressor promotes resistance to CDK4/6 inhibitors via the hippo pathway. *Cancer Cell* 34: 893–905.e898.

75 McCartney, A., Bonechi, M., De Luca, F. et al. (2020). Plasma thymidine kinase activity as a biomarker in patients with luminal metastatic breast cancer treated with palbociclib within the TREnd trial. *Clin. Cancer Res.* 26: 2131–2139.

76 Herrera-Abreu, M.T., Palafox, M., Asghar, U. et al. (2016). Early adaptation and acquired resistance to CDK4/6 inhibition in estrogen receptor-positive breast cancer. *Cancer Res.* 76: 2301–2313.

77 Al-Lazikani, B., Banerji, U., and Workman, P. (2012). Combinatorial drug therapy for cancer in the post-genomic era. *Nat. Biotechnol.* 30: 679–692.

78 Bayat Mokhtari, R., Homayouni, T.S., Baluch, N. et al. (2017). Combination therapy in combating cancer. *Oncotarget* 8: 38022–38043.

79 Rampioni Vinciguerra, G.L., Sonego, M., Segatto, I. et al. (2022). CDK4/6 inhibitors in combination therapies: better in company than alone: a mini review. *Front. Oncol.* 12: 891580.

80 Papadimitriou, M.C., Pazaiti, A., Iliakopoulos, K. et al. (2022). Resistance to CDK4/6 inhibition: mechanisms and strategies to overcome a therapeutic problem in the treatment of hormone receptor-positive metastatic

breast cancer. *Biochim. Biophys. Acta, Mol. Cell Res.* 1869: 119346.

81 Braal, C.L., Jongbloed, E.M., Wilting, S.M. et al. (2021). Inhibiting CDK4/6 in breast cancer with palbociclib, ribociclib, and abemaciclib: similarities and differences. *Drugs* 81: 317–331.

82 Patel, P., Asbach, B., Shteyn, E. et al. (2015). Brk/Protein tyrosine kinase 6 phosphorylates p27KIP1, regulating the activity of cyclin D–cyclin-dependent kinase 4. *Mol. Cell. Biol.* 35: 1506–1522.

83 Rampioni Vinciguerra, G.L., Dall'Acqua, A., Segatto, I. et al. (2021). p27kip1 expression and phosphorylation dictate palbociclib sensitivity in KRAS-mutated colorectal cancer. *Cell Death Dis.* 12: 951.

84 Wander, S.A., Cohen, O., Gong, X. et al. (2020). The genomic landscape of intrinsic and acquired resistance to cyclin-dependent kinase 4/6 inhibitors in patients with hormone receptor–positive metastatic breast cancer. *Cancer Discov.* 10: 1174–1193.

85 Lee, M.S., Helms, T.L., Feng, N. et al. (2016). Efficacy of the combination of MEK and CDK4/6 inhibitors in vitro and in vivo in KRAS mutant colorectal cancer models. *Oncotarget* 7: 39595–39608.

86 Ziemke, E.K., Dosch, J.S., Maust, J.D. et al. (2016). Sensitivity of KRAS-mutant colorectal cancers to combination therapy that cotargets MEK and CDK4/6. *Clin. Cancer Res.* 22: 405–414.

87 Iyengar, M., O'Hayer, P., Cole, A. et al. (2018). CDK4/6 inhibition as maintenance and combination therapy for high grade serous ovarian cancer. *Oncotarget* 9: 15658–15672.

88 Teo, Z.L., Versaci, S., Dushyanthen, S. et al. (2017). Combined CDK4/6 and PI3Kα inhibition is synergistic and immunogenic in triple-negative breast cancer. *Cancer Res.* 77: 6340–6352.

89 Michaloglou, C., Crafter, C., Siersbaek, R. et al. (2018). Combined inhibition of mTOR and CDK4/6 is required for optimal blockade of E2F function and long-term growth inhibition in estrogen receptor–positive breast cancer. *Mol. Cancer Ther.* 17: 908–920.

90 Mao, P., Cohen, O., Kowalski, K.J. et al. (2020). Acquired FGFR and FGF alterations confer resistance to estrogen receptor (ER) targeted therapy in ER+ metastatic breast cancer. *Clin. Cancer Res.* 26: 5974–5989.

91 Jaaks, P., Coker, E.A., Vis, D.J. et al. (2022). Effective drug combinations in breast, colon and pancreatic cancer cells. *Nature* 603: 166–173.

92 Finn, R.S., Dering, J., Conklin, D. et al. (2009). PD 0332991, a selective cyclin D kinase 4/6 inhibitor, preferentially inhibits proliferation of luminal estrogen receptor-positive human breast cancer cell lines in vitro. *Breast Cancer Res.* 11: R77.

93 Gao, H., Korn, J.M., Ferretti, S. et al. (2015). High-throughput screening using patient-derived tumor xenografts to predict clinical trial drug response. *Nat. Med.* 21: 1318–1325.

94 Garutti, M., Targato, G., Buriolla, S. et al. (2021). CDK4/6 inhibitors in melanoma: a comprehensive review. *Cells* 10: 1334.

95 Ascierto, P.A., Bechter, O., Wolter, P. et al. (2017). A phase Ib/II dose-escalation study evaluating triple combination therapy with a BRAF (encorafenib), MEK (binimetinib), and CDK 4/6 (ribociclib) inhibitor in patients (Pts) with BRAF V600-mutant solid tumors and melanoma. *J. Clin. Oncol.* 35: 9518–9518.

96 Franco, J., Witkiewicz, A.K., and Knudsen, E.S. (2014). CDK4/6 inhibitors have potent activity in combination with pathway selective therapeutic agents in models of pancreatic cancer. *Oncotarget* 5: 6512–6525.

97 Sharma, P., Goswami, S., Raychaudhuri, D. et al. (2023). Immune checkpoint therapy – current perspectives and future directions. *Cell* 186: 1652–1669.

98 Goel, S., DeCristo, M.J., Watt, A.C. et al. (2017). CDK4/6 inhibition triggers anti-tumour immunity. *Nature* 548: 471–475.

99 Uzhachenko, R.V., Bharti, V., Ouyang, Z. et al. (2021). Metabolic modulation by CDK4/6 inhibitor promotes chemokine-mediated recruitment of T cells into mammary tumors. *Cell Rep.* 35: 108944.

100 Heckler, M., Ali, L.R., Clancy-Thompson, E. et al. (2021). Inhibition of CDK4/6 promotes CD8 T-cell memory formation. *Cancer Discov.* 11: 2564–2581.

101 Zhang, J., Bu, X., Wang, H. et al. (2018). Cyclin D-CDK4 kinase destabilizes PD-L1 via cullin 3-SPOP to control cancer immune surveillance. *Nature* 553: 91–95.

102 Bisi, J.E., Sorrentino, J.A., Jordan, J.L. et al. (2017). Preclinical development of G1T38: a novel, potent and selective inhibitor of cyclin dependent kinases 4/6 for use as an oral antineoplastic in patients with CDK4/6 sensitive tumors. *Oncotarget* 8: 42343–42358.

103 Xu, B., Zhang, Q., Zhang, P. et al. (2021). Dalpiciclib or placebo plus fulvestrant in hormone receptor-positive and HER2-negative advanced breast cancer: a randomized, phase 3 trial. *Nat. Med.* 27: 1904–1909.

104 Long, F., He, Y., Fu, H. et al. (2019). Preclinical characterization of SHR6390, a novel CDK 4/6 inhibitor, in vitro and in human tumor xenograft models. *Cancer Sci.* 110: 1420–1430.

105 Zhang, P., Zhang, Q., Tong, Z. et al. (2023). Dalpiciclib plus letrozole or anastrozole versus placebo plus letrozole or anastrozole as first-line treatment in patients

with hormone receptor-positive, HER2-negative advanced breast cancer (DAWNA-2): a multicentre, randomised, double-blind, placebo-controlled, phase 3 trial. *Lancet Oncol.* 24: 646–657.

106 Zhang, J., Liu, R., Gao, S. et al. (2023). BPI-16350, a novel promising CDK4/6 inhibitor for HR+/HER2- metastatic breast cancer (MBC): results from a phase I study. *J. Clin. Oncol.* 41: e13051–e13051.

107 Freeman-Cook, K.D., Hoffman, R.L., Behenna, D.C. et al. (2021). Discovery of PF-06873600, a CDK2/4/6 inhibitor for the treatment of cancer. *J. Med. Chem.* 64: 9056–9077.

108 Dietrich, C., Trub, A., Ahn, A. et al. (2023). INX-315, a selective CDK2 inhibitor, induces cell cycle arrest and senescence in solid tumors. *Cancer Discov.* 14 (3): 446–467.

109 Patel, M.R., Juric, D., Henick, B.S. et al. (2023). BLU-222, an oral, potent, and selective CDK2 inhibitor, in patients with advanced solid tumors: Phase 1 monotherapy dose escalation. *J. Clin. Oncol.* 41: 3095–3095.

110 Said, M.A., Abdelrahman, M.A., Abourehab, M.A.S. et al. (2022). A patent review of anticancer CDK2 inhibitors (2017-present). *Expert Opin. Ther. Pat.* 32: 885–898.

111 Sokolsky, A., Winterton, S., Kennedy, K. et al. (2022). Discovery of 5,7-dihydro-6H-pyrrolo[2,3-d] pyrimidin-6-ones as highly selective CDK2 Inhibitors. *ACS Med. Chem. Lett.* 13: 1797–1804.

112 Yap, T.A., Elhaddad, A.M., Grisham, R.N. et al. (2023). First-in-human phase 1/2a study of a potent and novel CDK2 selective inhibitor PF-07104091 in patients (pts) with advanced solid tumors, enriched for CDK4/6 inhibitor resistant HR+/HER2- breast cancer. *J. Clin. Oncol.* 41: 3010–3010.

113 Bekes, M., Langley, D.R., and Crews, C.M. (2022). PROTAC targeted protein degraders: the past is prologue. *Nat. Rev. Drug Discov.* 21: 181–200.

114 Chirnomas, D., Hornberger, K.R., and Crews, C.M. (2023). Protein degraders enter the clinic – a new approach to cancer therapy. *Nat. Rev. Clin. Oncol.* 20: 265–278.

115 Rana, S., Mallareddy, J.R., Singh, S. et al. (2021). Inhibitors, PROTACs and molecular glues as diverse therapeutic modalities to target cyclin-dependent kinase. *Cancer* 13: 5506.

23

Tapentadol, a Clinically Proven Analgesic with Two Mechanisms

Thomas Christoph, Helmut Buschmann, Norbert Handler, and Michal Poznik

23.1 Introduction

Tapentadol was invented at the German pharmaceutical company Grünenthal in the late 1980s. The Grünenthal team started by analyzing the chemistry and activity of tramadol, which had been invented at the same company in 1962 [1].

Tapentadol, 3-[(2R,3R)-1-(dimethylamino)-2-methylpentan-3-yl]phenol or phenol, 3-[(1R,2R)-3-(dimethylamino)-1-ethyl-2-methylpropyl]- (9CI, ACI) is after more than 15 years on the market still an innovative and centrally acting synthetic analgesic with a dual mechanism of action. The commercial and scientific interest is also related to the number of patents and scientific publications. Up to now (June 2024), more than 1000 scientific publications and more than 500 patents are published and filed (Figure 23.1), with an average of 100 publications and 25–30 patents each year since 2012:

Tapentadol has two chiral centers and can exist in four stereoisomeric forms, which are shown in Figure 23.2:

The (+)-(1S,2S)-enantiomer of tapentadol (laboratory code BN-202) is an extraordinarily strong μ-opioid receptor (MOR) agonist. The respective epimers of tapentadol and its (+) enantiomer have no relevant biological activity.

In 2022, generic tapentadol was registered in European member states according to the national registers of authorized medicines [2]. This entry in the generic market marked another important milestone in the development and life-cycle management of this fascinating drug, which began decades ago. By 2022, tapentadol is a well-established member of a new opioid class, combining agonistic efficacy at the MOR and reuptake inhibitory efficacy for norepinephrine (NRI), a class of drugs meanwhile called MOR-NRI. The chemical structure of tapentadol was designed by purpose to create exactly such a combination in a single molecule. Tapentadol reflects the lessons learned with the weak opioid analgesic tramadol, which carries not only MOR and NRI but also, in addition, inhibits reuptake of serotonin (SRI). Thereafter, it took a long data-driven journey to prove the hypotheses that these two mechanisms enable treatment of moderate to severe pain conditions and are specifically beneficial in the treatment of hard-to-treat chronic pain indications such as neuropathic pain. Step by step, preclinical data paved the way toward clinical development, and clinical studies confirmed many of the promises suggested in preclinical experiments. Key aspects of this translational approach will be discussed in this article. At the same time, Chemistry, Manufacturing, and Controls (CMC) activities were developed to meet regulatory requirements for market authorization. Tapentadol reached market authorization in 2008 (FDA) and 2010 (EMA) and by then was ready to face clinical reality checks in day-to-day clinical practice.

23.2 The Discovery of Tapentadol – From Morphine and Tramadol to the Discovery of Tapentadol

Since the isolation of morphine two centuries ago, numerous derivatives have been synthesized, with the primary aim of separating analgesic effects from unwanted effects, such as nausea and emesis, constipation, respiratory depression, addiction, and dependence. An alternative approach to improving the therapeutic usefulness of opioid analgesics is to combine MOR activation with an additional mechanism of action aimed at enhancing analgesic efficacy. Such an additional analgesic mechanism may result in an opiate-sparing effect, thereby attenuating the unwanted effects of MOR activation [1].

The concept of combined mechanisms of action was realized for the first time with the introduction of tramadol, a compound that produces MOR activation and inhibition of SRI (5-HT) and NRI reuptake.

Polypharmacology: Strategies for Multi-Target Drug Discovery, First Edition. Edited by Jens-Uwe Peters.

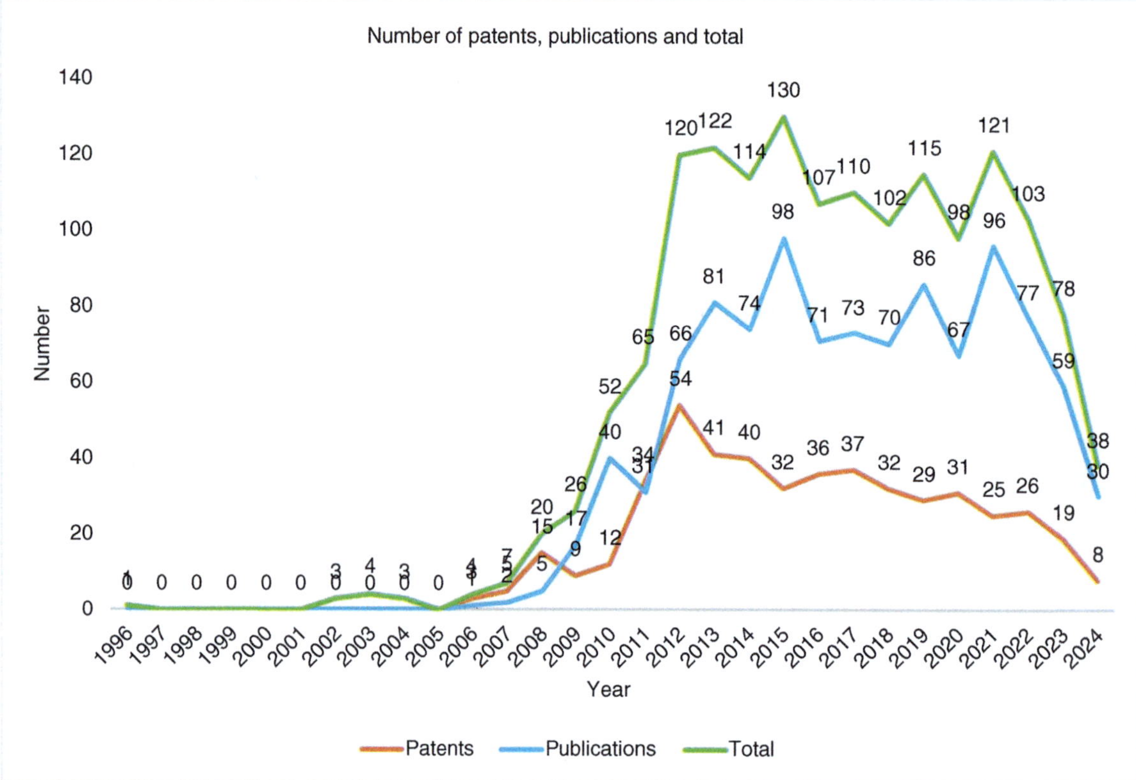

Figure 23.1 Scientific publications and patents registered in the SciFinder data base between 1996 (first patent covering of tapentadol) and June 2024 (https://scifinder-n.cas.org/, assessed on June 14, 2024).

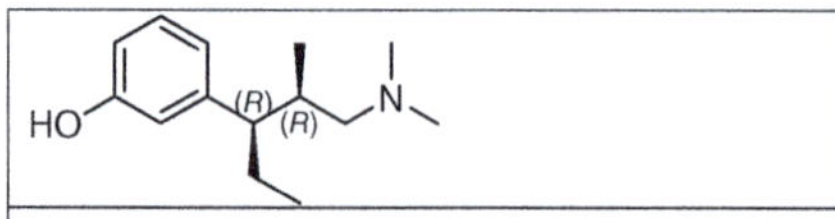	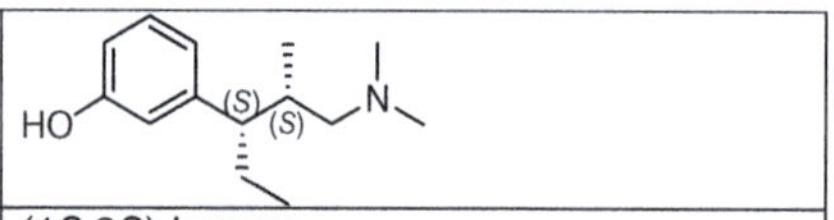
(1*R*,2*R*)-isomer Tapentadol, 3-[(1*R*,2*R*)-3-(dimethylamino)-1-ethyl-2-methylpropyl]-phenol CAS No.: 175591-23-8	(1*S*,2*S*)-isomer 3-[(1*S*,2*S*)-3-(dimethylamino)-1-ethyl-2-methylpropyl]-phenol, (+)-enantiomer of tapentadol CAS No.: 454221-04-6
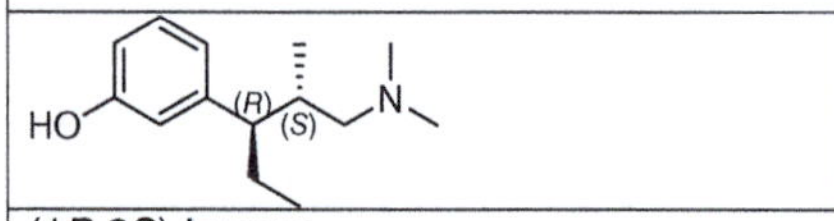	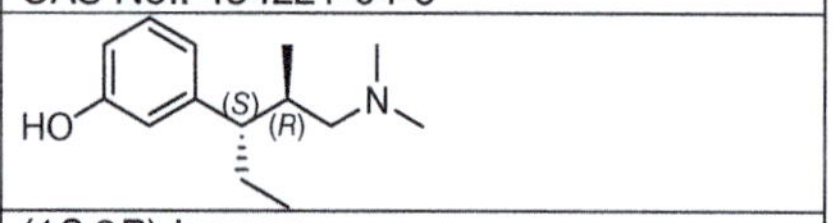
(1*R*,2*S*)-isomer 3-[(1*R*,2*S*)-3-(dimethylamino)-1-ethyl-2-methylpropyl]-phenol, diastereomer of tapentadol CAS No.: 953400-56-1	(1*S*,2*R*)-isomer 3-[(1*S*,2*R*)-3-(dimethylamino)-1-ethyl-2-methylpropyl]-phenol, diastereomer of tapentadol CAS No.: 953400-57-2

Figure 23.2 The four stereoisomers of 3-[1-(dimethylamino)-2-methylpentan-3-yl]phenol.

Tapentadol is shown in Figure 23.3 together with morphine and tramadol to show the structural relationships.

Tramadol was first synthesized in 1962 by Flick at Grünenthal and has been available for pain treatment in Germany since 1977. Immediate-release preparations of tramadol were introduced in the United Kingdom in 1994, in the United States in 1995, and subsequently in many other countries. Nowadays, several once-daily, sustained-release formulations of tramadol have been marketed in several countries all over the world. The drug has a wide range of applications, including treatment for restless leg syndrome and fibromyalgia [1].

Tramadol is generally considered a lower-risk opioid option for the treatment of moderate to severe pain. Tramadol binds weakly to κ- and δ-opioid receptors and to the MOR with 6000-fold less affinity than

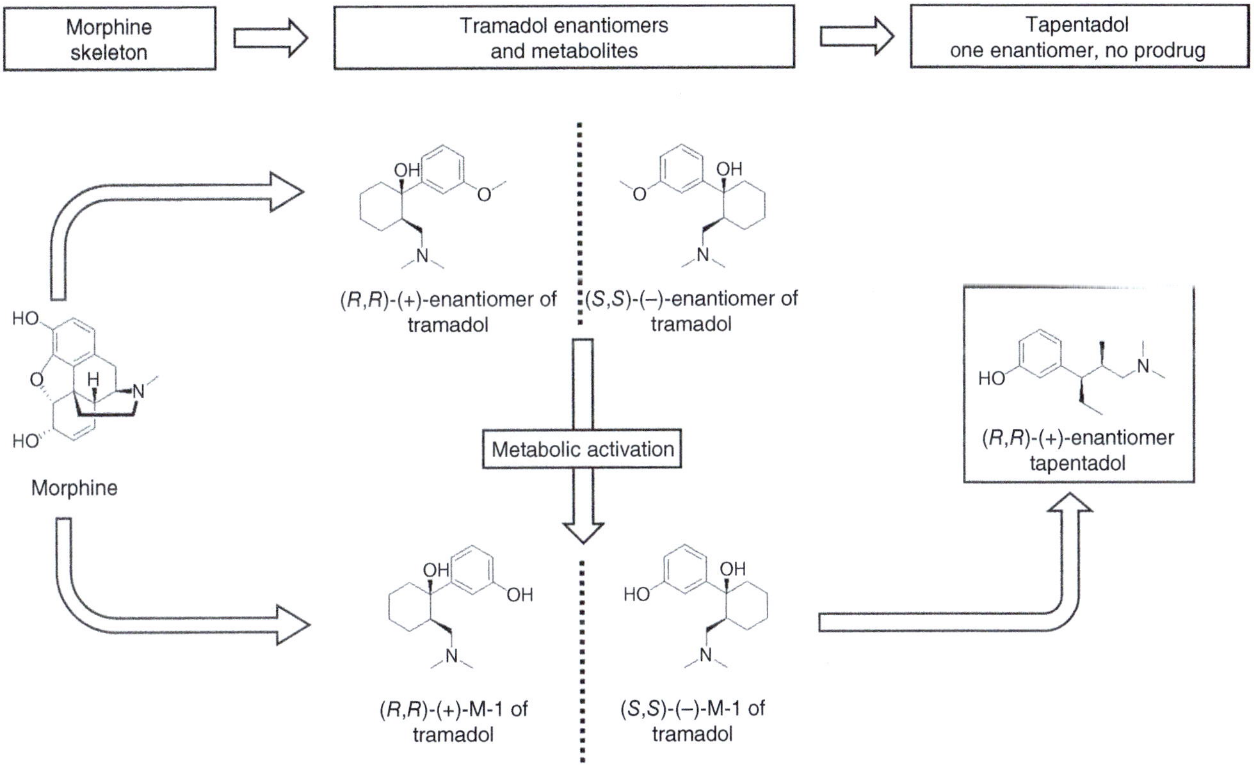

Figure 23.3 Chemical structure of tapentadol, morphine, and tramadol (the absolute stereochemistry of the (*S*,*S*)-(-)-M-1 metabolite of tramadol corresponds to the stereochemistry (-)-(*R*,*R*)-(-)-tapentadol, absolute chirality remains identical, and descriptors have changed according to the priority rules of the Cahn-Ingold-Prelog nomenclature for chiral descriptors)

morphine [3]. It is considered a Step 2 option on the World Health Organization's pain ladder and has about 1/10th of the potency of morphine.

Tramadol differs from other traditional opioid medications in that it does not just act as a μ-opioid agonist but also affects monoamines by modulating the effects of neurotransmitters involved in the modulation of pain, such as SRI and NRI, which activate descending pain inhibitory pathways [4]. Tramadol's effects on SRI and NRI mimic the effects of other SNRI antidepressants such as duloxetine and venlafaxine.

Tramadol exists as a racemic mixture consisting of two pharmacologically active enantiomers that both contribute to its analgesic property through different mechanisms and are also themselves metabolized into active metabolites: (+)-tramadol and its primary metabolite (+)-O-desmethyl-tramadol (M1) are agonists of the MOR, while (+)-tramadol inhibits SRI reuptake and (−)-tramadol inhibits NRI reuptake. These pathways are complementary and synergistic, improving tramadol's ability to modulate the perception of and response to pain [1S,25]. The (1R,2R)-(+)-enantiomer of tramadol is approximately four times more potent than the (1S,2S)-(−)-enantiomer in terms of MOR affinity and 5-HT reuptake. In animal models, racemic M-1 is up to six times more potent than tramadol in producing analgesia and 200 times more potent in μ-opioid binding [5].

The fact that these four compounds (the enantiomers of tramadol and the enantiomeric metabolites M-1) have different pharmacological properties makes tramadol an atypical opioid with a complex mechanism of action [6, 7]. This is the reason why tramadol is used as a racemate, despite known different physiological effects of the enantiomers because the racemate has a better pharmacological profile than the single enantiomers in terms of efficacy and tolerability in animals and humans.

However, the relative contribution of the different mechanisms of action to the overall analgesic effect changes over time. As the parent molecule is metabolized, the contribution of 5-HT and NE reuptake inhibition is reduced, whereas the contribution of MOR agonism increases, resulting in a complex time- and metabolism-dependent pattern of pharmacological activities. Because tramadol is mainly metabolized via cytochrome P450 2D6 (CYP2D6), which is polymorphic in humans, approximately 5–15% of the white population are "poor metabolizers" of tramadol and do not experience satisfactory analgesia with standard doses.

Tramadol has also been shown to affect a number of other pain modulators within the central nervous system as well as non-neuronal inflammatory markers and immune mediators [6–10]. Due to the broad spectrum of targets involved in pain and inflammation, it's not surprising that the evidence has shown that tramadol is effective for a number of pain types, including neuropathic pain, postoperative pain, lower back pain, as well as pain associated with labor, osteoarthritis, fibromyalgia, and cancer. Due to its SNRI activity, tramadol also has anxiolytic, antidepressant, and anti-shivering effects, which are all frequently found as comorbidities with pain [8]. Based on elucidation of the favorable aspects of the multimodal analgesic mechanism of action of tramadol, a research program was initiated in the late 1980s at Grünenthal to design a new class of opioid/nonopioid analgesics that would combine MOR agonism and NRI reuptake inhibition, with minimal serotonergic activity. It was also desired that both activities would reside in a single parent molecule that does not require metabolic activation via the cytochrome P450 system. Increasing evidence demonstrates that many pharmaceutically important compounds elicit their effects through binding to multiple targets. Tapentadol is an example of a rationally designed multiple ligand starting from the tramadol structure and its understanding of the complex pharmacological activity [1].

The discovery of tapentadol involved multiple steps of chemical innovations to become a unique drug, which may be summarized as follows in the following Figure 23.4.

23.3 Pharmacokinetics of Tapentadol

After oral administration, 32% of the drug is absorbed in humans [12]. It is widely distributed in the body. The plasma protein binding is low and amounts to approximately 20%.

Tapentadol is mainly metabolized by phase II pathways, and only a small amount is metabolized by Phase I oxidative pathways; thus, drug metabolism mediated by the cytochrome P450 system is of less importance than phase II conjugation [13]. The major pathway of tapentadol metabolism is conjugation with glucuronic acid to produce glucuronides [13, 14]. After oral administration, approximately 70% of the drug, of which 55% is the O-glucuronide metabolite and 15% is the sulfate metabolite, is excreted in urine. About 3% of the dose was excreted in urine as the unchanged parent drug [13].

Tapentadol can also undergo CYP2C9- and CYP2C19-mediated demethylation to form N-desmethyl tapentadol,

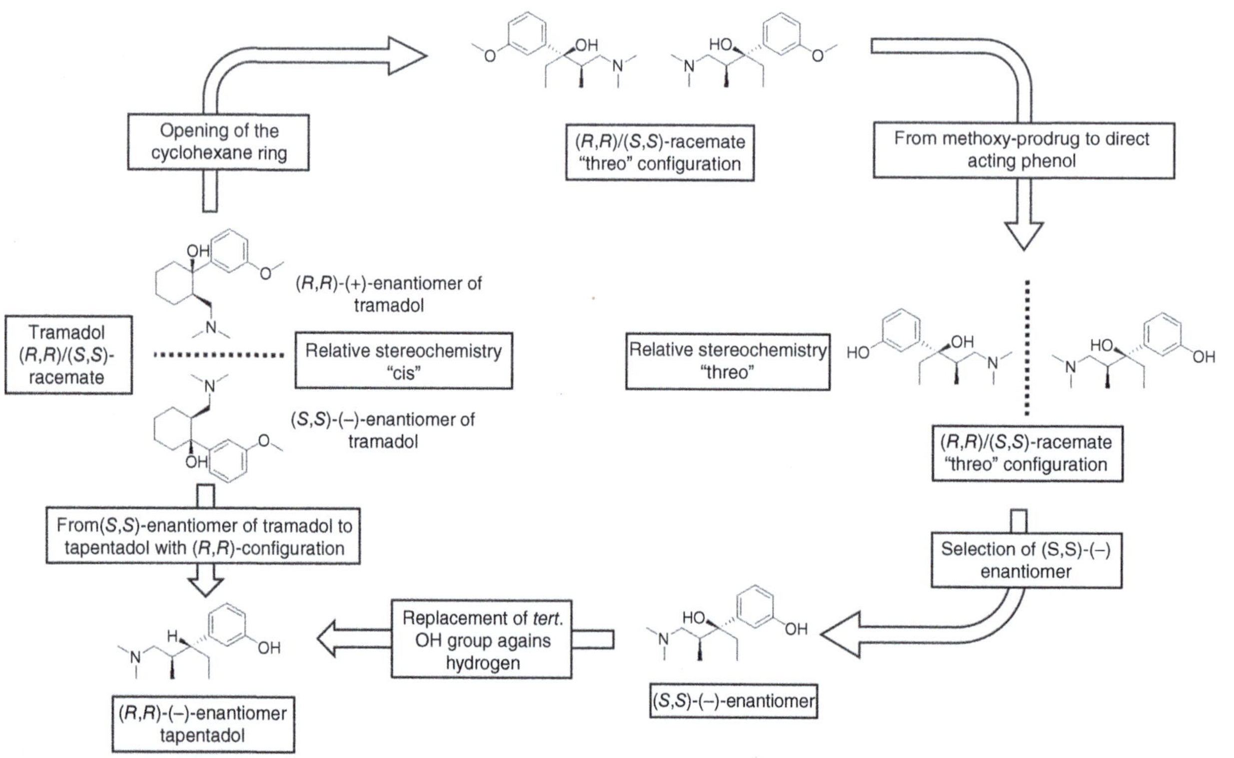

Figure 23.4 Chemical innovations introduced in the development of tapentadol starting from the tramadol structure. Adapted from Raffa et al. [11].

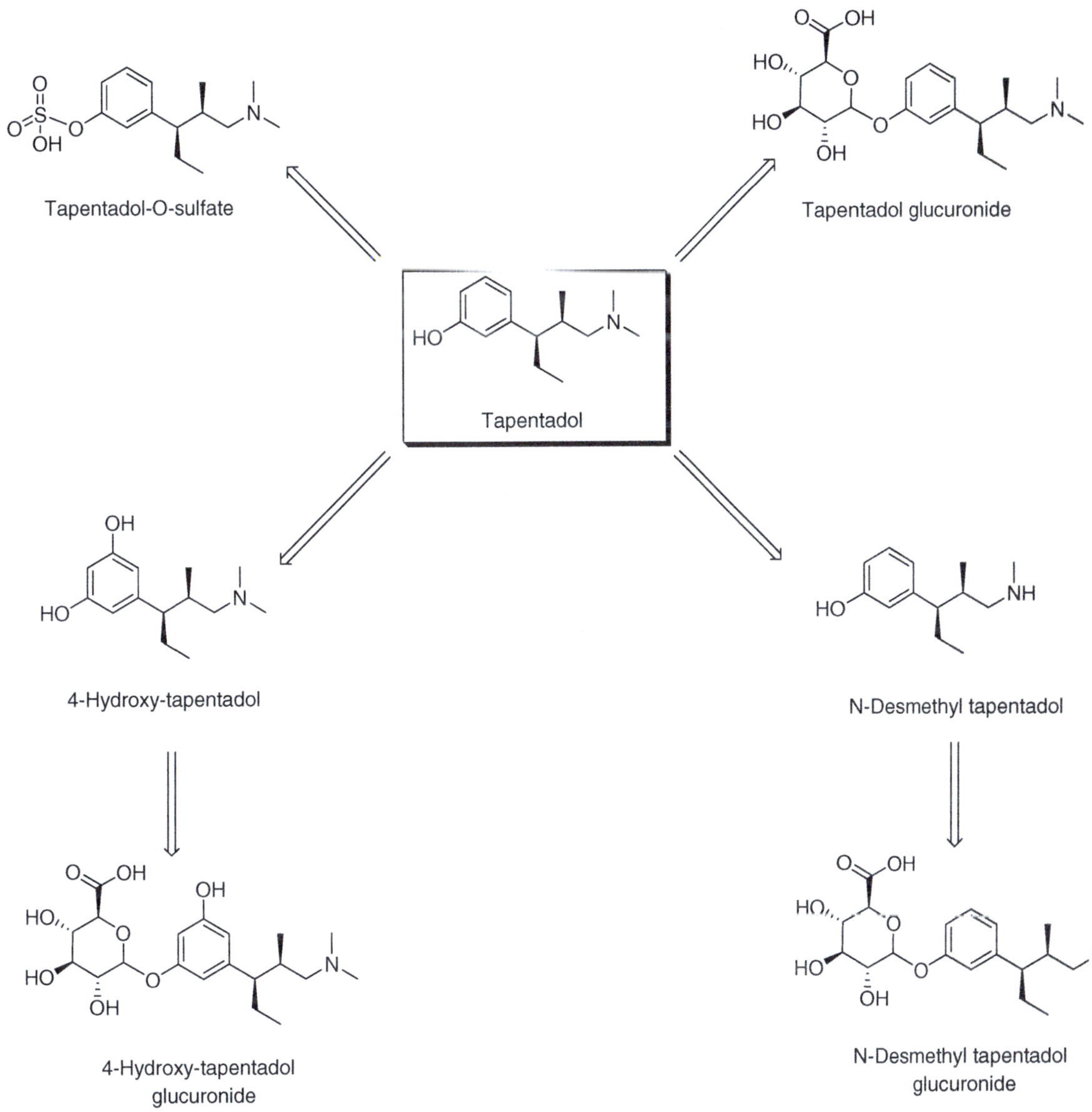

Figure 23.5 Major metabolic pathways of tapentadol. Adapted from Drugbank [12].

which accounts for 13% of the dose. About 2% of tapentadol can also undergo CYP2D6-mediated hydroxylation to form hydroxy tapentadol. N-desmethyl tapentadol and hydroxy tapentadol can further be glucuronidated.

The activity of tapentadol is independent of metabolic activation and resides in a single enantiomer that readily crosses the blood–brain barrier; hence, tapentadol displays a rapid onset of action after administration. The half-life is four hour and the peak effect is attained after one hour. The duration of action is four to six hours. The drug undergoes extensive first-pass hepatic metabolism, that is, 97%.

Drug–drug interaction studies to date have documented no induction of CYP1A2, CYP2C9, or CYP3A4 activity; in addition, these studies found no significant inhibition of CYP1A2, CYP2A6, CYP2C9, CYP2C19, CYP2E1, or CYP3A4 activity. CYP2D6 activity was inhibited, but to a degree that did not appear to be clinically relevant [15].

The major metabolic pathways are shown in the following Figure 23.5 [12].

23.4 The Polymorphic Forms of Tapentadol Hydrochloride

Polymorphism refers to the ability of crystalline material to exist in more than one solid or crystalline form. It is believed that 50–80% of all pharmaceuticals exist in at least two polymorphic forms. Polymorphic forms can potentially be characterized by techniques such as powder X-ray diffraction (PXRD), vibrational spectroscopy, including

IR and Raman spectroscopy, and solid-state NMR spectroscopy (ssNMR) [16]. Pharmaceutical regulators, such as the United States FDA and the European Medicines Agency, require screening of polymorphs during new pharmaceutical development to uncover polymorphic forms and evidence as to whether the choice of polymorph will affect the pharmaceutical efficacy or safety [17]. This is because pharmaceutical polymorphs often differ in their physicochemical properties, such as solubility, dissolution rate, stability, hygroscopicity, wettability, hardness, and optical and elec. properties [17, 18].

European patents EP1612203 [19] and EP 1799633 [20] disclose polymorphic Forms A and B of tapentadol hydrochloride, and characterize them by PXRD, infrared spectroscopy (IR), Raman spectroscopy, and crystal structure analysis.

In the meantime, several methods are disclosed to prepare either Form A or Form B. It is disclosed that the crystalline Form A has the same pharmacological activity than Form B but is stable under ambient conditions, while Form B is descripted to be meta-stable only [1]. US patent publication US2010272815A1 [21] discloses a process for the preparation of amorphous tapentadol hydrochloride by dissolving tapentadol hydrochloride in a solvent, such as water, organic solvent, or mixture thereof, and amorphous tapentadol hydrochloride is isolated by the removal of the solvent by vacuum drying or spray drying.

Chinese patent publication CN102924303A [22] discloses a process for the preparation of tapentadol hydrochloride Form C. According to the process disclosed in this Chinese patent publication, tapentadol hydrochloride Form C is prepared by reacting tapentadol-free base with hydrochloric acid/ethanol and using acetone as solvent. The existence of this so-called Form C is doubted by several experts.

US Patent No. US 6,248,737 ("the '737 patent") [23] discloses tapentadol hydrochloride and sets forth a method of synthesizing it in Example 25 [24]. The '737 patent indicates that tapentadol hydrochloride was crystallized, but it does not describe the resulting crystal structure or discuss whether tapentadol hydrochloride is polymorphic. Later investigation would show that faithful reproduction of Example 25 results in what is now known as Form B of tapentadol hydrochloride.

Years after the discovery of the compound, scientists at Grünenthal discovered that tapentadol hydrochloride was polymorphic and eventually identified and characterized two forms, later called Form A and Form B. Forms A and B are enantiotropic polymorphs, with Form A being the thermodynamically more stable form at room temperature. The discovery of Form A led to the filing of a patent application, which resulted in the issuance of US Patent No. 7,994,364 ("the '364 patent") [25].

At the time of the application of the '364 patent, several things were known in the art: the structure of tapentadol hydrochloride; the synthesis method of Example 25 for manufacturing tapentadol hydrochloride; the concept of polymorph screening; and Food and Drug Administration (FDA) guidance concerning polymorph screening [24]. On this basis, the patent challengers argued, among other things, that Form A of tapentadol hydrochloride was rendered obvious by virtue of the disclosure of Example 25 of the '737 patent, plus general information regarding polymorph screening [26].

The US Federal Circuit Court considered Grünenthal's polymorph patent for tapentadol hydrochloride for treating severe pain. Polymorph Form A was held to be non-obvious and patentable in the light of an earlier patent relating to polymorph Form B.

The '364 patent states that "Crystalline Form A … has the same pharmacological activity as Form B but is more stable under ambient conditions. It can be advantageously used as [an] active ingredient in pharmaceutical compositions." The patent challenger contended that the '364 patent lacks utility because the '364 patent's disclosure purportedly is vague; fails to provide benefit; and does not demonstrate that Form A is superior to Form B. The Federal Circuit explained that the '364 patent's disclosure concretely discloses the practical benefit of Form A as an analgesic and, further, explained that the patente need not prove that Form A has superior stability over Form B for purposes of determining utility [24].

But in an interesting contrast in China, Grünenthal's patent for tapentadol hydrochloride was held invalid, the case turning on the issue of insufficient data appearing in the patent specification [18].

23.5 Pharmaceutical Salts of Tapentadol

Tapentadol was introduced to the market as the hydrochloride form. After losing exclusivity beside the hydrochloride, also the phosphate and tartrate salts are available as generic brands for tapentadol.

Beside the most commercially relevant pharmaceutical salt forms of tapentadol, in total more than 170 patents have been published describing and claiming specific crystalline salts of tapentadol or specific polymorphic forms thereof. Beside pharmaceutically acceptable salt forms, also cocrystals of tapentadol-free base and tapentadol hydrochloride are described. Some of the scientifically interesting pharmaceutical salt forms as well as cocrystals of tapentadol-free base and tapentadol hydrochloride are summarized in the following Table 23.1.

Table 23.1 Pharmaceutical and crystalline salt forms of tapentadol as well as cocrystals of tapentadol-free base and tapentadol hydrochloride.

Pharmaceutical salts of tapentadol	
Tapentadol salt CAS No. (stoichiometric ratio)	**Acidic component (chemical structure, CAS No. Of free acid)**
Hydrochloride (1:1) 175591-09-0	HCl
Nitrate (1:1) 1356394-24-5	HNO_3
Acetate (1:1) 1356394-82-5	CH_3COOH
Sulfate (1:1) 1356394-26-7	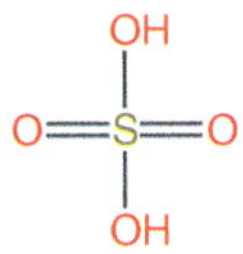
Ethanesulfonate (1:1) 1356394-38-1	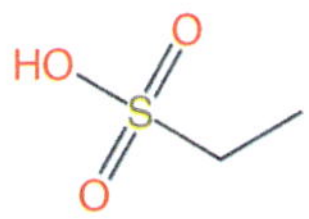 594-45-6
Phosphate 1356394-30-3 (1:1)	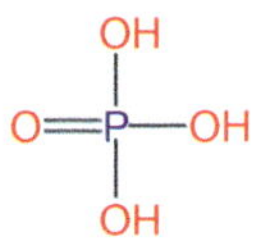 7664-38-2
Besylate 2467572-37-6 (1:1)	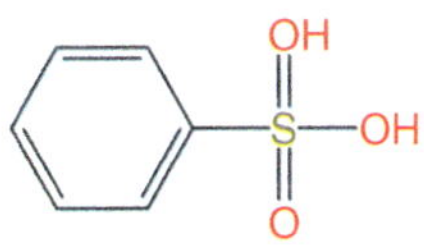 98-11-3
Tosylate 1356394-37-0 (1:1)	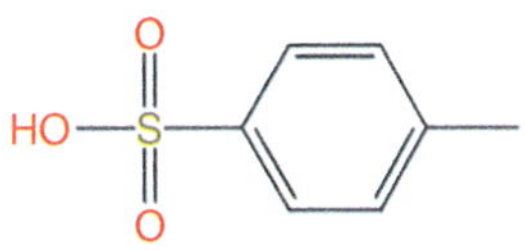 104-15-4
Benzoate 1356394-39-2 (1:1)	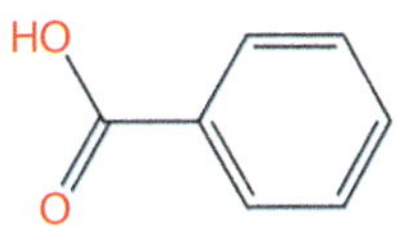
2-Acetyloxy benzoate 1443156-44-2 (1:1)	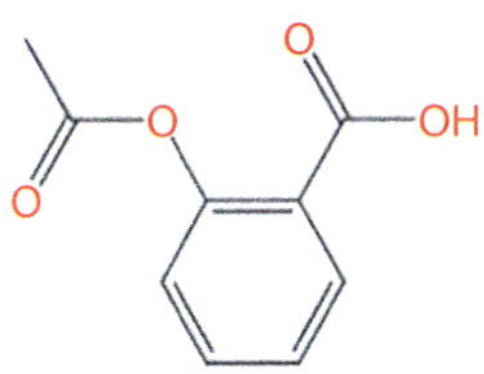 50-78-2

(Continued)

Table 23.1 (Continued)

Pharmaceutical salts of tapentadol	
Tapentadol salt CAS No. (stoichiometric ratio)	**Acidic component (chemical structure, CAS No. Of free acid)**
Nicotinate 1356394-25-6 (1:1)	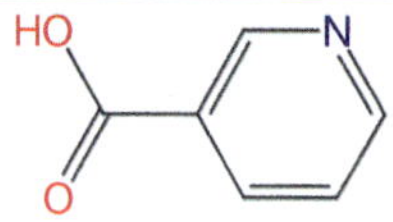 59-67-6
Salicylate 1277097-35-4 (1:1)	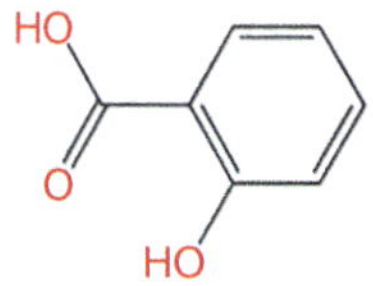 69-72-7
2,2-Dihydroxybenzoate 1356394-62-1 (1:1)	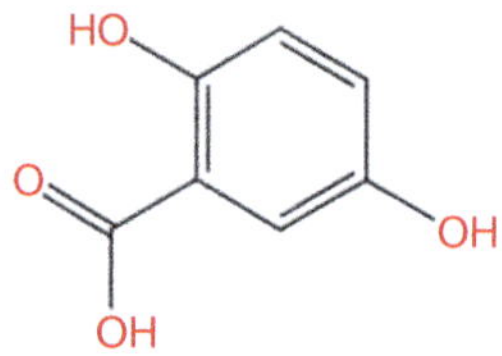 490-79-9
2-Aminobenzoate 1443157-13-8 (1:1)	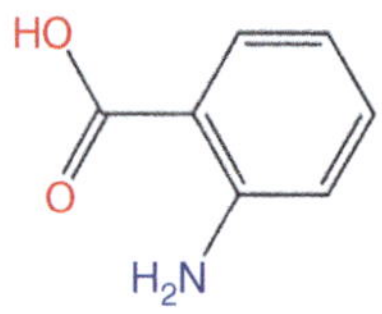 118-92-3
1-Hydroxy-2-naphthoate 1356394-21-2 (1:1)	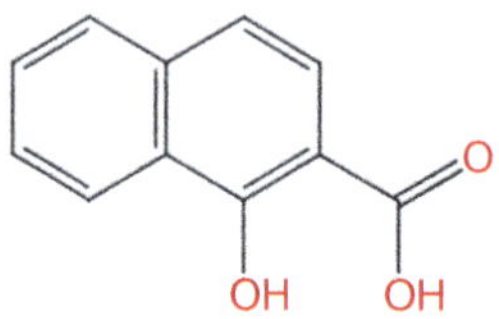 86-48-6
Naphthalene-1,5-disulfonate 1356394-52-9 (1:?)	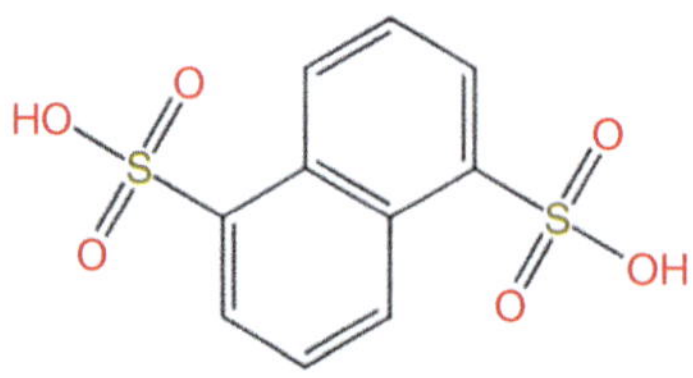 81-04-9
DL-Mandelate 1356394-95-0 (1:1)	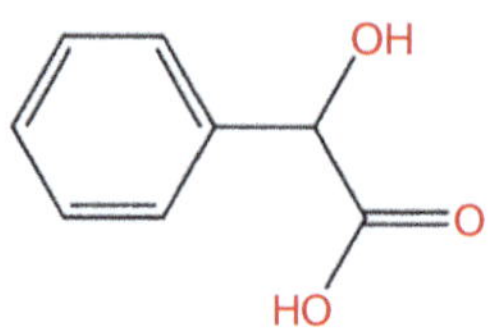 90-64-2

Table 23.1 (Continued)

Pharmaceutical salts of tapentadol	
Tapentadol salt CAS No. (stoichiometric ratio)	**Acidic component (chemical structure, CAS No. Of free acid)**
S-(+)-Mandelate 1356394-42-7 (1:1)	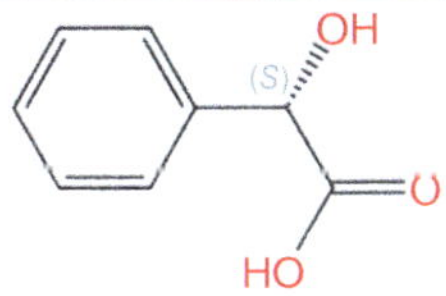 17199-29-0
Cinnamate 1356394-40-5 (1:1)	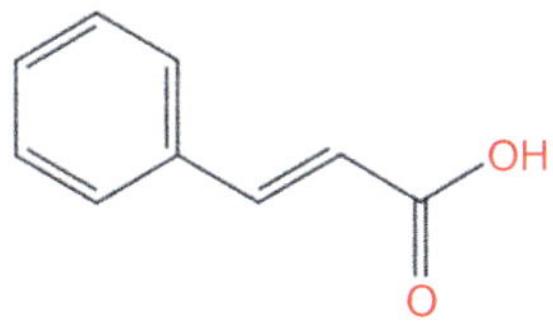 621-82-9
Oxalate 175591-23-8 (1:1) 1356394-83-6 (1:?)	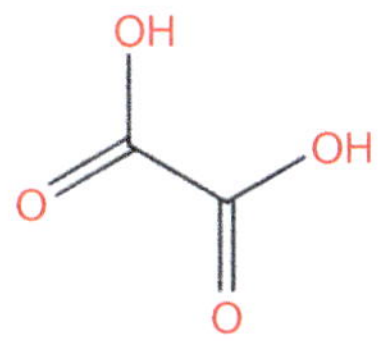 144-62-7
Malonate 1356394-28-9 (1:1)	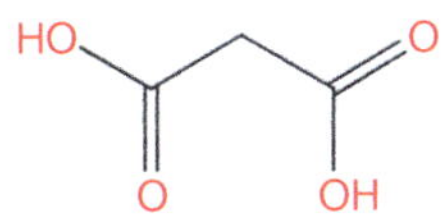 141-82-2
Butanedioic acid salt 1443156-39-5 (1:1) 1356394-85-8 (1:?)	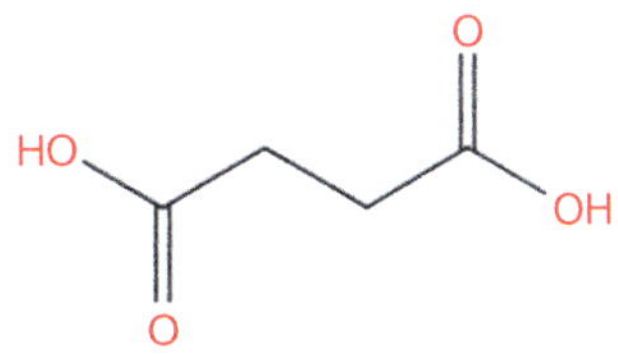 110-15-6
D,L-Malate 1443156-40-8 (1:1) 1356394-98-3 (1:?) 1277097-33-2 (1:2)	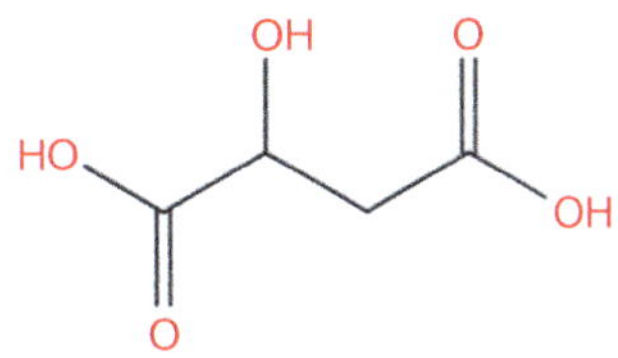 6915-15-7
D-Malate 1356394-36-9 (1:?)	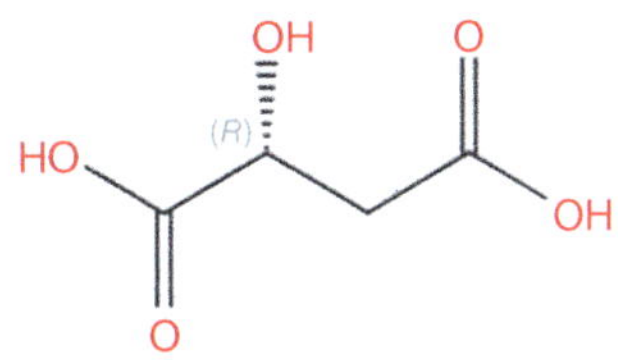 636-61-3

(Continued)

Table 23.1 (Continued)

Pharmaceutical salts of tapentadol	
Tapentadol salt CAS No. (stoichiometric ratio)	**Acidic component (chemical structure, CAS No. Of free acid)**
D,L-Lactate 1356394-93-8 (1:1)	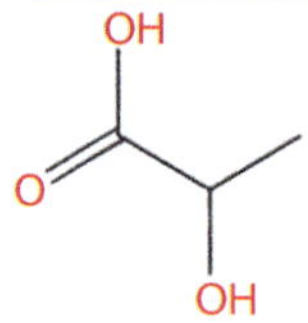 50-21-5
L-(+)-Lactate 1356394-41-6 (1:1)	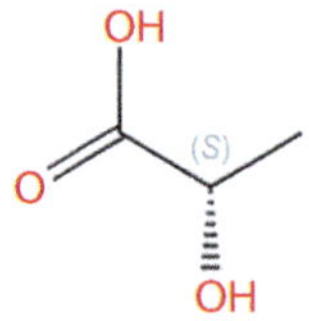 79-33-4
L-Lysinate 1356394-34-7 (1:1)	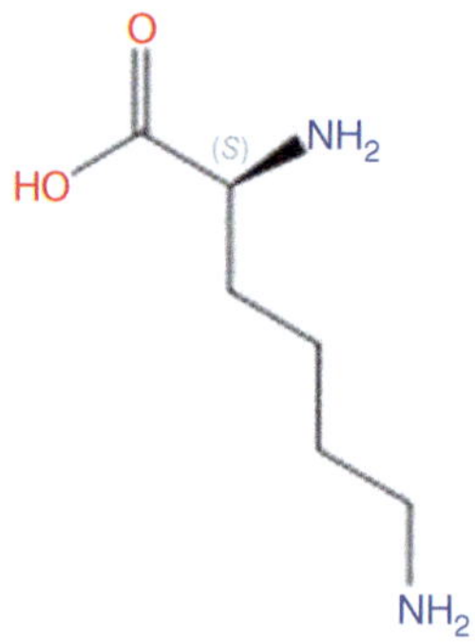 56-87-1
L-Aspartate 1443156-46-4 (1:1)	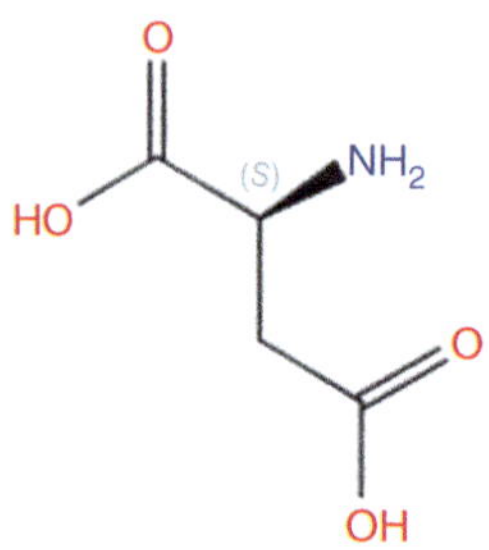 56-84-8
L-Glutamate 1443156-43-1 (1:1)	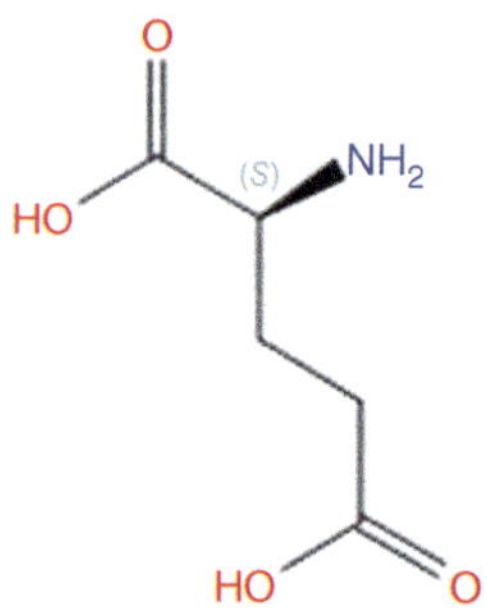 56-86-0

Table 23.1 (Continued)

Pharmaceutical salts of tapentadol	
Tapentadol salt CAS No. (stoichiometric ratio)	**Acidic component (chemical structure, CAS No. Of free acid)**
Hexanoate 1356394-56-3 (1:1)	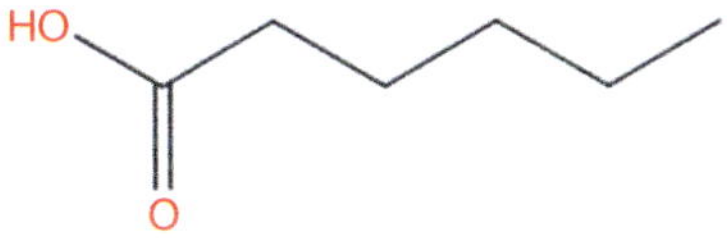 142-62-1
Octanoate 1356394-65-4 (1:1)	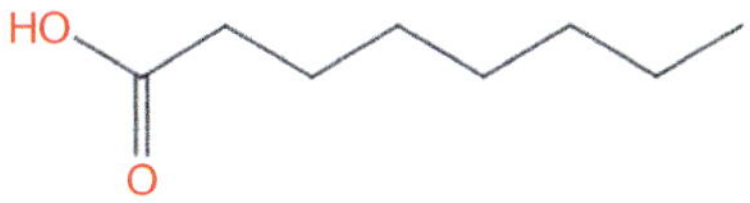 124-07-2
Decanoate 1356394-69-8 (1:1)	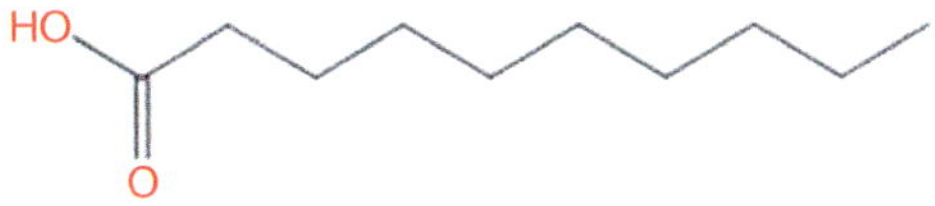 334-48-5
Sebacate 1356394-20-1 (1:1)	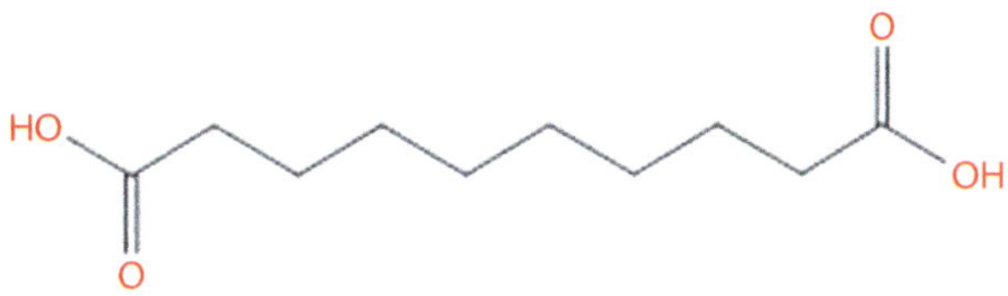 111-20-6
Hexadecanoate 1356394-50-7(1:1)	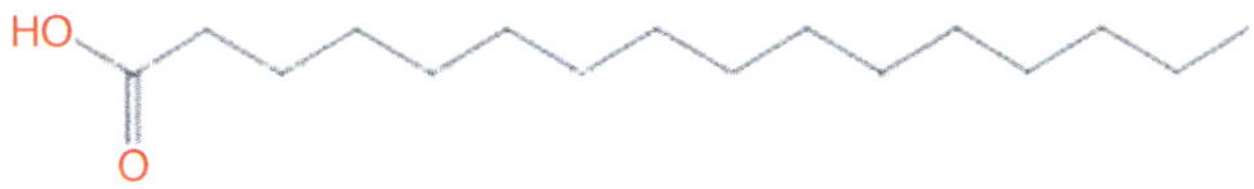 57-10-3
Octadecanoate 1356394-60-9 (1:1)	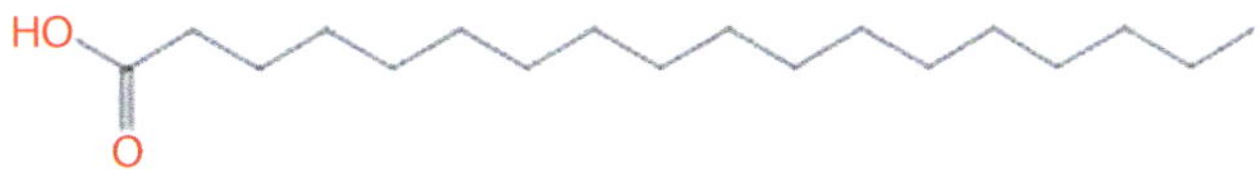 57-11-4
Oleate 1356394-58-5 (1:1)	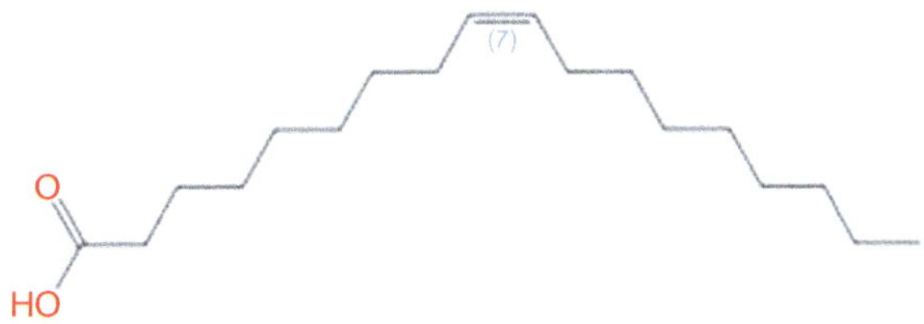 112-80-1
D-Gluconate 1356394-46-1 (1:1)	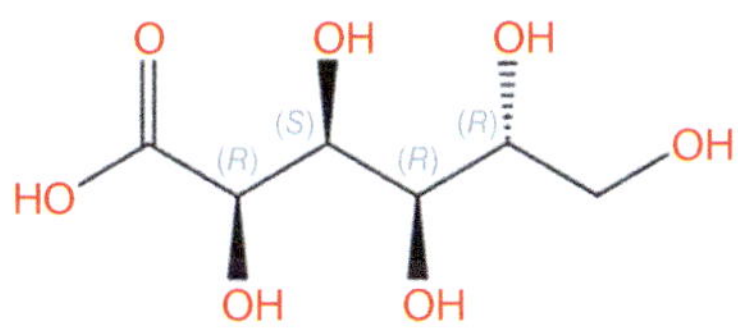 526-95-4

(Continued)

Table 23.1 (Continued)

Pharmaceutical salts of tapentadol	
Tapentadol salt CAS No. (stoichiometric ratio)	**Acidic component (chemical structure, CAS No. Of free acid)**
L-(+)-Tartrate 1443156-41-9 (1:1) 1356394-31-4 (1:?)	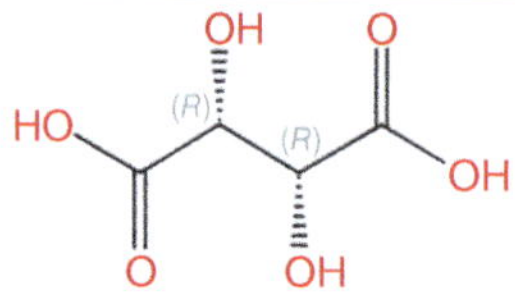 87-69-4
D-(-)-Dibenzoyltartrate 1313429-31-0 (1:1) 1277097-32-1 (1:2) 1404478-83-6 (1:?)	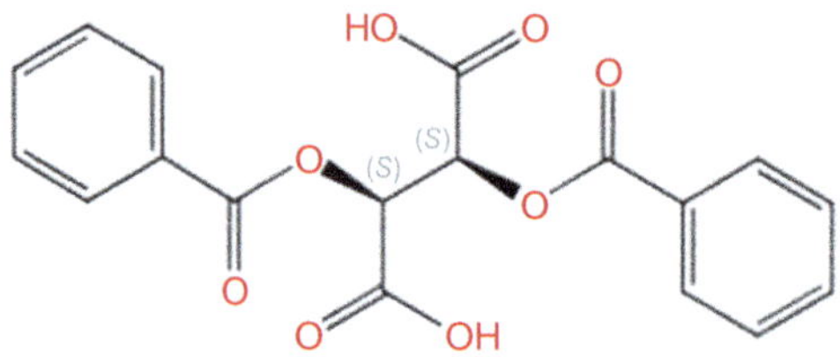 17026-42-5
L-(+)-Dibenzoyltartrate 1277097-31-0 (1:2)	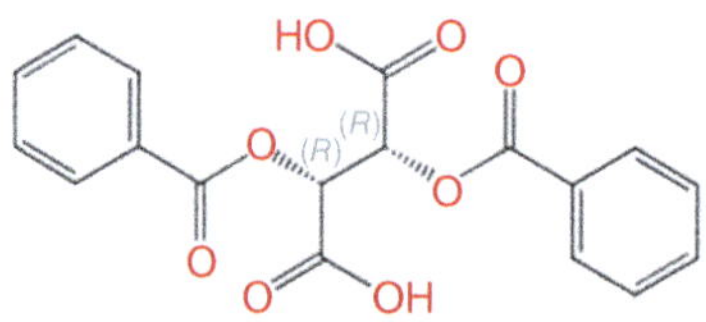 2743-38-6
Mucate 1356394-75-6 (1:?)	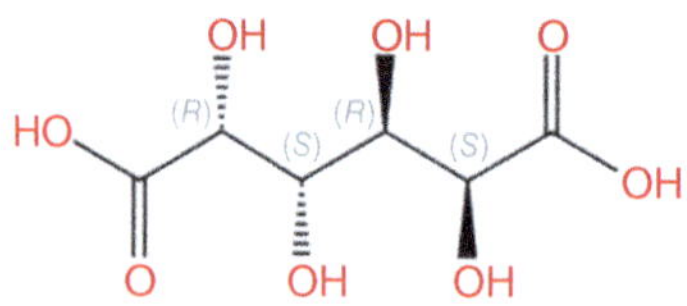 526-99-8
Maleate 1443156-37-3 (1:1) 1356394-87-0 (1:?) 1277097-34-3 (1:2)	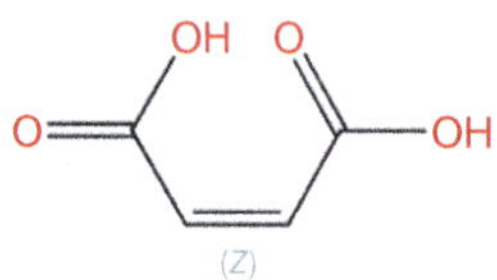 110-16-7
Fumarate 1443156-36-2 (1:1) 2468015-11-2 (1:?) 1356394-27-8 (2:1)	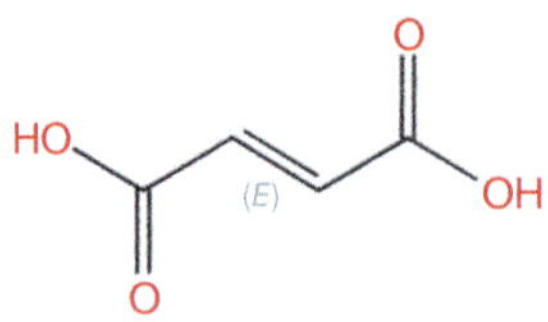 110-17-8
(3R)-1,2-Dithiolane-3-pentanoate 1443156-45-3 (1:1)	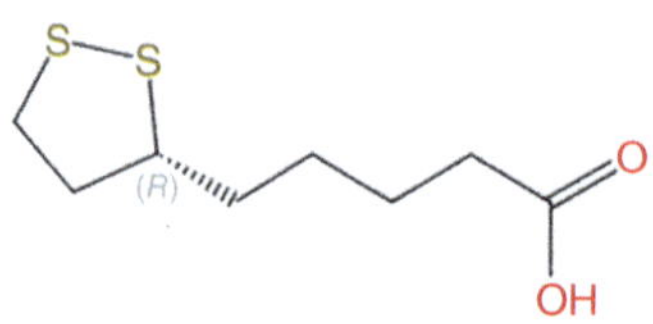 1200-22-2

Table 23.1 (Continued)

Pharmaceutical salts of tapentadol	
Tapentadol salt CAS No. (stoichiometric ratio)	**Acidic component (chemical structure, CAS No. Of free acid)**
Hippurate 1356394-90-5 (1:1)	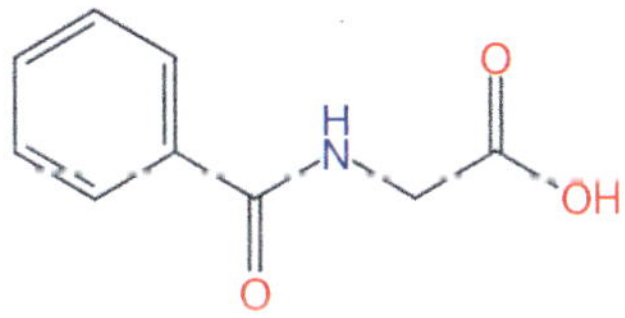 495-69-2
Orotate 1356394-72-3 (1:1)	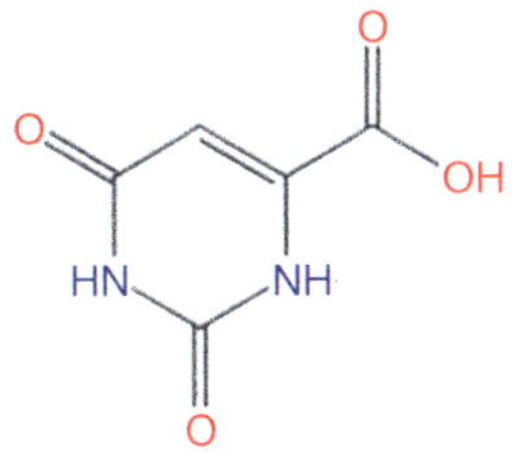 65-86-1
Acesulfamate 1356394-81-4 (1:1)	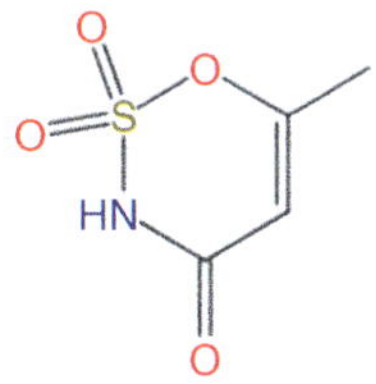 33665-90-6
Citrate 1443156-42-0 (1:1) 1356395-02-2 (1:?)	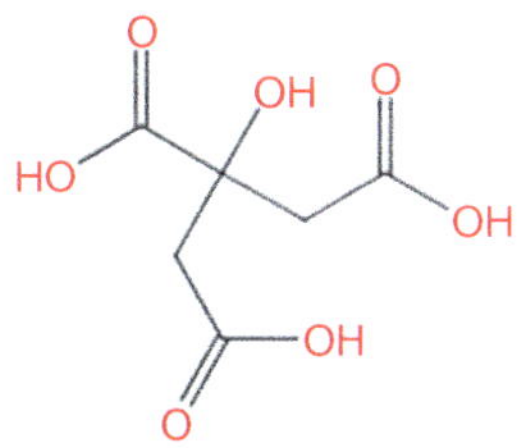 77-92-9
Saccharinate 1356394-29-0 (1:1)	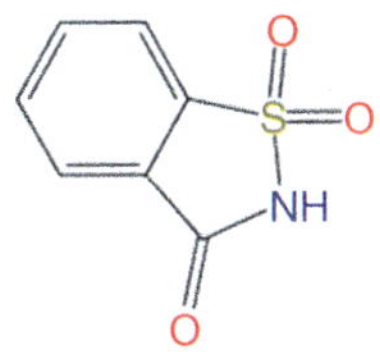 81-07-2
L-(+)-Ascorbate 1356394-48-3 (1:1)	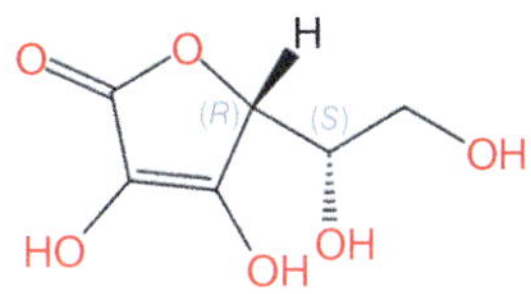 50-81-7

(Continued)

Table 23.1 (Continued)

Pharmaceutical salts of tapentadol	
Tapentadol salt CAS No. (stoichiometric ratio)	**Acidic component (chemical structure, CAS No. Of free acid)**
Embonate 1356394-23-4 (1:1)	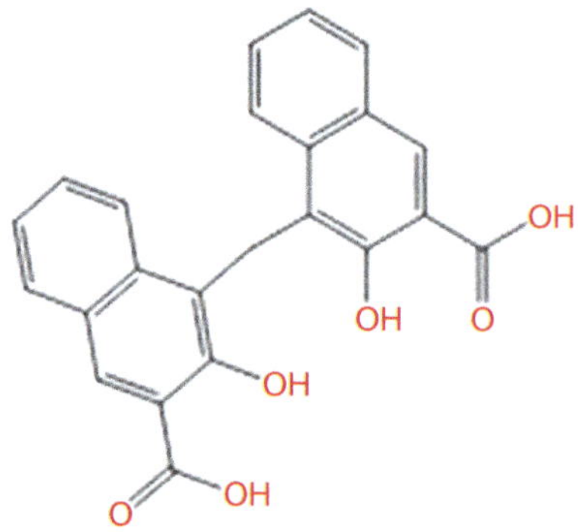 130-85-8
(+)-Camphor-10-sulfonate 1356394-44-9 (1:1)	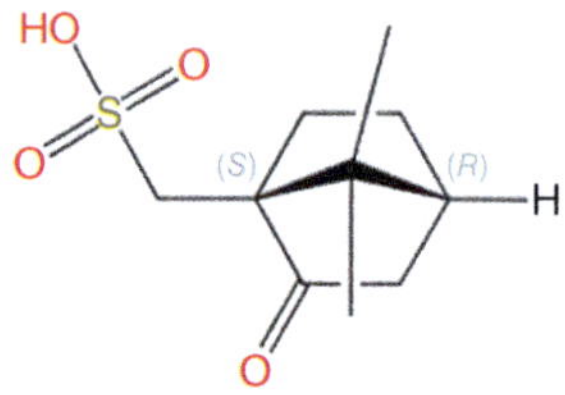 3144-16-9
(-)-Camphor-10-sulfonate 1277097-30-9 (1:1)	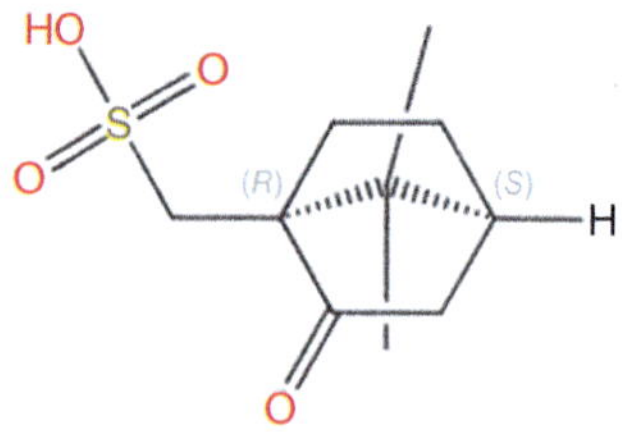 35963-20-3

Cocrystals of tapentadol-free base	
Components of cocrystal CAS No. (stoichiometric ratio)	**Cocrystal former (chemical structure, CAS No. of cocrystal former)**
Tapentadol and Paracetamol 1355078-25-9 (1:1)	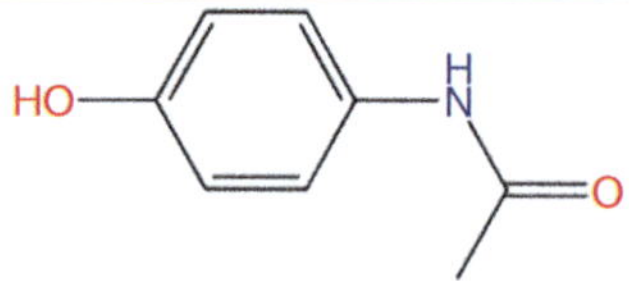 103-90-2
Tapentadol and 1-Butanone, 1-(4-fluorophenyl)-4-(4-methyl-1-piperidinyl)- 1240784-42-2 (1:1)	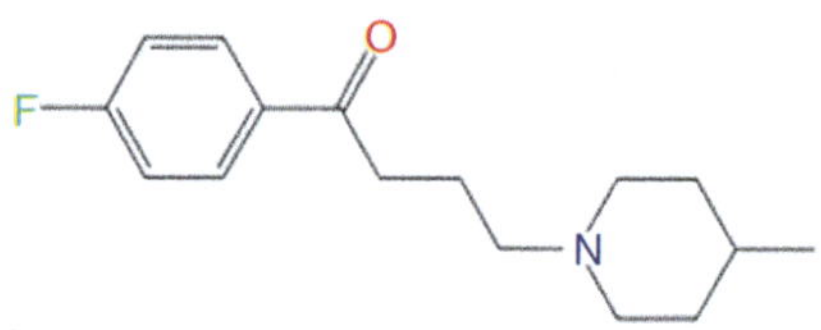 3575-80-2

Table 23.1 (Continued)

Cocrystals of tapentadol-free base	
Components of cocrystal CAS No. (stoichiometric ratio)	**Cocrystal former (chemical structure, CAS No. of cocrystal former)**
Tapentadol and *N*,*N*,α-trimethyl-10*H*-phenothiazine-10-ethanamine 1240784-28-4 (1:1)	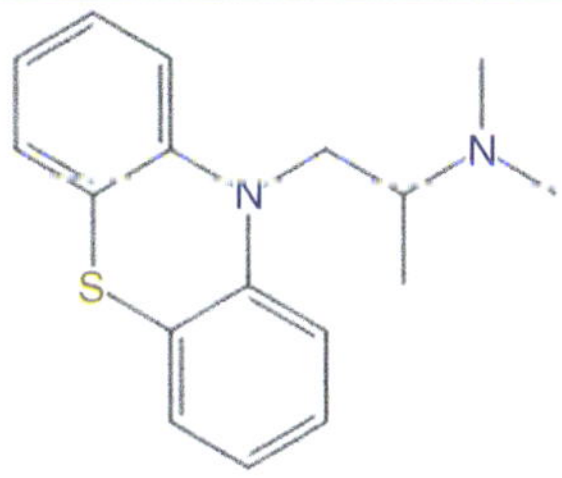 60-87-7
Tapentadol and *N*,*N*-dimethyl-10*H*-pyrido[3,2-*b*][1,4] benzothiazine-10-propanamine 1240784-38-6 (1:1)	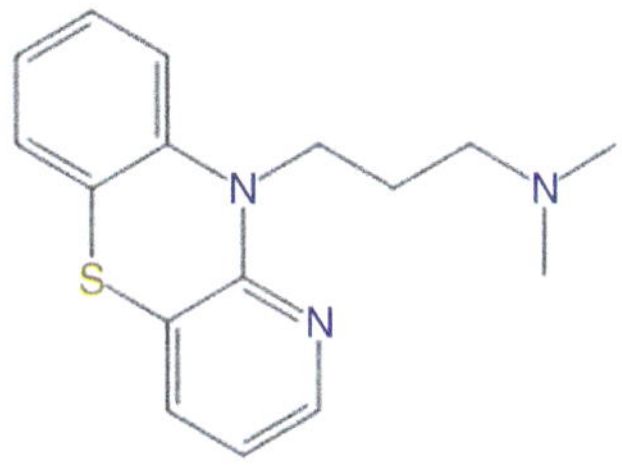 303-69-5
Tapentadol and 1-Butanone, 1-(4-fluorophenyl)-4-(4-methyl-1-piperidinyl)- 1240784-42-2 (1:1)	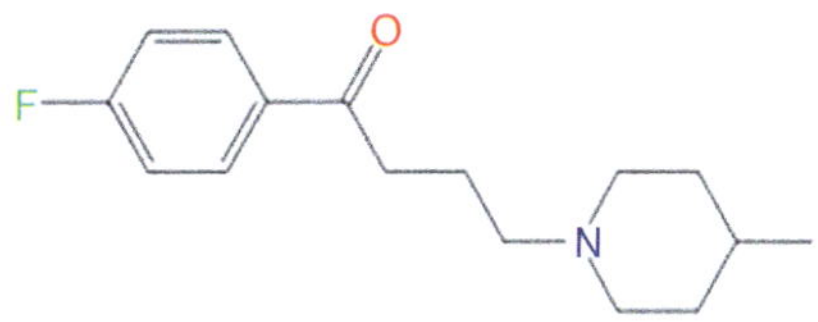 3575-80-2
Tapentadol and (β*R*)-2-methoxy-*N*,*N*,β-trimethyl-10*H*-phenothiazine-10-propanamine 1240784-31-9 (1:1)	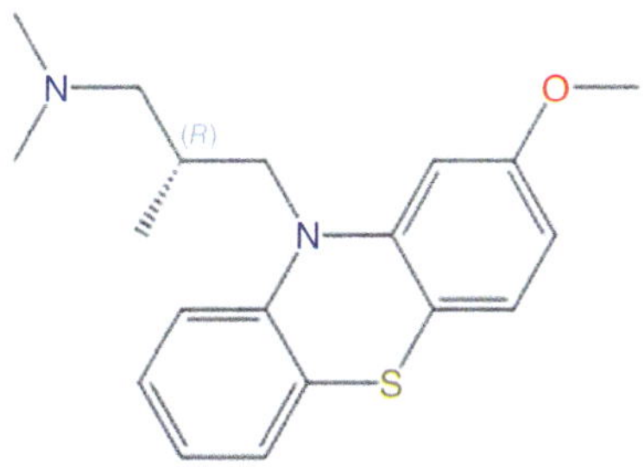 60-99-1
Tapentadol and (3*Z*)-3-(2-chloro-9*H*-thioxanthen-9-ylidene)-*N*,*N*-dimethyl-1-propanamine 1240784-35-3 (1:1)	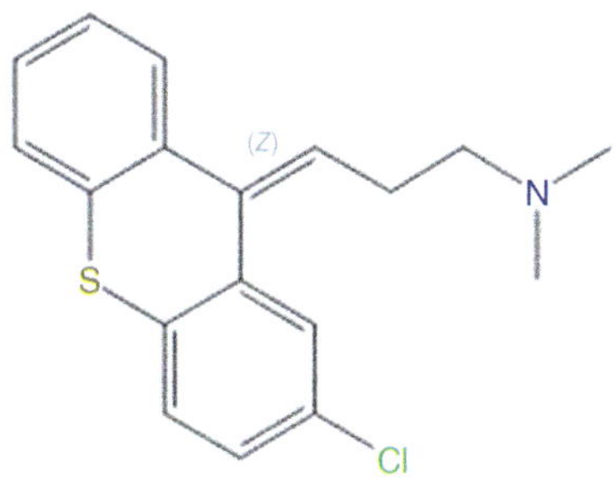 113-59-7

(Continued)

Table 23.1 (Continued)

Cocrystals of tapentadol hydrochloride	
Components of cocrystal CAS No. (stoichiometric ratio)	**Cocrystal former (chemical structure, CAS No. of cocrystal former)**
Tapentadol hydrochloride and Paracetamol 1355078-24-8 (1:1:1)	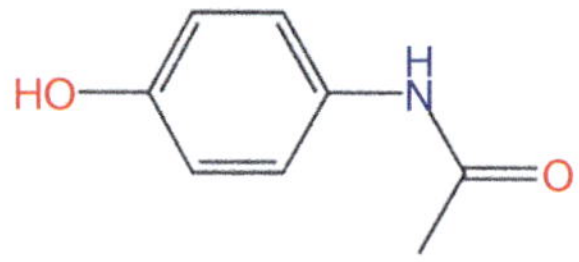 103-90-2
Tapentadol hydrochloride and Naproxen 1157853-66-1 (1:1:1)	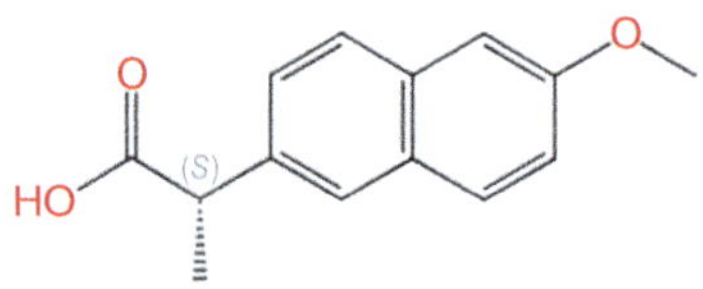 22204-53-1
Tapentadol hydrochloride and Gabapentin 1157853-69-4 (1:1:1)	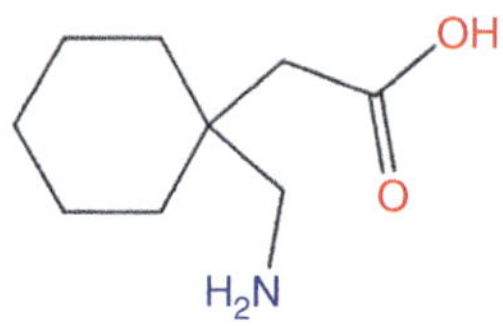 60142-96-3
Tapentadol Hydrochloride and (+)-Tramadol hydrochloride 1157853-65-0 (1:1:1:1)	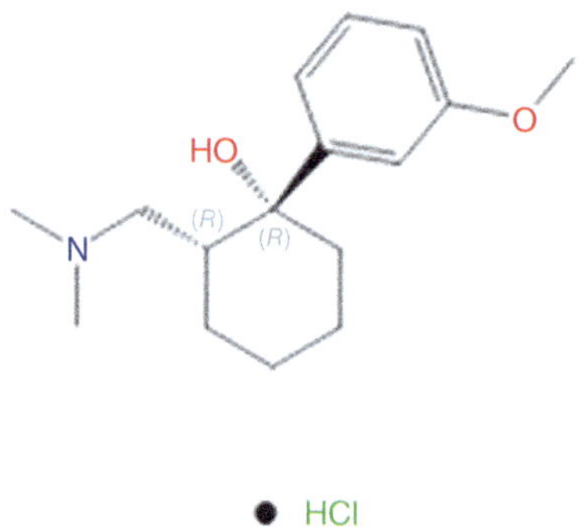 148229-78-1

23.6 Synthesis Routes to Tapentadol Hydrochloride

In the original patent [27], tapentadol hydrochloride is synthesized by ether cleavage (e.g., conc. HBr) from its corresponding methyl ether **1**. Compound **1** can be obtained by several methods, for example, starting from diethyl ketone, which is transformed under standard Mannich conditions to racemic **2**. The Grignard reaction with 3-bromoanisole and subsequent separation of diastereomers (crystallization as HCl salt from 2-buta-none with TMSCl/H_2O) and enantiomers (HPLC) gives (1*S*,2*S*)-(−)-enantiomer **3**. Removal of the hydroxy group can be achieved diastereoselectively (under retention) by chlorination with thionyl chloride to yield **4**, followed by treatment with zinc borohydride (from $ZnCl_2$ and $NaBH_4$), leading to tapentadol hydrochloride (Figure 23.6).

During process development, several alternative pathways were developed to reduce the cost of goods for the commercial phase. After market authorization, several patents were published describing improved and alternative synthetic approaches to tapentadol. The most advanced approaches, based on stereoselective syntheses, appeared only after 2010. Graczyk and coworkers [28] described in their review from 2017 the different phases of innovation applied for synthetic pathways to manufacture tapentadol hydrochloride. Due to the small number of the relevant peer-reviewed publications, this review makes

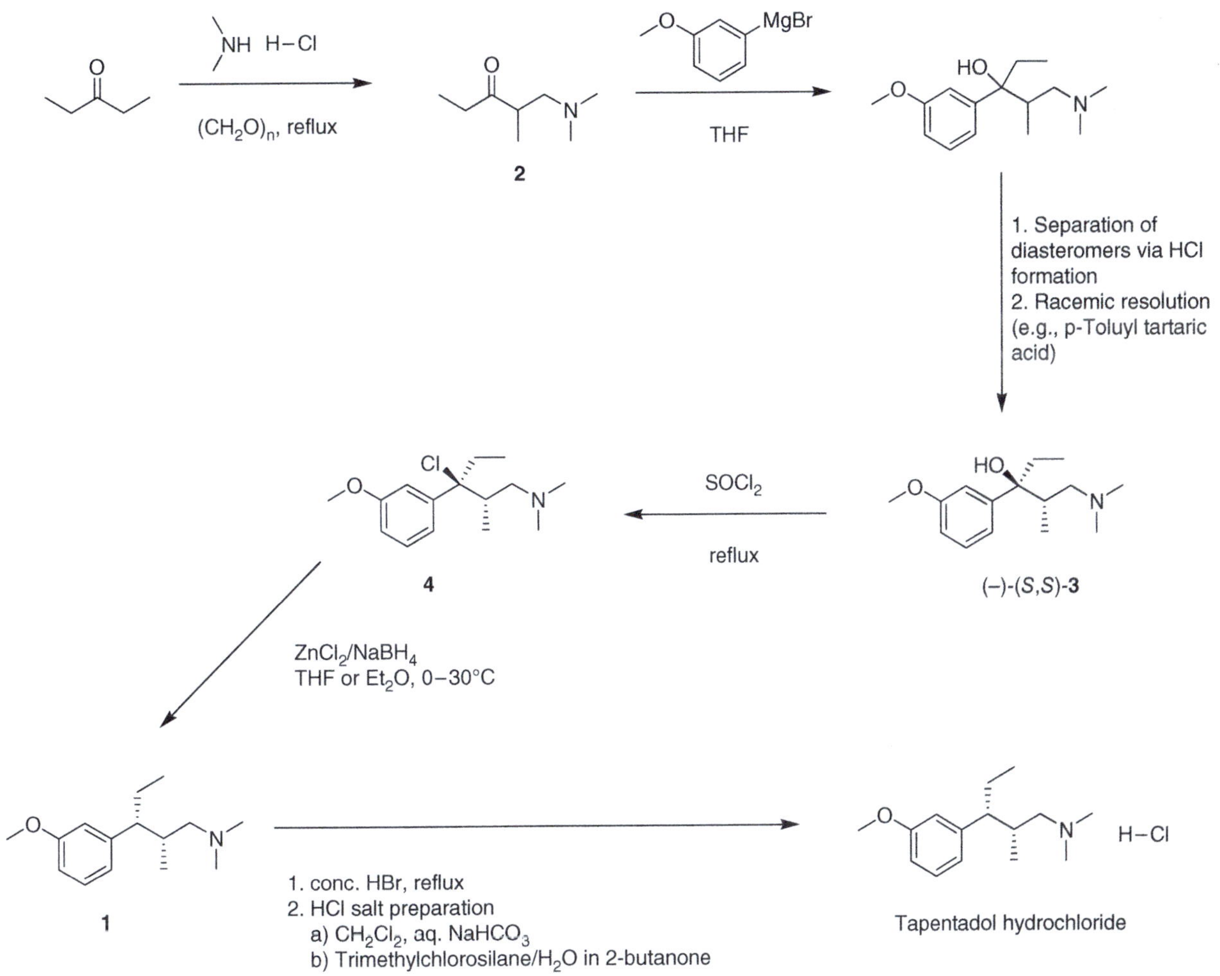

Figure 23.6 The synthesis of tapentadol hydrochloride as described in the first patent. Adapted from Hell et al. [27].

extensive use of the patent literature, bearing in mind that the disclosed procedures may not always be reproducible in another lab.

Based on the example of tapentadol, the authors have shown in their review that the synthetic approaches have been developed in four phases, which are characterized by the following criteria (Figure 23.7).

The authors also recognized that a considerable number of the recent stereoselective syntheses of tapentadol have been devised in China. This demonstrates that this area of the world is becoming increasingly important in terms of innovative input to industrial chemistry.

The synthetic approaches to tapentadol can be assigned to four distinct phases, whereas these phases partially overlap in time.

- Phase 1 includes the synthesis of racemic material with the desired relative stereochemistry of the two stereogenic centers. This initial phase 1 is characterized by the synthesis of the racemic material with the desired relative stereochemistry of the two stereogenic centers determined by the chemical pathways used to construct the backbone, followed by separation of the enantiomers by diastereomeric salt formation by chiral acids or chiral chromatography and additional inversion of configuration, if needed. Some further non-stereoselective syntheses also appeared at this early stage.
- Phase 2 comprises the synthesis of racemic precursors with a single stereogenic center, followed by resolution of the racemate by different racemic resolution techniques and final installation of the second stereogenic center. This can be followed by an inversion of configuration, if needed.
- Phase 3 is characterized by a synthesis of homochiral intermediates using optically active substrates (chiral building blocks) or chiral auxiliary strategies, which includes the establishment of the chiral auxiliary, a chiral induction to establish the chiral center, and finally followed by a final cleavage.

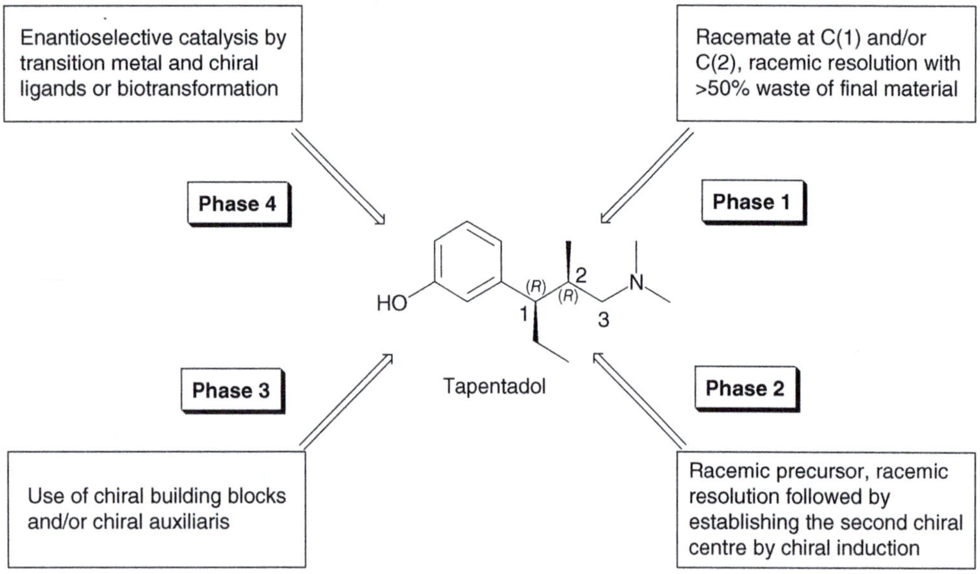

Figure 23.7 The development of different synthetic approaches to tapentadol hydrochloride over time according to Grazyk *et al.* [28]/with permission of Elsevier.

- Phase 4 includes the synthesis of a homochiral intermediate using stereoselective catalysis. In this last phase, several asymmetric syntheses of tapentadol have appeared.

Finally, the following conclusions were drawn: The case of tapentadol also shows that standard non-stereoselective approaches dominated the first 10 years since the discovery of this drug. Growing advances in process chemistry made it possible to devise the recent stereoselective routes. The application of a stereoselective syntheses is time-consuming. More transformation steps normally have a negative impact on the cost of goods. Future synthetic approaches to Tapentadol and other homochiral drugs with two or more chiral centers are likely to see a growing use of organocatalysts and/or biotransformation steps, which allow significant shortening of syntheses with the consequence of reduced competitive commercial prices of the drug substances [1].

23.7 The Pharmacological Profile of Tapentadol as a Multiple Ligand for the Treatment of Several Types of Pain

23.7.1 Preclinical Development

Based on its *in vitro* and *in vivo* profiles, tapentadol qualified for IND-enabling studies and further development. Non-clinical toxicity was assessed according to the regulatory guidelines and acknowledged by the authorities, resulting in EMA and FDA marketing authorizations. Notably, tolerability was clearly improved when compared to the standard opioid morphine in classical models for the assessment of central and peripheral opioid-typic side effects. Acute administration demonstrated less induction of emesis in the ferret. Comparing opioid-induced constipation in the gastro-intestinal charcoal transit test in mice with analgesic dose range revealed a therapeutic window for tapentadol but not for morphine. Symptoms of physical dependence were strongly detected for subacute treatment with morphine but only moderately for tapentadol after spontaneous withdrawal and naloxone-induced precipitation in rats [29, 30]. Likewise, development of tolerance was delayed when comparing daily administration of equianalgesic doses of tapentadol compared to morphine in a rat model of neuropathic pain [31].

Basic pharmacokinetic assessment of absorption, distribution, metabolism, and excretion (ADME) characterized tapentadol as reasonably suited for drug development [29]. A favorable metabolic profile was detected as compared to tramadol and other opioids with active metabolites. Tapentadol is eliminated mainly via phase II metabolism as direct glucuronidation by uridine diphosphate glucuronyl transferase (UGT) isoforms UGT1A6, UGT1A9, and UGT2B7. Tapentadol-*O*-glucuronide represents the major metabolite in all species investigated, including rodents and mans. Only a minor fraction is metabolized by phase I cytochrome P450 (CYP) enzymes CAP2B6, CYP2C8, CYP2C9, CYP2C19, and CYP2D6 resulting in the generation of *N*-demethyl tapentadol (M2) and hydroxy-tapentadol (M1). This feature reduces the risk of drug–drug interactions between tapentadol and CYP

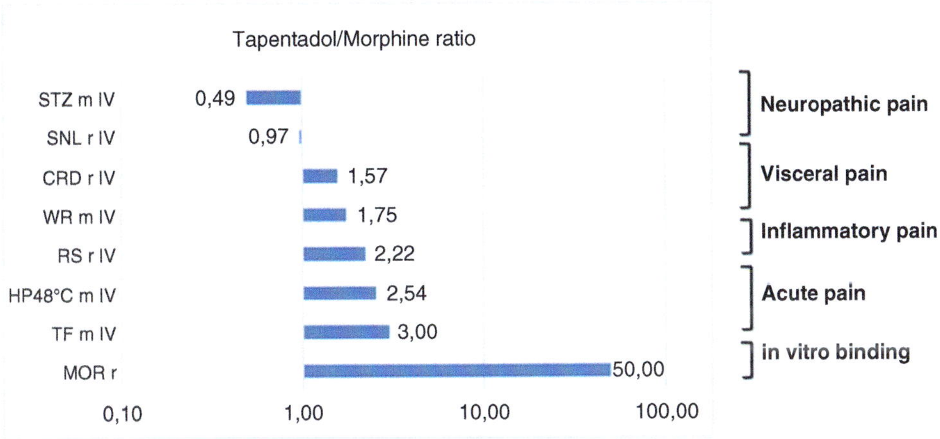

Figure 23.8 Rodent potency ratios of tapentadol/morphine for *in vitro* (rat MOR binding affinity according to [31] and *in vivo* data (ED50 after IV administration according to [29, 33, 34]. m mouse, r rat, IV intravenous, CRD colo-rectal distension, HP hot plate, MOR μ-opioid receptor, RS randall sellito, SNL spinal nerve ligation, STZ streptozotozin, TF tail flick.

inhibitors and/or CYP inducers, which is prominent in opioids prone to CYP-based phase I metabolism such as fentanyl, buprenorphine, oxycodone, tilidine, or tramadol. Tapentadol-*O*-glucuronide and eight other metabolites of tapentadol were analyzed for their potential contribution to analgesic efficacy. While sub-μmolar *in vitro* efficacies (MOR binding or NA uptake inhibition) were detected in 3/9 metabolites (5/9 had K_i values below 10 μM), none of them was able to induce anti-nociception in the mouse tail flick test, a standard model for acute nociceptive pain and opioid-typical efficacy. Anti-nociceptive activity was detected for 5/9 metabolites in the mouse writhing test, a test, which can detect weak analgesic principles. However, none of the metabolites with *in vitro* activity reached relevant plasma steady state concentrations in man. Thus, tapentadol can be considered being devoid of active metabolites [32].

23.7.2 Efficacy

MOR agonism and NA reuptake inhibition, the two active principles of tapentadol, have earlier been shown to contribute to analgesia. Since tapentadol is the first drug selectively combining both mechanisms in a single molecule, it was important to analyze the contribution and interaction of MOR agonism and NA reuptake inhibition in basic settings of pain transmission *in vitro* and *in vivo*. These data can give important insights into potential translation to human pain indications. The opioid activity of tapentadol was assessed in several experimental paradigms. Tapentadol binds with sub-μM affinity to rodent (K_i 0.1 μM) and human MOR (K_i 0.16 μM), while it shows only minor affinity to DOR and KOR and no affinity to NOP. Functionally, tapentadol is characterized as an MOR agonist in the electrically induced twitch response of the guinea pig ileum and in a GTPgS assay on human recombinant MOR. In the latter assay on recombinant human MOR, morphine shows full (i.e., 100%) efficacy, while tapentadol (EC_{50} 0.67 μM) reaches an efficacy of 88%. In terms of binding affinity, tapentadol is up to 50-fold weaker when compared to morphine (Figure 23.8) [29, 31]. NA reuptake inhibition was quantified in functional assays for rats and humans. In both species, tapentadol showed sub-μM inhibition of the NA reuptake transporter with K_i values of 0.48 μM and 0.75 μM for rat and human, respectively [31, 35]. When comparing this inhibitory potency with the selective NA reuptake inhibitor desipramine, tapentadol shows around 500-fold weaker inhibition [29]. When comparing MOR agonism and NA reuptake inhibition *in vitro*, tapentadol comes with similar potencies and a less than fivefold difference (rat 4.8-fold and human 4.7-fold), thus showing a balanced profile *in vitro*.

Assessment of a drug's mechanism of action *in vitro* usually is an isolated and highly specific experimental undertaking. In the case of tapentadol, this revealed a balanced MOR-NRI profile. This profile needs to be translated into the next level of complexity, i.e., to answer the question whether the balanced MOR-NRI profile of tapentadol also pays into analgesic *in vivo* efficacy. The mammalian pain transmitting system is organized in several complex layers, and analgesic intervention is possible based on the mechanism of a given drug as well as based on its site of action (for review, see ref. 36, 37).

MOR agonism is well known as an analgesic principle and can exert its action via numerous receptors both in the CNS and in the periphery. In the spinal cord, activation of MOR expressed both pre- and post-synaptically can inhibit afferent pain transmission to the CNS. On the other hand,

MOR activation in the brainstem is essential in controlling pain transmission by descending inhibitory pathways. An easy and reliable way to prove MOR-based efficacy is the combination with MOR antagonism such as naloxone or the use of genetically engineered knock-out mice lacking MOR (OPRM1 KO mice). Tapentadol, when administered systemically, readily crosses the blood–brain barrier and is therefore available in all compartments (peripheral and central), which are important for an analgesic net effect. Likewise, systemically administered naloxone reaches both central and peripheral compartments, and general genetic deletion of MOR, as in the OPRM1 KO mice used by Kögel *et al.* [38], mimics an overall unavailability of the receptor activated by the MOR-part of tapentadol. Thus, the combination of tapentadol (IV) and naloxone (IV) in a mouse model of acute chemically induced pain (phenylquinone writhing model) clearly demonstrated an opioid contribution of tapentadol in acute anti-nociceptive efficacy [31]. Furthermore, the anti-nociceptive efficacy of tapentadol (IP) in a mouse model of heat pain (hot plate) was clearly diminished when comparing normal wild-type mice with OPRM1-KO mice [38]. Both experiments were controlled with the standard opioid morphine, which only carries a single mechanism (i.e., MOR agonism). Interestingly, the efficacy of morphine was diminished more clearly in both pain models, which already suggests that the second (non-opioid) mechanism of tapentadol contributes to the anti-nociceptive efficacy. A more focused assessment of the peripheral site of action can be done by local administration of the agonist (tapentadol) or the antagonist (naloxone) and by local measurement of specific neuronal responses. In a rat study, systemically administered tapentadol (SC) ameliorated neuronal excitations in the spinal cord of naïve rats, and this was inhibited by local (spinal) combination with naloxone [39]. Thus, regarding MOR agonism, *in vivo* data clearly demonstrated an anti-nociceptive MOR-dependent efficacy of tapentadol in rodent models of acute nociception.

Reuptake inhibition of norepinephrine (NRI) can be demonstrated *in vivo* by means of microdialysis. Extracellular NA is cleared by reuptake, and NRI results in increased NA concentration. Systemic administration of tapentadol (IP) resulted in a relevant and dose-dependent increase in NA concentrations in the hippocampus of freely moving rats [31] and the spinal cord of anesthetized rats [40]. Lack of NA increase in control groups treated with equi-analgesic doses of morphine suggests that the pure brainstem-based MOR agonism-derived NA increase via descending neurons, which is also known for morphine, is not sufficient. Instead, the NRI mechanism of tapentadol is likely to be responsible for the dose-dependent increase in NA concentration. While an increase in NA concentration can be rated as a direct effect of NRI, antagonistic studies can indirectly corroborate these findings. Dose-dependent anti-nociceptive efficacy of tapentadol (IV) in naïve rats was shifted to the right by pretreatment with yohimbine (IP), a selective a2-antagonist [41]. Likewise, yohimbine diminished the anti-nociceptive effect of tapentadol in a rat model of persistent pain (formalin test, phase II) [29]. Atipamezole, another selective a2-antagonist, was shown to inhibit the effect of tapentadol (SC) when administered locally to the spinal cord [39]. Direct and indirect evidence therefore suggests that tapentadol's second non-opioidergic mechanism (NRI) can contribute to anti-nociceptive efficacy in rodent models of acute and persistent nociception.

Thus, the first step from *in vitro* paradigms to an *in vivo* mammalian system confirmed the functional efficacy of both mechanisms in naïve pain-transmitting pathways. Building on these findings, the next obvious question is how efficacy develops in pathological pain conditions such as inflammatory, visceral, or neuropathic pain. Earlier data with tramadol gave first hints that tapentadol might play an important role in the treatment of neuropathic pain. With MOR agonism and NRI, tapentadol partly shares the mechanisms responsible for the analgesic efficacy of tramadol. When comparing the potencies and efficacies of tramadol with morphine (a prototypic pure MOR agonist), it turned out that tramadol has a better potency ratio in neuropathic pain models than in nociceptive rodent pain models [42]. Hence, a similar comparison seems worthwhile between tapentadol and morphine. Anti-nociceptive, anti-hyperalgesic, and anti-allodynic potencies (ED50 values) after IV administration of both compounds generated in mouse and rat models of acute nociception, inflammatory, visceral, and neuropathic pain [29, 33, 34] were used and compared to the *in vitro* ratio of MOR binding in the [31]. Figure 23.8 shows the tapentadol/morphine ratios in descending rank order. The *in vitro* data ratio shows the 50-fold stronger MOR binding affinity of morphine. Likewise, the largest potency ratio toward morphine is reached in models of acute nociception. Models of inflammatory and visceral pain result in a more balanced potency ratio, while a clear tendency toward tapentadol is seen in models of neuropathic pain.

Indeed, as in the case of tramadol, tapentadol demonstrates stronger relative potency as compared to morphine in models of neuropathic pain. This is further corroborated when comparing antagonistic studies in rat acute nociception with neuropathic hypersensitivity [41]. Naloxone and yohimbine were shown to rightward-shift the ED50 curves of tapentadol in the low-intensity tail flick (acute nociception and heat stimulus) as well as in the spinal nerve ligation (neuropathic hypersensitivity and mechanical stimulus). However, although the test conditions were kept constant (IV administration of tapentadol, IP administration of antagonists, same doses of antagonists), the degrees by which these shifts occurred were opposite. The anti-nociceptive effect of tapentadol was more prone toward MOR agonism (6.4-fold shift by naloxone and 1.7-fold shift by yohimbine). In contrast, the anti-hypersensitive effect of tapentadol in neuropathic pain was more prone toward NRI (2.5-fold shift by naloxone and 4.5-fold shift by yohimbine). The recently introduced concept of μ-load can be applied to these data, stating that the MOR agonistic contribution of tapentadol is in the range of 54% for the anti-nociceptive effect and around 36% for the anti-hypersensitive effect [43]. In contrast, morphine would contribute a μ-load of 100% based on its pure MOR-agonistic mechanism of action. Nevertheless, one might criticize that these data were generated with different (thermal vs mechanical) readouts. Of note, assessment in a mouse model of diabetic hypersensitivity compared anti-nociceptive with anti-hyperalgesic potency and efficacy of tapentadol and morphine by means of exactly the same read-out, i.e., the number of nocifensive reactions over two minutes on a hot plate [33]. Figure 23.9 reflects the outcome of this study, demonstrating the selective anti-hyperalgesic efficacy of tapentadol, while morphine reaches full anti-hyperalgesic efficacy only at a dose that is also anti-nociceptive.

Overall, these data suggest that tapentadol is more potent in conditions of nerve injury reflecting neuropathic pain. Beside the comparison of naïve and pathological conditions of the pain transmitting system, another way of looking at the dual MOR-NRI mechanism of action is by means of isobolographic analysis, a method, which was introduced in [44, 45] and widely applied by Tallarida and others (for review, see [46]). Classically, isobolographic analysis is based on interaction studies with two independent principles (i.e., drugs), which are combined alone and in equi-effective combination ratios. The experimentally determined potency is compared with the theoretically additive combination calculated based on the potencies of both drugs when tested alone. An example of combined MOR agonism and monoaminergic reuptake inhibition is given by [35]. The experimental combination of morphine and desipramine clearly is more potent than expected based on the potencies generated with the two drugs alone in the phenylquinone writhing model in mice. This is a clear example of synergistic interaction. In the case of tapentadol, however, the situation is more complex as the combination of the two mechanisms is realized in a single molecule. This makes the classical isobolographic assessment impossible. Nevertheless, a combination of *in vivo* efficacy data [41], tapentadol's corresponding brain concentrations, and its *in vitro* dissociation constants for MOR and NRI were used to assess the intrinsic interaction between both mechanisms. The receptor fractional occupancies were determined and analyzed. This analysis strongly supports a synergistic interaction between MOR and NRI in rodent models of acute nociception and neuropathic pain [47]. As outlined earlier, MOR agonism and NRI can interfere at several sites of the pain-transmitting nervous system to exert an analgesic efficacy, such as the brainstem and the spinal cord. Based on the potent anti-hyperalgesia (Figure 23.8) and the selectivity of this anti-hyperalgesia after systemic administration of tapentadol (Figure 23.9), the mouse model of STZ-induced diabetic neuropathy was used to further elaborate on the site of action [48]. Comparing local administration on the spinal and supraspinal levels suggests the contribution of both sites to anti-hyperalgesia. Interestingly, when comparing with morphine, the supraspinal potency of tapentadol after ICV administration very well reflects the 50-fold *in vitro* ratio (Figure 23.8). The ICV ED50 values are 0.180 mg/animal for tapentadol and 0.004 mg/animal for morphine, resulting in a ratio of 45. In contrast, spinal IT administration results in IT ED50 values of 0.420 mg/animal for tapentadol and 0.547 mg/animal for morphine, resulting in a balanced ratio of around 1. Both drugs induced spinal release of norepinephrine as demonstrated by antagonistic inhibition of the anti-hyperalgesic efficacy by spinal administration of yohimbine. However, the shift in potency ratio from 45 to 1 is most likely

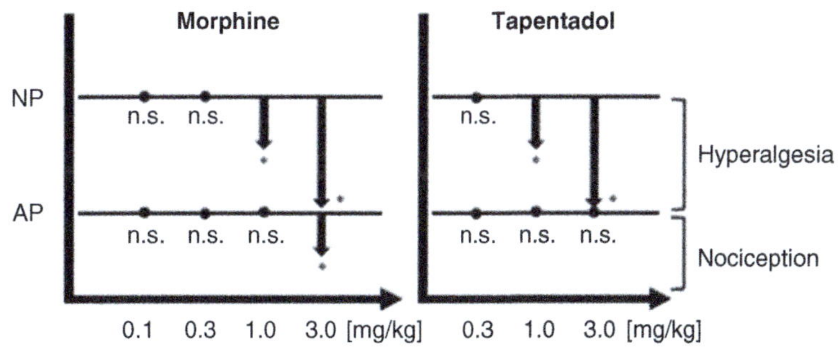

Figure 23.9 Selective anti-hyperalgesic efficacy of tapentadol in a mouse model of diabetic heat hyperalgesia. Christoph et al. [33]/with permission of Elsevier. Inhibitory effect of tapentadol or morphine on heat hyperalgesia in diabetic mice (neuropathic pain) and heat nociception in non-diabetic controls (acute pain) at the time of maximal efficacy. *Significant effect, n.s. = not significant, AP = acute pain, NP = neuropathic pain.

due to the NRI component of tapentadol, which is missing in the selective MOR agonist morphine. Finally, an isobolographic analysis on the combination of spinal (IT) and supra-spinal (ICV) tapentadol was performed in the same mouse model and clearly revealed a site-site synergism [48]. Thus, detailed analysis of anti-nociceptive and anti-neuropathic efficacy suggests (i) intrinsic synergistic activity after systemic administration both in rodent models of acute and neuropathic pain, (ii) a strong contribution of the spinal cord level to the anti-neuropathic efficacy, and (iii) spinal/supra-spinal synergy by combining tapentadol's efficacy on different levels of the neuraxis.

Beside behavioral studies in rodent models of neuropathic pain, tapentadol also showed clear mechanistic evidence of MOR agonism and NRI when assessed by means of *in vivo* electrophysiology. In rats with spinal nerve ligation, systemic tapentadol (SC) inhibits evoked responses of spinal cord neurons, and these effects were antagonized by naloxone and atipamezole, suggesting that MOR and NRI contribute to the inhibition of pain transmission [39]. The right central nucleus of the amygdala (CeA) is an important central control station for pain transmission and plays an important role in the control of the emotional-affective dimension of pain. Systemic tapentadol (SC) was effective in CeA neurons in rats with spinal nerve ligation, which again was subject to antagonistic interaction with naloxone and yohimbine [49]. Furthermore, efficacy of tapentadol was demonstrated in another central nucleus, the locus coeruleus (LC), which plays an important role in pain modulation. MOR agonism and NRI were demonstrated by means of a2- and MOR-antagonists [50]. The same experimental setup in a rat model of diabetic neuropathy revealed a shift of tapentadol's efficacy from MOR agonism to NRI [51], reflecting data obtained in behavioral pharmacology (see above). Finally, an elegant study in rats with spinal nerve ligation could demonstrate efficacy of tapentadol on restoring diffuse noxious inhibitory control (DNIC), which reflects brain stem-driven descending inhibition and is reduced in neuropathic pain states. Again, tapentadol's efficacy could be attributed to MOR agonism and NRI by means of appropriate antagonists [52]. Thus, beside striking evidence in behavioral pharmacology, electrophysiological data strongly support the efficacy of tapentadol on the spinal and supraspinal level as well as on descending circuits.

Conditioned pain modulation (CPM) is an experimental measure of endogenous pain modulation using a distant painful conditioning stimulus to affect a test stimulus. Thus, CPM can be seen as the human counterpart of DNIC [53] and offers a unique opportunity to build a translational bridge between rodents and humans, focusing on the dual mechanistic aspects of tapentadol. Chronic pain states lead to a reduction in descending inhibition, as reflected by reduced and/or absent DNIC in SNL rats [52] and CPM in patients with painful diabetic polyneuropathy (DPN) [54]. In this randomized controlled experimental study, DPN patients lacked CPM at baseline. Treatment with tapentadol for four weeks resulted in significant pain reduction, which was accompanied by a strong increase in CPM, suggesting that the analgesic efficacy in these neuropathic pain patients was driven by leveraging the recovery of NRI-dependent descending pain inhibition [54].

The final translational proof of a new drug has to be demonstrated in randomized controlled trials in the intended target patient population. Tapentadol was tested in an immediate release (IR) and a prolonged release (PR) formulation in several pain indications comprising phase 3 clinical trials in acute post-operative pain, chronic low back, joint and osteoarthritis pain, neuropathic pain, and cancer pain (see www.clinicaltrials.gov).

Efficacy of tapentadol IR was demonstrated in two postoperative phase 3 studies for acute pain after bunionectomy using 50, 75, or 100 mg every 4–6 hours for 72 hours [55] and 50 or 75 mg every 4–6 hours for 72 hours [56]. Tapentadol was compared to placebo and oxycodone IR and demonstrated clear improvements in pain intensity. Another phase 3 study used tapentadol IR at 50 or 75 mg every four to six hours in patients with end-stage joint disease [57]. Efficacy in these patients awaiting knee replacement surgery was demonstrated for tapentadol IR and the active comparator oxycodone IR as compared to placebo. Further evidence on efficacy with tapentadol IR came out of a 90-day safety study with flexible doses of 50 or 100 mg every four to six hours (up to 600 mg/day) for up to 90 days [58]. In patients with lower back pain or osteoarthritis of the hip or knee, tapentadol IR showed pain relief that was similar to that of the comparator oxycodone IR. In addition, patients with moderate to severe acute lower back pain experienced relevant pain relief after treatment with tapentadol IR (50, 75, or 100 mg) or oxycodone IR every four to six hours as needed for pain up to 10 days [59]. Thus, tapentadol IR demonstrated convincing efficacy in acute pain of different origins (for review see [60]).

Efficacy of tapentadol PR was demonstrated in numerous clinical studies and has been reviewed extensively [61, 62], including low back pain [63], neuropathic pain [64], osteoarthritis [65], and cancer pain [66]. While it is out of the scope of the current article to discuss all clinical trials performed with tapentadol PR, it is worthwhile to mention some studies, which are of interest in terms of translational value. The dual mode of action qualifies tapentadol as a potential treatment option for chronic neuropathic pain, and this view is supported by the preclinical data described above. Two pivotal phase 3 studies in chronic low back pain (NCT00449176) and painful DPN (NCT00455520) demonstrated efficacy of tapentadol PR (100–250 mg bid) in a 12 weeks maintenance period after three weeks of titration [61, 67, 68]. Beside nociceptive pain, chronic low back pain often comes with symptoms of neuropathic pain [69]. Both pain types have separate origins, which argues for the use of separate mechanisms for analgesic treatment [70]. Thus, the efficacy of tapentadol in chronic low back pain translates from the preclinical efficacy data. A phase 3b study in chronic low back pain patients with neuropathic pain (NCT01352741) compared tapentadol PR (500 mg/day) with a combination of tapentadol PR (300 mg/day) and pregabalin (300 mg/day) over eight weeks after poor response in three weeks of titration to tapentadol (300 mg/day). Tapentadol PR demonstrated efficacy both when given alone or as combination with pregabalin, and tapentadol PR alone was noninferior to the combination. Tolerability was improved in the tapentadol PR group over the combination group [71]. Treatment of patients responding well to tapentadol PR 300 mg/day in the titration phase was continued in an eight week open-label continuation arm [72]. Although the open-label nature of the study and the selective enrichment for responders are clear limitations, the positive effects of tapentadol in these patients might well reflect a population with effective restoration of descending inhibitory control. While the general finding of tapentadol's efficacy in neuropathic pain supports the theory generated based on preclinical data, the phase 3b combination data do not translate from preclinical data demonstrating synergistic interaction of both drugs in SNL rats [34], which might be due to missing face validity in the animal model, which is dominated by a pure nerve lesion. Beside efficacy in the gold standard of randomized controlled studies, it is of relevance and interest how a drug is used in daily practice. The German Pain eRegistry is a national web-based pain registry developed on behalf of the German Pain Association and was used for a noninterventional retrospective cohort study comparing tapentadol PR with standard oral opioid PR drugs (morphine, oxycodone, and hydromorphone) over 12 weeks in patients with chronic low back pain who did not receive sufficient pain relief by use of WHO-I/II standard analgesics. These real-word data clearly corroborate the results from randomized controlled studies demonstrating efficacy and tolerability of tapentadol PR, which was better than that of standard opioid drugs in this pain indication. Of note, this also was seen with respect to the neuropathic pain component in these patients [73]. While chronic low back pain frequently comprises nociceptive and neuropathic pain, painful DPN is clearly a neuropathic pain. In a randomized-withdrawal, placebo-controlled, phase 3 study (NCT00455520), patients with painful DPN were titrated over an open-label period of three weeks to their optimal dose of tapentadol PR (100–250 mg bid). Responding patients were randomized and treated for 12 weeks with placebo or the optimal dose of tapentadol PR. Tapentadol was well tolerated and resulted in a significant and clinically relevant reduction in pain [67]. This efficacy is in line with the functional assessment of CPM in painful DPN described above [54] and supports a reasonable translational bridge from preclinical findings in rodents with diabetes-induced hypersensitivity [33, 48].

23.7.3 Adverse Effects

Opioid agonists have been in clinical use for decades and come with specific and well-known peripheral and central adverse effects, such as opioid-induced constipation, respiratory depression, emesis, and euphoria. In contrast, adverse effects generated by selective NRIs are less well-described. Clinically used antidepressants such as the tricyclic antidepressants comprise NRI as one mechanism of action among others. Antimuscarinic efficacy is common in these antidepressants, and anticholinergic adverse effects are most disturbing in patients [74]. There is no selective NRI compound in clinical use. However, nisoxetine is a highly selective NRI and is frequently used as a reference compound for this mechanism of action. It was stopped during development as an antidepressant, but clinical phase 1 data have been published suggesting that no NRI-specific (hyper-adrenergic) adverse effects occurred over a treatment period of seven days [75]. The assessment of tolerability and adverse effects of a MOR-NRI compound like tapentadol therefore needs to focus on opioid-specific adverse effects and on potentially harmful interactions of both mechanisms.

Preclinical data compare the adverse effects of tapentadol to the pure MOR agonist morphine. Overall, opioid-specific adverse effects occur under the treatment of both compounds, but compared to the relative anti-nociceptive and anti-hypersensitive efficacy, tapentadol shows a more favorable profile. Gastrointestinal transit time of orally administered charcoal is dose-dependently delayed by MOR agonists, reflecting inhibition of gastrointestinal motility and hence opioid-induced constipation. Another measure related to opioid-induced constipation is the inhibition of PGE2-induced diarrhea. In mice, tapentadol showed inhibition of charcoal transit and PGE2-induced diarrhea with an ED25 of 28.1 mg/kg IP and an ED50 of 10.3 mg/kg IP, respectively. In contrast, morphine delayed charcoal transit with an ED25 of 5.9 mg/kg IP and PGE2-induced diarrhea with an ED50 of 1.1 mg/kg IP [29]. The tapentadol/morphine ratios are 4.8 and 9.4 for opioid-induced constipation and anti-diarrhea, respectively. These ratios are higher as compared to the ratios in efficacy models of pain (Figure 23.8) and suggest that the gastrointestinal adverse effects are driven by the MOR agonism of both compounds. Two sets of efficacy data can be compared to these data based on species and route of administration. The mouse hot plate assay results in ED50 values of 11.8 and 4.7 mg/kg IP for tapentadol and morphine, respectively [29], and the inhibition of diabetic heat hypersensitivity in mice was quantified with ED50 values of 0.274 and 1.107 mg/kg IP for tapentadol and morphine, respectively [48]. The tapentadol/morphine ratios are 2.51 and 0.25 for hot plate and diabetic mice, respectively, and corroborate the findings after IV administration (Figure 23.8). In any ratio of adverse effect over efficacy, tapentadol shows a more favorable profile when compared to morphine. These findings suggest that the intrinsic synergistic interaction of tapentadol seen in rodent models of efficacy might not be found in rodent models of adverse effects. A more detailed analysis of the intrinsic type of MOR-NRI interaction was performed in the gastrointestinal charcoal transit model in rats [76]. Based on the experimental and mathematical demonstration of intrinsic synergy in acute nociception and neuropathic hypersensitivity [47], a similar setup was analyzed with respect to opioid-induced constipation. The data generated with SC tapentadol in the absence or presence of naloxone (1 mg/kg IP) or yohimbine (0.25 mg/kg IP) suggest an additive interaction of MOR agonism and NRI [76]. While in line with the findings in corresponding mouse models of adverse effects (see above), this is in sharp contrast to the synergistic interaction of both mechanisms in rodent models of efficacy discussed earlier.

Nausea and vomiting occur under acute treatment conditions with opioids, and tolerance develops in chronic treatment. Acute treatment with opioids induces retches and vomits in the ferret and is used as an experimental model of emesis. Tapentadol up to a high dose of 31.6 mg/kg IP induces weak signs of emesis with vomiting and retching only in up to 20% and 40% of the treated animals, respectively. In contrast, morphine-induced vomiting and retching in 100% of the treated animals at low doses of 0.5 and 0.25 mg/kg SC, respectively [29].

Respiratory depression is another well-described adverse effect of MOR agonists and can lead to death upon overdose of potent opioids. As for the gastrointestinal effects, tapentadol shows a comparable opioid-specific profile with less potency than morphine [29]. Effects were assessed in conscious rats by means of spontaneous respiratory frequency, plethysmography, arterial pCO_2 in blood gas analysis, and CO_2-induced stimulation of respiratory frequency. The latter model allows for determination of ED50 values, which were calculated as 0.71 and 0.23 mg/kg IV for tapentadol and morphine, respectively, resulting in a tapentadol/morphine ratio of 3.1. Sub-acute treatment reveals tolerance development against respiratory depressants within 22 days for both tapentadol and morphine [29]. Notably, this is in contrast to the efficacy upon sub-acute treatment in a rat model of neuropathic pain, where tolerance development is clearly delayed for tapentadol compared to morphine [31].

In summary, preclinical data suggest better tolerability of tapentadol compared with MOR agonists. The adverse effect profile of tapentadol resembles that of classical opioids, but equianalgesic doses evoke less adverse effects. In contrast to an increased efficacy based on synergistic interaction of MOR agonism and NRI, the occurrence and intensity of the typical opioid adverse effects are ameliorated. Finally, tolerance development against central adverse effects such as respiratory depression or emesis is comparable to MOR agonists supporting the profile of increased analgesia with lower adverse effects.

As expected from the outcome of the preclinical studies, opioid-typic adverse effects of gastrointestinal (e.g., constipation, nausea, and vomiting), central (e.g., dizziness, somnolence, and headache), or peripheral somatic (e.g., pruritus) origin were detected in clinical studies with tapentadol.

Tapentadol IR, when used in different pain settings with analgesic efficacy comparable to oxycodone IR, demonstrated an opioid-typic adverse effect profile with better gastrointestinal tolerability than oxycodone IR [55–59]. Furthermore, a direct comparison of equianalgesic doses of tapentadol IR with oxycodone IR in end-stage joint disease for 14 days, followed by 28 days of treatment with PR formulations of both drugs, resulted in improved gastrointestinal function, including constipation, nausea, and vomiting, in patients receiving tapentadol (NCT00784277, [77]. Safety and tolerability of tapentadol PR was evaluated in a pooled analysis of four phase 3 studies on moderate to severe chronic osteoarthritis or low back pain (NCT00361504, [78]; NCT00421928, NCT00449176, NCT00486811, [61]) with large total numbers of patients in the placebo (n = 993), tapentadol PR (n = 1874), and oxycodone PR (n = 1224) groups [79]. Incidences of gastrointestinal/nervous system treatment-emergent adverse effects (TEAEs) were 65.4%/45.1%, 47.3%/42.6%, and 26.5%/22.5% for oxycodone PR, tapentadol PR, and placebo, respectively. Premature discontinuation was larger in the oxycodone PR group (62.3%) as compared to tapentadol PR (48.4%) and placebo (40.7%). The incidences of gastrointestinal/nervous system TEAEs leading to discontinuation were 24.1%/16.3%, 8.3%/7.9%, and 2.1%/1.4% for oxycodone PR, tapentadol PR, and placebo, respectively [79]. Thus, equianalgesic treatment doses of tapentadol result in an improved adverse effect profile when compared to oxycodone, reflected by better tolerability with respect to gastrointestinal and nervous system TEAEs, irrespective of the formulation (IR or PR).

23.8 Summary

Tapentadol is a fascinating centrally acting synthetic analgesic with a dual mechanism of action in one single molecule without any metabolic activation. Tapentadol exerts its analgesic effects without a pharmacologically active metabolite.

Tapentadol is a selective MOR agonist: it binds to MOR with an affinity greater than or equal to 10-fold affinity compared to delta- and kappa-opioid receptors. Tapentadol also inhibits norepinephrine reuptake, thereby increasing norepinephrine levels and activating alpha-2 receptors to promote analgesia.

Tapentadol is effective in treating moderate to severe pain (WHO step 3). In clinical trials, tapentadol has been shown to be equiefficacious to oxycodone with fewer gastrointestinal adverse effects.

Polymorphic Forms A and B of tapentadol hydrochloride are known and characterized beside an amorphous form. The crystalline Form A has the same pharmacological activity than Form B but is stable under ambient conditions, while Form B is descripted to be meta-stable only.

Tapentadol was introduced as the hydrochloride salt to the pharmaceutical market. Due to the loss of exclusivity, several generic brands are on the market also using the phosphate and tartrate salt of tapentadol. A tremendous number of acceptable pharmaceutical salts is described mainly in patents. Beside salt forms, also cocrystals of the tramadol-free base or the hydrochloride with cocrystal formers are described. Also, other analgesics are used as cocrystal formers.

References

1 Buschmann, H. (2012). Tapentadol – from morphine and tramadol to the discovery of tapentadol. Chapter 12. In: *Analogue-based Drug Discovery III* (ed. J. Fischer, C.R. Ganellin, and D.P. Rotella). Weinheim: Wiley-VCH Verlag.

2 European Medicines Agency (2023). National registers of authorized medicines. https://european-union.europa.eu/index_en (accessed 23 September 2024).

3 Barakat, A. (2019). Revisiting tramadol: a multi-modal agent for pain management. *CNS Drugs* 33 (5): 481–501.

4 Reimann, W. and Schneider, F. (1998). Induction of 5-hydroxytryptamine release by tramadol. *Eur. J. Pharmacol.* 349 (2–3, 203): 199.

5 FDA Approved Drug Products (2019). Ultram (tramadol hydrochloride) oral tablets, 2003, revised 2019 by Janssen Pharmaceutical Companies. https://www.accessdata.fda.gov/drugsatfda_docs/label/2019/020281s045lbl.pdf (accessed 23 September 2024).

6 Raffa, R.B., Friderichs, E., Reimann, W. et al. (1992). Opioid and nonopioid components independently contribute to the mechanism of action of tramadol, an "atypical" opioid analgesic. *J. Pharmacol. Exp. Ther.* 260: 275–285.

7 Raffa, R.B., Friderichs, E., Reimann, W. et al. (1993). Complementary and synergistic antinociceptive interaction between the enantiomers of tramadol. *J. Pharmacol. Exp. Ther.* 267: 331–340.

8 Valle, M., Garrido, M.J., Pavo, J.M. et al. (2000). Pharmacokinetic pharmacodynamic modeling of the antinociceptive effects of main active metabolites of tramadol, (1)-O-desmethyltramadol and (2)-Odesmethyltramadol, in rats. *J. Pharmacol. Exp. Ther.* 293: 646–653.

9 Casy, A.F. and Parfitt, R.T. (1986). *Opioid Analgesics – Chemistry and Receptors*. New York: Plenum Press.

10 Hu, Y., Stumpfe, D., and Bajorath, J. (2011). Lessons learned from molecular scaffold analysis. *J. Chem. Inf. Model.* 51: 1742–1753.

11 Raffa, R.B., Buschmann, H., Christoph, T. et al. (2012). Mechanistic and functional differentiation of tapentadol and tramadol. *Expert. Opin. Pharmacother.* 13 (10): 1437–1449.

12 Drugbank. Tapentadol. https://go.drugbank.com/drugs/DB06204 (assessed 14 June 2024).

13 FDA Approved Drug Products (2023). NUCYNTA (tapentadol) tablets for oral use C-II initial U.S. (revised July 2023). https://www.accessdata.fda.gov/drugsatfda_docs/label/2023/022304s024lbl.pdf (accessed 23 September 2024).

14 Barbosa, J., Faria, J., Queirós, O. et al. (2016). Comparative metabolism of tramadol and tapentadol: a toxicological perspective. *Drug Metab. Rev.* 48 (4): 577–592.

15 Wade, W.E. and Spruill, W.J. (2009). Tapentadol hydrochloride: a centrally acting oral analgesic. *Clin. Ther.* 31: 2804–2818.

16 Lischke, D. (2021). Standards and trends in solid-state characterization techniques: infrared (IR) spectroscopy.

Chapter 4. In: *Solid State Development and Processing of Pharmaceutical Molecules – Salts Cocrystals, and Polymorphism* (ed. M. Gruß), 151. Weinheim: Wiley-VCH GmbH.
17 Gruß, M. (2021). Aspects for developing and processing solid forms. Chapter 1. In: *Solid State Development and Processing of Pharmaceutical Molecules - Salts, Cocrystals, and Polymorphism* (ed. M. Gruß), 1–43. Weinheim: Wiley-VCH GmbH.
18 Rolls, C. and Onishi., J. (2020). Patenting polymorphs: a complex process. *Chem. Aust.* Sep-Nov: 34–35.
19 Fischer, A., Buschmann, H., Gruss, M., and Lischke, D. (2004). Crystalline forms of (-)-(1R,2R)-3-(3-dimethylamino-1-ethyl-2-methylpropyl)-phenol hydrochloride. EP1612203B1, https://patents.google.com/patent/EP1612203B1/en (accessed 23 September 2024).
20 Rajan, S. T., Kumar, M. K., Venkatesh, M., and Praneeth Kumar, A. (2021). Crystalline forms of tapentadol salts and process for preparation thereof. EP3377049B1. https://patents.google.com/patent/EP3377049B1/en (accessed 23 September 2024).
21 Devjibhai, K. M., Prabhurao, B. S., Daulatrao, A. V., Sharadchandra, P. N., and Vikas, A. D. (2010). Amorphous form of tapentadol hydrochloride. US2010272815A1. https://worldwide.espacenet.com/patent/search/family/042992362/publication/US2010272815A1?q=US2010%2F0272815 (accessed 23 September 2024).
22 Ziao, H., Yonghai, Z., Degang, L., and Xiaoxiang, L. (2012). Tapentadol hydrochloride crystal form C, and preparation method and application thereof. CN102924303A. https://worldwide.espacenet.com/patent/search/family/047639266/publication/CN102924303A?q=CN102924303A (accessed 23 September 2024).
23 Buschmann, H., Strassburger, W., and Frieserichs, E. (1995). 1-phenyl-3-dimethylaminopropane compounds with a pharmacological effects. US6248737B1. https://worldwide.espacenet.com/patent/search/family/006524045/publication/US6248737B1?q=US6248737 (accessed 23 September 2024).
24 McIntyre, K. (2021). Polymorphs and patents – the US perspective. Chapter 9.1. In: *Solid State Development and Processing of Pharmaceutical Molecules - Salts, Cocrystals, and Polymorphism*, vol. 2021 (ed. M. Gruß). Weinheim: Wiley-VCH GmbH.
25 Fischer, A., Buschmann, H., Gruss, M., and Lischke, D. (2011). Crystalline forms of (-)-(1R,2R)-3-(3-dimethylamino-1-ethyl-2-methylpropyl)-phenol hydrochloride. US7994364B2. https://worldwide.espacenet.com/patent/search/family/034925507/publication/US7994364B2?q=US7994364 (accessed 23 September 2024).
26 Byrn, S., Pfeiffer, R., Ganey, M. et al. (1995). Pharmaceutical solids: a strategic approach to regulatory considerations. *Pharm. Res.* 12 (7): 945–954.
27 Hell, W., Kegel, M., Akteries, B. et al. (2004). Method for the preparation of 3-substituted 3-aryl-butyl amine. https://patents.google.com/patent/WO2004108658A1/en (accessed 23 September 2024).
28 Graczyk, P.P., Zbrojkiewicz, O., and Nerdinger, S. (2017). Evolution of synthetic routes towards homochiral Tapentadol. *Tetrahedron Asymmetry* 28: 387–400.
29 Tzschentke, T.M., De Vry, J., Terlinden, R. et al. (2006). Tapentadol hydrochloride. *Drugs Future* 31 (12): 1053–1061.
30 Tzschentke, T.M., Jahnel, U., Kogel, B. et al. (2009). Tapentadol hydrochloride: a next-generation, centrally acting analgesic with two mechanisms of action in a single molecule. *Drugs Today* 45 (7): 483–496.
31 Tzschentke, T.M., Christoph, T., Kögel, B. et al. (2007). (-)-(1R,2R)-3-(3-dimethylamino-1-ethyl-2-methyl-propyl)-phenol hydrochloride (tapentadol HCl): a novel mu-opioid receptor agonist/norepinephrine reuptake inhibitor with broad-spectrum analgesic properties. *J. Pharmacol. Exp. Ther.* 323 (1): 265–276.
32 Terlinden, R., Kogel, B.Y., Englberger, W., and Tzschentke, T.M. (2010). In vitro and in vivo characterization of tapentadol metabolites. *Methods Find. Exp. Clin. Pharmacol.* 32 (1): 31–38.
33 Christoph, T., De Vry, J., and Tzschentke, T.M. (2010). Tapentadol, but not morphine, selectively inhibits disease-related thermal hyperalgesia in a mouse model of diabetic neuropathic pain. *Neurosci. Lett.* 470 (2): 91–94.
34 Christoph, T., De Vry, J., Schiene, K. et al. (2011). Synergistic antihypersensitive effects of pregabalin and tapentadol in a rat model of neuropathic pain. *Eur. J. Pharmacol.* 666 (1–3): 72–79.
35 Tzschentke, T.M., Christoph, T., and Kögel, B.Y. (2014). The mu-opioid receptor agonist/noradrenaline reuptake inhibition (MOR-NRI) concept in analgesia: the case of tapentadol. *CNS Drugs* 28 (4): 319–329.
36 Millan, M.J. (1999). The induction of pain: an integrative review. *Prog. Neurobiol.* 57 (1): 1–164.
37 Millan, M.J. (2002). Descending control of pain. *Prog. Neurobiol.* 66 (6): 355–474.
38 Kögel, B., De Vry, J., Tzschentke, T.M., and Christoph, T. (2011). The antinociceptive and antihyperalgesic effect of tapentadol is partially retained in OPRM1 (μ-opioid receptor) knockout mice. *Neurosci. Lett.* 491 (2): 104–107.

39 Bee, L.A., Bannister, K., Rahman, W., and Dickenson, A.H. (2011). Mu-opioid and noradrenergic α(2)-adrenoceptor contributions to the effects of tapentadol on spinal electrophysiological measures of nociception in nerve-injured rats. *Pain* 152 (1): 131–139.

40 Tzschentke, T.M., Folgering, J.H., Flik, G., and De Vry, J. (2012). Tapentadol increases levels of noradrenaline in the rat spinal cord as measured by in vivo microdialysis. *Neurosci. Lett.* 507 (2): 151–155.

41 Schröder, W., Vry, J.D., Tzschentke, T.M. et al. (2010). Differential contribution of opioid and noradrenergic mechanisms of tapentadol in rat models of nociceptive and neuropathic pain. *Eur. J. Pain* 14 (8): 814–821.

42 Christoph, T., Kögel, B., Strassburger, W., and Schug, S.A. (2007). Tramadol has a better potency ratio relative to morphine in neuropathic than in nociceptive pain models. *Drugs R D.* 8 (1): 51–57.

43 Raffa, R.B., Elling, C., and Tzschentke, T.M. (2018). Does 'strong analgesic' equal 'strong opioid'? Tapentadol and the concept of 'μ-load'. *Adv. Ther.* 35 (10): 1471–1484.

44 Loewe, S. (1953). The problem of synergism and antagonism of combined drugs. *Arzneimittelforschung* 3: 285–290.

45 Loewe, S. (1957). Antagonisms and antagonists. *Pharmacol. Rev.* 9: 237–242.

46 Tallarida, R.J. (2001). Drug synergism: its detection and applications. *J. Pharmacol. Exp. Ther.* 298: 865–872.

47 Schröder, W., Tzschentke, T.M., Terlinden, R. et al. (2011). Synergistic interaction between the two mechanisms of action of tapentadol in analgesia. *J. Pharmacol. Exp. Ther.* 337 (1): 312–320.

48 Christoph, T., Schröder, W., Tallarida, R.J. et al. (2013). Spinal-supraspinal and intrinsic μ-opioid receptor agonist-norepinephrine reuptake inhibitor (MOR-NRI) synergy of tapentadol in diabetic heat hyperalgesia in mice. *J. Pharmacol. Exp. Ther.* 347 (3): 794–801.

49 Gonçalves, L., Friend, L.V., and Dickenson, A.H. (2015). The influence of μ-opioid and noradrenaline reuptake inhibition in the modulation of pain responsive neurones in the central amygdala by tapentadol in rats with neuropathy. *Eur. J. Pharmacol.* 749: 151–160.

50 Torres-Sanchez, S., Alba-Delgado, C., Llorca-Torralba, M. et al. (2013). Effect of tapentadol on neurons in the locus coeruleus. *Neuropharmacology* 72: 250–258.

51 Torres-Sanchez, S., Borges, G.D.S., Mico, J.A., and Berrocoso, E. (2018). Opioid and noradrenergic contributions of tapentadol to the inhibition of locus coeruleus neurons in the streptozotocin rat model of polyneuropathic pain. *Neuropharmacology* 135: 202–210.

52 Bannister, K., Patel, R., Goncalves, L. et al. (2015). Diffuse noxious inhibitory controls and nerve injury: restoring an imbalance between descending monoamine inhibitions and facilitations. *Pain* 156 (9): 1803–1811.

53 Yarnitsky, D. (2010). Conditioned pain modulation (the diffuse noxious inhibitory control-like effect): its relevance for acute and chronic pain states. *Curr. Opin. Anaesthesiol.* 23 (5): 611–615.

54 Niesters, M., Proto, P.L., Aarts, L. et al. (2014). Tapentadol potentiates descending pain inhibition in chronic pain patients with diabetic polyneuropathy. *Br. J. Anaesth.* 113 (1): 148–156.

55 Daniels, S.E., Upmalis, D., Okamoto, A. et al. (2009). A randomized, double-blind, phase III study comparing multiple doses of tapentadol IR, oxycodone IR, and placebo for postoperative (bunionectomy) pain. *Curr. Med. Res. Opin.* 25 (3): 765–776.

56 Daniels, S., Casson, E., Stegmann, J.U. et al. (2009). A randomized, double-blind, placebo-controlled phase 3 study of the relative efficacy and tolerability of tapentadol IR and oxycodone IR for acute pain. *Curr. Med. Res. Opin.* 25 (6): 1551–1561.

57 Hartrick, C., Van Hove, I., Stegmann, J.U. et al. (2009). Efficacy and tolerability of tapentadol immediate release and oxycodone HCl immediate release in patients awaiting primary joint replacement surgery for end-stage joint disease: a 10-day, phase III, randomized, double-blind, active- and placebo-controlled study. *Clin. Ther.* 31 (2): 260–271.

58 Hale, M., Upmalis, D., Okamoto, A. et al. (2009). Tolerability of tapentadol immediate release in patients with lower back pain or osteoarthritis of the hip or knee over 90 days: a randomized, double-blind study. *Curr. Med. Res. Opin.* 25 (5): 1095–1104.

59 Biondi, D., Xiang, J., Benson, C. et al. (2013). Tapentadol immediate release versus oxycodone immediate release for treatment of acute low back pain. *Pain Physician* 16 (3): E237–E246.

60 Afilalo, M., Stegmann, J.U., and Upmalis, D. (2010). Tapentadol immediate release: a new treatment option for acute pain management. *J. Pain Res.* 3: 1–9.

61 Afilalo, M. and Morlion, B. (2013). Efficacy of tapentadol ER for managing moderate to severe chronic pain. *Pain Physician* 16 (1): 27–40.

62 Deeks, E.D. (2018). Tapentadol prolonged release: a review in pain management. *Drugs* 78 (17): 1805–1816.

63 Coluzzi, F., Polati, E., Freo, U., and Grilli, M. (2019). Tapentadol: an effective option for the treatment of back pain. *J. Pain Res.* 12: 1521–1528.

64 Freo, U., Romualdi, P., and Kress, H.G. (2019). Tapentadol for neuropathic pain: a review of clinical studies. *J. Pain Res.* 12: 1537–1551.

65 Rinonapoli, G., Coaccioli, S., and Panella, L. (2019). Tapentadol in the treatment of osteoarthritis: pharmacological rationale and clinical evidence. *J. Pain Res.* 12: 1529–1536.

66 Kress, H.G. and Coluzzi, F. (2019). Tapentadol in the management of cancer pain: current evidence and future perspectives. *J. Pain Res.* 12: 1553–1560.

67 Schwartz, S., Etropolski, M., Shapiro, D.Y. et al. (2011). Safety and efficacy of tapentadol ER in patients with painful diabetic peripheral neuropathy: results of a randomized-withdrawal, placebo-controlled trial. *Curr. Med. Res. Opin.* 27 (1): 151–162.

68 Buynak, R., Shapiro, D.Y., Okamoto, A. et al. (2010). Efficacy and safety of tapentadol extended release for the management of chronic low back pain: results of a prospective, randomized, double-blind, placebo- and active-controlled Phase III study. *Expert. Opin. Pharmacother.* 11 (11): 1787–1804.

69 Freynhagen, R. and Baron, R. (2009). The evaluation of neuropathic components in low back pain. *Curr. Pain Headache Rep.* 13 (3): 185–190.

70 Costigan, M., Scholz, J., and Woolf, C.J. (2009). Neuropathic pain: a maladaptive response of the nervous system to damage. *Annu. Rev. Neurosci.* 32: 1–32.

71 Baron, R., Martin-Mola, E., Müller, M. et al. (2015). Effectiveness and safety of tapentadol prolonged release (PR) versus a combination of tapentadol PR and pregabalin for the management of severe, chronic low back pain with a neuropathic component: a randomized, double-blind, phase 3b study. *Pain Pract.* 15 (5): 455–470.

72 Baron, R., Kern, U., Müller, M. et al. (2015). Effectiveness and tolerability of a moderate dose of tapentadol prolonged release for managing severe, chronic low back pain with a neuropathic component: an open-label continuation arm of a randomized phase 3b study. *Pain Pract.* 15 (5): 471–486.

73 Überall, M.A., Elling, C., Eibl, C. et al. (2022). Tapentadol prolonged release in patients with chronic low back pain: real-world data from the German Pain eRegistry. *Pain Manag.* 12 (2): 211–227.

74 Reinert, J.P., Veronin, M.A., and Medina, C. (2023). Tricyclic antidepressants in nociceptive and neuropathic pain: a review of their analgesic properties in combination with opioids. *J. Pharm. Technol.* 39 (1): 35–40.

75 Lemberger, L., Terman, S., Rowe, H., and Billings, R. (1976). The effect of nisoxetine (Lilly compound 94939), a potential antidepressant, on biogenic amine uptake in man. *Br. J. Clin. Pharmacol.* 3 (2): 215–220.

76 Cowan, A., Raffa, R.B., Tallarida, C.S. et al. (2014). Lack of synergistic interaction between the two mechanisms of action of tapentadol in gastrointestinal transit. *Eur. J. Pain* 18 (8): 1148–1156.

77 Etropolski, M., Kelly, K., Okamoto, A., and Rauschkolb, C. (2011). Comparable efficacy and superior gastrointestinal tolerability (nausea, vomiting, constipation) of tapentadol compared with oxycodone hydrochloride. *Adv. Ther.* 28 (5): 401–417.

78 Wild, J.E., Grond, S., Kuperwasser, B. et al. (2010). Long-term safety and tolerability of tapentadol extended release for the management of chronic low back pain or osteoarthritis pain. *Pain Pract.* 10 (5): 416–427.

79 Etropolski, M., Kuperwasser, B., Flügel, M. et al. (2014). Safety and tolerability of tapentadol extended release in moderate to severe chronic osteoarthritis or low back pain management: pooled analysis of randomized controlled trials. *Adv. Ther.* 31 (6): 604–620.

24

Thalidomide – From a Banned Drug to Molecular Glues, PROTACs, and New Concepts in Drug Discovery

Junichi Yamamoto, Hiroshi Handa, and Yuki Yamaguchi

24.1 Introduction

Developed in the 1950s as a sedative and antiemetic, thalidomide's journey through the history of medicinal chemistry was tumultuous but transformative. It caused severe teratogenic effects, exemplified by characteristic limb defects in more than 10,000 newborns, which are recorded as one of the worst drug disasters in history. Consequently, thalidomide was withdrawn from the market in 1962, marking a turning point in drug regulation and safety standards [1].

However, the story of thalidomide has taken an unexpected turn. Subsequent studies showed that thalidomide had the potential to treat a variety of diseases, particularly erythema nodosum leprosum (ENL), a complication of leprosy [2] and multiple myeloma [3], leading to its re-approval by the U.S. Food and Drug Administration (FDA). This led to the development of thalidomide derivatives such as lenalidomide and pomalidomide, collectively known as immunomodulatory drugs (IMiDs), which are now an integral class of drugs in the treatment of multiple myeloma. The molecular mechanism of action of thalidomide was greatly advanced with the breakthrough identification of cereblon (CRBN) as its primary molecular target [4]. It acts as a substrate recognition receptor for the cullin 4 RING E3 ubiquitin ligase (CRL4), a protein complex that plays a crucial role in protein degradation within the cell. The binding of thalidomide to CRBN triggers a cascade of molecular events, leading to the proteasomal degradation of non-native substrates. This new class of drugs that exert their pharmacological effects by inducing the degradation of target proteins is collectively referred to as protein degraders and is expected to become a new modality in drug discovery. Thalidomide and its derivatives are the first clinically useful protein degraders.

This chapter delves into two key events where thalidomide significantly impacted medicinal chemistry: its notorious drug disaster and the emergence of a novel drug discovery approach, the protein degrader. It focuses on thalidomide's interaction with CRBN and the subsequent molecular-level pharmacological effects. Thalidomide's unique properties, with its multiple pharmacological actions, are also outlined in the context of polypharmacology.

24.2 Thalidomide History: From Tragedy to Therapeutic Revival

24.2.1 The Early Promise and Subsequent Disaster

Thalidomide emerged in the mid-twentieth century as a promising pharmaceutical innovation. Developed in 1954 and marketed as Contergan by the German pharmaceutical company Grunenthal in 1957, thalidomide was initially prescribed as a sedative and to treat morning sickness. Rodent studies conducted during its development reported no significant toxicity, leading to its widespread promotion in over 40 countries as a "completely safe drug with no side effects." The early years of thalidomide were marked by widespread use and confidence in its safety. The absence of immediate adverse effects in standard animal testing protocols reinforced the perception of thalidomide as a benign, even ideal, pharmaceutical solution for a range of ailments, particularly for pregnant women suffering from morning sickness. However, this initial promise was soon overshadowed by a series of alarming reports.

As the use of thalidomide grew, so did the reports of its devastating effects. The first troubling signs were reports of peripheral neuropathy in patients, a condition characterized by nerve damage resulting in numbness, weakness, and pain, typically in the hands and feet [5]. The severity of these symptoms varied, but they were undeniably related to thalidomide use. This was just the beginning of a much larger and tragic story that unfolded around this drug. Even more disturbing was the increasing number

Polypharmacology: Strategies for Multi-Target Drug Discovery, First Edition. Edited by Jens-Uwe Peters.

of miscarriages and reports of deformed newborns. These were not isolated incidents but part of a terrifying pattern directly linked to thalidomide. In a tragic twist, a drug marketed for its safety during pregnancy turned out to be teratogenic, causing severe birth defects. The spectrum of malformations observed in thalidomide embryopathy was broad, affecting multiple organ systems. The most visible and heartbreaking were limb deficiencies such as amelia (complete absence of limbs) or phocomelia (shortened limbs), which led to the term thalidomide babies. In addition, development of the face, eyes, inner ears, genitourinary and central nervous systems, and vital organs such as the heart, kidneys, and components of the digestive system are also disrupted [6].

The worldwide impact of thalidomide was catastrophic. More than 10,000 babies were born with deformities caused by the drug. This figure only scratches the surface of the true impact, however, as it does not include the countless miscarriages and stillbirths that also resulted from thalidomide exposure [1]. The teratogenic effects of the drug were officially recognized following reports by McBride and Lenz in 1961 and 1962, respectively [7, 8], leading to its withdrawal in all countries by the end of 1962. In the U.S., the history of thalidomide took a different turn. Dr. Frances Kathleen Oldham Kelsey, a pharmacologist and physician working for the U.S. FDA, was concerned about the insufficient data on thalidomide and refused to approve it. This decision was vindicated when birth defects in other countries were later linked to thalidomide. Her vigilant actions crucially prevented a comparable tragedy in the U.S. [9].

The thalidomide disaster remains one of the most significant and tragic events in pharmaceutical history. This event marked a turning point in drug regulation, paving the way for more stringent measures to ensure the safety of pharmaceutical products, particularly for vulnerable populations such as pregnant women. It still serves as a poignant reminder of the imperative need for drug safety and regulatory oversight in pharmaceutical development.

24.2.2 Rediscovery and Therapeutic Applications

Thalidomide's journey from pharmaceutical disaster to therapeutic revival is a remarkable story of scientific inquiry and medical innovation. A pivotal moment in the history of thalidomide occurred in 1965, when Israeli dermatologist Dr. Jacob Sheskin administered it to leprosy patients and observed an improvement in ENL, a severe inflammatory complication of leprosy [2]. This unexpected efficacy marked the beginning of thalidomide's therapeutic revival. Further research revealed its immunomodulatory and anti-inflammatory properties, making it a candidate for the treatment of several conditions, including systemic lupus erythematosus (SLE) [10], systemic aphthae [11], prurigo nodularis, Behçet's disease [12], and ulcers caused by HIV/AIDS [13]. In 1998, the U.S. FDA approved thalidomide for ENL treatment under strict regulations due to its teratogenic risks. An important finding regarding its immunomodulatory mechanism was reported in 1991, showing thalidomide's ability to suppress TNF-α, a key cytokine in inflammation [14]. This rediscovery sparked subsequent research, leading to the proposal of several molecular mechanisms, including degradation of *TNF-α* mRNA [15], regulation of TNF-α expression via NK-κB [16], and inhibition of TNF-α via alpha1-acid glycoprotein [17]. Thalidomide has also been reported to modulate various cytokines, including IL-2, IL-4, IL-6, IL-12, IFN-γ, NF-κB, and TNF-α [14, 16, 18, 19]. However, inflammatory responses involve complex immune cell interactions, and the molecular basis for thalidomide's immunomodulatory effects is not fully understood. Due to their immunomodulatory effects, thalidomide and its derivatives, lenalidomide and pomalidomide (Figure 24.1), are collectively referred to as IMiDs.

In addition, early studies revealed thalidomide's anti-angiogenic effects. Dr. Judah Folkman's pioneering work in the 1970s established the concept that angiogenesis is critical for cancer development [20]. He proposed that tumors require a blood supply to grow beyond a certain size, suggesting that inhibiting this blood supply could be a therapeutic strategy. The first significant report on thalidomide's anti-angiogenic action was published by Dr. Robert D'Amato and colleagues in 1994, leading to increased interest in thalidomide's potential in oncology [21]. Thalidomide exerts its anti-angiogenic effect by inhibiting the production of angiogenic cytokines, particularly VEGF and FGF [22]. These findings were critical in establishing thalidomide's role in the treatment of refractory multiple myeloma, where bone marrow neovascularization is significant. Clinical efficacy studies were expanded to include several hematologic malignancies and solid tumors, including renal cell carcinoma, glioblastoma, melanoma, and AIDS-related Kaposi's sarcoma [23–25]. Subsequent studies demonstrated its remarkable efficacy in multiple myeloma treatment, particularly in combination with dexamethasone. The initial recognition of this efficacy by Dr. Bart Barlogie and his team in 1999 [3] marked a significant milestone in the medical comeback of thalidomide.

This anti-angiogenic effect was considered one of the factors contributing to thalidomide's teratogenic effects, due to inadequate blood supply to developing tissues in affected organs. Another pharmacological effect of thalidomide thought to be responsible for teratogenicity is the generation of oxidative stress via reactive oxygen

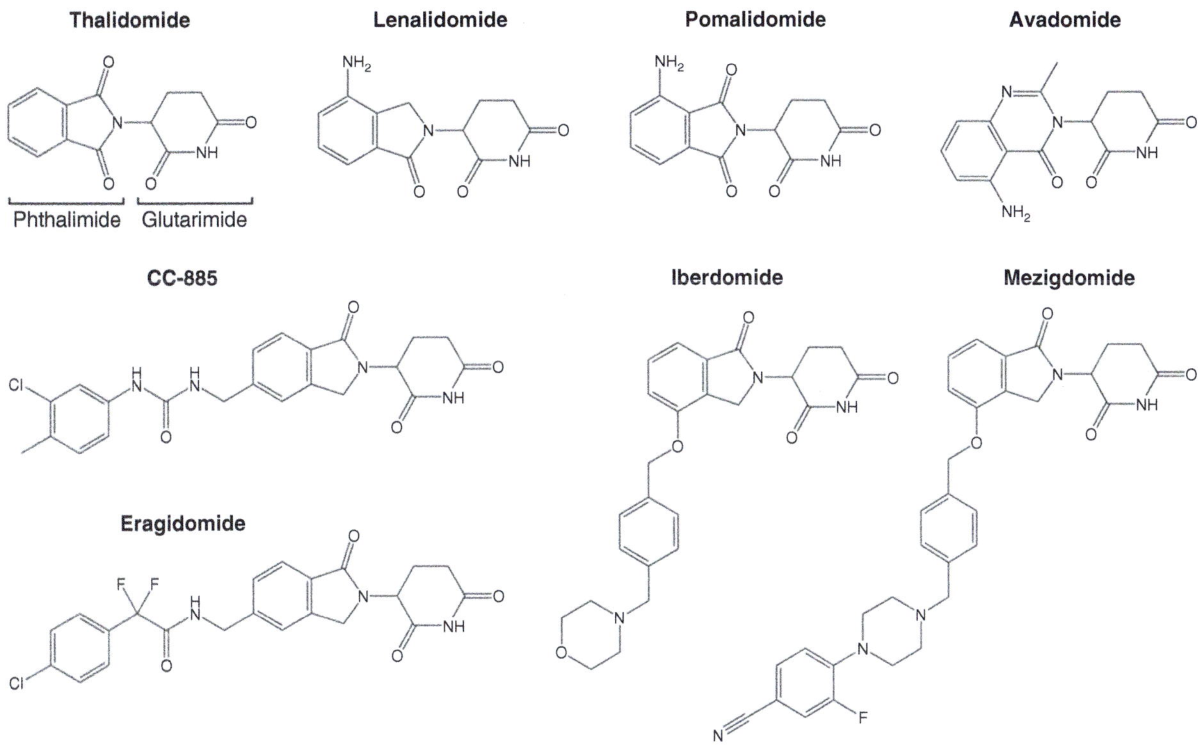

Figure 24.1 The structures of thalidomide and its derivatives. Thalidomide is composed of glutarimide and phthalimide. Thalidomide, lenalidomide, and pomalidomide are collectively referred to as IMiDs; all thalidomide derivatives targeting CRBN are collectively referred to as CELMoDs.

species (ROS), leading to cellular and tissue damage during embryonic development. Key experiments supporting this hypothesis involved the use of a scavenger of reactive oxygen intermediates, which prevented thalidomide-induced limb abnormalities in rabbits [26]. The oxidative stress hypothesis complements the anti-angiogenesis hypothesis because local ischemia due to the lack of blood vessels causes oxidative stress. Despite numerous studies on thalidomide teratogenicity, however, its exact mechanism remains unclear, in part due to the difficulty of studying it at the organismal level due to interspecies differences, such as the absence of side effects in rodents.

24.2.3 The Discovery of CRBN: Paving the Way for Targeted Protein Degradation

The identification of CRBN was a major breakthrough that paved the way to elucidate the molecular mechanism of its action, which had been a mystery for decades. This pivotal study identified CRBN as the primary target of thalidomide through affinity chromatography with thalidomide immobilized on nano-sized magnetic beads [4]. In the same study, animal experiments using zebrafish and chickens as model organisms demonstrated the involvement of CRBN in thalidomide teratogenicity [4]. Subsequent studies have further confirmed that the anti-myeloma effects of thalidomide and its derivatives, such as lenalidomide and pomalidomide, are also mediated by CRBN [27].

CRBN functions as a substrate recognition receptor for the CRL4 complex, forming with DDB1, RBX1, and Cullin 4 (CUL4A and CUL4B). Binding of thalidomide and its derivatives to CRBN results in the recruitment of non-native substrates to $CRL4^{CRBN}$ and subsequent proteasomal degradation, the major pathway for protein degradation in cells. Non-native substrates induced to degrade in the presence of such drugs are termed neosubstrates. The first reported CRBN neosubstrate-dependent on thalidomide derivatives are Ikaros (IKZF1) and Aiolos (IKZF3). These transcription factors play a crucial role in hematopoietic cell differentiation, and their degradation has been shown to mediate the therapeutic effect of thalidomide derivatives on multiple myeloma [28, 29]. Subsequent studies have identified a number of neosubstrates, including those that appear to mediate thalidomide teratogenicity [30–41]. Thus, thalidomide and its derivatives essentially hijack the $CRL4^{CRBN}$ E3 ubiquitin ligase complex, redirecting non-native substrates for protein degradation.

This new mode of action offers several advantages over the classical mechanism of action of small-molecule drugs, which typically involve binding to target enzymes and inhibiting their activities. First, targeting protein degradation allows for the targeting of proteins lacking a clear active site, such as transcription factors, overcoming limitations observed in conventional drugs, as evidenced by the drug-induced degradation of Ikaros and Aiolos. Such undruggable proteins constitute the majority of the human proteome. Second, the enzymatic mechanism of action involved in protein degradation, where a single compound molecule induces the degradation of multiple target protein molecules, is kinetically superior to stoichiometric binding and inhibition by conventional drugs. Thus, the elucidation of the mechanism of action of thalidomide and its derivatives has provided an excellent opportunity to demonstrate the utility of the new drug discovery modality.

24.3 Polypharmacology of Thalidomide and its Derivatives

Thalidomide and its derivatives serve as compelling examples of polypharmacology in drug discovery. Polypharmacology refers to the capability of a drug to affect multiple targets, pathways, or diseases. The primary target of thalidomide is CRBN, and binding of thalidomide to CRBN alters its substrate specificity, inducing drug-dependent targeted degradation of numerous downstream proteins, each of which contributes to the diverse pharmacological activities of thalidomide. Thus, it is the capability to modulate multiple downstream proteins that characterizes the polypharmacological nature of thalidomide and its derivatives (Figure 24.2).

As mentioned above, thalidomide exhibits multiple pharmacological effects, including immunomodulatory, anti-myeloma, and anti-angiogenesis effects, alongside its original hypnotic and teratogenic effects. Since the revival of thalidomide, its derivatives such as lenalidomide and pomalidomide have been developed and approved by the FDA; the former serves as a first-line treatment for multiple myeloma, while the latter is indicated for relapsed and refractory multiple myeloma. Lenalidomide also holds FDA approval for the treatment of myelodysplastic syndromes (MDS) with deletion of chromosome 5q (MDS 5q-), as well as for mantle and follicular lymphomas [42–44]. Subsequently, a number of thalidomide derivatives have been developed to enhance efficacy and broaden indications (Figure 24.1). The next-generation thalidomide derivatives, significantly expanded in structure, are collectively referred to as Cereblon E3 ligase-modulating drugs (CELMoDs), signifying their ability to modify the substrate specificity of CRBN. Interestingly, certain derivatives have been found to specifically target particular neosubstrates, raising the possibility that new medical properties may emerge through chemical modification of thalidomide. The mechanisms of action underlying the major pharmacological activities of thalidomide and its derivatives

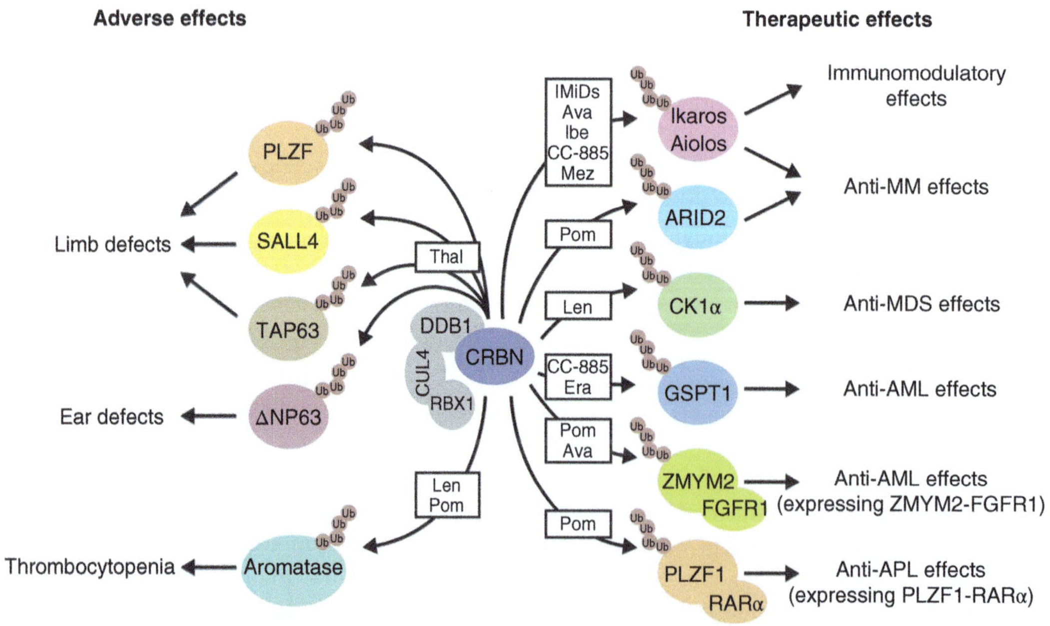

Figure 24.2 Polypharmacological properties of CELMoDs. CELMoDs exert their diverse pharmacological activities mediating the degradation of diverse neosubstrates. Abbreviations: Ava, avadomide; Era, eragidomide; Ibe, Iberdomide; IMiDs, immunomodulatory drugs; Len, lenalidomide; Mez, mezigdomide; Pom, pomalidomide; Thal, thalidomide; AML, acute myeloid leukemia; APL, acute promyelocytic leukemia; MDS, myelodysplastic syndromes; MM, multiple myeloma; Ub, ubiquitin.

are described below with a focus on the neosubstrates mediating these effects.

24.3.1 Anti-Myeloma Effects

Multiple myeloma, one of the most common blood cancers originating from plasma cells, was once untreatable; however, today it can be managed with thalidomide derivatives, proteasome inhibitors, and monoclonal antibodies, although it remains incurable. Given lenalidomide's demonstrated efficacy against multiple myeloma, the neosubstrates of CRBN were primarily investigated in the context of its therapeutic activity against this disease. In 2014, Ikaros and Aiolos were identified as neosubstrates for CRBN by two research groups [28, 29]. Both Ikaros and Aiolos belong to the Ikaros transcription factor family, which plays an important role in hematopoietic cell differentiation, and bind to thalidomide-bound CRBN in a similar molecular fashion. Ikaros and Aiolos regulate transcription of *IRF4* [45], which is essential for multiple myeloma survival and forms a self-regulatory circuit with the oncogene *MYC* [46]. Drug-induced degradation of Ikaros and Aiolos leads to decreased expression of both genes and thereby inhibits proliferation. The anti-myeloma effects of the drugs, thalidomide being the weakest and pomalidomide the strongest, correlate well with their degradative capacity of Ikaros and Aiolos.

Ikaros and Aiolos are the primary mediators of the anti-myeloma effects of lenalidomide and pomalidomide, prompting extensive research into compounds that can more effectively induce their degradation. Mezigdomide (Figure 24.1), also known as CC-92480, is one such CELMoD that induces the degradation of Ikaros and Aiolos much faster and at lower concentrations than lenalidomide or pomalidomide and is currently in phase III clinical trials for the treatment of relapsed refractory multiple myeloma [47]. The molecular basis for mezigdomide's strong potency in degrading Ikaros and Aiolos is described below.

Interestingly, from a polypharmacological perspective, another neosubstrate, ARID2, also undergoes degradation in the presence of pomalidomide and is important for survival in multiple myeloma [37]. Pomalidomide's ability to target multiple survival-related proteins in multiple myeloma may contribute to its superior efficacy and the suppression of drug resistance. Drug resistance often results from mutations or epigenetic changes in the pathways leading to downstream targets. Indeed, in mouse xenograft models, cell lines acquiring pomalidomide resistance exhibited significantly reduced CRBN expression, suggesting that they acquired resistance through direct loss of the primary target rather than downstream adaptation [48]. In patients with relapsed refractory multiple myeloma, downregulation of CRBN and expression of a CRBN mutant lacking the drug-binding pocket are frequently observed [49], posing a significant challenge in treatment.

24.3.2 Anti-MDS Effects

MDS are hematopoietic diseases characterized by specific chromosomal abnormalities that disrupt normal blood cell production in the bone marrow. Lenalidomide is approved for treating MDS with deletion of the long arm of chromosome 5 (MDS 5q-), unlike thalidomide and pomalidomide [42]. The identification of CK1α as a lenalidomide-dependent CRBN substrate involved in lenalidomide's therapeutic activity in MDS [30], following Ikaros and Aiolos, was a significant finding from a polypharmacological perspective. This finding represents the first report demonstrating that lenalidomide can target multiple substrates and that modification of thalidomide can lead to the acquisition of new targets. CK1α belongs to the serine/threonine kinase family, involved in a variety of cellular processes including the cell cycle, DNA repair, transcription, signal transduction, and cell polarity. The *CSNK1A1* gene encoding CK1α is located in the 5q region and is haploinsufficient in MDS 5q- patients, rendering the abnormal cells susceptible to drug-induced CK1α degradation [30]. This provides a compelling explanation for lenalidomide's particular effectiveness in MDS 5q-. The structural basis underlying lenalidomide's selective recognition of CK1α will be described later.

24.3.3 Anti-AML Effects

Several CELMoDs, though not currently approved, are in development for other cancers, particularly blood cancers. One such CELMoD, CC-885 (Figure 24.1), exhibits a broad anti-cancer spectrum, with particularly strong growth inhibition against acute myeloid leukemia (AML) [31]. Lenalidomide and pomalidomide have no direct anti-cancer activity against AML, indicating the involvement of unidentified neosubstrates. Research showed that CC-885 induces the proteasomal degradation of GSPT1 in a $CRL4^{CRBN}$-dependent manner. Its degradation is specific to CC-885 because the extended urea and chloro-methyl-phenyl moieties contribute to substrate recognition (Figure 24.1). GSPT1 is a component of the protein translation machinery that recognizes stop codons and is also involved in cell cycle regulation. The degradation of GSPT1 mediates the unique anti-AML effects of CC-885, as its anti-AML activity is suppressed by the expression of a GSPT1 mutant deficient in CRBN binding [31]. CC-885 induces the degradation of not only GSPT1 but also Ikaros, Aiolos, and CK1α; the lack of amino acid sequence homology among these neosubstrates indicates that primary structure is not the determining factor in which proteins are targeted by CELMoDs. Although CC-885 exhibits remarkably potent anti-AML activity *in vitro*, its clinical development has been suspended due to unmanageable toxicities resulting from the degradation of the multiple target proteins. Eragidomide (Figure 24.1), also known as CC-90009, was developed to address this issue. Eragidomide is a CELMoD with higher specificity for GSPT1, exhibiting minimal effect on the proteome beyond GSPT1 in an AML cell line, and is currently in phase II clinical trials for AML treatment [50].

Interestingly, several fusion proteins implicated in AML are reported to be targeted by CRBN in a CELMoD-dependent manner. In AML, certain chromosomal translocations commonly occur, resulting in the production of abnormal fusion proteins that may disrupt cell physiology and drive leukemia development. ZMYM2 and PLZF are newly discovered CELMoD-dependent CRBN neosubstrates [38–41]. A fusion protein composed of ZMYM2 and FGFR1, which is generated by chromosomal translocation of t(8;13)(p11;q12) and contributes to AML oncogenesis, is induced for degradation by pomalidomide or avadomide (CC-122) treatment [51]. Meanwhile, PLZF forms fusion proteins with RARα or ABL1, which are associated with acute promyelocytic leukemia (APL) and T-cell acute lymphoblastic leukemia (T-ALL), respectively [52, 53]. Pomalidomide induces the degradation of PLZF-RARα in a CRBN-dependent manner and inhibits proliferation in a model cell line expressing PLZF-RARα.

The evolution of CELMoDs, exemplified by CC-885 and eragidomide, marks a significant leap from earlier compounds such as thalidomide and lenalidomide, particularly in anti-AML effects. This advancement, coupled with the ability of pomalidomide and avadomide to degrade critical cancer driver fusion proteins in AML, underscores the expanding therapeutic spectrum of CELMoDs. This pivotal shift not only reinforces their established role in multiple myeloma but also indicates their potential applicability to a broader range of cancers.

24.3.4 Immunomodulatory Effects

The immunomodulatory effects have played a pivotal role in triggering the resurgence of thalidomide as a pharmaceutical product. These effects have established the clinical utility of IMiDs, including lenalidomide and pomalidomide, along with anti-myeloma properties. IMiDs, as their names suggest, have both immunostimulatory and anti-inflammatory activities, including activation of T cells, natural killer cells, and natural killer T cells; modulation of regulatory T cells; suppression of B cells and monocytes; and modulation of cytokine production depending on the immune cell type [18, 54–58]. For example, IMiD treatment suppresses TNF-α production in monocytes but enhances it in T cells, indicating cell type-specific regulation by IMiDs [14, 54]. The immunomodulatory effects of IMiDs arise from complex interactions involving such diverse immune cell responses and microenvironments. Therefore, understanding their mechanisms comprehensively is more challenging than understanding their direct, cell-autonomous anti-cancer effects. Although the full spectrum of immunomodulatory effects of IMiDs is not entirely clear, at least some of them are exerted through the degradation of neosubstrates via $CRL4^{CRBN}$.

IMiDs generally exhibit immunostimulatory effects on T cells, enhancing co-stimulation, which is a secondary stimulation required for full T-cell activation following antigen exposure, and promoting the production of IL-2, a crucial cytokine for T-cell proliferation and expansion [18, 56]. Lenalidomide and pomalidomide are much more potent than thalidomide in inducing T-cell proliferation and production of IL-2 and IFN-γ [18, 55]. In multiple myeloma, where immune exhaustion is common, the immunostimulatory effects of IMiDs are clinically important and complement their direct anti-myeloma effects [59]. The degradation of Ikaros and Aiolos by IMiDs appears to be responsible, at least in part, for the increased production of IL-2 in T cells. Furthermore, knockdown of Ikaros and Aiolos in $CD3^+$ T cells upregulates cytokines such as IL-4, IL-6, IL-10, IL-13, TNF-α, and IFN-γ, suggesting that drug-induced degradation of Ikaros and Aiolos contributes to increased production of these cytokines [60].

SLE is a chronic autoimmune disease that causes widespread inflammation and tissue damage, including joint pain, skin rashes, fatigue, and organ damage. Thalidomide and lenalidomide have been investigated for their potential in treating SLE, particularly in cases where traditional treatments are less effective. Although it is a small sample size study, they have been shown to be remarkably potent [61]. Interestingly, genetic variations of Ikaros and Aiolos are associated with SLE, and these genes are upregulated in patients with SLE [62, 63]. Iberdomide is a next-generation CELMoD (Figure 24.1) with a higher affinity for CRBN than lenalidomide and a greater capability for the degradation of Ikaros and Aiolos. Phase II clinical trials demonstrated the efficacy of Iberdomide in treating SLE. It is noteworthy that patients with a higher Aiolos expression appear to show a better response to the iberdomide treatment [64]. Although the mechanism of action of the therapeutic effect of CELMoDs on SLE remains unclear, these pieces of evidence suggest that it is partially through the degradation of Ikaros and Aiolos.

24.3.5 Teratogenic Effects

Teratogenicity is the most well-known side effect of thalidomide, and continued efforts have been made to understand its mechanism. The identification of CRBN as a primary target of thalidomide was a breakthrough toward this goal. FGF8 and FGF10 are growth factors expressed in the limb bud, specifically in the apical ectodermal ridge and the underlying mesenchyme, respectively, and are critical for the outgrowth and patterning of the limbs. Thalidomide has been reported to downregulate their expression in various model organisms [1]. Overexpression of a thalidomide-binding deficient mutant of CRBN restored *FGF8* and *FGF10* expression and suppressed limb malformations in thalidomide-treated zebrafish and chicken embryos [4], suggesting that thalidomide's teratogenic effects are also mediated by the degradation of unknown neosubstrates. Recent studies have reported several candidate neosubstrates responsible for thalidomide embryopathy.

In 2018, two groups reported SALL4 as a neosubstrate involved in thalidomide teratogenicity [33, 34]. SALL4 is a transcription factor that plays a crucial role in embryonic development. It cooperates with GLI3 to regulate the development of limb skeletal elements in mice [65]. Loss-of-function mutations in the *SALL4* gene are associated with Duane radial ray syndrome, Okihiro syndrome, and Holt-Oram syndrome [66–68]. Patients with these syndromes exhibit phenotypes similar to those of thalidomide-exposed babies, including limb malformations, heart defects, and ear defects. Some patients previously misdiagnosed with thalidomide embryopathy were later found to have the hereditary *SALL4* mutation [69]. While thalidomide induces the degradation of SALL4 in human embryonic stem cells [33] and fetal rabbits [34], a SALL4 mutant impaired in CRBN binding shows resistance to the degradation *in vitro*. During embryonic limb development, the lateral plate mesoderm (LPM) is a transient tissue that contributes to the formation of limb buds [70]. Thalidomide inhibits the differentiation of human-induced pluripotent stem cells into LPM-like cells. The introduction of a point mutation into the *SALL4* locus that confers resistance to drug-induced degradation reverses the inhibitory effect of thalidomide on differentiation, suggesting that SALL4 degradation mediates limb malformation induced by thalidomide treatment [71].

TAp63 and ΔNp63, isoforms encoded by the *TP63* gene, were also identified as neosubstrates involved in thalidomide teratogenicity [35]. TP63, a member of the p53 family of transcription factors, plays a crucial role in the development and differentiation of epithelial tissues [72]. It exists in multiple isoforms, including TAp63 and ΔNp63, each with distinct functions. TAp63 is involved in cochlear development and hearing, while ΔNp63 is essential for limb development, including the maintenance of the apical ectodermal ridge, a key structure in limb development [73]. Mutations in *TP63* cause a range of congenital malformations, most notably affecting the limbs and facial structures, such as ectrodactyly-ectodermal dysplasia-cleft lip/palate and limb-mammary syndrome [74, 75]. Thalidomide induces the degradation of both TAp63 and ΔNp63 via $CRL4^{CRBN}$. In zebrafish, TAp63 and ΔNp63 are associated with different teratogenic phenotypes. During early development, ΔNp63 plays a role in *fgf8* expression, and the overexpression of its non-degradable mutant suppresses thalidomide-induced fin malformation. Meanwhile, TAp63 is critical for the expression of *atoh1*, a key regulator of ear development, and the overexpression of its non-degradable mutant suppresses otic vesicle (ear) malformation [35].

PLZF is reported as another neosubstrate involved in thalidomide teratogenicity [39]. PLZF is a transcription factor involved in hematopoiesis, spermatogenesis, and anterior-to-posterior patterning during limb development [76]. PLZF knockout mice show impaired skeletal patterns in the limbs with no anteriorization [77]. PLZF has been reported to act as a pomalidomide-dependent neosubstrate in the context of its anti-AML effects in one study [40], as mentioned above. Another study demonstrated that thalidomide and its primary metabolite, 5-hydroxythalidomide, induce the degradation of PLZF in chick limb buds as well as in cultured cells [41]. In chickens, PLZF knockdown causes thalidomide-like limb defects, and its overexpression restores thalidomide-induced *Fgf8* and *Fgf10* downregulation and limb shortening. Meanwhile, there is no thalidomide-dependent degradation of Sall4 in chick limb buds, indicating the importance of PLZF in thalidomide teratogenicity in chickens [39].

Thalidomide exhibits teratogenic effects in humans, non-human primates, rabbits, chickens, and zebrafish, but not in rodents, including mice [1]. In particular, the absence of thalidomide teratogenicity in rodents is considered to be a reason for the disastrous drug incident [78]. The species specificity of thalidomide teratogenicity presents a significant challenge in comprehending the pharmacotoxicology of this drug class. As the studies described above used different model organisms, the differences in species-specific responses to thalidomide make it difficult to achieve an integrated understanding. To fill this gap in understanding, interspecies differences in the recognition of neosubstrates by CRBN and subsequent degradation have been intensively analyzed. In mice, Crbn does not recognize Ikaros or Aiolos even in the presence of IMiDs due to a single amino acid substitution in Crbn [79]. This substitution, I391 in mice (as opposed to V388 in humans), prevents neosubstrate access to mouse Crbn. When a humanized version of Crbn with the I391V substitution is expressed in mouse cells, it is sufficient to recreate drug-induced degradation of Ikaros, Aiolos, and Ck1α. Moreover, genetically modified mice expressing humanized Crbn (I391V) induce the degradation of Ikaros and CK1α when thalidomide and lenalidomide are administered. However, it is important to note that even with this substitution, these mice do not display limb malformations typically associated with thalidomide embryopathy, despite a high incidence of fetal lethality being observed upon exposure to thalidomide or lenalidomide [79]. Sall4 is not recognized by humanized Crbn in mice, unlike Ikaros and Aiolos, suggesting that species-specific amino acid substitutions in both CRBN and neosubstrates are responsible for the species-specificity of thalidomide teratogenicity [33]. However, since limb defects are observed in thalidomide-treated chick embryos despite the absence of SALL4 degradation, thalidomide teratogenic phenotypes may occur through different mechanisms in different species [39].

The identification of SALL4, p63, and PLZF as neosubstrates potentially associated with thalidomide teratogenicity is a significant advancement in the development of thalidomide-based drugs. These new findings are complementary to, not antithetical to, past research. In early studies, the leading hypotheses for thalidomide teratogenicity were angiogenesis inhibition and ROS production. SALL4 reportedly regulates the expression of VEGF, a critical factor in blood vessel formation, and is involved in angiogenesis. Knockdown of SALL4 in endothelial cells reduces VEGF-A expression and decreases cell proliferation, motility, and tube formation, which are all essential for angiogenesis [80]. In addition, there are reports that both SALL4 and p63 are involved in protection against ROS-induced damage [81, 82]. Thus, the recent identification of neosubstrates fills a gap in our previous understanding of the detailed mechanisms of thalidomide teratogenicity. However, the wide range of thalidomide embryopathy phenotypes suggests that known neosubstrates and their combinations may not fully explain thalidomide teratogenicity. Therefore, besides SALL4, p63, and PLZF, other unknown neosubstrates may be involved in thalidomide embryopathy. The multi-targeting profile of thalidomide likely underlies the wide array of its pharmacological and adverse effects.

24.3.6 Proplatelet Formation Inhibition

Thrombocytopenia, a significant reduction in platelet count, is a known side effect of IMiD treatment, often necessitating drug discontinuation [83]. A recent study has shown that the degradation of a neosubstrate mediates this side effect [36]. Platelet synthesis begins with the differentiation of hematopoietic stem cells in the bone marrow into megakaryocytes. Megakaryocytes undergo endomitosis, replicating their DNA without cell division and forming mature cells that extend long protrusions called proplatelets into bone marrow blood vessels. These proplatelets fragment into smaller units, forming platelets. The study identified CYP19A1 or aromatase, an enzyme crucial for estradiol biosynthesis, as a novel neosubstrate involved in thrombocytopenia. IMiD-induced degradation of aromatase via $CRL4^{CRBN}$ impairs estradiol autocrine signaling, which is important for proplatelet formation in megakaryocytes. Patients who developed thrombocytopenia due to IMiD treatment had lower levels of aromatase expression in isolated megakaryocytes compared to those who did not. *In vitro*, estradiol reversed the IMiD-induced inhibition of proplatelet formation. These findings contribute to overcoming one of the major side effects of IMiDs, thrombocytopenia.

24.4 Structural Understanding of the Mechanisms of Action of CELMoDs

24.4.1 The Interaction Between CRBN and Thalidomide

CRBN is a 441 or 442 amino acid protein with an N-terminal LON-like domain (LLD) and a C-terminal thalidomide-binding domain (TBD). The LLD is often further subdivided into an N-terminal domain (NTD) and an α-helical bundle domain (HBD). Several tryptophan residues located in the TBD are important for thalidomide binding. Early studies showed that overexpression of mutants carrying substitutions at critical residues in the TBD confers resistance to thalidomide teratogenicity in zebrafish and chicken models. In addition, thalidomide has a structure consisting of a glutarimide ring linked to a phthalimide ring (Figure 24.1), and structure-activity relationship analysis showed that the glutarimide moiety mainly contributes to CRBN binding [4]. These findings not only stimulated the development of thalidomide derivatives but also opened the way to a deeper exploration of its action mechanism.

In 2014, two groups determined the structure of the CRBN-DDB1 dimer containing CELMoDs by X-ray crystallography (Figure 24.3a). One group investigated the structures of human CRBN-DDB1 dimer bound to one of three CELMoDs (thalidomide, lenalidomide, and pomalidomide) [84], while the other group examined the same complexes using chicken CRBN [85], yielding results in general agreement. DDB1 is mainly composed of three

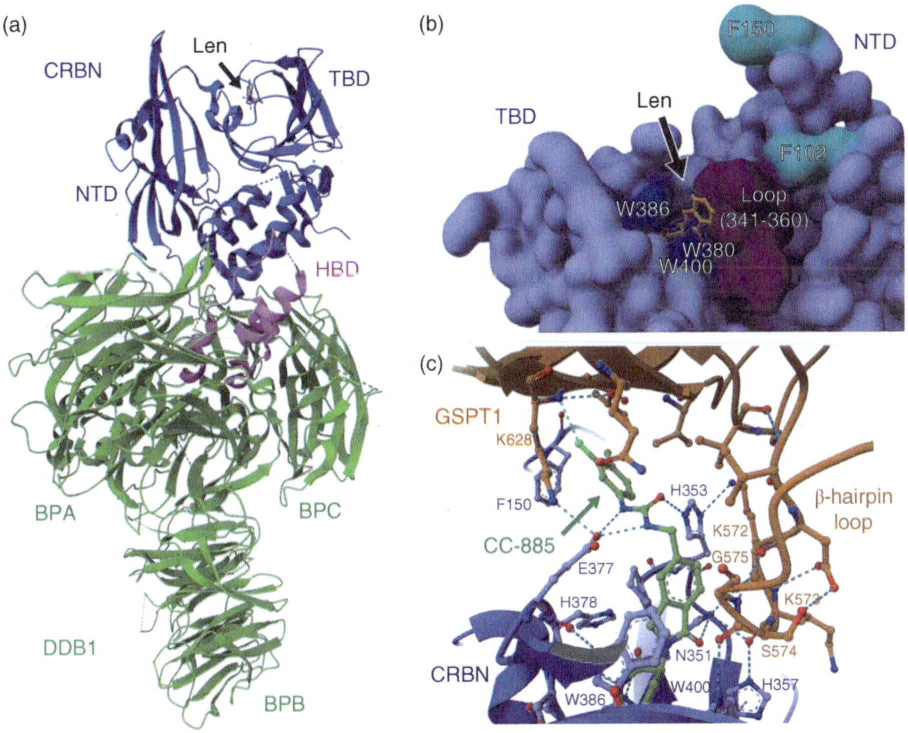

Figure 24.3 Structure of the complex containing CRBN and CELMoDs. (a) Three-dimensional structure of the complex consisting of CRBN (violet), DDB1 (green), and lenalidomide (brown). The α-helical bundle domain (HBD) of CRBN is highlighted in magenta (PDB: 4TZ4). (b) Close-up view of the molecular surface of thalidomide-binding pocket of CRBN. Amino acids comprising the pocket are highlighted in blue. In addition, the loop of residues 341-360 is highlighted in purple. F102 and F105, with which the extended functional moieties of the CELMoDs interact, are also highlighted in turquoise (PDB: 4TZ4). (c) Close-up view of the co-crystal structure of DDB1-CRBN-CC-885-GSPT1 (5HXB). Dotted lines indicate non-covalent interactions among CRBN (violet), CC-885 (green), and GSPT1 (brown). Abbreviations: Len, lenalidomide; NTD, N-terminal domain; HBD, α-helical bundle domain; TBD, thalidomide-binding domain; BPA, β-propeller A; BPB, β-propeller B; BPC, β-propeller C.

β-propeller domains (BPA, BPB, and BPC). The helices (residues 188–248) within the HBD of CRBN are situated in the cavity between the BPA and BPC of DDB1, facilitating binding between CRBN and DDB1. The TBD of CRBN, composed of a C4-type zinc finger motif (C323, C326, C391, and C394) and an antiparallel β-sheet core, is located on the opposite side of its DDB1-binding domain (Figure 24.3a). The glutarimide ring of CELMoDs fits into a highly conserved hydrophobic pocket consisting of three tryptophan residues (W380, W386, and W400) and one phenylalanine residue (F402) in the TBD (Figure 24.3b). The carbonyl and amide groups of glutarimide form hydrogen bonds with H378 and W380 of CRBN, respectively. Meanwhile, the phthalimide moiety is exposed on the surface of CRBN, supporting the idea that this side of the compounds contributes to neosubstrate binding.

24.4.2 Formation of the Ternary Complexes: CELMoDs Acting as Molecular Glue

The next step was to determine the structures of the ternary complexes (CRBN-CELMoDs-neosubstrates). In 2016, the X-ray crystal structure of DDB1-CRBN-lenalidomide-CK1α was determined [30]. CK1α binds to CRBN-lenalidomide via its β-hairpin loop (residues 35–41), with G40 being particularly important for binding and its mutation abolishing the binding. Amino acid residues around the drug-binding pocket of CRBN (e.g., N351, H357, and W400), along with the phthalimide moiety, provide an interaction surface to CK1α. Upon binding to CK1α, the phthalimide moiety of lenalidomide is repositioned to become proximal to E377 of CRBN. In the case of thalidomide and pomalidomide, however, the additional carbonyl group of these compounds causes a clash with the CRBN backbone. This provides an excellent explanation for the lenalidomide-specific recognition of CK1α.

In the same year, the structure of DDB1-CRBN-CC-885 was also determined (Figure 24.3c) [31]. Consistent with the CK1α study, GSPT1 also interacts with CRBN-CC885 via the β-hairpin loop structure (residues 571–575). Similar motifs are present in Ikaros and Aiolos in regions critical for CRBN binding (residues 146–152 for Ikaros). Mutation analysis revealed that substitution of the conserved glycine residue (G575 for GSPT1 and G151 for Ikaros) in the β-hairpin loop disrupts CRBN binding to these neosubstrates (Figure 24.3c). The β-hairpin loop motif responsible for CRBN binding shows no sequence homology among GSPT1, Ikaros, and CK1α, except for the glycine residue. Thus, it was concluded that the tertiary structure of the β-hairpin loop, rather than its specific primary structure, serves as a structural degron for these neosubstrates. This idea is supported by the observation that major interactions between the β-hairpin loop and CRBN are provided by the peptide backbone rather than the side chains. Within the complex, the key glycine residue of the β-hairpin loop adopts backbone angles that could only accommodate glycine [31]. This is why glycine mutants were used extensively in the neosubstrate studies presented in Section 24.3. CC-885 has a more extended structure than thalidomide and lenalidomide. Its urea and chloro-methyl-phenyl moiety of CC-885 is involved in binding to both CRBN (via F150 and E377) and GSPT1 (via K628) as shown in Figure 24.3c, which gives it specificity to GSPT1 [31]. The mechanism of action of CELMoDs is therefore called "molecular glue" because it physically mediates the interaction between CRBN and neosubstrates. Subsequently, the tertiary structures of other neosubstrates such as Ikaros and SALL4 bound to CRBN with CELMoDs were determined [86, 87]. These studies confirmed a common mechanism of action of CELMoDs as molecular glue.

24.4.3 Allosteric Regulation for Neosubstrate Recognition

Recent cryogenic electron microscopy analysis revealed a new aspect of neosubstrate recognition by CRBN [88]. The study provided evidence for allosteric regulation and conformational dynamics of CRBN upon binding to CELMoDs. The three-dimensional structure of CRBN, as determined by X-ray crystallography, shows a close alignment of the N-terminal NTD and the C-terminal TBD (closed conformation) [84, 85]. However, in the absence of CELMoDs, the two domains remain separated (open conformation). Binding of CELMoDs to CRBN causes a transition from an open to a closed conformation. In the closed conformation, a loop of residues 341–360, which was unstructured in the open conformation, closely contacts the CELMoDs and seems to facilitate interactions between the N- and C-terminal domains (Figure 24.3b). The N-terminal 63 amino acid residues appear to stabilize the closed conformation since their loss prevents the formation of the closed conformation [88].

This study provided two important insights for the development of CELMoDs. The first is that the efficiency of conversion to the closed conformation varies among compounds and appears to correlate with their ability to recognize neosubstrates. Approximately 20% of pomalidomide-bound CRBN adopts the closed conformation under saturating conditions, while it is ~50% for iberdomide, which is known to be more potent in inducing the degradation of Ikaros and Aiolos [88]. This is consistent with previous structural studies showing that the loop of residues 341–360 contributes to neosubstrate binding (Figure 24.3b). The second is that the conversion efficiency does not correlate with the binding affinity of the compound to CRBN. Indeed, a CRBN mutant lacking the N-terminus, which is not converted to the closed conformation, binds to pomalidomide as well as full-length CRBN. In contrast, mezigdomide, described above as a CELMoD that induces the degradation of Ikaros and Aiolos more potently than lenalidomide and pomalidomide, promotes the conversion to the closed conformation with nearly 100% efficiency. The extended benzonitrile moiety of mezigdomide interacts with F102 and F150 in the NTD, favoring the closed conformation (Figure 24.3b) [88]. Similar interactions between the extended moiety of CC-885 and the NTD have also been observed [31]. Thus, tethering the N- and C-terminal domains and stabilizing the closed conformation appears to be a common mechanism of action for very-high-potency CELMoDs.

24.4.4 Expanding Neosubstrates of $CRL4^{CRBN}$

The number of proteins reported as CRBN neosubstrate-dependent on CELMoDs is expanding. Neosubstrates such as Ikaros, Aiolos, and SALL4 belong to the C2H2-type zinc finger (C2H2-ZF) protein family, with their structural degrons located within a C2H2-ZF motif(s). These findings prompted a comprehensive search for additional CRBN neosubstrates containing C2H2-ZF motifs. The C2H2-ZF protein family, one of the largest transcription factors in eukaryotes, plays a crucial role in regulating key biological processes. A library containing more than 6500 C2H2-ZF motifs predicted from the human proteome was screened, leading to the identification of 11 ZF motifs targeted by CRBN in the presence of thalidomide, lenalidomide, or pomalidomide [87]. These motifs, which contain a β-hairpin loop that serves as a structural degron, share a flexible consensus sequence of CXXCG. Despite the presence of many ZF motifs matching this sequence in the library, only these 11 motifs were found to function as structural degrons. This specificity may be attributed to the requirement for highly complementary interactions between the neosubstrate and CRBN-CELMoD surfaces, limiting the number of C2H2-ZF motifs that actually function as neosubstrates. Nevertheless, it is noteworthy that CELMoDs are able to target, at least in part, transcription factors, an important class of disease-causing factors that are usually considered undruggable.

The polypharmacological properties of CELMoDs stem from the degradation of diverse neosubstrates based on the flexible substrate recognition by CRBN as described above. This flexibility is attributed to (1) the fact that the β-hairpin loop, a potential CRBN degron, is a common secondary structure and, more importantly, to (2) the fact that CRBN recognizes this loop in a sequence-independent manner. This mode of recognition, likely driven by interactions with the peptide backbone rather than the amino acid side chains [31], allows CRBN to target a range of substrates within certain structural contexts. Despite this breadth, specific amino acid combinations around the degron must be fulfilled to be recognized by CRBN, highlighting a crucial direction for future research. CELMoDs' ability to modulate a wide array of targets underscores their therapeutic potential, making the elucidation of CRBN's substrate specificity and its manipulation a key research focus. The expanding roster of neosubstrates will broaden the therapeutic applications of CELMoDs by uncovering unexpected disease targets. For instance, the degradation of PLZF and ZMYM2 by CELMoDs indicates potential pharmacological effects in AML, where their chimeric fusion proteins are common causative factors [38, 40, 41]. Such polypharmacological characteristics of CELMoDs could facilitate neosubstrate-based approaches to drug development and drug repositioning.

24.5 Challenges and Future Perspectives in the Development of CELMoDs

24.5.1 Advancing Drug Design: Molecular Glues and PROTACs

Over the last few decades, the pharmaceutical industry has heavily relied on small-molecule drugs that target active sites on proteins to inhibit their enzymatic functions. However, this approach is limited because it fails to address proteins without clear active sites, such as transcription factors, which constitute about 80% of the human proteome and are deemed undruggable. One of the groundbreaking alternatives to this challenge is the development of protein degraders. These compounds act by modifying the substrate specificity of an E3 ubiquitin ligase, facilitating the degradation of previously untouchable target proteins through the ubiquitin-proteasome system (UPS). This approach represents a novel therapeutic modality capable of targeting the undruggable proteome through a catalytic mechanism, where a single compound molecule can induce the degradation of multiple target protein molecules, a significant advancement over traditional stoichiometric inhibitors.

Molecular glue degraders are a type of protein degraders. This term originated from studies on the plant hormone auxin [89]. Thalidomide and its derivatives, such as lenalidomide and pomalidomide, are the only protein degraders currently used in clinical settings. They take advantage of the flexible substrate recognition of $CRL4^{CRBN}$ to trigger the degradation of diverse target proteins. Recent discoveries have expanded the family of molecular glue degraders beyond thalidomide derivatives to include sulfonamides (e.g., E7820, indisulam, and chloroquinoxaline sulfonamides) and cyclin-dependent kinase (CDK) inhibitors (e.g., CR8 and HQ461), which utilize different components of the UPS [90–93]. These developments suggest a diverse and expanding potential for molecular glue degraders in targeting a broad range of previously undruggable proteins.

PROteolysis-TArgeting Chimeras (PROTACs) represent a distinct class of protein degraders, consisting of two functional groups targeting an E3 ubiquitin ligase and a protein of interest (POI), connected by a spacer. This design allows for the targeted degradation of virtually any protein for which a binding small-molecule compound exists [94]. However, their tripartite structure results in larger molecular weights, potentially leading to unfavorable pharmacokinetics. Additionally, PROTACs are subject to the "hook effect," where high concentrations may promote the formation of binary complexes over the desired ternary complex, reducing degradation efficiency. Introduced in 2001 with an attempt to target methionine aminopeptidase MetAP-2 for ubiquitination [95], early PROTACs faced challenges such as membrane impermeability. The discovery of thalidomide's mechanism of action as molecular glue significantly accelerated PROTAC development. Various E3 ligases, including CRBN, VHL, IAP, and MDM2 [96–99], have been targeted by PROTACs, with CRBN being one of the common choices due to its effectiveness and versatility. CRBN-based PROTACs have successfully targeted a range of proteins, including BRD4, BCR-ABL, SIRT2, BTK, and CDK6, with the list of targets continually expanding [96, 100–103].

24.5.2 Physiological Functions of CRBN

Understanding the physiological function of CRBN is a crucial unresolved issue in CELMoD development. While CELMoDs bind to CRBN and facilitate the degradation of non-native substrates, they may also inhibit the degradation of native substrates. CRBN was initially discovered as a gene associated with intellectual disability, and research has shown that it plays regulatory roles in brain function, ion channels, and energy metabolism. The physiological functions of CRBN are thought to occur through its interaction with native substrates such as SLO1, MEIS2, CLC-1, glutamine synthetase, and AMPKα, which are targeted for ubiquitination and degradation in the drug-free state [85, 104–107]. For example, CRBN-mediated ubiquitination regulates SLO1, a component of the BK channels, affecting its endoplasmic retention and BK channel activity, which is essential for learning and memory. Intellectual abilities are impacted by a nonsense mutation in CRBN (E419X) that disrupts its interaction with BK channels. Furthermore, CRBN regulates energy metabolism through AMPK, a key regulator of energy balance that is ubiquitinated by $CRL4^{CRBN}$. In *Crbn* knockout mice, AMPK is constitutively activated, leading to reduced weight gain on a high-fat diet and to improved insulin sensitivity, highlighting CRBN's role as a negative AMPK modulator. Additionally, CRBN is involved in regulating glutamine homeostasis by controlling the degradation of glutamine synthetase in response to glutamine levels. Despite these insights, many aspects of CRBN's physiological functions and their implications for the clinical use of CELMoDs remain to be fully understood.

Recent findings have illuminated a fascinating aspect of CRBN's intrinsic substrate recognition, revealing a new role of cyclized asparagine and glutamine residues at the C-terminus, known as aspartimide and aminoglutarimide, respectively, in CRBN-mediated protein degradation [108, 109]. These studies embarked on identifying natural degron sequences, with a particular focus on the C-terminal cyclic imides. In one study, the authors utilized both human CRBN's TBD and its bacterial counterpart from *Magnetospirillum gryphiswaldense* (MsCI4) for affinity and crystallographic analysis, unveiling that these cyclized C-terminal residues insert into the drug-binding pocket in a manner that mirrors the binding of CELMoDs. Binding experiments employing synthetic peptides revealed that CRBN exhibits marked specificity toward peptides with C-terminal aspartimide and aminoglutarimide residues [109]. In another study, the authors demonstrated that the dipeptide motifs XcN and XcQ indeed serve as transferable degrons that engage CRBN in targeted protein degradation. They also presented evidence that naturally occurring proteins with C-terminal cyclic imides are targeted for degradation via $CRL4^{CRBN}$ [108]. These findings suggest a native, yet previously underappreciated, role for CRBN in protein homeostasis: recognizing and directing the UPS to target proteins modified with these cyclic imides for degradation. These proteins, which can result from non-enzymatic reactions or protein splicing, are indicative of molecular damage or aging [110]. Thus, the above findings also highlight the importance of further research to unravel the full biological implications of cyclized C-terminal amino acids.

24.5.3 Challenges of CELMoD Development From a Polypharmacological Perspective

Although CELMoDs have significant therapeutic potential, their development presents unique challenges from a polypharmacological perspective. Polypharmacological properties, i.e., the effects of a drug on multiple targets, represent a double-edged sword. On the one hand, they offer the promise of addressing complex diseases through multifaceted mechanisms of action. On the other hand, it complicates drug development due to the intricate balance required between efficacy and safety.

CELMoDs utilize CRBN's flexible substrate recognition to exert diverse pharmacological effects by degrading various neosubstrates. However, this polypharmacological nature also introduces complexity in drug development, particularly in predicting and managing potential side effects. The case of CC-885 illustrates this point well, highlighting difficulties in delineating the drug's safety profile due to its broad spectrum of action. Although CELMoDs offer therapeutic advantages by degrading multiple targets, they also carry the risk of off-target effects. The development of CELMoDs requires overcoming the challenges of predicting and identifying multiple targets and designing drugs that avoid degrading undesirable targets. However, the feasibility of this has been demonstrated by the development of eragidomide.

The development of CELMoDs requires the integration of comprehensive strategies, including proteomics, structural analysis, and computational modeling, to effectively navigate the polypharmacological landscape. Proteomics provides an overview of the cellular protein landscape, identifying potential neosubstrates that may be affected by the developing CELMoDs, facilitating the discovery of therapeutic targets and potential sources of toxicity. Analyzing the structures of the CRBN-CELMoD-neosubstrate complexes provides insight into the specific interactions that determine substrate specificity and degradation. Precise structural information is essential for optimal drug design of CELMoDs, as they require highly complementary interfaces between CRBN and neosubstrates. Computational modeling plays a pivotal role in the rational design of CELMoDs. By simulating the interactions among CRBN, CELMoDs, and potential neosubstrates, researchers can predict how modifications to the drug's structure might influence its binding affinity and specificity. In the future, it would be nice to be able to screen *in silico* targets for a specific compound from β-hairpin loop-containing proteins.

24.5.4 Thalidomide and Beyond: Pursuing Non-CRBN Pathways for Hypnotic Action

In this chapter, the focus has been on the pharmacological effects of thalidomide and CELMoDs related to CRBN. A recent report indicates that thalidomide's original pharmacological effect, its hypnotic action, does not involve CRBN [111]. To investigate the mechanism of hypnotic action, sleep patterns in mice were analyzed by examining the impact of thalidomide on electroencephalography/electromyography signals. Furthermore, investigations were carried out to understand thalidomide's influence on synaptic excitation and inhibition in the brain, as well as on gene expression within hypothalamic regions known to be sensitive to sedative drugs [112]. The findings indicate that thalidomide enhances non-rapid eye movement (REM) sleep duration in a dose-responsive manner without diminishing total REM sleep time, demonstrating its hypnotic potential. Additionally, thalidomide was shown to suppress excitatory synaptic transmission in the cortex and to stimulate Fos expression in vasopressin neurons within the supraoptic nucleus. This pattern mirrors that induced by traditional sedatives and some general anesthetics [113]. Intriguingly, the hypnotic effects of thalidomide, unlike its teratogenic and anti-myeloma properties, were replicated in mice. The absence of teratogenic and anti-myeloma effects in mice is attributed to species-specific amino acid variations in CRBN and neosubstrates that fail to trigger the degradation of neosubstrates [33, 78, 84]. Noteworthy is that the hypnotic effects were consistently observed in CRBN YW/AA knock-in mice, which harbor a mutated version of CRBN that cannot bind thalidomide yet maintains its ubiquitin ligase function, underscoring the notion that thalidomide's hypnotic effects do not rely on CRBN. These findings open the possibility of developing thalidomide-like hypnotic medications devoid of teratogenic risks.

24.6 Conclusions

The history of thalidomide illustrates the dual nature of drug discovery. It highlights how tragedy can lead to significant therapeutic advancements. This narrative serves as a cautionary tale about the importance of rigorous drug testing, particularly for teratogenic effects. It is also a testament to the resilience and innovative capacity of medical research. The re-emergence of thalidomide and its derivatives in modern medical science is not only just a story of its reuse as a beneficial drug but also a reflection of over a decade-long, extensive exploration into its mechanism of action, ignited by the identification of CRBN. This journey has revealed thalidomide's distinctive features, including its innovative mode of action as a molecular glue degrader and its polypharmacological nature, demonstrating flexibility in substrate recognition and the ability to act on a variety of neosubstrates simultaneously. These properties not only shed light on the drug's distinct mechanisms of action but also have significantly influenced drug discovery modalities, promising to advance current pharmacological strategies.

References

1. Vargesson, N. (2015). Thalidomide-induced teratogenesis: history and mechanisms. *Birth Defects Res. C Embryo Today* 105: 140–156.
2. Sheskin, J. (1965). Thalidomide in the treatment of lepra reactions. *Clin. Pharmacol. Ther.* 6: 303–306.
3. Singhal, S., Mehta, J., Desikan, R. et al. (1999). Antitumor activity of thalidomide in refractory multiple myeloma. *N. Engl. J. Med.* 341: 1565–1571.
4. Ito, T., Ando, H., Suzuki, T. et al. (2010). Identification of a primary target of thalidomide teratogenicity. *Science* 327: 1345–1350.
5. Mellin, G.W. and Katzenstein, M. (1962). The saga of thalidomide. Neuropathy to embryopathy, with case reports of congenital anomalies. *N. Engl. J. Med.* 267: 1184–1193.
6. Miller, M.T. and Strömland, K. (1999). Teratogen update: thalidomide: a review, with a focus on ocular findings and new potential uses. *Teratology* 60: 306–321.
7. McBride, W. (1961). Thalidomide and congenital malformations. *Lancet* 278: 1358.
8. Lenz, W., Pfeiffer, R.A., Kosenow, W., and Hayman, D.J. (1962). Thalidomide and congenital abnormalities. *Lancet* 279: 45–46.
9. Scheindlin, S. (2011). The courage of one's convictions: the due diligence of Frances Oldham Kelsey at the FDA. *Mol. Interv.* 11: 3–9.
10. Atra, E. and Sato, E.I. (1993). Treatment of the cutaneous lesions of systemic lupus erythematosus with thalidomide. *Clin. Exp. Rheumatol.* 11: 487–493.
11. Santis, H.R. (1991). Aphthous stomatitis and its management. *Curr. Opin. Dent.* 1: 763–768.
12. Saylan, T. and Saltik, I. (1982). Thalidomide in the treatment of Behcet's syndrome. *Arch. Dermatol.* 118: 536.

13 Louvel, D., Musso, S., Métivier, S. et al. (1996). Idiopathic esophageal ulceration associated with human immunodeficiency virus infection: efficacy of thalidomide treatment. *Endoscopy* 28: 471–472.

14 Sampaio, E.P., Sarno, E.N., Galilly, R. et al. (1991). Thalidomide selectively inhibits tumor necrosis factor alpha production by stimulated human monocytes. *J. Exp. Med.* 173: 699–703.

15 Moreira, A.L., Sampaio, E.P., Zmuidzinas, A. et al. (1993). Thalidomide exerts its inhibitory action on tumor necrosis factor alpha by enhancing mRNA degradation. *J. Exp. Med.* 177: 1675–1680.

16 Keifer, J.A., Guttridge, D.C., Ashburner, B.P., and Baldwin, A.S. Jr. (2001). Inhibition of NF-kappaB activity by thalidomide through suppression of IkappaB kinase activity. *J. Biol. Chem.* 276: 22382–22387.

17 Turk, B.E., Jiang, H., and Liu, J.O. (1996). Binding of thalidomide to alpha1-acid glycoprotein may be involved in its inhibition of tumor necrosis factor alpha production. *Proc. Natl. Acad. Sci. U.S.A* 93: 7552–7556.

18 Corral, L.G., Haslett, P.A., Muller, G.W. et al. (1999). Differential cytokine modulation and T cell activation by two distinct classes of thalidomide analogues that are potent inhibitors of TNF-alpha. *J. Immunol.* 163: 380–386.

19 Moller, D.R., Wysocka, M., Greenlee, B.M. et al. (1997). Inhibition of IL-12 production by thalidomide. *J. Immunol.* 159: 5157–5161.

20 Sherwood, L.M., Parris, E.E., and Folkman, J. (1971). Tumor angiogenesis: therapeutic implications. *N. Engl. J. Med.* 285: 1182–1186.

21 D'Amato, R.J., Loughnan, M.S., Flynn, E., and Folkman, J. (1994). Thalidomide is an inhibitor of angiogenesis. *Proc. Natl. Acad. Sci. U.S.A* 91: 4082–4085.

22 Kenyon, B.M., Browne, F., and D'Amato, R.J. (1997). Effects of thalidomide and related metabolites in a mouse corneal model of neovascularization. *Exp. Eye Res.* 64: 971–978.

23 Eisen, T., Boshoff, C., Mak, I. et al. (2000). Continuous low dose thalidomide: a phase II study in advanced melanoma, renal cell, ovarian and breast cancer. *Br. J. Cancer* 82: 812–817.

24 Fine, H.A., Figg, W.D., Jaeckle, K. et al. (2000). Phase II trial of the antiangiogenic agent thalidomide in patients with recurrent high-grade gliomas. *J. Clin. Oncol.* 18: 708–715.

25 Fife, K., Phillips, R.H., Howard, M.R. et al. (1998). Activity of thalidomide in AIDS-related Kaposi's sarcoma and correlation with HHV8 titre. *Int. J. STD AIDS* 9: 751–755.

26 Parman, T., Wiley, M.J., and Wells, P.G. (1999). Free radical-mediated oxidative DNA damage in the mechanism of thalidomide teratogenicity. *Nat. Med.* 5: 582–585.

27 Zhu, Y.X., Braggio, E., Shi, C.-X. et al. (2011). Cereblon expression is required for the antimyeloma activity of lenalidomide and pomalidomide. *Blood* 118: 4771–4779.

28 Krönke, J., Udeshi, N.D., Narla, A. et al. (2014). Lenalidomide causes selective degradation of IKZF1 and IKZF3 in multiple myeloma cells. *Science* 343: 301–305.

29 Lu, G., Middleton, R.E., Sun, H. et al. (2014). The myeloma drug lenalidomide promotes the cereblon-dependent destruction of Ikaros proteins. *Science* 343: 305–309.

30 Krönke, J., Fink, E.C., Hollenbach, P.W. et al. (2015). Lenalidomide induces ubiquitination and degradation of CK1α in Del(5Q) MDS. *Nature* 523: 183–188.

31 Matyskiela, M.E., Lu, G., Ito, T. et al. (2016). A novel cereblon modulator recruits GSPT1 to the CRL4(CRBN) ubiquitin ligase. *Nature* 535: 252–257.

32 An, J., Ponthier, C.M., Sack, R. et al. (2017). pSILAC mass spectrometry reveals ZFP91 as IMiD-dependent substrate of the $CRL4^{CRBN}$ ubiquitin ligase. *Nat. Commun.* 8: 15398.

33 Donovan, K.A., An, J., Nowak, R.P. et al. (2018). Thalidomide promotes degradation of SALL4, a transcription factor implicated in Duane Radial Ray Syndrome. *eLife* 7: e38430.

34 Matyskiela, M.E., Couto, S., Zheng, X. et al. (2018). SALL4 mediates teratogenicity as a thalidomide-dependent cereblon substrate. *Nat. Chem. Biol.* 14: 981–987.

35 Asatsuma-Okumura, T., Ando, H., De Simone, M. et al. (2019). p63 is a cereblon substrate involved in thalidomide teratogenicity. *Nat. Chem. Biol.* 15: 1077–1084.

36 Tochigi, T., Miyamoto, T., Hatakeyama, K. et al. (2020). Aromatase is a novel neosubstrate of cereblon responsible for immunomodulatory drug-induced thrombocytopenia. *Blood* 135: 2146–2158.

37 Yamamoto, J., Suwa, T., Murase, Y. et al. (2020). ARID2 is a pomalidomide-dependent $CRL4^{CRBN}$ substrate in multiple myeloma cells. *Nat. Chem. Biol.* 16: 1208–1217.

38 Shimizu, N., Asatsuma-Okumura, T., Yamamoto, J. et al. (2021). PLZF and its fusion proteins are pomalidomide-dependent CRBN neosubstrates. *Commun. Biol.* 4: 1277.

39 Yamanaka, S., Murai, H., Saito, D. et al. (2021). Thalidomide and its metabolite 5-hydroxythalidomide

induce teratogenicity via the cereblon neosubstrate PLZF. *EMBO J.* 40: e105375.

40 Renneville, A., Gasser, J.A., Grinshpun, D.E. et al. (2021). Avadomide induces degradation of ZMYM2 fusion oncoproteins in hematologic malignancies. *Blood Cancer Discov.* 2: 250–265.

41 Yamanaka, S., Horiuchi, Y., Matsuoka, S. et al. (2022). A proximity biotinylation-based approach to identify protein-E3 ligase interactions induced by protacs and molecular glues. *Nat. Commun.* 13: 183.

42 List, A., Kurtin, S., Roe, D.J. et al. (2005). Efficacy of lenalidomide in myelodysplastic syndromes. *N. Engl. J. Med.* 352: 549–557.

43 Habermann, T.M., Lossos, I.S., Justice, G. et al. (2009). Lenalidomide oral monotherapy produces a high response rate in patients with relapsed or refractory mantle cell lymphoma. *Br. J. Haematol.* 145: 344–349.

44 Ramsay, A.G., Clear, A.J., Kelly, G. et al. (2009). Follicular lymphoma cells induce T-cell immunologic synapse dysfunction that can be repaired with lenalidomide: implications for the tumor microenvironment and immunotherapy. *Blood* 114: 4713–4720.

45 Lopez-Girona, A., Heintel, D., Zhang, L. et al. (2011). Lenalidomide downregulates the cell survival factor, interferon regulatory factor-4, providing a potential mechanistic link for predicting response. *Br. J. Haematol.* 154: 325–336.

46 Shaffer, A.L., Emre, N.C., Lamy, L. et al. (2008). IRF4 addiction in multiple myeloma. *Nature* 454: 226–231.

47 Hansen, J.D., Correa, M., Nagy, M.A. et al. (2020). Discovery of CRBN E3 ligase modulator CC-92480 for the treatment of relapsed and refractory multiple myeloma. *J. Med. Chem.* 63: 6648–6676.

48 Ocio, E.M., Fernández-Lázaro, D., San-Segundo, L. et al. (2015). In vivo murine model of acquired resistance in myeloma reveals differential mechanisms for lenalidomide and pomalidomide in combination with dexamethasone. *Leukemia* 29: 705–714.

49 Franssen, L.E., Nijhof, I.S., Couto, S. et al. (2018). Cereblon loss and up-regulation of c-Myc are associated with lenalidomide resistance in multiple myeloma patients. *Haematologica* 103: e368–e371.

50 Surka, C., Jin, L., Mbong, N. et al. (2021). CC-90009, a novel cereblon E3 ligase modulator, targets acute myeloid leukemia blasts and leukemia stem cells. *Blood* 137: 661–677.

51 Xiao, S., Nalabolu, S.R., Aster, J.C. et al. (1998). FGFR1 is fused with a novel zinc-finger gene, ZNF198, in the T(8;13) leukaemia/lymphoma syndrome. *Nat. Genet.* 18: 84–87.

52 Melnick, A. and Licht, J.D. (1999). Deconstructing a disease: RARalpha, its fusion partners, and their roles in the pathogenesis of acute promyelocytic leukemia. *Blood* 93: 3167–3215.

53 Chen, B., Jiang, L., Zhong, M.-L. et al. (2018). Identification of fusion genes and characterization of transcriptome features in T-cell acute lymphoblastic leukemia. *Proc. Natl. Acad. Sci. U.S.A* 115: 373–378.

54 Haslett, P.A.J., Corral, L.G., Albert, M., and Kaplan, G. (1998). Thalidomide costimulates primary human T lymphocytes, preferentially inducing proliferation, cytokine production, and cytotoxic responses in the CD8+ subset. *J. Exp. Med.* 187: 1885–1892.

55 Davies, F.E., Raje, N., Hideshima, T. et al. (2001). Thalidomide and immunomodulatory derivatives augment natural killer cell cytotoxicity in multiple myeloma. *Blood* 98: 210–216.

56 Schafer, P.H., Gandhi, A.K., Loveland, M.A. et al. (2003). Enhancement of cytokine production and AP-1 transcriptional activity in T cells by thalidomide-related immunomodulatory drugs. *J. Pharmacol. Exp. Ther.* 305: 1222–1232.

57 Galustian, C., Meyer, B., Labarthe, M.-C. et al. (2008). The anti-cancer agents lenalidomide and pomalidomide inhibit the proliferation and function of T regulatory cells. *Cancer Immunol. Immunother.* 58: 1033–1045.

58 Busch, A., Zeh, D., Janzen, V. et al. (2014). Treatment with lenalidomide induces immunoactivating and counter-regulatory immunosuppressive changes in myeloma patients. *Clin. Exp. Immunol.* 177: 439–453.

59 Zelle-Rieser, C., Thangavadivel, S., Biedermann, R. et al. (2016). T cells in multiple myeloma display features of exhaustion and senescence at the tumor site. *J. Hematol. Oncol.* 9: 116.

60 Gandhi, A.K., Kang, J., Havens, C.G. et al. (2013). Immunomodulatory agents lenalidomide and pomalidomide co-stimulate T cells by inducing degradation of T cell repressors Ikaros and Aiolos via modulation of the E3 ubiquitin ligase complex CRL4(CRBN). *Br. J. Haematol.* 164: 811–821.

61 Yuki, E.F., Silva, C.A., Aikawa, N.E. et al. (2019). Thalidomide and lenalidomide for refractory systemic/cutaneous lupus erythematosus treatment: a narrative review of literature for clinical practice. *J. Clin. Rheumatol.* 27: 248–259.

62 Lessard, C.J., Adrianto, I., Ice, J.A. et al. (2012). Identification of IRF8, TMEM39A, and IKZF3-ZPBP2 as susceptibility loci for systemic lupus erythematosus in a large-scale multiracial replication study. *Am. J. Hum. Genet.* 90: 648–660.

63 Nakayama, Y., Kosek, J., Capone, L. et al. (2017). Aiolos overexpression in systemic lupus erythematosus B cell subtypes and BAFF-induced memory B cell

differentiation are reduced by CC-220 modulation of cereblon activity. *J. Immunol.* 199: 2388–2407.
64 Merrill, J.T., Werth, V.P., Furie, R. et al. (2022). Phase 2 trial of iberdomide in systemic lupus erythematosus, *New Engl. J. Med.* 386: 1034–1045.
65 Akiyama, R., Kawakami, H., Wong, J. et al. (2015). Sall4-Gli3 system in early limb progenitors is essential for the development of limb skeletal elements. *Proc. Natl. Acad. Sci. U.S.A* 112: 5075–5080.
66 Al-Baradie, R., Yamada, K., St. Hilaire, C. et al. (2002). Duane radial ray syndrome (Okihiro syndrome) maps to 20q13 and results from mutations in SALL4, a new member of the SAL family. *Am. J. Hum. Genet.* 71: 1195–1199.
67 Kohlhase, J. (2002). Okihiro syndrome is caused by SALL4 mutations. *Hum. Mol. Genet.* 11: 2979–2987.
68 Kohlhase, J. (2003). Mutations at the SALL4 locus on chromosome 20 result in a range of clinically overlapping phenotypes, including Okihiro syndrome, Holt-Oram syndrome, acro-renal-ocular syndrome, and patients previously reported to represent thalidomide embryopathy. *J. Med. Genet.* 40: 473–478.
69 Kohlhase, J. and Holmes, L.B. (2004). Mutations in SALL4 in malformed father and daughter postulated previously due to reflect mutagenesis by thalidomide. *Birth Defects Res. A Clin. Mol. Teratol.* 70: 550–551.
70 Tanaka, M. (2012). Molecular and evolutionary basis of limb field specification and limb initiation. *Develop. Growth Differ.* 55: 149–163.
71 Belair, D.G., Lu, G., Waller, L.E. et al. (2020). Thalidomide inhibits human iPSC mesendoderm differentiation by modulating CRBN-dependent degradation of SALL4. *Sci. Rep.* 10: 2864.
72 Yang, A., Kaghad, M., Caput, D., and McKeon, F. (2002). On the shoulders of giants: p63, p73 and the rise of p53. *Trends Genet.* 18: 90–95.
73 Mills, A.A., Zheng, B., Wang, X.-J. et al. (1999). p63 is a p53 homologue required for limb and epidermal morphogenesis. *Nature* 398: 708–713.
74 Restelli, M., Lopardo, T., Lo Iacono, N. et al. (2014). DLX5, FGF8 and the Pin1 isomerase control ΔNp63α protein stability during limb development: a regulatory loop at the basis of the SHFM and EEC congenital malformations. *Hum. Mol. Genet.* 23: 3830–3842.
75 van Bokhoven, H., Hamel, B.C.J., Bamshad, M. et al. (2001). p63 gene mutations in EEC syndrome, limb-mammary syndrome, and isolated split hand–split foot malformation suggest a genotype-phenotype correlation. *Am. J. Hum. Genet.* 69: 481–492.
76 Barna, M., Pandolfi, P.P., and Niswander, L. (2005). Gli3 and PLZF cooperate in proximal limb patterning at early stages of limb development. *Nature* 436: 277–281.
77 Barna, M., Hawe, N., Niswander, L., and Pandolfi, P.P. (2000). PLZF regulates limb and axial skeletal patterning. *Nat. Genet.* 25: 166–172.
78 Brent, R.L. (1964). Drug testing in animals for teratogenic effects: thalidomide in the pregnant rat. *J. Pediatr.* 64: 762–770.
79 Fink, E.C., McConkey, M., Adams, D.N. et al. (2018). Crbn I391V is sufficient to confer in vivo sensitivity to thalidomide and its derivatives in mice. *Blood* 132: 1535–1544.
80 Sun, J., Zhao, Z., Zhang, W. et al. (2020). Spalt-like protein 4 (SALL4) promotes angiogenesis by activating vascular endothelial growth factor A (VEGFA) signaling. *Med. Sci. Monit.* 26: e920851.
81 Huynh, D.L., Zhang, J.J., Chandimali, N. et al. (2018). SALL4 suppresses reactive oxygen species in pancreatic ductal adenocarcinoma phenotype via FoxM1/Prx III axis. *Biochem. Biophys. Res. Commun.* 503: 2248–2254.
82 Ellisen, L.W., Ramsayer, K.D., Johannessen, C.M. et al. (2002). REDD1, a developmentally regulated transcriptional target of p63 and p53, links p63 to regulation of reactive oxygen species. *Mol. Cell* 10: 995–1005.
83 Dimopoulos, M., Spencer, A., Attal, M. et al. (2007). Lenalidomide plus dexamethasone for relapsed or refractory multiple myeloma. *N. Engl. J. Med.* 357: 2123–2132.
84 Chamberlain, P.P., Lopez-Girona, A., Miller, K. et al. (2014). Structure of the human Cereblon–DDB1–lenalidomide complex reveals basis for responsiveness to thalidomide analogs. *Nat. Struct. Mol. Biol.* 21: 803–809.
85 Fischer, E.S., Böhm, K., Lydeard, J.R. et al. (2014). Structure of the DDB1–CRBN E3 ubiquitin ligase in complex with thalidomide. *Nature* 512: 49–53.
86 Matyskiela, M.E., Clayton, T., Zheng, X. et al. (2020). Crystal structure of the SALL4–pomalidomide–cereblon–DDB1 complex. *Nat. Struct. Mol. Biol.* 27: 319–322.
87 Sievers, Q.L., Petzold, G., Bunker, R.D. et al. (2018). Defining the human C2H2 zinc finger degrome targeted by thalidomide analogs through CRBN. *Science* 362: eaat0572.
88 Watson, E.R., Novick, S., Matyskiela, M.E. et al. (2022). Molecular glue CELMoD compounds are regulators of cereblon conformation. *Science* 378: 549–553.
89 Tan, X., Calderon-Villalobos, L.I., Sharon, M. et al. (2007). Mechanism of auxin perception by the TIR1 ubiquitin ligase. *Nature* 446: 640–645.
90 Han, T., Goralski, M., Gaskill, N. et al. (2017). Anticancer sulfonamides target splicing by inducing

RBM39 degradation via recruitment to DCAF15. *Science* 356: eaal3755.

91 Uehara, T., Minoshima, Y., Sagane, K. et al. (2017). Selective degradation of splicing factor CAPERα by anticancer sulfonamides. *Nat. Chem. Biol.* 13: 675–680.

92 Lv, L., Chen, P., Cao, L. et al. (2020). Discovery of a molecular glue promoting CDK12-DDB1 interaction to trigger cyclin K degradation. *eLife* 9: e59994.

93 Słabicki, M., Kozicka, Z., Petzold, G. et al. (2020). The CDK inhibitor CR8 acts as a molecular glue degrader that depletes cyclin K. *Nature* 585: 293–297.

94 Li, K. and Crews, C.M. (2022). PROTACs: past, present and future. *Chem. Soc. Rev.* 51: 5214–5236.

95 Sakamoto, K.M., Kim, K.B., Kumagai, A. et al. (2001). Protacs: chimeric molecules that target proteins to the Skp1–Cullin–F box complex for ubiquitination and degradation. *Proc. Natl. Acad. Sci.* 98: 8554–8559.

96 Winter, G.E., Buckley, D.L., Paulk, J. et al. (2015). Phthalimide conjugation as a strategy for in vivo target protein degradation. *Science* 348: 1376–1381.

97 Hon, W.-C., Wilson, M.I., Harlos, K. et al. (2002). Structural basis for the recognition of hydroxyproline in HIF-1 alpha by pVHL. *Nature* 417: 975–978.

98 Itoh, Y., Ishikawa, M., Naito, M., and Hashimoto, Y. (2010). Protein knockdown using methyl bestatin–ligand hybrid molecules: design and synthesis of inducers of ubiquitination-mediated degradation of cellular retinoic acid-binding proteins. *J. Am. Chem. Soc.* 132: 5820–5826.

99 Schneekloth, A.R., Pucheault, M., Tae, H.S., and Crews, C.M. (2008). Targeted intracellular protein degradation induced by a small molecule: en route to chemical proteomics. *Bioorg. Med. Chem. Lett.* 18: 5904–5908.

100 Lai, A.C., Toure, M., Hellerschmied, D. et al. (2015). Modular PROTAC design for the degradation of oncogenic BCR-ABL. *Angew. Chem. Int. Ed. Engl.* 55: 807–810.

101 Schiedel, M., Herp, D., Hammelmann, S. et al. (2017). Chemically induced degradation of Sirtuin 2 (Sirt2) by a proteolysis targeting Chimera (PROTAC) based on Sirtuin rearranging ligands (SirReals). *J. Med. Chem.* 61: 482–491.

102 Zorba, A., Nguyen, C., Xu, Y. et al. (2018). Delineating the role of cooperativity in the design of potent PROTACs for BTK. *Proc. Natl. Acad. Sci. U.S.A* 115: E7285–E7292.

103 Su, S., Yang, Z., Gao, H. et al. (2019). Potent and preferential degradation of CDK6 via proteolysis targeting chimera degraders. *J. Med. Chem.* 62: 7575–7582.

104 Liu, J., Ye, J., Zou, X. et al. (2014). CRL4A(CRBN) E3 ubiquitin ligase restricts BK channel activity and prevents epileptogenesis. *Nat. Commun.* 5: 3924.

105 Chen, Y.-A., Peng, Y.-J., Hu, M.-C. et al. (2015). The Cullin 4A/B-DDB1-Cereblon E3 ubiquitin ligase complex mediates the degradation of CLC-1 chloride channels. *Sci. Rep.* 5: 10667.

106 Van Nguyen, T., Lee, J.E., Sweredoski, M.J. et al. (2016). Glutamine triggers acetylation-dependent degradation of glutamine synthetase via the thalidomide receptor cereblon. *Mol. Cell* 61: 809–820.

107 Kwon, E., Li, X., Deng, Y. et al. (2019). AMPK is down-regulated by the CRL4A-CRBN Axis through the polyubiquitination of ampkα isoforms. *FASEB J.* 33: 6539–6550.

108 Ichikawa, S., Flaxman, H.A., Xu, W. et al. (2022). The E3 ligase adapter cereblon targets the C-terminal cyclic imide degron. *Nature* 610: 775–782.

109 Heim, C., Spring, A.-K., Kirchgäßner, S. et al. (2022). Identification and structural basis of C-terminal cyclic imides as natural degrons for cereblon. *Biochem. Biophys. Res. Commun.* 637: 66–72.

110 Friedrich, M.G., Wang, Z., Schey, K.L., and Truscott, R.J. (2019). Mechanism of protein cleavage at asparagine leading to protein–protein cross-links. *Biochem. J.* 476: 3817–3834.

111 Hirose, Y., Kitazono, T., Sezaki, M. et al. (2020). Hypnotic effect of thalidomide is independent of teratogenic ubiquitin/proteasome pathway. *Proc. Natl. Acad. Sci. U.S.A* 117: 23106–23112.

112 Jiang-Xie, L.-F., Yin, L., Zhao, S. et al. (2019). A common neuroendocrine substrate for diverse general anesthetics and sleep. *Neuron* 102: e4.

113 Lu, J., Nelson, L.E., Franks, N. et al. (2008). Role of endogenous sleep-wake and analgesic systems in anesthesia. *J. Comp. Neurol.* 508: 648–662.

25

The Polypharmacology of Cariprazine and its Implications to Clinical Indications

Attila Egyed, Dóra J. Kiss, and György M. Keserű

25.1 Introduction

Mental disorders like schizophrenia, bipolar disorders, and depression affect a significant percent of the world's population at some point in their lives, and they lead to a severe drop in the quality of life. Reflecting to this significant unmet medical need, offering effective treatment for these patients is paramount. Schizophrenia is characterized by positive (hallucinations, delusions, movement, and thought disorders), negative (social withdrawal, anhedonia, avolition, alogia, and blunted affect), and cognitive (memory problems, impaired sensory perception, and motivational issues) symptoms that vary heavily from patient to patient [1, 2]. Bipolar disorder disrupts the normal lifestyle of the patients as it is characterized by extreme shifts in mood, activity, and energy levels and results in mania and depressive episodes [3, 4]. Major depressive disorder (MDD) is one of the most common mental health problems causing the constant feeling of sadness, lack of interest and motivation, sleeping issues like insomnia or too much sleeping, anxiety, irritability, and weight issues [5]. Considering the complexity of these conditions and the variability of symptoms among patients, effective pharmacological intervention requires a broad selection of treatment options to find the best possible outcome for everyone. Novel drugs are introduced to the market to answer the unmet need by offering new ways to improve response rates, help treatment-resistant conditions [6–9], and minimize the adverse effects.

The malfunction of dopaminergic signaling pathways has been linked to severe diseases affecting the central nervous system; among others Parkinson's [10], drug addiction [11], schizophrenia [12], and MDD [13]. The dopamine receptors belong to the G-protein-coupled receptor (GPCR) family, characterized by seven transmembrane (TM) helixes that are the leading target for drug discovery, with over 25% of approved drugs targeting GPCRs [14]. The D_1 and D_5 receptors are grouped to the D1-like and the D_2, D_3, and D_4 receptors are grouped to the D_2-like subfamily. Traditionally, the D_2 receptors were major drug targets into the treatment of schizophrenia and other neurodegenerative disorders [15]. One major group of the first-generation antipsychotics, also termed typical antipsychotics like haloperidol, were D_2 receptor antagonists that are able to treat effectively the positive symptoms however are associated with strong extrapyramidal side effects and also hyperprolactinemia, treatment-induced or worsening of pre-existing negative, affective, or cognitive symptoms, tardive dyskinesia, and other cardiac and metabolic side effects [16–18] that may contribute to the shorter lifespan of the patients [19]. Second-generation antipsychotics are generally acting on the serotonin/dopamine receptors and were associated with a different side effect profile compared to the first-generation drugs while maintaining high efficacy against the positive symptoms. A main advantage of these drugs was linked to less extrapyramidal symptoms (EPS); therefore, they quickly overcome the first-generation options in the front line of treatment [20], although some studies have refuted this [21]. These drugs also induce only moderate improvement in cognition according to large clinical trials relative to first-generation antipsychotics [22] and could produce strong metabolic side effects like weight gain [23]. Clozapine, a prototypical second-generation antipsychotics, is considered a gold-standard in treatment-resistant schizophrenia [9], although the task proved to be quite challenging [24] and the efficacy of other options is also monitored closely [25–27]. Third-generation antipsychotics like cariprazine, aripiprazole, and brexpiprazole offer a unique polypharmacological profile with partial agonism on the D_2 and D_3 receptors and on the 5-HT_{1A} receptor [28, 29]. This sensitive regulatory profile is linked with their outstanding clinical outcome that opened the door for treatments of previously resistant cases [26, 30], increasing the possibility of avoiding severe adverse metabolic effects like weight gain [23] and hyperprolactinemia [31] while effectively

Polypharmacology: Strategies for Multi-Target Drug Discovery, First Edition. Edited by Jens-Uwe Peters.

improving the positive symptoms and, especially, in the case of cariprazine, also the negative symptoms[32–34].

Cariprazine is a D_3-preferring D_3 and D_2 partial agonist with activity on serotonin and other aminergic receptors that is marketed as Vraylar® in the USA and Reagila® in the EU [35–37]. The first FDA approval was focused to the treatment of schizophrenia and manic or mixed episodes of bipolar I disorder in 2015 [38]. From 2017, the cariprazine indications were expanded to the maintenance treatment of schizophrenia. After successful clinical trials, the FDA approved the drug in bipolar depression indication in 2019, while from 2022, the drug is also approved as an adjunctive treatment in MDD [39]. Although cariprazine was originally introduced as an antipsychotic agent, presently the drug is even more prescribed as an antidepressant. From a medicinal chemistry perspective, cariprazine has two major pharmacophores that are connected by a linker. The drug therefore can be classified as a bitopic compound that binds two different sites of the targeted GPCRs. In this chapter, we discuss the role of the pharmacophores in the binding affinity and selectivity as well as on the functional profile and link its polypharmacological profile to the preclinical and clinical outcomes.

25.2 Structure and Binding

The orthosteric binding pocket (OBP) of aminergic GPCRs is quite conserved and bundled by the extracellular part of TM helices 3, 5, 6, and 7 [40]. This binding pocket is historically identified as the binding site of the endogenous ligand, *e.g.*, dopamine in the case of dopaminergic receptors and serotonin for the 5-HT receptors, therefore providing a direct control on the functionality of the receptor. Naturally, this orthosteric site was the center of attention in the early stages of drug discovery; however, due to the very high sequency similarity between the GPCR families and subfamilies targeting this pocket, it often led to complex receptor profiles that caused adverse effects [41].

A significant number of allosteric sites identified in GPCRs, however, offer an opportunity to design drugs with improved selectivity and an optimal polypharmacological profile [42, 43]. Allosteric sites [44, 45] can be an extension of the OBP; like the secondary binding pocket (SBP) defined by the top part of TMs 1, 2, 3, and 7 and the ECL2 loop forming the extracellular vestibule in aminergic GPCRs or the deep pocket opened up by aripiprazole and cariprazine in the recent 5-HT_{2A} receptor structure [46]; or they can be spatially completely distinct from the OBP. These sites are located on several different areas of the receptors, including the extracellular site, the sodium site, the extrahelical site, and the intracellular side of the protein [47]. Allosteric modulators binding to either of these sites may modulate the effect of an orthosteric ligand by enhancing or decreasing its effect, resulting in positive (PAM), negative (NAM), or neutral allosteric modulators or may have a direct effect on the function. Targeting these less conserved allosteric sites benefits drug discovery efforts by allowing better control on the selectivity, affinity, and functional profile of the compounds and might offer a smoother regulation of the signaling pathways. Due to its favorable properties, a large number of compounds have been designed for these sites in the last few decades, with some of them already marketed and in use in the clinics [48, 49]. Allosteric sites, modulators, and regulators of GPCRs have been recently thoroughly reviewed [50].

Cariprazine [51] belongs to bitopic compounds that combine the benefits of targeting both the orthosteric site and a secondary pocket [47]. These compounds utilize the efficacy of targeting the OBP while aiming to achieve selectivity and fine control through the secondary site. Compared to the allosteric modulators, though they can exert their effect without the presence of the endogenous ligand that could be necessary in conditions that are characterized by the depletion of the endogenous substrate level, like Parkinsons's disease. These compounds have a primary pharmacophore (PP) that is able to bind to the OBP, and usually the careful choice of this moiety determines the functional outcome of the bitopic ligand, although it can be modulated from the secondary pocket as well [52]. The secondary pharmacophore (SP) targeting an allosteric site is connected to the PP with a carefully designed linker. The general scheme of a bitopic ligand and the structure of cariprazine highlighting the different parts of the compound are shown in Figure 25.1a,b,c, respectively.

Cariprazine shares its PP with aripiprazole, another third-generation antipsychotic, with both utilizing the 2,3-dichloro-phenylpiperazine moiety. Arylpiperazines are widespread among candidates targeting the dopamine receptors with their partial agonist profile on the dopamine D_3 and D_2 receptors. Partial agonists are getting more and more popular since they allow a refined control of the receptor acting as agonist or antagonist depending on the endogenous ligand level, sensitively responding to even small changes in its concentration. While the phenylpiperazine moiety is expected to occupy the orthosteric pocket, an ethylcyclohexyl linker fuses it to the SP that is a dimethylurea moiety in the case of cariprazine.

The experimental binding mode of cariprazine in D_3 receptors is still not available; however, we have investigated its potential binding pose (Figure 25.1c and d) both with molecular dynamics simulations and docking studies [53, 54]. The basic piperazine ring forms a hydrogen bond

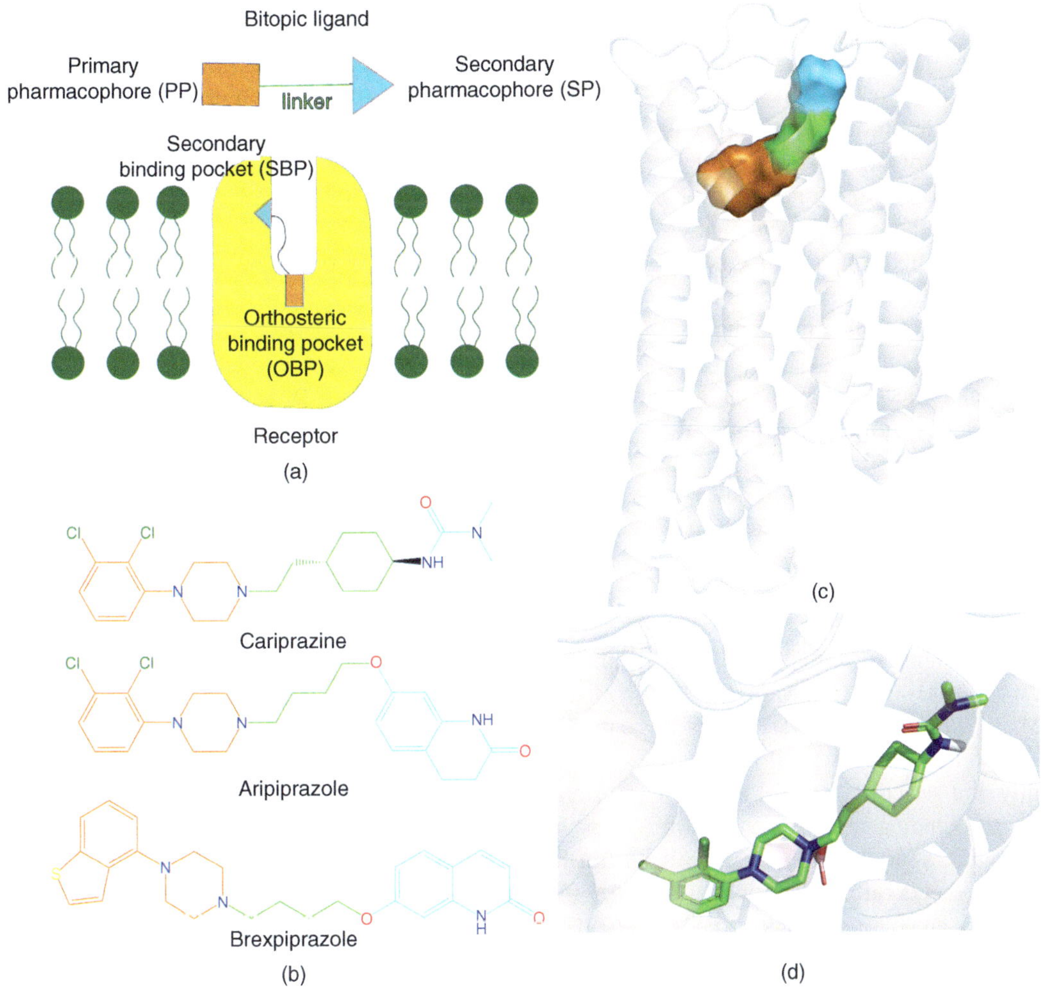

Figure 25.1 (a) The general concept of bitopic ligands that are built from a PP binding to the OBP linked to a SP targeting an allosteric site (Egyed et al. [47]/Frontiers Media S.A/CC BY 4.0). (b) The structure of cariprazine, aripiprazole, and brexpiprazole, with orange representing the PP, green the linker, and cyan the SP moiety. (c) The predicted binding mode of cariprazine in the dopamine D3 receptor based on computational modeling. Orange represents the orthosteric and cyan the secondary region, while the linker region is colored green. (d) Closeup of the binding pose of cariprazine in the hD3 receptor for better visuality, all carbon atoms of the ligand are colored green.

with the conserved Asp3.32 residue, while the SP is located in the extracellular vestibule, potentially forming polar interactions with TM2 and/or TM7 residues. The aromatic and hydrophobic residues like L2.64, V3.33, W6.48, F6.51, F6.52, and Y7.35 around the pocket and also the ECL2 loop significantly contribute to the stabilization of the ligand. In contrast to this canonical binding mode in a recent 5-HT_{2A} crystal structure, the ligand turns into an upside-down binding mode [46]. The secondary moiety occupies a deeper binding pocket penetrating between TM5 and TM6, reaching toward the membrane. The PP moiety still connects with Asp3.32 through a hydrogen bond; however, the orientation of the phenylpiperazine is reoriented, aligning toward the intracellular bundle of the receptor instead of turning toward TM5. Although no structures have been solved yet with cariprazine in the 5-HT_{1A} receptor, aripiprazole occupies the canonical binding mode, and cariprazine is expected to behave similarly.

25.3 The Role of the Primary and Secondary Pharmacophore in Binding and Selectivity

The structure of cariprazine can be separated to three parts, the PP, the linker, and the SP, as we show in Figure 25.1b. In this section, we examine the effect of the structural moieties one by one on binding affinity, receptor, and subtype selectivity, while in the next section (section 4), we present their impacts on the function response of the receptor. The PP moiety strongly determines the functional and receptor selectivity profiles of the compounds. However, further understanding on the role of the linker and SP associated with different OBPs in affinity, selectivity, and functional selectivity is critical for further drug development. The SP and the linker generally fine-tune the affinity and subtype selectivity profiles; however, in some cases, the functional profile changes when different linkers and SPs are used on the same PP. It should also be mentioned that not only the functional profile but the functional selectivity of the compound can change as well.

First, we discuss the receptor selectivity profiles of three third-generation antipsychotics, cariprazine, aripiprazole, and brexpiprazole. (Figure 25.1b and Table 25.1).

Since the residues at the orthosteric binding site are highly conserved within the GPCR subtypes, it is particularly difficult to achieve appropriate selectivity exclusively at that site. SBPs show greater variation in residues, and therefore the introduction of SP might help to achieve the desired receptor selectivity. While 1-(2,3-dichlorophenyl) piperazine shows only 197.4 nM affinity toward the D_3 receptor and 137.3 nM activity toward the D_2 receptor [57], cariprazine and aripiprazole have several orders of magnitude better affinity and selectivity values. Furthermore, when comparing the binding profiles of the two drugs, we can see that there are several differences, with aripiprazole having slightly better affinity at the 5-HT receptors, while cariprazine is much more potent at the D_3 receptor. These effects can be attributed to differences in the linker and SP. In the case of brexpiprazole, another PP, 1-(benzo[b]thiophen-4-yl)piperazine is utilized, but the SP is very similar to that of aripiprazole. This combination resulted in a more potent compound than aripiprazole at the 5-HT_{1A}, 5-HT_{2A}, 5-HT_{2C}, and 5-HT_7 receptors. If we examine the subtype selectivity of the three compounds on the dopamine receptors, we find that cariprazine shows D_3 receptor preference, aripiprazole is balanced, and brexpiprazole tends toward D_2 receptor preference. These differences may lead to different clinical effects and side effect profiles for the three drugs, as discussed in section 7. Previously, we published a computational protocol that can be used to design selective bitopic compounds. 1-(2-Methoxyphenyl)piperazine is a PP (pK_i D_2: 6.17; pK_i D_3: 5.65) with a weak preference toward the D_2 receptor; however, it can be modified by introducing a linker and SP to obtain a D_3 selective compound [52]. Several examples can be found from Newman and her group, where they have developed compounds with D_3 receptor selectivity from different reference compounds. In these structure–activity relationship (SAR) studies, there are several examples of how altering SP can affect dopamine receptor subtype selectivity [57–60]. They have also looked at the linker as part of their work. Comparing the binding profiles of the pure enantiomers and racemic mixtures of several chiral compounds, they concluded that the vestibule forming ECL2 may play an important role in binding on these receptors [60]. Furthermore, in another work, they have shown that the relative and absolute stereochemistry of the cyclopropyl linker they use can have a strong influence on the polypharmacological profile [61]. Juza and coworkers designed, synthesized, and evaluated a new series of 1,4-disubstituted aromatic piperazines inspired by aripiprazole and cariprazine using *in vitro* and *in vivo* models. These new compounds simultaneously modulate the D_2 receptor and the 5-HT_3 receptor. In behavioral tests using an acute amphetamine model of psychosis, one of the compounds was shown to ameliorate positive and negative effects of amphetamine, including hyperlocomotion, social impairment, and prepulse inhibition disorder. In all cases, the effect of the highest dose (10 mg/kg) was similar or greater than the effect of the reference dose of aripiprazole (1 mg/kg) [62]. Lee and co-workers synthesized N-phenylpiperazine analogues with 3-thiophene and 4-thiazolylphenyl SP, resulting in a compound with 450-fold selectivity for D_3. They first evaluated the substituent effect on the phenylpiperazine moiety. Substitution at most phenyl positions is well tolerated by both D_2 and D_3 receptors, with the most selective OBPs being substituted at 2nd and 4th or 5th positions. While high selectivity was not achieved from the PP itself, modification of the two selected SPs at the D_3 receptor caused both an improvement in selectivity and affinity. This study, together with other data, further confirms that the D_3 receptor accommodates a wide range of linkers and SBPs, whereas here in the D_2 receptor, binding affinity

Table 25.1 The binding affinity (K_i/nM) of cariprazine, aripiprazole, and brexpiprazole on selected GPCRs.

Receptor	Cariprazine [51]	Aripiprazole [55]	Brexpiprazole [56]
5-HT_{1A}	3	5.6	0.12
5-HT_{2A}	19	8.7	0.47
5-HT_{2B}	0.58	0.36	1.9
5-HT_{2C}	134	76	12–34
5-HT_7	111	103	3.7
α_{1A}	155	25.7	3.8
D_{2L}	0.49	0.74	0.30
D_{2S}	0.69	1.2	n.d.
D_3	0.085	1.0–9.7	1.1
H_1	23.2	25.1	19
mACh	>1000	n.d	>1000

was not significantly altered by the applied butyl-linker and the selected SPs, leading to K_i values well into the range of several hundred – thousand nM and above [63]. Stark's group has also investigated phenylpiperazine analogues, including unsubstituted phenylpiperazine and the 2-methoxy version, and in combination with aromatic SPs they have produced D_3-preferring compounds with limited selectivity. The length of the aliphatic linker was also varied, longer linkers generally resulted in better D_3 and D_2 affinity, while the worst results were obtained with a linker containing only two carbon atoms [64]. These studies confirmed that there is no universal linker and SP for the design of compounds, but that they have to be customized to the specific PP part. The primary aim of these SAR studies is to establish a suitable dopamine receptor profile with a certain level of D_2 activity needed for antipsychotic effects. But cariprazine has shown that a high level of D_3 affinity combined with affinities at other receptors like the 5-HT_{1A} and $5HT_{2A}$ might be beneficial for depressive symptoms and mood regulation.

25.4 Cariprazine-Functional Profile, Polypharmacology, and Functional Selectivity

While binding affinity data is crucial in characterizing ligands, it does not reveal the functional activity and selectivity of receptors, which is key to assessing the benefits and risks of any given compound. Functional selectivity causes significantly different signal transduction across receptors. Recently, this phenomenon has been widely recognized as having implications for understanding the mechanism of action of certain drugs and is therefore important for drug discovery. As mentioned above, the functional profile is primarily influenced by PP, but we have shown in a previous work that it can also be tuned by the SP [52]. The antagonist 3-(piperazin-1-yl)-5-(trifluoromethyl)benzonitrile was transformed into a weak partial agonist by the introduction of a new SP. In this chapter, we focus on the third generation of antipsychotics, including the previously mentioned cariprazine, aripiprazole, and brexpiprazole, which are D_2 receptor partial agonists. As we have seen in the previous chapter that the binding profiles of these compounds are different, we now compare their functional profiles (Tables 25.2 and 25.3).

The three marketed drugs share several properties, including their high affinity partial agonism at D_2, D_3, and $5HT_{1A}$ receptors combined with very low affinity antagonism at muscarinic and histamine receptors [29] These similarities are probably due to the similar chemical structures shown in Figure 25.1c. The small differences between individual partial agonists are due to the variation in intrinsic activity and affinity for different receptors.

Table 25.2 Functional activity of cariprazine, brexpiprazole and aripiprazole on selected aminergic receptors (D_2, D_3, $5HT_{1A}$, $5HT_{2A}$, $5HT_{2B}$).

Receptor	Cariprazine [65]	Brexpiprazole [56]	Aripiprazole [56]
D_2	EC_{50} 1.0 nM E_{max} 78% partial agonist	EC_{50} 4.0 nM E_{max} 43% partial agonist	EC_{50} 5.6 nM E_{max} 61% partial agonist
D_3	EC_{50} 2.7 nM E_{max} 78% partial agonist	EC_{50} 2.8 nM E_{max} 15% partial agonist	EC_{50} 5.9 nM E_{max} 28% partial agonist
$5HT_{1A}$	EC_{50} 158 nM E_{max} 95% partial agonist	EC_{50} 0.49 nM E_{max} 60% partial agonist	EC_{50} 2.1 nM E_{max} 73% partial agonist
$5HT_{2A}$	n.d. antagonist [46]	IC_{50} 6.5 nM antagonist	EC_{50} 48 nM IA 12.7% partial agonist [55]
$5HT_{2B}$	IC_{50} 3.2 nM antagonist	IC_{50} 14 nM antagonist	EC_{50} 11 nM inverse agonist [55]

Table 25.3 Broader functional receptor profile of cariprazine, brexpiprazole and aripiprazole.

Site	Cariprazine [51]	Brexpiprazole [56]	Aripiprazole [55]
5-HT_{1A}	Partial agonist	Partial agonist	Partial agonist
5-HT_{2A}	Antagonist	Antagonist	Partial agonist
5-HT_{2B}	Antagonist	Antagonist	Inverse agonist
5-HT_{2C}	Inverse agonist	Antagonist	Partial agonist
5-HT_7	Antagonist	Antagonist	Partial agonist
α_{1A}	Antagonist	Antagonist	Antagonist
D_{2L}	Partial agonist	Partial agonist	Partial agonist
D_{2S}	Partial agonist	n.d.	Partial agonist
D_3	Partial agonist	Partial agonist	Partial agonist
H_1	Antagonist	Antagonist	Antagonist

Aripiprazole is likely to have the highest affinity for D_2 receptors [66], but cariprazine and brexpiprazole have less intrinsic activity than aripiprazole [67]. Cariprazine has the strongest affinity for D_3 [51] and clinical data suggests its importance in the treatment of both negative symptoms [67] and cognitive symptoms [68]. The binding affinity of cariprazine is not only higher for the D_3 receptor than for the D_2 receptor, but it is three orders of magnitude higher than that of dopamine itself. In terms of serotonergic $5HT_{1A}$ partial agonism and affinity for $5HT_{1A}$ receptors, brexpiprazole has a significantly higher affinity than either aripiprazole or cariprazine. 5-HT_{1A}R is associated with an increase in dopamine release in the mesocortical pathway, which accounts for possible antidepressant effects. For 5-HT_{2A}R antagonism, brexpiprazole has the strongest affinity. Antagonism of 5-HT_{2A}R allows greater dopamine activity to exist in the nigrostriatal system, which lowers EPS [29]. More recently, 5-HT_{2A} partial agonists such as aripiprazole have been associated with antidepressive effects in animal models [69–71].

The lack of significant histamine H_1-antagonism means that these drugs showed limited ability to induce sedation and weight gain. The absence of an anticholinergic effect is of particular interest. Some patients may have a high anticholinergic load due to other medications such as antidepressants, antihistamines, or gastroenterological drugs, and the anticholinergic effect of second-generation antipsychotics may further increase this, with negative effects on cognition. At alpha-1 receptors, brexpiprazole has the strongest affinity, while cariprazine has the lowest. The low activity of alpha-1 receptors indicates that they are less prone to cause postural hypotension [72]. Cariprazine has the greatest affinity for D3R and the weakest affinity for serotonine receptors (5-HT_{1A}R, 5-HT_{2A}R, 5-HT_{2C}R, and 5-HT_7). This suggests that cariprazine has antipsychotic and antidepressant activity through modulation of DA and serotonin tones, respectively, and furthermore, it might have anti-anhedonic, wakefulness, and anxiolytic effects as well as a role in improved cognition.

25.5 *In Vivo* Profile of Cariprazine

Before entering the market in 2015, cariprazine went through a thorough pharmacological characterization in different animal models to test its effectiveness and tolerance. The first *in vivo* tests mainly assessed the efficacy of the compound in improving the symptoms associated with schizophrenia. Over the functional characterization, the effect of cariprazine on the neurotransmitter levels may shed light on the receptor involvement in the mechanism of action [51]. In contrast to first- and second-generation antipsychotics, cariprazine and aripiprazole dose-dependently increased the dopamine turnover lasting over 8h in mouse striatum and olfactory tubercles, with the higher effect in the latter. The observed elevated 3,4-dihydroxyphenyl acetic acid (DOPAC) concentration in the frontal cortex is in agreement with D_2 receptor antagonism. In contrast, the serotonin turnover decreased moderately at every timepoint for all three regions investigated. To further assess its partial agonist nature, the drugs were investigated in the γ-butyrolactone (GBL) model *in vivo*. Both cariprazine and aripiprazole partially reduced the GBL-induced DOPA accumulation in mouse striatum if given before GBL while fully antagonizing the effect of a full agonist-evoked reduction of the GBL-induced increase in DOPA accumulation. In the mice reserpine model, cariprazine again confirmed its dopamine receptor partial agonist activity. These experiments support the unique ability of cariprazine and aripiprazole to behave as either an agonist or antagonist depending on the assay format and dopamine level.

Before the assessment of the compound in animal models, its oral bioavailability and blood–brain barrier penetration were confirmed in rats. The preferential binding of cariprazine to the D_3 receptors was confirmed *in vivo* in a displacement assay, while its close analogue aripiprazole does not show this tendency [73]. The D_2-linked antipsychotic properties were evaluated in apomorphine-induced climbing, conditioned avoidance response, novelty and stimulant-induced motor activity, and spontaneous motor activity tests in rats and mice with all confirming its antipsychotic effect. In the apomorphine-induced climbing in mice, cariprazine has increased potency compared to olanzapine and aripiprazole, however, it was found to be less potent compared to risperidone, while in rats, cariprazine blocked amphetamine-induced hypermotility with higher potency than aripiprazole and olanzapine and comparable potency to risperidone. A conditioned avoidance test confirmed the potential of cariprazine to treat the positive symptoms of schizophrenia with comparable potency to risperidone. Cariprazine blocked potently both the NMDA antagonists MK-801 and PCP-induced hyperactivity in mice and rats, respectively. The increased potency compared to aripiprazole and olanzapine of the MK-801-induced behavior inhibition was linked to its higher D_3 affinity, while risperidone likely exerts its potent effect through the blockage of the 5-HT_{2A} receptors. Cariprazine, similarly to other D_2 partial agonists, reduces the novelty-induced mobility in rats; however, only partially that might be linked with its D_3 partial agonism attenuating some of the D_2 antagonist effect. Cariprazine has not induced catalepsy even in concentrations up to 100-fold its conditioned avoidance response ED_{50}, while other antipsychotics induce this behavior in lower doses that have already prognosed potential to avoid EPS in clinic. Outside the D_2 and D_3 partial agonism, the compound showed antagonism on the 5-HT_{2A} and 5-HT_{2B} receptors and partial agonism on 5-HT_{1A} receptors in functional assays [51], although newer measurements suggest partial agonism on the 5-HT_{2A} receptor [46, 74]. The preferential occupancy of the D_2 and D_3 receptors compared to the 5-HT_{1A} receptor was confirmed in monkey brain with positron emission tomography [75]; nevertheless, the occupancy found at serotonergic receptors might contribute to some therapeutic effect. Serotonergic effects were assessed in DOI-induced head twitch and lower lip retraction tests [73]. In line with its 5-HT_{1A} partial agonism, cariprazine was active in the lower lip retraction test similar to aripiprazole and dose-dependently reduced the DOI-induced head twitch in connection with antagonizing the 5-HT_{2A} receptor. This result is also in line with the newly suggested weak partial agonism, as the compound may act as an antagonist based on the circumstances. Cariprazine antimanic potential was assessed in an ouabain-induced hyperactivity model in rats, confirming its beneficial effects, and it was linked to the differences of the D_2-mediated G-protein and β-arrestin signaling induced by the ligand [76].

Schizophrenia is a highly complex condition, and outside the positive symptoms that were generally the first targets of treatment, the quality of life is strongly influenced by the negative and cognitive symptoms. Therefore, aiming to improve those symptoms significantly contributes to easing the burden of the patients.

Cariprazine, in contrast to the investigated other antipsychotics, enhanced cognition assessed by the scopolamine-induced learning impairment paradigm in the water-labyrinth test that was linked to its preferential D_3 binding affinity [73]. In line with its lack of D_3 preference, aripiprazole only showed a slight improvement. The potential to exert a pro-cognitive effect was further assessed in mice with PCP-triggered cognitive deficit models [77]. Cariprazine pretreatment attenuated the severity of the PCP-induced disruption of social recognition and related memory, spatial working memory, and extradimensional attention set-shifting; however, the improvement was not observed in PCP-treated D_3 KO mice. The potential to improve attentional processing deficit, a symptom linked to schizophrenia, was assessed by the 5-choice serial reaction time task (5-CSRTT) deficit induced by PCP [78]. Both cariprazine and aripiprazole attenuated the effect of PCP. In lower doses, cariprazine reduced the increase of PCP-induced inappropriate answers; however, in higher doses, nonspecific response suppression was observed.

The influence of cariprazine on social behavior and cognition after PCP treatment was also assessed in rats [79]. PCP treatment caused a reduction in the ability of discrimination between novel and familiar objects, however, low and middle doses of cariprazine attenuated this effect. Locomotor activity was significantly reduced for both PCP and higher doses of cariprazine; however, this effect was absent for the lower dose of cariprazine (0.05 mg/kg). In the reversal learning task, the higher doses of cariprazine attenuated the PCP-induced impairment similarly to risperidone. The social behavior was characterized by the animals avoiding, following, and sniffing behavior and object exploration time. The PCP-induced alterations in the avoiding behavior and object exploration time were reversed by cariprazine and risperidone; however, the latter did not show significant improvement on the following and sniffing behavior. Cariprazine significantly improved the deficit caused by the PCP treatment in the following behavior in the 0.05 mg/kg dose. The highest dose of cariprazine (0.25 mg/kg) reduced both line crossing and object exploration time, likely connected to a mild seductive effect. Recent experiments in rats also confirmed its benefits in learning, memory and cognition that were linked to its D_3 and 5-HT_{1A} partial agonism [80]. These results were indicative of cariprazine's ability to treat the negative and cognitive symptoms of schizophrenia.

Microdialysis and simultaneous monitoring of locomotor activity were applied to assess the effect of cariprazine on neurotransmitter levels in the prefrontal cortex in the rat phencyclidine schizophrenia model [81]. While PCP treatment increased the extracellular levels of dopamine, noradrenaline, serotonin, and glutamate, it did not affect the GABA levels. Both cariprazine and aripiprazole are able to reverse the effect of the PCP treatment by decreasing the neurotransmitter levels dose-dependently and cariprazine was found to be 5-fold more potent compared to aripiprazole. A schizophrenia model in rats was developed by neonatal PCP administration combined with social isolation rearing of rats that resulted in animals with behavior analogues to the positive, negative, and cognitive schizophrenia symptoms [82]. Cariprazine and aripiprazole reversed the effect of the PCP treatment in locomotor hyperactivity and improved the novel object recognition impairment. While with cariprazine treatment in a 0.3 mg/kg dose the animals showed some improvement in social interactions, but neither drug affected the fear-motivated learning and memory nor the ultrasonic vocalization deficits.

While the exact pathomechanism is still not completely elaborated behind the impairment in these mood and cognition tests, they were linked to the disfunction of the dopaminergic regulation [83–90]. Therefore, cariprazine's unique D_3-preferring partial agonist profile that is able to both agonize and antagonize the receptor based on the endogenous ligand concentration, could be behind its beneficial *in vivo* profile. Furthermore, 5-HT_{1A} receptors were also linked to the restoration of PCP-induced deficits [91, 92]; therefore, mediation through the 5-HT_{1A} receptors could also play a role.

In bipolar I disorder, manic and depressive periods are interchanging cyclically. While some symptoms overlap in each section with schizophrenia, modeling bipolar disorder in rats was quite challenging. Cariprazine was assessed in the only model available that monitors the ouabain-evoked epileptiform spikes and loss of activity in rat hippocampal slices [93]. In this model, the bipolar disorder is modeled as alternating hyperactivity and hypoactivity periods scattered in normal activity. Higher cariprazine concentrations effectively delayed the onset of the epiletform bursts and activity loss similar to lithium, attenuating the effect of the sodium pump blocker ouabain. Although this model has its limitations and further proof is required to confirm that the ouabain-induced behavior is related to bipolar disorder, the experiments implicated a potential usage of cariprazine in bipolar disorder that was later confirmed in clinical settings [94, 95].

Addiction was linked to the dopaminergic system; therefore, compounds targeting these receptors may show anti-abuse potential. Cariprazine, aripiprazole, and bifeprunox were investigated in cocaine self-administration tests with focusing on both the continuous self-administration regimen and the relapse-preventing potential in rats [96]. All three compounds successfully attenuate cocaine seeking and also moderate relapse.

Cariprazine has first been marketed in schizophrenia indication; therefore, early investigations mainly aimed to evaluate the antipsychotic and antimanic effects. However, further studies revealed anhedonic and anxiolytic effects. First, cariprazine together with aripiprazole and imipramine was investigated in the chronic mild stress-induced anhedonia model in rats [97]. While chronic stress reduces the sucrose intake, all three drugs significantly reversed this effect. Cariprazine showed greater potency compared to aripiprazole with ED_{50} of 0.052 mg/kg and 4.4 mg/kg, respectively, and showed antidepressant-like behavior in doses that were also therapeutically relevant for the antipsychotic effect. This behavior is relevant in the treatment of the negative symptoms of schizophrenia as well as other conditions characterized by depressive episodes like bipolar I and MDD. To further understand the mechanism of action behind the antianhedonic effect, chronic unpredictable stress-induced anhedonic-like behavior was assessed in wild-type as well as D_3 knockout mice [98]. While cariprazine attenuated the effect of the stress similarly to aripiprazole and imipramine in the wild-type animal, it could not be observed in the knockout mice, suggesting that this effect is at least partially exerted through the D_3 receptors. However, no further studies dig deeper behind the role of any other potential receptors like 5-HT_{1A} or 5-HT_{2A} that are also major targets of antidepressants. Next, cariprazine was investigated in the novelty-induced hypophagia test, which evaluates anxiolytic behavior. It reduced drinking latency in wild-type mice showing anxiolytic behavior, the role of the D_3 receptor, however, could not be clarified. Furthermore, while aripiprazole was effective in this test, its antidepressant-like response was combined with a general increase of total fluid intake linked to polydipsia or polyuria, a potential side effect of antidepressants not observed for cariprazine. Finally, cariprazine modulated the sleep architecture in rats by reducing the rapid eye movement sleep time and enhancing its latency that could be linked to its antidepressant effect, while its influence on the gamma band frequency oscillations might be connected to its antipsychotic effect [99].

In conclusion, cariprazine showed antipsychotic, antimanic, antianhedonic, and anxiolytic potential while also improving cognitive deficits in rodent models that is beneficial in the treatment of the positive, negative, and cognitive symptoms of schizophrenia. More recently, cariprazine showed preferential behavior in the prenatal valproic acid exposure model of autism spectrum disorder as well [100]. The effect of cariprazine is comparable to aripiprazole, with generally even more potency observed for improving the negative and cognitive deficits. The unique D_3 preferring D_3 and D_2 partial agonist profile of cariprazine was often linked to its beneficial pharmacological effects and efficacy against cognitive and negative symptoms. In fact, cariprazine was often found to be more potent in these models than its closest analogue aripiprazole without any D_3 preference [101]. Monitoring different dopaminergic receptor levels across several brain regions in rats [102], cariprazine - similar to other antipsychotics – upregulated the D_2 receptor levels in the medial prefrontal cortex, nucleus accumbens, caudate putamen, and the hippocampus. The D_4 receptor levels in the nucleus accumbens, caudate putamen, and hippocampus and uniquely dose-dependently, the D_3 levels of the olfactory tubercle, island of Calleja, and nucleus accumbens shell, were upregulated. Fluorescent-labeled cariprazine binding to the island of Calleja, a part of the brain-rich in D_3 receptors, was also confirmed with the superresolution microscopy protocol PharmacoSTORM [54]. On the 5-HT_{1A}, NMDA and glutamate receptors, both cariprazine and aripiprazole behaved similarly; however, aripiprazole has less effect on the D_3 receptors as compared to cariprazine but selectively reduced the 5-HT_{2A} receptor levels [103]. Long-term administration upregulated the 5-HT_{1A} neurotransmission in the hippocampus and increased locus coeruleus noradrenergic neuron activity [104]. The serotonergic 5-HT_{1A} and 5-HT_{2A} receptors' involvement in the actions of cariprazine were further confirmed in rat brains *in vivo* by electrophysiological recordings [105].

In a recent study, cariprazine reversed the effect of a dopamine D_3 agonist on social behavior and re-increased the extracellular dopamine level in rats [106]. Cariprazine occupancy on the D_3 receptor is unique among the third-generation antipsychotics, as confirmed in monkey brains and rats [75, 107]. [^{11}C] cariprazine brain uptake and distribution was also investigated with positron emission tomography in the brain of cynomolgus monkeys. This study confirmed its availability in the primate brain and binding to D_2 and D_3 receptors, with the highest uptake in the caudate and putamen and lower uptake in the cortex, thalamus, and cerebellum [108]. PET studies also confirmed the preferential binding to the D_3 over the D_2 receptors in lower doses in human patients with schizophrenia [109]. While cariprazine has a fairly short half-life of 1-3 days, it has two active metabolites, desmethyl cariprazine (DCAR) and didesmethyl cariprazine (DDCAR), the latter having a half-life of 2–3 weeks, strongly suggesting that metabolites could contribute to the clinical outcome [65]. The metabolites are mostly behaving similar to the parent compound, they influence the rat striatal dopamine and cortical serotonin turnover similar to cariprazine; however, the latter displayed more potency toward D_2 receptor occupancy. While DDCAR did produce an antipsychotic and precognitive effect in several *in vivo* tests, it was generally less potent than cariprazine; nevertheless, it could contribute to the observed clinical profile.

25.6 Cariprazine in Clinical Practice

Since cariprazine was first approved by the FDA for the treatment of schizophrenia and the manic episodes in bipolar disorder, the usage of this drug spread quickly and appeared in many off-label indications as well. In line with the expansion of its indication profile, the number of clinical studies involving cariprazine surged a lot in recent years; therefore, we do not aim to provide a full, comprehensive review of all available studies but will focus on the most important findings.

Cariprazine was first introduced as an antipsychotic agent, and therefore a large amount of data pooled up in connection with its effect in schizophrenic patients. The Phase II and Phase III clinical trials and post hoc analysis of their data in patients with acute exacerbation of schizophrenia confirmed the efficacy of cariprazine by evaluating the improvement in the Positive and Negative Syndrome Scale (PANSS) and Global Impressions-Severity of Illness (CGI-S) scores [110–116]. The safety of the drug was assessed by monitoring the adverse effects and other parameters like clinical laboratory values, vital signs, electrocardiogram, Columbia-Suicide Rating scale, and EPS scales. The most common adverse effects during the trials were akathisia and extrapyramidal disorder, but symptoms like tremor, insomnia, sedation, somnolence, nausea, headache, and constipation were also observed, although usually at only mild or moderate levels. The most serious adverse effects were linked to the worsening of the schizophrenia symptoms themselves both in the placebo and cariprazine treatment groups. Metabolic changes were usually comparable to placebo with no increase in prolactin levels, however, cariprazine induced somewhat greater mean increases in weight compared to placebo. Cardiac parameters also remained normal for most cariprazine groups, with some increase in diastolic blood pressure observed only in the higher dose (9–12 mg/day) group. The safety and tolerability of cariprazine were also assessed in long-term treatment 48 weeks open-label studies [117–119]. Overall the side effect profile was quite similar to the outcome of the previous shorter trials, with the most common adverse effects being usually extrapyramidal effects like akathisia, tremor, restlessness, and extrapyramidal disorder with insomnia, weight gain, and headache also among the higher frequency outcomes. The mean prolactin [31], low and high-density lipoprotein levels decreased in the groups; however, the observed small metabolic changes were not dose-dependent. No clinically significant differences in the cardiovascular parameters like blood pressure were detected. In a post-hoc analysis, cariprazine was also found to improve hostility monitored by the PNASS hostility item [110]. Based on the short- and long-term trials and a relapse trial, cariprazine was found to be safe and tolerable within the FDA-approved dose ranges (1.5–6 mg/day). Recently, potential implications in pediatric care were also reported [120]. Cariprazine also proved to be beneficial in treatment-resistant schizophrenia [26, 121].

While the positive symptoms of schizophrenia are significantly improved by all generations of antipsychotics, the negative and cognitive symptoms remained a burden on the patients for a long time, strongly lowering their and their family's quality of life [122]. The negative symptoms may be grouped into five large categories, namely, alogia, anhedonia, asociality, avolition, and blunted affect [123].

There are several potential pharmacological options to achieve improvement in the negative symptoms [124]. Targeting the glutaminergic system mainly through the NMDA receptors was successfully pursued usually in adjunctive therapy with compounds like glycine, D-serine, and sarcosine to antipsychotics [125]. Serotonergic and trace amine-associated receptor 1 treatments like 5-HT_{2A} modulators could also lead to some improvement [126, 127]. Cholinergic and muscarinic pathways were also targeted; however, in most trials, the effect on the negative symptoms was not outstanding. Nevertheless, the combination of xanomeline and trospium just proved to be effective in Phase III clinical studies; if approved by the FDA, it will open a new door with a completely new pharmacological profile compared to other antipsychotics, characterized by M_1/M_4 agonism without targeting the

D_2 receptors [128]. Reductions in the PANSS score were observed in every subscale, including the positive, negative, and Marder negative factors; however, no large-scale clinical data is accumulated yet about its effect on negative symptoms.

The dysfunction of the dopaminergic system and the dopamine D_3 receptors were linked to the pathogenesis of the negative symptoms of schizophrenia [129]. Cariprazine is a D_3-preferring D_3 and D_2 partial agonist, and already the preclinical *in vivo* experiments prognosed a potential to improve the negative symptoms, and this was thoroughly investigated in the clinic [33, 34, 130–136] due to the significant unmet medical need in this area. While most of the studies compare the efficacy to placebo, the effect of cariprazine on negative symptoms was assessed in monotherapy against risperidone. Cariprazine was found to improve both positive and negative symptoms in case studies. It improved the PANSS-FSNS score more compared to risperidone in monotherapy, with the adverse effect profile similar as described before for the non-negative symptoms' specific trials and in post-hoc analysis even with different analysis methods [134]. The observed improvement was broadscale across the individual items of the negative symptoms targeting several core domains, although anhedonia is harder to characterize with the PANSS item scores [131]. We note here that network analysis of patients with acute exacerbation of schizophrenia and predominant negative symptoms also shed light to differences between the two groups' conditions, suggesting distinctive structures of the negative symptoms [137].

The other most investigated indication of cariprazine are the different variations of the bipolar I disorder, as the FDA approved the drug first in the manic than also in the depressive stages of the disease. In line with that, a large amount of trial [138–145] and real-life case studies [146] data and post hoc analysis/review [147–156] has been accumulated over the years. During the trials, usually the Montgomery-Åsberg Depression Rating Scale (MADRS), Clinical Global Impression-Severity (CGI-S), and Young Mania rating scale (YMRS) scores were monitored depending on if the depressive or manic episodes were followed, with some trials also following functional outcome as measured on the FAST score change. Cariprazine significantly improved these scores compared to placebo; however, in some studies, the significance was dose-dependent. In response/remission analysis, cariprazine also induced improvement compared to placebo [150]. In conclusion, cariprazine was found to be effective both in bipolar depression and the manic phase of the disorder, although there were also counter opinions [157] suggesting cariprazine and aripiprazole's effects are too weak in improving the depression and mania and also criticized the scales the trials measure the outcome. The adverse effect profile of cariprazine was favorable with the most common side effects similar to what was observed with schizophrenic patients. Mainly EPS-related symptoms appeared, the most common being akathisia [158] and Parkinsonism, but also gastrointestinal symptoms and constipation were detected, although usually the adverse effects were mild and moderate. Even though treatment-emergent affective switches were not reported for cariprazine during the trials, this adverse effect appeared for some patients [159] while problems with the eye like cataracts were identified through the analysis of the FAERS database [160]. Newer studies also suggest the potential of cariprazine in the treatment of comorbid bipolar disorder and substance use disorder; however, more clinical research is needed to further support this hypothesis [161, 162].

Recently, cariprazine has been approved as adjunctive therapy for MDD, and at present its use is even more widespread as an antidepressant than as an antipsychotic, with the majority of prescriptions utilizing its benefits against depression. Although usage for the depressive part of bipolar disorder was originally off-label, its benefits were later contributed to the start of bipolar depression and MDD trials. While there is still less data accumulated connected to this indication, several trials and reviews cover this topic [163–169]. The major outcome was that cariprazine was able to decrease the MADRS score significantly. Other scales, like the HAMD-17 and CGI-S, and also the MADRS response confirmed that cariprazine outperforms the placebo; however, these changes sometimes did not reach significance. Monitoring the safety and tolerability showed similar adverse effect profiles as previously, with the most common adverse effects being akathisia, nausea, headaches, sleeplessness, and somnolence, however, no significant difference was observed compared to the placebo groups in serious adverse effects and suicidal ideation. In case studies, cariprazine improved several patients with unipolar depression who were not responsive to general antidepressant and adjunctive aripiprazole, brexpiprazole, or risperidone treatment [170].

While many reviews now focus on the efficacy, tolerability, and treatment options of the general palette of indications of cariprazine [68, 171–174], newer off-label indications like autism spectrum disorder [175], Huntington disease [176], and Wernicke–Korsakoff syndrome [177] are also reported [178]. Psychiatric conditions are complex, with a large individual variance in response rates. As with other antipsychotics, while for many cariprazine finally brings relief from the burden of severe symptoms, some patients do not react well to cariprazine and have to be switched from the drug to other treatment options [179].

25.7 Conclusions

Cariprazine is a D_3-preferring D_3 and D_2 partial agonist that was introduced to the market as an antipsychotic agent beneficial in the treatment of schizophrenia and mania in bipolar I disorder. Further clinical studies confirmed its potential as an antidepressant both in bipolar depression and as an adjunctive therapy in MDD. While multiple treatment options are available against positive symptoms, efficient management of negative and cognitive symptoms is more challenging. In addition, depressive symptoms affect approximately 5% of the adult population; however, the depression also affects children [180]. Still, the majority of the depressive patients, especially in low- and medium-income countries never receive treatments even though the burden on the patients and the severe decrease in the quality of life could lead to suicide that is among the leading causes of death in the younger generation [181]. Not surprisingly, any novel treatment option that offers a beneficial side effect profile while improving negative, cognitive, and depressive symptoms often to those who were resistant to previous medication is a huge step ahead in improving the quality of life for many people.

Cariprazine belongs to the third-generation antipsychotics that are characterized by partial agonism on the dopaminergic receptors versus antagonism that was typical to many first- and second-generation antipsychotics. Cariprazine, however, is unique even among the third-generation drugs as it prefers the D_3 receptor versus the D_2 receptor and basically achieves full occupancy on the D_3 receptor that was not observed for aripiprazole [75, 107].

While we could not cover the full *in vivo* and clinical profile of other antipsychotics, in a recent review, the pharmacodynamic and clinical profile of four third-generation antipsychotic drugs, cariprazine, aripiprazole, brexpiprazole, and lumateperone was compared [72]. Another review focused on the comparison of first-, second-, and third-generation antipsychotics [29]. Our basic objective in this work was connecting the multitarget polypharmacological profile of cariprazine to its clinical effects. Cariprazine and aripiprazole share the same OBP, while aripiprazole and brexpiprazole are equipped with almost the same SBP moiety. Lumateperone, the fourth drug, is also a bitopic compound, however, structurally distinct from the other three. The pharmacological profile of these drugs is very similar [51, 182], with the main difference that lumateperone antagonizes the D_1 and D_4 receptors, the serotonin reuptake transporter (SERT), and the alpha 2 adrenergic receptor, while the other three drugs are partial agonists on the D_3 receptor and $5\text{-}HT_{1A}$ receptor and antagonize the $5\text{-}HT_{2C}$ and $5\text{-}HT_7$ receptors. All third-generation drugs target the D_2, $5\text{-}HT_{2A}$, H_1, M_1, and α_1 receptors, though with different affinity. Interestingly, lumateperone has partial agonist activity at presynaptic and antagonistic activity at postsynaptic dopamine D_2 receptors. We have to note here that while most antipsychotics were considered antagonists on the $5\text{-}HT_{2A}$ receptor, recently BRET assays propose partial agonism on these receptors without inducing hallucinations, as proved by their clinical profiles. Since these compounds can act as antagonists based on the level of the endogenous neurotransmitter, they may provide all the benefits of $5\text{-}HT_{2A}$ antagonists like antidepressant, anti-anxiety, and pro-circadian properties.

The complex nature of the symptoms and the polypharmacological profile of the drugs harden the clarification of individual receptors in the treatment and adverse effect profiles. While all drugs have been approved for the treatment of schizophrenia, aripiprazole is indicated in the manic and mixed episodes and maintenance monotherapy of bipolar disorder, while cariprazine is indicated in both the manic and mixed phases and also in bipolar depression. Lumateperone is approved in the depressive phases of bipolar I and II disorder. Aripiprazole, brexpiprazole, and cariprazine are all approved as adjuncts in MDD; however, lumateperone is only approved in bipolar depression. Furthermore, aripiprazole is also approved in other indications like Tourette syndrome and specific cases of autism.

The D_1 receptors have been linked to cognition [183, 184], although many other receptors like D_3 and D_4 have been indicated in memory, learning, and cognition as well [84, 185, 186]. From the four third-generation drugs, lumateperone has the highest affinity against D_1 receptors, while cariprazine basically does not bind to this receptor subtype; nevertheless, its cognitive effects are likely mediated through other receptors. Antagonism on the D_2 receptor was associated to the improvement in psychotic symptoms. Already first-generation antipsychotics like haloperidol were D_2 antagonists; however, this pharmacology is also associated with EPS side effects. Second-generation antipsychotics tried to alleviate this by dual targeting the $5\text{-}HT_{2A}$ receptors, while the third generation of the compounds shows partial agonism that allows a controlled regulation of the receptors depending on the endogenous dopamine level. Brexpiprazole, aripiprazole, and cariprazine show comparable, high-binding affinity, while lumateperone is slightly lower toward this receptor, in line with their antipsychotic efficacy. Interestingly, cariprazine was associated with the most EPS related adverse effects like extrapyramidal disorder and akathisia but no hyperprolactemia, while for aripiprazole, EPS except akathisia and Parkinsonism were not so common. As expected, lumateperone with the weakest D_2 profile has a lower chance of akathisia and also fairly low rates for other EPS symptoms. Cariprazine shows the most affinity toward the D_3 receptor that was indicated in its anhedonic, anxiolytic, and pro-cognitive properties. The D_4 receptor was linked to antipsychotic effects, including mediating the negative symptoms; however, only brexpiprazole exerts high affinity toward it.

5-HT_{1A} partial agonism and 5-HT_{2A} antagonism were associated with antidepressant effects, while the latter also played a role in mediating EPS-related adverse effects [105, 126, 187–189]. Except for lumateperone, the other drugs exert potent partial agonism on the 5-HT_{1A} receptor that might be an important feature behind their antidepressant effect. Notably, cariprazine showing the highest rate of extrapyramidal side effects is the weakest on the 5-HT_{2A} receptor; nevertheless, that might still contribute to its antidepressant effect. Cariprazine, aripiprazole, and brexpiprazole are all 5-HT_{2B} antagonist that are crucial to avoid cardiac side effects [190]. The role of the other serotonergic receptors is less well clarified, though they were linked to depression [191], however cariprazine shows fairly low affinity toward them, and therefore their role is less likely to contribute to its therapeutic effects. However, they might be more relevant in the antidepressive effect of aripiprazole and brexpiprazole. Interestingly, the major antidepressive target SERT is only mediated by lumateperone [192].

Out of the four drugs, cariprazine was linked with the smallest effect on body weight in the clinic, while brexpiprazole, showing the highest H_1 receptor affinity that was previously indicated in obesity, is associated with the highest rate of increased appetite and weight gain [193]. Cariprazine does not show significant binding affinity toward muscarinic receptors, while recent Phase III trials confirmed the antipsychotic activity of an M_1/M_4 agonist [128]. In contrast to lumateperone, sedation was not found to be a characteristic adverse effect of cariprazine that might be linked to its weaker affinity toward the α1 receptors.

The clinical trials and real-life experience confirmed cariprazine benefits in the treatment of psychiatric diseases, including different forms of depression that may improve the quality of life for patients suffering from treatment-resistant forms of the disorders. Similar to other antipsychotics, there are no drug without adverse effects; however, cariprazine was found to be generally safe and tolerable in the FDA-approved dose range with an acceptable side effect profile. Despite the wide range of clinically proven indications, its mechanism of action is still not fully clarified. Elucidating and separating the important signaling pathways behind its antipsychotic and antidepressant therapeutic effects, especially focusing on its unique D_3 preferring profile, might aid the development of a newer generation of drugs.

References

1 Jauhar, S., Johnstone, M., and McKenna, P.J. (2022). Schizophrenia. *Lancet* 399 (10323): 473–486.

2 McCutcheon, R.A., Reis Marques, T., and Howes, O.D. (2020). Schizophrenia—an Overview. *JAMA Psychiatry* 77 (2): 201–210.

3 Bauer, M.S. (2022). Bipolar disorder. *Ann. Intern. Med.* 175 (7): ITC97–ITC112.

4 Nierenberg, A.A., Agustini, B., Köhler-Forsberg, O. et al. (2023). Diagnosis and treatment of bipolar disorder: a review. *JAMA* 330 (14): 1370–1380.

5 Marx, W., Penninx, B.W.J.H., Solmi, M. et al. (2023). Major depressive disorder. *Nat. Rev. Dis. Primers* 9 (1): 1–21.

6 Voineskos, D., Daskalakis, Z.J., and Blumberger, D.M. (2020). Management of treatment-resistant depression: challenges and strategies. *NDT.* 21 (16): 221–234.

7 Breeksema, J.J., Niemeijer, A., Krediet, E. et al. (2024). Patient perspectives and experiences with psilocybin treatment for treatment-resistant depression: a qualitative study. *Sci. Rep.* 14 (1): 2929.

8 Kumar, R., Nuñez, N.A., Joshi, N. et al. (2024). Metabolomic biomarkers for (R, S)-ketamine and (S)-ketamine in treatment-resistant depression and healthy controls: a systematic review. *Bipolar Disord.* 24: 321–330.

9 Kane, J.M., Schoretsanitis, G., Rubio, J.M., and Correll, C.U. (2024). Clozapine in treatment-resistant schizophrenia: Reflections from the Hallmark US clinical trial and beyond. *Schizophr. Res.* 268: 9–13.

10 Vallano, A., Fernandez-Duenas, V., Pedros, C. et al. An update on adenosine A2A receptors as drug target in Parkinson's disease. *CNS Neurol. Disord. Drug Targets* 10 (6): 659–669.

11 Galaj, E., Ewing, S., and Ranaldi, R. (2018). Dopamine D1 and D3 receptor polypharmacology as a potential treatment approach for substance use disorder. *Neurosci. Biobehav. Rev.* 89: 13–28.

12 Roth, B.L., Sheffler, D.J., and Kroeze, W.K. (2004). Magic shotguns versus magic bullets: selectively non-selective drugs for mood disorders and schizophrenia. *Nat. Rev. Drug Discov.* 3 (4): 353–359.

13 Belujon, P. and Grace, A.A. (2017). Dopamine system dysregulation in major depressive disorders. *Int. J. Neuropsychopharmacol.* 20 (12): 1036–1046.

14 Hauser, A.S., Attwood, M.M., Rask-Andersen, M. et al. (2017). Trends in GPCR drug discovery: new agents, targets and indications. *Nat. Rev. Drug Discov.* 16 (12): 829–842.

15 Beaulieu, J.M. and Gainetdinov, R.R. (2011). The physiology, signaling, and pharmacology of dopamine receptors. *Pharmacol. Rev.* 63 (1): 182–217.

16 Leucht, S., Corves, C., Arbter, D. et al. (2009). Second-generation versus first-generation antipsychotic drugs for schizophrenia: a meta-analysis. *Lancet* 373 (9657): 31–41.

17 Shad, M.U. (2023). Seventy years of antipsychotic development: a critical review. *Biomedicine* 11 (1): 130.

18 Vardar, M.K., Ceylan, M.E., and Ünsalver, B.Ö. (2020). Assesment of risk factors for tardive dyskinesia. *Psychopharmacol. Bull.* 50 (3): 36–46.

19 Lobo, M.C., Whitehurst, T.S., Kaar, S.J., and Howes, O.D. (2022). New and emerging treatments for schizophrenia: a narrative review of their pharmacology, efficacy and side effect profile relative to established antipsychotics. *Neurosci. Biobehav. Rev.* 132: 324–361.

20 Dibben, C.R.M., Khandaker, G.M., Underwood, B.R. et al. (2016). First-generation antipsychotics: not gone but forgotten. *BJPsych Bulletin.* 40 (2): 93–96.

21 Peluso, M.J., Lewis, S.W., Barnes, T.R.E., and Jones, P.B. (2012). Extrapyramidal motor side-effects of first- and second-generation antipsychotic drugs. *Br. J. Psychiatry* 200 (5): 387–392.

22 Hill, S.K., Bishop, J.R., Palumbo, D., and Sweeney, J.A. (2010). Effect of second-generation antipsychotics on cognition: current issues and future challenges. *Expert. Rev. Neurother.* 10 (1): 43–57.

23 Pillinger, T., McCutcheon, R.A., Vano, L. et al. (2020). Comparative effects of 18 antipsychotics on metabolic function in patients with schizophrenia, predictors of metabolic dysregulation, and association with psychopathology: a systematic review and network meta-analysis. *Lancet Psychiatry* 7 (1): 64–77.

24 Elkis, H. and Buckley, P.F. (2016). Treatment-resistant schizophrenia. *Psychiatr. Clin. N. Am.* 39 (2): 239–265.

25 Lin, T.C. and Lin, C.H. (2023). Schizophrenia patients discharged on clozapine plus long-acting injectable antipsychotics from a public psychiatric hospital in Taiwan, 2006–2021. *Int. J. Neuropsychopharmacol.* 26 (11): 808–816.

26 Boydstun, C., Lynch, S., and DiGenova, P. (2023). Cariprazine: an augmentation strategy for treatment-resistant schizophrenia with pro-cognitive and anti-hostility effects. *Int. Clin. Psychopharmacol.* 38 (5): 361.

27 Malik, R. and Zaidi, M. (2023). A review of current strategies in treatment-resistant schizophrenia: considering paliperidone palmitate lai with and without olanzapine as an alternative to clozapine. *CNS Spectr.* 28 (2): 257–257.

28 Mailman, R.B. and Murthy, V. (2010). Third generation antipsychotic drugs: partial agonism or receptor functional selectivity? *Curr. Pharm. Des.* 16 (5): 488–501.

29 Taylor, D., Chithiramohan, R., Grewal, J. et al. (2023). Dopamine partial agonists: a discrete class of antipsychotics. *Int. J. Psychiatry Clin. Pract.* 27 (3): 272–284.

30 Vasiliu, O. (2022). Third-generation antipsychotics in patients with schizophrenia and non-responsivity or intolerance to clozapine regimen: what is the evidence? *Front. Psych.* 29 (13): 1069432.

31 Culpepper, L., Vieta, E., Kelly, D.L. et al. (2022). Minimal effects of cariprazine on prolactin levels in bipolar disorder and schizophrenia. *Neuropsychiatr. Dis. Treat.* 18: 995–1011.

32 Demjaha, A., Iacoponi, E., Hansen, L. et al. (2022). Cariprazine as a treatment for negative psychotic symptoms in first-episode psychosis: case series. *BJPsych Open.* 8 (3): e88.

33 Earley, W., Guo, H., Daniel, D. et al. (2019). Efficacy of cariprazine on negative symptoms in patients with acute schizophrenia: a post hoc analysis of pooled data. *Schizophr. Res.* 204: 282–288.

34 Németh, G., Laszlovszky, I., Czobor, P. et al. (2017). Cariprazine versus risperidone monotherapy for treatment of predominant negative symptoms in patients with schizophrenia: a randomised, double-blind, controlled trial. *Lancet* 389 (10074): 1103–1113.

35 De Deurwaerdère, P. (2016). Cariprazine: new dopamine biased agonist for neuropsychiatric disorders. *Drugs Today* 52 (2): 97–110.

36 VRAYLAR (2024). VRAYLAR® (cariprazine) | official HCP site. https://www.vraylarhcp.com/ (accessed 25 September 2024).

37 Reagila (2024). European medicines agency. https://www.ema.europa.eu/en/medicines/human/EPAR/reagila

38 Drugs.com (2024). Vraylar (cariprazine) FDA approval history. https://www.drugs.com/history/vraylar.html (accessed 25 September 2024).

39 Newswire MP (2024). MultiVu. U.S. FDA approves VRAYLAR® (cariprazine) as an adjunctive treatment for major depressive disorder. https://www.multivu.com/players/English/9107351-vraylar-cariprazine-fda-approval-major-depressive-disorder/ (accessed 25 September 2024).

40 Chan, H.C.S., Li, Y., Dahoun, T. et al. (2019). New binding sites, new opportunities for GPCR drug discovery. *Trends Biochem. Sci.* 44 (4): 312–330.

41 Lee, S.M., Booe, J.M., and Pioszak, A.A. (2015). Structural insights into ligand recognition and selectivity for classes A, B, and C GPCRs. *Eur. J. Pharmacol.* 15 (763): 196–205.

42 Chan, H.C.S., Li, Y., Dahoun, T. et al. (2019). New binding sites, new opportunities for GPCR drug discovery. *Trends Biochem. Sci.* 44 (4): 312–330.

43 Congreve, M., Oswald, C., and Marshall, F.H. (2017). Applying structure-based drug design approaches to allosteric modulators of GPCRs. *Trends Pharmacol. Sci.* 38 (9): 837–847.

44 Christopoulos, A., Changeux, J.P., Catterall, W.A. et al. (2014). International union of basic and clinical pharmacology. XC. Multisite pharmacology: recommendations for the nomenclature of receptor allosterism and allosteric ligands. *Pharmacol. Rev.* 66 (4): 918–947.

45 Christopoulos, A. (2014). Advances in G protein-coupled receptor allostery: from function to structure. *Mol. Pharmacol.* 86 (5): 463–478.

46 Chen, Z., Fan, L., Wang, H. et al. (2022). Structure-based design of a novel third-generation antipsychotic drug lead with potential antidepressant properties. *Nat. Neurosci.* 25 (1): 39–49.

47 Egyed, A., Kiss, D.J., and Keserű, G.M. (2022). The impact of the secondary binding pocket on the pharmacology of class A GPCRs. *Front. Pharmacol.* 13: 847788.

48 Dorr, P., Westby, M., Dobbs, S. et al. (2005). Maraviroc (UK-427,857), a potent, orally bioavailable, and selective small-molecule inhibitor of chemokine receptor CCR5 with broad-spectrum anti-human immunodeficiency virus type 1 activity. *Antimicrob. Agents Chemother.* 49 (11): 4721–4732.

49 Casadó, V. and Casadó-Anguera, V. (2023). What are the current trends in G protein-coupled receptor targeted drug discovery? *Expert Opin. Drug Discov.* 18 (8): 815–820.

50 Shpakov, A.O. (2023). Allosteric sites and allosteric regulators of G protein-coupled receptors: gray cardinals of signal transduction. *J. Evol. Biochem. Physiol.* 59 (1): S1–S106.

51 Kiss, B., Horváth, A., Némethy, Z. et al. (2010). Cariprazine (RGH-188), a dopamine D(3) receptor-preferring, D(3)/D(2) dopamine receptor antagonist-partial agonist antipsychotic candidate: in vitro and neurochemical profile. *J. Pharmacol. Exp. Ther.* 333 (1): 328–340.

52 Egyed, A., Domány-Kovács, K., Koványi, B. et al. (2020). Controlling receptor function from the extracellular vestibule of G-protein coupled receptors. *Chem. Commun.* 56 (91): 14167–14170.

53 Frank, A., Kiss, D.J., Keserű, G.M., and Stark, H. (2018). Binding kinetics of cariprazine and aripiprazole at the dopamine D3 receptor. *Sci. Rep.* 8: 12509.

54 Prokop, S., Ábrányi-Balogh, P., Barti, B. et al. (2021). PharmacoSTORM nanoscale pharmacology reveals cariprazine binding on Islands of Calleja granule cells. *Nat. Commun.* 12 (1): 6505.

55 Shapiro, D.A., Renock, S., Arrington, E. et al. (2003). Aripiprazole, a novel atypical antipsychotic drug with a unique and robust pharmacology. *Neuropsychopharmacology* 28 (8): 1400–1411.

56 Maeda, K., Sugino, H., Akazawa, H. et al. (2014 Sep 1). Brexpiprazole I: in vitro and in vivo characterization of a novel serotonin-dopamine activity modulator. *J. Pharmacol. Exp. Ther.* 350 (3): 589–604.

57 Newman, A.H., Beuming, T., Banala, A.K. et al. (2012). Molecular determinants of selectivity and efficacy at the dopamine D3 receptor. *J. Med. Chem.* 55 (15): 6689–6699.

58 Kumar, V., Bonifazi, A., Ellenberger, M.P. et al. (2016). Highly selective dopamine D3 receptor (D3R) antagonists and partial agonists based on eticlopride and the D3R crystal structure: new leads for opioid dependence treatment. *J. Med. Chem.* 59 (16): 7634–7650.

59 Gogarnoiu, E.S., Vogt, C.D., Sanchez, J. et al. (2023). Dopamine D3/D2 receptor ligands based on cariprazine for the treatment of psychostimulant use disorders that may be dual diagnosed with affective disorders. *J. Med. Chem.* 66 (3): 1809–1834.

60 Newman, A.H., Grundt, P., Cyriac, G. et al. (2009). N-(4-(4-(2,3-Dichloro- or 2-methoxyphenyl) piperazin-1-yl)butyl)heterobiarylcarboxamides with functionalized linking chains as high affinity and enantioselective D3 receptor antagonists. *J. Med. Chem.* 52 (8): 2559–2570.

61 Battiti, F.O., Cemaj, S.L., Guerrero, A.M. et al. (2019). The significance of chirality in drug design and synthesis of bitopic ligands as D3 receptor (D3R) selective agonists. *J. Med. Chem.* 62 (13): 6287–6314.

62 Juza, R., Vojtechova, I., Stefkova-Mazochova, K. et al. (2022). Novel D2/5-HT receptor modulators related to cariprazine with potential implication to schizophrenia treatment. *Eur. J. Med. Chem.* (232): 114193.

63 Lee, B., Taylor, M., Griffin, S.A. et al. (2021). Evaluation of substituted N-phenylpiperazine analogs as D3 vs. D2 dopamine receptor subtype selective ligands. *Molecules* 26 (11): 3182.

64 Schübler, M., Sadek, B., Kottke, T. et al. (2017). Synthesis, molecular properties estimations, and

dual dopamine D2 and D3 receptor activities of benzothiazole-based ligands. *Front. Chem.* 5: 64.

65 Kiss, B., Némethy, Z., Fazekas, K. et al. (2019 Sep). Preclinical pharmacodynamic and pharmacokinetic characterization of the major metabolites of cariprazine. *Drug Des. Devel. Ther.* 16 (13): 3229–3248.

66 Tadori, Y., Forbes, R.A., McQuade, R.D., and Kikuchi, T. (2011 Oct 15). In vitro pharmacology of aripiprazole, its metabolite and experimental dopamine partial agonists at human dopamine D2 and D3 receptors. *Eur. J. Pharmacol.* 668 (3): 355–365.

67 Keks, N., Hope, J., Schwartz, D. et al. (2020). Comparative tolerability of dopamine D2/3 receptor partial agonists for schizophrenia. *CNS Drugs* 34 (5): 473–507.

68 McIntyre, R.S., Daniel, D.G., Vieta, E. et al. (2023). The efficacy of cariprazine on cognition: a post hoc analysis from phase II/III clinical trials in bipolar mania, bipolar depression, and schizophrenia. *CNS Spectr.* 28 (3): 319–330.

69 Cameron, L.P., Tombari, R.J., Lu, J. et al. (2021). A non-hallucinogenic psychedelic analogue with therapeutic potential. *Nature* 589 (7842): 474–479.

70 Vargas, M.V., Dunlap, L.E., Dong, C. et al. (2023). Psychedelics promote neuroplasticity through the activation of intracellular 5-HT2A receptors. *Science* 379 (6633): 700–706.

71 Kaplan, A.L., Confair, D.N., Kim, K. et al. (2022 Oct). Bespoke library docking for 5-HT2A receptor agonists with antidepressant activity. *Nature* 610 (7932): 582–591.

72 Munayco Maldonado, G.M. and Schwartz, T.L. (2024). Differentiating the third generation of antipsychotics: a focus on lumateperone's similarities and differences. *Int. Clin. Psychopharmacol.* 39 (1): 4.

73 Gyertyán, I., Kiss, B., Sághy, K. et al. (2011). Cariprazine (RGH-188), a potent D3/D2 dopamine receptor partial agonist, binds to dopamine D3 receptors *in vivo* and shows antipsychotic-like and procognitive effects in rodents. *Neurochem. Int.* 59 (6): 925–935.

74 Gaitonde, S., Shahraki, A., Pastor, A. et al. (2023). Pharmacological fingerprint of antipsychotic drugs at the serotonin 5-HT2A receptor. *Mol. Psychiatry* 29 (9): 2753–2764.

75 Seneca, N., Finnema, S.J., Laszlovszky, I. et al. (2011). Occupancy of dopamine D_2 and D_3 and serotonin 5-HT_1A receptors by the novel antipsychotic drug candidate, cariprazine (RGH-188), in monkey brain measured using positron emission tomography. *Psychopharmacology* 218 (3): 579–587.

76 Gao, Y., Peterson, S., Masri, B. et al. (2015). Cariprazine exerts antimanic properties and interferes with dopamine D2 receptor β-arrestin interactions. *Pharmacol. Res. Perspect.* 3 (1): e00073.

77 Zimnisky, R., Chang, G., Gyertyán, I. et al. (2013). Cariprazine, a dopamine D(3)-receptor-preferring partial agonist, blocks phencyclidine-induced impairments of working memory, attention set-shifting, and recognition memory in the mouse. *Psychopharmacology* 226 (1): 91–100.

78 Barnes, S.A., Young, J.W., Markou, A. et al. (2018). The effects of cariprazine and aripiprazole on PCP-induced deficits on attention assessed in the 5-choice serial reaction time task. *Psychopharmacology* 235 (5): 1403–1414.

79 Neill, J.C., Grayson, B., Kiss, B. et al. (2016). Effects of cariprazine, a novel antipsychotic, on cognitive deficit and negative symptoms in a rodent model of schizophrenia symptomatology. *Eur. Neuropsychopharmacol.* 26 (1): 3–14.

80 Zlatanova, H.I., Georgieva-Kotetarova, M.T., Vilmosh, N.B., and Kandilarov, I.K. (2022). Evaluation of the effect of cariprazine on memory and cognition in experimental rodent models. *Int. J. Environ. Res. Public Health* 19 (22): 14748.

81 Kehr, J., Yoshitake, T., Ichinose, F. et al. (2018). Effects of cariprazine on extracellular levels of glutamate, GABA, dopamine, noradrenaline and serotonin in the medial prefrontal cortex in the rat phencyclidine model of schizophrenia studied by microdialysis and simultaneous recordings of locomotor activity. *Psychopharmacology* 235 (5): 1593–1607.

82 Watson, D.J.G., King, M.V., Gyertyán, I. et al. (2016). The dopamine D_3-preferring D_2/D_3 dopamine receptor partial agonist, cariprazine, reverses behavioural changes in a rat neurodevelopmental model for schizophrenia. *Eur. Neuropsychopharmacol.* 26 (2): 208–224.

83 Gross, G., Wicke, K., and Drescher, K.U. (2013). Dopamine D3 receptor antagonism – still a therapeutic option for the treatment of schizophrenia. *Naunyn Schmiedeberg's Arch. Pharmacol.* 386 (2): 155–166.

84 Nakajima, S., Gerretsen, P., Takeuchi, H. et al. (2013). The potential role of dopamine D3 receptor neurotransmission in cognition. *Eur. Neuropsychopharmacol.* 23 (8): 799–813.

85 Black, K.J., Hershey, T., Koller, J.M. et al. (2002). A possible substrate for dopamine-related changes in mood and behavior: Prefrontal and limbic effects of a D3-preferring dopamine agonist. *Proc. Natl. Acad. Sci. U.S.A* 99 (26): 17113–17118.

86 Millan, M.J., Svenningsson, P., Ashby, C.R. Jr., et al. (2008). S33138 N-4-[2-[(3aS,9bR)-8-cyano-1,3a,4,9b-tetrahydro[1]-benzopyrano[3,4-c] pyrrol-2(3H)-yl-ethyl]

phenylacetamide], a preferential dopamine D3 versus D2 receptor antagonist and potential antipsychotic agent. II. A neurochemical, electrophysiological and behavioral characterization in vivo. *J. Pharmacol. Exp. Ther.* 324 (2): 600–611.
87 Gross, G. and Drescher, K. (2012). The role of dopamine D 3 receptors in antipsychotic activity and cognitive functions. *Handb. Exp. Pharmacol.* 213: 167–210.
88 Gyertyán, I., Sághy, K., Laszy, J. et al. (2008). Subnanomolar dopamine D3 receptor antagonism coupled to moderate D2 affinity results in favourable antipsychotic-like activity in rodent models: II. Behavioural characterisation of RG-15. Naunyn-Schmiedeberg's. *Arch. Pharmacol.* 378 (5): 529–539.
89 Joyce, J.N. and Millan, M.J. (2005). Dopamine D3 receptor antagonists as therapeutic agents. *Drug Discov. Today* 10 (13): 917–925.
90 Sokoloff, P., Diaz, J., Le Foll, B. et al. (2006). The dopamine D3 receptor: a therapeutic target for the treatment of neuropsychiatric disorders. *CNS Neurol. Disord. Drug Targets* 5 (1): 25–43.
91 Snigdha, S. and Neill, J.C. (2008). Improvement of phencyclidine-induced social behaviour deficits in rats: Involvement of 5-HT1A receptors. *Behav. Brain Res.* 191 (1): 26–31.
92 McLean, S.L., Woolley, M.L., Thomas, D., and Neill, J.C. (2009). Role of 5-HT receptor mechanisms in sub-chronic PCP-induced reversal learning deficits in the rat. *Psychopharmacology* 206 (3): 403–414.
93 El-Mallakh, R.S., Payne, R.S., Schurr, A. et al. (2015). Cariprazine delays ouabain-evoked epileptiform spikes and loss of activity in rat hippocampal slices. *Psychiatry Res.* 229 (1–2): 370–373.
94 Pinto, J.V., Saraf, G., Vigo, D. et al. (2020). Cariprazine in the treatment of bipolar disorder: a systematic review and meta-analysis. *Bipolar Disord.* 22 (4): 360–371.
95 Do, A., Keramatian, K., Schaffer, A., and Yatham, L. (2021). Cariprazine in the treatment of bipolar disorder: within and beyond clinical trials. *Front. Psych.* 12: 769897.
96 Román, V., Gyertyán, I., Sághy, K. et al. (2013). Cariprazine (RGH-188), a D_3-preferring dopamine D_3/D_2 receptor partial agonist antipsychotic candidate demonstrates anti-abuse potential in rats. *Psychopharmacology* 226 (2): 285–293.
97 Papp, M., Gruca, P., Lasoń-Tyburkiewicz, M. et al. (2014). Attenuation of anhedonia by cariprazine in the chronic mild stress model of depression. *Behav. Pharmacol.* 25 (5–6): 567–574.
98 Duric, V., Banasr, M., Franklin, T. et al. (2017). Cariprazine exhibits anxiolytic and dopamine D3 receptor-dependent antidepressant effects in the chronic stress model. *Int. J. Neuropsychopharmacol.* 20 (10): 788–796.
99 Nyitrai, G., Kiss, B., Farkas, B. et al. (2021). Cariprazine modulates sleep architecture in rats. *J. Psychopharmacol.* 35 (3): 303–310.
100 Román, V., Adham, N., Foley, A.G. et al. (2021). Cariprazine alleviates core behavioral deficits in the prenatal valproic acid exposure model of autism spectrum disorder. *Psychopharmacology* 238 (9): 2381–2392.
101 Adham, N., Gyertyán, I., and Kiss, B. (2014). Cariprazine demonstrates greater potency than aripiprazole in animal models of psychosis, cognitive impairment, and negative symptoms. *Eur. Neuropsychopharmacol.* 24: S233–S234.
102 Choi, Y.K., Adham, N., Kiss, B. et al. (2014). Long-term effects of cariprazine exposure on dopamine receptor subtypes. *CNS Spectr.* 19 (3): 268–277.
103 Choi, Y.K., Adham, N., Kiss, B. et al. (2017). Long-term effects of aripiprazole exposure on monoaminergic and glutamatergic receptor subtypes: comparison with cariprazine. *CNS Spectr.* 22 (6): 484–494.
104 El Mansari, M., Ebrahimzadeh, M., Hamati, R. et al. (2020). Long-term administration of cariprazine increases locus coeruleus noradrenergic neurons activity and serotonin1A receptor neurotransmission in the hippocampus. *J. Psychopharmacol.* 34 (10): 1143–1154.
105 Herman, A., Mansari, M.E., Adham, N. et al. (2018). Involvement of 5-HT1A and 5-HT2A receptors but not α2-adrenoceptors in the acute electrophysiological effects of cariprazine in the rat brain in vivo. *Mol. Pharmacol.* 94 (6): 1363–1370.
106 Kehr, J., Wang, F.H., Ichinose, F. et al. (2022). Preferential effects of cariprazine on counteracting the disruption of social interaction and decrease in extracellular dopamine levels induced by the dopamine D3 receptor agonist, PD-128907 in rats: implications for the treatment of negative and depressive symptoms of psychiatric disorders. *Front. Psych.* 12: 801641.
107 Adham, N., Gyertyán, I., Laszlovszky, I., and Kiss, B. (2014). EPA-0579 – at antipsychotic-like effective doses, cariprazine displays potent dopamine D3 and D2 receptor occupancy in vivo and efficacy across animal models. *Eur. Psychiatry* 29: 1.
108 Tóth, M., Varrone, A., Steiger, C. et al. (2013). Brain uptake and distribution of the dopamine D3 /D2 receptor partial agonist [11 C]cariprazine: an in vivo positron emission tomography study in nonhuman primates. *Synapse* 67 (5): 258–264.

109 Girgis, R.R., Slifstein, M., D'Souza, D. et al. (2016). Preferential binding to dopamine D3 over D2 receptors by cariprazine in patients with schizophrenia using PET with the D3/D2 receptor ligand [11C]-(+)-PHNO. *Psychopharmacology* 233 (19): 3503–3512.

110 Durgam, S., Starace, A., Li, D. et al. (2014). An evaluation of the safety and efficacy of cariprazine in patients with acute exacerbation of schizophrenia: a phase II, randomized clinical trial. *Schizophr. Res.* 152 (2): 450–457.

111 Kane, J.M., Zukin, S., Wang, Y. et al. (2015). Efficacy and safety of cariprazine in acute exacerbation of schizophrenia: results from an international, phase III clinical trial. *J. Clin. Psychopharmacol.* 35 (4): 367.

112 Durgam, S., Cutler, A.J., Lu, K. et al. (2015). Cariprazine in acute exacerbation of schizophrenia: a fixed-dose, phase 3, randomized, double-blind, placebo- and active-controlled trial. *J. Clin. Psychiatry* 76 (12): 2310.

113 Durgam, S., Earley, W., Li, R. et al. (2016). Long-term cariprazine treatment for the prevention of relapse in patients with schizophrenia: A randomized, double-blind, placebo-controlled trial. *Schizophr. Res.* 176 (2): 264–271.

114 Earley, W., Durgam, S., Lu, K. et al. (2017). Safety and tolerability of cariprazine in patients with acute exacerbation of schizophrenia: a pooled analysis of four phase II/III randomized, double-blind, placebo-controlled studies. *Int. Clin. Psychopharmacol.* 32 (6): 319.

115 Barabássy, Á., Sebe, B., Acsai, K. et al. (2021). Safety and tolerability of cariprazine in patients with schizophrenia: a pooled analysis of eight phase II/III studies. *NDT.* 17: 957–970.

116 Falkai, P., Dombi, Z.B., Acsai, K. et al. (2023). The efficacy and safety of cariprazine in the early and late stage of schizophrenia: a post hoc analysis of three randomized, placebo-controlled trials. *CNS Spectr.* 28 (1): 104–111.

117 Durgam, S., Greenberg, W.M., Li, D. et al. (2017). Safety and tolerability of cariprazine in the long-term treatment of schizophrenia: results from a 48-week, single-arm, open-label extension study. *Psychopharmacology* 234 (2): 199–209.

118 Nasrallah, H.A., Earley, W., Cutler, A.J. et al. (2017). The safety and tolerability of cariprazine in long-term treatment of schizophrenia: a post hoc pooled analysis. *BMC Psychiatry* 17 (1): 305.

119 Cutler, A.J., Durgam, S., Wang, Y. et al. (2018). Evaluation of the long-term safety and tolerability of cariprazine in patients with schizophrenia: results from a 1-year open-label study. *CNS Spectr.* 23 (1): 39–50.

120 Riccobene, T., Riesenberg, R., Yeung, P.P. et al. (2022). Pharmacokinetics, safety, and tolerability of cariprazine in pediatric patients with bipolar I disorder or schizophrenia. *J. Child Adolesc. Psychopharmacol.* 32 (8): 434–443.

121 Karoline, R.D., Sekar, S., Charitha, S.A., and Jacob, J.A. (2023). Cariprazine augmentation in a patient with clozapine-resistant schizophrenia. Industrial. *Psychiatry J.* 32 (Suppl 1): S279.

122 Barabassy, A., Szatmári, B., Laszlovszky, I. et al. (2018). Negative symptoms of schizophrenia: constructs, burden, and management. In: *Psychotic Disorders – An Update*. IntechOpen https://www.intechopen.com/chapters/58847 (accessed 26 February 2024).

123 Galderisi, S., Mucci, A., Buchanan, R.W., and Arango, C. (2018). Negative symptoms of schizophrenia: new developments and unanswered research questions. *Lancet Psychiatry* 5 (8): 664–677.

124 Kantrowitz, J.T. (2021). How do we address treating the negative symptoms of schizophrenia pharmacologically? *Expert. Opin. Pharmacother.* 22 (14): 1811–1813.

125 Goh, K.K., Wu, T.H., Chen, C.H., and Lu, M.L. (2021). Efficacy of N-methyl-D-aspartate receptor modulator augmentation in schizophrenia: a meta-analysis of randomised, placebo-controlled trials. *J. Psychopharmacol.* 35 (3): 236–252.

126 Kantrowitz, J.T. (2020). Targeting serotonin 5-HT2A receptors to better treat schizophrenia: rationale and current approaches. *CNS Drugs* 34 (9): 947–959.

127 Davidson, M., Saoud, J., Staner, C. et al. (2017). Efficacy and safety of MIN-101: A 12-week randomized, double-blind, placebo-controlled trial of a new drug in development for the treatment of negative symptoms in schizophrenia. *Am. J. Psychiatry* 174 (12): 1195–1202.

128 Kaul, I., Sawchak, S., Correll, C.U. et al. (2024). Efficacy and safety of the muscarinic receptor agonist KarXT (xanomeline–trospium) in schizophrenia (EMERGENT-2) in the USA: results from a randomised, double-blind, placebo-controlled, flexible-dose phase 3 trial. *Lancet* 403 (10422): 160–170.

129 Collo, G., Mucci, A., Giordano, G.M. et al. (2020). Negative symptoms of schizophrenia and dopaminergic transmission: translational models and perspectives opened by iPSC techniques. *Front. Neurosci.* 14: https://www.frontiersin.org/journals/neuroscience/articles/10.3389/fnins.2020.00632 (accessed 26 February 2024).

130 Laszlovszky, I., Barabássy, Á., and Németh, G. (2021). Cariprazine, a broad-spectrum antipsychotic for the treatment of schizophrenia: pharmacology, efficacy, and safety. *Adv. Ther.* 38 (7): 3652–3673.

131 Fleischhacker, W., Galderisi, S., Laszlovszky, I. et al. (2019). The efficacy of cariprazine in negative symptoms of schizophrenia: Post hoc analyses of PANSS individual items and PANSS-derived factors. *Eur. Psychiatry* 58: 1–9.

132 Leucht, S., Dombi, Z.B., Szabó, P. et al. (2023 Nov). Single trajectory treatment response for predominant negative symptoms: Post-hoc analysis of a clinical trial with cariprazine and risperidone. *Schizophr. Res.* 1 (261): 24–30.

133 Rana, T. and André, V. (2023). From evidence to clinical experience: three cases of community patients with negative symptoms of schizophrenia and treated with cariprazine. *Minerva Psychiat.* 64, 64 (2, 2): 270, 270–271, 1.

134 Czobor, P., Sebe, B., Acsai, K. et al. (2022). What is the minimum clinically important change in negative symptoms of schizophrenia? PANSS based post-hoc analyses of a phase III clinical trial. *Front. Psych.* 13: 816339. https://www.frontiersin.org/journals/psychiatry/articles/10.3389/fpsyt.2022.816339 (accessed 26 February 2024).

135 Németh, G., Dombi, Z.B., Laszlovszky, I., and Barabássy, Á. (2022). Addressing negative symptoms of schizophrenia pharmacologically with cariprazine: evidence from clinical trials, a real-world study, and clinical cases. *Expert. Opin. Pharmacother.* 23 (12): 1467–1468.

136 Krause, M., Zhu, Y., Huhn, M. et al. (2018). Antipsychotic drugs for patients with schizophrenia and predominant or prominent negative symptoms: a systematic review and meta-analysis. *Eur. Arch. Psychiatry Clin. Neurosci.* 268 (7): 625–639.

137 Demyttenaere, K., Leenaerts, N., Acsai, K. et al. (2022). Disentangling the symptoms of schizophrenia: network analysis in acute phase patients and in patients with predominant negative symptoms. *Eur. Psychiatry* 65 (1): e18.

138 Durgam, S., Earley, W., Lipschitz, A. et al. (2016). An 8-week randomized, double-blind, placebo-controlled evaluation of the safety and efficacy of cariprazine in patients with bipolar I depression. *Am. J. Psychiatry* 173 (3): 271–281.

139 Sachs, G.S., Greenberg, W.M., Starace, A. et al. (2015). Cariprazine in the treatment of acute mania in bipolar i disorder: a double-blind, placebo-controlled, phase III trial. *J. Affect. Disord.* 174: 296–302.

140 Calabrese, J.R., Keck, P.E., Starace, A. et al. (2014). Efficacy and safety of low- and high-dose cariprazine in acute and mixed mania associated with Bipolar I disorder: a double-blind, placebo-controlled study. *J. Clin. Psychiatry* 76 (3): 6179.

141 Durgam, S., Starace, A., Li, D. et al. (2015). The efficacy and tolerability of cariprazine in acute mania associated with bipolar I disorder: a phase II trial. *Bipolar Disord.* 17 (1): 63–75.

142 Earley, W., Burgess, M.V., Rekeda, L. et al. (2019). Cariprazine treatment of bipolar depression: a randomized double-blind placebo-controlled phase 3 study. *Am. J. Psychiatry* 176 (6): 439–448.

143 Ketter, T.A., Sachs, G.S., Durgam, S. et al. (2018). The safety and tolerability of cariprazine in patients with manic or mixed episodes associated with bipolar I disorder: a 16-week open-label study. *J. Affect. Disord.* 225: 350–356.

144 Yatham, L.N., Vieta, E., and Earley, W. (2020). Evaluation of cariprazine in the treatment of bipolar I and II depression: a randomized, double-blind, placebo-controlled, phase 2 trial. *Int. Clin. Psychopharmacol.* 35 (3): 147.

145 Earley, W.R., Burgess, M.V., Khan, B. et al. (2020). Efficacy and safety of cariprazine in bipolar I depression: A double-blind, placebo-controlled phase 3 study. *Bipolar Disord.* 22 (4): 372–384.

146 Dyrmishi, E., De Pieri, M., Ferrari, M. et al. (2022). Case report: long-acting oral cariprazine. *Front. Psych.* 13: https://www.frontiersin.org/journals/psychiatry/articles/10.3389/fpsyt.2022.876003 (accessed 26 February 2024).

147 Jain, R., McIntyre, R.S., Cutler, A.J. et al. (2024). Efficacy of cariprazine in patients with bipolar depression and higher or lower levels of baseline anxiety: a pooled post hoc analysis. *Int. Clin. Psychopharmacol.* 39 (2): 82.

148 Earley, W., Durgam, S., Lu, K. et al. (2017). Tolerability of cariprazine in the treatment of acute bipolar I mania: a pooled post hoc analysis of 3 phase II/III studies. *J. Affect. Disord.* 215: 205–212.

149 McIntyre, R.S., Masand, P.S., Earley, W., and Patel, M. (2019). Cariprazine for the treatment of bipolar mania with mixed features: a post hoc pooled analysis of 3 trials. *J. Affect. Disord.* 257: 600–606.

150 Earley, W., Durgam, S., Lu, K. et al. (2018). Clinically relevant response and remission outcomes in cariprazine-treated patients with bipolar I disorder. *J. Affect. Disord.* 226: 239–244.

151 Tohen, M. (2021). Cariprazine as a treatment option for depressive episodes associated with bipolar 1

disorder in adults: an evidence-based review of recent data. *Drug Des. Devel. Ther.* 15: 2005–2012.

152 Patel, M., Jain, R., Tohen, M. et al. (2021). Efficacy of cariprazine in bipolar I depression across patient characteristics: a post hoc analysis of pooled randomized, placebo-controlled studies. *Int. Clin. Psychopharmacol.* 36 (2): 76–83.

153 Citrome, L. (2019). Cariprazine for bipolar depression: What is the number needed to treat, number needed to harm and likelihood to be helped or harmed? *Int. J. Clin. Pract.* 73 (10): e13397.

154 Findlay, L.J., El-Mallakh, P.L., and El-Mallakh, R.S. (2017). Cariprazine for the Treatment of Bipolar Disorder. *Perspect. Psychiatr. Care* 53 (3): 148–155.

155 Vieta, E., Calabrese, J.R., Whelan, J. et al. (2021). The efficacy of cariprazine on function in patients with bipolar depression: a post hoc analysis of a randomized controlled trial. *Curr. Med. Res. Opin.* 37 (9): 1635–1643.

156 Yildiz, A., Siafis, S., Mavridis, D. et al. (2023). Comparative efficacy and tolerability of pharmacological interventions for acute bipolar depression in adults: a systematic review and network meta-analysis. *Lancet Psychiatry* 10 (9): 693–705.

157 Rossiter, J.R. (2022). Aripiprazole and cariprazine for bipolar disorder? *Open J. Med. Psychol.* 11 (4): 254–273.

158 Citrome, L., Yatham, L.N., Patel, M.D. et al. (2021). Cariprazine and akathisia, restlessness, and extrapyramidal symptoms in patients with bipolar depression. *J. Affect. Disord.* 288: 191–198.

159 Pons-Cabrera, M.T., Palacios-Garrán, R., Tardón-Senabre, L. et al. (2022). Cariprazine-induced mania: a case series report. *Bipolar Disord.* 24 (4): 457–460.

160 Zhu, H., Qu, Y., Du, Z. et al. (2024). Mining and analysis of adverse event signals of Cariprazine based on the real-world data of FAERS database. *J. Affect. Disord.* 347: 45–50.

161 Grunze, H., Csehi, R., Born, C., and Barabássy, Á. (2021). Reducing addiction in bipolar disorder via hacking the dopaminergic system. *Front. Psych.* 12: https://www.frontiersin.org/journals/psychiatry/articles/10.3389/fpsyt.2021.803208 (accessed 26 February 2024).

162 Grunze, H. (2023). The role of the D3 dopamine receptor and its partial agonist cariprazine in patients with schizophrenia and substance use disorder. *Expert. Opin. Pharmacother.* 24 (18): 1985–1992.

163 Xie, M., Qiu, Y., Wang, M. et al. (2024). Adjunctive cariprazine as a novel effective strategy for treating major depressive disorder: a systematic review and meta-analysis. *J. Psychiatr. Res.* (172): 71–80.

164 Durgam, S., Earley, W., Guo, H. et al. (2016). Efficacy and safety of adjunctive cariprazine in inadequate responders to antidepressants: a randomized, double-blind, placebo-controlled study in adult patients with major depressive disorder. *J. Clin. Psychiatry* 77 (3): 6112.

165 Fava, M., Durgam, S., Earley, W. et al. (2018). Efficacy of adjunctive low-dose cariprazine in major depressive disorder: a randomized, double-blind, placebo-controlled trial. *Int. Clin. Psychopharmacol.* 33 (6): 312.

166 Earley, W.R., Guo, H., Németh, G. et al. (2018). Cariprazine augmentation to antidepressant therapy in major depressive disorder: results of a randomized, double-blind, placebo-controlled trial. *Psychopharmacol. Bull.* 48 (4): 62–80.

167 Vieta, E., Earley, W.R., Burgess, M.V. et al. (2019). Long-term safety and tolerability of cariprazine as adjunctive therapy in major depressive disorder. *Int. Clin. Psychopharmacol.* 34 (2): 76.

168 Sachs, G.S., Yeung, P.P., Rekeda, L. et al. (2023). Adjunctive cariprazine for the treatment of patients with major depressive disorder: a randomized, double-blind, placebo-controlled phase 3 study. *Am. J. Psychiatry* 180 (3): 241–251.

169 Riesenberg, R., Yeung, P.P., Rekeda, L. et al. (2023). Cariprazine for the adjunctive treatment of major depressive disorder in patients with inadequate response to antidepressant therapy: results of a randomized, double-blind, placebo-controlled study. *J. Clin. Psychiatry* 84 (5): 48439.

170 Pessina, E., Martini, A., Raffone, F., and Martiadis, V. (2023). Cariprazine augmentation in patients with treatment resistant unipolar depression who failed to respond to previous atypical antipsychotic add-on. A case-series. *Front. Psych.* 14: https://www.frontiersin.org/journals/psychiatry/articles/10.3389/fpsyt.2023.1299368 (accessed 26 February 2024).

171 Durgam, S., Earley, W., Lu, K. et al. (2017). Global improvement with cariprazine in the treatment of bipolar I disorder and schizophrenia: a pooled post hoc analysis. *Int. J. Clin. Pract.* 71 (12): e13037.

172 Németh, G. and Csehi, R. (2022). Editorial: novel antipsychotics within and beyond clinical trials: The treatment of overlapping psychiatric disorders with D3-D2 partial agonists. *Front. Psych.* 13: 1038627.

173 Garnock-Jones, K.P. (2017). Cariprazine: a review in schizophrenia. *CNS Drugs* 31 (6): 513–525.

174 Fagiolini, A., Alcalá, J.Á., Aubel, T. et al. (2020). Treating schizophrenia with cariprazine: from clinical research to clinical practice. Real world experiences

and recommendations from an International Panel. *Ann. General Psychiatry* 19 (1): 55.

175 Yeung, P.P., Johnson, K.A., Riesenberg, R. et al. (2023). Cariprazine in pediatric patients with autism spectrum disorder: results of a pharmacokinetic, safety and tolerability study. *J. Child Adolesc. Psychopharmacol.* 33 (6): 232–242.

176 Molnar, M.J., Molnar, V., Fedor, M. et al. (2022). Improving mood and cognitive symptoms in Huntington's disease with cariprazine treatment. *Front. Psych.* 12: 825532.

177 Jimoh, I.J., Sebe, B., Balicza, P. et al. (2020). Wernicke–Korsakoff syndrome associated with mtDNA disease. *Ther. Adv. Neurol. Disord.* 13: 1756286420938972.

178 Csehi, R., Dombi, Z.B., Sebe, B., and Molnár, M.J. (2022). Real-life clinical experience with cariprazine: a systematic review of case studies. *Front. Psych.* 13: 827744.

179 Heck, J., Seifert, J., Stichtenoth, D.O. et al. (2021). A case series of serious and unexpected adverse drug reactions under treatment with cariprazine. *Clinical Case Rep.* 9 (5): e04084.

180 Depressive Disorder (Depression) (2024). [Internet]. https://www.who.int/news-room/fact-sheets/detail/depression (accessed 25 September 2024).

181 Evans-Lacko, S., Aguilar-Gaxiola, S., Al-Hamzawi, A. et al. (2018). Socio-economic variations in the mental health treatment gap for people with anxiety, mood, and substance use disorders: results from the WHO World Mental Health (WMH) surveys. *Psychol. Med.* 48 (9): 1560–1571.

182 Stahl, S.M. (2021). *Stahl's Essential Psychopharmacology: Neuroscientific Basis and Practical Applications*, 5the. Cambridge: Cambridge University Press https://www.cambridge.org/core/books/stahls-essential-psychopharmacology/1231F597AE3471AE53B1CC9AFDA34B32 (accessed 25 September 2024).

183 Jones-Tabah, J., Mohammad, H., Paulus, E.G. et al. (2022). The signaling and pharmacology of the dopamine D1 receptor. *Front. Cell. Neurosci.* 15: 806618. https://www.frontiersin.org/articles/10.3389/fncel.2021.806618 (accessed 25 September 2024).

184 Yang, Y., Lee, S.M., Imamura, F. et al. (2021). D1 dopamine receptors intrinsic activity and functional selectivity affect working memory in prefrontal cortex. *Mol. Psychiatry* 26 (2): 645–655.

185 Woolley, M.L., Waters, K.A., Reavill, C. et al. (2008). Selective dopamine D4 receptor agonist (A-412997) improves cognitive performance and stimulates motor activity without influencing reward-related behaviour in rat. *Behav. Pharmacol.* 19 (8): 765–776.

186 Huang, M., Kwon, S., He, W., and Meltzer, H.Y. (2017). Neurochemical arguments for the use of dopamine D4 receptor stimulation to improve cognitive impairment associated with schizophrenia. *Pharmacol. Biochem. Behav.* 157: 16–23.

187 Celada, P., Puig, M.V., Amargós-Bosch, M. et al. (2004). The therapeutic role of 5-HT1A and 5-HT2A receptors in depression. *J. Psychiatry Neurosci.* 29 (4): 252–265.

188 Robinson, D.S. (1991). Antidepressant efficacy of 5-HT1A partial agonist drugs. In: *5-Hydroxytryptamine in Psychiatry: A Spectrum of Ideas* [Internet] (ed. M. Sandler, A. Coppen, and S. Harnett). Oxford University Press https://doi.org/10.1093/acprof:oso/9780192620118.003.0013.

189 Smith, A.L.W., Harmer, C.J., Cowen, P.J., and Murphy, S.E. (2023). The serotonin 1A (5-HT1A) receptor as a pharmacological target in depression. *CNS Drugs* 37 (7): 571–585.

190 McIntyre, R.S. (2023). Serotonin 5-HT2B receptor agonism and valvular heart disease: implications for the development of psilocybin and related agents. *Expert Opin. Drug Saf.* 22 (10): 881–883.

191 Millan, M.J. (2005). Serotonin 5-HT2C receptors as a target for the treatment of depressive and anxious states: focus on novel therapeutic strategies. *Therapie* 60 (5): 441–460.

192 White, K.J., Walline, C.C., and Barker, E.L. (2005). Serotonin transporters: Implications for antidepressant drug development. *AAPS J.* 7 (2): E421–E433.

193 Ratliff, J., Barber, J., Palmese, L. et al. (2010). Association of prescription H1 antihistamine use with obesity: results from the national health and nutrition examination survey. *Obesity* 18 (12): 2398–2400.

26

Multi-Targeted Antivirals

A Novel Drug Design Strategy for Potent SARS-CoV-2 Inhibitors

Bing Ye, Letian Song, Meehyein Kim, Shenghua Gao, Peng Zhan, and Xinyong Liu

Since the outbreak of COVID-19, the scientific community has been committed to developing preventive and therapeutic agents as countermeasures. Although vaccination has reduced infection and mortality rates, current vaccines are often sensitive to the mutations in the spike protein of SARS-CoV-2, making immune escape a concern [1, 2]. As supplements of vaccine, other treatment options are still needed. Novel antivirals against specific and less mutable targets may be more successful. The genome of SARS-CoV-2 is close to 30 kb in size, contains 14 open reading frames (ORFs), and encodes 29 viral proteins [3]. Approximately two-thirds of the 5′ end of the SARS-CoV-2 genome encodes two overlapping polyproteins: pp1a and pp1ab. These two polyproteins are digested by main protease (M^{pro}) and papain-like protease (PL^{pro}) into 16 non-structural proteins (nsp), which are essential for viral replication and transcription [4]. Functional biomolecules encoded by viruses or host cells established a synergistic and continuous life cycle of SARS-CoV-2. Therefore, a multi-target pharmacological approach against key targets associated with SARS-CoV-2 infection has been developed nowadays, which aims to reduce the mutation escape caused by single-targeting therapy. This part summarizes compounds which exert synergistic effects on different stages of virus entry and replication, achieving more potential antiviral effects.

26.1 Multi-Target Inhibitors Targeting Both SARS-CoV-2 and Host Proteins

As commonly acknowledged, viruses exploit signal pathways and functional proteins of host cells to complete their life cycles. Developing inhibitors that can simultaneously target both the virus and host cells holds potential for reducing drug dosage, enhancing the efficacy through different inhibition pathways, and minimizing adverse reactions associated with taking multiple drugs.

26.1.1 Multi-Target Inhibitors Targeting SARS-CoV-2 Protease and Host Targets

26.1.1.1 Targeting SARS-CoV-2 M^{pro} and Host Cathepsin L

SARS-CoV-2 M^{pro} is a cysteine protease, functioning through the catalytic dyad formed by Cys145 and His41. The M^{pro} active site can be divided into several sub-pockets, among which S1, S1', S2, and S4 are well-defined cavities corresponding to sidechains of substrate polypeptide (Figure 26.1). Previous studies have demonstrated that M^{pro} is highly conserved across coronaviruses and plays an indispensable role in the viral replication cycle, making it an ideal target for drug development [5]. As a cysteine protease, M^{pro} can be reversibly inhibited by lots of peptide-like covalent inhibitors that react with Cys145 to form a covalent hemiacetal intermediate through aldehyde, α-ketonamide, or other functional groups. The nonspecific reaction of ketone towards thiol or hydroxyl groups also enable peptidomimetic M^{pro} inhibitors easier to interact with cysteine and serine proteases in host cells, particularly those crucial for SARS-CoV-2 infection [6].

During the invasion of SARS-CoV-2 into host cells, several host proteases, including serine protease transmembrane protease serine 2 (TMPRSS2) and cathepsin, facilitate the activation of SARS-CoV-2 spike protein for viral packaging and ACE2 interaction [7]. Cathepsin B (CatB) and cathepsin L (CatL) are a class of endosomal cysteine proteases that mediate the cleavage of S protein within the acidic endosomal/lysosomal compartments, thereby playing an essential step for viral entry into host cells [8, 9].

Polypharmacology: Strategies for Multi-Target Drug Discovery, First Edition. Edited by Jens-Uwe Peters.

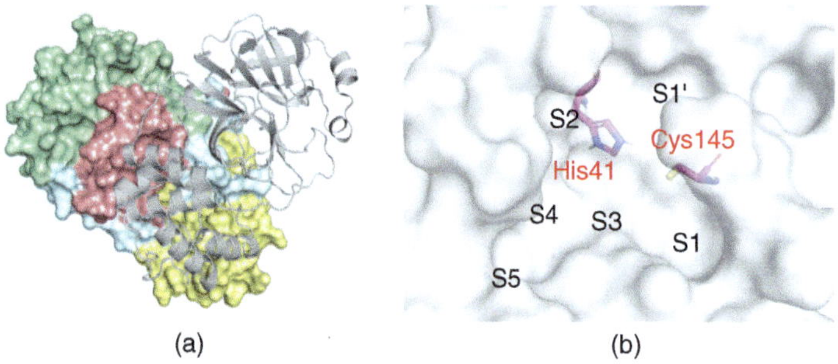

Figure 26.1 (a) Domains of M^{pro} dimer. Domain I (green), domain II (red), and domain III (yellow); (b) sub-sites and catalytic dyad in M^{pro} active center.

To date, research about multi-inhibitors targeting host proteases and viral proteases has greatly advanced. Several inhibitors have been developed to target viral proteases as well as human CatL, effectively acting on the invasion stages of virus.

Upon the emergence of SARS-CoV-2, Ma et al. identified two potent human cathepsin inhibitors [10], calpain inhibitor II (**1**) and XII (**2**), which exhibited significant inhibition against SARS-CoV-2 M^{pro} activity with IC_{50} values of 0.97 μM and 0.45 μM, respectively. The EC_{50} values of anti-SARS-CoV-2 were determined to be 2.07 μM and 0.49 μM. Unlike previous M^{pro} inhibitors, the two inhibitors are assembled with lipophilic P1 groups (Figure 26.2). As an alternative to glutamine in the S1 pocket of M^{pro}, they maintained hydrogen bonding with His163 through the sulfur atom of methionine and the nitrogen atom of pyridine. This pioneering research has paved the way for discovering peptidomimetic multiple-target inhibitors against SARS-CoV-2.

Olgotrelvir (**3**) is an orally administered inhibitor targeting M^{pro} and CatL [11]. Its activity against SARS-CoV-2 is enhanced through synergistic dual inhibition of viral polyprotein processing and cell entry. The active form of olgotrelvir, AC1115 (**4**), exhibited potent antiviral activity against all tested SARS-CoV-2 variants (Figure 26.3a). Crystal structures revealed that AC1115 displayed a

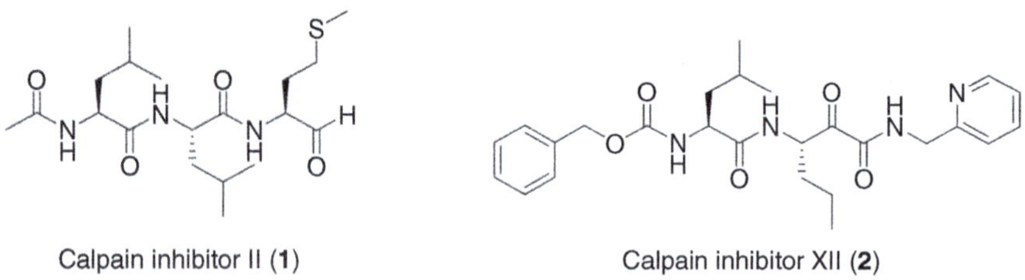

Figure 26.2 Chemical structure of calpain inhibitors II (**1**) and XII (**2**).

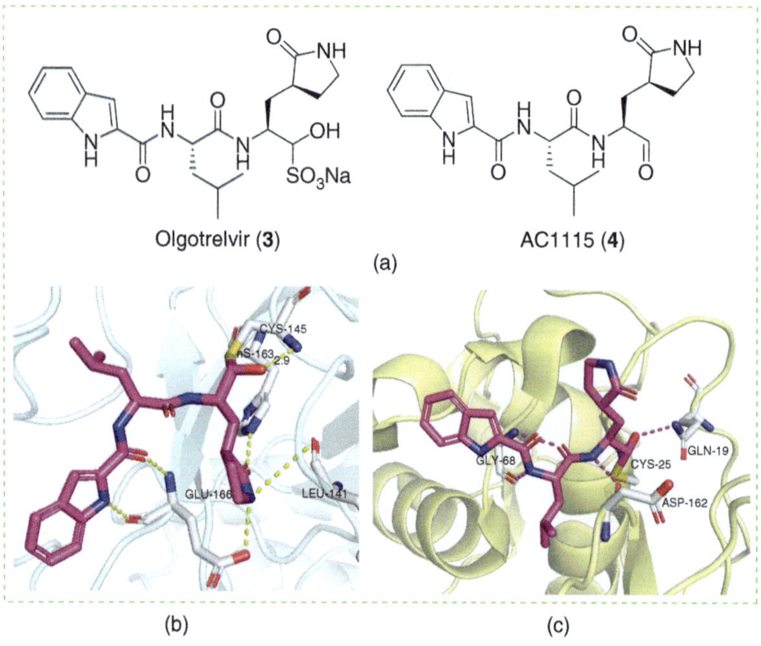

Figure 26.3 (a) Olgotrelvir and its active form AC1115; (b) Co-crystal of AC1115 covalently binding in the SARS-CoV-2 M^{pro} (PDB ID: 8UAB); (c) Co-crystal of AC1115 covalently binding in the human cathesin L (PDB ID: 8UAC).

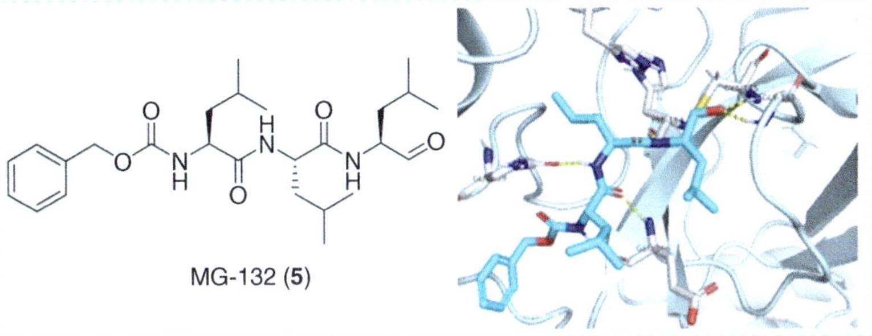

Figure 26.4 The structure of MG-132 (**5**) and its binding mode with SARS-CoV-2 M^pro^ (PDB ID: 7NF5).

complementary shape to the active sites of M^pro^ and CatL (Figure 26.3b,c). In the K18-hACE2 transgenic mouse model infected with SARS-CoV-2, olgotrelvir significantly reduced viral load in the lungs, as well as attenuating cytokine release and mitigating lung pathology severity. The exceptional antiviral activity and substantial plasma exposure made it a promising standalone candidate for the treatment of COVID-19. Moreover, its dual-targeting property demonstrates efficacy in inhibiting the nirmatrelvir-resistant M^pro^ E166 mutant.

Compound MG-132 (**5**), described by Elisa's team, is a peptidomimetic inhibitor also with an aldehyde warhead which exhibits well suitability for M^pro^ active sites forming a covalent bond with Cys145 independent of reducing agents and crystallization conditions [12]. The docking mode of MG-132 with CatL showed its ability to form effective covalent interactions with cysteine residues. MG-132 exhibited nanomolar inhibition against CatL (IC_{50} = 0.15 nM) and reversible micromolar inhibition against M^pro^ (IC_{50} = 3.8 μM, preincubated for 30 min), while the residence time is prolonged. Structural and biochemical analysis suggested a preference of M^pro^ for leucine at the P2 position (Figure 26.4), however, it can also accommodate other hydrophobic residues. Conversely, CatL has a stronger preference towards aromatic residues at P2 position. Therefore, the selectivity of MG-132 was improved by introducing a leucine residue modification at P2. Additionally, it is worth considering the replacement of aldehyde-based warheads with alternative reversible covalent warheads.

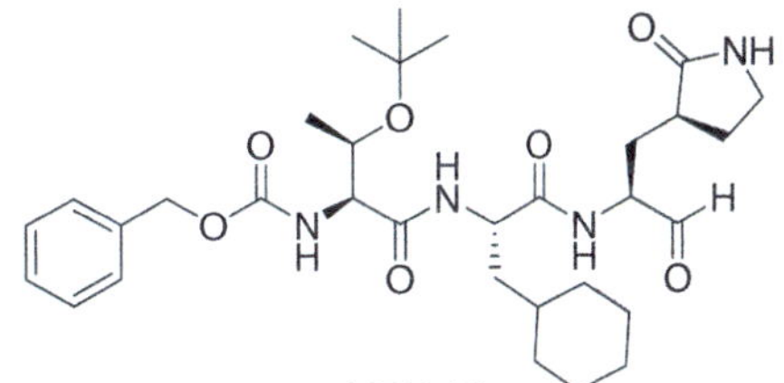

MPI8 (**6**)

Figure 26.5 The structure of MPI8 (**6**).

Another aldehyde-based inhibitor MPI8 (**6**, Figure 26.5) with similar structure exhibited the most potent effect on M^pro^ (IC_{50} = 105 nM) and outstanding selectivity towards CatL (IC_{50} = 1 nM) [13]. High antiviral potency of MPI8 was observed in Vero E6 cells (EC_{50} = 30 nM), and its pronounced selectivity for CatL effectively reduced the potential toxicity to host cells.

Peptide-like inhibitors SM141 (**7**) and SM142 (**8**) (Figure 26.6a) bind to the catalytic site of M^pro^ in a typical mode like other peptidic inhibitors [14]. The unique acrylate warhead, however, forms a thioether bond with Cys145, while the γ-lactam glutamine mimetic occupies the S1 pocket of the substrate-binding site (Figure 26.6b). Interestingly, the corresponding EC_{50} values (8.2 and 14.7 nM)

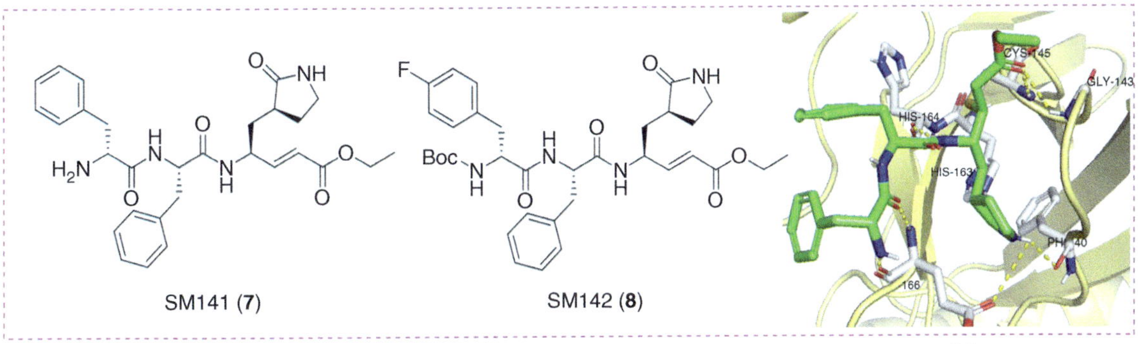

(a) (b)

Figure 26.6 (a) Chemical structure of SM141 (**7**) and SM142 (**8**). (b) Co-crystal and interactions between SM141 with SARS-CoV-2 M^pro^ (PDB ID: 7MBO).

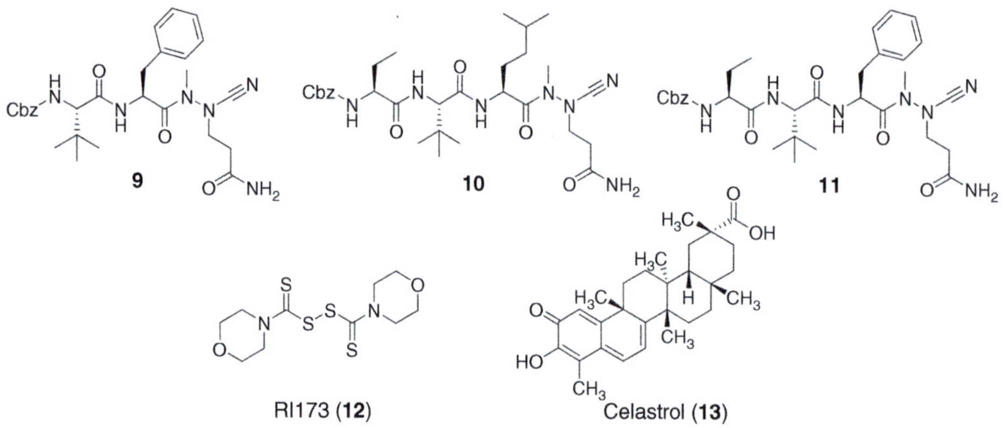

Figure 26.7 Chemical structures of compounds **9**, **10**, **11**, RI173 (**12**) and Celastrol (**13**).

are more than 100-fold lower than their IC_{50} values (0.9 and 1.8 μM) of M^{pro} inhibition, suggesting potential targeting to additional sites in A549-hACE2 cells beyond M^{pro}. Subsequent studies have confirmed the inhibition of SM141 and SM142 on CatL. Therefore, treatment with SM141 and SM142 can block both viral replication and entry into host cells, leading to a significant reduction in SARS-CoV-2 infection. In K18-ACE2 transgenic mice infected with SARS-CoV-2, intranasal and intraperitoneal administration of them resulted in reduced viral replication, decreased lung viral load, and improved survival.

With an improved high-throughput screening system, Müller's team successfully identified several M^{pro} inhibitors, including a class of azapeptide nitriles [15]. The cyano group acted as a covalent warhead targeting M^{pro} Cys145. To enhance the non-covalent interaction with M^{pro}, the researchers assembled glutamine sidechains to these azapeptides. The representative compounds **9**, **10**, and **11** (Figure 26.7) demonstrated strong inhibitory effects on SARS-CoV-2 M^{pro}, along with effective anti-cathepsin L activity, exhibiting K_i values of 53 nM, 22 nM, and 120 nM respectively. Therefore, these cysteine inhibitors with nitronitrile warheads can effectively target various stages of virus replication.

Moreover, disulfiram exhibits covalent binding affinity towards both M^{pro} and PL^{pro} [16]. Ittipats' group identified a range of thiuram disulfide or dithiobis-(thioformate) analogs which target SARS-CoV-2 M^{pro}, PL^{pro}, and human CatL [17]. Notably, the inhibitory effectiveness of these compounds has been improved compared to disulfiram. The most remarkable compound RI173 (**12**, Figure 26.7) exhibited potent antiviral activity without exerting cytotoxic effects on host cells, whereas disulfiram and other compounds within the series showed significant toxicity at antiviral concentrations. In addition to direct inhibition of these cysteine proteases, RI173 also attenuates the release of inflammatory cytokines by inhibiting PDK1, gasdermin D, and phosphoglycerate dehydrogenase, thereby mitigating acute lung injury and reducing mortality caused by SARS-CoV-2. As a promising candidate for the treatment of SARS-CoV-2 or its related variants, RI173 warrants further development.

Celastrol (**13**, Figure 26.7) was isolated from *Tripterygium wilfordii*. In 2022, Carlos' team further demonstrated the inhibitory effects of this pentacyclic triterpene on three cysteine proteases essential for the entry and replication of SARS-CoV-2 (PL^{pro}, M^{pro}, and CatL) [18]. Time-dependent inhibition experiments have shown that celastrol is an irreversible inhibitor, effectively suppressing virus proliferation *in vitro* (EC_{50} = 0.22 μM, Vero cells) and attenuating IL-6 secretion in infected Caco-2 cells at concentrations below 1 μM.

26.1.1.2 Targeting SARS-CoV-2 M^{pro} and Spike Protein-ACE2 Interactions

The interaction of SARS-CoV-2 spike protein with human cell surface receptor angiotensin-converting enzyme 2 (ACE2) is a pivotal stage for viral entry. Targeting protein-protein interactions (PPIs) to develop inhibitors represents an effective therapeutic strategy for blocking viral infection.

Filomena's group synthesized a constrained dipeptide isoscere (**14**, Figure 26.8) which inserted into the short peptide epitope EDLFYQ of ACE2 α1 helix domain to disrupt the ACE2/spike interaction at a micromolar level (IC_{50} = 20 ± 5 μM) [19]. Considering the presence of a glutamine in both the epitope sequence EDLFYQ of ACE2 and the substrate recognizing motif of M^{pro}, compound **14** has been confirmed to exhibit a certain inhibitory effect on M^{pro} in the FRET assay (IC_{50} = 127 ± 76 μM).

Li and colleagues synthesized a series of biaryl amide compounds and tested their activities against SARS-CoV-2 [20]. Representative compound **15** (Figure 26.8) showed broad-spectrum activity against both α-coronavirus

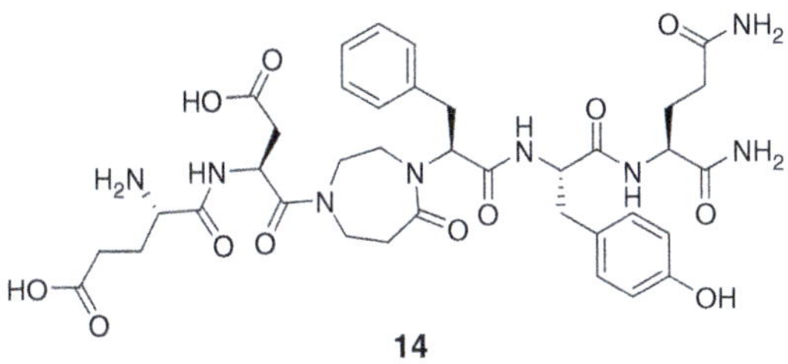

14

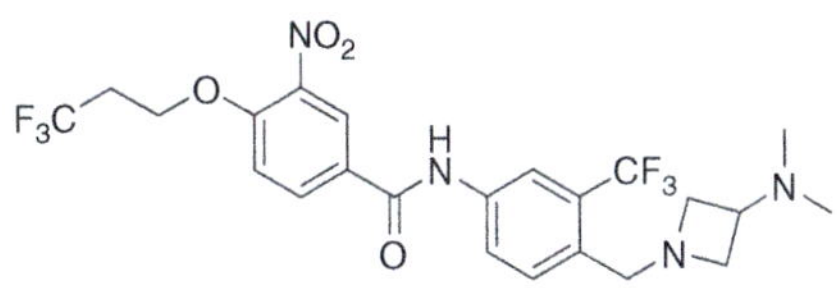

15

Figure 26.8 Chemical structures of compounds **14** and **15**.

(HCoV-229E) and β-coronavirus with low toxicity, demonstrating an EC_{50} value of 2.13 μM. **15** showed a high predicted affinity to human TMPRSS2, owing to its structural resemblance to the classical TMPRSS2 inhibitor camostat. SPR and enzyme inhibition assays confirmed its modest affinity. In subsequent investigations, **15** also had a certain affinity towards M^{pro} in SPR and effectively inhibited M^{pro} activity at an IC_{50} of 38.27 μM. This is a promising lead for multitarget therapy against SARS-CoV-2, but further optimization is required.

26.1.1.3 Targeting SARS-CoV-2 M^{pro} and Host Protein Neuropilin-1 (NRP1)

Neuropilin-1 (NRP1) is a relatively conserved host protein receptor which mediates the entry of SARS-CoV-2 into host cells for replication [21]. The b1 domain of NRP1 interacts with and is activated by the arginine-rich region (amino acid sequence RRAR) located at the C-terminus of the SARS-CoV-2 S1 protein, thereby promoting viral invasion and infection. The expression of the NRP1 gene was found to be upregulated in respiratory and olfactory epithelial cells of COVID-19 patients, which is closely related to adverse pulmonary and neurological manifestations. Therefore, blocking NRP1 represents a therapeutic approach to reduce SARS-CoV-2 infection and multisystemic diseases [22]. However, currently available NRP1 inhibitors have limited antiviral effects against COVID-19.

Yin et al. identified four cyclic peptides (**16-19**) that non-covalently target both NRP1 and M^{pro} using virtual screening step by step (Table 26.1) [23]. Among them, MN-2 (**17**) showed highly potent affinity towards NRP1 ($K_d = 12.3 \pm 1.2$ nM) and M^{pro} ($K_d = 18.2 \pm 1.9$ nM), demonstrating a remarkable enhancement of approximately 74-fold and 3,478-fold stronger than the positive single target inhibitors EG3287 ($K_{d\,(NRP1)}$ = 912.4 ± 10.6 nM) and Peptide-21 ($K_{d\,(M^{pro})}$ = 63.3 ± 4.8 μM). Furthermore, MN-2 exhibited remarkable antiviral activity against the Omicron BA.2.75 variant without inducing noticeable cytotoxic effects.

26.1.1.4 Targeting SARS-CoV-2 M^{pro} and Human Phospholipase

Pathobiological biology studies have revealed a close association between human lipid metabolism enzymes and the severity of SARS-CoV-2 infection [24], with phospholipase A2 (PLA2) playing a pivotal role in the inflammatory response [25]. Its potent inhibitor, GK241 (**20**, Figure 26.9), is effective in the treatment of inflammatory diseases. The research conducted by Maria's group in 2022 has confirmed that the 2-oxamide inhibitor GK241 exhibits a moderate inhibitory effect on M^{pro} with an IC_{50} value of 24 μM [26]. Overall, it is considered a potential dual-target agent against PLA2 and SARS-CoV-2 M^{pro} for the treatment of SARS-CoV-2, reducing infection and inflammation reaction caused by COVID-19.

Table 26.1 Sequences and bioactivities of MN-1, MN-2, MN-3, and MN-4.

Peptide	Sequence	M^{pro}, K_d [nM]	NRP1, K_d [nM]
MN-1	(CF)^{D}LQFAVLRHRRKQ	53.4	95.6
(**16**)	(C)^{D}GGGRPRRGR		
MN-2	(CF)^{D}LQYAVLKQKRRE	18.2	12.3
(**17**)	(C)^{D}GGGRPPRRAR		
MN-3	(CI)^{D}LQLAVFKKRQRK	205.5	110.4
(**18**)	(C)^{D}GGGRGERPPR		
MN-4	(CW)^{D}LQFAVLREKKQA	116.2	15.1
(**19**)	(C)^{D}GGRPARPAR		

26.1.1.5 Targeting SARS-CoV-2 M^{pro} and Human Inflammatory Factors

The interaction of Licorice-saponin A3 (**21**, Figure 26.10) with multiple pathways related to COVID-19 has drawn great interest from researchers [27]. A3 showed high affinity with the spike RBD region of SARS-CoV-2 omicron variants (RBD IC_{50} = 4.36–22.98 μM) and potent antiviral

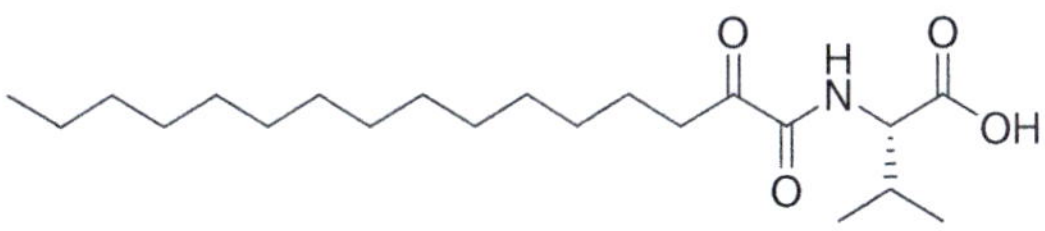

GK241 (**20**)

Figure 26.9 Chemical structure of compound GK241 (**20**).

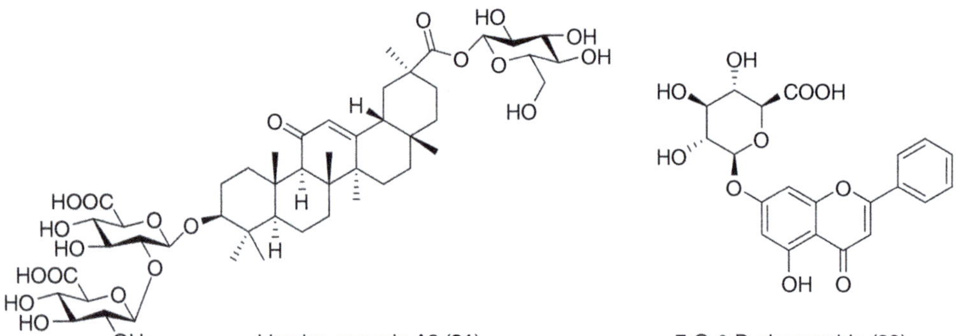

Figure 26.10 Chemical structure of compound licorice-saponin A3 (**21**) and 7-O-β-D-glucuronide (**22**).

activity (Omicron BA.2 EC_{50} = 1.02 μM). The inhibition of RBD is predominantly attributed to A3's interaction with conserved Y453 residue. In the presence of A3, host inflammation is additionally suppressed through modulation of the JNK and p38 MAPK pathways. Application of A3 could synergistically enhance antivirus and anti-inflammation effects, thereby offering undeniable clinical benefits for COVID-19 treatment.

Similarly, the natural product chrysin 7-O-β-D-glucuronide (**22**, Figure 26.10) was also reported to inhibit multiple targets of SARS-CoV-2 [28]. A total of 25 flavonoids derived from *Scutellaria baicalensis Georgi* were screened by molecular docking and enzymatic activity tests against SARS-CoV-2 M^{pro}/PL^{pro}. Only chrysin 7-O-β-D-glucuronide had significant activity towards both proteases (SARS-CoV-2 M^{pro} IC_{50} = 1.82 μM, SARS-CoV-2 PL^{pro} IC_{50} = 2.54 μM), which also inhibited SARS-CoV-2 on Vero E6 cells with an EC_{50} value of 8.72 μM. On the other hand, chrysin 7-O-β-D-glucuronide reduced the levels of interleukin (IL)-1β and IL-6, therefore demonstrating its anti-inflammatory effects.

26.1.2 Multi-Target Inhibitors of Host Cell for Virus Entry

SARS-CoV-2 has been demonstrated to exploit alternative pathways for host cell invasion. The preference for entry pathways varies among viral variants and host cell types, necessitating the simultaneous blockade of both distinct routes [29]. To achieve this object, a bifunctional compound **25** merging TMPRSS2 inhibitor nafamostat (**23**) and cathepsin L/B inhibitor K777 (**24**) was designed and synthesized (Figure 26.11) [30]. In accordance with design theory, **25** potently inhibited TMPRSS2, CatB, and CatL, meanwhile showing 64 nM EC_{50} value towards SARS-CoV-2 Omicron BA.2.

The design of peptide-like covalent inhibitors targeting both CatL and calpain I was based on shared characteristics

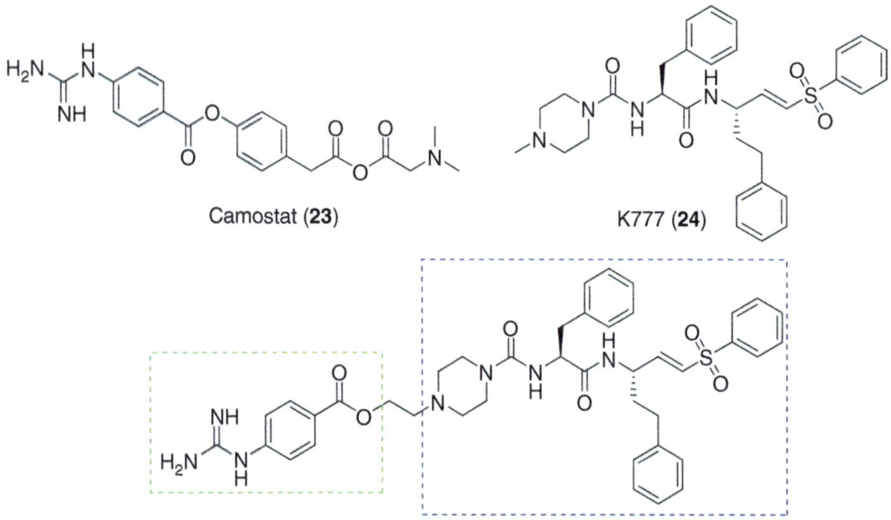

Figure 26.11 Chemical structures of camostat (**23**), K777 (**24**) and the bifunctional compound **25**.

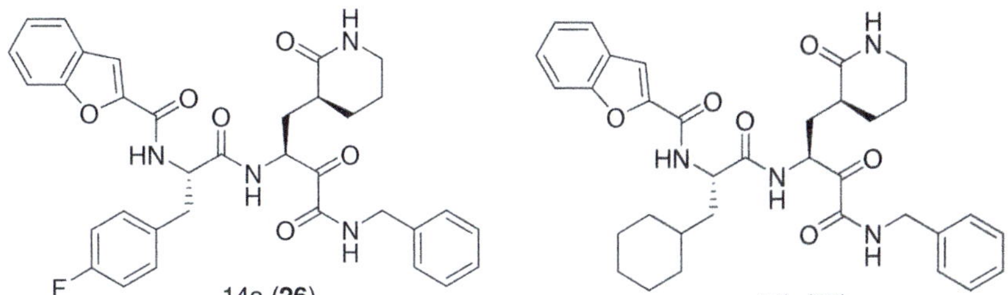
14a (**26**) 14b (**27**)

Figure 26.12 The structures of 14a (**26**) and 14b (**27**).

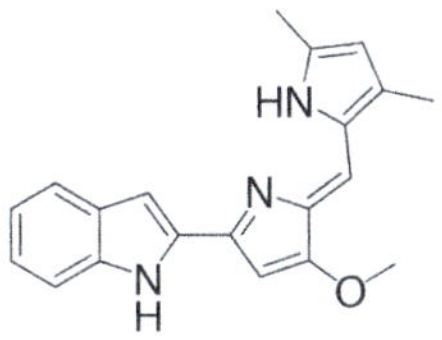
Obatoclax (**28**)

Figure 26.13 Chemical structure of obatoclax (**28**).

of these two host proteases, aiming to effectively block virus invasion and alleviate cellular inflammatory response. Compounds 14a (**26**) and 14b (**27**) (Figure 26.12) inhibited CatL with IC_{50} values of 3.34 nM and 6.88 nM, and also had activity against CatL with IC_{50} values of 375.1 nM and 347.6 nM, respectively [31]. In cellular systems, 14a (EC_{50} = 51.7 nM) and 14b (EC_{50} < 40 nM) inhibited SARS-CoV-2 wild type strains at nanomolar level. Time-of-addition experiments confirmed that 14a and 14b inhibited SARS-CoV-2 specifically by blocking virus entry. Both compounds effectively reduced viral replication in K18-hACE2 mice model. Although being closely resemble to typical peptide M^{pro} inhibitors in structure, 14a and 14b only had moderate inhibition of SARS-CoV-2 M^{pro}, proving their antiviral activity was mainly contributed by CatL and calpain I inhibition.

The Bcl-2 inhibitor obatoclax also impeded SARS-CoV-2 entry through multiple targets. Obatoclax (**28**, Figure 26.13) was proved to disrupt the endocytosis of SARS-CoV-2 by interfering endosomal acidification [32]. Obatoclax at 330 nM concentration is able to inhibit over 90% pseudo-SARS-CoV-2 entry in Calu-3-hACE2, Calu-3-hACE2, and A549-hACE2 cell lines. Furthermore, treatment with obatoclax in Calu3-hACE2 and A549-hACE2 cells significantly decreased the expression of mature CTSL.

26.2 Multi-Target Inhibitors Directly Targeting SARS-CoV-2

26.2.1 Multiple-Targeting Inhibitors Towards SARS-CoV-2 RdRp

Nucleoside analogues (NA) are extensively investigated for their broad biological activities, and are classical virus RdRp (RNA dependent RNA polymerase) inhibitors. Some of the widely-known nucleosides have been extensively studied and developed in SARS-CoV-2 treatment [33].

The nsp12 protein of coronaviruses does not only function as RdRp [34], but also bears a Nidovirus RdRp-Associated Nucleotidyltransferase (NiRAN) domain [35]. It has been proven to facilitate the covalent transfer of nucleoside monophosphate (NMP) to viral cofactor proteins nsp7, nsp8, and nsp9. Furthermore, NiRAN has also been proposed to get involved in the viral RNA capping as a guanyltransferase.

The nucleoside analog, AT-527 (**29**), exhibited an EC_{90} value of 0.47 μM against SARS-CoV-2 in human airway epithelial (HAE) cells [36]. This compound exhibited favorable DMPK file and *in vivo* therapeutic effect, having previously progressed phase III clinical trials for the treatment of COVID-19 [37]. As a prodrug, AT-527 is converted into its corresponding triphosphate AT-9010 (**30**, Figure 26.14a) after entering host cells. AT-9010 effectively binds to the substrate entry site of SARS CoV-2 RdRp and causes immediate chain termination. Ashleigh reported the complex structure between SARS-CoV-2 nsp12-nsp7-nsp8-RNA, and AT-9010, suggests that this compound exhibits dual inhibitory mechanisms by binding to two distinct functional sites on nsp12. [38]. Importantly, AT-9010 is incorporated at the 3' end of RNA product strand within the RdRp active site. The presence of its modified ribose group hinders proper alignment of the incoming NTP, leading to immediate termination of the RNA synthesis (Figure 26.12b). For another aspect, AT-9010 is also observed to bind to the NiRAN site (Figure 26.12c) of nsp12, which is located in close proximity to the N-terminal domain. Due to its flipping direction, the guanine base of AT-9010 occupies a previously unnoticed cavity. AT-9010 demonstrates higher affinity towards the NiRAN site compared with any other natural nucleotides, effectively inhibiting nucleotide transferase activity.

Favipiravir (**31**) has a broad-spectrum inhibition against RNA viruses. While it has demonstrated effective suppression of virus replication in hamster models during the SARS-CoV-2 outbreak [39, 40], its efficacy appears to be comparatively lower in non-human primate and human subjects [41–43]. This molecule is converted

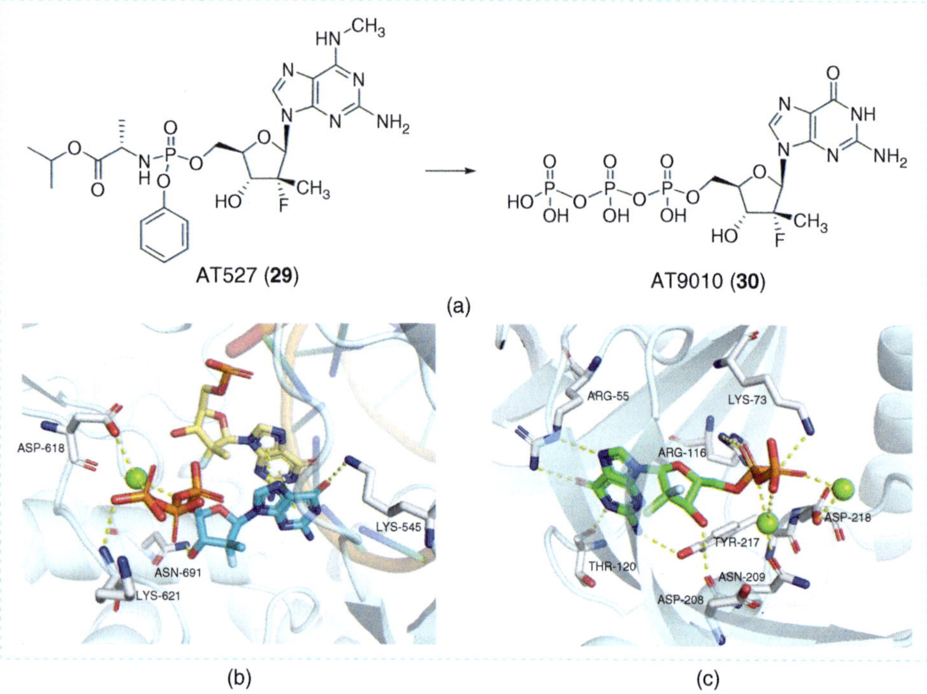

Figure 26.14 (a) Structure of AT-527 (**29**) and its active metabolite AT-9010 (**30**); (b) Binding mode of AT-9010 in SARS-CoV-2 nsp12 RdRp catalytic center; (c) Binding mode of AT-9010 in SARS-CoV-2 nsp12 NiRAN site (PDB ID: 7ED5).

to ribonucleoside form (**32**) and could be effectively recognized by virus polymerase. Favipiravir ribonucleoside inhibited viral RNA replication and nucleotide incorporation at 10 μM [44], albeit displaying relatively weak antiviral activity in vero cells (EC_{50} = 118.3 μM, CC_{50} > 500 μM, CPE assay) [45]. Structural biology studies have indicated that favipiravir can be incorporated into elongating RNA chains of SARS-CoV-2 genome. In subsequent replication, favipiravir could mimic either A or G by flexibly rotating its carboxamide bond and pairing with different base groups, thereby functioning as a mutagenic nucleoside (Figure 26.15) [46]. The mutagenic rate of SARS-CoV-2 was found to be increased threefold by favipiravir treatment, as demonstrated in another study, through the induction of G-A and C-U conversions [45]. These above-limit mutations in the genome are detrimental for virus replication and survival. Collectively, favipiravir has multifaceted inhibitory effect on coronavirus RNA synthesis; however, its therapeutic efficacy is limited, rendering it unsuitable as a first-line candidate drug.

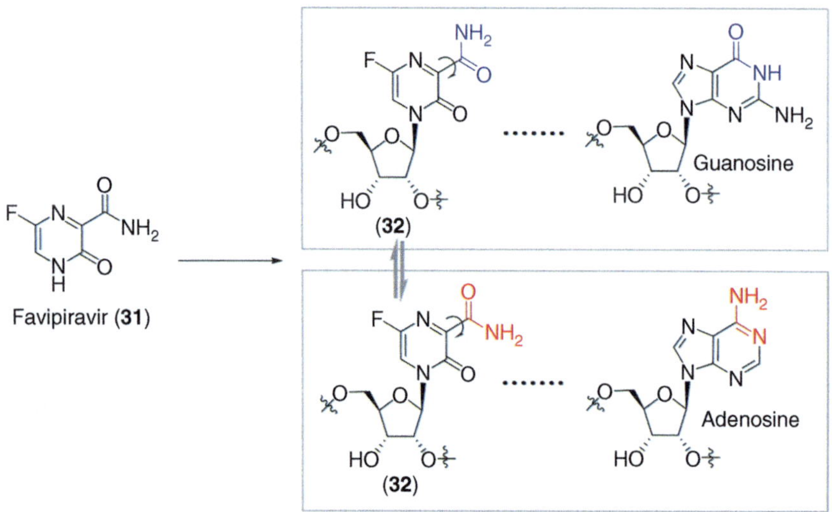

Figure 26.15 The structures of favipiravir (**31**), favipiravir nucleoside (**32**) and its mutagenic mechanism.

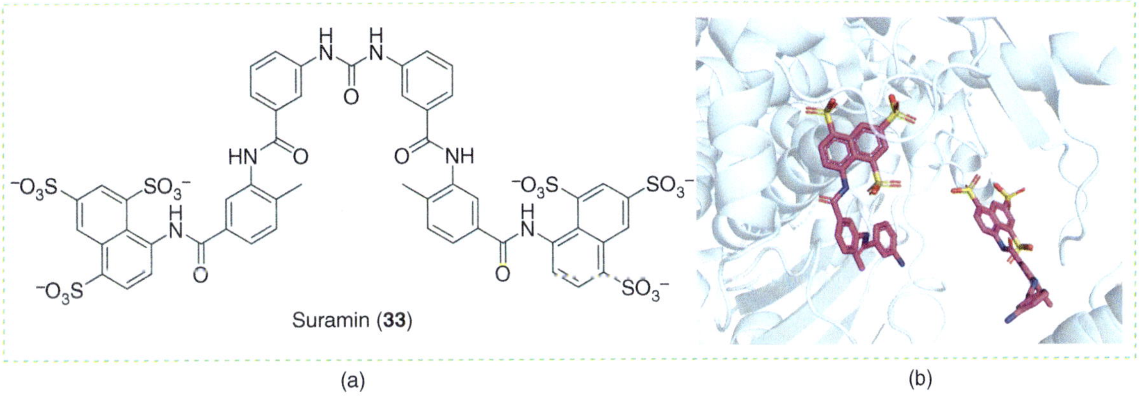

Figure 26.16 (a) The structure of suramin (**33**). (b) Binding mode of a part of suramin molecules at SARS-CoV-2 nsp12 catalytic center (PDB ID: 7D4F).

Suramin (**33**, Figure 26.16a) is a century-old drug molecule with diverse applications, including parasitic disease, viral infection, and cancer [47]. Salgado-Benvindo *et al.* first demonstrated the anti-SARS-CoV-2 potency of suramin in cell culture. Its remarkable inhibition of RdRp replication activity (IC_{50} = 0.26 μM) was then reported [48]. Co-crystal structure suggested that two suramin molecules inserted at different sites of nsp12, both contacting with conserved amnio acids critical for RdRp function (Figure 26.16b) [49]. In M^{pro} inhibition screening of approved drugs, suramin showed moderate activity towards M^{pro} in a non-competitive manner (IC_{50} = 6.3–6.5 μM) [50]. Further studies conducted by Kwon et al. revealed that suramin could also act on the spike protein of SARS-CoV-2. In the SPR assay, suramin binds to the RBD of the SARS-CoV-2 spike with an K_d value of 151 nM [51]. Given its approved status as a drug with an established safety profile and extensive clinical application history, suramin is a promising clinical candidate in the treatment of COVID-19.

SARS-CoV-2 nsp14 and nsp16 play crucial roles as methyltransferases in the processing of viral RNA. They are responsible for adding methyl groups to form functional cap structures in RNA [52]. Nsp14 also involves in RNA proofreading and the excision process to prove replication fidelity, while nsp10 acts as a cofactor for both nsp14 and nsp16. Through analyzing protein-protein interactions between nsp10 and nsp14/nsp16, a series of peptides were designed to inhibit the two functional proteins by disrupting nsp10 interactions between nsp14/nsp16 [53]. Yet their effects have only been studied *in sillico*.

Besides this example, small-molecule inhibitors also demonstrated clear nsp14/nsp16 dual inhibition activity. From a library of mimics of MTase inhibitor SAH (S-adenosylhomocysteine), potent, selective, and broad-spectrum nsp14 inhibitors were identified through radiometric assay: SS148 (**34**) with an IC_{50} value of 0.07 μM and WZ16 (**35**) with an IC_{50} value of 0.19 μM (Figure 26.17). [54]. A range of identified active nucleoside analogs targeting the nsp14 N7-methyltransferase domain were repurposed for inhibiting nsp16 2'-O-methyltransferase activity. Interestingly, SS148 and WZ16 also inhibited nsp16 with IC_{50} values of 1.2 μM and 3.4 μM, respectively. These two compounds competitively bind to the active site of nsp16, blocking its methylation function by competing with the natural substrate S-Adenosyl-L-Methionine (SAM). Interestingly, under increased concentrations of RNA, WZ16 showed decreased IC_{50} values, indicating its RNA-dependent inhibition mechanism. In contrast, IC_{50} of SS148 remained unaffected by RNA concentration [55]. Crystal studies of both compounds complexed with nsp16-nsp10 provided further possibilities of designing

Figure 26.17 The structures of SS148 (**34**) and WZ16 (**35**).

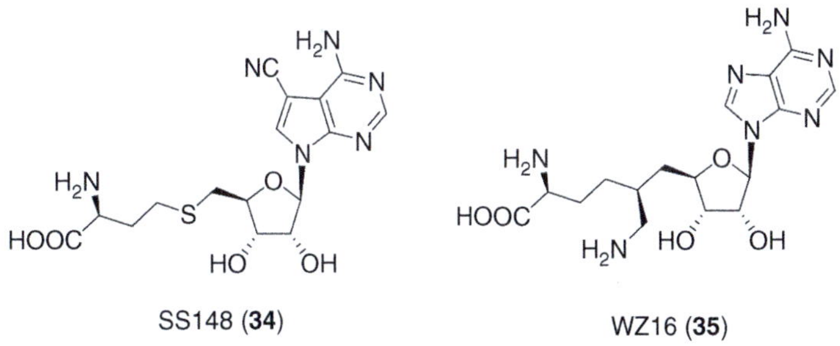

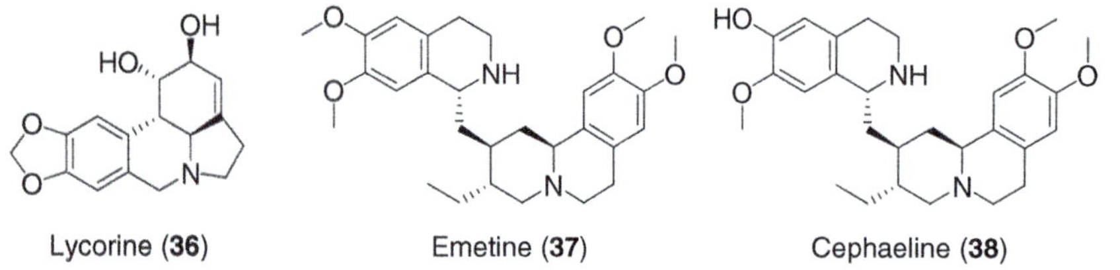

Figure 26.18 The structures of lycorine (**36**), emetine (**37**), and cephaeline (**38**).

future nsp16/nsp14 inhibitors based on their structural characteristics. [56].

To effectively inhibit key proteins related to SARS-CoV-2 RNA replication, Ren et al. conducted a multi-targeting virtual screening through the DrugBank and MedChemExpress databases [57]. The host ribosome, SARS-CoV-2 RNA, RdRp, and nucleoprotein were chosen as protein models for the docking study. The alkaloid compounds, namely lycorine (**36**), emetine (**37**), and cephaeline (**38**) were identified in this study (Figure 26.18). These compounds simultaneously interact with the four targets above, as evidenced by SPR assays and docking simulations. Antiviral activity and toxicity assays were performed in Vero E6 cells to evaluate the efficacy of lycorine (EC_{50} = 0.44 μM, CC_{50} > 1000 μM), emetine (EC_{50} = 0.0077 μM, CC_{50} = 2.17 μM), and cephaeline (EC_{50} = 0.012 μM, CC_{50} = 49.05 μM) indicated that they have therapeutic potential against SARS-CoV-2 infection.

26.2.2 Multiple-Targeting Inhibitors Towards SARS-CoV-2 Proteases

As a representative compound with benzo-selenazolone scaffold, ebselen (**39**) has been found out to inhibit both M^{pro} (IC_{50} = 0.67 μM) and PL^{pro} of SARS-CoV-2 (IC_{50} = 2.02 μM) [58–60]. An irreversible, covalent mechanism between ebselen and M^{pro} Cys145 was confirmed by co-crystal study. The situation with PL^{pro} is more intricate: while ebselen may form a covalent bond with Cys111 of PL^{pro}, zinc injection into PL^{pro} by ebselen was also observed through fluorescence assay and mass spectrometry [61]. In a screening conducted by Hannah's group, the inhibitory effect of ebselen on the nuclease activity of nsp14-nsp10 (IC_{50} = 3.3 μM) was observed. The researchers postulated that ebselen exerts its inhibitory action on Nsp14-nsp10 through zinc ejection, as zinc finger motifs are essential for enzymatic functions of the complex [62].

Modifications based on ebselen scaffold were subsequently pursued to discover more potent analogues. Multiple series of ebselen derivatives were synthesized and evaluated, resulting in some compounds exhibiting enhanced protease inhibition potency and antiviral effects. In 2023, Mikolaj's team screened 34 ebselen derivatives for SARS-CoV-2 PL^{pro} and M^{pro} inhibitors [63]. Representative compounds **40** (M^{pro} IC_{50} = 0.05 μM, PL^{pro} IC_{50} = 0.58 μM) and **41** (M^{pro} IC_{50} = 0.015 μM, PL^{pro} IC_{50} = 2.07 μM) (Figure 26.19) exhibit potent inhibitory effects on both proteases, surpassing the activity of ebselen. Additionally, they demonstrate inhibition of N7-methyltransferase activity related to SARS-CoV-2 nsp14 protein involved in viral RNA cap modification (**40** IC_{50} = 0.35 μM, **41** IC_{50} = 3.38 μM). Similar ebselen/ebsulfur derivatives 6c (**42**) / 7e (**43**) (Figure 26.19) were also reported by the Wandong group [64]. With additional groups on the N-phenyl ring, the two compounds **42** (M^{pro} 78.2% inhibition at 0.5 μM, PL^{pro} 43% inhibition at 0.05 μM) and **43** (M^{pro} IC_{50} = 0.44 μM, PL^{pro} IC_{50} = 0.063 μM) had improved enzyme inhibition activity. Notably, these compounds exhibited measurable antiviral activity in cellular models. However, their *in vivo* DMPK evaluation and assessment of antiviral activity in animal models are yet to be conducted.

The overall findings suggest that ebselen derivatives have a great potential to function as multi-target inhibitors of

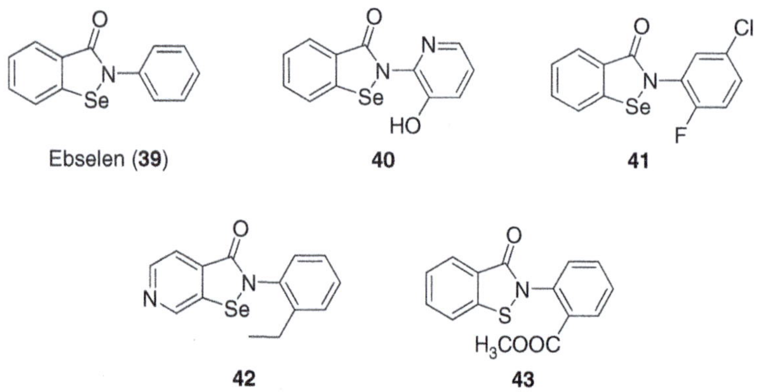

Figure 26.19 The structures of ebselen (**39**) and its derivatives **40**–**43**.

SARS-CoV-2. Importantly, phenyl group modification is an effective way to optimize the selectivity of these compounds towards M^{pro}, PL^{pro}, and nsp14.

Lai's team developed a novel series of dual-target inhibitors using a structure-based co-pharmacophore screening method [65]. With quinone-like warhead, representative compound LY-1 (**44**) could bind to catalytic cysteine residues of both SARS-CoV-2 M^{pro} and PL^{pro}. LY1-O (43), the hydrolyzed open-ring form of LY1, reaches chemical equilibration with LY1 (Figure 26.18a). Both molecules can bind to the active center of M^{pro}/PL^{pro} by Michael addition with cysteine. The 2-(piperazin-1-yl)aniline moiety of LY-1 could be flexibly reoriented to fit into the active center of both proteases (Figure 26.20b). Enzyme kinetics experiments revealed a 710 $M^{-1}\,s^{-1}$ of k_{inact}/K_i value between LY1 and M^{pro}, and 500 $M^{-1}\,s^{-1}$ between LY1 and PL^{pro}. LY-1 exhibited an EC_{50} value of 3.9 μM, demonstrating high specificity towards target enzymes and low toxicity in animal models.

Similarly, C10 (**46**), a non-covalent M^{pro} inhibitor based on 2-chloroquinoline and designed initially by considering common features of catalytic site structure [66], was further modified to incorporate an imidine moiety into azetidinone, resulting in the development of dual inhibitors of both M^{pro} (Ki = 0.82 μM) and PL^{pro} (K_i = 0.35 μM). Native mass confirmed C11 (**47**) binded to M^{pro} non-covalently. Moreover, this compound showed low toxicity across multiple cell lines (CC_{50} = 195.79 μM, Vero E6).

Guided by structural biology information, a class of indole-based compounds was developed through *in silico* design, followed by antiviral assay screening and additional *in vitro* enzymatic assays [67]. Compound **48** was identified as a covalent inhibitor of M^{pro} (IC_{50} = 1.72 μM) and PL^{pro} (IC_{50} = 0.67 μM). The compound also demonstrated significant antiviral activity in Vero E6 cells (EC_{50} = 0.44 μM), positioning it as a promising lead candidate for M^{pro} inhibitor discovery. However, the introduction of a chloro-acetamide warhead resulted in ineligible cytotoxicity (CC_{50} = 38.67 μM, Vero E6).

Large-scale screening enables the identification of inhibitors that target multiple proteases of SARS-CoV-2. Originally purposed for discovering M^{pro} inhibitors, Santos *et al.* virtually screened a unique naphthoquinone library, and one of the top candidates, compound **49**, inhibited M^{pro} at an IC_{50} of 1.9 μM [68]. Surprisingly, in subsequent PL^{pro} activity screening, this compound also exhibited potent inhibition of PL^{pro} at a low-micromolar level (IC_{50} = 3.1 μM). Dilution experiment confirmed **49** as an irreversible covalent inhibitor however, further investigation is required to elucidate its underlying mechanism. In another attempt, a high-throughput screening of 2560 compounds was conducted to evaluate their activities against both M^{pro} and PL^{pro}. Cetylpyridinium chloride (**50**) exhibited micromolar-level IC_{50} values against both M^{pro} and PL^{pro} [69]. Considering its significant inhibition of SARS-CoV-2 (EC_{50} = 0.62 μM, Caco-2) and MERS-CoV (EC_{50} = 0.69 μM, BHK-21) [70, 71], the addition of **50** into mouthwash and nasal sprays may be helpful to prevent coronavirus infection. Zang *et al.* conducted a high-throughput assay of M^{pro} and PL^{pro} on 1.8 million compounds. Among the active candidates, 3a (**51**, M^{pro} IC_{50} = 4.37 μM, PL^{pro} IC_{50} = 102.5 μM), 3b (**52**, M^{pro} IC_{50} = 1.53 μM, PL^{pro} IC_{50} = 7.3 μM) and 3c (**53**, M^{pro} IC_{50} = 9.67 μM, PL^{pro} IC_{50} = 43.9 μM) showed moderate inhibition of the both proteases [72].

Ayoup *et al.* incorporated the 1,2,4-oxadiazole structure into validated PLpro inhibitors as a replacement for their conventional scaffolds and acquired several dual inhibitors towards SARS-CoV-2 PL^{pro}. Compound **54** exhibited a well-balanced inhibitory potential (PL^{pro} IC_{50} = 7.20 μM, spike RBD IC_{50} = 8.67 μM) and favorable safety profile [73] (Figure 26.21).

As SARS-CoV-2 M^{pro} inhibitors with significant *in vitro* antiviral activities, baicalein (**55**, IC_{50} = 0.94 μM, EC_{50} = 2.94 μM, CC_{50} > 200 μM) and its glycoside baicalin (**56**, IC_{50} = 6.41 μM, EC_{50} = 27.87 μM, CC_{50} > 200 μM) were initially identified from Chinese medicine preparation (Figure 26.22a). Co-crystal structure revealed baicalein's

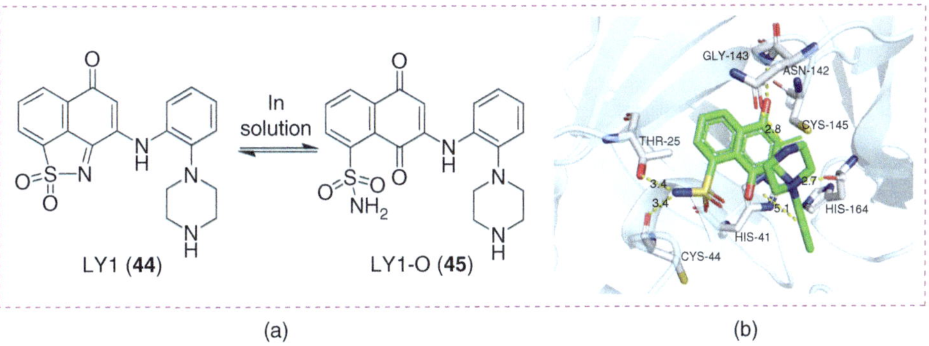

Figure 26.20 (a) The structure of LY1 (**44**) and LY1-O (**45**). (b) Covalent binding mode of LY-1 with M^{pro} (PDB ID: 7V1T).

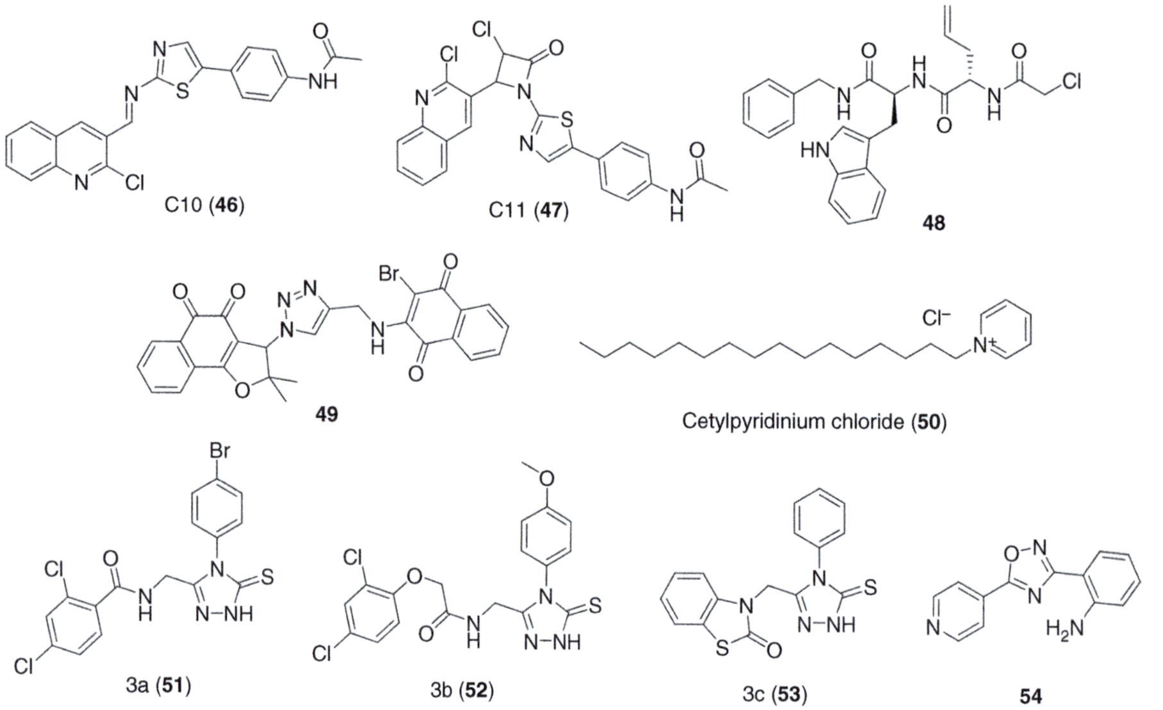

Figure 26.21 The structures of compounds **46**–**54** [66–73].

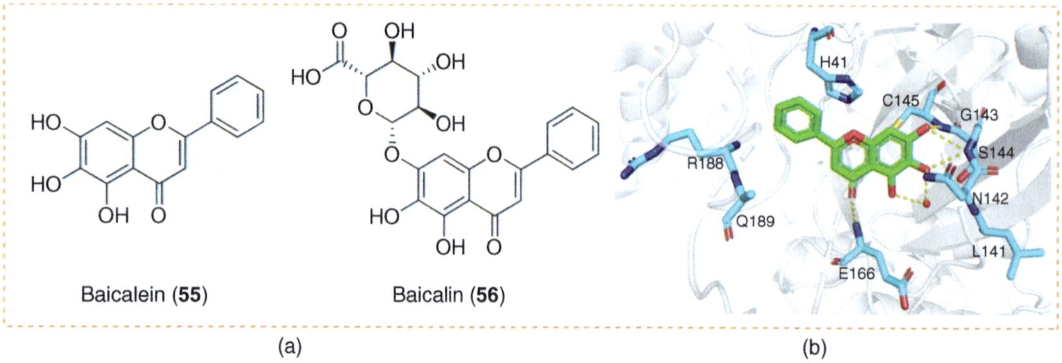

Figure 26.22 (a) The structures of baicalein (**55**) and baicalin (**56**); (b) Binding mode of baicalein in M^{pro} active center (PDB ID: 6M2N).

non-covalent binding mode at the active center of M^{pro} (Figure 26.22b). Multiple interactions with key residues of M^{pro} confer it with a high affinity [74]. Furthermore, baicalein and baicalin have been demonstrated to significantly inhibit SARS-CoV-2 RdRp replication activity at concentrations > 2.5 μM in biochemical and cellular experiments proving them as potent RdRp inhibitors [75]. High activity and wide source of baicalein render it a potent candidate in the early stages of M^{pro} inhibitor discovery.

Discovery of dual-target SARS-CoV-2 inhibitors is greatly facilitated by computational simulations. Murugan *et al.* conducted a screening of the drugbank database to identify possibly active inhibitors, towards SARS-CoV-2 M^{pro}, PL^{pro}, and RdRp simultaneously. Based on the screening results, a few approved drugs, including Tadalafil (**57**) and Lonafarnib (**58**) (Figure 26.23), exhibited potential multi-target inhibition towards the aforementioned three targets [76].

26.2.3 Multi-Targeting SARS-CoV-2 Cell Entry Inhibitors Targeting Spike Proteins

Clofazimine (**59**, Figure 26.24) has demonstrated broad-spectrum inhibitory activity against coronaviruses, as evidenced by viral-load reduction assays. It exhibits potential inhibitory effects on SARS-CoV-2 (EC_{50} = 0.31 μM) and MERS-CoV (EC_{50} = 1.48 μM), indicating its ability to effectively inhibit the replication of these

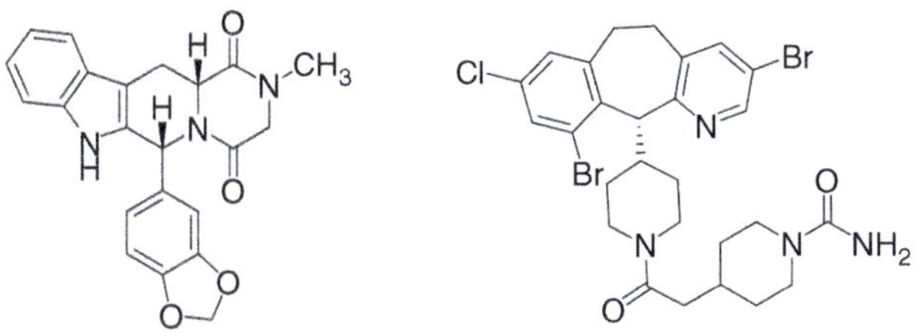

Tadalafil (**57**) Lonafarnib (**58**)

Figure 26.23 The structures of Tadalafil (**57**) and Lonafarnib (**58**).

viruses [77]. Cell membrane fusion assays characterized that clofazimine can reduce the fusion between virus spike protein and target cells, implied its inhibition of the viral invasion into host cells [78]. Researchers also found that clofazimine inhibited the unwinding activity of nsp13, in situations which double-stranded DNA or double-stranded RNA is the substrate. Subsequent experiments have proved that clofazimine reduced viral loads in the lung and shed the virus in the feces, while also mitigating inflammation associated with viral infection. Collectively, clofazimine disrupts multiple stages in the replication cycle of SARS-CoV-2, thereby demonstrating potential as a dual-target inhibitor for combating COVID-19.

Inspired by the membrane fusion inhibition activity of clofazimine in SARS-CoV-2 and the rabies virus, 32 clofazimine analogs were prepared to evaluate their antivirus effects [79]. Among them, 15f (**60**, Figure 26.24) is potent against both the rabies virus and pseudo-SARS-CoV-2. 15f gives the EC_{50} value of 14.6 μM and CC_{50} of 88.9 μM. Moreover, it demonstrates a high safety profile with an LD_{50} in mice exceeding 800 mg/kg. In the SPR assay, 15f binds with spike protein with a K_d value of 2.52 μM, which surpasses that of clofazimine (K_d = 3.82 μM). Considering that clofazimine has already shown nsp13 inhibition activity [77], docking results suggest that 15f could be a more potent nsp13 inhibitor acting through multiple mechanisms.

In addition to small-molecule compounds, peptides have been identified as potent dual inhibitors of SARS-CoV-2 cell entry. Through an integrated approach combining virtual screening and molecular dynamics simulations targeting spike RBD and M^{pro}, four peptides MR-1 to MR-4 (**61–64**, Table 26.2) showed IC_{50} values of M^{pro} and binding affinity to spike RBD at nanomolar level. In cell-based assays, these four peptides were found to be non-toxic to 293T/A549 cells up to 50 μM concentration, and inhibited over 50% virus replication under 5 μM in 293T cells [80].

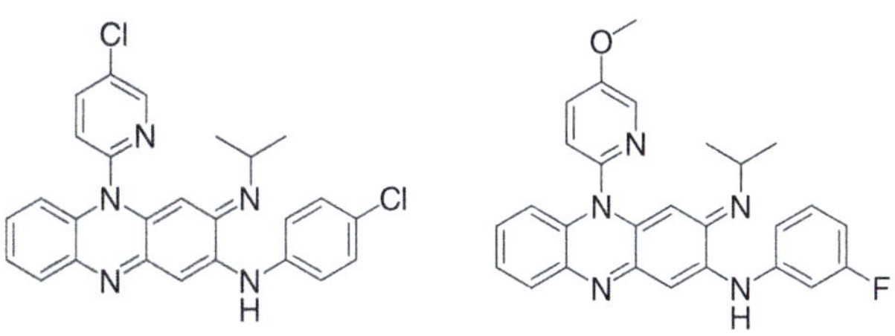

Clofazimine (**59**) 15f (**60**)

Figure 26.24 The structures of clofazimine (**59**) and 15f (**60**).

Table 26.2 Sequences and bioactivities of MR-1, MR-2, MR-3, and MR-4.

Peptide	Sequence	M^{pro}, IC_{50} [nM]	RBD, K_d [nM]
MR-1 (**61**)	(CF)^{D}LQYAVLQAKKRE(C)^{D}GGG DKYNHEAEDLYYQSSLASFN	28.2	14.4
MR-2 (**62**)	(CY)^{D}LQFAVLKHRRRE(C)^{D}GGG QQRTYLDKFNHEAEDLFYQRAL	19.4	39.2
MR-3 (**63**)	(CY)^{D}LQYAVLRHKRKE(C)^{D}GGG ERKTYADKFNHEAEDLYYQSKL	22.1	34.2
MR-4 (**64**)	(CY)^{D}LQLAVLAHRRAE(C)^{D}GGG DKFNHEAEDLLYQKSLAKYD	35.6	27.1

26.3 Summary and Prospect

Evidence on the therapeutic potential of multi-target inhibitors is growing even stronger in the post-pandemic era. Multi-target inhibitors are an attractive therapeutic approach that holds promise for generating novel chemical probes for antiviral treatment [81, 82]. Currently, the predominant strategies employed for the discovery of multi-target SARS-CoV-2 inhibitors encompass drug repurposing, structure-based screening, pharmacophore fusion, and computational approaches. However, a significant challenge persists in developing multi-target SARS-CoV-2 inhibitors, particularly when dealing with structurally diverse targets exhibiting low structural similarity. The current challenge lies in maintaining a delicate balance among the activities of multiple targets to achieve optimal antiviral efficacy. Conversely, it would be intriguing to reassess the targets and mechanisms of action of previous single-targeting inhibitors against SARS-CoV-2, with the aim of identifying potential multi-targeting inhibitors.

Despite significant efforts, most multi-target inhibitors described here have not progressed to pre-clinical studies. Therefore, it is imperative for chemists to optimize these molecules in order to obtain desirable pharmacological properties and ensure safety.

Acknowledgments

We gratefully acknowledge financial support from the Major Basic Research Project of Shandong Provincial Natural Science Foundation (No. ZR2021ZD17), the Guangdong Basic and Applied Basic Research Foundation (No. 2021A1515110740), the Young Scientists Fund of the National Natural Science Foundation of China (No. 22307067), the Key Research and Development Program, Ministry of Science and Technology of the People's Republic of China (Grant No. 2023YFC2606500), the Shandong Laboratory Program (SYS202205). We acknowledge the Korea Health Industry Development Institute (KHIDI; HI22C2067 to M. K.) funded by the Korean government (MHW) and also by an intramural fund from Korea Research Institute of Chemical Technology (KRICT; KK2432-10 to M.K.).

References

1 Choi, J.Y. and Smith, D.M. (2022). SARS-CoV-2 variants of concern. *Yonsei Med. J.* 62: 961–968.

2 Feikin, D.R., Higdon, M.M., Abu-Raddad, L.J. et al. (2022). Duration of effectiveness of vaccines against SARS-CoV-2 infection and COVID-19 disease: results of a systematic review and meta-regression. *Lancet* 399: 924–944.

3 Yang, H. and Rao, Z. (2021). Structural biology of SARS-CoV-2 and implications for therapeutic development. *Nat. Rev. Microbiol.* 19: 685–700.

4 Chakraborty, C., Bhattacharya, M., Mallick, B. et al. (2021). SARS-CoV-2 protein drug targets landscape: a potential pharmacological insight view for the new drug development. *Expert. Rev. Clin. Pharmacol.* 14: 225–238.

5 Tan, B., Joyce, R., Tan, H. et al. (2023). SARS-CoV-2 main protease drug design, assay development, and drug resistance studies. *Acc. Chem. Res.* 56: 157–168.

6 Citarella, A., Scala, A., Piperno, A., and Micale, N. (2021). SARS-CoV-2 M^{pro}: A potential target for peptidomimetics and small-molecule inhibitors. *Biomol. Ther.* 11 (4): 607.

7 Jackson, C.B., Farzan, M., Chen, B., and Choe, H. (2022). Mechanisms of SARS-CoV-2 entry into cells. *Nat. Rev. Mol. Cell Biol.* 23: 3–20.

8 Muralidar, S., Gopal, G., and Visaga Ambi, S. (2021). Targeting the viral-entry facilitators of SARS-CoV-2 as a therapeutic strategy in COVID-19. *J. Med. Virol.* 93: 5260–5276.

9 Xiu, S., Dick, A., Ju, H. et al. (2020). Inhibitors of SARS-CoV-2 entry: current and future opportunities. *J. Med. Chem.* 63: 12256–12274.

10 Ma, C., Sacco, M.D., Hurst, B. et al. (2020). Boceprevir, GC-376, and calpain inhibitors II, XII inhibit SARS-CoV-2 viral replication by targeting the viral main protease. *Cell Res.* 30: 678–692.

11 Mao, L., Shaabani, N., Zhang, X. et al. (2024). Olgotrelvir, a dual inhibitor of SARS-CoV-2 Mpro and cathepsin L, as a standalone antiviral oral intervention candidate for COVID-19. *Med.* 5: 42–61.

12 Costanzi, E., Kuzikov, M., Esposito, F. et al. (2021). Structural and biochemical analysis of the dual inhibition of MG-132 against SARS-CoV-2 main protease (Mpro/3CLpro) and human cathepsin-L. *Int. J. Mol. Sci.* 22: 11779.

13 Ma, X.R., Alugubelli, Y.R., Ma, Y. et al. (2022). MPI8 is potent against SARS-CoV-2 by inhibiting dually and selectively the SARS-CoV-2 main protease and the host cathepsin L. *ChemMedChem* 17: e202100456.

14 Mondal, S., Chen, Y., Lockbaum, G.J. et al. (2022). Dual inhibitors of main protease (MPro) and cathepsin L as potent antivirals against SARS-CoV2. *J. Am. Chem. Soc.* 144: 21035–21045.

15 Breidenbach, J., Lemke, C., Pillaiyar, T. et al. (2021). Targeting the main protease of SARS-CoV-2: from the establishment of high throughput screening to the design of tailored inhibitors. *Angew. Chem. Int. Ed. Engl.* 60: 10423–10429.

16 Ma, C., Hu, Y., Townsend, J.A. et al. (2020). *ACS Pharmacol. Transl. Sci.* 3: 1265–1277.

17 Meewan, I., Kattoula, J., Kattoula, J.Y. et al. (2022). Discovery of triple inhibitors of both SARS-CoV-2 proteases and human cathepsin L. *Pharmaceuticals* 15: 744.

18 Fuzo, C.A., Martins, R.B., Fraga-Silva, T.F.C. et al. (2022). Celastrol: a lead compound that inhibits SARS-CoV-2 replication, the activity of viral and human cysteine proteases, and virus-induced IL-6 secretion. *Drug Dev. Res.* 83: 1623–1640.

19 Tedesco, F., Calugi, L., Lenci, E., and zTrabocchi, A. (2022). Peptidomimetic small-molecule inhibitors of 3CLPro activity and spike-ACE2 interaction: Toward dual-action molecules against coronavirus infections. *J. Org. Chem.* 87: 12041–12051.

20 Li, J., Zhong, X., Li, H. et al. (2024). Design, synthesis and biological evaluation of biaryl amide derivatives against SARS-CoV-2 with dual-target mechanism. *Eur. J. Med. Chem.* 264: 115978.

21 Cantuti-Castelvetri, L., Ojha, R., Pedro, L.D. et al. (2020). *Science* 370: 856–860.

22 Daly, J.L., Simonetti, B., Klein, K. et al. (2020). Neuropilin-1 is a host factor for SARS-CoV-2 infection. *Science* 370: 861–865.
23 Yin, S., Mei, S., Li, Z. et al. (2022). Non-covalent cyclic peptides simultaneously targeting Mpro and NRP1 are highly effective against Omicron BA.2.75. *Front. Pharmacol.* 13: 1037993.
24 Casari, I., Manfredi, M., Metharom, P., and Falasca, M. (2021). Dissecting lipid metabolism alterations in SARS-CoV-2. *Prog. Lipid Res.* 82: 101092.
25 Snider, J.M., You, J.K., Wang, X. et al. (2021). Group IIA secreted phospholipase A2 is associated with the pathobiology leading to COVID-19 mortality. *J. Clin. Invest.* 131: e149236.
26 Theodoropoulou, M.A., Koutoulogenis, G.S., Zhang, L. et al. (2022). Identification of a dual inhibitor of secreted phospholipase A2 (GIIA sPLA2) and SARS-CoV-2 main protease. *Pharmaceuticals (Basel)* 15: 961.
27 Yi, Y., Li, W., Liu, K. et al. (2023). Licorice-saponin A3 is a broad-spectrum inhibitor for COVID-19 by targeting viral spike and anti-inflammation. *J. Pharm, Anal.* 14: 115–127.
28 Yi, Y., Yu, R., Xue, H. et al. (2024). Chrysin 7-O-β-D-glucuronide, a dual inhibitor of SARS-CoV-2 3CLpro and PLpro, for the prevention and treatment of COVID-19. *Int. J. Antimicrob. Agents* 63: 107039.
29 Hashimoto, R., Sakamoto, A., Deguchi, S. et al. (2021). Dual inhibition of TMPRSS2 and Cathepsin B prevents SARS-CoV-2 infection in iPS cells. *Mol. Ther. Nucleic Acids* 26: 1107–1114.
30 Wang, H., Yang, Q., Liu, X. et al. (2023). Structure-based discovery of dual pathway inhibitors for SARS-CoV-2 entry. *Nat. Commun.* 14: 7574.
31 Xie, X., Lan, Q., Zhao, J. et al. (2024). Structure-based design of pan-coronavirus inhibitors targeting host cathepsin L and calpain-1. *Signal Transduct. Target. Ther.* 9: 54.
32 Mao, B., Le-Trilling, V.T.K., Wang, K. et al. (2022). Obatoclax inhibits SARS-CoV-2 entry by altered endosomal acidification and impaired cathepsin and furin activity in vitro. *Emerg. Microb. Infect.* 11: 483–497.
33 Xu, T. and Zhang, L. (2023). Current understanding of nucleoside analogs inhibiting the SARS-CoV-2 RNA-dependent RNA polymerase. *Comput. Struct. Biotechnol. J.* 21: 4385–4394.
34 Wang, Q., Wu, J., Wang, H. et al. (2020). Structural basis for rna replication by the sars-cov-2 polymerase. *Cell* 182: 417–428.
35 Park, G.J., Osinski, A., Hernandez, G. et al. (2022). The mechanism of RNA capping by SARS-CoV-2. *Nature* 609: 793–800.
36 Good, S.S., Westover, J., Jung, K.H. et al. (2021). AT-527, a double prodrug of a guanosine nucleotide analog, is a potent inhibitor of SARS-CoV-2 in vitro and a promising oral antiviral for treatment of COVID-19. *Antimicrob. Agents Chemother.* 65: e02479–e02420.
37 Horga, A., Saenz, R., Yilmaz, G. et al. (2023). Oral bemnifosbuvir (AT-527) vs placebo in patients with mild-to-moderate COVID-19 in an outpatient setting (MORNINGSKY). *Future Virol.* 10: 2217/fvl-2023-0115.
38 Shannon, A., Fattorini, V., Sama, B. et al. (2023). A dual mechanism of action of AT-527 against SARS-CoV-2 polymerase. *Nat. Commun.* 13: 621.
39 Driouich, J.S., Cochin, M., Lingas, G. et al. (2021). Favipiravir antiviral efficacy against SARS-CoV-2 in a hamster model. *Nat. Commun.* 12: 1735.
40 Kaptein, S.J.F., Jacobs, S., Langendries, L. et al. (2020). Favipiravir at high doses has potent antiviral activity in SARS-CoV-2-infected hamsters, whereas hydroxychloroquine lacks activity. *Proc. Natl. Acad. Sci. USA* 117: 26955–26965.
41 Marlin, R., Desjardins, D., Contreras, V. et al. (2022). Antiviral efficacy of favipiravir against Zika and SARS-CoV-2 viruses in non-human primates. *Nat. Commun.* 13: 5108.
42 Boretti, A. (2020). Favipiravir use for SARS CoV-2 infection. *Pharmacol. Rep.* 72: 1542–1552.
43 Solaymani-Dodaran, M., Ghanei, M., Bagheri, M. et al. (2021). Safety and efficacy of Favipiravir in moderate to severe SARS-CoV-2 pneumonia. *Int. Immunopharmacol.* 95: 107522.
44 Naydenova, K., Muir, K.W., Wu, L.F. et al. (2021). Structure of the SARS-CoV-2 RNA-dependent RNA polymerase in the presence of favipiravir-RTP. *Proc. Natl. Acad. Sci. USA* 118: e2021946118.
45 Shannon, A., Selisko, B., Le, N.T. et al. (2020). Rapid incorporation of Favipiravir by the fast and permissive viral RNA polymerase complex results in SARS-CoV-2 lethal mutagenesis. *Nat. Commun.* 11: 4682.
46 Hadj Hassine, I., Ben M'hadheb, M., and Menéndez-Arias, L. (2022). Lethal mutagenesis of RNA viruses and approved drugs with antiviral mutagenic activity. *Viruses* 14: 841.
47 Wiedemar, N., Hauser, D.A., and Mäser, P. (2020). 100 years of suramin. *Antimicrob. Agents Chemother.* 64: e01168–e01119.
48 Salgado-Benvindo, C., Thaler, M., Tas, A. et al. (2020). Suramin inhibits SARS-CoV-2 infection in cell culture by interfering with early steps of the replication cycle. *Antimicrob. Agents Chemother.* 64: e00900–e00920.
49 Yin, W., Luan, X., Li, Z. et al. (2021). Structural basis for inhibition of the SARS-CoV-2 RNA polymerase by suramin. *Nat. Struct. Mol. Biol.* 28: 319–325.

50 Zhu, W., Xu, M., Chen, C.Z. et al. (2020). Identification of SARS-CoV-2 3CL protease inhibitors by a quantitative high-throughput screening. *ACS Pharmacol. Transl. Sci.* 3: 1008–1016.

51 Kwon, P.S., Xu, S., Oh, H. et al. (2023). Suramin binds and inhibits infection of SARS-CoV-2 through both spike protein-heparan sulfate and ACE2 receptor interactions. *Commun. Biol.* 6: 387.

52 Samrat, S.K., Bashir, Q., Zhang, R. et al. (2023). A universal fluorescence polarization high throughput screening assay to target the SAM-binding sites of SARS-CoV-2 and other viral methyltransferases. *Emerg. Microb. Infect.* 12: 2204164.

53 Arabi-Jeshvaghani, F., Javadi-Zarnaghi, F., and Ganjalikhany, M.R. (2023). Analysis of critical protein-protein interactions of SARS-CoV-2 capping and proofreading molecular machineries towards designing dual target inhibitory peptides. *Sci. Rep.* 13: 350.

54 Devkota, K., Schapira, M., Perveen, S. et al. (2021). Probing the SAM binding site of SARS-CoV-2 Nsp14 in vitro using SAM competitive inhibitors guides developing selective bisubstrate inhibitors. *SLAS Discov.* 26: 1200–1211.

55 Li, F., Ghiabi, P., Hajian, T. et al. (2023). SS148 and WZ16 inhibit the activities of nsp10-nsp16 complexes from all seven human pathogenic coronaviruses. *Biochim. Biophys. Acta Gen. Subj.* 1867: 130319.

56 Klima, M., Khalili Yazdi, A., Li, F. et al. (2022). Crystal structure of SARS-CoV-2 nsp10-nsp16 in complex with small molecule inhibitors, SS148 and WZ16. *Protein Sci.* 31: e4395.

57 Ren, P.X., Shang, W.J., Yin, W.C. et al. (2022). A multi-targeting drug design strategy for identifying potent anti-SARS-CoV-2 inhibitors. *Acta Pharmacol. Sin.* 43: 483–493.

58 Jin, Z., Du, X., Xu, Y. et al. (2020). Structure of Mpro from SARS-CoV-2 and discovery of its inhibitors. *Nature* 582: 289–293.

59 Weglarz-Tomczak, E., Tomczak, J.M., Talma, M. et al. (2021). Identification of ebselen and its analogues as potent covalent inhibitors of papain-like protease from SARS-CoV-2. *Sci. Rep.* 11: 3640.

60 Amporndanai, K., Meng, X., Shang, W. et al. (2021). Inhibition mechanism of SARS-CoV-2 main protease by ebselen and its derivatives. *Nat. Commun.* 12: 3061.

61 Sargsyan, K., Lin, C.C., Chen, T. et al. (2020). Multi-targeting of functional cysteines in multiple conserved SARS-CoV-2 domains by clinically safe Zn-ejectors. *Chem. Sci.* 11: 9904–9909.

62 Baddock, H.T., Brolih, S., Yosaatmadja, Y. et al. (2022). Characterization of the SARS-CoV-2 ExoN (nsp14ExoN-nsp10) complex: implications for its role in viral genome stability and inhibitor identification. *Nucleic Acids Res.* 50: 1484–1500.

63 Zmudzinski, M., Rut, W., Olech, K. et al. (2023). Ebselen derivatives inhibit SARS-CoV-2 replication by inhibition of its essential proteins: PLpro and Mpro proteases, and nsp14 guanine N7-methyltransferase. *Sci. Rep.* 13: 9161.

64 Liu, W., Wang, J., Wang, S. et al. (2023). Discovery of new non-covalent and covalent inhibitors targeting SARS-CoV-2 papain-like protease and main protease. *Bioorg. Chem.* 140: 106830.

65 Yu, W., Zhao, Y., Ye, H. et al. (2022). Structure-based design of a dual-targeted covalent inhibitor against papain-like and main proteases of SARS-CoV-2. *J. Med. Chem.* 65: 16252–16267.

66 Kattula, B., Reddi, B., Jangam, A. et al. (2023). Development of 2-chloroquinoline based heterocyclic frameworks as dual inhibitors of SARS-CoV-2 MPro and PLPro. *Int. J. Biol. Macromol.* 242: 124772.

67 Di Sarno, V., Lauro, G., Musella, S. et al. (2021). Identification of a dual acting SARS-CoV-2 proteases inhibitor through in silico design and step-by-step biological characterization. *Eur. J. Med. Chem.* 226: 113863.

68 Santos, L.H., Kronenberger, T., Almeida, R.G. et al. (2022). Structure-based identification of naphthoquinones and derivatives as novel inhibitors of main protease Mpro and Papain-like protease PLpro of SARS-CoV-2. *J. Chem. Inf. Model.* 62: 6553–6573.

69 Puhl, A.C., Godoy, A.S., Noske, G.D. et al. (2023). Discovery of PLpro and Mpro inhibitors for SARS-CoV-2. *ACS Omega* 8: 22603–22612.

70 Ellinger, B., Bojkova, D., Zaliani, A. et al. (2021). A SARS-CoV-2 cytopathicity dataset generated by high-content screening of a large drug repurposing collection. *Sci. Data* 8: 70.

71 Shen, L., Niu, J., Wang, C. et al. (2019). High-throughput screening and identification of potent broad-spectrum inhibitors of coronaviruses. *J. Virol.* 93: e00023–e00019.

72 Zang, Y., Su, M., Wang, Q. et al. (2022). High-throughput screening of SARS-CoV-2 main and papain-like protease inhibitors. *Protein Cell* 14: 17–27.

73 Ayoup, M.S., ElShafey, M.M., Abdel-Hamid, H. et al. (2023). Repurposing 1,2,4-oxadiazoles as SARS-CoV-2 PLpro inhibitors and investigation of their possible viral entry blockade potential. *Eur. J. Med. Chem.* 252: 115272.

74 Su, H.X., Yao, S., Zhao, W.F. et al. (2020). Anti-SARS-CoV-2 activities in vitro of Shuanghuanglian preparations and bioactive ingredients. *Acta Pharmacol. Sin.* 41: 1167–1177.

75 Zandi, K., Musall, K., Oo, A. et al. (2021). Baicalein and baicalin inhibit SARS-CoV-2 RNA-dependent-RNA polymerase. *Microorganisms* 9: 893.

76 Murugan, N.A., Kumar, S., Jeyakanthan, J., and Srivastava, V. (2020). Searching for target-specific and multi-targeting organics for Covid-19 in the Drugbank database with a double scoring approach. *Sci. Rep.* 10: 19125.

77 Riva, L., Yuan, S., Yin, X. et al. (2020). Discovery of SARS-CoV-2 antiviral drugs through large-scale compound repurposing. *Nature* 586: 113–119.

78 Yuan, S., Yin, X., Meng, X. et al. (2021). Clofazimine broadly inhibits coronaviruses including SARS-CoV-2. *Nature* 593: 418–423.

79 Zhang, X., Shi, Y., Guo, Z. et al. (2022). Clofazimine derivatives as potent broad-spectrum antiviral agents with dual-target mechanism. *Eur. J. Med. Chem.* 234: 114209.

80 Xu, Z., Zou, Y., Gao, X. et al. (2022). Dual-targeting cyclic peptides of receptor-binding domain (RBD) and main protease (Mpro) as potential drug leads for the treatment of SARS-CoV-2 infection. *Front. Pharmacol.* 13: 1041331.

81 Zhan, P. and Liu, X. (2009). Designed multiple ligands: an emerging anti-HIV drug discovery paradigm. *Curr. Pharm. Des.* 15: 1893–1917.

82 Zhan, P. and Liu, X. (2013). Rationally designed multi-target anti-HIV agents. *Curr. Pharm. Des.* 20: 1743–1758.

27

Multi-target Antimalarials as a Strategy to Reduce Resistance Risk

Lauren B. Coulson and Kelly Chibale

27.1 Introduction

Malaria, an ancient disease caused by the protozoan *Plasmodium* and transmitted via the female anopheles mosquito, has had a devastating effect on global health for centuries. Despite the discovery of effective treatments and the implementation of a range of other control measures, malaria remains a serious public health burden with over 249 million reported cases and an estimate of 608,000 associated deaths in 2022, of which over 95% occurred in sub-Saharan Africa [1]. Malaria affects infants and pregnant women most severely, posing additional challenges for drug development. Notably, 76% of malaria deaths were in children under the age of 5 in 2022 [1]. Hopes of disease eradication by 2030 have dwindled in recent years, largely due to drug and insecticide resistance, in addition to logistical challenges and environmental factors.

Plasmodium, unicellular eukaryotes with ~5000 genes, have co-evolved with humans for centuries and have a remarkable ability to evade the human immune system and antimalarial drugs. Human malaria is caused by five *Plasmodium* species. *Plasmodium falciparum* is the most prevalent in Africa and causes the most lethal form of the disease, while *Plasmodium vivax*, more common in Southeast Asia, causes a relapsing form of the disease, where parasites go into a dormant state in the liver for long periods of time making the disease more difficult to treat. The *Plasmodium* parasite has a complex life cycle, which includes human liver, human blood, and mosquito stages. During a blood meal, parasites in the form of sporozoites are transferred from the saliva of an infected mosquito into the human bloodstream. They make their way to the liver, where they replicate, maturing into schizonts that rapture to release daughter merozoites (ex-erythrocytic schizogony). The liver stages are asymptomatic and are targeted by prophylactics for malaria prevention. The merozoites released from the liver enter the bloodstream and infect red blood cells. Within the red blood cells, the parasites undergo rapid asexual replication (erythrocytic schizogony), maturing from ring-stage trophozoites into schizonts that rupture releasing merozoites into the bloodstream that go on to infect other red blood cells. A small proportion of blood-stage parasites undergo sexual differentiation forming gametocytes, the transmissible form of the parasite. The gametocytes are taken up by a mosquito. Within the mosquito, the male and female gametes fuse, forming zygotes, which become ookinetes. The ookinetes migrate to the mosquito midgut and pass through the midgut wall, forming oocysts. The oocysts divide to form sporozoites that migrate to the salivary glands of the mosquito and continue the transmission cycle to the next human host.

The exceptionally rapid increase in parasite load during asexual blood-stage replication is not only responsible for malaria symptoms, including high fevers, but also provides the ideal environment for the evolution and selection of drug-resistant parasites. Consequently, drug resistance has been reported in the field for almost all clinically used drugs. Malaria drugs are now developed as combination therapies to slow the development of parasite resistance, although eventual resistance to the individual partner drugs remains inevitable.

The frontline treatments for uncomplicated malaria are artemisinin-based combination therapies (ACTs) with the most used being artemether–lumefantrine. Artemether–lumefantrine remains clinically effective in Africa, although partial resistance to artemisinin, resulting in a delayed parasite death phenotype, is prevalent in Southeast Asia and has been reported in East Africa [2–4]. Parasites with a decreased susceptibility to the partner drug lumefantrine as well as dihydroartemisinin have also been identified in Uganda raising concerns about how long the current frontline ACTs will remain effective [5].

Identification of drugs with novel targets and distinct resistance mechanisms from currently used antimalarials is critical. Since antimalarials are predominantly used in

Polypharmacology: Strategies for Multi-Target Drug Discovery, First Edition. Edited by Jens-Uwe Peters.

vulnerable population groups including pregnant women and young children, and drug adherence is crucial to slow down the emergence of resistance; drug safety and tolerability are also paramount. The drug development pipeline must be continually replenished to contend with existing and emerging drug resistance.

Until recently, most malaria drugs have been developed using a phenotypic-based drug discovery approach relying mostly on screening against asexual blood-stage parasites in culture for hit identification and compound optimization. While this approach has led to the development of novel antimalarials, many active compounds target the same parasite pathways, and it does not allow for rational drug optimization. Target deconvolution studies for phenotypic hits have led to the identification of novel malaria targets enabling target-based drug discovery for malaria [6–8]. Information on the mode of action allows for rational optimization of compounds for potency against the target and selectivity relative to key human off-targets where these exist. However, a compound that is optimized to exert its therapeutic effect by binding to a single *Plasmodium* protein target may be more easily evaded by the parasite through single-point mutations in the binding site or other compensatory mechanisms. In contrast, many of the compounds that have low resistance propensity hit nonprotein immutable targets such as the hemozoin pathway or likely act through multiple pathways making target deconvolution challenging. Rational target-based optimization of compounds to reduce resistance risk is still in its infancy. Designed polypharmacology, where a compound is optimized to exert the desired therapeutic effect via two or more targets (or target binding sites), has the potential to reduce the resistance risk associated with an individual target, as well as improve compound efficacy [9–11].

27.2 Next-generation Antimalarials

The requirements for next-generation antimalarial drugs are high, and funding for drug development is limited in this field. Consequently, drug discovery programs need to be carefully prioritized, with only the most promising compounds being progressed into clinical development. In recent years, there has been a notable improvement to the malaria drug pipeline with an increase in the number of new compounds entering preclinical and early clinical development [7]. Despite this, attrition due to safety concerns, resistance risk, and efficacy is inevitable, warranting continued efforts to maintain the pipeline. In addition to oral drugs to treat uncomplicated malaria, drugs for malaria prevention and transmission blocking are also development goals. Various stakeholders, including clinicians, scientists, and public health professionals, have put together criteria for new drugs. These have been formalized by the Medicines for Malaria Venture (MMV) in the form of target product profiles (TPPs) and target candidate profiles (TCPs) [12]. MMV is a public–private product development partnership responsible for coordinating the discovery, development, and delivery of new antimalarial drugs, with the focus being to maximize the impact and safeguard the efficacy of current and new antimalarials (https://www.mmv.org/). The TPPs outline the ideal requirements for new products that will offer advantages over the current malaria therapies, with TPP-1 focusing on case management, particularly the treatment of uncomplicated malaria, and TPP-2 focusing on chemoprotection for migratory populations. The requirements of individual drug molecules that will be used in a combination product to meet the TPPs are described by the TCPs as follows: **TCP-1:** molecules that clear asexual blood-stage parasitemia; **TCP-2**: profile retired; **TCP-3**: molecules with activity against hypnozoites, the dormant liver form of the parasite that causes a relapsing form of disease (mainly for *Plasmodium vivax*); **TCP-4**: molecules with activity against liver-stage schizonts; **TCP-5**: molecules that block transmission by targeting the sexual forms of the parasite; and **TCP-6**: molecules that block transmission by targeting the insect vector (endectocides).

The key challenges for drug development can be summarized as follows: (i) multidrug resistance, (ii) the need for cheap and simplified drug regimens suitable for resource-limited settings, and (iii) good safety and tolerability in vulnerable populations including pregnant women, infants, and patients with co-infections or suffering from malnutrition. New molecules entering clinical development for the treatment of malaria ideally need to be cheap and chemically stable, exhibit high potency and rapid parasite clearance, have a long half-life enabling single-dose cure, be devoid of fetal toxicity-associated risk, and have a low risk of resistance.

27.3 Resistance Risk as a Criterion for the Prioritization of New Molecules and Targets

27.3.1 Assessing the Resistance Risk of Compounds in Development

Even though molecules will ultimately be administrated in combination, the resistance propensity of individual molecules serves as a key criterion for compound prioritization and progression [13]. A standardized workflow has been established to assess resistance risk at the preclinical development stage using a combination of *in vitro*, *in vivo*, and *ex vivo* resistance studies. Resistance propensity can be a characteristic of both the chemical class and mechanism of action/target binding site. *In vitro* evolution followed by whole-genome sequencing of resistant parasites is routinely carried out to identify resistance mechanisms and has proved useful for target identification [14]. Single-point mutations and/or copy number variations are regularly observed for the target gene or genes within related pathways. Mutations also regularly occur in multidrug-resistant genes such as *pfmdr1* or *pfcrt*, which are associated with drug efflux away from the site of action.

The *in vitro* minimum inoculum for resistance (MIR) is used to quantify resistance risk. The MIR refers to the minimum number of asexual blood-stage parasites required to yield resistance when exposed to drug pressure at a defined drug concentration (typically $3 \times EC_{50}$ or $3 \times EC_{90}$) within a standard timeframe (typically 60 days) [13]. A MIR $\leq 10^7$ is considered high risk and will typically lead to de-prioritization of the chemical series and the associated target; an MIR between 10^7 and 10^9 is considered moderate risk, and attempts may be made to identify other analogs from the series with an improved MIR; and an MIR $> 10^9$ is indicative of low risk and is now considered a requirement for candidate selection. In addition to the MIR, the shift in effective drug concentration (EC_{50}) is also important as some mutations yield low-grade resistance, meaning that drugs will remain effective in a clinical setting, while other mutations result in high-grade resistance that will likely result in complete treatment failure. The fitness of the resistant parasites, although more difficult to quantify, will also have an impact on the rate that resistance emerges and spreads in the field. At the candidate selection phase, *ex vivo* resistance studies using field isolates are carried out to check for preexisting resistance in the field. Parasite recrudescence and presence of resistant mutants are also assessed following drug efficacy studies in mouse models of malaria infection, where NOD/SCID/IL2rγnull (NSG) mice are ingrafted with human red blood cells infected with *P. falciparum* parasites [15]. Finally, parasite recrudescence is also monitored in human volunteer infection studies. These infection models are used to predict the likelihood and speed at which resistance leading to treatment failure will emerge in the field [16].

27.3.2 Resistance Risk for Target-based Drug Discovery Programs

In vitro evolution studies coupled with whole-genome sequencing have led to the identification and subsequent phenotypic validation of a range of promising protein drug targets essential for parasite development enabling target-based drug discovery programs for malaria [6, 14]. Biochemical assays coupled with structural information provide powerful information to optimize compounds for potency, selectively, and physicochemical properties. However, inherent in the method used to identify the target, the target can mutate to decrease or compensate for drug binding, and consequently, inhibitors typically show high-to-moderate risk of resistance. Consequently, there are several drug candidates with novel protein targets that are in the patient exploratory phase of development that display good efficacy but carry a high resistance risk. Cabamiquine, a *Plasmodium*-specific eukaryotic elongation factor 2 (eEF2) inhibitor, was effective in a human volunteer infection study and shows potent TCP-1, TCP-4, and TCP-5 activity although it readily selected for resistance *in vitro* (MIR $\sim 10^7$), *in vivo*, and in a human volunteer infection study, with a total of 11 mutations identified within eEF2 across these assays [16]. To manage this risk, it has been paired with pyronaridine, a hemozoin formation inhibitor with a low propensity for resistance, for combination studies [17]. Similarly, cipargamin, an inhibitor of the *Plasmodium* cation adenosine triphosphatase (ATPase) ATP4, has a high propensity for resistance [18, 19] and will need to be paired with a compound with low resistance risk to safeguard its efficacy. The promising dihydroorotate dehydrogenase (DHODH) inhibitor DSM265 with an exceptionally long half-life reached phase II clinical trials but also has a resistance liability (*in vitro* MIR between 10^5 and 10^7) and clinical development has been stalled based on this and other challenges [20, 21].

It is plausible that the MIR for a particular target and chemical class could be optimized through medicinal chemistry. However, the throughput of MIR studies is low making it challenging to generate the data required to understand the molecular basis for resistance propensity and guide optimization. MIRs have been reported to shift significantly for a given chemical class/target. While this may result from changes to the binding site interactions, it is difficult to distinguish this from the implication of new/additional targets without extensive mode of action studies. Efforts to understand and improve the propensity for resistance of DHODH inhibitors have been reported, and structure–activity relationship (SAR) analysis has identified compounds within the DSM265 series with lower *in vitro* resistance risk based on binding mode and/or physicochemical properties [22, 23].

27.3.3 Irresistible Compounds

For some molecules, it is more challenging or not possible to generate resistance *in vitro*. A hypermutable *P. falciparum* line (deoxyribonucleic acid [DNA] polymerase δ mutant parasite line) has been engineered to increase the chances of generating mutations *in vitro* [24]. Compounds that fail to generate resistance *in vitro* after multiple attempts carried out under a range of conditions are referred to as "irresistibles" [6]. These are promising molecules for drug discovery since they have a lower risk of generating resistance in the field, but there are other significant challenges associated with these compounds. Target identification is generally more challenging, requiring alternative approaches. Many of these compounds may act at least partially through the inhibition of hemozoin formation and/or polypharmacology.

Hemozoin formation is essential to parasite survival during asexual blood-stage development [25]. Parasites break down hemoglobin as a source of amino acids. This leads to the release of toxic-free heme that the parasite converts to hemozoin, an inert crystalline form of heme. This heme detoxification process is essential to parasite survival within the red blood cell, and the inhibition of this process leads to rapid parasite death. Hemozoin formation takes place within the parasite's digestive vacuole [25]. Hemozoin inhibitors typically bind to the interface of the hemozoin crystals and/or free heme interfering with crystal packing [26]. Many clinically used antimalarials including chloroquine and mefloquine inhibit hemozoin biocrystallization. Their potent cellular activity relies on drug accumulation within the digestive vacuole driven by pH trapping due to the low pH of the digestive vacuole relative to the cytosol. Hemozoin is an excellent immutable nonprotein target that is absent in the host, and hemozoin inhibitors typically show a low risk of resistance *in vitro*. However, clinically acquired resistance, caused by mutations that reduce drug accumulation in the digestive vacuole, has diminished the efficacy of many of these drugs in the field. For example, chloroquine resistance is primarily mediated through mutations in the "chloroquine resistance transporter" (*Pf*CRT) [27]. *Pf*CRT-mediated resistance is compound- or chemical-class-specific, so cross-resistance is not necessarily seen for novel chemical classes of hemozoin inhibitors. Consequently, novel hemozoin inhibitors that do not display cross-resistance with clinically used antimalarials remain of interest for TCP-1 but will not lead to transmission blocking unless additional modes of action are involved.

The high bar for resistance risk (MIR > 10^9 or ideally irresistible) set out my MMV is likely to bias the portfolio away from target-based drug discovery programs and toward hemozoin inhibitors or other irresistible compounds with unknown targets. To allow for structure-enabled drug discovery focused on novel protein targets with the potential to yield multistage active antimalarials meeting the resistance guidelines, new strategies to reduce the risk of drug resistance need to be explored. Some strategies could include medicinal chemistry optimization to improve the MIR, multi-target inhibitors, and covalent inhibitors. However, the benefits of these approaches still need to be conclusively demonstrated.

27.4 Polypharmacology in Malaria Drug Discovery

Target-based drug discovery typically focuses on optimizing compounds for potency against a single validated target, while reducing activity against related off-targets. For multigenetic and infectious diseases, the benefits of modulating multiple pathways simultaneously to achieve efficacy and overcome resistance are becoming increasingly evident [28]. In many cases, this is achieved serendipitously, particularly for phenotypic drug discovery programs where the mode of action is unknown or is not used to drive medicinal chemistry optimization. It is likely that many antimalarials, particularly those classified as "irresistible" or "target-less" [6], exert their effect through multiple mechanisms or polypharmacology. While mode of action studies focus on identifying the primary target, the extent to which secondary parasite or host targets may contribute to antiplasmodium activity *in vivo* or in the clinic is poorly understood.

In addition to reduced resistance risk, polypharmacology offers other potential advantages including improved efficacy due to additive or synergistic effects resulting from the simultaneous modification of multiple pathways or multiple proteins in the same pathway. These attributes have the potential to reduce the treatment costs and pill burden associated with complex combination therapies.

Designed polypharmacology for malaria, where a single molecule is intentionally optimized to modulate multiple targets, is still in its infancy. However, there are encouraging examples of compounds interacting with more than one target/binding site, displaying good efficacy and reduced resistance risk. Designed polypharmacology requires the identification of promising target pairs or sets, based on i) pharmacophore matching, i.e., targets with similar binding sites that are amenable to modulation by a single compound, and ii) parasite biology, i.e., targets for which simultaneous modulation leads to favorable synergistic effects and reduced resistance risk. While many compounds show binding to multiple proteins *in vitro*, achieving therapeutically relevant levels of inhibition and balancing potency for all intended targets in a cellular context may be challenging.

27.4.1 Dual Inhibition of *Plasmodium* Plasmepsin X and Plasmepsin IX

The plasmepsins, a family of 10 aspartic proteases in *Plasmodium*, play key roles in a wide range of biological processes, and several of these proteins have shown promise as antimalarial targets [29]. A phenotypic screen of an aspartyl protease inhibitor library identified hits with potent antiplasmodium asexual blood-stage activity [9]. Subsequent medicinal chemistry optimization led to the identification of WM382 (Figure 27.1) [30]. Resistance selection with WM4 and related hit compound WM5 led to resistant parasites with an amplification event on chromosome 8 resulting in a copy number variation for plasmepsin X (PMX) and an associated 13- to 14-fold increase in EC_{50} [9]. In contrast, WM382 did not select for resistance *in vitro* and displayed a lower level of cross-resistance (fourfold increase in EC_{50}) to parasites resistant to WM4 and WM5, suggesting that this compound has a higher barrier to resistance. Target engagement studies showed that all three compounds directly inhibited PMX [9]. A cellular thermal shift assay (CETSA), allowing for a global unbiased assessment of protein binding, showed that WM382 stabilized both PMX and plasmepsin IX (PMIX), while WM4 only stabilized PMX. Dual PMX and PMIX targeting by WM382 was further confirmed using PMX and PMIX conditional knockdown parasite lines. Reducing expression levels of either PMX or PMIX led to an increase in parasite sensitivity to WM382. In contrast, only knockdown of PMX led to increased sensitivity to WM4. Encouragingly, WM384 achieved *in vivo* efficacy in a humanized mouse model of *P. falciparum* infection and displayed transmission-blocking and liver-stage activity [9].

PMX and PMIX are essential proteases that process proteins required to disrupt host cell membranes for parasite invasion and mass egress [31, 32]. An earlier study also identified PMX and PMIX as the molecular targets of the hydroxyl-ethyl-amine–based inhibitor, referred to as 49c (Figure 27.1), that also possesses potent multistage antimalarial activity [31]. While PMX and PMIX are very distinct proteins with different substrates and subcellular localization patterns, they have very similar active sites, making them amenable to modulation by a single molecule. High-resolution X-ray structure determination of WM4 and WM384 in complex with PMX, coupled with enzyme kinetic analysis for PMX and PMIX inhibition, has provided insight into the mechanistic details of inhibitor binding and the requirements for dual PMX/PMIX inhibition [33]. These studies demonstrate that the dual targeting of PMX/PMIX is a promising avenue for the development of antimalarials with a reduced resistance risk.

27.4.2 Dual-Site Inhibitors of the Cytochrome bc_1 Complex

The mitochondrial respiratory chain relies on a series of distinct membrane protein complexes located in the inner mitochondrial membrane that serve as catalysts. Ubiquinone and cytochrome c serve as the links between the complexes, enabling electron transport. The mitochondrial respiratory transport chain cytochrome bc_1 complex (complex III) is a well-established drug target and is the molecular target of the clinically used antimalarial atovaquone (Figure 27.2) [34, 35]. In eukaryotes, the bc_1 complex is an oligomeric protein localized in the inner mitochondrial membrane. The subunits cytochrome *b*, cytochrome c_1, and the Rieske iron–sulfur protein ([2Fe2S] ISP) form the catalytic core of the complex [36, 37]. The complex catalyzes the transfer of electrons from the low-potential substrate ubiquinol to high-potential cytochrome c via a Q-cycle mechanism. This contributes to the generation of a proton gradient that serves as a proton-motive force required for ATP synthesis and other cellular functions. The *bc1* complex has two distinct

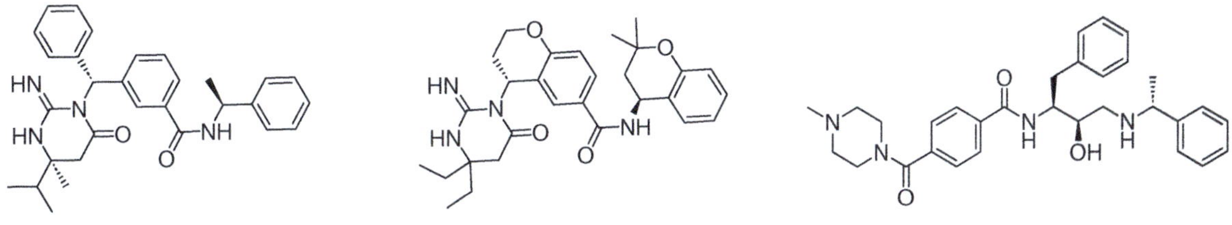

PMX inhibitor WM4
P. falciparum ABS EC_{50} = 7.4 nM
*Pf*PMX K_i = 0.4 nM
*Pf*PMIX K_i = 306 nM
(a)

Dual PMX/PMIX inhibitor WM382
P. falciparum ABS EC_{50} = 0.4 nM
*Pf*PMX K_i = 0.01 nM
*Pf*PMIX K_i = 0.5 nM
(b)

Dual PMX/PMIX inhibitor 49c
P. falciparum ABS EC_{50} = 0.6 nM
(c)

Figure 27.1 *Plasmodium* plasmepsin X and IX inhibitors. K_i values were determined using an *in vitro* FRET-based assay for *Pf*PMX and *Pf*PMIX cleavage of peptide substrates. ABS; asexual blood-stage.

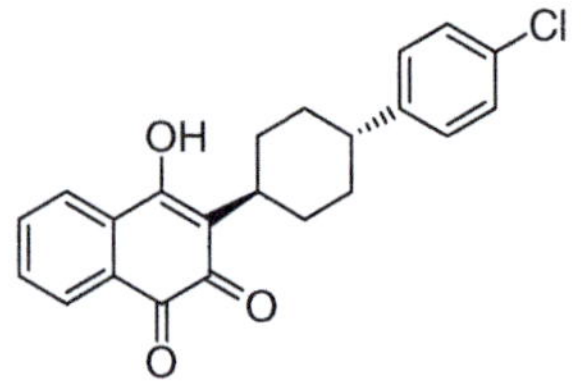

Atovaquone
*Pf*Cyt*Bc*$_1$ Q$_o$ site inhibitor

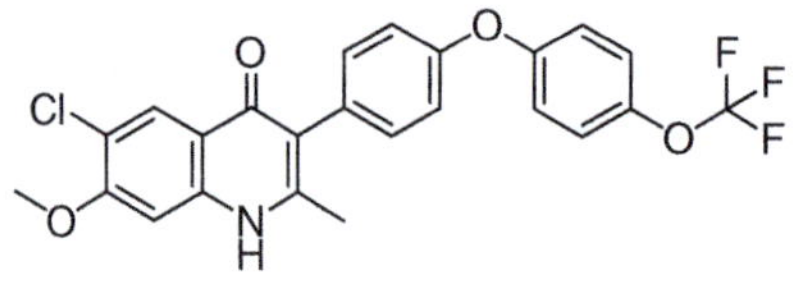

ELQ-300
*Pf*Cyt*Bc*$_1$ Q$_i$ site inhibitor

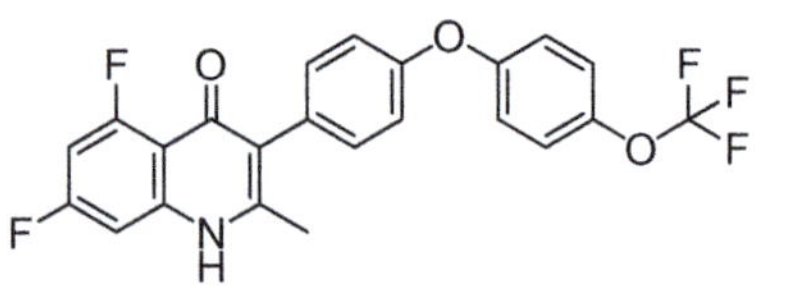

ELQ-400
*Pf*Cyt*Bc*$_1$ dual Q$_o$/Q$_i$ site inhibitor

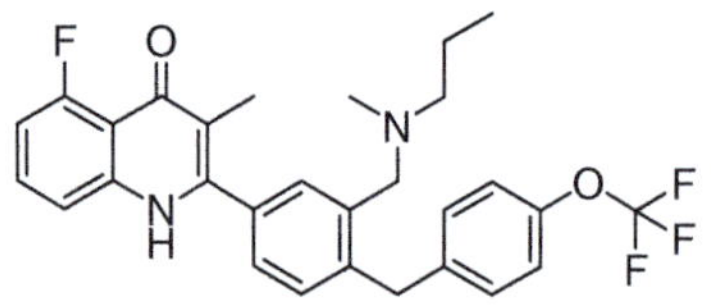

RYL-581
*Pf*Cyt*Bc*$_1$ dual Q$_o$/Q$_i$ site and *Pf*NDH2 inhibitor

Figure 27.2 Inhibitors of the *Plasmodium* mitochondrial respiratory transport chain cytochrome *bc*$_1$ complex.

quinone-binding sites located on opposite sides of the membrane within cytochrome *b*, the quinol oxidation site Q$_o$, and the quinone reduction site Q$_i$. The *Plasmodium* cytochrome *bc*$_1$ complex is a highly vulnerable target in liver, asexual blood, and sexual stage parasites [36, 38, 39], with the potential to yield antimalarials suitable for prophylaxis, treatment, and transmission-blocking strategies. Since ATP is mainly synthesized via glycolysis in asexual blood-stage parasites, the main role of the *bc*$_1$ complex in the blood stage is to regenerate ubiquinone, which serves as an electron acceptor for the *P. falciparum* DHODH (*Pf*DHODH) [40].

Inhibitors of the *bc*$_1$ complex, such as atovaquone (Q$_o$) and endochin-like quinolone-300 (ELQ-300) (Q$_i$) (Figure 27.2), bind to either the Q$_o$ binding site or the Q$_i$ binding site. High-grade parasite resistance to atovaquone, the only cytochrome *bc*$_1$ inhibitor in clinical use for malaria, emerges rapidly *in vitro* (MIR $\sim 10^{5-6}$) and in the clinic [41–44]. Due to the high resistance risk, atovaquone is given in combination with proguanil under the trade name Malarone™, but the high cost of this combination limits its use in resource-limited settings. The most common mutation associated with Malarone™ treatment failure, leading to atovaquone-resistant parasites, is a Y268S mutation in the Q$_o$ binding site of a cytochrome *bc*$_1$ [45, 46]. This mutation results in an increase in EC_{50} of several hundred-fold relative to drug-sensitive parasites [41].

These challenges have led to efforts to develop cheaper cytochrome *bc*$_1$ inhibitors, able to overcome atovaquone resistance, and reduce resistance risk. Notably, many developmental compounds do not show cross-resistance with atovaquone. The endochin-like quinolones (ELQs) are potent *Plasmodium* cytochrome *bc*$_1$ inhibitors and do not show cross-resistance with atovaquone or other Q$_o$ site inhibitors [47]. ELQ-300 has been shown to bind to the distinct Q$_i$ site, with a I22L mutation within this site resulting in ELQ-300 resistance [48]. The 4(1H)-quinolone scaffold is compatible with the inhibition of either cytochrome *bc*$_1$ sites, and minor modifications to substituents can influence ELQs specificity for the Q$_o$ or Q$_i$ binding sites [48]. Combining a Q$_o$ site inhibitor with a Q$_i$ site inhibitor has been proposed as a potential combination strategy for malaria. Atovaquone in combination with ELQ-300 was synergistic with a curative combined dose of 1 mg/kg in a *P. yoelii* murine model of malaria infection [49].

ELQ-400 is a closely related analog to ELQ-300 with potent multistage antiplasmodium activity (Figure 27.2). Interestingly, this compound was effective against both Q$_o$ site (atovaquone) and Q$_i$ site (ELQ-300) inhibitor-resistant parasites, indicating an alternative/additional mode of action/binding [10]. Further studies in yeast concluded that ELQ-400 targets both the Q$_o$ and Q$_i$ sites [10]. *In vitro* resistance studies, carried out in a Dd2 hypermutable line, suggested that ELQ-400 has a lower propensity for resistance and may yield lower-grade resistance than atovaquone. The V259L mutation that was identified in ELQ-400 resistance parasites generated *in vitro* mapped to the Q$_o$ site but only resulted in a 10-fold reduction in antiplasmodium activity. ELQ-400 shows potent multistage activity with single-dose liver-stage, blood-stage, and transmission-blocking activity in *in vivo* models of malaria infection [50]. While cytochrome *bc*$_1$ inhibitors generally display slow-killing kinetics, ELQ-400 treatment led to a rapid reduction in parasitemia *in vivo*. This dual-site inhibitory activity offers an attractive approach to overcome the resistance liability associated with *Plasmodium* cytochrome *bc*$_1$ inhibitors.

Obtaining selectivity over the closely related mammalian cytochrome bc_1 complex remains a challenge for drug discovery. Cytochrome bc_1 inhibitors such as atovaquone and ELQ-300 show good selectivity for *Plasmodium* cytochrome bc_1 over the human orthologue [51, 52], but clinical development of the pyridones developed by GlaxoSmithKline (GSK) such as the Q_i site inhibitor GSK932121 was halted due to toxicity concerns, potentially related to off-target mammalian cytochrome bc_1 inhibition [53]. Further development of ELQ-400 is also hampered by a poor target selectivity window (*Pf*Cytbc_1 IC_{50} = 6.7 nM; *Hu*Cytbc_1 IC_{50} = 108 nM), although ELQ-400 did not show evidence of cytotoxicity [50]. The challenge for the development of dual Q_o/Q_i site inhibitors will be to achieve selectivity margins comparable to atovaquone and ELQ-300, while potently inhibiting both the *Plasmodium* bc_1 Q_o and Q_i sites.

In a related study, a structure-based approach was used to identify inhibitors targeting multiple enzymes within the *P. falciparum* mitochondrial respiratory chain [54]. The complexes of the respiratory chain are amenable to designed polypharmacology since they share a common quinone-binding site. RYL-581, a compound with potent antiplasmodium activity, was shown to bind to the allosteric site of type II nicotinamide adenine dinucleotide (NADH) dehydrogenase (NDH2) and the Q_o and Q_i sites of cytochrome bc_1 [54]. The *Plasmodium* and human respiratory chains differ significantly. The homodimer *Plasmodium* NDH2 (*Pf*NDH2) serves as an alternative to the multi-subunit complex I in the human respiratory chain. *Pf*NDH2 is not essential for asexual blood-stage development, and the asexual blood-stage antiplasmodium activity of several *Pf*NDH2 inhibitors is now attributed to the inhibition of the cytochrome bc_1 complex [55]. However, *Pf*NDH2 is important for the mosquito stages of the parasites, so inhibition of *Pf*NDH2 could contribute to transmission blocking [56]. *In vitro* MIR values have not been reported for RYL-581, but it would be interesting to see whether this inhibitor shows a reduced risk of resistance compared to atovaquone and ELQ-300 as was observed for the dual Q_o/Q_i site bc_1 complex inhibitor ELQ-400. Targeting multiple proteins in the respiratory chain through designed polypharmacology is an attractive approach for malaria drug discovery, although selectivity relative to related human off-targets will need to be carefully monitored.

27.4.3 Multi-target Kinase Inhibitors for Malaria

Plasmodium kinases have emerged as a promising, albeit challenging, target class for drug discovery [11, 57]. These enzymes are highly druggable, and many are essential for multiple stages of parasite development [58, 59]. The *P. falciparum* kinome is comprised of ~100 kinases, many of which can be classified into eukaryotic kinase groups [60]. The *Plasmodium* kinome is divergent from the human kinome, in terms of both overall composition (e.g., there are no tyrosine kinases in *Plasmodium*) and the sequence of individual kinases relative to their human orthologues. The generally conserved nature of the kinase adenosine triphosphate (ATP)-binding site, while posing challenges for compound selectivity relative to human kinase off-targets, makes kinases an ideal target class for exploring the potential benefits of designed polypharmacology, particularly with regard to reducing the resistance risk and overcoming other liabilities associated with individual kinase targets [11, 57].

The benefits of multi-kinase inhibitors have been best illustrated in the oncology field. Multi-kinase anticancer drugs exert their therapeutic effect by modulating networks of targets, improving efficacy, and limiting compensation by other pathways. There are many examples of clinically approved anticancer drugs that intentionally or unintentionally inhibit multiple kinases, where it is now appreciated that the secondary or additional kinase targets play a key role in clinical efficacy of the drug and reduce the risk of resistance [61]. Well-known examples include sorafenib (rapidly accelerated fibrosarcoma kinase B [B-Raf], vascular endothelial growth factor receptor [VEGFR], platelet-derived growth factor receptor [PDGFR], tyrosine-protein kinase KIT [c-Kit], and Fms-like tyrosine kinase-3 [Flt-3] inhibitor) used for the treatment of kidney and liver cancers and lapatinib (epidermal growth factor receptor [EGFR/ErbB1] and human epidermal growth factor receptor 2 [HER2/ErbB2] inhibitor) used for the treatment of HER2-positive advanced breast cancer [28].

Acquired resistance to kinase inhibitors is a major problem for cancer treatment. Due to the inherent instability of malignant cells, resistance mutations typically within the catalytic domain of the kinase target inevitably arise rendering the inhibitor ineffective. In this respect, multi-kinase inhibitors can offer a distinct advantage.

Similarly, resistance propensity is a key concern for kinase-targeted malaria drug discovery programs. *Plasmodium* phosphatidylinositol 4-kinase kinase beta (PI4Kβ), cyclin-dependent-like protein kinase 3 (CLK3), and cyclic guanosine monophosphate (cGMP)-dependent protein kinase (PKG) are well-validated *Plasmodium* kinase targets, with the potential to yield antimalarials with prophylactic liver-stage, asexual blood-stage, and transmission-blocking activity. However, both PI4Kβ and CLK3 inhibitors are associated with a moderate-to-high resistance risk, with inhibitors readily leading to resistance-conferring mutation in the kinase domain. Inhibitors for a range of other potential *Plasmodium* targets have been reported [62–64], although in most of these cases inhibition of the kinase in question has not conclusively been confirmed as the primary driver of the compound's phenotypic antiplasmodium activity, so further phenotypic target validation is required.

27.4.3.1 *Plasmodium* Phosphatidylinositol 4-kinase Kinase Beta (PI4Kβ) Inhibitors

Plasmodium PI4Kβ was first reported as a promising antimalarial target when it was identified as the primary target of a novel class of imidazopyrazine/pyridine compounds (KDU691 and KAI407) that displayed potent multistage antiplasmodium activity [65]. *In vitro* resistance selections with these compounds led to recrudescent parasites with mutations in the kinase domain of PI4Kβ, and these mutations were confirmed to confer drug resistance. Potent inhibition of the PI4Kβ was subsequently confirmed in an *in vitro* kinase inhibition assay using recombinant *P. vivax* PI4Kβ. PI4Kβ was also identified as the primary target of the 2-aminopyridine MMV390048 (Figure 27.3a) that progressed to phase 2a clinical trials for malaria [17, 66–68]. Despite good efficacy in the clinic, the clinical development of MMV390048 was halted due to data that emerged from embryofetal toxicity studies in rodents identifying a potential teratogenicity risk [69]. Although MMV390048 showed good selectivity for *Plasmodium* PI4Kβ relative to related human kinases *in vitro*, it has been speculated that off-target inhibition of mammalian kinases PI4Kβ, MAP4K4, and/or MINK1 may have caused the teratogenicity observed in rats [69]. Notably, several backup compounds with distinct chemotypes show different preclinical toxicity and human kinase inhibition profiles to MMV390048, so alternative chemotypes may still yield a novel antimalarial targeting PI4Kβ.

In addition to concerns around the safety of PI4Kβ inhibitors, the increased stringency of the resistance risk criteria set out by MMV poses an additional challenge for the development of PI4Kβ inhibitors. MMV390048 and other related PI4Kβ inhibitors show a moderate-to-high risk of resistance with an *in vitro* MIR between 10^6 and 10^7 [66, 70]. Resistant parasites generated *in vitro* show an increase in EC_{50} between threefold and 22-fold compared to wild-type parasites, with single-point mutations in the PI4Kβ kinase domain, small GTPase Rab11a known to interact with PI4Kβ, and/or PI4Kβ copy number variations. Notably, more recently PI4Kβ inhibitors showing a significantly reduced risk of resistance based on *in vitro* MIR studies have been identified, but it is not yet clear whether this is a result of secondary targets or a resistant refractory binding mode (unpublished).

27.4.3.2 *Plasmodium* Cyclin-Dependent-Like Protein Kinase 3 (CLK3) Inhibitors

Plasmodium CLK3 is another promising kinase target that is being pursued for the development of novel antimalarials [71, 72]. CLKs regulate messenger ribonucleic acid (mRNA) splicing. *Plasmodium* has four CLKs that are highly conserved across *Plasmodium* species.

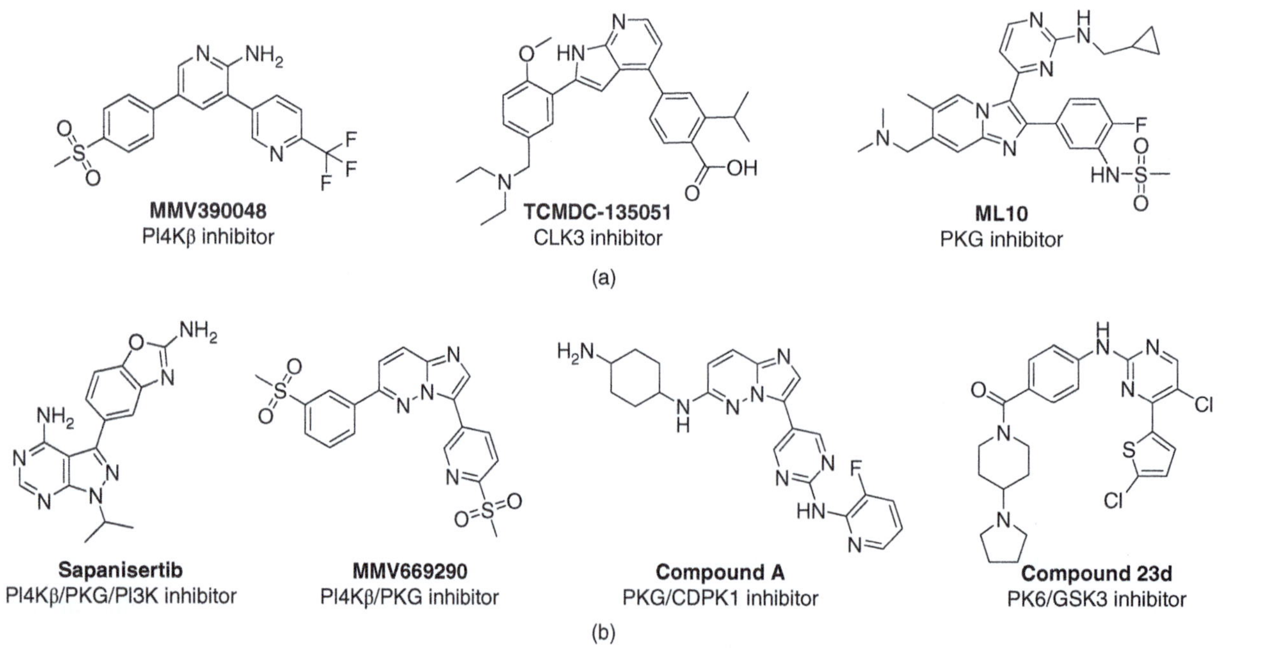

Figure 27.3 (a) Compounds targeting validated *Plasmodium* kinase targets PI4Kβ, CLK3, and PKG. (b) Compounds shown to inhibit multiple *Plasmodium* kinases.

The inhibitor TCMDC-135051 (Figure 27.3a), identified from a target-based screen against recombinant *P. falciparum* CLK3 (*Pf*CLK3), displayed multistage antiplasmodium activity and *in vivo* efficacy. *In vitro* resistance selections gave rise to CLK3-mediated resistance, with a mutation identified in the CLK3 kinase domain as well as a mutation in the putative RNA-processing protein ubiquitin-specific peptidase-39 (PfUSP39), yielding parasites that were fourfold to 11-fold less susceptible to TCMDC-135051 [71].

A related kinase, CLK1, is also essential for asexual blood-stage development and is of interest as a drug target for malaria although medicinal chemistry optimization programs against this target have not yet been reported [71].

27.4.3.3 *Plasmodium* cGMP-dependent Protein Kinase (PKG) Inhibitors

In contrast to PI4Kβ and CLK3, PKG has been reported to be refractory to resistance [73]. To date, resistance mutations have not been reported in PKG. *In vitro* resistance selections with the imidazopyridine ML10 and the trisubstituted imidazole MMV030084 (Figure 27.3a), two potent PKG inhibitors for which mode of action studies have clearly identified PKG as the primary efficacious target in asexual blood-stage parasites, did not lead to mutations in PKG [73, 74]. Reported efforts to generate resistance to ML10 *in vitro* were unsuccessful, and mutations in tyrosine kinase-like protein 3 (TKL3) resulted in low-level resistance to MMV030084. Conditional knockdown of PKG, but not other putative targets, resulted in increased parasite sensitivity to MMV030084, shown by a distinct leftward shift in EC_{50} [73].

Plasmodium PKG inhibitors typically extend into the so-called back pocket of PKG, a small hydrophobic pocket adjacent to the ATP-binding site that is accessible to bumped kinase inhibitors due to the presence of a small gatekeeper residue, that can be exploited for selectivity [74]. Engineered parasites with a PKG gatekeeper mutation (*P. falciparum* PKG [*Pf*PKG] T618Q) are resistant to PKG inhibitors such as ML10 and serve as a useful chemogenetic tool for demonstrating on-target PKG whole-cell activity [74, 75]. Notably, mutations to the gatekeeper residue in human kinase targets such as EGFR are major mediators of acquired resistance to cancer treatment [76]. It is therefore somewhat surprising that a *Plasmodium* PKG gatekeeper mutation has not been selected for using PKG inhibitors. Whether the apparent refractory nature of PKG to resistance is an inherent feature of the target, possibly due to restraints of the ATP-binding site, or whether it is a result of inhibition of secondary targets such as calcium-dependent protein kinase 1 (CDPK1), is not yet clear. The low risk of resistance observed for PKG inhibitors to date is attractive for drug discovery, although further work is required to confirm whether this will hold true *in vivo* and in the clinic.

27.4.3.4 Multi-target Kinase Inhibitors

Due to the conserved nature of the kinase ATP-binding site, ATP-competitive kinase inhibitors typically bind to multiple kinases as is observed in chemoproteomic studies (e.g., Kinobead studies, pull-down assays, or CETSA) aimed at identifying potential targets from whole-cell lysates [66, 73, 75, 77]. Despite this, the primary efficacious kinase target responsible for antiplasmodium activity in asexual blood-stage parasites can often be identified through resistance selections and/or conditional knockdown of the putative kinase targets [73, 77]. The potential role of the secondary targets in modulating resistance risk or contributing to compound efficacy at different stages of the parasite life cycle is often overlooked and more difficult to demonstrate.

Several kinase inhibitors have been shown to potently inhibit multiple *Plasmodium* kinases using *in vitro* kinase assays and/or chemoproteomic experiments [75, 77–81]. A few examples are shown in Figure 27.3b. These inhibitors may serve as good starting points for designed polypharmacology. Since it is significantly more challenging to optimize a compound for two *Plasmodium* kinase targets simultaneously, while maintaining good selectivity relative to human kinases, the benefits of multi-kinase activity need to be clearly demonstrated to justify this approach. Balancing inhibitory activity against *Plasmodium* kinase target pairs that are amenable to selective inhibition by a single compound with the potential to show additive or synergistic effects is not trivial. The relative potency against individual targets will need to be fine-tuned to fully achieve the desired benefits of dual inhibition. This will be affected by a range of factors in a cellular context including differences in target expression levels and patterns, cellular localization, target vulnerability, and ATP competition.

Sapanisertib, a human "mammalian target of rapamycin" (mTOR) inhibitor, was shown to inhibit multiple (kinase and non-kinase) targets to varying degrees with potent inhibition of *Plasmodium* PI4Kβ (IC_{50} = 4 nM), PKG (IC_{50} = 20 nM), and PI3K (IC_{50} = 56 nM) clearly noteworthy [77] (Figure 27.4b). Despite showing low nanomolar affinity against PKG and PI3K, it was clear from *in vitro* resistance selection and conditional knockdown experiments that PI4Kβ is the main driver of asexual blood-stage activity. To observe the benefits of polypharmacology that would be expected from the simultaneous inhibition of PI4Kβ and PKG such as reduced risk of resistance, further optimization will be required to fine-tune the relative potencies for PI4Kβ and PKG, taking into consideration the relative K_m values for ATP as well as target vulnerability [11]. Imidazopyridazines with potent antiplasmodium activity have also been reported to inhibit

Plasmodium PI4Kβ and PKG as well as CDPK1 in some cases [79, 80] (e.g., MMV669290 and Compound A; Figure 27.3b). *Plasmodium* CDPK1 was initially considered essential for asexual blood-stage development based on knockout experiments, but subsequent studies showed that CDPK1 inhibitors with potent antiplasmodium activity had additional modes of action driving whole-cell activity [80]. Furthermore, it was shown that PKG can compensate for reduced CDPK1 activity and that knockdown of CDPK1 has little effect on asexual blood-stage parasite viability [73, 82]. However, kinase inhibitors targeting both CDPK1 and PKG could be an effective strategy. CDPK1 plays a critical role in gametogenesis and mosquito stages, so inhibition of CDPK1 could also contribute to transmission blocking [83].

A high-throughput *in vitro* screen of nearly 30 000 compounds against recombinant *Pf*CLK1 and *Pf*CLK3 identified compounds with selective CLK1 and CLK3 inhibitory activity as well as compounds displaying potent dual CLK1/CLK3 inhibition [71]. While the reported study focused on using the selective CLK3 inhibitor TCMDC-135051 for CLK3 target validation, the data from this screen could provide a good starting point for the development of dual CLK1/CLK3 inhibitors. Other reports on multi-kinase inhibitors include dual *P. falciparum* protein kinase 6 (*Pf*PK6)/glycogen synthase kinase 3 (*Pf*GSK3) inhibitors [81] (e.g., Compound 23d; Figure 27.3b), although conclusive phenotypic validation of these targets is still lacking.

ATP-competitive kinase inhibitors are typically flat molecules and regularly show *in vitro* inhibition of beta-hematin formation, the synthetic form of hemozoin. This can translate into hemozoin inhibition and a corresponding increase in toxic-free heme in the parasite if the compound accumulates to sufficiently high concentrations within the acidic digestive vacuole, as is typically seen for basic compounds. A number of human kinase inhibitors show potent hemozoin inhibition, including human tyrosine kinase inhibitors lapatinib and nilotinib [58]. It is therefore plausible to develop dual *Plasmodium* kinase/hemozoin inhibitors and reports of *Plasmodium* [91] kinase inhibitors with hemozoin activity have started to emerge. This provides a unique opportunity to use polypharmacology to overcome the resistance risk associated with kinase inhibitors, while addressing the lack of multistage activity limiting the utility of hemozoin inhibitors for transmission blocking.

Plasmodium parasites modulate host erythrocyte signaling pathways to establish infection [84]. Human kinases that are activated during parasite infection have been proposed as potential targets for host-directed therapy [84–87]. While kinase-targeted drug discovery programs typically focus on improving selectivity for the *Plasmodium* kinase target relative to human kinase off-targets, simultaneously targeting *Plasmodium* kinases and human kinases that serve as host-dependency factors is another possible strategy for designed polypharmacology that could serve to reduce resistance propensity.

27.5 Concluding Remarks and the Way Forward

The malaria research community has made significant progress validating novel *Plasmodium* drug targets and developing the necessary tools to support targeted-based malaria drug discovery programs [6, 73, 77, 88–90]. The high resistance propensity typically associated with compounds optimized for potency against a single-protein target is problematic. New strategies are required to address resistance risk, prolong the life span of new antimalarials, and reduce the need for complex combination therapies. While designing compounds against multiple targets is far more complex, the potential benefits of this approach are substantial. Advances in artificial intelligence and machine learning tools, coupled with new lab-based technologies including gene editing and 'omics methods for *Plasmodium,* are enabling more complex approaches to rational drug design. Designed polypharmacology requires (i) the identification of suitable target pairs or sets, (ii) a strategy to optimize compounds against multiple *Plasmodium* targets while maintaining selectivity relative to related human proteins, and (iii) experimental methods to show that multi-target engagement in a cellular context leads to the desired pharmacological properties. While reports of multi-target inhibitors for malaria are starting to emerge, more work needs to be done to clearly demonstrate the advantage of dual or multi-target compounds for a particular target pair or set.

References

1 WHO (2023). World malaria report.

2 Uwimana, A., Umulisa, N., Venkatesan, M. et al. (2021). Association of *Plasmodium falciparum* kelch13 R561H genotypes with delayed parasite clearance in Rwanda: an open-label, single-arm, multicentre, therapeutic efficacy study. *Lancet Infect. Dis.* 21: 1120–1128.

3 Uwimana, A., Legrand, E., Stokes, B.H. et al. (2020). Emergence and clonal expansion of in

vitro artemisinin-resistant *Plasmodium falciparum* kelch13 R561H mutant parasites in Rwanda. *Nat. Med.* 26: 1602–1608.

4 Balikagala, B., Fukuda, N., Ikeda, M. et al. (2021). Evidence of artemisinin-resistant malaria in Africa. *N. Engl. J. Med.* 385: 1163–1171.

5 Tumwebaze, P.K., Conrad, M.D., Okitwi, M. et al. (2022). Decreased susceptibility of *Plasmodium falciparum* to both dihydroartemisinin and lumefantrine in northern Uganda. *Nat. Commun.* 13: 6353.

6 Yang, T., Ottilie, S., Istvan, E.S. et al. (2021). MalDA, accelerating malaria drug discovery. *Trends Parasitol.* 37: 493–507.

7 Siqueira-Neto, J.L., Wicht, K.J., Chibale, K. et al. (2023). Antimalarial drug discovery: progress and approaches. *Nat. Rev. Drug Discov.* 22: 807–826.

8 Guerra, F. and Winzeler, E.A. (2022). New targets for antimalarial drug discovery. *Curr. Opin. Microbiol.* 70: 102220.

9 Favuzza, P., de Lera Ruiz, M., Thompson, J.K. et al. (2020). Dual plasmepsin-targeting antimalarial agents disrupt multiple stages of the malaria parasite life cycle. *Cell Host Microbe* 27: 642–658.

10 Song, Z., Iorga, B.I., Mounkoro, P. et al. (2018). The antimalarial compound ELQ-400 is an unusual inhibitor of the bc1 complex, targeting both Qo and Qi sites. *FEBS Lett.* 592: 1346–1356.

11 Mogwera, K.S.P., Chibale, K., and Arendse, L.B. (2023). Developing kinase inhibitors for malaria: an opportunity or liability? *Trends Parasitol.* 39: 720–731.

12 Burrows, J.N., Duparc, S., Gutteridge, W.E. et al. (2017). New developments in anti-malarial target candidate and product profiles. *Malar. J.* 16: 26.

13 Duffey, M., Blasco, B., Burrows, J.N. et al. (2021). Assessing risks of *Plasmodium falciparum* resistance to select next-generation antimalarials. *Trends Parasitol.* 37: 709–721.

14 Cowell, A.N. and Winzeler, E.A. (2019). Advances in omics-based methods to identify novel targets for malaria and other parasitic protozoan infections. *Genome Med.* 11: 63.

15 Angulo-Barturen, I., Jiménez-Díaz, M.B., Mulet, T. et al. (2008). A murine model of falciparum-malaria by *in vivo* selection of competent strains in non-myelodepleted mice engrafted with human erythrocytes. *PLoS ONE* 3: e2252.

16 Stadler, E., Maiga, M., Friedrich, L. et al. (2023). Propensity of selecting mutant parasites for the antimalarial drug cabamiquine. *Nat. Commun.* 14: 5205.

17 McCarthy, J.S., Yalkinoglu, Ö., Odedra, A. et al. (2021). Safety, pharmacokinetics, and antimalarial activity of the novel plasmodium eukaryotic translation elongation factor 2 inhibitor M5717: a first-in-human, randomised, placebo-controlled, double-blind, single ascending dose study and volunteer infection study. *Lancet Infect. Dis.* 21: 1713–1724.

18 Qiu, D., Pei, J.V., Rosling, J.E.O. et al. (2022). A G358S mutation in the *Plasmodium falciparum* Na^+ pump PfATP4 confers clinically-relevant resistance to cipargamin. *Nat. Commun.* 13: 5746.

19 Schmitt, E.K., Ndayisaba, G., Yeka, A. et al. (2022). Efficacy of cipargamin (KAE609) in a randomized, phase II dose-escalation study in adults in sub-Saharan Africa with uncomplicated *Plasmodium falciparum* malaria. *Clin. Infect. Dis.* 74: 1831–1839.

20 Phillips, M.A., Lotharius, J., Marsh, K. et al. (2015). A long-duration dihydroorotate dehydrogenase inhibitor (DSM265) for prevention and treatment of malaria. *Sci. Transl. Med.* 7: 296ra111-296ra111.

21 Collins Katharine, A., Rückle, T., Elliott, S. et al. (2019). DSM265 at 400 milligrams clears asexual stage parasites but not mature gametocytes from the blood of healthy subjects experimentally infected with *Plasmodium falciparum*. *Antimicrob. Agents Chemother.* 63: e01837–e01818. https://doi.org/10.1128/AAC.01837-18.

22 White, J., Dhingra, S.K., Deng, X. et al. (2019). Identification and mechanistic understanding of dihydroorotate dehydrogenase point mutations in *Plasmodium falciparum* that confer in vitro resistance to the clinical candidate DSM265. *ACS Infect. Dis.* 5: 90–101.

23 Palmer, M.J., Deng, X., Watts, S. et al. (2021). Potent antimalarials with development potential identified by structure-guided computational optimization of a pyrrole-based dihydroorotate dehydrogenase inhibitor series. *J. Med. Chem.* 64: 6085–6136.

24 Kümpornsin, K., Kochakarn, T., Yeo, T. et al. (2023). Generation of a mutator parasite to drive resistome discovery in *Plasmodium falciparum*. *Nat. Commun.* 14: 3059.

25 Goldberg, D.E., Slater, A.F., Cerami, A., and Henderson, G.B. (1990). Hemoglobin degradation in the malaria parasite *Plasmodium falciparum*: an ordered process in a unique organelle. *Proc. Natl. Acad. Sci.* 87: 2931–2935.

26 Combrinck, J.M., Mabotha, T.E., Ncokazi, K.K. et al. (2013). Insights into the role of heme in the mechanism of action of antimalarials. *ACS Chem. Biol.* 8: 133–137.

27 Sidhu, A.B., Verdier-Pinard, D., and Fidock, D.A. (2002). Chloroquine resistance in *Plasmodium falciparum* malaria parasites conferred by pfcrt mutations. *Science* 298: 210–213.

28 Peters, J.-U. (2013). Polypharmacology – foe or friend? *J. Med. Chem.* 56: 8955–8971.

29 Nasamu, A.S., Polino, A.J., Istvan, E.S., and Goldberg, D.E. (2020). Malaria parasite plasmepsins: More than

just plain old degradative pepsins. *J. Biol. Chem.* 295: 8425–8441.

30 de Lera Ruiz, M., Favuzza, P., Guo, Z. et al. (2022). The invention of WM382, a highly potent PMIX/X dual inhibitor toward the treatment of malaria. *ACS Med. Chem. Lett.* 13: 1745–1754.

31 Pino, P., Caldelari, R., Mukherjee, B. et al. (2017). A multistage antimalarial targets the plasmepsins IX and X essential for invasion and egress. *Science* 358: 522–528.

32 Nasamu, A.S., Glushakova, S., Russo, I. et al. (2017). Plasmepsins IX and X are essential and druggable mediators of malaria parasite egress and invasion. *Science* 358: 518–522.

33 Hodder, A.N., Christensen, J., Scally, S. et al. (2022). Basis for drug selectivity of plasmepsin IX and X inhibition in *Plasmodium falciparum* and *vivax*. *Structure* 30: 947–961.

34 Srivastava, I.K., Morrisey, J.M., Darrouzet, E. et al. (1999). Resistance mutations reveal the atovaquone-binding domain of cytochrome *b* in malaria parasites. *Mol. Microbiol.* 33: 704–711.

35 Siregar, J.E., Kurisu, G., Kobayashi, T. et al. (2015). Direct evidence for the atovaquone action on the *Plasmodium* cytochrome *bc1* complex. *Parasitol. Int.* 64: 295–300.

36 Barton, V., Fisher, N., Biagini, G.A. et al. (2010). Inhibiting *Plasmodium* cytochrome *bc1*: a complex issue. *Curr. Opin. Chem. Biol.* 14: 440–446.

37 Solmaz, S.R. and Hunte, C. (2008). Structure of complex III with bound cytochrome c in reduced state and definition of a minimal core interface for electron transfer. *J. Biol. Chem.* 283: 17542–17549.

38 Barata, L., Houzé, P., Boutbibe, K. et al. (2016). In vitro analysis of the interaction between atovaquone and proguanil against liver stage malaria parasites. *Antimicrob. Agents Chemother.* 60: 4333–4335.

39 Goodman, C.D., Siregar, J.E., Mollard, V. et al. (2016). Parasites resistant to the antimalarial atovaquone fail to transmit by mosquitoes. *Science* 352: 349–353.

40 Painter, H.J., Morrisey, J.M., Mather, M.W., and Vaidya, A.B. (2007). Specific role of mitochondrial electron transport in blood-stage *Plasmodium falciparum*. *Nature* 446: 88–91.

41 Korsinczky, M., Chen, N., Kotecka, B. et al. (2000). Mutations in *Plasmodium falciparum* cytochrome *b* that are associated with atovaquone resistance are located at a putative drug-binding site. *Antimicrob. Agents Chemother.* 44: 2100–2108.

42 Chiodini, P.L., Conlon, C.P., Hutchinson, D.B.A. et al. (1995). Evaluation of atovaquone in the treatment of patients with uncomplicated *Plasmodium falciparum* malaria. *J. Antimicrob. Chemother.* 36: 1073–1078.

43 Fisher, N., Abd Majid, R., Antoine, T. et al. (2012). Cytochrome *b* mutation Y268S conferring atovaquone resistance phenotype in malaria parasite results in reduced parasite *bc1* catalytic turnover and protein expression. *J. Biol. Chem.* 287: 9731–9741.

44 Looareesuwan, S., Viravan, C., Webster, H.K. et al. (1996). Clinical studies of atovaquone, alone or in combination with other antimalarial drugs, for treatment of acute uncomplicated malaria in Thailand. *Am. J. Trop. Med. Hyg.* 54: 62–66.

45 Musset, L., Bouchaud, O., Matheron, S. et al. (2006). Clinical atovaquone-proguanil resistance of *Plasmodium falciparum* associated with cytochrome *b* codon 268 mutations. *Microbes Infect.* 8: 2599–2604.

46 Berry, A., Senescau, A., Lelièvre, J. et al. (2006). Prevalence of *Plasmodium falciparum* cytochrome *b* gene mutations in isolates imported from Africa, and implications for atovaquone resistance. *Trans. R. Soc. Trop. Med. Hyg.* 100: 986–988.

47 Nilsen, A., Miley, G.P., Forquer, I.P. et al. (2014). Discovery, synthesis, and optimization of antimalarial 4(1H)-quinolone-3-diarylethers. *J. Med. Chem.* 57: 3818–3834.

48 Stickles, A.M., de Almeida, M.J., Morrisey, J.M. et al. (2015). Subtle changes in endochin-like quinolone structure alter the site of inhibition within the cytochrome *bc1* complex of *Plasmodium falciparum*. *Antimicrob. Agents Chemother.* 59: 1977–1982.

49 Stickles, A.M., Smilkstein, M.J., Morrisey, J.M. et al. (2016). Atovaquone and ELQ-300 combination therapy as a novel dual-site cytochrome *bc1* inhibition strategy for malaria. *Antimicrob. Agents Chemother.* 60: 4853–4859.

50 Stickles, A.M., Ting, L.M., Morrisey, J.M. et al. (2015). Inhibition of cytochrome bc1 as a strategy for single-dose, multi-stage antimalarial therapy. *Am. J. Trop. Med. Hyg.* 92: 1195–1201.

51 Fry, M. and Pudney, M. (1992). Site of action of the antimalarial hydroxynaphthoquinone, 2-[trans-4-(4'-chlorophenyl) cyclohexyl]-3-hydroxy-1,4-naphthoquinone (566C80). *Biochem. Pharmacol.* 43: 1545–1553.

52 Nilsen, A., LaCrue, A.N., White, K.L. et al. (2013). Quinolone-3-diarylethers: a new class of antimalarial drug. *Sci. Transl. Med.* 5: 177ra137-177ra137.

53 Bueno, J.M., Herreros, E., Angulo-Barturen, I. et al. (2012). Exploration of 4(1H)-pyridones as a novel family of potent antimalarial inhibitors of the plasmodial cytochrome *bc1*. *Future Med. Chem.* 4: 2311–2323.

54 Yang, Y., Tang, T., Li, X. et al. (2021). Design, synthesis, and biological evaluation of multiple targeting antimalarials. *Acta Pharm. Sin. B* 11: 2900–2913.

55 Ke, H., Ganesan, S.M., Dass, S. et al. (2019). Mitochondrial type II NADH dehydrogenase of *Plasmodium falciparum* (PfNDH2) is dispensable in the asexual blood stages. *PLoS ONE* 14: e0214023.

56 Boysen, K.E. and Matuschewski, K. (2011). Arrested oocyst maturation in *Plasmodium* parasites lacking Type II NADH:ubiquinone dehydrogenase. *J. Biol. Chem.* 286: 32661–32671.

57 Arendse, L.B., Wyllie, S., Chibale, K., and Gilbert, I.H. (2021). *Plasmodium* kinases as potential drug targets for malaria: challenges and opportunities. *ACS Infect. Dis.* 7: 518–534.

58 Tewari, R., Straschil, U., Bateman, A. et al. (2010). The systematic functional analysis of *Plasmodium* protein kinases identifies essential regulators of mosquito transmission. *Cell Host Microbe* 8: 377–387.

59 Solyakov, L., Halbert, J., Alam, M.M. et al. (2011). Global kinomic and phospho-proteomic analyses of the human malaria parasite *Plasmodium falciparum*. *Nat. Commun.* 2: 565.

60 Adderley, J. and Doerig, C. (2022). Comparative analysis of the kinomes of *Plasmodium falciparum*, *Plasmodium vivax* and their host Homo sapiens. *BMC Genom.* 23: 237.

61 Roskoski, R. Jr. (2024). Properties of FDA-approved small molecule protein kinase inhibitors: a 2024 update. *Pharmacol. Res.* 200: 107059.

62 Ong, H.W., Adderley, J., Tobin, A.B. et al. (2023). Parasite and host kinases as targets for antimalarials. *Expert Opin. Ther. Targets* 27: 151–169.

63 Gonzalez-Cabrera, D., Horatscheck, A., Wilson, C.R. et al. (2018). Plasmodial kinase inhibitors: license to cure? *J. Med. Chem.* 61: 8061–8077.

64 Moolman, C., Sluis, R.V., Beteck, R.M., and Legoabe, L.J. (2020). An update on development of small-molecule plasmodial kinase inhibitors. *Molecules* 25: 5182.

65 McNamara, C.W., Lee, M.C.S., Lim, C.S. et al. (2013). Targeting *Plasmodium* PI(4)K to eliminate malaria. *Nature* 504: 248–253.

66 Paquet, T., Le Manach, C., Cabrera, D.G. et al. (2017). Antimalarial efficacy of MMV390048, an inhibitor of *Plasmodium* phosphatidylinositol 4-kinase. *Sci. Transl. Med.* 9: eaad9735.

67 Mohammed, R., Asres, M.S., Gudina, E.K. et al. (2023). Efficacy, safety, tolerability, and pharmacokinetics of MMV390048 in acute uncomplicated malaria. *Am. J. Trop. Med. Hyg.* 108: 81–84.

68 Sinxadi, P., Donini, C., Johnstone, H. et al. (2020). Safety, tolerability, pharmacokinetics, and antimalarial activity of the novel plasmodium phosphatidylinositol 4-kinase inhibitor MMV390048 in healthy volunteers. *Antimicrob. Agents Chemother.* 64: e01896-19.

69 Demarta-Gatsi, C., Donini, C., Duffy, J. et al. (2022). Malarial PI4K inhibitor induced diaphragmatic hernias in rat: potential link with mammalian kinase inhibition. *Birth Defects Res.* 114: 487–498.

70 Brunschwig, C., Lawrence, N., Taylor, D. et al. (2018). UCT943, a next-generation *Plasmodium falciparum* PI4K inhibitor preclinical candidate for the treatment of malaria. *Antimicrob. Agents Chemother.* 62: e00012–e00018.

71 Alam, M.M., Sanchez-Azqueta, A., Janha, O. et al. (2019). Validation of the protein kinase PfCLK3 as a multistage cross-species malarial drug target. *Science* 365: eaau1682.

72 Mahindra, A., Janha, O., Mapesa, K. et al. (2020). Development of potent PfCLK3 inhibitors based on TCMDC-135051 as a new class of antimalarials. *J. Med. Chem.* 63: 9300–9315.

73 Vanaerschot, M., Murithi, J.M., Pasaje, C.F.A. et al. (2020). Inhibition of resistance-refractory *P. falciparum* kinase PKG delivers prophylactic, blood stage, and transmission-blocking antiplasmodial activity. *Cell Chem. Biol.* 27: 806–816.e808.

74 Baker, D.A., Stewart, L.B., Large, J.M. et al. (2017). A potent series targeting the malarial cGMP-dependent protein kinase clears infection and blocks transmission. *Nat. Commun.* 8: 430.

75 Penzo, M., de Las Heras-Dueña, L., Mata-Cantero, L. et al. (2019). High-throughput screening of the *Plasmodium falciparum* cGMP-dependent protein kinase identified a thiazole scaffold which kills erythrocytic and sexual stage parasites. *Sci. Rep.* 9: 1–13.

76 Roskoski, R. (2023). Properties of FDA-approved small molecule protein kinase inhibitors: a 2023 update. *Pharmacol. Res.* 187: 106552.

77 Arendse, L.B., Murithi, J.M., Qahash, T. et al. (2022). The anticancer human mTOR inhibitor sapanisertib potently inhibits multiple *Plasmodium* kinases and life cycle stages. *Sci. Transl. Med.* 14: eabo7219.

78 Ong, H.W., de Silva, C., Avalani, K. et al. (2023). Characterization of 2,4-dianilinopyrimidines against five *P. falciparum* kinases PfARK1, PfARK3, PfNEK3, PfPK9, and PfPKB. *ACS Med. Chem. Lett.* 14: 1774–1784.

79 Cheuka, P.M., Centani, L., Arendse, L.B. et al. (2021). New amidated 3,6-diphenylated imidazopyridazines with potent antiplasmodium activity are dual inhibitors of *Plasmodium* phosphatidylinositol-4-kinase and

cGMP-dependent protein kinase. *ACS Infect. Dis.* 7: 34–46.

80 Green, J.L., Moon, R.W., Whalley, D. et al. (2015). Imidazopyridazine inhibitors of Plasmodium falciparum calcium-dependent protein kinase 1 also target cyclic GMP-dependent protein kinase and heat shock protein 90 to kill the parasite at different stages of intracellular development. *Antimicrob. Agents Chemother.* 60: 1464–1475.

81 Galal, K.A., Truong, A., Kwarcinski, F. et al. (2022). Identification of novel 2,4,5-trisubstituted pyrimidines as potent dual inhibitors of Plasmodial PfGSK3/PfPK6 with activity against blood stage parasites *in vitro*. *J. Med. Chem.* 65: 13172–13197.

82 Bansal, A., Ojo, K.K., Mu, J. et al. (2016). Reduced activity of mutant calcium-dependent protein kinase 1 is compensated in *Plasmodium falciparum* through the action of protein kinase G. *MBio* 7: e02011–e02016.

83 Bansal, A., Molina-Cruz, A., Brzostowski, J. et al. (2018). PfCDPK1 is critical for malaria parasite gametogenesis and mosquito infection. *Proc. Natl. Acad. Sci. USA* 115: 774–779.

84 Adderley, J.D., John von Freyend, S., Jackson, S.A. et al. (2020). Analysis of erythrocyte signalling pathways during *Plasmodium falciparum* infection identifies targets for host-directed antimalarial intervention. *Nat. Commun.* 11: 4015.

85 Rashidi, S., Mansouri, R., Ali-Hassanzadeh, M. et al. (2021). The host mTOR pathway and parasitic diseases pathogenesis. *Parasitol. Res.* 120: 1151–1166.

86 Wei, L., Adderley, J., Leroy, D. et al. (2021). Host-directed therapy, an untapped opportunity for antimalarial intervention. *Cell Rep. Med.* 2: 100423.

87 Adderley, J. and Grau, G.E. (2023). Host-directed therapies for malaria: possible applications and lessons from other indications. *Curr. Opin. Microbiol.* 71: 102228.

88 Adjalley, S. and Lee, M.C.S. (2022). CRISPR/Cas9 editing of the *Plasmodium falciparum* genome. In: *Malaria Immunology: Targeting the Surface of Infected Erythrocytes* (ed. A.T.R. Jensen and L. Hviid), 221–239. New York, NY, Springer.

89 Summers, R.L., Pasaje, C.F.A., Pisco, J.P. et al. (2022). Chemogenomics identifies acetyl-coenzyme A synthetase as a target for malaria treatment and prevention. *Cell Chem. Biol.* 29: 191–201.

90 Istvan, E.S., Das, S., Bhatnagar, S. et al. (2019). Plasmodium Niemann-Pick type C1-related protein is a druggable target required for parasite membrane homeostasis. *elife* 8: e40529.

91 Dziwornu, G.A., Seanego, D., Fienberg, S. et al. (2024). 2,8-Disubstituted-1, 5-naphthyridines as dual inhibitors of *Plasmodium falciparum* phosphatidylinositol-4-kinase and hemozoin formation with *in vivo* efficacy. *Journal of Medicinal Chemistry* 67: 11401–11420.

28

Multi-target Compounds for Tuberculosis

Giovanni Stelitano, Mario Cocorullo, and Laurent R. Chiarelli

28.1 Tuberculosis and the Problem of Antimicrobial Resistance

Tuberculosis (TB) is still among the most frequent chronic infectious diseases and the second leading cause of death from a single infectious agent after COVID-19, thus representing an important threat to public health [1]. According to the WHO annual report, in 2022, TB caused an estimated 1.30 million deaths, with about 7.5 million people that were newly diagnosed with TB, the highest number since the first WHO global TB monitoring in 1995 [2].

The main issue in treating TB is the spread of strains of *Mycobacterium tuberculosis* (*Mtb*), the etiological agent of TB that are resistant to currently available drugs. Indeed, the standard treatment regimen, consisting of 2 months' administration of isoniazid, rifampicin, pyrazinamide, and ethambutol, followed by 4 months of treatment with isoniazid and rifampicin, ensures a successful outcome in about 85% of patients [3]. Nevertheless, in 2022, about 450000 people developed drug-resistant TB (DR-TB), which includes RR-TB (resistant to rifampicin) and MDR-TB (resistant to both rifampicin and isoniazid). These resistant forms, together with the spread of pre-XDR (extensively drug-resistant, further resistant to any fluoroquinolone) and XDR-TB (further resistant to any fluoroquinolone drug plus at least bedaquiline (BDQ) or linezolid), represent a major challenge for the treatment [2]. Indeed, considering that the treatment is successful in only 60% of MDR-TB and 39% of XDR-TB cases and that regimens for DR-TB are more expensive, less effective, and more toxic, there is an urgent need to develop new anti-TB drugs.

Mtb is an aerobic obligate infectious microorganism that was identified for the first time by Robert Koch in 1882 by analyzing the sputum of TB-infected patients through a new staining method [4]. *Mtb* is transmitted through aerosol droplets since it mainly infects lungs, but it can become extrapulmonary, spreading to other organs via blood and lymphatic circulation or direct pathways [5].

This peculiar microorganism shows features shared with both Gram-positive and Gram-negative bacteria [6]. The *Mtb* cell wall is an example of this complexity, being composed by an outside bilayer of very long-chain mycolic acids and glycolipids that resemble to Gram-negative cell wall and an inner multilayer of peptidoglycan covalently linked to arabinogalactan similarly to Gram-positive bacteria [6, 7]. This thick lipophilic structure, which impairs the crossing of many molecules, including drugs, is *Mtb's* first defense and an exquisite example of intrinsic resistance, along with efflux systems and detoxifying enzymes.

Mtb possesses five types of efflux systems, classified by sequence homology: the ATP-binding cassette (ABC) and the major facilitator superfamilies (MFS), the multidrug and toxic compound extrusion family (MATE), the resistance-nodulation-cell-division superfamily (RND), and the small multidrug resistance family (SMR) [8]. Each family is associated to a specific multi-protein pump complex and utilizes a different energy source to export the substrate outside the cell through a specific pool of carrier proteins [9].

The ABC superfamily transporters, for instance, are the most abundant and are implicated in *Mtb* resistance to chloramphenicol, macrolides, tetracyclines, and other drugs [10]. These transporters show a dimeric structure with an intracellular and an extracellular gate in opposite states, whose closing and opening are regulated by the binding and hydrolysis of ATP in a specific site located at cytoplasmic region [11].

MFS transporters are associated with resistance to a wide number of drugs and other toxic compounds [10]. The protonation of specific transporter residues followed by the binding with the carrier induces the switch of state of this transporter, allowing the final release of the complex drug carrier outside of the cell [12].

RND pumps are generally characterized by a heterotrimeric structure composed by an inner membrane protein and an outer membrane protein connected by a

Polypharmacology: Strategies for Multi-Target Drug Discovery, First Edition. Edited by Jens-Uwe Peters.

periplasmic membrane fusion protein [13]. The pump can switch cyclically among three states: loose (L), tight (T), and open (O). This consequential change of conformation allows the binding with the substrate (during the L state), its movement along the pump (T), and its release outside the cell (O) [8]. RND pumps such as MmpS5/MmpL5 are associated with *Mtb* resistance to BDQ and clofazimine [14].

The only SMR family efflux pump reported to date is Mmr encoded by the gene *Rv3065*, which is associated with resistance to different drugs, including isoniazid, fluoroquinolones, macrolides, ethidium bromide, acriflavine, the pump inhibitor verapamil, and other small molecules [15, 16]. Finally, the *Mbt* genome codifies for only one MATE family exporter, DinF, whose function has not been fully clarified yet [17].

Another important cause of bacterial intrinsic resistance mechanism is represented by detoxifying enzymes. Transferases, for instance, are a class of enzymes that can inactivate a drug by adding chemical moieties in its structure, which alter the physico-chemical properties, hence the interaction with the target. Some examples are the aminoglycoside acetyltransferase, Eis, which inactivates kanamycin and amikacin by acetylating the free amine group of the drugs [18, 19], or the arylamine N-acetyltransferase NAT, which exploits the same mechanism of action to inactivate isoniazid [19]. Other enzymes may hydrolyze specific bonds producing inactive metabolites, such as the β-lactamase BlaC, capable of inactivating a plethora of β-lactams by hydrolyzing the β-lactam ring [20].

In addition to intrinsic resistance mechanisms, acquired resistance is also present as gene mutations, while no examples of exogen DNA acquisition have been reported in *Mtb* until now. Acquired resistance often comes with a fitness cost, such as a slower growth rate compared to the wild type, but compensatory mechanisms may overcome this problem [21]. A single point mutation (SNP) on a gene is enough to produce an amino acid change in the target, eventually altering the bond with the drug. Similarly, an amino acid change in the activating enzyme could interfere with the activation of prodrugs. Both cases have been reported in *Mtb*. For instance, 35 different mutations conferring resistance to rifampicin have been identified in the gene *rpoB* encoding the RNA polymerase β subunit [22], located in the region corresponding to the rifampicin binding site [23]. By contrast, the most predominant resistance mechanism observed for isoniazid relies on mutations in the catalase-peroxidase KatG enzyme, which is responsible for the activation of this prodrug [24].

The treatment of drug-susceptible TB with first-line drugs has led to the development of DR-TB due to the evolutionary development of *Mtb* and the redundant treatment regimens [25]. The treatment of DR-TB remains difficult because of the long duration and complicated drug regimens that, in case of MDR-TB and XDR-TB, frequently require the administration of multiple drugs, which affect the treatment response [26]. Therapy adherence is crucial in the TB regimen, as inadequate treatment can increase the probability of treatment failure, leading to several issues such as disease recurrence and the development of drug resistance. However, therapy is often not always adhered to, for several reasons, including the absence of medicine or the high cost of the treatment [27]. Moreover, several factors also related to the abovementioned intrinsic characteristics of *Mtb* limit drug discovery attempts to find highly effective drugs with low side effects for the treatment of DR-TB. Indeed, high-throughput screening has been associated with a high failure rate in drug discovery and development for this disease, as lead candidates usually are not active against whole-cell *Mtb* [28]. The spread of resistant pathogens frequently makes even modern antimicrobials unresponsive. Therefore, finding new antibiotics is not the only solution to manage antibiotic resistance. The diverse behavior of TB in the various stages of replication and in the dormant state requires a strong synergy between potential drug therapies and alternative strategies to discover new putative drug regimens that address many drug targets associated with *Mtb* metabolism [29].

All the described features involved in antimicrobial resistance (AMR) of *Mtb* are major obstacles to the development of new drugs against this pathogen, resulting in the need to develop new strategies.

28.2 Polypharmacology to Fight *M. tuberculosis* Antimicrobial Resistance

Polypharmacology, which refers to a multitarget strategy, represents an attractive route in the path to eradicate TB and counteract the spread of MDR and XDR *Mtb* strains [30]. Polypharmacology is in contraposition with classical polypharmacy, that uses different co-administered drugs, each with high affinity for a single target, and which represents the current approach in TB. However, as previously pointed out, AMR has become a public health issue, resulting in the emergence of multi-drug-resistant bacteria that are no longer treatable with any of the conventional classes of antibiotics. For a long time, antibiotics with a specific protein target have been highly demanded in drug design. The idea that single-target drugs are the most effective antibiotics is based on the assumption that a high affinity and selectivity for the target is associated with reduced risk of side effects. This approach has primarily led to a biased approach to antibiotic discovery, in which

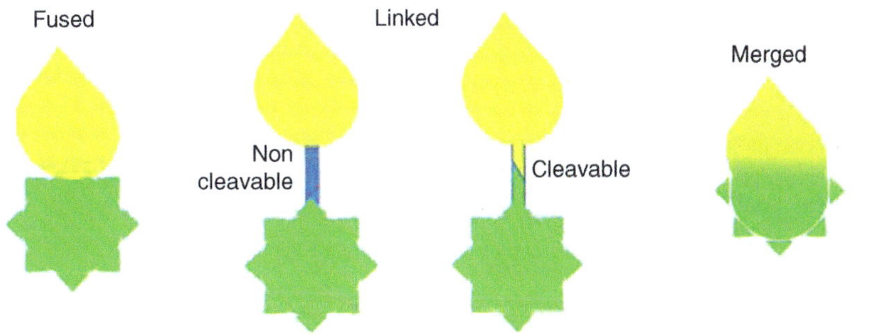

Figure 28.1 Possible strategy for the rational design of multitargeting compounds by combining different pharmacophoric moieties: in fused compounds, the pharmacophores are directly connected together, in linked compounds, the moieties are attached through a linker that may be cleavable or not, and in merged compounds, two parts of the compounds overlap and can be merged.

single protein targets are investigated, while compounds with multiple targets are thought to be nonspecific and not worthy of being developed. Nevertheless, antibiotics with a long history of clinical efficacy rarely target a single molecule. In fact, even naturally occurring antibiotics often do not have a single strict target [31].

In the fight against MDR- and XDR-TB strains, drug resistance mechanisms are a significant concern, considering that a pathogen often needs only a single base mutation to develop resistance to an antimicrobial. In this context, the development of a single pharmaceutical molecule that acts on multiple targets is one possible strategy to circumvent the problem [32]. Furthermore, polypharmacology may present additional advantages: for instance, a single multitargeting molecule should conceivably have potential toxicity lower than a mix of several compounds and likely without the issue of drug–drug interactions [33].

Multitargeting compounds can be classified according to their mechanism of action and the different levels at which they act. For instance, it has been proposed to distinguish them into multi-effective, multifunctional, and multitarget [31].

Multi-effectives are compounds that have a single target involved in several pathways, all of which are affected by the inhibition. This is the case with antibiotics that interfere with the integrity of the cytoplasmic membrane, such as Cyclic β-sheet peptides. Indeed, these antimicrobial peptides by altering membrane fluidity and integrity, impair the function of several processes, including cell division and cell wall biosynthesis, as well as membrane potential, thus ATP synthesis [34].

Multifunctional compounds, besides their antimicrobial activity, have an additional unrelated activity that results in useful for the treatment; these compounds usually exhibit anti-inflammatory or immunomodulatory properties [35].

Finally, actual multitarget compounds can act through series inhibition or parallel inhibition. The first case refers to compounds that target related proteins belonging to the same pathway, such as β-lactams that can inhibit several penicillin-binding proteins [36] or D-cycloserine, which is an inhibitor of two enzymes involved in peptidoglycan synthesis, namely, alanine racemase and D-Ala-D-Ala ligase [37].

By contrast, in parallel inhibition, the compounds act on two or more different targets, not related and belonging to distinct pathways. This type of inhibition has the advantage of being the one associated with the lowest rate of resistance. In fact, to develop resistance in this case, several mutations in several genes are required; moreover, the bacteria have to cope with more stresses caused by the different activities of the molecule, thus having a reduced adaptability [31].

Given the great potential of multitarget compounds, the possibility of modifying and optimizing the scaffolds has also been studied, as well as the possibility to project a potential multitarget drug through rational design [30]. In this context, multitargeting compounds can be distinguished into fused, linked, or merged pharmacophores, depending on how the different molecules are joined together to obtain a compound that preserves their activities [38]. Fused pharmacophores are obtained by joining small molecules with a bond that can be cleavable or not. In linked pharmacophores, the molecules are bound together by either a stable or cleavable linker. Finally, merged pharmacophores can be obtained if a portion of two different molecules overlaps, resulting in their integration (Figure 28.1). Clearly, in all cases, the different moieties should be combined in order to maintain the correct molecular geometry, charge distribution, and all the features necessary to interact with the different targets.

28.3 Multitarget Compounds Against TB

Considering that one of the major problems related to TB is the spread of multiresistance, there has been a growing interest in polypharmacology in recent years, with several multitargeting antitubercular compounds that have emerged [39].

Actually, many of these compounds have been identified by phenotypic screening campaigns or by biochemical

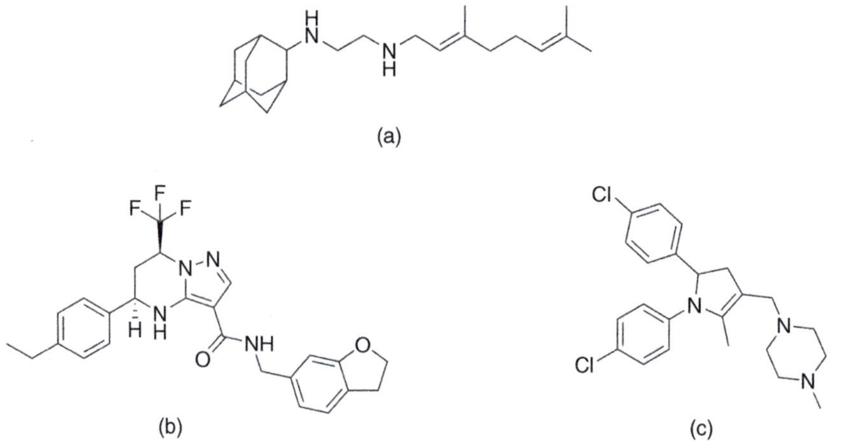

Figure 28.2 Chemical structure of the multitargeting inhibitors of MmpL3: (a) the ethylenediamine compound SQ109; (b) the tetrahydropyrazolo[1,5-*a*]pyrimidine-3-carboxamide GSK951A; and (c) the pyrrole derivative BM212.

screens against specific targets, and in retrospect, they were found to have multiple targets. A clear example of this type of compound is represented by the MmpL3 inhibitors. The mycolic acid transporter MmpL3 was classified among the "promiscuous target", targets that are frequently found in phenotypic screening campaigns to be inhibited by different scaffolds [40].

Among these inhibitors, the ethylenediamine compound SQ109 is the most advanced in the TB drug pipeline, being in Phase II clinical trials (Figure 28.2a). Interestingly, MmpL3 was proposed as the target on the basis that *Mtb*-resistant mutants, isolated from cultures in the presence of several SQ109 analogues, showed mutations in the *mmpL3* gene and cross-resistance to SQ109 itself. Nevertheless, the level of resistance to SQ109 of these strains was relatively low. This observation, together with the fact that no spontaneous-resistant mutants have ever been obtained with SQ109, has suggested that this compound might have multiple targets [41]. For instance, it has been suggested that the compound has an indirect effect on MmpL3, hypothesizing that it disrupts the pH gradient or the membrane potential, leading to the disruption of the proton relay necessary for MmpL3 activity.

This hypothesis was effectively confirmed by both *in vitro* and whole cell assays, which demonstrated that SQ109 and analogues act as uncouplers, thus affecting the proton motive force. [32]. Moreover, this study demonstrated that the compound was also able to bind and inhibit the two enzymes MenA and MenG, which are involved in menaquinone biosynthesis, thus affecting electron transport and consequently inhibiting respiration and ATP biosynthesis [32]. The direct involvement of MmpL3 was however confirmed by crystallographic studies, which showed that SQ109, as well as other inhibitors, binds inside the proton-translocation channel of the protein [42].

In this view, SQ109 could be considered a network multitarget compound, involving series inhibition by affecting energetic pathways through the inhibition of MenA and MenG and the disruption of proton motif force, but also parallel inhibition by directly and indirectly interfering with MmpL3 activity, thus with mycolates transport [32].

Interestingly, several other MmpL3 inhibitors show the characteristic of being multitarget compounds, such as the tetrahydropyrazolo[1,5-*a*]pyrimidine-3-carboxamide (THPP, Figure 28.2b), which also inhibits the enoyl-coenzyme A (CoA) hydratase EchA6, an enzyme involved in the biosynthesis of mycolates [43].

Similarly, the pyrrole derivative BM212 (Figure 28.2c), besides inhibiting MmpL3, was demonstrated to act as an uncoupler [44], as well as to bind the transcriptional repressor EthR2 [45]. The latter protein represses the expression of EthA2, an enzyme involved in the bioactivation of the antitubercular drug ethionamide (ETH), and has been exploited for the development of ETH boosters, compounds useful to circumvent the innate or acquired resistance of *Mtb* to ETH [46].

Actually, ETH can also be bioactivated by two other monooxygenases, EthA and MymA, whose expression is repressed by the transcriptional regulators, EthR and VirS, respectively; this evidence has led to the development of multi-target compounds capable of interacting with these regulators [47]. Moreover, a recent phenotypic-drug screening performed with *Mtb*-infected macrophages disclosed a series of oxadiazole compounds with dual activity, showing anti-virulence properties by inhibiting the ESX-1 secretion system and ETH boost activity. Interestingly, for some compounds, the ETH boost was achieved by inhibiting EthR, while for others by binding VirS [48]; this observation demonstrates how pharmacological pleiotropy can be modulated even by small chemical modifications of compounds.

In recent years, several other examples of multitarget compounds inhibiting a wide variety of metabolic pathways have been discovered by phenotypic screening, including inhibitors of the pyrimidine and coenzyme A biosynthesis [49], of lipids and cell wall metabolism [50], of efflux mechanisms and peptidoglycan biosynthesis [51], of folate and nucleotide biosynthesis [52], and of protein homeostasis and virulence factors [53].

Clinically relevant antitubercular drugs mainly target cell wall synthesis, cell membrane function, and protein and nucleic acid biosynthesis; recently, the introduction of BDQ, which targets the membrane-bound F_1F_0 ATP synthase, has made energy generation another interesting target for drug development [54]. This evidence was further confirmed by the discovery of the imidazopyridine amide compound Q203, which targets the respiratory cytochrome bc1 complex [55] and is currently in Phase II clinical trials, as well as by the previously described MmpL3 inhibitor SQ109. Thus, several efforts have been made in the search of novel inhibitors of this essential pathway, which in several cases afforded multitargeting compounds.

This is the case of the amilorides, a drug class originally identified as potassium-sparing diuretics that have been investigated as potential inhibitors of the *Mtb* F_1F_0-ATP synthase [56]. Among these, the derivative HM2-16F, previously optimized to remove the diuretic activity, showed an antitubercular activity comparable to that of BDQ, but the inhibition of the F1Fo-ATP synthase was found to be relatively poor. Interestingly, HM2-16F that has displayed synergic effects with Q203, was demonstrated to inhibit the cytochrome bd, proving to be a multitargeting compound [56]. Moreover, a further optimization campaign of amiloride compounds disclosed the 6-substituted 5-(N,N-hexamethylene)amiloride derivatives as more potent antituberculars [57]. Once again, the compounds were demonstrated to act as multi-targeting bioenergetic inhibitors, inhibiting the F_1F_0-ATP synthase and the catalytic site of the succinate dehydrogenase and promoting the production of reactive oxygen species. Moreover, the best compound of this class, BB2-50F, was demonstrated to have rapid bactericidal activity, not only against replicating mycobacteria but also against hypoxic non-replicating *Mtb*, demonstrating the efficacy of inhibiting multiple mechanisms within the bioenergetics pathway [57].

Although the most advanced multitargeting compounds in the TB pipeline are derived from phenotypic screenings, many studies are also underway to specifically project and design compounds with multiple inhibitory effects. However, the rational design of a multitargeting inhibitor could be a very challenging task, particularly if it aims to involve structurally divergent enzymes.

One successful strategy could be to combine the selection of common pharmacophores for the two potential targets with molecular docking studies [58]. For instance, this approach was successfully used to design common inhibitors of the two enzymes FabG4 and HtdX involved in *Mtb* fatty acid metabolism [59], as well as in the selection of inhibitors of the two aminoacyl-tRNA synthetases LeuRS and MetRS [60].

Thanks to bioinformatic development, different tools are now available to design or identify potential multitargeting compounds or putative scaffolds to be implemented for multitargeting. Anyway, the rising number of those tools and their different features require a precise working design and a careful setting.

Different examples of *in silico* identification of putative multitarget compounds and scaffolds, specifically against *Mtb*, are reported in the literature. The steps for the identification of a multitarget scaffold against different enzymes cooperating in the same biosynthetic pathway have been described by Kumari and colleagues, focusing on the Mur enzymes (MurA-MurF), pivotal players in the biosynthesis of the peptidoglycan component of *Mtb* cell wall [61]. They used the available crystal structure of Mur enzymes to build the missing models, exploiting the homology of conserved regions. All the inhibitors reported in the literature were analyzed by three methods - Murcko scaffold, plain ring system, and ring with atomic no substitutions - to generate scaffold models for multitargeting. Those scaffolds have been evaluated by docking analysis against the active sites of both crystal structures and enzyme models [62]. From these results, the 1-[(2R,5S)-5-{[(2S)-oxolan-2-yloxy]methyl}oxolan-2-yl]-1,2,3,4-tetrahydropyrimidine-2,4-dione scaffold was predicted to establish the major number and best interactions with different amino acids within many Mur active sites. From more than a hundred derivatives of this scaffold, it has been predicted that the top fifteen may establish interactions with all the enzymes considered, so they have been further analyzed by molecular dynamic (MD) simulation and free binding energy calculation. One of these derivatives called HPT was revealed to be a promising hit compound since its high stability and binding affinity for Mur enzymes [62].

Similarly, it is theoretically possible identifying multitarget inhibitors starting from putative molecules that interact with the enoyl-acyl carrier protein reductase InhA, the dihydrodipicolinate reductase DapB, and the N-acetyl-glucosamine-1-phosphate uridyltransferase GlmU [63]. The virtual screening of the Asinex database, filtered through the 2D-QSAR model for the identification of DapB putative inhibitors and through the structure-based pharmacophore model for InhA and GlmU inhibitors, identified 784 molecules as hit compounds to be refined. A docking analysis of those molecules confirmed the plausibility of the identified hits against the respective enzyme models. Successively, these hits have

Table 28.1 AI models used in drug discovery. Adapted from [68].

Phase of drug discovery	Aim	AI models	Reference
Target identification	Protein–protein interaction analysis	FD/DCA	[69]
	Protein folding study	CNN	[70]
Hit identification	Drug repurposing	Network pathology	[71]
	Virtual screening	SVM, AAE	[70]
	Activity scoring	SVM. RF, 3D graph CNN	[72]
Lead Identification	*De novo* drug design	Deep learning, VAE, AAE	[73]
	QSAR	Machine learning methods, DNN	[74]
Lead optimization	ADME[a)] and toxicity analysis	CNN, neural network	[75]

a) ADME: absorption, distribution, metabolism, and excretion.

been evaluated *in silico* against the models of 33 *Mtb* druggable targets that have been selected by the Open Source Drug Discovery (OSDD) consortium through a protein interactome approach [64]. The poses obtained with three different docking protocols, Glide, GOLD, and AutoDock Vina, were compared for similarity and revealed a selection of 110 polypharmacological hits. To improve this selection, some filters have been applied, such as the prediction of membrane permeability by the QSAR models [65] and putative cytotoxicity analysis by the Free ADMETox filtering (FAF)-drugs [66]. Finally, 59 molecules have been identified as putative *Mtb* multitarget inhibitors, of which 9 of them were scored in the top 100 inhibitors for 10 different enzymes [63]. A similar workflow was performed against 20 selected *Mtb* putative targets among the enzymes involved in different metabolisms. Starting from a pool of 978 available molecules, 4 candidates have been identified as possible polypharmacological active inhibitors of all the targets [67]. Anyway, the above-described studies were developed without *in vitro* experiments to prove the reliability of predictions, so results remain to be confirmed.

Artificial intelligence (AI) can be integrated into the ongoing methodologies in the attempt to obtain predictive results by exploiting a different path [68]. This difference stands in the concept that AI is not based on the quantomechanic equations that are generally used by the predictive tools above described. It recognizes common features of the molecules, based on the docking results or the MD simulations used for its training, to predict possible inhibitors among an unstudied set. Different AI models are available, and their choice depends on the analysis to be performed, as summarized in Table 28.1.

MetRS and LeuRS, the two aminoacyl-tRNA synthases for methionine and leucine, respectively, have been studied with the support of AI. Upon training the AI with 242 known inhibitors of both enzymes, a database of 100.000 small molecules was analyzed to identify some clusters of pharmacophores with putative dual-targeting capability through the webtool PharmaGist [76]. The results were validated by PharmDeveloper, another tool for pharmacophore analysis. Preliminary data enlightened 143 pharmacophore models divided into eight clusters according to their properties. Only 6 of these models showed dual-targeting properties against both synthases. For the following studies, the best-fitting 4 models have been selected to identify putative dual-target inhibitors among a database of 100.000 small molecules, leading to the identification of 156 putative multitargeting compounds. The reliability of this prediction stands in five molecules classified in the top 30 best-scored compounds that showed a significant inhibition effect *in vitro* against both enzymes [76]. The most active compound, the 3-(3-chloro-4- methoxy-phenyl)-5-[3-(4-fluorophenyl)-[1,2,4]oxadiazol-5-yl]-3H-[1,2,3]triazol-4-ylamine, showing an IC_{50} of 13 μM against LeuRS and of 13.8 μM against MetRS, is confirmed to be an interesting hit compound for further chemical and biological optimization [76].

Similarly, the AI model "artificial neural network" was integrated as a single-hidden layer using the module nnet from R [77] to implement the analysis of LeuRS and MetRS inhibitors in another work. In this case, the AI was trained with known inhibitors of those enzymes, then used to predict putative dual-inhibitors among the Otava dataset of 250.000 compounds. The analysis resulted in 200 compounds, of which six showed inhibition activity against both enzymes with IC_{50}s in the μM range of concentrations [77].

Many studies based on the described methodologies and strategies are been published in the past few years, and it is expected that their number will increase rapidly.

28.4 Multitarget Compounds Against TB-HIV Co-infection

HIV is the causative agent of acquired immunodeficiency syndrome (AIDS), which is still a global health problem of a magnitude that has never been reached before [78]. A combination of drugs acting on different targets is commonly used in the current treatment of AIDS [79]. These drugs target the essential HIV enzymes reverse transcriptase (RT), integrase (IN), and protease (PR) [80], and the recommended treatment consists of a combination of two nucleoside reverse transcriptase inhibitors (NRTIs) together with a third interfering nucleotide transfer inhibitor (INSTI) [81].

The infection and destruction of the CD4+ lymphocytes lead to the strong immunosuppression characteristic of AIDS and therefore facilitate the emergence of secondary infections. The main obstacle is the latent viral reservoir. The virus has the ability to survive in macrophages, and this event constitutes a huge challenge for the further development of a successful virus elimination strategy. Moreover, HIV-infected monocytes and macrophages play an important role in initiating inflammation, leading to co-morbidity in people living with HIV [82, 83].

Tuberculosis is the leading cause of death among people living with HIV (PLHIV); for instance, in 2019, an estimated 30% of AIDS-related deaths were caused by TB [84]. Moreover, PLHIV are about 18 times more likely to develop active TB and, even when on antiretroviral therapy, are 3 times more likely to die.

Mtb is capable of infecting different organs, leading to extrapulmonary TB. This type of TB primarily affects PLHIV, who are more susceptible to the spread of the pathogen due to the immunosuppression that favors the dissemination of *Mtb*, accelerating HIV replication and placing patients at high risk of co-infection [85]. Several factors may explain the interplay between the two diseases. First, the depletion of macrophages may delay the recruitment of this type of cell. Moreover, necrotic cell death is associated with increased *Mtb*, suggesting that phagocytosis of mycobacteria may limit their growth [86, 87]. In addition, macrophages in HIV infected individuals manifest an activated phenotype, M2, as opposed to M1. This is significant since this alteration may reduce the activity of the enzyme nitric oxide synthase, leading to decreased *Mtb*-induced macrophage apoptosis [88].

The management of the HIV-TB co-infection involves prolonged treatment with the use of drug combinations for both pathological conditions at the same time. The most significant therapeutic difficulties are the occurrence of drug–drug interactions, overlapping toxicity, non-adherence to treatment, and consequent cases of resistance [89].

In recent therapeutic strategies, the use of molecules that synergistically target different targets is being explored. The development of multitarget molecules represents a potential solution to overcome the limitations associated with current therapies for HIV-TB co-infection. The advantages that multitarget compounds may offer are more predictable pharmacokinetics, a reduction of potential drug–drug interactions, and increased patient tolerability [90].

This strategy involves the use of molecules with activity against both HIV and *Mtb* through the use of molecular hybridization and multi-target approaches [80, 91]. Molecular hybridization is becoming an important method in the development of drugs for the treatment of complex diseases and is one of the main tools in the development of bioactive compounds for the treatment of multifactorial diseases [92].

The first example of an HIV-*Mtb* multitarget compound is represented by the coumarin compound Calanolide A and its related pyranocoumarin [93]. These compounds, already investigated as RT inhibitors, have been shown to possess significant efficacy against *Mtb*, including strains resistant to the most common antimycobacterial drugs isoniazid, ethambutol, rifampicin, and streptomycin [93].

As previously described, molecular hybridization is a useful method for drug development for the treatment of complex diseases, and particularly for the treatment of multifactorial diseases [92]. In this context, one of the first works evaluated potential multitargeting molecules obtained through esterification of the antiretroviral drug zidovudine (ZDV) with the antimicrobial drugs isoniazid (Figure 28.3a), ciprofloxacin (Figure 28.3b), norfloxacin, and acetic acid from pyrazinamide (Figure 28.3c). These compounds displayed potent inhibitory activity against both HIV and *Mtb*, showing a good efficacy against HIV-1 in T4 lymphocytes, with a high selectivity index, and also being able to inhibit *Mtb* replication by 99% [94].

In continuation with this study, ZVD was linked to further different fluoroquinolones (lomefloxacin, gatifloxacin, sparfloxacin, enrofloxacin, levofloxacin, and ofloxacin), resulting in a series of compounds with promising activity against HIV and *Mtb* and also active against a panel of different pathogenic bacteria [95].

The hybridization of ZDV has recently resumed in a study in which the anti-HIV pharmacophoric part of ZDV was linked through a 1,2,3-triazol spacer to azaaurone derivatives (Figure 28.3d) previously shown to display antimycobacterial activity [96].

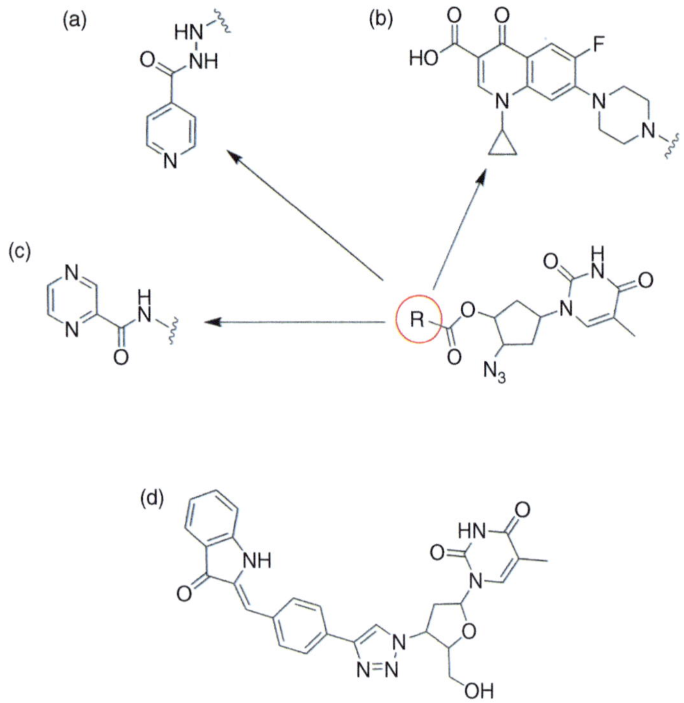

Figure 28.3 Chemical structure of the molecular hybrids of ZDV with the antitubercular drugs isoniazid (a), ciprofloxacin (b), pyrazinamide (c), and with the azaaurone compounds.

A similar approach was used with the non-nucleoside RT inhibitor nevirapine (NVP), which was hybridized with isoniazid, pirazinamide, and fluoroquinolones. Of these, a hybrid of NVP and isoniazid was particularly noteworthy, being highly active against HIV, with a good selectivity index and exhibiting excellent inhibition of *Mtb* growth [97].

The strategy of combining antimicrobial with antiretroviral compounds has also been exploited for inhibitors of other HIV enzymes. For instance, tetracycline and its derivatives, such as oxytetracycline, doxycycline, and minocycline, are able to inhibit the HIV-1 IN enzyme. A number of novel tetracycline derivatives containing fluoroquinolones were synthesized and tested against HIV-1 and *Mtb*. The best derivatives showed inhibitory activity against the two reactions catalyzed by HIV-1 IN, the 3′ processing and the strand transfer, with good potency in inhibiting virus replication as well as a very promising activity against *Mtb*, displaying a higher potency than that of the evaluated standards [98].

Another strategy proposed to target both HIV and Mtb is the use of metal-based complexes, for which antimicrobial and antiviral activity have been well documented [99, 100]. In this context, the use of three generations of gallium compounds has shown that they may be a good starting point for the development of potential new drugs. Indeed, gallium resembles iron and has been reported to potentially inhibit the growth and metabolism of some microorganisms, such as *Mtb*. In particular, it was found that most Fe-dependent proteins cannot distinguish Fe from Ga, and unlike Fe, Ga bound to the active site is able to inactivate the enzymes [101]. For instance, Ga tetraphenyl porphyrin nanoparticles were able to inhibit the growth of both mycobacterium and HIV in a macrophage coinfection model. Moreover, this formulation demonstrated to have prolonged activity, being effective up to 15 days after a single compound loading [102]. Similarly, promising results have been reported for platinum(II) and Pd(II) complexes [103] and ruthenium(II) complexes [104].

Designing inhibitors that target two completely different organisms, such as mycobacteria and HIV, remains anyway a very difficult task. One possibility could be the identification of a common privileged scaffold already reported as potential inhibitors of both targets. In this context, it has been shown that piperidine could represent a useful scaffold for the development of multitargeting compounds. A series of 3,5-bis(furan-2-ylmethylidene)piperidin-4-substituted imines have been evaluated against *Mtb* and HIV, yielding compounds with a good activity against *Mtb*, a moderate inhibition of HIV, as well as a low in vitro toxicity. In silico molecular docking suggested that the best hits inhibit the enoyl-ACP-reductase of *Mtb* and the RT of HIV, but experimental proof has not yet been reported [105].

Another emerging and promising scaffold for drug development is niclosamide, which has the potential to treat bacterial and viral infections. Indeed, this compound has already been shown to inhibit *Mtb* growth and appears to be active against both *Mtb* and HIV at the same time.

Niclosamide, assayed in human macrophages co-infected with mycobacteria and HIV-1, reduced the replication of both pathogens by more than 50%, with no significant cell death observed due to drug toxicity, confirming promising a dual activity for the treatment of HIV-TB co-infection [106].

28.5 Conclusions

The emergence of the spread of multi-drug and extensively-drug resistant M. TB strains leads to the development of novel strategies to fight the so-called resistance–resistant drugs. Among these approaches, polypharmacology has shown that it can make an important contribution, as evidenced by the fact that at least two molecules in clinical trials are actually multitargeting compounds. Despite numerous studies in this direction, there is still much work to be done. A great help is provided by the ongoing development of computer models and advances in AI. Thanks to these, the development of new effective molecules to implement the treatment of TB and *Mtb*-HIV co-infections is expected to increase in the near future.

References

1 Bu, Q., Qiang, R., Fang, L., and Peng; X., Zhang; H., Cheng; H. (2023). Global trends in the incidence rates of MDR and XDR tuberculosis: Findings from the global burden of disease study 2019. *Front. Pharmacol.* 14: 1156249.

2 Bagcchi, S. (2023). WHO's global tuberculosis report 2022. *Lancet. Microbe* 4: e20.

3 Rich, M.L., Khan, U., Zeng, C. et al. (2023). Outcomes of WHO-conforming, longer, all-oral multidrug-resistant TB regimens and analysis implications. *Int. J. Tuberc. Lung Dis.* 27: 451–457.

4 Koch, R., Brock, T.D., and Fred, E.B. (1982). The etiology of tuberculosis. *Rev. Infect. Dis.* 4: 1270–1274.

5 Pezzella, A.T. (2019). History of pulmonary tuberculosis. *Thorac. Surg. Clin.* 29: 1–17.

6 Fu, L.M. and Fu-Liu, C.S. (2002). Is *Mycobacterium tuberculosis* a closer relative to Gram-positive or Gram-negative bacterial pathogens? *Tuberculosis* 82: 85–90.

7 Lucarelli, A.P., Buroni, S., Pasca, M.R. et al. (2010). *Mycobacterium tuberculosis* phosphoribosylpyrophosphate synthetase: biochemical features of a crucial enzyme for mycobacterial cell wall biosynthesis. *PLoS One* 5: e15494.

8 Laws, M., Jin, P., and Rahman, K.M. (2022). Efflux pumps in *Mycobacterium tuberculosis* and their inhibition to tackle antimicrobial resistance. *Trends Microbiol.* 30: 57–68.

9 Pule, C.M., Sampson, S.L., Warren, R.M. et al. (2016). Efflux pump inhibitors: targeting mycobacterial efflux systems to enhance TB therapy. *J. Antimicrob. Chemother.* 71: 17–26.

10 Machado, D., Lecorche, E., Mougari, F. et al. (2018). Insights on *Mycobacterium leprae* efflux pumps and their implications in drug resistance and virulence. *Front. Microbiol.* 9: 3072.

11 Rempel, S., Gati, C., Nijland, M. et al. (2020). A mycobacterial ABC transporter mediates the uptake of hydrophilic compounds. *Nature* 580: 409–412.

12 Huang, Y., Lemieux, M.J., Song, J. et al. (2003). Structure and mechanism of the glycerol-3-phosphate transporter from Escherichia coli. *Science* 301: 616–620.

13 Venter, H., Mowla, R., Ohene-Agyei, T., and Ma, S. (2015). RND-type drug efflux pumps from Gram-negative bacteria: molecular mechanism and inhibition. *Front. Microbiol.* 6: 377.

14 Briffotaux, J., Huang, W., Wang, X., and Gicquel, B. (2017). MmpS5/MmpL5 as an efflux pump in Mycobacterium species. *Tuberculosis (Edinb.)* 107: 13–19.

15 Sharma, A., Gupta, V.K., and Pathania, R. (2019). Efflux pump inhibitors for bacterial pathogens: From bench to bedside. *Indian J. Med. Res.* 149: 129–145.

16 Rodrigues, L., Villellas, C., Bailo, R. et al. (2013). Role of the Mmr efflux pump in drug resistance in *Mycobacterium tuberculosis*. *Antimicrob. Agents Chemother.* 57: 751–757.

17 Mishra, M.N. and Daniels, L. (2013). Characterization of the MSMEG_2631 gene (mmp) encoding a multidrug and toxic compound extrusion (MATE) family protein in *Mycobacterium smegmatis* and exploration of its polyspecific nature using biolog phenotype microarray. *J. Bacteriol.* 195: 1610–1621.

18 Zaunbrecher, M.A., Sikes, R.D. Jr., Metchock, B. et al. (2009). Overexpression of the chromosomally encoded aminoglycoside acetyltransferase eis confers kanamycin resistance in *Mycobacterium tuberculosis*. *Proc. Natl. Acad. Sci. USA* 106: 20004–20009.

19 Kashyap, A., Singh, P.K., and Silakari, O. (2018). Mechanistic investigation of resistance via drug-inactivating enzymes in *Mycobacterium tuberculosis*. *Drug Metab. Rev.* 50: 448–465.

20 Tremblay, L.W., Fan, F., and Blanchard, J.S. (2010). Biochemical and structural characterization of *Mycobacterium tuberculosis* beta-lactamase with the carbapenems ertapenem and doripenem. *Biochemistry* 49: 3766–3773.

21 Nimmo, C., Millard, J., Faulkner, V. et al. (2022). Evolution of *Mycobacterium tuberculosis* drug resistance in the genomic era. *Front. Cell. Infect. Microbiol.* 12: 954074.

22 Ramaswamy, S. and Musser, J.M. (1998). Molecular genetic basis of antimicrobial agent resistance in *Mycobacterium tuberculosis*: 1998 update. *Tuber. Lung Dis.* 79: 3–29.

23 Musser, J.M. (1995). Antimicrobial agent resistance in mycobacteria: molecular genetic insights. *Clin. Microbiol. Rev.* 8: 496–514.

24 Unissa, A.N., Subbian, S., Hanna, L.E., and Selvakumar, N. (2016). Overview on mechanisms of isoniazid action and resistance in *Mycobacterium tuberculosis*. *Infect. Genet. Evol.* 45: 474–492.

25 Koch, A., Cox, H., and Mizrahi, V. (2018). Drug-resistant tuberculosis: challenges and opportunities for diagnosis and treatment. *Curr. Opin. Pharmacol.* 42: 7–15.

26 Kurz, S.G., Furin, J.J., and Bark, C.M. (2016). Drug-resistant tuberculosis: challenges and progress. *Infect. Dis. Clin. N. Am.* 30: 509–522.

27 Alipanah, N., Jarlsberg, L., Miller, C. et al. (2018). Adherence interventions and outcomes of tuberculosis treatment: A systematic review and meta-analysis of trials and observational studies. *PLoS Med.* 15: e1002595.

28 Machado, D., Girardini, M., Viveiros, M., and Pieroni, M. (2018). Challenging the drug-likeness dogma for new drug discovery in tuberculosis. *Front. Microbiol.* 9: 1367.

29 Stephanie, F., Saragih, M., and Tambunan, U.S.F. (2021). Recent progress and challenges for drug-resistant tuberculosis treatment. *Pharmaceutics* 13: 592.

30 Stelitano, G., Sammartino, J.C., and Chiarelli, L.R. (2020). Multitargeting compounds: a promising strategy to overcome multi-drug resistant tuberculosis. *Molecules* 25: 1239.

31 Gray, D.A. and Wenzel, M. (2020). Multitarget approaches against multiresistant superbugs. *ACS Infect. Dis.* 6: 1346–1365.

32 Li, K., Schurig-Briccio, L.A., Feng, X. et al. (2014). Multitarget drug discovery for tuberculosis and other infectious diseases. *J. Med. Chem.* 57: 3126–3139.

33 Anighoro, A., Bajorath, J., and Rastelli, G. (2014). Polypharmacology: challenges and opportunities in drug discovery. *J. Med. Chem.* 57: 7874–7887.

34 Wenzel, M., Rautenbach, M., Vosloo, J.A. et al. (2018). The multifaceted antibacterial mechanisms of the pioneering peptide antibiotics tyrocidine and gramicidin S. *MBio* 9: e00802–e00818.

35 Tauber, S.C. and Nau, R. (2008). Immunomodulatory properties of antibiotics, *Curr Mol. Pharmacol. Ther.* 1: 68–79.

36 Kocaoglu, O. and Carlson, E.E. (2015). Profiling of β-lactam selectivity for penicillin-binding proteins in *Escherichia coli* strain DC2. *Antimicrob. Agents Chemother.* 59: 2785–2790.

37 Noda, M., Kawahara, Y., Ichikawa, A. et al. (2004). Self-protection mechanism in D-cycloserine-producing Streptomyces lavendulae. Gene cloning, characterization, and kinetics of its alanine racemase and D-alanyl-D-alanine ligase, which are target enzymes of D-cycloserine. *J. Biol. Chem.* 279: 46143–46152.

38 Morphy, R. and Rankovic, Z. (2005). Designed multiple ligands. An emerging drug discovery paradigm. *J. Med. Chem.* 48: 6523–6543.

39 Hazra, S., Hazarika, R., and Patra, S. (2023). Multitargeting: an alternative approach to tackle multidrug resistance in tuberculosis. *Curr. Drug Targets* 24: 751–775.

40 Degiacomi, G., Belardinelli, J.M., Pasca, M.R. et al. (2020). Promiscuous targets for antitubercular drug discovery: the paradigm of DprE1 and MmpL3. *Appl. Sci.* 10: 623.

41 La Rosa, V., Poce, G., Canseco, J.O. et al. (2012). MmpL3 is the cellular target of the antitubercular pyrrole derivative BM212. *Antimicrob. Agents Chemother.* 56: 324–331.

42 Zhang, B., Li, J., Yang, X. et al. (2019). Crystal Structures of membrane transporter MmpL3, an anti-TB drug target. *Cell* 176: 636–648.e13.

43 Cox, J.A., Abrahams, K.A., Alemparte, C. et al. (2016). THPP target assignment reveals EchA6 as an essential fatty acid shuttle in mycobacteria. *Nat. Microbiol.* 1: 15006.

44 Li, W., Stevens, C.M., Pandya, A.N. et al. (2019). Direct Inhibition of MmpL3 by Novel Antitubercular Compounds. *ACS Infect Dis.* 5: 1001–1012.

45 Moorey, A.R., Cabanillas, A., Batt, S.M. et al. (2021). The multi-target aspect of an MmpL3 inhibitor: The BM212 series of compounds bind EthR2, a transcriptional regulator of ethionamide activation. *Cell Surf.* 7: 100068.

46 Blondiaux, N., Moune, M., Desroses, M. et al. (2017). Reversion of antibiotic resistance in *Mycobacterium*

tuberculosis by spiroisoxazoline SMARt-420. *Science* 355: 1206–1211.

47 Flipo, M., Frita, R., Bourotte, M. et al. (2022). The small-molecule SMARt751 reverses *Mycobacterium tuberculosis* resistance to ethionamide in acute and chronic mouse models of tuberculosis. *Sci. Transl. Med.* 14: eaaz6280.

48 Gries, R., Chhen, J., van Gumpel, E. et al. (2024). Discovery of dual-active ethionamide boosters inhibiting the *Mycobacterium tuberculosis* ESX-1 secretion system. *Cell Chem. Biol.* S2451-9456 (23): 00436–00431.

49 Chiarelli, L.R., Mori, G., Orena, B.S. et al. (2018). A multitarget approach to drug discovery inhibiting *Mycobacterium tuberculosis* PyrG and PanK. *Sci. Rep.* 8: 3187.

50 Nguyen, P.C., Delorme, V., Bénarouche, A. et al. (2018). Oxadiazolone derivatives, new promising multi-target inhibitors against *M. tuberculosis*. *Bioorg. Chem.* 81: 414–424.

51 Kaur, P., Potluri, V., Ahuja, V.K. et al. (2021). A multi-targeting pre-clinical candidate against drug-resistant tuberculosis. *Tuberculosis* 129: 102104.

52 Hajian, B., Scocchera, E., Shoen, C. et al. (2019). Drugging the folate pathway in *Mycobacterium tuberculosis*: the role of multi-targeting agents. *Cell Chem. Biol.* 26: 781–791.

53 Washburn, A., Abdeen, S., Ovechkina, Y. et al. (2019). Dual-targeting GroEL/ES chaperonin and protein tyrosine phosphatase B (PtpB) inhibitors: a polypharmacology strategy for treating *Mycobacterium tuberculosis* infections. *Bioorg. Med. Chem. Lett.* 29: 1665–1672.

54 Hards, K. and Cook, G.M. (2018). Targeting bacterial energetics to produce new antimicrobials. *Drug Resist. Updat.* 36: 1–12.

55 Pethe, K., Bifani, P., Jang, J. et al. (2013). Discovery of Q203, a potent clinical candidate for the treatment of tuberculosis. *Nat. Med.* 19: 1157–1160.

56 Hards, K., Cheung, C.Y., Waller, N. et al. (2022). An amiloride derivative is active against the F_1F_0-ATP synthase and cytochrome bd oxidase of *Mycobacterium tuberculosis*. *Commun. Biol.* 5: 166.

57 Adolph, C., Cheung, C.Y., McNeil, M.B. et al. (2024). A dual-targeting succinate dehydrogenase and F_1F_0-ATP synthase inhibitor rapidly sterilizes replicating and non-replicating *Mycobacterium tuberculosis*. *CellChem-Biol* S2451-9456 (23): 00432–00434.

58 Wei, D., Jiang, X., Zhou, L. et al. (2008). Discovery of multitarget inhibitors by combining molecular docking with common pharmacophore matching. *J. Med. Chem.* 51: 7882–7888.

59 Banerjee, D.R., Biswas, R., Das, A.K., and Basak, A. (2015). Design, synthesis and characterization of dual inhibitors against new targets FabG4 and HtdX of *Mycobacterium tuberculosis*. *Eur. J. Med. Chem.* 100: 223–234.

60 Kovalenko, O.P., Volynets, G.P., Rybak, M.Y. et al. (2019). Dual-target inhibitors of mycobacterial aminoacyl-tRNA synthetases among *N*-benzylidene-*N'*-thiazol-2-yl-hydrazines. *MedChemComm* 10: 2161–2169.

61 Shinde, Y., Ahmad, I., Surana, S., and Patel, H. (2021). The Mur enzymes chink in the armour of *Mycobacterium tuberculosis* cell wall. *Eur. J. Med. Chem.* 222: 113568.

62 Kumari, M., Waseem, M., and Subbarao, N. (2023). Discovery of multi-target mur enzymes inhibitors with anti-mycobacterial activity through a scaffold approach. *J. Biomol. Struct. Dyn.* 41: 2878–2899.

63 Janardhan, S., John, L., Prasanthi, M. et al. (2017). A QSAR and molecular modelling study towards new lead finding: polypharmacological approach to *Mycobacterium tuberculosis*. *SAR QSAR Environ. Res.* 28: 815–832.

64 Vashisht, R., Mondal, A.K., Jain, A. et al. (2012). Crowd sourcing a new paradigm for interactome driven drug target identification in *Mycobacterium tuberculosis*. *PLoS One* 7: e39808.

65 Janardhan, S., Vivek, M.R., and Sastry, G.N. (2016). Modeling the permeability of drug-like molecules through the cell wall of *Mycobacterium tuberculosis*: an analogue based approach. *Mol. BioSyst.* 12: 3377–3384.

66 Lagorce, D., Sperandio, O., Galons, H. et al. (2008). FAF-Drugs2: free ADME/tox filtering tool to assist drug discovery and chemical biology projects. *BMC Bioinformatics* 9: 396.

67 Madugula, S.S., Nagamani, S., Jamir, E. et al. (2022). Drug repositioning for anti-tuberculosis drugs: an in silico polypharmacology approach. *Mol. Divers.* 26: 1675–1695.

68 Sahu, A., Mishra, J., and Kushwaha, N. (2022). Artificial intelligence (AI) in drugs and pharmaceuticals. *Comb. Chem. High Throughput Screen.* 25: 1818–1837.

69 Bai, F., Morcos, F., Cheng, R.R. et al. (2016). Elucidating the druggable interface of protein-protein interactions using fragment docking and coevolutionary analysis. *Proc. Natl. Acad. Sci. USA* 113: E8051–E8058.

70 Wang, S., Sun, S., Li, Z. et al. (2017). Accurate De Novo prediction of protein contact map by ultra-deep learning model. *PLoS Comput. Biol.* 13: e1005324.

71 Luo, Y., Zhao, X., Zhou, J. et al. (2017). A network integration approach for drug-target interaction

prediction and computational drug repositioning from heterogeneous information. *Nat. Commun.* 8: 573.
72 Kadurin, A., Aliper, A., Kazennov, A. et al. (2017). The cornucopia of meaningful leads: Applying deep adversarial autoencoders for new molecule development in oncology. *Oncotarget* 8: 10883–10890.
73 Kadurin, A., Nikolenko, S., Khrabrov, K. et al. (2017). druGAN: an advanced generative adversarial autoencoder model for de novo generation of new molecules with desired molecular properties in silico. *Mol. Pharm.* 14: 3098–3104.
74 Ma, J., Sheridan, R.P., Liaw, A. et al. (2015). Deep neural nets as a method for quantitative structure-activity relationships. *J. Chem. Inf. Model.* 55: 263–274.
75 Kearnes, S., Goldman, B., and Pande, V. (2016). Modeling industrial ADMET data with multitask networks. arXiv. 1606, 08793v3.
76 Volynets, G.P., Starosyla, S.A., Rybak, M.Y. et al. (2019). Dual-targeted hit identification using pharmacophore screening. *J. Comput. Aided Mol. Des.* 33: 955–964.
77 Volynets, G.P., Usenko, M.O., Gudzera, O.I. et al. (2022). Identification of dual-targeted *Mycobacterium tuberculosis* aminoacyl-tRNA synthetase inhibitors using machine learning. *Future Med. Chem.* 14: 1223–1237.
78 Broder, S. (2010). The development of antiretroviral therapy and its impact on the HIV-1/AIDS pandemic. *Antivir. Res.* 85: 1–18.
79 Phanuphak, N. and Gulick, R.M. (2020). HIV treatment and prevention 2019: current standards of care. *Curr. Opin. HIV AIDS* 15: 4–12.
80 Bianco, M.D.C.A.D., Inacio Leite, D., Branco, S.C., and F., Boechat, N., Uliassi, E., Bolognesi, M. L., Bastos, M. M. (2022). The use of zidovudine pharmacophore in multi-target-directed ligands for AIDS therapy. *Molecules* 27: 8502.
81 Pau, A.K. and George, J.M. (2014). Antiretroviral therapy: current drugs. *Infect. Dis. Clin. N. Am.* 28: 371–402.
82 Joseph, J., Daley, W., Lawrence, D. et al. (2022). Role of macrophages in HIV pathogenesis and cure: NIH perspectives. *J. Leukoc. Biol.* 112: 1233–1243.
83 Waight, E., Zhang, C., Mathews, S. et al. (2022). Animal models for studies of HIV-1 brain reservoirs. *J. Leukoc. Biol.* 112: 1285–1295.
84 Litvinjenko, S., Magwood, O., Wu, S., and Wei, X. (2023). Burden of tuberculosis among vulnerable populations worldwide: an overview of systematic reviews. *Lancet Infect. Dis.* 23: 1395–1407.
85 Rosas-Taraco, A.G., Arce-Mendoza, A.Y., Caballero-Olín, G., and Salinas-Carmona, M.C. (2006). *Mycobacterium tuberculosis* upregulates coreceptors CCR5 and CXCR4 while HIV modulates CD14 favoring concurrent infection. *AIDS Research Hum. Retroviruses* 22: 45–51.
86 Geldmacher, C., Schuetz, A., Ngwenyama, N. et al. (2008). Early depletion of *Mycobacterium tuberculosis*-specific T helper 1 cell responses after HIV-1 infection. *J. Infect. Dis.* 198: 1590–1598.
87 Cambier, C.J., Takaki, K.K., Larson, R.P. et al. (2014). Mycobacteria manipulate macrophage recruitment through coordinated use of membrane lipids. *Nature* 505: 218–222.
88 Walter, N.D., de Jong, B.C., Garcia, B.J. et al. (2016). Adaptation of *Mycobacterium tuberculosis* to Impaired host immunity in HIV-infected patients. *J. Infect. Dis.* 214: 1205–1211.
89 McIlleron, H., Meintjes, G., Burman, W.J., and Maartens, G. (2007). Complications of antiretroviral therapy in patients with tuberculosis: drug interactions, toxicity, and immune reconstitution inflammatory syndrome. *J. Infect. Dis.* 196 (Suppl 1): S63–S75.
90 Bolognesi, M.L. (2013). Polypharmacology in a single drug: multitarget drugs. *Curr. Med. Chem.* 20: 1639–1645.
91 Leite, D.I., de Castro Bazan Moura, S., da Conceição Avelino Dias, M., and Costa, C.C.P. (2023). A review of the development of multitarget molecules against HIV-TB coinfection pathogens. *Molecules* 28: 3342.
92 Fraga, C.A. (2009). Drug hybridization strategies: before or after lead identification? *Expert Opin. Drug Discov.* 4: 605–609.
93 Xu, Z.Q., Barrow, W.W., Suling, W.J. et al. (2004). Anti-HIV natural product (+)-calanolide A is active against both drug-susceptible and drug-resistant strains of *Mycobacterium tuberculosis*. *Bioorg. Med. Chem.* 12: 1199–1207.
94 Sriram, D., Srichakravarthy, N., Bal, T.R., and Yogeeswari, P. (2005). Synthesis of zidovudine prodrugs with broad-spectrum chemotherapeutic properties for the effective treatment of HIV/AIDS. *Biomed. Pharmacother.* 59: 452–455.
95 Senthilkumar, P., Long, J., Swetha, R. et al. (2009). Synthesis of zidovudine derivatives with anti-HIV-1 and antibacterial activities. *Nucleosides Nucleotides Nucleic Acids* 28: 89–102.
96 Leite, D.I., Campaniço, A., Costa, P.A.G. et al. (2024). New azaaurone derivatives as potential multitarget agents in HIV-TB coinfection. *Arch. Pharm.* 357: e2300560.
97 Sriram, D., Srichakravarthy, N., Bal, T.R., and Yogeeswari, P. (2005). Nevirapine derivatives with

broad-spectrum chemotherapeutic properties for the effective treatment of HIV/AIDS. *Biomed. Pharmacother.* 59: 456–459.

98 Sriram, D., Yogeeswari, P., Senchani, G., and Banerjee, D. (2007). Newer tetracycline derivatives: synthesis, anti-HIV, antimycobacterial activities and inhibition of HIV-1 integrase. *Bioorg. Med. Chem. Lett.* 17: 2372–2375.

99 Che, C.M. and Siu, F.M. (2010). Metal complexes in medicine with a focus on enzyme inhibition. *Curr. Opin. Chem. Biol.* 14: 255–261.

100 Sharma, B., Shukla, S., Rattan, R. et al. (2022). Antimicrobial agents based on metal complexes: present situation and future prospects. *Int. J. Biomater.* 8 (2022): 6819080.

101 Stelitano, G., Cocorullo, M., Mori, M. et al. (2023). Iron acquisition and metabolism as a promising target for antimicrobials (bottlenecks and opportunities): where do we stand? *Int. J. Mol. Sci.* 24: 6181.

102 Narayanasamy, P., Switzer, B.L., and Britigan, B.E. (2015). Prolonged-acting, multi-targeting gallium nanoparticles potently inhibit growth of both HIV and mycobacteria in co-infected human macrophages. *Sci. Rep.* 6: 8824.

103 Gama, N.H., Elkhadir, A.Y., Gordhan, B.G. et al. (2016). Activity of phosphino palladium(II) and platinum(II) complexes against HIV-1 and *Mycobacterium tuberculosis*. *Biometals* 29: 637–650.

104 Yufanyi, D.M., Abbo, H.S., Titinchi, S.J.J., and Neville, T. (2020). Platinum(II) and ruthenium(II) complexes in medicine: antimycobacterial and anti-HIV activities. *Coord. Chem. Rev.* 414: 213285.

105 Kumar, A., Revathi, R., Sriram, D. et al. (2019). Targeting HIV-TB coinfection by developing novel piperidin-4-substituted imines: Design, synthesis, in vitro and in silico studies. *Arch. Pharm.* 352: e1800358.

106 Fan, X., Xu, J., Files, M. et al. (2019). Dual activity of niclosamide to suppress replication of integrated HIV-1 and *Mycobacterium tuberculosis* (Beijing). *Tuberculosis (Edinb.)* 116S: S28–S33.

29

Dual-acting HIV Inhibitors

María-José Camarasa, Ana-Rosa San-Félix, and Sonia de Castro

29.1 Introduction

Acquired Immune Deficiency Syndrome (AIDS), caused by human immunodeficiency virus (HIV) infection, is one of the great pandemics that humanity has suffered and still is present nowadays. Since it began, back in 1982, there have been around 84 million infections and more than 40 million AIDS-related deaths worldwide. It is estimated that 38 million people are currently infected with HIV. There are 1.5 million people infected each year and about one million deaths annually [1–3].

Thanks to the advances in knowledge about HIV and AIDS epidemics in the management, prevention, and treatment with antiretroviral therapy (ART) [4–6], today HIV infection/AIDS can be considered a chronic manageable disease. HIV-infected patients have a life expectancy very similar to that of healthy people. However, this is not the case for poor-income countries. It is estimated that 15% of HIV-infected people do not have access to medications and are untreated [4, 6]. Today there is no cure for this disease, and although current treatments allow the infected people to keep the viral load at practically undetectable levels and are no longer infectious [4], they do not eliminate the virus, so it is essential to maintain lifelong care and drug treatments [6]. A great number of asymptomatic people living with HIV (PLHIV) survive and get older but require lifelong care and treatment, and they are at high risk to develop chronic co-diseases and complications [7, 8].

HIV rapidly mutates and escapes the action of the immune system. On the other hand, the virus infects the T lymphocytes (CD4+), an infection that destroys these cells, causing immunodeficiency and, as a consequence of a weakening and vulnerable immune system, the infected people are very susceptible to opportunistic infections caused by other pathogens such as viruses, bacteria, fungal, protozoa, parasites, etc. [7, 9]. Opportunistic infections are more frequent and with higher severity among immunocompromised people, such as PLHIV [9]. These infections are responsible of more than 90% of all HIV-related deaths [9–11]. Remarkably, immunosuppression predisposes HIV-infected children to several opportunistic infections and is the cause of HIV-related morbidity and mortality worldwide [11]. Co-infections have a major impact in the course of the HIV disease. The term co-infection indicates that a person is living/infected with two or more pathogens at the same time. This is a real health threaten that puts in a serious health risk at millions worldwide. Even the HIV-infected people with good treatments are at high risk of suffering co-morbidities.

Treatment of the co-infected patients requires lifelong combination therapy with different agents. An alternative for the treatment of these patients is the use of dual drugs to tackle these complex infections and also drug resistance [11, 12]. Dual drugs combine, in a single compound, two different pharmacological activities at a similar efficacious dose; for example, in the case of HIV and another viral co-infection, the dual drug would be active against both viruses blocking their replication. This kind of dual drug may prevent drug–drug interactions and/or the emergence of viral resistance.

This book chapter is focused on dual drugs with concomitant activity against HIV replication and that of other viruses such as hepatitis B and C viruses and enterovirus A71 (EV-A71).

29.2 HIV and Hepatitis Viruses Co-infections

In PLHIV, co-infections are very common with hepatitis viruses being a significant cause of morbidity and mortality [6, 13–15]. These co-infected patients have a bigger probability to progress to liver cirrhosis and to develop hepatocellular carcinoma [14, 16–18]. The major co-infections, of clinical concern, with hepatitis viruses in PLHIV are those with hepatitis B, C, and D [6]. HIV and hepatitis

Polypharmacology: Strategies for Multi-Target Drug Discovery, First Edition. Edited by Jens-Uwe Peters.

viruses share the same transmission routes, which explains the frequent occurrence of co-infections. These routes are: exposure to infected blood and other body fluids, unprotected sexual intercourse, sharing contaminated needles (among injecting drug users), and perinatal transmission from an infected mother to the newborn during childbirth [18–20]. Although transmission mother-to-child has been controlled in developing countries, it is still a serious health concern in poor countries [6, 18].

The latest WHO estimates indicate that 7.4% of individuals living with HIV are co-infected with HBV [21], and around 1.5 million (4%) are co-infected with the HCV worldwide. Moreover, triple HIV/HBV/HCV viral infection has also been reported in PLHIV [22]. Co-infection with both viruses (HIV and HBV or HCV) increases the risk of liver damage and mortality and decreases the CD4 cell count, lowering the immune response to HIV [23, 24]. Co-infection with HIV and HCV increases the risk of transmitting HCV through childbirth (a co-infected mother has a bigger chance to transmit HCV to the baby than if she is only infected by HCV). On the other hand, sexual transmission of HCV in co-infected people is greatly increased with respect to only HCV-infected people.

The management of these comorbid conditions requires a careful balance between ART for HIV and antiviral treatments for HBV or HCV. The simultaneous presence of both viruses greatly complicates the treatment regimens of the patients, due to possible drug–drug interactions, suboptimal therapeutic outcomes, and an elevated risk of liver-related morbidity and mortality. In developing countries, the availability of effective drugs against HBV and HCV and the vaccination against HBV has reduced the clinical burden of HIV/hepatitis co-infections to have them more or less controlled [14, 25]. But due to the migration of population from highly endemic of hepatitis infection areas not covered by the HBV vaccine and the high risk of HCV reinfections, viral hepatitis infections in people living with HIV are a major health concern.

The development of drugs with dual activity against both HBV and HIV viruses, or HCV and HIV, may represent a convenient and efficient therapeutic approach to reduce the complexity associated with the use of two individual drugs and to improve patient adherence to treatment, which is a critical factor in the long-term success of managing chronic viral infections.

29.2.1 Compounds with Dual Activity Against HIV and HBV Viruses

HBV infection, that primarily affects the liver, is a major global health concern, with millions of people being chronically infected and at risk of developing severe liver complications. According to the latest estimates from the World Health Organization, it was estimated that there were 316 million people living with chronic HBV in 2019 [21, 26].

In the life cycle of both viruses (HBV and HIV), there is a reverse transcriptase (RT) enzyme that plays a key role in the viral replication, the synthesis of the viral DNA from the proviral RNA (in the case of HIV). In the case of HBV (a DNA virus), the virus replicates its DNA genome by reverse transcription through an RNA intermediate. The presence of the key RT in HIV and HBV viruses represents a unique opportunity to develop antivirals with dual activity against both viruses.

Currently, there are three drugs (emtricitabine, lamivudine, and tenofovir (TFV)) and two prodrugs (tenofovir disoproxil fumarate (TDF) and tenofovir alafenamide (TAF)) (Figure 29.1) approved for use as dual drugs against HIV and HBV replication that act as RT inhibitors [25, 27, 28].

All of the above mentioned drugs are 2′,3′-dideoxynucleosides (emtricitabine and lamivudine) or acyclic nucleosides (TFV) that share a similar mechanism of action. These molecules need to be phosphorylated to their corresponding triphosphates by cellular kinases to interact with RT. The triphosphates, which compete with the natural substrates, are incorporated into the growing viral DNA sequence. Once incorporated, and due to the lack of a hydroxyl group at the 3′-position, they act as chain terminators, thus halting the expansion of the nascent DNA chain. In this process, the bottleneck for emtricitabine and

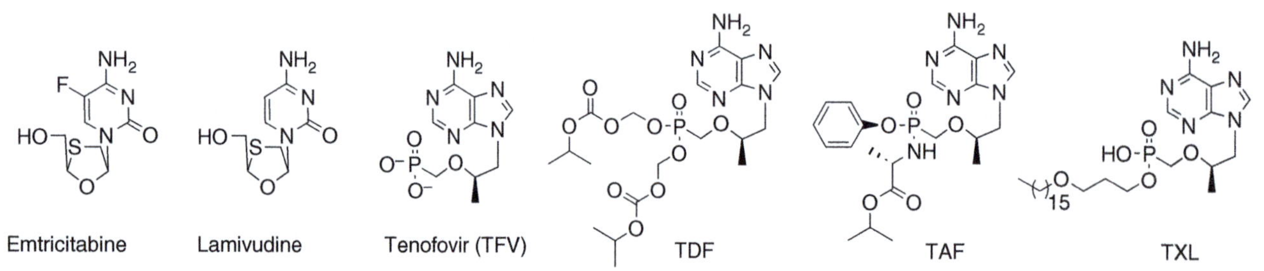

Figure 29.1 Structures of reverse transcriptase inhibitors.

lamivudine is the first phosphorylation, due to the high selectivity of the kinase involved in this initial step. TFV was designed to bypass this first phosphorylation step. This compound and its analogues bear a phosphonate group in their structures and therefore require only two phosphorylation steps for activation. Furthermore, the substitution of the phosphate group (P–O) by a phosphonate group (P–C), as in TFV, prevents hydrolysis to the parent compound. On the other hand, the absence of a cyclic structure in TFV contributes to the conformational flexibility of the compound, thus enabling the interaction with the different enzymes involved in its activation [28]. However, due to the presence of the phosphonate group, TFV, has low membrane permeability and low oral bioavailability derived from the dianionic character of this group under physiological conditions. To enhance membrane permeability and oral bioavailability, two prodrugs of TFV were developed. The first was TDF. TDF is a diester derivative of TFV, which shows efficient absorption through the cell membrane. Once absorbed, the esterases cleave the ester linkage liberating TFV (Figure 29.2).

In 2015 [29], the prodrug of TFV, TAF, was approved for the treatment of HIV infection and in 2016 for the treatment of hepatitis B-infected patients [29]. This prodrug, based on the ProTide technology developed by Professor Chris McGuigan [30], is an aryloxy phosphoramidate derivative of TFV. ProTide TAF showed good anti-HIV activity, higher solubility, *in vivo* stability, and fewer side effects when compared to TFV and TDF [31]. The superior plasma stability of TAF, compared to TDF, results in more efficient uptake by hepatocytes, showing lower plasma concentrations compared to TDF. Consequently, the circulating concentration of TFV is 90% lower after administering a 25 mg dose of TAF than after a 300 mg dose of TDF [32] (Figure 29.1). Although TAF has a significantly better safety and solubility profile than TDF, a single dose of TAF is transformed into TFV by up to 17% in *in vivo* studies [33]. In an attempt to enhance the stability of TFV prodrugs and mitigate their side effects, Painter described in 2007 a new prodrug of TFV, FV exalidex (TXL) (Figure 29.1), by conjugating a lipid moiety to TFV. TXL is stable in plasma and is released by the action of phospholipase C (PLC). However, clinical pharmacokinetic studies indicated a substantial hepatic extraction, and currently, this family of compounds continues to be investigated only as an antiviral agent against the HBV [34].

Despite the advances in the development of dual drugs against HIV and HBV co-infection, the emergence of resistant virus strains and the side effects of existing medications make it necessary to discover new drugs with dual HIV/HBV activity. Over the years, several dual compounds exhibiting activity against both viruses have been investigated (see below).

The use of Toll-like receptor (TLR) agonists as targets for the discovery of dual HIV/HBV inhibitors has been studied. TLRs in humans are involved in the recognition of different pathogens [35]. In the context of HIV, one of these receptors, TLR-7, has also been associated with the reversal of latency [35]. Latency is a critical factor in the HIV life cycle as it enables the virus to persist in a dormant state within the host cells, thus evading detection by the immune system, which represents a significant obstacle to its complete eradication. The elimination of HIV latency is essential to achieve a complete cure since it would disrupt the viral reservoirs, allowing effective targeting and clearance of all infected cells. This hopefully will bring us closer to the total eradication of the virus.

Multiple studies [36, 37] suggest that agonists of this receptor (TLR-7) may trigger the reversal of latency by promoting the release of type I interferon and activating interferon-stimulated genes without causing global immune activation. On the other hand, TRL-7 activation plays a crucial role in hepatitis B infection due to the ability to effectively induce interferons, proinflammatory cytokines, and chemokines in both hepatocytes and non-parenchymal cells [38].

Taking into consideration the involvement of TLR-7 receptors in both infections (HIV and HBV), Gilead developed vesatolimod (GS-9620) (Figure 29.3) as a dual HBV/HIV inhibitor. GS-9620 demonstrated inhibition of HBV replication in animal models with chronic infection and induced a temporary activation of HIV expression in

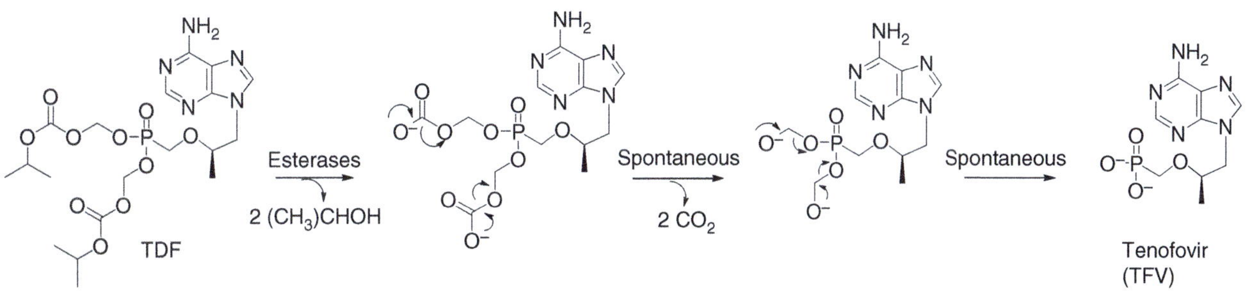

Figure 29.2 Metabolism of TDF to release TFV by esterases.

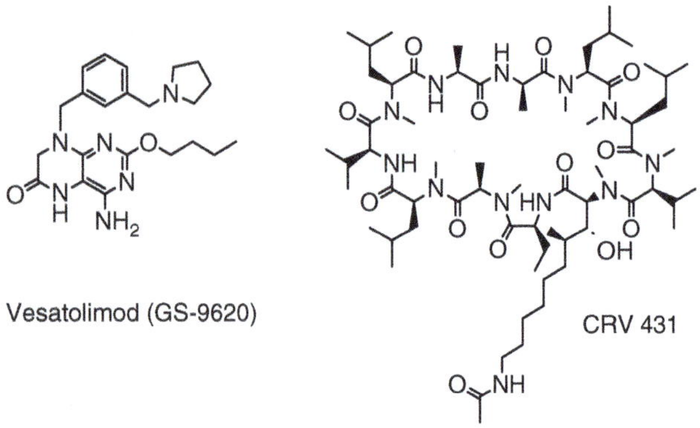

Figure 29.3 Structure of vesatolimod and of cyclophilin inhibitor **CRV431.**

cultured latently infected peripheral blood mononuclear cells (PBMCs) in patients with suppressed viral loads [39]. Currently, this compound is in clinical trials for HIV in combination with therapeutic vaccines (NCT02858401), one of them in phase II [40–42].

In the case of HBV, different clinical trials (phase II) were completed. These studies indicated the safety and good tolerance of vesatolimod in individuals with chronic hepatitis B. Nevertheless, the treatment did not lead to a notable reduction in hepatitis B surface antigen (HBsAg) or HBV DNA of clinical significance. Additionally, the reports mentioned an increase in T cells and natural killer responses, but these immunological changes did not translate into substantial improvements in the clinical outcome of the patients [43, 44].

Another therapeutic target that has been studied in the search for dual HIV/HBV inhibitors is the family of cyclophilins, in particular cyclophilin A (CypA). These proteins belong to a group known as immunophilins that possess peptidyl-prolyl *cis-trans* isomerase activity and are widely expressed in almost all human tissues. Cyclophilins stabilize the *cis-trans* transition state and accelerate isomerization, a process considered important not only in protein folding but also during the assembly of multidomain proteins [45]. Furthermore, these proteins participate in various intracellular signaling pathways, such as oxidative stress, mitochondrial dysfunction, cell migration, and apoptosis. The fact that they act as regulators of intercellular communication has increasingly attracted attention due to their implications in viral infections. CypA been associated with promoting or inhibiting the replication and infection of other viruses such as hepatitis C, influenza, or EV-A71, which makes it a widely studied therapeutic target [46].

In the case of HIV, it has been suggested that CypA is involved in several key processes of viral infection [46]. This protein binds to the virus capsid and modulates capsid uncoating, thus facilitating infection. Additionally, it enhances the efficiency of reverse transcription and nuclear import [47]. CypA not only plays a crucial role in the HIV life cycle but also in various stages of the HBV life cycle. In the cytoplasm, HBsAg interacts with CypA, triggering the extracellular release of CypA. Interestingly, it seems that the presence of CypA also promotes the secretion of HBsAg in return, as evidenced by the reduction in secreted HBsAg levels when CypA inhibitors are employed [48]. This released CypA exerts a potent chemotactic effect on inflammatory cells, thereby exacerbating liver inflammation, which could potentially contribute to the pathogenicity of HBV [49].

In 2015, Gallay described a novel cyclophilin inhibitor, CPI-431-32 or CRV431 [50] (Figure 29.3). This compound targets CypA. CRV431 shows broad-spectrum antiviral activity against HIV, HBV, HDV, and HCV. The compound disrupts the interactions between CypA and viral proteins. Specifically, it prevents the interaction between CypA and the HIV-1 capsid, effectively inhibiting HIV replication, including that of drug-resistant strains [50]. Moreover, in the case of HBV, CRV431 reduces the presence of different viral proteins and viral DNA. This reduction is achieved by blocking the interaction between HBV X protein (a pleiotropic regulatory protein) or HBsAg and CypA, or by inhibiting viral entry via the vital receptor sodium taurocholate cotransporting polypeptide [51].

Metabolism studies indicate that CRV431 is primarily metabolized by the cytochrome P450 enzyme [52], while nucleotides are metabolized predominantly in the kidneys. Consequently, potential drug–drug interactions between CRV431 and nucleotide drugs are speculated to be minimal. A phase I clinical trial (NCT03596697) demonstrated that CRV431 is safe and well tolerated, with single doses up to 525 mg and multiple ascending doses up to 375 mg for 28 days [53].

Despite the fact that extensive research has been conducted to develop new dual HIV/HBV antivirals that act through a mechanism of action distinct from that of RT inhibition, only RT inhibitors have demonstrated sufficient efficacy and safety for commercialization so far. On the other hand, the renal toxicity observed in some of the drugs in clinical use, especially TDF/TAF, underscores the need to continue searching for novel compounds with different mechanisms of action.

29.2.2 Compounds with Dual Activity Against HIV and HCV Viruses

HIV and HCV are two chronic infections that affect millions all over the world. It has been reported that around 5 million HIV-infected people are also co-infected with

HCV (approximately one third) [54–56]. HIV-1 infection has a negative impact on HCV infection [54–58]. This co-infection provoques a higher risk of hepatic fibrosis and, as a result, the liver disease is more aggressive [59], putting the patients on an increased risk of a liver-related death [57]. HCV infection is the leading co-infection in the late stage of HIV-infected patients, resulting, very frequently, into liver cirrhosis and finally to death [19, 58, 60]. Combination therapy is the regimen currently used to treat these co-infected people [60]. Although there have been great advances in the treatment of HCV-infected people with direct-acting antivirals [19, 61, 62], that even resulted in a feasible HCV erradication, the treatment of HIV-1/HCV co-infected individuals remains a real challenge. As mentioned above, these co-infected patients develop serious liver diseases such as cirrhosis, liver fibrosis, or liver cancer more easily than HCV mono-infected patients.

One strategy that may be useful to cope with these problems would be the discovery of compounds that diminish the infection and replication of both HIV and HCV. However, there are very few examples in the literature, and very few compounds under investigation, of potential dual drugs with concomitant activities against both viruses (HIV and HCV). Some of them are presented below.

Cali[4]arene derivatives were reported to block HIV and HCV viral replication in the low micromolar range [63]. In the same publication, it was determined that substitutions at the lower edge of these molecules with four alkoxy or hydroxyl groups stabilize a cone conformation of the molecules. Moreover, aromatic rings substituted with two carboxylic groups are essential to keep anti-HCV, while for the anti-HIV activity, the presence of aromatic spacers at the top of the cone is required [12, 63]. These combined substitutions in the molecules are essential for the potent dual HIV/HCV activity. Compound **1** (Figure 29.4) is an example of this family of dual HIV/HCV inhibitors. It inhibits the replication of both viruses with values of EC_{50s} in the low micromolar range [63, 64].

Several ring-expanded nucleosides, as compounds **2** and **3** (Figure 29.4), of different configurations at the anomeric carbon, were reported in 2014 [65] to behave as dual inhibitors of HIV and HCV. These compounds possess an imidazol[4,5-e]diazepine ring substituted at the C-6 position with a long alkyl chain. The authors described previously (2008) [66] that compounds **2** and **3** bearing 17 methylenes were good inhibitors of HIV (low micromolar IC_{50}). They also reported [67] that this family of compounds inhibited the NPAses and helicases encoded by several flaviviruses (the family to which belongs the HCV). Therefore, these compounds were explored to determine their potential as dual inhibitors of HIV and HCV. From this study, it was concluded that a long chain of a minimum of 18 carbon atoms is required for dual activity against HIV and HCV. The compounds behave as inhibitors of the viruses both in enzymatic and *in vitro* antiviral assays. The activity profile of the compounds is close and comparable to that of interferon-alfa (treatment of reference in combination with ribavirin approved by the FDA to treat HCV infection), but they showed lower toxicity. Both compounds represent promising leads as dual inhibitors of HIV and HCV.

In 2021, Kobayasi [68] described dual inhibitors of HIV and HCV based on substituted fullerenes. Water-soluble fullerenes have been reported to show diverse pharmacological properties [69–71]. Notably, they show inhibitory effects against several viral enzymes, including HCV and HIV polymerases and proteases [72–75], among others.

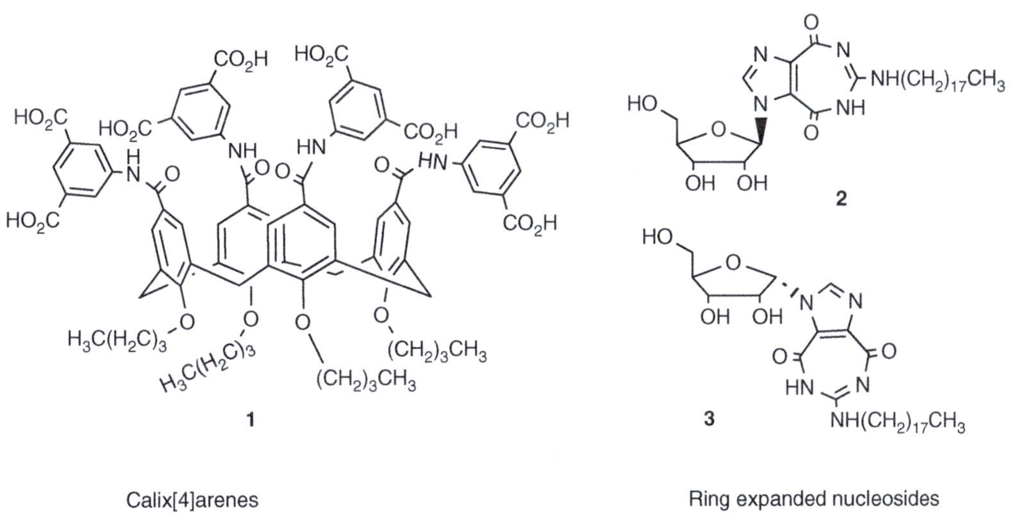

Figure 29.4 Structures of calix[4]arenes and ring-expanded nucleosides.

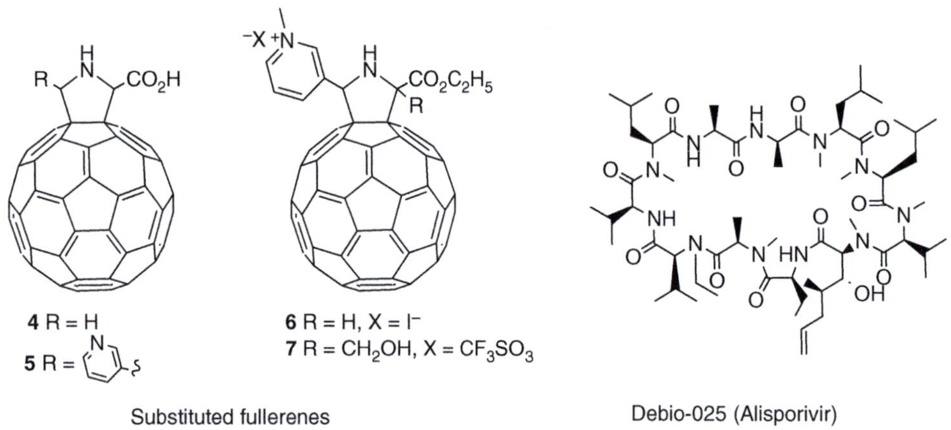

Figure 29.5 Structures of substituted fullerenes and cyclophilin inhibitors.

The authors previously reported that proline-type fullerene **4** and pyridinium-type analogues **5**, **6**, and **7** (Figure 29.5) inhibited HIV RT in a similar fashion [74]. In addition, it was further reported the inhibitory activity of proline substituted **4** against HCV NS3/4A serine protease and NS5B polymerase [72]. These facts promoted the authors to study the antiviral activity of substituted fullerenes, in particular the pyridinium-type fullerenes, as dual inhibitors of HIV/HCV that may eventually be useful to treat co-infected individuals. They found that these derivatives acted as dual inhibitors of HIV and HCV, targeting HIV-RT, HIV protease (HIV-Pr), and HCV NS5B polymerase, with IC_{50} values well-balanced in the low μM range [68]. Most of the compounds were non-cytotoxic. They also determined the cellular uptake of the compounds, concluding that the best compounds were the pyridinium-ethyl ester-type compounds [68]. The effect of these compounds in HIV or HCV-infected cells is currently under study.

As already mentioned above, cyclophilin inhibitors demonstrated good antiviral activities against HIV-1 replication and also against HCV replication, individually. Therefore, it was considered interesting to explore the activity of this kind of inhibitor in the case of HIV/HCV co-infection.

Debio-025 (Alisporivir) (Figure 29.5) is an oral cyclophilin inhibitor with dual activity against HIV-1 and HCV [76–78]. The compound shows good anti-HIV-1 and potent anti-HCV (low nanomolar range) *in vitro* activities [76].

In a clinical phase I study carried out with HIV-1 asymptomatically infected volunteers treated with alisporivir (at 50, 400, and 1200 mg/day for 10 days), it was observed to have a good antiviral effect (viral reduction <1 log 10). A second double-blind randomized placebo controlled study was carried out to determine the antiviral effect, safety, and pharmacological profile in both HIV-1/HCV co-infected and HIV-1-monoinfected people upon treatment with Debio-025 monotherapy for 15 days [76]. It was observed a direct anti-HIV-1 effect (1 log 10 reduction) and a good anti-HCV effect (a viral reduction among 2 to 3.6 log 10 reduction). However, the HIV viral load reduction was lower than the reduction of HCV viral load. It should be pointed out that normally these co-infected patients respond poorly to anti-HCV treatment. From the study, it was concluded that Debio-025 (administered at 1200 mg twice a day/daily for 14 days) has a noticeable antiviral effect against several genotypes of HCV. Moreover, no viral rebound was observed on the treatment, which suggests a high barrier for the selection of resistance by the drug.

As already mentioned, CRV431 (Figure 29.3), an analogue of cyclosporin A, is another cyclophilin inhibitor that has also been described as an efficient inhibitor of hepatitis B and hepatitis C replication by blocking important virus-host cyclophilin interactions [50]. CRV431 inhibits simultaneously both HIV-1 and HCV replication. The compound blocks the interactions of both the CypA and HIV-1 capsid and of CypA and HCV NS5A [50]. In particular, the compound disrupts the interactions of CypA with the viral proteins HIV-1 p24 capsid [79, 80] and HCV NS5A [50]. Moreover, CRV431 inhibits, efficiently, HIV-1 and HCV drug resistant variants [50]. When compared with Alisporivir, CRV431 is antivirally more efficient.

Taking into consideration all the above mentioned, CRV431 can be considered a potent and attractive dual inhibitor for the inhibition of HIV-1/HCV co-infection that deserves further development.

29.3 Compounds with Dual Activity Against HIV and EV-A71

EVs are non-enveloped single-stranded RNA viruses that belong to the picornaviridae family. They infect people worldwide and are transmitted through droplets, the fecal-oral route, or by direct contact [81]. The non-polio EVs are associated with the emergence of different diseases worldwide [82]. There have been described more than 100 different serotypes (all of them infect humans) that are clustered into 4 species : EV-A, -B, -C, and D.

Enteroviral infections (especially the non-polio enteroviruses) are attracting increasing attention in sub-Saharan countries since they mainly affect the pediatric population [9, 82–84]. Moreover, it has been detected within PLHIV, which are immunocompromised [85]. As mentioned above, the HIV-infected patients are more vulnerable to infections by other pathogens [84]. Co-infections with EVs are reported to worsen and speed the progression of HIV-infected individuals to the AIDS [86].

EV-A71 causes hand, foot, and mouth disease and other serious health problems in children under 5 years of age. The virus is a non-polio neurotropic EV with pandemic potential. There are no antivirals approved to prevent or treat EV-A71 infections. Effective antivirals are urgently needed for prophylaxis or treatment of patients with severe EV-A71 infections. In 2015, San Félix described a novel and unique family of dendrimers, containing from 9-18 tryptophan (Trp) units on the periphery, which showed a marked dual activity against HIV-1 and EV-A71 replication [87]. The prototype of this family **8** (Figure 29.6), with 12 Trp units on the periphery, linked through trivalent spacer arms to a central pentaerythritol core, showed potent dual anti-HIV activity (EC_{50}: 2.3 μM; CC_{50} > 100 μM) and EV-A71 activity (EC_{50}: 0.29 μM; CC_{50}: 30 μM). Interestingly, compound **8** showed extraordinary antiviral activity *in vitro* (low nanomolar/high picomolar range) against circulating EV-A71 clinical isolates [88]. This compound binds to the surface HIV-1 gp120 and gp41 proteins and blocks the attachment and internalization of EV-A71 [87, 89, 90]. Further SAR studies within this family of compounds indicated that aromatic amino acids, such as Trp and Tyr, are required for dual anti-HIV and anti-EV-A71 activities. The presence of free carboxylic acids on the periphery is also crucial for the activity [91]. On the other hand, only compounds with tetravalent branched arms showed similar anti-HIV and anti-EV-A71 activity (low micromolar) than that of the prototype **8** (with trivalent branched arms) [92]. The new compounds (with tetravalent branched arms) also showed low nanomolar activity against a representative EV-A71 clinical isolate. Some compounds of this series interact with the viral envelopes of HIV-1, EV-A71, and HSV-2 and thus prevent virus attachment to the host cell. Interestingly, these compounds even gained significant antiviral activity against new pathogens like Herpes simplex virus type 2 (HSV-2), adenovirus-2, human coronavirus, and respiratory syncytial virus [92]. Some compounds of this series interact with the viral envelopes of HIV-1, EV-A71, and HSV-2 and thus prevent virus attachment to the host cell.

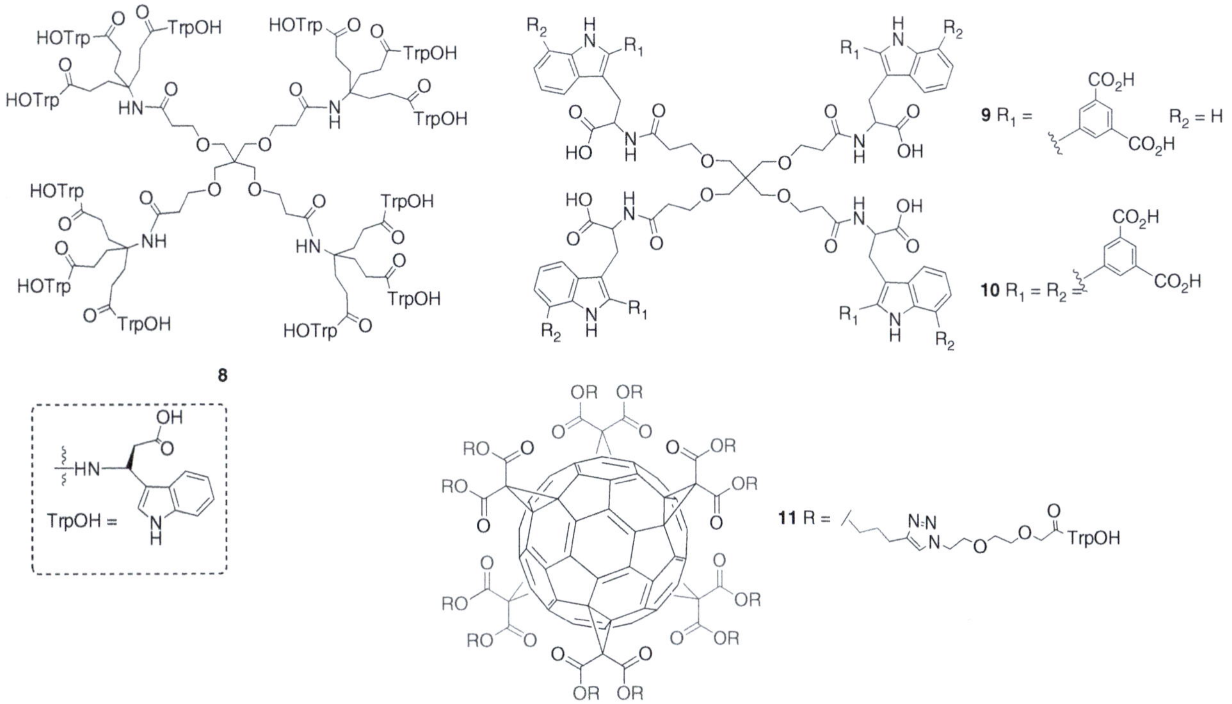

Figure 29.6 Structures of HIV/EV-A71 dual inhibitors.

For drug development purposes, compounds of reduced size were prepared [93]. The new prototype **9** (Figure 29.6) contains a pentaerythritol central scaffold substituted with four Trp units bearing an extra isophthalic acid moiety at the C-2 position of each indole ring, was 100-fold more potent inhibitor against HIV-1 replication than **8**, and also was a potent inhibitor of EV-A71 replication, inhibiting not only the laboratory BrCr strain but also different EV-A71 clinical isolates from different genogroups and geographic origins. This new half-size prototype **9** is non-toxic for different types of host cells. Further optimization studies led to the C2/C7 doubly arylated tetrapodal compound (**10**) as the most potent entry inhibitor of this series synthesized so far (EC_{50} HIV-1: 0.006 μM; EC_{50} EV-A71: 0.040 μM; SI >16000) [94].

From the different modifications carried out in the prototype **9**, it was concluded that both the Trp and the isophthalic acid moiety were the basic structural elements responsible for the antiviral activity (pharmacophore) [95].

In order to complete SAR studies within this family of compounds, in 2021, the synthesis and antiviral activity of 3D hexa-adducts of [60]fullerene peripherally decorated with twelve Trp or tyrosine (Tyr) residues was reported [96]. These substituted fullerenes, whose prototype is **11** (Figure 29.6), were much more potent against HIV-1 and equally active against EV-A71 than the previously described dendrimer **8**, which possesses the same number of Trp/Tyr residues on the periphery but linked to a smaller and more flexible pentaerythritol core. This demonstrated the relevance of the globular 3D presentation of the peripheral groups (Trp/Tyr), as well as the length of the spacer connecting them to the central core, to bind with the viral envelopes, particularly in the case of HIV-1. It was hypothesized that the fullerene core of these compounds helps to distribute the peripheral aromatic rings (indole/phenol) and negative charges in a spherical fashion that can be optimal for binding to the globular HIV-1 envelope.

Finally, biodistribution experiments in animal models carried out with a fluorescently labeled (Alexa Fluor 647) derivative of this series of dendrimers [97] demonstrate that this family of compounds accumulates preferably in the intestines and lungs, which are the primary replication sites for EV-A71 and locations of HIV-associated pathogenesis. Therefore, this family of compounds may prevent the spread of both viruses to other tissues and organs. As entry inhibitors, they have the advantage, with respect to other classes of antivirals, that act outside the host cell, making them good candidates for development as pre- and post-exposure prophylaxis, microbicides, and therapeutics.

Acknowledgement

We authors would like to dedicate this book chapter in Memoriam of Dr. Sonsoles Velázquez, an excellent Medicinal Chemistry researcher and a better person.

References

1 UNAIDS 2023: UNAIDS global aids update. https://thepath.unaids.org/wp-content/themes/unaids2023/assets/files/2023_report.pdf. (accessed 31 January 2024).

2 UNAIDS data 2023. https://www.unaids.org/sites/default/files/media_asset/data-book-2023_en.pdf https://www.unaids.org/es/resources/documents/2022/in-danger-global-aids-update-summary (accessed 8 February 2024).

3 World Health Organization (2023). HIV-AIDS: https://www.who.int/health-topics/hiv-aids#tab=tab_1. World AIDS day 2023. https://isid.org/world-aids-day-2023/ (accessed 31 January 2024).

4 De Cock, K.M., Jaffe, H.W., and Curran, J.W. (2021). Reflections on 40 years of AIDS. *Emerg. Infect. Dis.* 27: 1553–1560.

5 Gandhi, R.T., Bedimo, R., Hoy, J. et al. (2023). Antiretroviral drugs for treatment and prevention of HIV infection in adults: 2022 Recommendations of the International Antiviral Society-USA Panel. *JAMA* 329: 63–84.

6 Soriano, V., Moreno-Torres, V., de Mendoza, C. et al. (2023). Viral hepatitis in persons living with HIV in the post.COVID era. *AIDS Rev.* 25: 1–13.

7 Chang, C.C., Crane, M.C., Zhou, J.L. et al. (2013). HIV and co-infections. *Immunol. Rev.* 254: 114–142.

8 WHO (Global HIV Programme): chronic comorbidities & coinfections among PLHIV. https://www.who.int/teams/global-hiv-hepatitis-and-stis-programmes/hiv/treatment/chronic-comorbidities-and-coinfections (accessed 22 January 2024).

9 Mekonnen, G.B., Birhane, B.M., Engdaw, M.T. et al. (2023). Predictors of a high incidence of opportunistic infections among HIV-infected children receiving antiretroviral therapy at Amhara regional state comprehensive specialized hospitals, Ethiopia: a multicenter institutionbased retrospective follow-up study. *Front. Pediatr.* 11: 1107321–1107336.

10 Candiani, T., Pinto, J., Cardoso, C.A.A. et al. (2007). Impact of highly active antiretroviral therapy (HAART) on the incidence of opportunistic infections,

hospitalizations and mortality among children and adolescents living with HIV/AIDS in Belo Horizonte, minas Gerais state, Brazil. *Cad. Saude Publica* 23 (Suppl 3): S414–S423.

11 Boyd, A.T., Oboho, I., Paulin, H. et al. (2000). Addressing advanced HIV disease and mortality in global HIV programming. *AIDS Res. Ther.* 17: 40–47.

12 de Castro, S. and Camarasa, M.-J. (2018). Polypharmacology in HIV inhibition: can a drug with simultaneous action against two relevant targets be an alternative to combination therapy? *Eur. J. Med. Chem.* 150: 206e227.

13 Castellares, C., Barreiro, P., Martín-Carbonero, L. et al. (2008). Liver cirrhosis in HIV-infected patients: prevalence, etiology and clinical outcome. *J. Viral Hepat.* 15: 165–172.

14 Luetkemeyer, A.F. and Wyles, D. (2021). CROI 2021: viral hepatitis and other forms of liver injury impacting on people with HIV. *Top. Antivir. Me.* 29: 379–385.

15 Kenfack-Momo, R., Kenmoe, S., Takuissu, G.R. et al. (2022). Epidemiology of hepatitis B virus and/or hepatitis C virus infections among people living with human immunodeficiency virus in Africa: a systematic review and meta-analysis. *PLoS ONE* 31, 17(5): e0269250.

16 García-Samaniego, J., Rodríguez, M., Berenguer, J. et al. (2001). Hepatocellular carcinoma in HIV-infected patients with chronic hepatitis C. *Am. J. Gastroenterol.* 96: 179–183.

17 Soriano, V., Martín-Carbonero, L., García-Samaniego, J., and Puoti, M. (2001). Mortality due to chronic viral liver disease among patients infected with HIV. *Clin. Infect. Dis.* 33: 1793–1795.

18 Gobran, S.T., Ancuta, P., and Shoukry, N.H. (2021). A tale of two viruses: immunological insights into HCV/HIV coinfection. *Front. Immunol.* 12: 726419.

19 Platt, L., Easterbrook, P., Gower, E. et al. (2016). Prevalence and burden of HCV co-infection in people living with HIV: a global systematic review and meta-analysis. *Lancet Infect. Dis.* 16: 797–808.

20 Jin, F., Matthews, G.V., and Grulich, A.E. (2017). Sexual transmission of Hepatitis C Virus among gay and bisexual men: a systematic review. *Sex. Health* 14: 28–41.

21 World Health Organization (WHO) (2023). Hepatitis B. https://www.who.int/news-room/fact-sheets/detail/hepatitis-b (accessed 23 January 2024).

22 Maida, I., Ríos, M., Pérez-Saleme, L. et al. (2008). Profile of patients triply infected with HIV and the hepatitis B and C viruses in the HAART era. *AIDS Res. Hum. Retrovir.* 24: 679–683.

23 Nikolopoulos, G.K., Paraskevis, D., Hatzitheodorou, E. et al. (2009). Impact of hepatitis B virus infection on the progression of AIDS and mortality in HIV-infected individuals: a cohort study and meta-analysis. *Clin. Infect. Dis.* 48: 1763–1771.

24 Sun, H.Y., Sheng, W.H., Tsai, M.S. et al. (2014). Hepatitis B virus coinfection in human immunodeficiency virus-infected patients: a review. *World J. Gastroenterol.* 20: 14598–14614.

25 Naing, C., Poovorawan, Y., and Tong, K.S. (2018). Comparative effectiveness of anti-viral drugs with dual activity for treating hepatitis B and HIV co-infected patients: a network meta-analysis. *BMC Infect. Dis.* 18: 564–575.

26 GBD 2019 Hepatitis B Collaborators (2022). Global, regional and national burden of hepatitis B, 1990-2019: a systematic analysis for the global burden of disease study 2019. *Lancet Gastroenterol. Hepatol.* 7: 796–829.

27 Cheng, Z., Lin, P., and Cheng, N. (2021). HBV/HIV coinfection: impact on the development and clinical treatment of liver diseases. *Front. Med.* 8: 713981–713996.

28 De Clercq, E. and Holy, A. (2005). Case history: Acyclic nucleoside phosphonates: a key class of antiviral drugs. *Nat. Rev. Drug Discov.* 4: 928–940.

29 Department of Health & Human Services (2016). FDA approves vemlidy (tenofovir alafenamide) for chronic hepatitis B in adults. https://web.archive.org/web/20191011163514/https://www.hhs.gov/hepatitis/blog/2016/11/21/fda-approves-vemlidy-tenofovir-alafenamide-for-chronic-hepatitis-b-in-adults.html (accessed 23 January 2024).

30 McGuigan, C., Pathirana, R.N., Balzarini, J., and De Clercq, E. (1993). Intracellular delivery of bioactive AZT nucleotides by aryl phosphate derivatives of AZT. *J. Med. Chem.* 36: 1048–1052.

31 De Clercq, E. (2016). Tenofovir alafenamide (TAF) as the successor of tenofovir disoproxil fumarate (TDF). *Biochem. Pharmacol.* 119: 1–7.

32 Tamuzi, J.L., Tshimwanga, J.L., Bulabula, A.N.H., and Muyaya, L.M. (2018). Tenofovir Alafenamide versus Tenofovir Disoproxil Fumarate: systematic review and meta-analysis. *Int. J. Pul. Res. Sci.* 2: 555600–555616.

33 Lee, W.A., He, G.X., Eisenberg, E. et al. (2005). Selective intracellular activation of a novel prodrug of the human immunodeficiency virus reverse transcriptase inhibitor tenofovir leads to preferential distribution and accumulation in lymphatic tissue. *Antimicrob. Agents Chemother.* 49: 1898–1906.

34 Pribut, N., D'Erasmo, M., Dasari, M. et al. (2021). ω-Functionalized lipid prodrugs of HIV NtRTI Tenofovir with enhanced pharmacokinetic properties. *J. Med. Chem.* 64: 12917–12937.

35 Janssens, S. and Beyaert, R. (2003). Role of toll-like receptors in pathogen recognition. *Clin. Microbiol. Rev.* 16 (4): 637–646.

36 Tsai, A., Irrinki, A., Kaur, J. et al. (2017). Toll-like receptor 7 agonist GS-9620 induces HIV expression and HIV-specific immunity in cells from HIV-infected individuals on suppressive antiretroviral therapy. *J. Virol.* 91: e02166–e02116.

37 Macedo, A.B., Novis, C.L., De Assis, C.M. et al. (2018). Dual TLR2 and TLR7 agonists as HIV latency-reversing agents. *JCI Insight* 3 (19): e122673.

38 Kayesh, M.E.H., Kohara, M., and Tsukiyama-Kohara, K. (2021). Toll-Like Receptor response to hepatitis b virus infection and potential of TLR agonists as immunomodulators for treating chronic hepatitis b: an overview. *Int. J. Mol. Sci.* 22 (19): 10462.

39 Bam, R.A., Hansen, D., Irrinki, A. et al. (2016). TLR7 agonist GS-9620 is a potent inhibitor of acute HIV-1 infection in human peripheral blood mononuclear cells. *Antimicrob. Agents Chemother.* 61: e01369–e01316.

40 Ventura, J.D., Nkolola, J.P., Chandrashekar, A. et al. (2022). Therapeutic efficacy of an Ad26/MVA vaccine with SIV gp140 protein and Vesatolimod in ART-suppressed rhesus macaques. *Vaccine* 7: 53–61.

41 Bailón, L., Llano, A., Cedeño, S. et al. (2022). Safety, immunogenicity and effect on viral rebound of HTI vaccines in early treated HIV-1 infection: a randomized, placebo-controlled phase 1 trial. *Nat. Med.* 28: 2611–2621.

42 Akbari, E., Seyedinkhorasani, M., and Bolhassani, A. (2023). Conserved multiepitope vaccine constructs: a potent HIV-1 therapeutic vaccine in clinical trials. *Braz. J. Infect. Dis.* 27: 102774–102789.

43 Omer, I., Abuthiyab, N., Al-Omari, B. et al. (2023). Efficacy and safety of vesatolimod in chronic hepatitis B: a systematic review and meta-analysis. *Medicine* 102 (16): e33609.

44 Woong Kim, S.W., Yoon, J.S., Lee, M., and Cho, Y. (2022). Toward a complete cure for chronic hepatitis B: novel therapeutic targets for hepatitis B virus. *Clin. Mol. Hepatol.* 28: 17–30.

45 Wang, P. and Heitman, J. (2005). The cyclophilins. *Genome Biol.* 6: 226–232.

46 Liao, Y., Luo, D., Peng, K., and Zeng, Y. (2021). Cyclophilin a: a key player for etiological agent infection. *Appl. Microbiol. Biotechnol.* 105: 1365–1377.

47 Bose, S., Mathur, M., Bates, P. et al. (2003). Requirement for cyclophilin A for the replication of vesicular stomatitis virus New Jersey serotype. *J. Gen. Virol.* 84: 1687–1699.

48 Phillips, S., Chokshi, S., Chatterji, U. et al. (2015). Alisporivir inhibition of hepatocyte cyclophilins reduces HBV replication and hepatitis B surface antigen production. *Gastroenterology* 148: 403–414.

49 Tian, X., Zhao, C., Zhu, H. et al. (2010). Hepatitis B Virus (HBV) surface antigen interacts with and promotes cyclophilin a secretion: possible link to pathogenesis of HBV infection. *J. Virol.* 84: https://doi.org/10.1128/jvi.02555-09.

50 Gallay, P.A., Bobardt, M., Chatterji, U. et al. (2015). The novel cyclophilin inhibitor CPI-431-32 concurrently blocks HCV and HIV-1 infections via a similar mechanism of action. *PLoS ONE* 10 (8): e0134707.

51 BIOSPACE: contraVir's cyclophilin inhibitor CRV431 potently inhibits essential pathway in hepatitis B. https://www.biospace.com/article/releases/contravir-s-cyclophilin-inhibitor-crv431-potently-inhibits-essential-pathway-in-hepatitis-b-/ (accessed 30 January 2023).

52 Trepanier, D.J., Ure, D.R., and Foster, R.T. (2017). In vitro phase I metabolism of CRV431, a novel oral drug candidate for chronic Hepatitis B. *Pharmaceutics.* 9: 51–64.

53 National Library of Medicine (NIH) (2021). A Study in healthy volunteers. https://clinicaltrials.gov/study/NCT03596697 (accessed 31 January 2024).

54 Hernandez, M.D. and Sherman, K.E. (2011). HIV/hepatitis C coinfection natural history and disease progression. *Curr. Opin. HIV AIDS* 6: 478–482.

55 Kim, A.Y. and Chung, R.T. (2009). Coinfection with HIV-1 and HCV—a one-two punch. *Gastroenterology* 137: 795–814.

56 Sherman, K.E. (2004). HCV and HIV: a tale of two viruses. *Rev. Gastroenterol. Disord.* 4 (Suppl 1): S48–S54.

57 Smit, C., van den Berg, C., Geskus, R. et al. (2008). Risk of hepatitis-related mortality increased among hepatitis C virus/HIV-coinfected drug users compared with drug users infected only with hepatitis C virus: a 20-year prospective study. *J. Acquir. Immune Defic. Syndr.* 47: 221–225.

58 Shepard, C.W., Finelli, L., and Alter, M.J. (2005). Global epidemiology of hepatitis C virus infection. *Lancet Infect. Dis.* 5: 558–567.

59 Operskalski, E.A. and Kovacs, A. (2011). HIV/HCV co-infection: pathogenesis, clinical complications, treat-ment, and new therapeutic technologies. *Curr. HIV/AIDS Rep.* 8: 12–22.

60 Dienstag, J.L. and McHutchison, J.G. (2006). American gastroenterological association technical review on the management of hepatitis C. *Gastroenterology* 130: 231–264.

61 van Santen, D., Sacks-Davis, R., Stewart, A. et al. (2010). Treatment as prevention effect of direct-acting antivirals on primary hepatitis C virus incidence: findings from a

multinational cohort between 2010 and 2019. *E. Clinical Med.* 56: 101810–101822.

62 Lodi, S., Klein, M., Rauch, A. et al. (2022). Sustained virological response after treatment with direct antiviral agents in individuals with HIV and hepatitis C co-infection. *J. Int. AIDS Soc.* 25: e26048.

63 Tsou, L.K., Dutschman, G.E., Gullen, E.M. et al. (2010). Discovery of a synthetic dual inhibitor of HIV and HCV infection based on tetrabutoxy-calix[4]arene scaffold. *Bioorg. Med. Chem. Lett.* 20: 2137–2139.

64 Sun, J., Wang, D.A., Jain, R.K. et al. (2005). Inhibiting angiogenesis and tumorigenesis by a synthetic molecule that blocks binding of both VEGF and PDGF to their receptors. *Oncogene* 24: 4701e4709.

65 Zhang, N., Zhang, P., Baier, A. et al. (2014). Dual inhibition of HCV and HIV by ring-expanded nucleosides containing the 5:7-fused imidazo[4,5-e][1,3]diazepine ring system. *In vitro* results and implications. *Bioorg. Med. Chem. Lett.* 24: 1154–1157.

66 Yedavalli, V.S.R.K., Zhang, N., Cai, H. et al. (2008). Ring expanded nucleoside analogues inhibit RNA helicase and intracellular human immunodeficiency virus type 1 replication. *J. Med. Chem.* 51: 5043–5051.

67 Zhang, N., Chen, H.M., Koch, V. et al. (2003). Ring-expanded ("fat") nucleoside and nucleotide analogues exhibit potent in vitro activity against flaviviridae NTPases/helicases, including those of the west nile virus, hepatitis C virus, and Japanese encephalitis virus. *J. Med. Chem.* 46: 4149–4164.

68 Kobayashi, T., Yasuno, T., Takahashi, K. et al. (2021). Novel pyridinium-type fullerene derivatives as multitargeting inhibitors of HIV-1 reverse transcriptase, HIV-1 protease, and HCV NS5B polymerase. *Bioorg. Med. Chem. Lett.* 49: 128267–112872.

69 Nishizawa, C., Hashimoto, N., Yokoo, S. et al. (2009). Pyrrolidinium-type fullerene derivatives-induced apoptosis by the generation of reactive oxygen species in HL-60 cells. *Free Radic. Res.* 43: 1240–1247.

70 Mashino, T., Nishikawa, D., Takahashi, K. et al. (2003). Antibacterial and antiproliferative activity of cationic fullerene derivatives. *Bioorg. Med. Chem. Lett.* 13: 4395–4397.

71 Kobzar, O.L., Trush, V.V., Tanchuk, V.Y. et al. (2014). Fullerene derivatives as a new class of inhibitors of protein tyrosine phosphatases. *Bioorg. Med. Chem. Lett.* 24: 3175–3179.

72 Kataoka, H., Ohe, T., Takahashi, K. et al. (2016). Novel fullerene derivatives as dual inhibitors of hepatitis C virus NS5B polymerase and NS3/4A protease. *Bioorg. Med. Chem. Lett.* 26: 4565–4567.

73 Mashino, T., Shimotohno, K., Ikegami, N. et al. (2005). Human immunodeficiency virus-reverse transcriptase inhibition and hepatitis C virus RNA-dependent RNA polymerase inhibition activities of fullerene derivatives. *Bioorg. Med. Chem. Lett.* 15: 1107–1109.

74 Yasuno, T., Ohe, T., Takahashi, K. et al. (2015). The human immunodeficiency virus-reverse transcriptase inhibition activity of novel pyridine/ pyridinium-type fullerene derivatives. *Bioorg. Med. Chem. Lett.* 25: 3226–3229.

75 Yasuno, T., Ohe, T., Kataoka, H. et al. (2021). Fullerene derivatives as dual inhibitors of HIV-1 reverse transcriptase and protease. *Bioorg. Med. Chem. Lett.* 31: 127675–127680.

76 Flisiak, R., Horban, A., Gallay, P. et al. (2008). The cyclophilin inhibitor Debio-025 shows potent anti–hepatitis C effect in patients coinfected with hepatitis C and Human Immunodeficiency Virus. *Hepatology* 47: 817–826.

77 Inoue, K., Umehara, T., Ruegg, U.T. et al. (2007). Evaluation of a cyclophilin inhibitor in hepatitis C virus-infected chimeric mice in vivo. *Hepatology* 45: 921–928.

78 Paeshuyse, J., Kaul, A., De Clercq, E. et al. (2006). The non-immunosuppressive cyclosporin DEBIO-025 is a potent inhibitor of hepatitis C virus replication in vitro. *Hepatology* 43: 761–770.

79 Hilditch, L. and Towers, G.J. (2014). A model for cofactor use during HIV-1 reverse transcription and nuclear entry. *Curr. Opin. Virol.* 4: 32–36.

80 Marini, B., Kertesz-Farkas, A., Ali, H. et al. (2015). Nuclear architecture dictates HIV-1 integration site selection. *Nature* 521: 227–231.

81 Wells, A.I. and Coyne, C.B. (2019). Enteroviruses: a gut-wrenching game of entry, detection, and evasion. *Viruses* 11: 460–480.

82 Di Cristanziano, V., Weimer, K., Böttcher, S. et al. (2020). Molecular characterization and clinical cescription of non-polio enteroviruses detected in stool samples from HIV-positive and HIV-negative adults in Ghana. *Viruses* 12: 221–233.

83 Chakraborty, R., Iturriza-Gómara, M., Musoke, R. et al. (2004). An epidemic of enterovirus 71 infection among HIV-1-infected orphans. *AIDS* 18: 1968–1979.

84 Liste, M.B., Natera, I., Suarez, J.A. et al. (2000). Enteric virus infections and diarrhea in healthy and human immunodeficiency virus-infected children. *J. Clin. Microbiol.* 2873–2877.

85 Alemu, G., Alelign, D., and Abossie, A. (2018). Prevalence of opportunistic intestinal parasites and associated factors among HIV patients while receiving ART at Arba Minch hospital in southern Ethiopia: a cross-sectional study. *Ethiop. J. Health Sci.* 28: 147–156.

86 Gedle, D., Kumera, G., Eshete, T. et al. (2017). Intestinal parasitic infections and its association with undernutrition and CD4 T cell levels among HIV/AIDS patients on HAART in Butajira, Ethiopia. *J. Health Popul. Nutr.* 36: 15–25.

87 Rivero-Buceta, E., Doyagüez, E.G., Colomer, I. et al. (2015). Tryptophan dendrimers that inhibit HIV replication, prevent virus entry and bind to the HIV envelope glycoproteins Gp120 and Gp41. *Eur. J. Med. Chem.* 106: 34–43.

88 Rivero-Buceta, E., Sun, L., Martínez-Gualda, B. et al. (2016). Optimization of a class of tryptophan dendrimers that inhibit HIV replication leads to a selective, specific, and low-nanomolar inhibitor of clinical isolates of enterovirus A71. *Antimicrob. Agents Chemother.* 60: 5064–5067.

89 Sun, L., Lee, H., Thibaut, H.J. et al. (2019). Viral engagement with host receptors blocked by a novel class of tryptophan dendrimers that targets the 5-fold-axis of the enterovirus-A 71 capsi. *PLoS Pathog.* 15: 1–20.

90 Ren, J., Wang, X., Hu, Z. et al. (2013). Picornavirus uncoating intermediate captured in atomic detail. *Nat. Commun.* 4: 1929–1936.

91 Martínez-Gualda, B., Sun, L., Rivero-Buceta, E. et al. (2017). Structure-activity relationship studies on a Trp dendrimer with dual activities against HIV and enterovirus A71, modifications on the amino acid. *Antivir. Res.* 139: 32–40.

92 Martínez-Gualda, B., Sun, L., Martí-Marí, O. et al. (2019). Modifications in the branched arms of a class of dual inhibitors of HIV and EV71 replication expand their antiviral spectrum. *Antivir. Res.* 168: 210–214.

93 Martínez-Gualda, B., Sun, L., Martí-Marí, O. et al. (2020). Scaffold simplification strategy leads to a novel generation of dual human immunodeficiency and enterovirus-A71 entry inhibitors. *J. Med. Chem.* 63: 349–368.

94 Martí-Marí, O., Martínez-Gualda, B., de la Puente-Secades, S. et al. (2021). Double arylation of the indole side chain of tri- and tetrapodal tryptophan derivatives renders highly potent HIV-1 and EV-A71 entry inhibitors. *J. Med. Chem.* 64: 10027–10046.

95 Martí-Marí, O., Abdelnabi, R., Schols, D. et al. (2023). Insertion of an amphipathic linker in a tetrapodal tryptophan derivative leads to a novel and highly potent entry inhibitor of enterovirus A71 clinical isolates. *Int. J. Mol. Sci.* 24: 3539–1564.

96 Ruiz-Santaquiteria, M., Illescas, B.M., Abdelnabi, R. et al. (2021). Multivalent tryptophan- and tyrosine-containing [60]fullerene hexa-adducts as dual HIV and EV71 entry inhibitors. *Chem. Eur. J.* 27: 10700–10710.

97 Martí-Marí, O., Martínez-Gualda, B., Fernández-Barahona, I. et al. (2022). Organotropic dendrons with high potency as HIV-1, HIV-2 and EV-A71 cell entry inhibitors. *Eur. J. Med. Chem.* 237: 114414–114431.

30

Multi-kinase Inhibitors for the Treatment of Pancreatic Cancer

Paul Dent and Andrew Poklepovic

Our studies over the past 15 years in pancreatic cancer and other gastrointestinal (GI) tumor cell types have shown that combinations of multi-kinase inhibitors and histone deacetylase (HDAC) inhibitors can kill tumor cells *in vitro*, suppress tumor growth in mouse models, and prolong the survival of patients. Additionally, we have assessed the evolutionary changes in signaling, which occurred as tumor cells became resistant to the initial two-drug treatment, leading to the rational development of three-drug combinations, which cause more killing and prevent the primary evolutionary mechanisms that maintain survival.

Our initial studies in this field combined the multi-kinase inhibitors sorafenib and regorafenib with HDAC inhibitors such as vorinostat and sodium valproate in GI and genitourinary (GU) tumor cells (Figure 30.1) [1–6]. The mechanistic conclusions of several manuscripts were that the drugs interacted to increase the expression of the death receptor CD95, also known as the FAS receptor, and its ligand FAS-L. The drug combination caused reactive oxygen species (ROS)-dependent tyrosine phosphorylation of CD95, i.e., ROS-dependent phosphatase inhibition, which promoted its trimerization, leading to activation of pro-caspase 8, followed by active caspase 8, causing the cleavage of the toxic BH3 domain protein BID. Cleavage of BID resulted in mitochondrial dysfunction and ultimately tumor cell death. In all cells, except hepatoma, we observed increased levels of autophagosomes, which played an active role in mediating tumor cell killing alongside the actions of the death receptor CD95. Reduced signaling through the ERK1/2 pathway permitted expression of the toxic BH3 domain protein BIM to rise, and endoplasmic reticulum stress (ER) signaling by reducing translation caused the levels of cytoprotective mitochondrial proteins with short half-lives, BCL-XL and MCL-1, to decline, which collectively further enhanced mitochondrial dysfunction. The key pathway for causing cell death was from the death receptor CD95, as overexpression of the caspase 8/10 inhibitor FLICE-inhibitory protein – short (FLIP-s) abolished the synergy of killing between the drugs in colony formation assays.

Based on these data, a phase I trial was opened to treat unresectable liver cancer patients combining sorafenib and vorinostat (NCT01075113) [7]. Sixteen patients were treated, and thirteen patients were subsequently evaluated for any responses. Eleven of the thirteen patients required dose reductions or omissions for non-dose-limiting toxicity. Ten patients (77%) had a response of stable disease (SD). The median treatment duration was 4.7 months. One patient with SD was on treatment for 29.9 months, and another patient, also with SD, was on treatment for 18.7 months. Another SD patient electively stopped therapy after 15 months and remained without evidence of progression three years later. Three out of thirteen patients had prolonged disease control; however, in HCC patients, the addition of the HDAC inhibitor vorinostat to sorafenib resulted in multiple dose-limiting toxicities, which precluded defining a recommended phase II dose.

While this clinical trial was proceeding, additional work was performed to define whether the [sorafenib plus vorinostat] combination could interact with other standard of care modalities in the treatment of GI tumors, in particular ionizing radiation, to control tumor growth [5]. *In vitro*, the drug combination of [sorafenib plus vorinostat], more so than the individual agents, radio sensitized pancreatic cancer cells. In mouse tumor models, the addition of radiation to the drug combination caused a greater growth delay in tumors compared to either individual treatment. Based on these findings, a second phase I trial was opened focused on stage I-III pancreatic cancer patients (NCT02349867). At present, the results of this completed study have not been published, though a poster was presented at the 2021 ASCO meeting [8]. Adult patients with borderline resectable and unresectable and lymph node-positive pancreatic cancer were enrolled. Patients had to have completed at least eight

Polypharmacology: Strategies for Multi-Target Drug Discovery, First Edition. Edited by Jens-Uwe Peters.

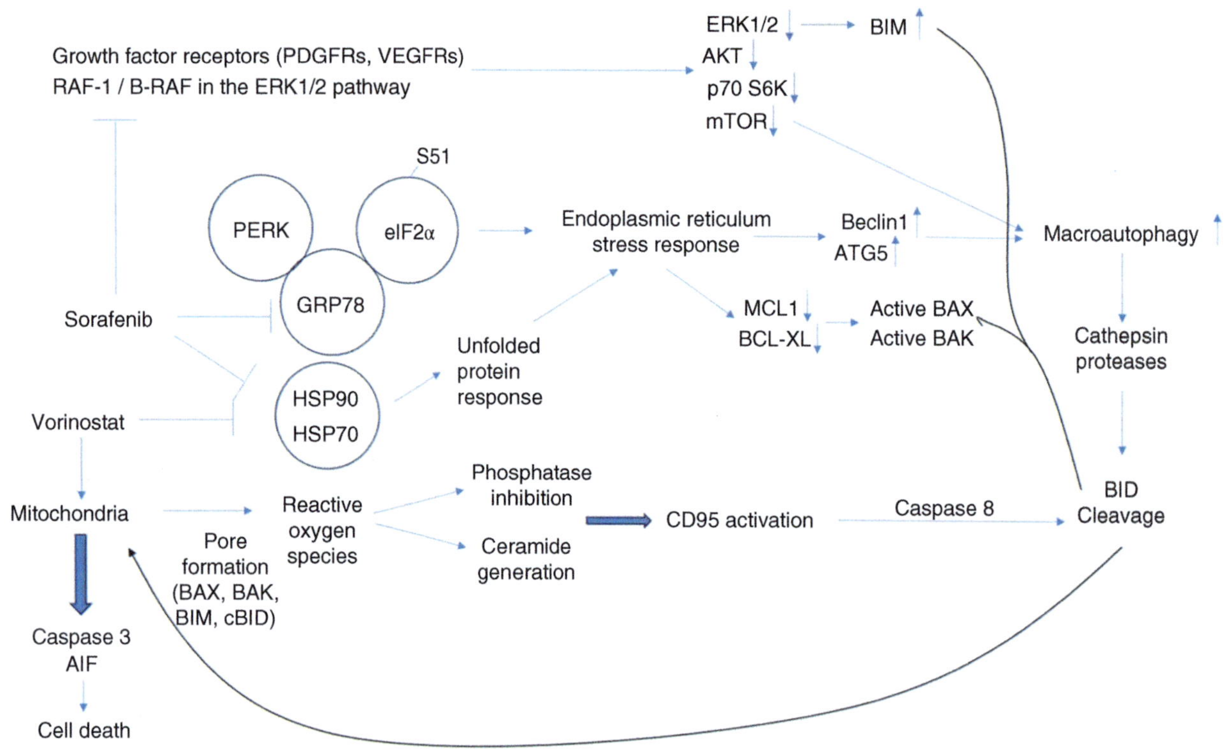

Figure 30.1 Mechanisms by which the multi-kinase inhibitor sorafenib and HDAC inhibitors interact to kill tumor cells: Sorafenib inhibits the activities of HSP90 and HSP70 family chaperones. Sorafenib also inhibits class III receptor tyrosine kinases and RAF-1/B-RAF in the ERK1/2 MAP kinase pathway. Reduced ERK1/2 activity permits BIM to become stabilized. Because sorafenib inhibits GRP78, PERK becomes activated and phosphorylates eIF2α serine 51. This inactivates eIF2α, and for the majority of genes, transcription is halted, e.g., MCL1 and BCL-XL, but for the autophagy proteins Beclin1 and ATG5, their levels are enhanced. Vorinostat acts to increase the acetylation of chaperones, which acts to reduce their activities and is a key nodal interaction between both drugs. This further enhances ER stress signaling and autophagy and facilitates activation of BAX and BAK. Autophagic flux ultimately causes the release of cathepsin proteases into the cytosol. Vorinostat enhances the levels of ROS in cells, which leads to CD95 activation and downstream, to that of caspase 8. Caspase 8 and cathepsins cleave and activate BID. BAX, BAK, BIM, and cleaved BID, together with reduced levels of MCL1 and BCL-XL, cause pore formation and mitochondrial dysfunction. Killing is mediated by both caspase 3 and by AIF.

weeks of neoadjuvant chemotherapy prior to [sorafenib plus vorinostat] and intensity-modulated radiotherapy. Twenty-two patients were enrolled on the trial, and twenty-one were treated. Thirteen patients became eligible for surgery, and nine of these had R0 Whipple resections. Encouragingly, anti-tumor activity was observed across multiple drug dose levels within the trial.

The development of drug resistance is the bane of medical oncologists, and understanding how tumor cells become resistant to drugs or to drug combinations represents the cutting edge of developmental cancer therapeutics. Hence, we performed studies to understand the mechanisms by which cells become resistant to [sorafenib plus vorinostat] and whether we could rationally include a third agent to block the evolution of resistance and simultaneously enhance tumor cell killing.

The treatment of established pancreatic tumors with [sorafenib plus vorinostat] evolved the tumor cells to activate the receptors ERBB1 (the EGF receptor), ERBB2 (HER2), ERBB3, c-MET, and the intracellular kinase AKT [9]. These evolutionary events occurred regardless of the specific point mutation in mutant KRAS in each cell line. For reasons to be explained later, we chose the multi-kinase inhibitor neratinib to block signaling by the activated ERBB family receptors. Neratinib significantly enhanced the anti-tumor efficacy of [sorafenib plus vorinostat] *in vitro* and *in vivo*. Compared to [sorafenib plus vorinostat] as an agent or to neratinib as a single agent, the three-drug combination in pancreatic cancer cells further enhanced the phosphorylation of eIF2α (ER stress) and NFκB, the expression of Beclin1 and ATG5 (macroautophagy) and CD95 (death receptor signaling); and suppressed the levels of β-catenin (epithelial-mesenchymal transition). The knockdown of Beclin1, ATG5, CD95, eIF2α, or NFκB suppressed cell killing by the three-drug combination, whereas knockdown of β-catenin enhanced killing. The three-drug combination further increased autophagosome formation, and macroautophagy and cell killing were suppressed when we transfected cells to express an activated mammalian-target-of-rapamycin (mTOR) protein or when either Beclin1 or ATG5 were knocked down. A significant portion of the killing mechanism also required CD95 signaling, and knockdown of NFκB prevented the three-drug combination from increasing CD95 expression.

Unrelated work in breast cancer resulted in the laboratory working with the irreversible pan-ERBB receptor inhibitors afatinib and neratinib (Figure 30.2) [10]. Further studies explored the biology of neratinib, alone and combined with HDAC inhibitors, in pancreatic cancer cells. One of our initial observations was that neratinib not only inhibited the catalytic activity of ERBB receptors, but it also caused their internalization and degradation via LC3-associated phagocytosis (LAP) and macroautophagy [11–19]. Based on the concept that receptors and transducing small GTP-binding proteins, e.g., RAS proteins, co-localize as a component transducing of signals away from the plasma membrane, we went on to discover that neratinib could cause KRAS and NRAS proteins to mislocalize away from the plasma membrane within 30 min [9, 19, 20]. For RAS proteins to mediate signaling away from the plasma membrane, they must form short-lived, ~1 second, nanoclusters of 4-6 RAS proteins. Neratinib, within 5-10 min, disrupted both KRAS V12G nanoclustering and reduced the amount of RAF-1 that co-precipitated with KRAS V12G – GFP. In parallel to these events, neratinib reduced the amount of phosphatidyl serine (PS) in the plasma membrane. PS plays a key role both in maintaining KRAS in the plasma membrane and for its ability to form nanoclusters [21–23]. The precise mechanisms by which neratinib prevents RAS nanoclustering and RAS/RAF association are as yet to be elucidated.

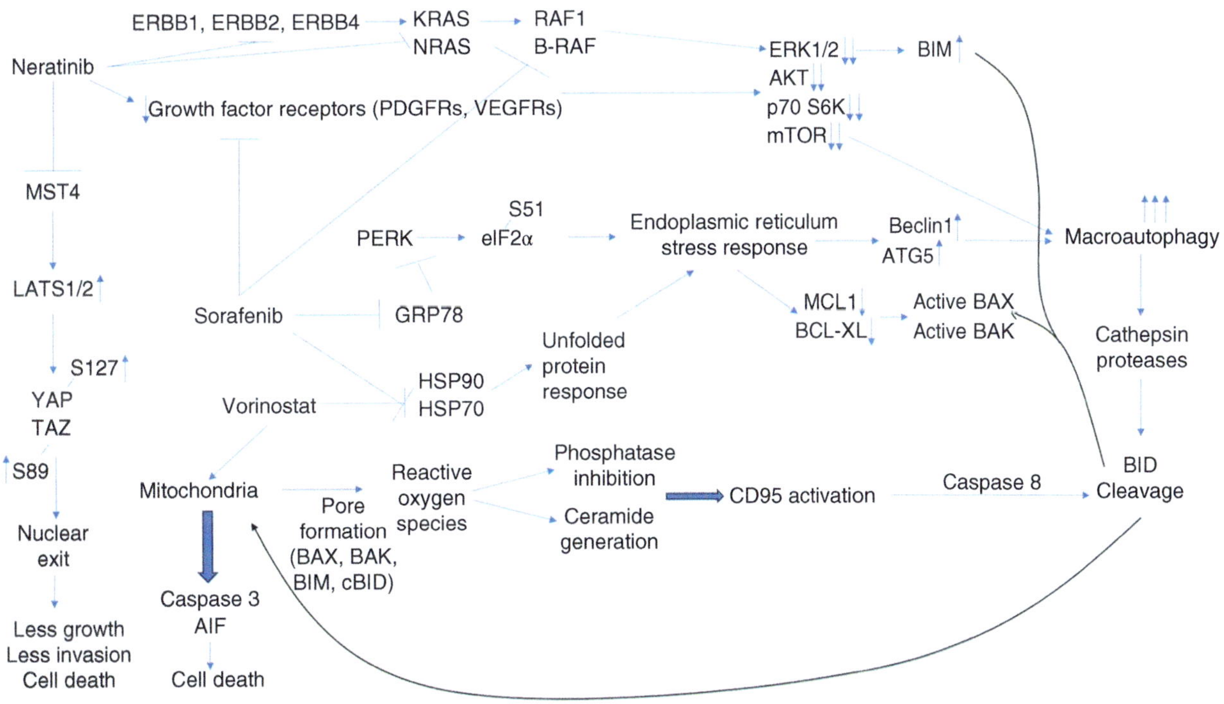

Figure 30.2 Mechanisms by which the multi-kinase inhibitor neratinib enhances the efficacy of [sorafenib plus HDAC inhibitors] to kill tumor cells: Our findings demonstrated that tumor cells that become resistant to [sorafenib plus vorinostat] have activated ERBB1, ERBB2, ERBB3, and ERBB4. Neratinib was developed to irreversibly inhibit ERBB2 and was subsequently shown to also act upon ERBB1 and ERBB4. Because of these effects, the transphosphorylation of ERBB3 in a cell-type specific fashion can be reduced. Growth factor receptors are proposed to be organized in lipid rafts, and we found that in addition to catalytic inhibition, neratinib could cause the internalization of ERBB family receptors and, in a cell-type-specific fashion, reduce the protein levels of other receptors such as c-MET and PDGFRb. This led us to discover that neratinib could cause the mislocalization of KRAS and NRAS proteins from the plasma membrane to intracellular sites where they were degraded. Although neratinib was developed to catalytically inhibit tyrosine kinases, it also inhibits serine/threonine kinases, notably the MAP4K MST4. Neratinib both catalytically inhibits MST4 and causes its degradation by autophagy. This results in activation of LATS1/2 and increased phosphorylation of YAP and TAZ (Hippo pathway). Phosphorylated YAP and TAZ exit the nucleus and no longer function as co-transcription factors. YAP and TAZ can compensate for the loss of mutant KRAS in pancreatic cancer cells, and co-transcription factors regulate the activities of genes coding for cytokines and growth factors. This as a secondary effect results in tumor cells that grow more slowly, that lose their ability to invade, and that cannot survive anoikis.

In our initial *in vivo* work, neratinib interacted with the HDAC inhibitor sodium valproate to suppress tumor growth using the mouse TNBC-equivalent model 4T1 [14]. Several of our *in vitro* observations were found in drug-treated tumors *in vivo* using immunohistochemistry approaches. For example, the levels of ERBB1 (EGF receptor) and ERBB2 (HER2) after a three-day drug exposure remained very low compared to vehicle control-treated tumors two weeks later. Similar immunohistochemical data were observed for wild-type KRAS expression, which remained significantly lower. The drug-treated tumor cells expressed less programed death ligand 1 (PD-L1) and more major histocompatibility complex (MHCA), and tumors contained more CD8+ cytotoxic T cells, and this collectively resulted in the tumors responding more efficaciously to immunotherapeutic treatment with an anti-PD1 antibody. These studies resulted in a new phase I trial in collaboration with Puma Biotechnology: "Neratinib + Valproate in Advanced Solid Tumors, w/Expansion Cohort in Ras-Mutated Ca" (NCT03919292).

As of 2024, checkpoint-inhibitory immunotherapy antibodies have not shown particular efficacy in pancreatic adenocarcinoma. Our preclinical studies and subsequently a phase I trial demonstrated the safety and efficacy of combinatorial radiochemotherapy plus surgery, including the combination of sorafenib and vorinostat (NCT01075113). The lethality of [sorafenib plus vorinostat] was enhanced by the standard of care drug gemcitabine [11, 12, 24]. Exposure of pancreatic tumor cells to [sorafenib plus vorinostat] reduced the expression of β-catenin, ERBB1, BCL-XL, and MCL-1, and the phosphorylation of AKT T308, AKT S473, GSK3 S9/21, mTORC1, and mTORC2. The drug combination increased the expression of Beclin1 and the phosphorylation of eIF2α S51. The drug combination rapidly reduced the levels of multiple HDAC proteins that were directly associated with the previously noted changes in tumor cell biology, as well as with alterations in the expression of biomarkers predictive for a response to checkpoint inhibitor antibodies, i.e., reduced PD-L1 expression and elevated MHCA expression. *In vivo* studies using the pancreatic PAN02 tumor model in its syngeneic mouse demonstrated that an anti-PD-1 antibody had no impact on pancreatic tumor growth, whereas a transient exposure to [sorafenib plus vorinostat] significantly suppressed growth [24]. The combination of [sorafenib plus vorinostat] together with an anti-PD-1 antibody caused a significant further reduction in tumor growth compared to the drug combination alone. Tumors transiently exposed to drugs three weeks earlier were found to contain elevated levels of CD8+ cells, M1 macrophages, and natural killer cells. Drug exposure plus an anti-PD-1 antibody further significantly enhanced the levels of these immune cells in the tumor. These findings argue that our [sorafenib plus vorinostat] combination could turn pancreatic tumors "hot."

Another outcome of our neratinib studies in pancreatic cancer came from manuscripts arguing that neratinib was, proposed in chemical biology assays to be an inhibitor of Ste20 serine/threonine kinases (MAP4Ks), in particular mammalian STE20-like protein kinase 4 (MST4) [17, 19]. This finding likely explains why neratinib caused diarrhea in patients, as MST4 is essential for a functioning GI brush border. MST4 also plays an essential role in regulating the rigidity of the plasma membrane by regulating the function of ezrin/radixin/moesin (ERM) proteins. In performing agnostic signaling studies on proteins that interact with MST4, we discovered that neratinib caused dephosphorylation of ezrin and plasma membrane flaccidity and increased the phosphorylation of large tumor suppressor kinase (LATS1/2), and of the targets of LATS1/2, the co-transcription factors (Yes-associated protein) YAP and its paralog TAZ. Neratinib caused the majority of YAP and TAZ to translocate from the nucleus into the cytosol and reduced YAP/TAZ protein levels. LATS1/2, YAP, and TAZ are distal components of the Hippo pathway [17, 19, 20].

Kinases liminal to the plasma membrane, below the level of GTP-binding proteins, are MAP4Ks (MAP4Ks1-7 and MST1-4). MAP4Ks are core upstream signaling components of the Hippo pathway. The Hippo pathway controls organ size and tissue homeostasis by regulating apoptosis and cell proliferation. MST1-4, MAP4K1/2/3, and MAP4K4/6/7 can all phosphorylate and activate the kinase large tumor suppressor, LATS1/2. Activated LATS1/2 phosphorylate and caused cytoplasmic sequestration of the transcriptional co-activators YAP and TAZ. YAP is an oncogene that is overexpressed in many tumors and is known to synergize with RAS to promote tumor growth, invasion, and therapeutic resistance in pancreatic cancer. YAP and TAZ partner with transcriptional enhancer factor TEF-1 (TEAD) as the nuclear effectors of the Hippo pathway. Inactivation of YAP and TAZ has a complex myriad of downstream global anti-tumor effects in a wide variety of tumor cell types, particularly with the reduced production of cytokines and growth factors. Collectively, our findings with neratinib argue that in pancreatic cancer cells as a single agent, the drug has the potential to block signaling by mutant KRAS, block oncogenic signaling by the Hippo pathway, and suppress evolutionary survival signaling from ERBB family growth factor receptors.

In conclusion, our initial studies at using multi-kinase inhibitors in GI tumors and pancreatic cancer resulted in two clinical trials where the combination of [sorafenib plus vorinostat] exhibited significant anti-tumor effects, including nine pancreatic cancer patients receiving R0 Whipple resections. A key evolutionary mechanism by which pancreatic cancer cells become resistant to [sorafenib plus vorinostat] was evolving to activate members of the ERBB receptor family. Neratinib blocked these evolutionary activations and further enhanced tumor control by [sorafenib plus vorinostat].

Acknowledgements

The authors wish to thank Dr. Daniel Von Hoff for his advice during our studies and his proofreading of this manuscript.

References

1 Zhang, G., Park, M.A., Mitchell, C. et al. (2008). Vorinostat and sorafenib synergistically kill tumor cells via FLIP suppression and CD95 activation. *Clin. Cancer Res.* 14: 5385–5399.
2 Park, M.A., Zhang, G., Martin, A.P. et al. (2008). Vorinostat and sorafenib increase ER stress, autophagy and apoptosis via ceramide-dependent CD95 and PERK activation. *Cancer Biol. Ther.* 7: 1648–1662.
3 Walker, T., Mitchell, C., Park, M.A. et al. (2009). Sorafenib and vorinostat kill colon cancer cells by CD95-dependent and -independent mechanisms. *Mol. Pharmacol.* 76: 342–355.
4 Martin, A.P., Park, M.A., Mitchell, C. et al. (2009). BCL-2 family inhibitors enhance histone deacetylase inhibitor and sorafenib lethality via autophagy and overcome blockade of the extrinsic pathway to facilitate killing. *Mol. Pharmacol.* 76: 327–341.
5 Park, M.A., Mitchell, C., Zhang, G. et al. (2010). Vorinostat and sorafenib increase CD95 activation in gastrointestinal tumor cells through a Ca(2+)-de novo ceramide-PP2A-reactive oxygen species-dependent signaling pathway. *Cancer Res.* 70: 6313–6324.
6 Park, M.A., Reinehr, R., Häussinger, D. et al. (2010). Sorafenib activates CD95 and promotes autophagy and cell death via Src family kinases in gastrointestinal tumor cells. *Mol. Cancer Ther.* 9: 2220–2231.
7 Gordon, S.W., McGuire, W.P. 3rd, Shafer, D.A. et al. (2019). Phase I study of sorafenib and vorinostat in advanced hepatocellular carcinoma. *Am. J. Clin. Oncol.* 42: 649–654.
8 Poklepovic, A., Fields, E.C., Bandyopadhyay, D. et al. A phase 1 study of neoadjuvant chemotherapy followed by concurrent chemoradiation with gemcitabine; sorafenib; and vorinostat in pancreatic cancer. *J. Clin. Oncol.* 39 (15_suppl): e16268.
9 Booth, L., Poklepovic, A., and Dent, P. (2020). Neratinib decreases pro-survival responses of [sorafenib + vorinostat] in pancreatic cancer. *Biochem. Pharmacol.* 178: 114067.
10 Tavallai, M., Booth, L., Roberts, J.L. et al. (2016). Rationally repurposing ruxolitinib (Jakafi (®)) as a solid tumor therapeutic. *Front. Oncol.* 6: 142.
11 Booth, L., Roberts, J.L., Poklepovic, A. et al. (2017). HDAC inhibitors enhance neratinib activity and when combined enhance the actions of an anti-PD-1 immunomodulatory antibody *in vivo*. *Oncotarget* 8: 90262–90277.
12 Booth, L., Roberts, J.L., Poklepovic, A. et al. (2018). The levels of mutant K-RAS and mutant N-RAS are rapidly reduced in a Beclin1/ATG5 -dependent fashion by the irreversible ERBB1/2/4 inhibitor neratinib. *Cancer Biol. Ther.* 19: 132–137.
13 Booth, L., Roberts, J.L., Samuel, P. et al. (2018). The irreversible ERBB1/2/4 inhibitor neratinib interacts with the PARP1 inhibitor niraparib to kill ovarian cancer cells. *Cancer Biol. Ther.* 19: 525–533.
14 Booth, L., Roberts, J.L., Rais, R. et al. [Neratinib + Valproate] exposure permanently reduces ERBB1 and RAS expression in 4T1 mammary tumors and enhances M1 macrophage infiltration. *Oncotarget* 9: 6062–6074.
15 Booth, L., Roberts, J.L., Rais, R. et al. (2019). Palbociclib augments Neratinib killing of tumor cells that is further enhanced by HDAC inhibition. *Cancer Biol. Ther.* 20: 157–168.
16 Booth, L., Roberts, J.L., Sander, C. et al. (2019). Neratinib and entinostat combine to rapidly reduce the expression of K-RAS, N-RAS, Gαq and Gα11 and kill uveal melanoma cells. *Cancer Biol. Ther.* 20: 700–710.
17 Dent, P., Booth, L., Roberts, J.L. et al. (2019). Neratinib inhibits Hippo/YAP signaling, reduces mutant K-RAS expression, and kills pancreatic and blood cancer cells. *Oncogene* 38: 5890–5904.
18 Dent, P., Booth, L., Poklepovic, A. et al. (2020). Enhanced signaling via ERBB3/PI3K plays a compensatory survival role in pancreatic tumor cells exposed to [neratinib + valproate]. *Cell. Signal.* 68: 109525.
19 Dent, P., Booth, L., Poklepovic, A. et al. (2020). Neratinib degrades MST4 via autophagy that reduces membrane stiffness and is essential for the inactivation of PI3K, ERK1/2, and YAP/TAZ signaling. *J. Cell. Physiol.* 235: 7889–7899.
20 Dent, P., Booth, L., Poklepovic, A. et al. (2021). Osimertinib-resistant NSCLC cells activate ERBB2 and YAP/TAZ and are killed by neratinib. *Biochem. Pharmacol.* 190: 114642.

21 Morstein, J., Shrestha, R., Van, Q.N. et al. (2023). Direct modulators of K-Ras-membrane interactions. *ACS Chem. Biol.* 18: 2082–2093.

22 Simanshu, D.K., Philips, M.R., and Hancock, J.F. (2023). Consensus on the RAS dimerization hypothesis: strong evidence for lipid-mediated clustering but not for G-domain-mediated interactions. *Mol. Cell* 83: 1210–1215.

23 Kattan, W.E., Liu, J., Montufar-Solis, D. et al. (2021). Components of the phosphatidylserine endoplasmic reticulum to plasma membrane transport mechanism as targets for KRAS inhibition in pancreatic cancer. *Proc. Natl. Acad. Sci. USA* 118: e2114126118.

24 Booth, L., Roberts, J.L., Poklepovic, A., and Dent, P. (2019). Prior exposure of pancreatic tumors to [sorafenib + vorinostat] enhances the efficacy of an anti-PD-1 antibody. *Cancer Biol. Ther.* 20: 109–121.

Index

Polypharmacology: Strategies for Multi-Target Drug Discovery, First Edition. Edited by Jens-Uwe Peters.

q

r

s

t

u

v

w

y

z